Lecture Notes in Artificial Intelligence 9621

Subseries of Lecture Notes in Computer Science

LNAI Series Editors

Randy Goebel
  University of Alberta, Edmonton, Canada
Yuzuru Tanaka
  Hokkaido University, Sapporo, Japan
Wolfgang Wahlster
  DFKI and Saarland University, Saarbrücken, Germany

LNAI Founding Series Editor

Joerg Siekmann
  DFKI and Saarland University, Saarbrücken, Germany

More information about this series at http://www.springer.com/series/1244

Ngoc Thanh Nguyen · Bogdan Trawiński
Hamido Fujita · Tzung-Pei Hong (Eds.)

Intelligent Information and Database Systems

8th Asian Conference, ACIIDS 2016
Da Nang, Vietnam, March 14–16, 2016
Proceedings, Part I

 Springer

Editors
Ngoc Thanh Nguyen
Wrocław University of Technology
Wrocław
Poland

Bogdan Trawiński
Wrocław University of Technology
Wrocław
Poland

Hamido Fujita
Iwate Prefectural University
Takizawa
Japan

Tzung-Pei Hong
National University of Kaohsiung
Kaohsiung
Taiwan

ISSN 0302-9743   ISSN 1611-3349 (electronic)
Lecture Notes in Artificial Intelligence
ISBN 978-3-662-49380-9   ISBN 978-3-662-49381-6 (eBook)
DOI 10.1007/978-3-662-49381-6

Library of Congress Control Number: 2016930675

LNCS Sublibrary: SL7 – Artificial Intelligence

Printed on acid-free paper

This Springer imprint is published by SpringerNature
The registered company is Springer-Verlag GmbH Berlin Heidelberg

Preface

ACIIDS 2016 was the eighth event of the series of international scientific conferences for research and applications in the field of intelligent information and database systems. The aim of ACIIDS 2016 was to provide an internationally respected forum for scientific research in the technologies and applications of intelligent information and database systems. ACIIDS 2016 was co-organized by the Vietnam–Korea Friendship Information Technology College (Vietnam) and Wrocław University of Technology (Poland) in co-operation with IEEE SMC Technical Committee on Computational Collective Intelligence, Bina Nusantara University (Indonesia), Ton Duc Thang University (Vietnam), and Quang Binh University (Vietnam). It took place in Da Nang (Vietnam) during March 14–16, 2016.

The ACIIDS conference series is well established. The first two events, ACIIDS 2009 and ACIIDS 2010, took place in Dong Hoi City and Hue City in Vietnam, respectively. The third event, ACIIDS 2011, took place in Daegu (Korea), while the fourth event, ACIIDS 2012, took place in Kaohsiung (Taiwan). The fifth event, ACIIDS 2013, was held in Kuala Lumpur in Malaysia, while the sixth event, ACIIDS 2014, was held in Bangkok, Thailand. The last event, ACIIDS 2015, took place in Bali (Indonesia).

We received papers from 36 countries all over the world. Each paper was peer reviewed by at least two members of the international Program Committee and international reviewer board. Only 153 papers with the highest quality were selected for oral presentation and publication in the two-volume proceedings of ACIIDS 2016.

Papers included in the proceedings cover the following topics: knowledge engineering and the Semantic Web, social networks and recommender systems, text processing and information retrieval, database systems and software engineering, intelligent information systems, decision support and control systems, machine learning and data mining, computer vision techniques, intelligent big data exploitation, cloud and network computing, multiple model approach to machine learning, advanced data mining techniques and applications, computational intelligence in data mining for complex problems, collective intelligence for service innovation, technology opportunity, e-learning and fuzzy intelligent systems, analysis of image, video, and motion data in life sciences, real-world applications in engineering and technology, ontology-based software development, intelligent and context systems, modeling and optimization techniques in information systems, database systems, and industrial systems, smart pattern processing for sports, and intelligent services for smart cities.

Accepted and presented papers highlight the new trends and challenges of intelligent information and database systems. The presenters showed how new research could lead to novel and innovative applications. We hope you will find these results useful and inspiring for your future research.

We would like to extend our heartfelt thanks to Mr. Jarosław Gowin, the Deputy Prime Minister of the Republic of Poland and Minister of Science and Higher Education for his support and honorary patronage over the conference.

We would like to express our sincere thanks to the honorary chairs, Mr. Minh Hong Nguyen (Deputy Minister of Information and Communications, Vietnam), and Prof. Tadeusz Więckowski (Rector of the Wrocław University of Technology, Poland) for their support.

Our special thanks go to the program chairs, special session chairs, organizing chairs, publicity chairs, liaison chairs, and local Organizing Committee for their work for the conference. We sincerely thank all members of the international Program Committee for their valuable efforts in the review process, which helped us to guarantee the highest quality of the selected papers for the conference. We cordially thank the organizers and chairs of special sessions, who essentially contributed to the success of the conference.

We also would like to express our thanks to the keynote speakers (Prof. Tzung-Pei Hong, Prof. Saeid Nahavandi, Prof. Jun Wang, and Prof. Piotr Wierzchoń) for their interesting and informative talks of world-class standard.

We cordially thank our main sponsors, Vietnam–Korea Friendship Information Technology College (Vietnam), Wrocław University of Technology (Poland), IEEE SMC Technical Committee on Computational Collective Intelligence, Bina Nusantara University (Indonesia), Ton Duc Thang University (Vietnam), and Quang Binh University (Vietnam). Our special thanks are due to Springer for publishing the proceedings, and all our other sponsors for their kind support.

We wish to thank the members of the Organizing Committee for their very substantial work and the members of the local Organizing Committee for their excellent work.

We cordially thank all the authors for their valuable contributions and other participants of this conference. The conference would not have been possible without their input.

Thanks are also due to many experts who contributed to making the event a success.

March 2016

Ngoc Thanh Nguyen
Bogdan Trawiński
Hamido Fujita
Tzung-Pei Hong

Conference Organization

Honorary Chairs

Minh Hong Nguyen   Deputy Minister of Information and Communications, Vietnam

Tadeusz Więckowski   Rector of Wrocław University of Technology, Poland

General Chairs

Ngoc Thanh Nguyen   Wrocław University of Technology, Poland

Bao Hung Hoang   Vietnam–Korea Friendship Information Technology College, Vietnam

Program Chairs

Bogdan Trawiński   Wrocław University of Technology, Poland

Tzung-Pei Hong   National University of Kaohsiung, Taiwan

Hamido Fujita   Iwate Prefectural University, Japan

Duc Dung Nguyen   IOIT – Vietnamese Academy of Science and Technology, Vietnam

Special Session Chairs

Dariusz Król   Wrocław University of Technology, Poland

Lech Madeyski   Wrocław University of Technology, Poland

Bay Vo   Ton Duc Thang University, Vietnam

Publicity Chairs

Khanh Van Hoang Thi   Vietnam–Korea Friendship Information Technology College, Vietnam

Adrianna Kozierkiewicz-Hetmańska   Wrocław University of Technology, Poland

Liaison Chairs

Ford Lumban Gaol   Bina Nusantara University, Indonesia

Tan Hanh   Posts and Telecommunications Institute of Technology, Vietnam

Mong-Fong Horng	National Kaohsiung University of Applied Sciences, Taiwan
Jason J. Jung	Chung-Ang University, Korea
Ali Selamat	Universiti Teknologi Malaysia, Malaysia

Organizing Chairs

| The Son Tran | Vietnam–Korea Friendship Information Technology College, Vietnam |
| Elżbieta Kukla | Wrocław University of Technology, Poland |

Local Organizing Committee

Marcin Maleszka	Wrocław University of Technology, Poland
Zbigniew Telec	Wrocław University of Technology, Poland
Bernadetta Maleszka	Wrocław University of Technology, Poland
Marcin Pietranik	Wrocław University of Technology, Poland
Nguyen Quang Vu	Vietnam–Korea Friendship Information Technology College, Vietnam
Ngo Viet Phuong	Vietnam–Korea Friendship Information Technology College, Vietnam
Le Tu Thanh	Vietnam–Korea Friendship Information Technology College, Vietnam

Webmaster

| Jarosław Bernacki | Wrocław University of Technology, Poland |

Steering Committee

Ngoc Thanh Nguyen (Chair)	Wrocław University of Technology, Poland
Longbing Cao	University of Technology Sydney, Australia
Suphamit Chittayasothorn	King Mongkut's Institute of Technology Ladkrabang, Thailand
Ford Lumban Gaol	Bina Nusantara University, Indonesia
Tu Bao Ho	Japan Advanced Institute of Science and Technology, Japan
Tzung-Pei Hong	National University of Kaohsiung, Taiwan
Dosam Hwang	Yeungnam University, Korea
Lakhmi C. Jain	University of South Australia, Australia
Geun-Sik Jo	Inha University, Korea
Jason J. Jung	Chung-Ang University, Korea
Hoai An Le-Thi	University of Lorraine, France

Toyoaki Nishida   Kyoto University, Japan
Leszek Rutkowski   Technical University of Czestochowa, Poland
Ali Selamat   Universiti Teknologi Malaysia, Malaysia

Keynote Speakers

Tzung-Pei Hong   National University of Kaohsiung, Taiwan
Saeid Nahavandi   Deakin University, Victoria, Australia
Jun Wang   City University of Hong Kong, Hong Kong, SAR China
Piotr Wierzchoń   Adam Mickiewicz University in Poznań, Poland

Special Sessions Organizers

1. Multiple Model Approach to Machine Learning (MMAML 20165)

Tomasz Kajdanowicz   Wrocław University of Technology, Poland
Edwin Lughofer   Johannes Kepler University Linz, Austria
Bogdan Trawiński   Wrocław University of Technology, Poland

2. Workshop on Real-World Applications in Engineering and Technology (RWAET 2016)

Pandian Vasant   Universiti Teknologi PETRONAS, Malaysia
Vish Kallimani   Universiti Teknologi PETRONAS, Malaysia
Sujan Chowdhury   Universiti Teknologi PETRONAS, Malaysia

3. Special Session on Intelligent Services for Smart Cities (IS4SC 2016)

David Camacho   Universidad Autónoma de Madrid, Spain
Jason J. Jung   Chung-Ang University, Korea
Paulo Novais   University of Minho, Portugal
Salvatore Venticinque Seconda Università Degli Studi di Napoli, Italy

4. Special Session on Ontology-Based Software Development (OSD 2016)

Zbigniew Huzar   Wrocław University of Technology, Poland
Bogumiła Hnatkowska   Wrocław University of Technology, Poland

5. Special Session on Intelligent Big Data Exploitation (IBDE 2016)

Gottfried Vossen   University of Münster, Germany
Stuart Dillon   University of Waikato, New Zealand

6. Special Session on Intelligent and Context Systems (ICxS 2016)

Maciej Huk   Wrocław University of Technology, Poland
Jan Kwiatkowski   Wrocław University of Technology, Poland
Anita Pinheiro   Halmstad University, Sweden
  Sant'Anna

7. Special Session on Analysis of Image, Video, and Motion Data in Life Sciences (IVMLS 2016)

Kondrad Wojciechowski	Polish–Japanese Academy of Information Technology, Poland
Marek Kulbacki	Polish–Japanese Academy of Information Technology, Poland
Jakub Segen	Gest3D, USA
Andrzej Polański	Silesian University of Technology, Poland

8. Special Session on Collective Intelligence for Service Innovation, Technology Opportunity, E-Learning, and Fuzzy Intelligent Systems (CISTEF 2016)

Chao-Fu Hong	Aletheia University, Taiwan
Tzu-Fu Chiu	Aletheia University, Taiwan
Kuo-Sui Lin	Aletheia University, Taiwan

9. Special Session on Advanced Data Mining Techniques and Applications (ADMTA 2016)

Bay Vo	Ton Duc Thang University, Vietnam
Tzung-Pei Hong	National University of Kaohsiung, Taiwan
Bac Le	University of Science, VNU-HCM, Vietnam

10. Special Session on Modeling and Optimization Techniques in Information Systems, Database Systems, and Industrial Systems (MOT 2016)

Le Thi Hoai An	University of Lorraine, France
Pham Dinh Tao	National Institute for Applied Science – Rouen, France

11. Special Session on Computational Intelligence in Data Mining for Complex Problems (CIDMCP 2016)

Habiba Drias	University of Science and Technology USTHB Algiers, Algeria
Gabriella Pasi	University of Milano-Bicocca, Italy

12. Special Session on Smart Pattern Processing for Sports (SP2S 2016)

S.M.N. Arosha Senanayake	Universiti Brunei Darussalam, Brunei
Chu Kiong Loo	University of Malaya, Malaysia

International Program Committee

Ajith Abraham	Machine Intelligence Research Labs, USA
Muhammad Abulaish	Jamia Millia Islamia, India
El-Houssaine Aghezzaf	Ghent University, Belgium
Waseem Ahmad	International College of Auckland, New Zealand
Jesus Alcala-Fdez	University of Granada, Spain

Haider Alsabbagh	Basra University, Iraq
Ahmad Taher Azar	Benha University, Egypt
Le Hoai Bac	University of Science, VNU-HCM, Vietnam
Amelia Badica	University of Craiova, Romania
Costin Badica	University of Craiova, Romania
Emili Balaguer-Ballester	Bournemouth University, UK
Zbigniew Banaszak	Warsaw University of Technology, Poland
Dariusz Barbucha	Gdynia Maritime University, Poland
John Batubara	Bina Nusantara University, Indonesia
Ramazan Bayindir	Gazi University, Turkey
Maumita Bhattacharya	Charles Sturt University, Australia
Jacek Błażewicz	Poznań University of Technology, Poland
Veera Boonjing	King Mongkut's Institute of Technology Ladkrabang, Thailand
Mariusz Boryczka	University of Silesia, Poland
Urszula Boryczka	University of Silesia, Poland
Abdelhamid Bouchachia	Bournemouth University, UK
Zouhaier Brahmia	University of Sfax, Tunisia
Stephane Bressan	National University of Singapore, Singapore
Peter Brida	University of Zilina, Slovakia
Andrej Brodnik	University of Ljubljana, Slovenia
Grazyna Brzykcy	Poznań University of Technology, Poland
The Duy Bui	University of Engineering and Technology, VNU Hanoi, Vietnam
Robert Burduk	Wrocław University of Technology, Poland
David Camacho	Universidad Autonoma de Madrid, Spain
Frantisek Capkovic	Institute of Informatics, Slovak Academy of Sciences, Slovakia
Leopoldo Eduardo Cardenas-Barron	Tecnologico de Monterrey, Mexico
Oscar Castillo	Tijuana Institute of Technology, Mexico
Dariusz Ceglarek	Poznań School of Banking, Poland
Rituparna Chaki	University of Calcutta, India
Kit Yan Chan	Curtin University, Australia
Somchai Chatvichienchai	University of Nagasaki, Japan
Meng Chang Chen	Academia Sinica, Taiwan
Rung-Ching Chen	Chaoyang University of Technology, Taiwan
Shyi-Ming Chen	National Taiwan University of Science and Technology, Taiwan
Yi-Ping Phoebe Chen	La Trobe University, Australia
Yiu-Ming Cheung	Hong Kong Baptist University, Hong Kong, SAR China
Suphamit Chittayasothorn	King Mongkut's Institute of Technology Ladkrabang, Thailand
Tzu-Fu Chiu	Aletheia University, Taiwan
Sung-Bae Cho	Yonsei University, Korea

Kazimierz Choroś	Wrocław University of Technology, Poland
Kun-Ta Chuang	National Cheng Kung University, Taiwan
Robert Cierniak	Czestochowa University of Technology, Poland
Dorian Cojocaru	University of Craiova, Romania
Phan Cong-Vinh	NTT University, Vietnam
Jose Alfredo Ferreira Costa	UFRN – Federal University of Rio Grande do Norte, Brazil
Keeley Crockett	Manchester Metropolitan University, UK
Bogusław Cyganek	AGH University of Science and Technology, Poland
Ireneusz Czarnowski	Gdynia Maritime University, Poland
Piotr Czekalski	Silesian University of Technology, Poland
Paul Davidsson	Malmo University, Sweden
Roberto De Virgilio	Università degli Studi Roma Tre, Italy
Phuc Do	Vietnam National University, HCMC, Vietnam
Tien V. Do	Budapest University of Technology and Economics, Hungary
Grzegorz Dobrowolski	AGH University of Science and Technology, Poland
Rafał Doroz	University of Silesia, Poland
Habiba Drias	University of Science and Technology Houari Boumediene, Algeria
El-Sayed M. El-Alfy	King Fahd University of Petroleum and Minerals, Saudi Arabia
Irraivan Elamvazuthi	Universiti Teknologi PETRONAS, Malaysia
Rim Faiz	University of Carthage, Tunisia
Victor Falea	Alexandru Ioan Cuza University of Iasi, Romania
Thomas Fober	University of Marburg, Germany
Simon Fong	University of Macau, Macau
Dariusz Frejlichowski	West Pomeranian University of Technology, Poland
Hamido Fujita	Iwate Prefectural University, Japan
Mohamed Gaber	Robert Gordon University, UK
Patrick Gallinari	LIP6 – University of Paris 6, France
Junbin Gao	Charles Sturt University, Australia
Dariusz Gąsior	Wrocław University of Technology, Poland
Janusz Getta	University of Wollongong, Australia
Dejan Gjorgjevikj	Ss. Cyril and Methodius University in Skopje, Macedonia
Daniela Godoy	ISISTAN Research Institute, Argentina
Gergo Gombos	Eotvos Lorand University, Hungary
Fernando Gomide	University of Campinas, Brazil
Antonio Gonzalez-Pardo	Bilbao Center for Applied Mathematics, Spain
Janis Grundspenkis	Riga Technical University, Latvia
Claudio Gutierrez	Universidad de Chile, Chile
Sung Ho Ha	Kyungpook National University, Korea
Sajjad Haider	Institute of Business Administration, Karachi, Pakistan
Marcin Hajdul	Institute of Logistics and Warehousing, Poland
Pei-Yi Hao	National Kaohsiung University of Applied Sciences, Taiwan

Harisno	Bina Nusantara University, Indonesia
Tutut Herawan	University of Malaya, Malaysia
Marcin Hernes	Wrocław University of Economics, Poland
Bogumiła Hnatkowska	Wrocław University of Technology, Poland
Huu Hanh Hoang	Hue University, Vietnam
Jaakko Hollmen	Aalto University School of Science, Finland
Tzung-Pei Hong	National University of Kaohsiung, Taiwan
Mong-Fong Horng	National Kaohsiung University of Applied Sciences, Taiwan
Chia-Ling Hsu	Tamkang University, Taiwan
Jen-Wei Huang	National Cheng Kung University, Taiwan
Maciej Huk	Wrocław University of Technology, Poland
Zbigniew Huzar	Wrocław University of Technology, Poland
Dosam Hwang	Yeungnam University, Korea
Roliana Ibrahim	Universiti Teknologi Malaysia, Malaysia
Dmitry Ignatov	National Research University Higher School of Economics, Russia
Lazaros Iliadis	Democritus University of Thrace, Greece
Hazra Imran	Athabasca University, Canada
Agnieszka Indyka-Piasecka	Wrocław University of Technology, Poland
Mirjana Ivanovic	University of Novi Sad, Serbia
Sanjay Jain	National University of Singapore, Singapore
Chuleerat Jaruskulchai	Kasetsart University, Thailand
Khalid Jebari	LCS Rabat, Morocco
Joanna Jędrzejowicz	University of Gdańsk, Poland
Piotr Jędrzejowicz	Gdynia Maritime University, Poland
Janusz Jeżewski	Institute of Medical Technology and Equipment ITAM, Poland
Geun-Sik Jo	Inha University, Korea
Kang-Hyun Jo	University of Ulsan, Korea
Jason J. Jung	Chung-Ang University, Korea
Janusz Kacprzyk	Systems Research Institute, Polish Academy of Sciences, Poland
Tomasz Kajdanowicz	Wrocław University of Technology, Poland
Nadjet Kamel	Ferhat Abbas University of Setif, Algeria
Hung-Yu Kao	National Cheng Kung University, Taiwan
Mehmet Hakan Karaata	Kuwait University, Kuwait
Nikola Kasabov	Auckland University of Technology, New Zealand
Arkadiusz Kawa	Poznań University of Economics, Poland
Muhammad Khurram Khan	King Saud University, Saudi Arabia
Pan-Koo Kim	Chosun University, Korea
Yong Seog Kim	Utah State University, USA
Attila Kiss	Eotvos Lorand University, Hungary
Jerzy Klamka	Silesian University of Technology, Poland

Frank Klawonn	Ostfalia University of Applied Sciences, Germany
Goran Klepac	Raiffeisen Bank, Croatia
Joanna Kołodziej	Cracow University of Technology, Poland
Marek Kopel	Wrocław University of Technology, Poland
Józef Korbicz	University of Zielona Góra, Poland
Jacek Koronacki	Institute of Computer Science, Polish Academy of Sciences, Poland
Raymond Kosala	Bina Nusantara University, Indonesia
Leszek Koszałka	Wrocław University of Technology, Poland
Jan Kozak	University of Silesia, Poland
Adrianna Kozierkiewicz-Hetmańska	Wrocław University of Technology, Poland
Bartosz Krawczyk	Wrocław University of Technology, Poland
Ondrej Krejcar	University of Hradec Kralove, Czech Republic
Dalia Kriksciuniene	Vilnius University, Lithuania
Marzena Kryszkiewicz	Warsaw University of Technology, Poland
Adam Krzyzak	Concordia University, Canada
Elżbieta Kukla	Wrocław University of Technology, Poland
Marek Kulbacki	Polish–Japanese Academy of Information Technology, Poland
Marek Kurzyński	Wrocław University of Technology, Poland
Kazuhiro Kuwabara	Ritsumeikan University, Japan
Halina Kwaśnicka	Wrocław University of Technology, Poland
Mark Last	Ben-Gurion University of the Negev, Israel
Anabel Latham	Manchester Metropolitan University, UK
Hoai An Le Thi	Université de Lorraine, France
Chang-Hwan Lee	DongGuk University, Korea
Kun Chang Lee	Sungkyunkwan University, Korea
Yue-Shi Lee	Ming Chuan University, Taiwan
Philippe Lenca	Telecom Bretagne, France
Chunshien Li	National Central University, Taiwan
Jiuyong Li	University of South Australia, Australia
Rita Yi Man Li	Hong Kong Shue Yan University, Hong Kong, SAR China
Horst Lichter	RWTH Aachen University, Germany
Sebastian Link	University of Auckland, New Zealand
Igor Litvinchev	Nuevo Leon State University, Mexico
Lian Liu	University of Kentucky, USA
Rey-Long Liu	Tzu Chi University, Taiwan
Heitor Silverio Lopes	UTFPR, Federal University of Technology, Parana, Brazil
Edwin Lughofer	Johannes Kepler University Linz, Austria
Lech Madeyski	Wrocław University of Technology, Poland
Nezam Mahdavi-Amiri	Sharif University of Technology, Iran
Bernadetta Maleszka	Wrocław University of Technology, Poland
Marcin Maleszka	Wrocław University of Technology, Poland
Yannis Manolopoulos	Aristotle University of Thessaloniki, Greece

Konstantinos Margaritis	University of Macedonia, Greece
Francesco Masulli	University of Genoa, Italy
Mustafa Mat Deris	Universiti Tun Hussein Onn Malaysia, Malaysia
Tamas Matuszka	Eotvos Lorand University, Hungary
Joao Mendes-Moreira	University of Porto, Portugal
Gerardo Mendez	Instituto Tecnologico de Nuevo Leon, Mexico
Hector D. Menendez	University College London, UK
Jacek Mercik	Wrocław School of Banking, Poland
Radosław Michalski	Wrocław University of Technology, Poland
Peter Mikulecky	University of Hradec Kralove, Czech Republic
Marek Miłosz	Lublin University of Technology, Poland
Yang-Sae Moon	Kangwon National University, Korea
Leo Mrsic	IN2data Ltd Data Science Company, Croatia
Grzegorz J. Nalepa	AGH University of Science and Technology, Poland
Mahyuddin K.M. Nasution	Universitas Sumatera Utara, Indonesia
Prospero Naval	University of the Philippines, Philippines
Richi Nayak	Queensland University of Technology, Australia
Fulufhelo Nelwamondo	Council for Scientific and Industrial Research, South Africa
Dieu Ngoc Vo	Ho Chi Minh City University of Technology, Vietnam
Huu-Tuan Nguyen	Vietnam Maritime University, Vietnam
Loan T.T. Nguyen	Broadcasting College II, Vietnam
Thai-Nghe Nguyen	Can Tho University, Vietnam
Vinh Nguyen	University of Melbourne, Australia
Toyoaki Nishida	Kyoto University, Japan
Yusuke Nojima	Osaka Prefecture University, Japan
Mariusz Nowostawski	Norwegian University of Science and Technology, Norway
Alberto Nunez	Universidad Complutense de Madrid, Spain
Manuel Nunez	Universidad Complutense de Madrid, Spain
Richard Jayadi Oentaryo	Singapore Management University, Singapore
Kouzou Ohara	Aoyama Gakuin University, Japan
Tomasz Orczyk	University of Silesia, Poland
Shingo Otsuka	Kanagawa Institute of Technology, Japan
Marcin Paprzycki	Systems Research Institute, Polish Academy of Sciences, Poland
Jakub Peksiński	West Pomeranian University of Technology, Poland
Danilo Pelusi	University of Teramo, Italy
Xuan Hau Pham	Quang Binh University, Vietnam
Tao Pham Dinh	National Institute for Applied Sciences, France
Xuan-Hieu Phan	Vietnam National University, Hanoi, Vietnam
Maciej Piasecki	Wrocław University of Technology, Poland
Dariusz Pierzchała	Military University of Technology, Poland
Marcin Pietranik	Wrocław University of Technology, Poland
Elvira Popescu	University of Craiova, Romania

Piotr Porwik	University of Silesia, Poland
Bhanu Prasad	Florida A&M University, USA
Paulo Quaresma	Universidade de Evora, Portugal
R. Rajesh	Central University of Bihar, India
Mohammad Rashedur Rahman	North South University, Bangladesh
Ewa Ratajczak-Ropel	Gdynia Maritime University, Poland
Patricia Riddle	University of Auckland, New Zealand
Sebastian A. Rios	University of Chile, Chile
Miguel Angel Romero Orth	Universidad de Chile, Chile
Manuel Roveri	Politecnico di Milano, Italy
Przemysław Różewski	West Pomeranian University of Technology, Poland
Leszek Rutkowski	Czestochowa University of Technology, Poland
Alexander Ryjov	Lomonosov Moscow State University, Russia
Virgilijus Sakalauskas	Vilnius University, Lithuania
Daniel Sanchez	University of Granada, Spain
Cesar Sanin	University of Newcastle, Australia
Moamar Sayed-Mouchaweh	Ecole des Mines de Douai, France
Juergen Schmidhuber	Swiss AI Lab IDSIA, Switzerland
Bjorn Schuller	University of Passau, Germany
Jakub Segen	Gest3D, USA
Ali Selamat	Universiti Teknologi Malaysia, Malaysia
S.M.N. Arosha Senanayake	Universiti Brunei Darussalam, Brunei
Natalya Shakhovska	Lviv Polytechnic National University, Ukraine
Alexei Sharpanskykh	Delft University of Technology, The Netherlands
Andrzej Siemiński	Wrocław University of Technology, Poland
Guillermo R. Simari	Universidad Nacional del Sur, Argentina
Dragan Simic	University of Novi Sad, Serbia
Andrzej Skowron	Warsaw University, Poland
Adam Słowik	Koszalin University of Technology, Poland
Vladimir Sobeslav	University of Hradec Kralove, Czech Republic
Kulwadee Somboonviwat	University of Electro-Communications, Japan
Zenon A. Sosnowski	Białystok University of Technology, Poland
Jerzy Stefanowski	Poznań University of Technology, Poland
Serge Stinckwich	University of Caen-Lower Normandy, France
Stanimir Stoyanov	University of Plovdiv Paisii Hilendarski, Bulgaria
Suharjito	Bina Nusantara University, Indonesia
Jerzy Świątek	Wrocław University of Technology, Poland
Paweł Świątek	Wrocław University of Technology, Poland
Andrzej Świerniak	Silesian University of Technology, Poland
Edward Szczerbicki	University of Newcastle, Australia
Julian Szymański	Gdańsk University of Technology, Poland

Ryszard Tadeusiewicz	AGH University of Science and Technology, Poland
Yasufumi Takama	Tokyo Metropolitan University, Japan
Zbigniew Telec	Wrocław University of Technology, Poland
Krzysztof Tokarz	Silesian University of Technology, Poland
Behcet Ugur Toreyin	Cankaya University, Turkey
Bogdan Trawiński	Wrocław University of Technology, Poland
Krzysztof Trawiński	European Centre for Soft Computing, Spain
Hong-Linh Truong	Vienna University of Technology, Austria
Ualsher Tukeyev	Al-Farabi Kazakh National University, Kazakhstan
Olgierd Unold	Wrocław University of Technology, Poland
Pandian Vasant	Universiti Teknologi PETRONAS, Malaysia
Joost Vennekens	Katholieke Universiteit Leuven, Belgium
Jorgen Villadsen	Technical University of Denmark, Denmark
Maria Virvou	University of Piraeus, Greece
Bay Vo	Ton Duc Thang University, Vietnam
Gottfried Vossen	ERCIS Münster, Germany
M. Abdullah-Al Wadud	King Saud University, Saudi Arabia
Can Wang	CSIRO, Australia
Lipo Wang	Nanyang Technological University, Singapore
Yongkun Wang	University of Tokyo, Japan
Izabela Wierzbowska	Gdynia Maritime University, Poland
Konrad Wojciechowski	Silesian University of Technology, Poland
Michal Woźniak	Wrocław University of Technology, Poland
Krzysztof Wróbel	University of Silesia, Poland
Marian Wysocki	Rzeszow University of Technology, Poland
Guandong Xu	University of Technology Sydney, Australia
Xin-She Yang	Middlesex University, UK
Lina Yao	University of Adelaide, Australia
Shi-Jim Yen	National Dong Hwa University, Taiwan
Lean Yu	Chinese Academy of Sciences, AMSS, China
Sławomir Zadrożny	Systems Research Institute, Polish Academy of Sciences, Poland
Drago Zagar	University of Osijek, Croatia
Danuta Zakrzewska	Łódź University of Technology, Poland
Constantin-Bala Zamfirescu	Lucian Blaga University of Sibiu, Romania
Katerina Zdravkova	Ss. Cyril and Methodius University in Skopje, Macedonia
Vesna Zeljkovic	Lincoln University, USA
Aleksander Zgrzywa	Wrocław University of Technology, Poland
Zhongwei Zhang	University of Southern Queensland, Australia
Zhi-Hua Zhou	Nanjing University, China
Zhandos Zhumanov	Al-Farabi Kazakh National University, Kazakhstan
Maciej Zięba	Wrocław University of Technology, Poland

Program Committees of Special Sessions

Multiple Model Approach to Machine Learning (MMAML 2016)

Emili Balaguer-Ballester	Bournemouth University, UK
Urszula Boryczka	University of Silesia, Poland
Abdelhamid Bouchachia	Bournemouth University, UK
Robert Burduk	Wrocław University of Technology, Poland
Oscar Castillo	Tijuana Institute of Technology, Mexico
Dariusz Ceglarek	Poznań High School of Banking, Poland
Manuel Chica	European Centre for Soft Computing, Spain
Rung-Ching Chen	Chaoyang University of Technology, Taiwan
Suphamit Chittayasothorn	King Mongkut's Institute of Technology Ladkrabang, Thailand
José Alfredo F. Costa	Federal University (UFRN), Brazil
Bogusław Cyganek	AGH University of Science and Technology, Poland
Ireneusz Czarnowski	Gdynia Maritime University, Poland
Patrick Gallinari	Pierre et Marie Curie University, France
Fernando Gomide	State University of Campinas, Brazil
Tzung-Pei Hong	National University of Kaohsiung, Taiwan
Roliana Ibrahim	Universiti Teknologi Malaysia, Malaysia
Konrad Jackowski	Wrocław University of Technology, Poland
Piotr Jędrzejowicz	Gdynia Maritime University, Poland
Tomasz Kajdanowicz	Wrocław University of Technology, Poland
Yong Seog Kim	Utah State University, USA
Bartosz Krawczyk	Wrocław University of Technology, Poland
Elżbieta Kukla	Wrocław University of Technology, Poland
Kun Chang Lee	Sungkyunkwan University, Korea
Edwin Lughofer	Johannes Kepler University Linz, Austria
Hector Quintian	University of Salamanca, Spain
Andrzej Sieminski	Wrocław University of Technology, Poland
Dragan Simic	University of Novi Sad, Serbia
Adam Słowik	Koszalin University of Technology, Poland
Kulwadee Somboonviwat	University of Electro-Communications, Japan
Zbigniew Telec	Wrocław University of Technology, Poland
Bogdan Trawiński	Wrocław University of Technology, Poland
Krzysztof Trawiński	European Centre for Soft Computing, Spain
Olgierd Unold	Wrocław University of Technology, Poland
Pandian Vasant	University Technology Petronas, Malaysia
Michał Woźniak	Wrocław University of Technology, Poland
Zhongwei Zhang	University of Southern Queensland, Australia
Zhi-Hua Zhou	Nanjing University, China

Workshop on Real-World Applications in Engineering and Technology (RWAET 2016)

Vassili Kolokoltsov	University of Warwick, UK
Gerhard-Wilhelm Weber	METU, Turkey
Junzo Watada	Waseda University, Japan
Morteza Kalaji	Universiti Teknologi PETRONAS, Malaysia
Kwon-Hee Lee	Dong-A University, South Korea
Igor Litvinchev	Nuevo Leon State University, Mexico
Mohammad Abdullah-Al-Wadud	King Saud University, Saudi Arabia
Subhash Kamal	Universiti Teknologi PETRONAS, Malaysia
Vo Ngoc Dieu	HCMC University of Technology, Vietnam
Gerardo Maximiliano Mendez	Instituto Tecnologico de Nuevo Leon, Mexico
Leopoldo Eduardo Cárdenas Barrón	Tecnológico de Monterrey, Mexico
Hassan Soliemani	Universiti Teknologi PETRONAS, Malaysia
Denis Sidorov	Irkutsk State University, Russia
Weerakorn Ongsakul	Asian Institute of Technology, Thailand
Goran Klepac	Raiffeisen Bank Austria, Croatia
Jean Leveque	Universiti Teknologi PETRONAS, Malaysia
Herman Mawengkang	The University of Sumatera Utara, Indonesia
Igor Tyukhov	Moscow State University of Mechanical Engineering, Russia
Hayato Ohwada	Tokyo University of Science, Japan
Joga Setiawan	Universiti Teknologi PETRONAS, Malaysia
Ugo Fiore	Federico II University, Italy
Leo Mrsic	University College for Law and Finance Effectus Zagreb, Croatia
Nguyen Trung Thang	Ton Duc Thang University, Vietnam
Nikolai Voropai	Energy Systems Institute, Russia
Shiferaw Jufar	Universiti Teknologi PETRONAS, Malaysia
Xueguan Song	Dalian University of Technology, China
Ruhul A. Sarker	UNSW, Australia
Vipul Sharma	Lovely Professional University, India

Special Session on Intelligent Services for Smart Cities (IS4SC 2016)

Antonio Gonzalez	Basque Center for Applied Mathematics-TECNALIA, Spain
Jason J. Jung	Chung-Ang University, Korea
Pankoo Kim	Chosun University, Korea
Héctor D. Menéndez	University College London, UK
Xuan Hau Pham	Quang Binh University, Vietnam
Javier del Ser	Tecnalia, Spain

Special Session on Ontology-Based Software Development (OSD 2016)

Jose Maria Alvarez-Rodríguez	Universidad Carlos III de Madrid, Spain
Veera Boonjing	King Mongkut's Institute of Technology Ladkrabang, Bangkok Thailand
Somchai Chatvichienchai	University of Nagasaki, Japan
Rung-Ching Chen	Chaoyang University of Technology, Taiwan
Mustafa Mat Deris	Universiti Tun Hussein Onn, Malaysia
Iwona Dubielewicz	Wrocław University of Technology, Poland
Mirjana Ivanovic	University of Novi Sad, Serbia
Mark Last	Ben-Gurion University of the Negev, Israel
Adam Pease	R&D Manager at IPsoft, Hong Kong Polytechnic University, Hong Kong, SAR China
Sławomir Zadrożny	Systems Research Institute, Polish Academy of Sciences, Poland

Special Session on Intelligent Big Data Exploitation (IBDE 2016)

Alfredo Cuzzocrea	University of Trieste, Italy
Ernesto Damiani	University of Milan, Italy
Stuart Dillon	University of Waikato, New Zealand
Stefan Gruner	University of Pretoria, South Africa
Birgit Hofreiter	Technical University of Vienna, Austria
Christopher Holland	Manchester Business School, UK
Alexander Löser	Beuth University of Applied Sciences, Berlin, Germany
Ute Masermann	Decadis AG, Koblenz, Germany
Tadeusz Morzy	Technical University of Poznań, Poland
Florian Stahl	ERCIS, University of Münster, Germany
Heike Trautmann	ERCIS, University of Münster, Germany
Gottfried Vossen	ERCIS, University of Münster, Germany

Special Session on Intelligent and Context Systems (ICxS 2016)

Qiangfu Zhao	University of Aizu, Japan
Goutam Chakraborty	Iwate Prefectural University, Japan
Anita Sant'Anna	Halmstad University, Sweden
Michael Spratling	University of London, UK
Anna Fabijańska	Łódź University of Technology, Poland
Józef Korbicz	University of Zielona Góra, Poland
Jerzy Świątek	Wrocław University of Technology, Poland
Maciej Piasecki	Wrocław University of Technology, Poland
Michał Kędziora	Wrocław University of Technology, Poland
Nguyen Thanh Binh	Ho Chi Minh City University of Technology, Vietnam
Quan Thanh Tho	Ho Chi Minh City University of Technology, Vietnam
Ha Manh Tran	Ho Chi Minh City International University, Vietnam
Nguyen Khang Pham	Can Tho University, Vietnam

Nguyen Thai-Nghe	Can Tho University, Vietnam
Pedro Medeiros	University Nova of Lisbon, Portugal
Jan Kwiatkowski	Wrocław University of Technology, Poland
Maciej Huk	Wrocław University of Technology, Poland
Emilio Luque	University Autonoma of Barcelona, Spain
Dolores Rexachs	University Autonoma of Barcelona, Spain
Philip Moore	Lanzhou University, China
Norbert Jankowski	Nicholas Copernicus University, Poland
Bartlett W. Mel	University of Southern California, USA
Gregory Hager	Johns Hopkins University, USA
Shimon Ullman	Weizmann Institute of Science, Israel
Santosh S. Venkatesh	University of Pennsylvania, USA
Garrison W. Cottrell	University of California, USA
Xiao-Ping Zhang	Ryerson University, Canada
Grażyna Suchacka	Opole University, Poland
Ryszard Tadeusiewicz	AGH, Poland
William Dally	Stanford University, UK
Marek Wróblewski	University of Stuttgart, Germany
Wen Gao	Peking University, China
Elan Barenholtz	Florida Atlantic University, USA

Special Session on Analysis of Image, Video, and Motion Data in Life Sciences (IVMLS 2016)

Artur Bąk	Polish–Japanese Academy of Information Technology, Poland
Leszek Chmielewski	Warsaw University of Life Sciences, Poland
Aldona Barbara Drabik	Polish–Japanese Academy of Information Technology, Poland
Marcin Fojcik	Sogn og Fjordane University College, Norway
Adam Gudyś	Silesian University of Technology, Poland
Celina Imielińska	Vesalius Technologies LLC, USA
Henryk Josiński	Silesian University of Technology, Poland
Ryszard Klempous	Wrocław University of Technology, Poland
Ryszard Kozera	The University of Life Sciences – SGGW, Poland
Julita Kulbacka	Wrocław Medical University, Poland
Marek Kulbacki	Polish–Japanese Academy of Information Technology, Poland
Aleksander Nawrat	Silesian University of Technology, Poland
Jerzy Paweł Nowacki	Polish–Japanese Academy of Information Technology, Poland
Eric Petajan	LiveClips LLC, USA
Andrzej Polański	Silesian University of Technology, Poland
Joanna Rossowska	Polish Academy of Sciences, Institute of Immunology and Experimental Therapy, Poland
Jakub Segen	Gest3D LLC, USA

Aleksander Sieroń	Medical University of Silesia, Poland
Michał Staniszewski	Polish–Japanese Academy of Information Technology, Poland
Adam Świtoński	Silesian University of Technology, Poland
Agnieszka Szczęsna	Silesian University of Technology, Poland
Kamil Wereszczyński	Polish–Japanese Academy of Information Technology, Poland
Konrad Wojciechowski	Polish–Japanese Academy of Information Technology, Poland
Sławomir Wojciechowski	Polish–Japanese Academy of Information Technology, Poland

Special Session on Collective Intelligence for Service Innovation, Technology Opportunity, E-Learning, and Fuzzy Intelligent Systems (CISTEF 2016)

Chang, Ya-Fung	Tamkang University, Taiwan
Chen, Chi-Min	Aletheia University, Taiwan
Chiu, Chih-Chung	Aletheia University, Taiwan
Chiu, Kuan-Shiu	Aletheia University, Taiwan
Chiu, Tzu-Fu	Aletheia University, Taiwan
Chou, Chen-Huei	College of Charleston, USA
Hsu, Chia-Ling	Tamkang University, Taiwan
Hsu, Fang-Cheng	Aletheia University, Taiwan
Lin, Kuo-Sui	Aletheia University, Taiwan
Lin, Min-Huei	Aletheia University, Taiwan
Lin, Yuh-Chang	Aletheia University, Taiwan
Maeno, Yoshiharu	NEC Corporation, Japan
Sun, Pen-Choug	Aletheia University, Taiwan
Wang, Henry	Chinese Academy of Sciences, China
Wang, Leuo-Hong	Aletheia University, Taiwan
Yang, Feng-Sueng	Aletheia University, Taiwan
Yang, Hsiao-Fang	National Chengchi University, Taiwan
Yang, Ming-Chien	Aletheia University, Taiwan

Special Session on Advanced Data Mining Techniques and Applications (ADMTA 2016)

Bay Vo	Ton Duc Thang University, Vietnam
Tzung-Pei Hong	National University of Kaohsiung, Taiwan
Bac Le	University of Science, VNU-HCM, Vietnam
Chun-Hao Chen	Tamkang University, Taiwan
Chun-Wei Lin	Harbin Institute of Technology Shenzhen Graduate School, China
Wen-Yang Lin	National University of Kaohsiung, Taiwan
Yeong-Chyi Lee	Cheng Shiu University, Taiwan
Le Hoang Son	University of Science, Ha Noi, Vietnam
Le Hoang Thai	University of Science, Ho Chi Minh City, Vietnam

Vo Thi Ngoc Chau	Ho Chi Minh City University of Technology, Ho Chi Minh City, Vietnam
Van Vo	Ho Chi Minh University of Industry, Ho Chi Minh City, Vietnam

Special Session on Modeling and Optimization Techniques in Information Systems, Database Systems, and Industrial Systems (MOT 2016)

Le Thi Hoai An	University of Lorraine, France
Pham Dinh Tao	INSA-Rouen, France
Pham Duc Truong	University of Cardiff, UK
El-Houssaine Aghezzaf	University of Gent, Belgium
Azeddine Beghadi	University of Paris 13, France
Raymond Bisdorff	Université du Luxembourg
Jin-Kao Hao	University of Angers, France
Van-Dat Cung	INPG, France
Joaquim Judice	University Coimbra, Portugal
Amédéo Napoli	LORIA, France
Yann Germeur	LORIA, France
Conan-Guez Brieu	University of Lorraine, France
Gely Alain	University of Lorraine, France
Le Hoai Minh	University of Luxembourg, Luxembourg
Vo Xuan Thanh	University of Lorraine, France
Do Thanh Nghi	Can Tho University, Vietnam
Ibrahima Sakho	University of Lorraine, France

Special Session on Computational Intelligence in Data Mining for Complex Problems (CIDMCP 2016)

Sid Ahmed Benraouane	University of Minnesota, USA
Maria Gini	University of Minnesota, USA
Imed Kacem	University of Lorraine, France
Nadjet Kamel	University of Sétif, Algeria
Mehmed M. Kantardzic	University of Louisville, USA
Saroj Kaushik	Indian Institute of Technology Delhi, India
Samir Kechid	USTHB, Algeria
In-Young Ko	KAIST, Korea
Qin Lv	University of Colorado at Boulder, USA
Ana Maria Madureira	Polytechnic of Porto, Portugal
Brahim Medjahed	University of Michigan, USA
Madjid Merabti	Liverpool John Moores University, UK
Farid Meziane	University of Salford, UK
Erich J. Neuhold	Vienna University, Austria
Mourad Oussalah	Nantes University, France
Myeong Cheol Park	KAIST, Korea
Nelishia Pillay	University of KwaZulu-Natal, South Africa
Kalai Anand Ratnam	APUTI, Malaysia

Jae Jeung Rho	KAIST, Korea
Houari Sahraoui	University of Montreal, Canada
Lakhdar Sais	University of Artois, France
Thouraya Tebibel	ESI, Algeria
Farouk Yalaoui	Troyes University of Technology, France
Ning Zhong Maebashi	Institute of Technology, Japan

Special Session on Smart Pattern Processing for Sports (SP2S 2016)

Minoru Sasaki	Gifu University, Japan
Michael Yu Wang	National University of Singapore, Singapore
Eran Edirisinghe	Loughborough University, UK
Ajith Abraham	MIR Labs, USA
James F. Peters	University of Manitoba, Canada
Sergio Velastin	Kingston University, UK
Tadashi Ishihara	Fukushima University, Japan
Darwin Gouwanda	Monash University, Malaysia
Aaron Leung	The Hong Kong Polytechnic University, Hong Kong, SAR China
Toshiyo Tamura	Osaka Electro-Communication University, Japan
T. Nandha Kumaar	University of Nottingham, Malaysia
James Goh Cho Hong	National University Singapore, Singapore
Tsuyoshi Takagi	Kyushu University, Japan
William C. Rose	University of Delaware, USA

Contents – Part I

Text Processing and Information Retrieval

Database Systems and Software Engineering

Intelligent Information Systems

Decision Support and Control Systems

Machine Learning and Data Mining

Computer Vision Techniques

Contents – Part II

Advanced Data Mining Techniques and Applications

Computational Intelligence in Data Mining for Complex Problems

Collective Intelligence for Service Innovation, Technology Opportunity, E-Learning and Fuzzy Intelligent Systems

Analysis of Image, Video and Motion Data in Life Sciences

Real World Applications in Engineering and Technology

Ontology-Based Software Development

Intelligent and Context Systems

Modelling and Optimization Techniques in Information Systems, Database Systems and Industrial Systems

Smart Pattern Processing for Sports

Intelligent Services for Smart Cities

Knowledge Engineering
and Semantic Web

A Novel Approach to Multimedia Ontology Engineering for Automated Reasoning over Audiovisual LOD Datasets

Leslie F. Sikos$^{(\boxtimes)}$

Flinders University, Adelaide, Australia
siko0007@flinders.edu.au

Abstract. Multimedia reasoning, which is suitable for, among others, multimedia content analysis and high-level video scene interpretation, relies on the formal and comprehensive conceptualization of the represented knowledge domain. However, most multimedia ontologies are not exhaustive in terms of role definitions, and do not incorporate complex role inclusions and role interdependencies. In fact, most multimedia ontologies do not have a role box at all, and implement only a basic subset of the available logical constructors. Consequently, their application in multimedia reasoning is limited. To address the above issues, VidOnt, the very first multimedia ontology with $\mathcal{SROIQ}^{(\mathcal{D})}$ expressivity and a DL-safe ruleset has been introduced for next-generation multimedia reasoning. In contrast to the common practice, the formal grounding has been set in one of the most expressive description logics, and the ontology validated with industry-leading reasoners, namely HermiT and FaCT++. This paper also presents best practices for developing multimedia ontologies, based on my ontology engineering approach.

Keywords: Ontology · OWL · MPEG-7 · Video metadata · Video retrieval · Linked Open Data · Knowledge representation

1 Introduction to Multimedia Reasoning

Description logics (DL), which are formal knowledge representation languages, have been used in logic-based models for multimedia retrieval and reasoning since the 1990s [1]. They are suitable for the expressive formalization of multimedia contents and the semantic refinement of video segmentation [2]. DL-based knowledge representations, such as OWL ontologies, can serve as the basis for multimedia content analysis [3], event detection [4], high-level video scene interpretation [5], abductive reasoning to differentiate between similar concepts in image sequence interpretation [6], and constructing high-level media descriptors [7], particularly if the ontology contains not only terminological and assertional axioms (that form a knowledge base), but also a role box and a ruleset. Ontology rules make video content understanding possible and improve the quality of structured annotations of concepts and predicates [8]. Natural language processing algorithms can be used to curate the represented video concepts while preserving provenance data, and assist to achieve consistency in multimedia ontologies [9].

© Springer-Verlag Berlin Heidelberg 2016
N.T. Nguyen et al. (Eds.): ACIIDS 2016, Part I, LNAI 9621, pp. 3–12, 2016.
DOI: 10.1007/978-3-662-49381-6_1

In contrast to ontologies of other knowledge domains, video ontologies need a specific set of motion events to represent spatial changes of video scenes, which are characterized by subconcepts, multiple interpretations, and ambiguity [10]. Research results in structured video annotations are particularly promising for constrained videos, where the knowledge domain is known, such as medical videos, news videos, tennis videos, and soccer videos [11].

In spite of the benefits of multimedia reasoning in video scene interpretation and understanding, most multimedia ontologies lack the expressivity and constructors necessary for complex inference tasks [12]. To address the reasoning limitations of multimedia ontologies, the VidOnt ontology has been introduced, which exploits all mathematical constructors of the underlying expressive description logic, and features a role box and a ruleset missing from previous multimedia ontologies for automated scene interpretation and video understanding [13]. VidOnt is suitable for the knowledge representation and lightweight annotation of objects and actors depicted in videos, providing technical, licensing, and general metadata as structured data, as well as for multimedia reasoning and Linked Open Data (LOD) interlinking.

2 Formalism with Description Logics

The majority of web ontologies written in the Web Ontology Language (OWL) are implementations of a description logic [14]. Description logics are decidable fragments of first-order logic (FOL): DL concepts are equivalent to FOL unary predicates, DL roles to FOL binary predicates, DL individuals to FOL constants, DL concept expressions to FOL formulae with one free variable, role expressions to FOL formulae with two free variables, and so on. Description logics are more efficient in decision problems than first-order predicate logic (which uses predicates and quantified variables over non-logical objects) and more expressive than propositional logic (which uses declarative propositions and does not use quantifiers). A description logic can efficiently model concepts, roles, individuals, and their relationships.

Definition 1 (Concept). The concept C of an ontology is defined as a pair that can be expressed as $C = (X^C, Y^C)$, wherein $X^C \subseteq X$ is a set of attributes describing the concept, and $Y^C \subseteq Y$ is the domain of the attributes, $Y^C = \bigcup_{x \in X^c} Y_x$

Definition 2 (Role). A role is either $R \in N_R$, an inverse role R^- with $R \in N_R$, or a universal role U^1.

A core modeling concept of a description logic is the *axiom*, which is a logical statement about the relation between roles and/or concepts.

Definition 3 (Axiom). An axiom is either

---

[1] In the $\mathcal{SROIQ}$ description logic. Many less expressive DLs do not provide inverse roles, and no other ontology supports the universal role, which has been introduced in $\mathcal{SROIQ}$.

- a general concept inclusion of the form $A \sqsubseteq B$ for concepts A and B, or
- an individual assertion of one of the forms $a : C$, $(a, b) : R$, $(a, b) : \neg R$, $a = b$ or $a \neq b$ for individuals a, b and a role R, or
- a role assertion of one of the forms $R \sqsubseteq S$, $R_1 \circ \ldots \circ R_n \sqsubseteq S$, $\mathrm{Asy}(R)$, $\mathrm{Ref}(R)$, $\mathrm{Irr}(R)$, $\mathrm{Dis}(R, S)$ for roles R, R_i, S.

After determining the domain and scope of the ontology, and potential term reuse from external ontologies, the terms of the knowledge domain are enumerated, followed by the creation of the class hierarchy, the concept and predicate definitions, their relationships, and individuals. Both the first-order logic and the description logic syntax correspond to OWL, so axioms written in either syntax can be translated to the desired OWL serialization, such as Turtle, as demonstrated in Table 1.

Table 1. Description logic to OWL 2 DL translation examples

DL Axiom	Turtle syntax
$a \approx b$	`a owl:sameAs b .`
$a \not\approx b$	`a owl:differentFrom b .`
$C \sqsubseteq D$	`C rdfs:subClassOf D .`
$C(a)$	`a rdf:type C .`
$R(a, b)$	`a R b .`
$R^-(a, b)$	`b R a .`

The data model of the VidOnt ontology has been formalized in the very expressive yet decidable $\mathcal{SROIQ}^{(\mathcal{D})}$ description logic, which exploits all constructors of OWL 2 DL from concept constructors to complex role inclusion axioms, as will be discussed in the following sections.

Definition 4 ($\mathcal{SROIQ}$ ontology). A $\mathcal{SROIQ}$ ontology is a set $\mathcal{O}$ of axioms including $\varrho \sqsubseteq \mathbf{R}$ complex role inclusions, $\mathrm{Dis}(S_1, S_2)$ disjoint roles, $C \sqsubseteq D$ concept inclusions, $C(a)$ concept assertions, and $R(a, b)$ role assertions, wherein ϱ is a role chain, $R_{(i)}$ and $S_{(i)}$ are roles, C and D are concepts, and a, b individuals, such that the set of all role inclusion axioms in $\mathcal{O}$ are $\prec$ -regular for some regular order $\prec$ on roles.

2.1 Concept Constructors

The $\mathcal{SROIQ}$ description logic supports a wide range of concept expression constructors, including concept assertion, conjunction, disjunction, complement, top concept, bottom concept, role restrictions (existential and universal restrictions), number restrictions (at-least and at-most restrictions), local reflexivity, and nominals.

Definition 5 ($\mathcal{SROIQ}$ Concept Expression). A set of $\mathcal{SROIQ}$ concept expressions is defined as $\mathbf{C} ::= N_C \mid (\mathbf{C} \sqcap \mathbf{C}) \mid (\mathbf{C} \sqcup \mathbf{C}) \mid \neg\mathbf{C} \mid \top \mid \bot \mid \exists\mathbf{R}.\mathbf{C} \mid \forall\mathbf{R}.\mathbf{C} \mid \geqslant n\mathbf{R}.\mathbf{C} \mid \leqslant n\mathbf{R}.\mathbf{C} \mid \exists\mathbf{R}.Self \mid \{N_I\}$, wherein $\mathbf{C}$ represents concepts, $\mathbf{R}$ is a set of roles, and n is a non-negative integer.

2.2   Axioms

VidOnt defines terminological, assertional, and relational axioms. As you will see, constructors not exploited in previously released multimedia ontologies, in particular the role box axioms, significantly extend the application potential in data integration, knowledge management, and multimedia reasoning.

2.2.1   TBox Axioms

The concepts and roles of VidOnt have been defined in a hierarchy incorporating de facto standard structured definitions, and can be deployed in fully-featured knowledge representations in an RDF serialization, such as Turtle or RDF/XML, or as lightweight markup annotations in HTML5 Microdata, JSON-LD, or RDFa. Terminological knowledge is included in VidOnt by defining the relationship of classes and properties as subclass axioms and subproperty axioms, respectively, and specifying domains and ranges for the properties. The TBox axioms leverage constructors such as subclass relationships ($\sqsubseteq$), equivalence ($\equiv$), conjunction ($\sqcap$), and disjunction ($\sqcup$), negation ($\neg$), property restrictions ($\forall$, $\exists$), tautology ($\top$), and contradiction ($\bot$).

Definition 6 (TBox). A TBox $\mathcal{T}$ is a finite collection of concept inclusion axioms in the form $C \sqsubseteq D$ and concept equivalence axioms in the form $C \equiv D$, wherein C and D are concepts.

For example, TBox axioms can express that live action is a movie type, or narrators are equivalent to lectors, as shown in Table 2.

Table 2. Expressing terminological knowledge with TBox axioms

DL Syntax	Turtle syntax
liveAction $\sqsubseteq$ Movie	`:liveAction rdfs:subClassOf :Movie .`
remakeOf $\sqsubseteq$ basedOn	`:remakeOf rdfs:subPropertyOf :basedOn .`
Narrator $\equiv$ Lector	`:Narrator owl:equivalentClass :Lector .`

2.2.2   ABox Axioms

Individuals and their relationships are represented using ABox axioms.

Definition 7 (ABox). An ABox $\mathcal{A}$ is a finite collection of axioms of the form $x{:}D$, $\langle x, y \rangle{:}R$, where x and y are individual names, D is a concept, and R is a role. An individual assertion can be

- a concept assertion, $C(a)$
- a role assertion, $R(a, b)$, or a negated role assertion, $\neg R(a, b)$
- an equality statement, $a \approx b$
- an inequality statement, $a \not\approx b$

wherein $a, b \in N_I$ individual names, $C \in \mathbf{C}$ a concept expression, and $R \in \mathbf{R}$ a role, each of which is demonstrated in Table 3.

Table 3. Asserting individuals with ABox axioms

DL Syntax	Turtle syntax
computerAnimation(Zambezia)	`:Zambezia a :computerAnimation .`
directedBy(Unforgiven, ClintEastwood)	`:Unforgiven :directedBy :ClintEastwood .`
房仕龍 ≈ JackieChan	`:房仕龍 owl:sameIndividualAs :JackieChan .`
RobinWilliams ≢ RobbieWilliams	`:RobinWilliams owl:differentFrom :RobbieWilliams .`

2.2.3 RBox Axioms

Most multimedia ontologies define terminological and assertional axioms only, which form a knowledge base only, rather than a fully-featured ontology.

Definition 8 (Knowledge Base). A DL knowledge base κ is a pair $\langle \mathcal{T}, \mathcal{A} \rangle$ where

- $\mathcal{T}$ is a set of terminological axioms (Tbox)
- $\mathcal{A}$ is a set of assertional axioms (Abox)

Beyond Abox and TBox axioms, $\mathcal{SROIQ}$ also supports *role box* (*RBox*) axioms to collect all statements related to roles and the interdependencies between roles, which is particularly useful for multimedia reasoning.

Definition 9 (RBox). A rule box (RBox) $\mathcal{R}$ is a role hierarchy, a finite collection of generalized role inclusion axioms of the form $R \sqsubseteq S$, role equivalence axioms in the form $R \equiv S$, complex role inclusions in the form $R_1 \circ R_2 \sqsubseteq S$, and role disjointness declarations in the form Dis(R, S), wherein R and S are roles, and transitivity axioms of the form $R^+ \sqsubseteq R$, wherein R^+ is a set of transitive roles.

Some examples for role box axioms are shown in Table 4.

Table 4. Modeling relationships between roles with RBox axioms

DL Syntax	Turtle syntax
starredIn $\circ$ starredIn $\sqsubseteq$ co-starred	`:co-starred owl:propertyChainAxiom (:starredIn :starredIn) .`
Dis(parentOf, childOf)	`:x a owl:AllDisjointProperties ; owl:members (:parentOf :childOf) .`
basedOn $\circ$ basedOn $\sqsubseteq$ basedOn	`:basedOn a owl:TransitiveProperty .`

2.3 DL-Safe Ruleset

While $\mathcal{SROIQ}^{(\mathcal{D})}$, the description logic of OWL 2 DL, is very expressive, it can only express axioms of a certain tree structure, because OWL 2 DL corresponds to a decidable subset of first-order predicate logic. There are decidable rule-based formalisms, such as function-free Horn rules, which are not restricted in this regard.

Definition 10 (Rule). A *rule R* is given as $H \leftarrow B_1, \ldots, B_n(n \geq 0)$, wherein $H, B_1, \ldots, B_n$ are atoms, H is called the *head* (conclusion or consequent) and $B_1, \ldots, B_n$ the *body* (premise or antecedent).

While some OWL 2 axioms correspond to rules, such as class inclusion and property inclusion, some classes can be decomposed as rules, and property chain axioms provide rule-like axioms, there are rules that cannot be expressed in OWL 2 rules. For example, a rule head with two variables cannot be represented as a subclass axiom, or a rule body that contains a class expression cannot be described by a subproperty axiom. To add the additional expressivity of rules to OWL 2 DL, ontologies can be extended with *SWRL*[2] rules which, however, make ontologies undecidable. The solution is to apply *DL-safe rules*, wherein each variable must occur in a non-DL-atom in the rule body [15], i.e., DL-safe rules can be considered SWRL rules restricted to known individuals. DL-safe rules are very expressive and decidable at the same time.

Definition 11 (DL-safe rule). Let *KB* be a $\mathcal{SROIQ}^{(\mathcal{D})}$ knowledge base, and let N_P be a set of predicate symbols such that $N_C \cup N_{R_a} \cup N_{R_c} \subseteq N_P$. A DL-atom is an atom of the form $A(s)$, where $A \in N_C$, or of the form $R(s, t)$, where $R \in N_{R_a} \cup N_{R_c}$. A rule R is called DL-safe if each variable in R occurs in a non-DL-atom in the rule body.

As an example, assume we have axioms to define award-winning actors (1–4).

$$\text{AwardWinnerActor} \equiv \text{won}.\exists\text{Award} \tag{1}$$

$$\text{Actor}(a), \ \text{Actor}(b), \ \text{Actor}(c) \tag{2}$$

$$\text{Award}(d) \tag{3}$$

$$\text{won}(a, \ d) \tag{4}$$

Based on the axioms, a DL-safe rule can be written to infer new assertional axioms (5).

$$\text{AwardWinnerActor}(x) \leftarrow \text{won}(?x, \ ?y) \tag{5}$$

Using the above rule (5), reasoners can infer that actor *a* is an award winner (6).

$$\text{AwardWinnerActor}(a) \tag{6}$$

Without a DL-safe restriction containing special non-DL literals $O(x)$ and $O(y)$ in the rule body and the assertion of each individual, reasoners would assert that actors *a*, *b*, and *c* are award winners (7).

$$\text{AwardWinnerActor}(a), \ \text{AwardWinnerActor}(b), \ \text{AwardWinnerActor}(c) \tag{7}$$

---

[2] Semantic Web Rule Language.

3   Multimedia Reasoning

The feasibility and efficiency of automated reasoning relies on the accurate conceptualization and comprehensive description of relations between concepts, predicates, and individuals [16]. Advanced reasoning is infeasible without expressive constructors, most of which are not implemented in multimedia ontologies other than VidOnt. For example, the Visual Descriptor Ontology (VDO), which was published as an "ontology for multimedia reasoning" [17], has in fact very limited description logic expressivity (corresponding to $\mathcal{ALH}$) and reasoning potential. In the next sections we compare TBox and ABox reasoning supported by most ontologies to Rbox and rule-based reasoning not supported by any multimedia ontology except VidOnt.

3.1   Tableau-Based Consistency Checking

Most OWL-reasoners, such as FaCT++, Pellet, and RacerPro, are based on *tableau* algorithms. They attempt to construct a model that satisfies all axioms of an ontology to prove (un)satisfiability. Based on the ABox axioms, a set of elements is created, which is used to retrieve concept memberships and role assertions. Typically, the constructed intermediate model does not satisfy all TBox and RBox axioms, so the model is updated accordingly with each iteration. As a result, new concept memberships and role relationships might be generated. When a case distinction occurs, the algorithm might have to backtrack. If a state is reached where all axioms are satisfied, the ontology is considered satisfiable. OWL 2 reasoners, such as HermiT, usually use a tableau refinement based on the *hypertableau* and *hyperresolution* calculi to reduce the nondeterminism caused by general inclusion axioms [18].

To demonstrate integrity checking with reasoning, assume the following axioms:

$$\text{acts} \sqsubseteq \text{lives} \tag{8}$$

$$\text{canAct} \sqsubseteq \neg \text{DeadActor} \tag{9}$$

$$\text{Actor} \sqsubseteq \text{DeadActor} \sqcup \text{LivingActor} \tag{10}$$

$$\text{activeActor} \sqsubseteq \text{lives.Actor} \sqcap \forall \text{lives.canAct} \tag{11}$$

$$\text{activeActor}(a) \tag{12}$$

Based on the only ABox axiom (12), tableau-based reasoners would assume that a is an active actor, which would not satisfy the definition of living actors (11). Next, reasoners would introduce a new concept which logically corresponds to the Person concept. The connection between the individual (a) and the new concept (Person) is defined with the acts predicate. As a result, the definition of active actors (11) is now satisfied, however, other TBox axioms are invalidated (8 and 10). To address this issue, reasoners would introduce a lives connection between individual a and the Person concept. Finally, a case distinction is needed, because a person can be either dead (DeadActor) or alive (LivingActor). In the first case, (11) is violated because of the second part of its consequence. To address this issue, Person has to be marked with canAct, which in turn invalidates (9), meaning that Person must be ¬DeadActor. Because Person cannot be marked with both DeadActor and ¬DeadActor, the algorithm needs to backtrack. In the second case, Person is marked as LivingActor, which violates (11), so Person must be marked with canAct, which invalidates (9). Consequently, Person is marked as ¬DeadActor, which leads to a state with a knowledge representation model satisfying all axioms, upon which reasoners can conclude that the ontology is satisfiable.

3.2 RBox and Rule-Based Reasoning over Audiovisual Contents

Take a simplistic example which combines RBox reasoning with rule-based reasoning not supported by any other multimedia ontology but VidOnt, to infer statements that are not explicitly defined. Assume the following base ontology:

$$Actor(a),\ Actor(b),\ Actor(c),\ Actor(d) \tag{13}$$

$$Movie(m),\ Series(s),\ partOf(m,\ s) \tag{14}$$

$$partOf\ \circ\ starredIn\ \sqsubseteq\ co\text{-}starredWith \tag{15}$$

$$starredIn(a,\ m),\ starredIn(b,\ m),\ starredIn(c,\ m),\ starredIn(d,\ s) \tag{16}$$

Also assume the following rule:

$$starredIn(?x,\ m)\ \rightarrow\ co\text{-}starredWith(?x,\ d) \tag{17}$$

Based on the ABox and TBox axioms (13, 14, 16) and the DL-safe rule (17), reasoners can generate new object property assertions about the actors who co-starred with actor *d* (24–26):

$$co\text{-}starredWith(a,\ d),\ co\text{-}starredWith(b,\ d),\ co\text{-}starredWith(c,\ d) \tag{18}$$

Furthermore, based on the property chain axiom (15), it can be inferred that actors who starred in at least one part of a series appeared in the series (19):

$$starredIn(a,\ s),\ starredIn(b,\ s),\ starredIn(c,\ s) \tag{19}$$

The resulting axioms are automatically generated with full certainty, making the combination of complex role inclusion axioms and DL-safe rules suitable for big data implementations where manual annotation is not an option, for video cataloging to automatically generate new axioms through user or programmatic queries, and for knowledge discovery, such as identifying factors from medical videos that, when occur together, indicate a serious condition or disease.

4 Conclusions and Future Work

Multimedia ontology engineers often apply a bottom-up, top-down, or hybrid development method without mathematical grounding. The majority of mainstream domain-independent and domain-specific multimedia ontologies introduced in the past decade, with or without MPEG-7 alignment, lack complex role inclusion axioms and DL-safe rules, and are limited to terminological and assertional knowledge. Consequently, most multimedia ontologies are actually controlled vocabularies, taxonomies, or knowledge bases only, rather than fully-featured ontologies, and are not suitable for advanced multimedia reasoning. To address the above issues, concepts, roles, individuals, and relationships of the professional video production and broadcasting domains have been formally modeled using $\mathcal{SROIQ}^{(\mathcal{D})}$, one of the most expressive decidable description logics, and then the axioms translated into OWL 2. The vocabulary of the new ontology has been aligned with standards in a new concept and role hierarchy. To further improve expressivity, $\mathcal{SROIQ}^{(\mathcal{D})}$ has been combined with DL-safe rules, without sacrificing expressivity yet ensuring decidability by restricting rules to known individuals. There is ongoing work to extend this core ruleset further to reach an even higher level of reasoning power.

References

1. Meghini, C., Sebastiani, F., Straccia, U.: Reasoning about the Form and Content of Multimedia Objects. In: AAAI 1997 Spring Symposium on Intelligent Integration and Use of Text, Image, Video and Audio, pp. 89–94. AAAI Press, Menlo Park (1997)
2. Simou, N., Athanasiadis, T., Tzouvaras, V., Kollias, S.: Multimedia reasoning with f–SHIN. In: Second International Workshop on Semantic Media Adaptation and Personalization, IEEE (2007). doi:10.1109/SMAP.2007.40
3. Simou, N., Saathoff, C., Dasiopoulou, S., Spyrou, E., Voisine, N., Tzouvaras, V., Kompatsiaris, I., Avrithis, Y., Staab, S.: An ontology infrastructure for multimedia reasoning. In: Atzori, L., Giusto, D.D., Leonardi, R., Pereira, F. (eds.) VLBV 2005. LNCS, vol. 3893, pp. 51–60. Springer, Heidelberg (2006)
4. Town, C.: Ontological inference for image and video analysis. Mach. Vis. Appl. **17**(2), 94–115 (2006). doi:10.1007/s00138-006-0017-3
5. Gómez-Romero, J., Patricio, M.A., García, J., Molina, J.M.: Ontology-based context representation and reasoning for object tracking and scene interpretation in video. Expert Syst. Appl. **38**, 7494–7510 (2011). doi:10.1016/j.eswa.2010.12.118
6. Möller, R., Neumann, B.: Ontology-based reasoning techniques for multimedia interpretation and retrieval. In: Semantic Multimedia and Ontologies. Springer, London (2008). doi:10.1007/978-1-84800-076-6_3
7. Elleuch, N., Zarka, M., Ammar, A.B., Alimi, A.M.: A fuzzy ontology-based framework for reasoning in visual video content analysis and indexing. In: 11th International Workshop on Multimedia Data Mining (MDMKDD 2011), San Diego (2011). doi:10.1145/2237827.2237828

8. Jaimes, A., Tseng, B.L., Smith, J.R.: Modal keywords, ontologies, and reasoning for video understanding. In: Bakker, E.M., Lew, M.S., Huang, T.S., Sebe, N., Zhou, X.S. (eds.) Image and Video Retrieval. LNCS, vol. 2728, pp. 248–259. Springer, Heidelberg (2003). doi:10.1007/3-540-45113-7_25

9. Dasiopoulou, S., Heinecke, J., Saathoff, C., Strintzis, M.G.: Multimedia reasoning with natural language support. In: IEEE Sixth International Conference on Semantic Computing, pp. 413–420. IEEE (2007). doi:10.1109/ICSC.2007.28

10. D'Odorico, T., Bennett, B.: Automated reasoning on vague concepts using formal ontologies, with an application to event detection on video data. In: 11th International Symposium on Logical Formalizations of Commonsense Reasoning (COMMONSENSE 2013), Ayia Napa (2013)

11. Ballan, L., Bertini, M., Del Bimbo, A., Serra, G.: Semantic annotation of soccer videos by visual instance clustering and spatial/temporal reasoning in ontologies. Multimedia Tools Appl. **48**, 313–337 (2010). doi:10.1007/s11042-009-0342-4

12. Sikos, L.F., Powers, D.M.W.: Knowledge-driven video information retrieval with LOD: from semi-structured to structured video metadata. In: Eighth Workshop on Exploiting Semantic Annotations in Information Retrieval (ESAIR 2015), Melbourne (2015). doi:10.1145/2810133.2810141

13. VidOnt: The Video Ontology. http://vidont.org

14. Sikos, L.F.: Mastering Structured Data on the Semantic Web: From HTML5 Microdata to Linked Open Data. Apress Media, New York (2015). doi:10.1007/978-1-4842-1049-9

15. Motik, B., Sattler, U., Studer, R.: Query answering for OWL-DL with rules. J. Web Semant. **3**(1), 41–60 (2005). doi:10.1016/j.websem.2005.05.001

16. Hitzler, P., Krötzsch, M., Rudolph, S.: Foundations of Semantic Web Technologies. CRC Press, Boca Raton (2009)

17. Simou, N., Tzouvaras, V., Avrithis, Y., Stamou, G., Kollias, S.: A visual descriptor ontology for multimedia reasoning. In: 6th International Workshop on Image Analysis for Multimedia Interactive Services, Montreux (2005)

18. Glimm, B., Horrocks, I., Motik, B., Stoilos, G., Wang, Z.: HermiT: An OWL 2 Reasoner. J. Autom. Reasoning **53**(3), 245–269 (2014). doi:10.1007/s10817-014-9305-1

Finding Similar Clothes Based on Semantic Description for the Purpose of Fashion Recommender System

Dariusz Frejlichowski[1]([✉]), Piotr Czapiewski[1], and Radosław Hofman[2]

[1] Faculty of Computer Science and Information Technology,
West Pomeranian University of Technology, Zolnierska 49, 72-210 Szczecin, Poland
{dfrejlichowski,pczapiewski}@wi.zut.edu.pl
[2] FireFrog Media sp. z o.o., Jeleniogorska 16, 60-179 Poznań, Poland
radekh@fire-frog.pl

Abstract. The fashion domain has been one of the most growing areas of e-commerce, hence the issue of facilitating cloth searching in fashion-related websites becomes an important topic of research. The paper deals with measuring the similarity between items of clothing and between complete outfits, based on the semantic description prepared by users and experts according to a previously developed fashion ontology. Proposed approach deals with different types of attributes describing clothes and allows for calculating similarity between the whole outfits in a domain-aware manner. Exemplary results of experiments performed on real clothing datasets are presented.

Keywords: Clothes similarity · Object retrieval · Recommender systems

1 Introduction

Recently the fashion domain has been one of the most dynamically growing areas of e-commerce and social networking. Given the popularity of the topic, especially among women, number of potential customers is vast.

At the same time, buying clothes on-line poses a bigger problem, than in other areas of e-commerce. First, lack of physical contact with the merchandise is for many people a discouraging factor. Second, due to the great variety of clothing styles and certain difficulties in clearly and unambigously describing them, it is quite difficult to search for desirable clothes. Futhermore, in real-life shopping for clothes it is quite common to ask for advice of a shopping assistant.

Due to the above factors, the analysis and representation of clothes became an important topic of research within several subareas of computer science, including computer vision, knowledge representation, information retrieval and recommender systems. The method presented in this paper is intended to be incorporated in the fashion recommender system being currently under development. The ultimate goal of the system is to learn the individual style of the user

© Springer-Verlag Berlin Heidelberg 2016
N.T. Nguyen et al. (Eds.): ACIIDS 2016, Part I, LNAI 9621, pp. 13–22, 2016.
DOI: 10.1007/978-3-662-49381-6_2

(based on the outfits owned by them and on their clothing preferences expressed in other ways) and then to recommend some matching new clothes.

An important part of such system is comparing clothes in terms of fashion and visual characteristics. The proposed similarity measure aims at expressing the level of resemblance between two complete outfits (each consisting of any number of garments), and can be used in two basic usage scenarios:

– directly finding clothes similar to a given reference outfit;
– clustering outfits in order to recognize distinguishable styles.

The rest of the paper is organized as follows. Section 2 contains the review of relevant literature dealing with similar problems. Section 3 presents proposed method of representing clothing items and outfits. In Sect. 4 the proposed similarity measures is described, both for single clothing items and for complete outfits. Section 5 presents some experimental results. Section 6 concludes the paper.

2   Previous Works

Analysing various fashion-related data just recently became an active area of research. Most of the work has been done in the field of computer vision, focusing on analysing images containing clothes and outfits.

Zhang et al. [7] proposed a human-computer interaction system called a responsive mirror, intended to be used as an interactive tool supporting shopping in a real retail store. Cloth type and some attributes are automatically extracted from the image and similar clothes are looked for in a database. Di et al. [3] proposed a method to recognize clothes attributes from images, limiting the area of interest to one category of clothes (coats and jackets). A similarly limited approach can be found in [1], where only upper body clothes are analysed. The main focus in all the above research is on computer vision aspects and the semantic description of clothing is very limited. Also, none of the above deals with the whole outfits, only with single items of clothing.

The topic of similarity between items of clothing has been dealt with in [2,6]. In [2] the issue of similarity is taken only from computer vision perspective, no semantic description of clothing is considered. In [6] authors propose a tabular structure to describe cloth characteristics and then propose a procedure to calculate the similarity. Their proposed structure lacks the flexibility that can be achieved using the fashion ontology described in the next section of this paper. Also, both above mentioned papers deal only with single elements of clothing, no means of analysing complete outfits is proposed.

General similarity measures intended for usage in retrieval tasks are discussed in [4,5]. However they do not deal with domain-specific issues related to fashion datasets.

3   Semantic Description of Clothing

For the purpose of creating fashion recommender system, a lightweight fashion ontology has been developed. Then some clothing images have been collected and semantically annotated in accordance with the ontology.

3.1 Clothing Ontology

The four most important concepts of the developed fashion ontology are:

- cloth type (a class representing a type of clothing, e.g. trousers, shirt etc.);
- cloth attributes (describing a particular cloth item, e.g. sleeve length, collar type, colour etc.);
- cloth item (a single, particular instance of clothing, described by some attributes);
- cloth set (a complete outfit; a composition of clothes, intended to be worn together).

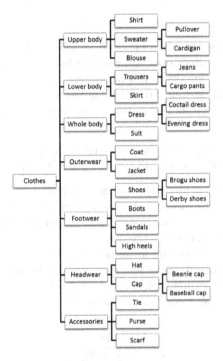

Fig. 1. Excerpt from the cloth type taxonomy defined within the fashion ontology.

A cloth type taxonomy has been defined (see Fig. 1), arranging all cloth types into a hierarchy, the first level of which corresponds to most general categories of clothing – 7 main categories were defined:

- upper body (e.g. shirt, sweater),
- lower body (e.g. jeans, skirt),
- whole body (e.g. dress, suit),
- footwear (e.g. boots, high-heels),
- headwear (e.g. hat, baseball cap),
- outerwear (e.g. coat, leather jacket),
- accessories (e.g. purse, tie, scarf).

Cloth attributes are defined on different levels – some are common to all cloth types (e.g. colour), some to only certain cloth categories (e.g. sleeve length and cut, dress cut, heels height etc.). The attributes are divided into to groups:

– fashion attributes – pertaining to particular characteristics of clothes in terms of shape or cut (e.g. sleeve length, dress style, heel type etc.);
– visual attributes – pertaining to the purely visual, non-fashion specific characteristics (colour, brightness, vividness, pattern).

Other concepts defined within the ontology cover different types of materials and possible usage contexts for a given outfit.

3.2   Clothing Dataset

The data under consideration is organized as follows. Each item of clothing is described using a set of attributes, appropriate for a given cloth type. However, none of the attributes is mandatory, hence different items can be described using different sets of attributes even if they belong to the same category and/or type.

An example of a semantic description of one item is given below:

– type – shirt;
– attributes – colour: sky blue; pattern: stripes; brightness: 0.8; vividness: 0.1; sleeve length: long; sleeve cut: slim; collar type: Italian; material: cotton.

Given the cloth type "shirt", the category "upper body" can be inferred.

The ontology contains mostly categorical attributes (e.g. sleeve length) and several real-valued ones (e.g. brightness). The colour attribute receives special treatment, as will be explained in the next section.

The most important concept forming the dataset is cloth set (or outfit), consisting of several cloth items of any type. No constraints have been imposed on the contents of a set. Some typical sets contain the following combinations of elements (in terms of the general cloth category):

Fig. 2. Exemplary outfit from the obtained fashion dataset, containing two reference items – a red dress and a denim jacket (Color figure online).

- 1x upper body, 1x lower body, 1x outerwear, 1x footwear (e.g. shirt, trousers, jacket, shoes);
- 1x whole body, 1x footwear, 2x accessories (e.g. dress, high-heels, purse, scarf);
- 2x upper body, 1x lower body, 1x headwear (e.g. shirt, pullover, jeans, hat).

Given the extreme variability of outfit composition in real life, the flexibility achieved using the above approach is crucial to properly represent fashion information. However, this level of flexibility poses a problem when trying to compare outfits or to search for an outfit similar to a given one.

4   Clothes Similarity

For the purpose of comparing clothes two similarity measures need to be defined. First, we introduce means to compare two single cloth items. Then, based on this, the measure of similarity for cloth sets (outfits) is defined.

4.1   Cloth Item Similarity

When comparing cloth items, two fundamental pieces of information need to be taken into consideration: cloth type and cloth attributes.

Cloth Types Similarity. First, we need to deal with cloth type. Two pieces of clothing of distinctively different types (e.g. shirt and shoes) must be considered completely different, regardless of their specific fashion attributes. Two items of the same type (e.g. two shirts) should be considered similar, with the value of similarity depending on the values of detailed cloth attributes. However, there are certain items of clothing, which could be considered similar, even if the types don't match exactly (e.g. shirt and blouse, jeans and trousers). Hence, we introduce a similarity index μ_t for two cloth types t_1 and t_2:

$$\mu_t(t_1, t_2) = \begin{cases} 1 & \text{if } t_1 = t_2, \\ 0.5 & \text{if } t_1 \text{ is ancestor of } t_2, \\ 0 & \text{otherwise.} \end{cases} \tag{1}$$

The relations within cloth types taxonomy are utilized in order to detect similar cloth types. More detailed ways of assigning similarity index based on the distance in hierarchy tree were tested. However, a simplified approach of assigning 0.5 value to types connected by ancestor/descendant relationship has been found to be sufficient and significantly faster and easier to implement.

Fashion Attributes Similarity. As for the cloth attributes, both categorical and real valued attributes must be uniformly taken care of. Bearing the above in mind, we introduce the following measure of similarity.

Let i denote an item of clothing, described by a set of attributes $p_1, p_2, ..., p_n$. Given two values v_1 and v_2 of a particular attribute p_i, we introduce a similarity index $\mu_a(p_i, v_1, v_2)$, defined differently for different types of attributes:

$$\mu_a(p_i, v_1, v_2) = \begin{cases} \mu_{ac}(v_1, v_2) \text{ if } p_i \text{ is a categorical attribute,} \\ \mu_{an}(v_1, v_2) \text{ if } p_i \text{ is a numerical attribute,} \end{cases} \quad (2)$$

where μ_{ac} and μ_{an} denote similarity indices for categorical and numerical (real-valued, normalised) attributes respectively:

$$\mu_{ac}(v_1, v_2) = \begin{cases} 1 \text{ if } v_1 = v_2, \\ 0 \text{ if } v_1 \neq v_2, \end{cases} \quad (3)$$

$$\mu_{an}(v_1, v_2) = 1 - |v_1 - v_2|. \quad (4)$$

Colour Values and Colour Sets Similarity. A special treatment is given to the colour attribute, for two reasons. First, the colour is described by categorical values (red, blue, yellow and so on), but the similarity between two colours can be calculated if the names are mapped into RGB values. Second, colour attribute can take several values for the same item (e.g. a shirt is red and white). In order to compare colours of two cloth items, two concepts need to be introduced: similarity between two colours and similarity between two sets of colours.

Assuming that colour v_i is described in RGB space as (r_i, g_i, b_i), the similarity index for two values of colour is defined as follows:

$$\mu_c(v_1, v_2) = 1 - \frac{\sqrt{w_r(r_1-r_2)^2 + w_g(g_1-g_2)^2 + w_b(b_1-b_2)^2}}{255}, \quad (5)$$

where w_r, w_g and w_b denote weights applied to particular RGB components. Given the characteristics of human perception, the following values are suggested:

$$w_r = 0.2989, \quad w_g = 0.587, \quad w_b = 0.114. \quad (6)$$

Assuming that two items of clothing i_1 and i_2 are described using two colour sets c_1 and c_2, the following procedure for determining similarity between c_1 and c_2 is performed:

1. for each colour pair $(v_i, v_j) \in c_1 \times c_2$ calculate $\mu_c(v_i, v_j)$;
2. for each colour $v_i \in c_1$ find the most similar colour $v_{1,i}^{sim} \in c_2$;
3. for each colour $v_j \in c_2$ find the most similar colour $v_{2,j}^{sim} \in c_1$;
4. calculate mean values μ_1 and μ_2 of similarity indexes found in two previous steps, according to the following formulas:

$$\mu_1 = \frac{1}{|c_1|} \sum_{i=1}^{|c_1|} v_{1,i}^{sim}, \quad \mu_2 = \frac{1}{|c_2|} \sum_{j=1}^{|c_2|} v_{2,j}^{sim}, \quad (7)$$

where $|c_i|$ denotes set cardinality;
5. take the smaller of μ_1 and μ_2 as the final similarity index.

Aggregated Cloth Items Similarity. The similarity measure between two cloth items is loosely based on Jaccard index, combined with the above similarity index for attributes, and defined as follows. For two items i_1 and i_2 being compared, let t_1 and t_2 denote cloth types, and P_1 and P_2 – the respective attribute sets. The total number of attributes and the number of common attributes is determined, then similarity indices are calculated for common attributes. The final measure of similarity between items is defined as:

$$s(i_1, i_2) = \frac{\mu_t(t_1, t_2) + \sum_{p \in P_1 \cap P_2} \mu_a(p, v_{p1}, v_{p2})}{1 + |P_1 \cup P_2|} \tag{8}$$

where v_{p1} and v_{p2} denote the value of p attribute for i_1 and i_2 respectively and $|\cdot|$ denotes set cardinality.

The above similarity measure takes into consideration both cloth type and specific fashion attributes of any type.

4.2   Cloth Set Similarity

Based on the cloth item similarity measure, the cloth set similarity measure has been developed. Let us first consider a composition of cloth set and possibility of comparing elements belonging to two sets. As mentioned before, no constraints exist as to what types of clothes can coexist within an outfit. Hence, two outfits being compared might consists of different number of items, belonging to different types, and even to different general clothing categories (see examples in Sect. 2). In the proposed approach we compare pairs of items from both sets belonging to the same general categories. For example, let's assume two sets consisting of:

– shirt, pants, shoes;
– sweater, jeans, hat.

In such a case the shirt would be compared to the sweater (category: upper body), the pants to the jeans (category: lower body), while shoes and hat have no match and are ignored in attributes comparison.

The final similarity measure consists of two components. The first one measures the similarity of composition in terms of types of clothing only. The second one masures the similarity of matching items in terms of detailed fashion attributes. Using weights the importance of both components may be controlled.

Let us introduce the following notation:

– o_1 and o_2 – two outfits (cloth sets) being compared;
– M – a set containing all matching (i_i, i_j) item pairs, where $i_i \in o_1$, $i_j \in o_2$, and both i_i and i_j belong to the same general category;
– C_{all} – a set containing all general categories present in any of o_1 and o_2;
– C_{common} – a set of common categories, present in both of o_1 and o_2.

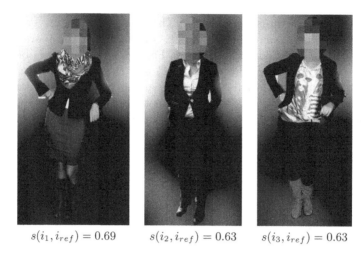

$$s(i_1, i_{ref}) = 0.69 \qquad s(i_2, i_{ref}) = 0.63 \qquad s(i_3, i_{ref}) = 0.63$$

Fig. 3. Examplary results of searching for single items with semantic attributes similar as the reference jacket i_{ref}. The values of similarity index $s(i_i, i_{ref})$ are given below each image.

Then the final similarity measure between outfits o_1 and o_2 can be defined as:

$$s(o_1, o_2) = w_1 \frac{|C_{common}|}{|C_{all}|} + w_2 \frac{\sum_{(i_i, i_j) \in M} s(i_i, i_j)}{|M|}, \qquad (9)$$

where $s(i_i, i_j)$ denotes similarity between items, introduced in the previous section, and $|\cdot|$ denotes a cardinality of set. By default it is assumed that:

$$w_1 = w_2 = 0.5.$$

However, the validity of these values depends on the dataset under examination, especially on the variance in outfit composition in terms of cloth types/categories and on the completeness of semantic attributes data.

5   Experimental Results

The proposed method for comparing outfits has been tested using a dataset collected from various fashion-related web sites, especially fashion blogs. The images depicting people wearing the outfits were collected and then annotated using the developed fashion ontology. Annotations were prepared by analysts based on users' descriptions and visual examination.

The test dataset contains 800 outfits (500 women, 300 men) of varying composition and characteristics. First, the similarity measure for single items was tested. Some clothing items were selected randomly and for each of them the most similar items were searched for. Exemplary results of finding items similar to the reference one (to the jacket visible on Fig. 2) are shown in Fig. 3.

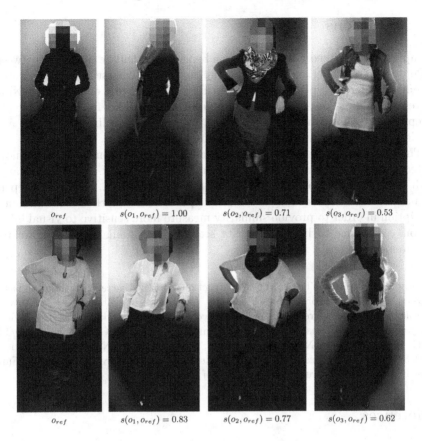

o_{ref} $s(o_1, o_{ref}) = 1.00$ $s(o_2, o_{ref}) = 0.71$ $s(o_3, o_{ref}) = 0.53$

o_{ref} $s(o_1, o_{ref}) = 0.83$ $s(o_2, o_{ref}) = 0.77$ $s(o_3, o_{ref}) = 0.62$

Fig. 4. Examplary results of searching for outfits with similar semantic attributes. The first image in each row is a reference outfit, the following three outfits are found to be similar.

Next, the similarity measure for complete outfits was tested. For each outfit in the dataset the most similar outfits were searched for. Some representative results are shown in Fig. 4.

The obtained results show, that the results of applying the proposed similarity measure to searching for similar single items is fully satisfactory. As for seeking similar outfits, the results are satisfactory in terms of fashion attributes (length, cut, types of clothes etc.). However, the similarity in terms of colours and visual patterns, as perceived by users, is not very good. However, this should be expected, given that when calculating the similarity index, the fashion attributes have more impact on the outcome, than visual attributes (colour, brightness, pattern).

Further research is currently in progress aimed at incorporating certain computer vision methods in the similarity measure, in order to find outfits which are similar both in terms of semantic description and subjectively perceived visual characteristics.

6   Conclusions

The approach described in this paper can be used to calculate similarity between single items of clothing and between complete outfits, consisting of any number of items of different types and characteristics. The semantic description used in the research, based on a developed fashion ontology, allows for great flexibility in representing cloth characteristics, both in terms of possible continuous and seamless extension, and dealing with incomplete description.

The experiments performed on real data gathered from various fashion-related social networking websites, confirmed the validity of the proposed approach and indicated possible areas of further enhancement. The next step in planned research will be integration of computer vision methods and semantic description in order to provide similarity measure more sensitive to visual information, and hence to better facilitate searching for similar and/or matching cloth items and outfits.

Acknowledgements. The project "Construction of innovative recommendation based on users styles system prototype: FireStyle" (original title: "Zbudowanie prototypu innowacyjnego systemu rekomendacji zgodnych ze stylami uytkownikw: FireStyle") is co-founded by European Union (project number: UDA-POIG.01.04.00-30-196/12, value: 14.949.474,00 PLN, EU contribution: 7.879.581,50 PLN, realization period: 01.2013–10.2014). European funds – for the development of innovative economy (Fundusze Europejskie – dla rozwoju innowacyjnej gospodarki).

References

1. Chen, H., Gallagher, A., Girod, B.: Describing clothing by semantic attributes. In: Fitzgibbon, A., Lazebnik, S., Perona, P., Sato, Y., Schmid, C. (eds.) ECCV 2012, Part III. LNCS, vol. 7574, pp. 609–623. Springer, Heidelberg (2012)
2. Chen, Q., Li, J., Liu, Z., Lu, G., Bi, X., Wang, B.: Measuring clothing image similarity with bundled features. Int. J. Cloth. Sci. Technol. **25**(2), 119–130 (2013)
3. Di, W., Wah, C., Bhardwaj, A., Piramuthu, R., Sundaresan, N.: Style finder: Fine-grained clothing style detection and retrieval. In: 2013 IEEE Conference on Computer Vision and Pattern Recognition Workshops (CVPRW), pp. 8–13. IEEE (2013)
4. Finnie, G., Sun, Z.: Similarity and metrics in case-based reasoning. Int. J. Intell. Syst. **17**(3), 273–287 (2002)
5. Liao, T.W., Zhang, Z., Mount, C.R.: Similarity measures for retrieval in case-based reasoning systems. Appl. Artif. Intell. **12**(4), 267–288 (1998)
6. Liu, Z., Wang, J., Chen, Q., Lu, G.: Clothing similarity computation based on tlac. Int. J. Cloth. Sci. Technol. **24**(4), 273–286 (2012)
7. Zhang, W., Begole, B., Chu, M., Liu, J., Yee, N.: Real-time clothes comparison based on multi-view vision. In: Second ACM/IEEE International Conference on Distributed Smart Cameras, ICDSC 2008, pp. 1–10. IEEE (2008)

An Influence Analysis of the Inconsistency Degree on the Quality of Collective Knowledge for Objective Case

Van Du Nguyen[✉] and Ngoc Thanh Nguyen

Department of Information Systems,
Faculty of Computer Science and Management,
Wroclaw University of Technology, Wroclaw, Poland
{van.du.nguyen, ngoc-thanh.nguyen}@pwr.edu.pl

Abstract. In collective knowledge determination, objective case is the case which the real knowledge state of a subject in the real world exists independently of the knowledge states given by autonomous units. With inconsistency we have in mind some conflicts between the knowledge states of a collective. Besides, the measure of the quality of collective knowledge is based on the distance from the collective knowledge to the real knowledge state. In this work we investigate the influence of the inconsistency degree of a collective on the quality of collective knowledge by increasing the number of collective members. Based on the Euclidean space, some criteria for adding members to a collective and simulating the real knowledge state of a subject in the real world are proposed. Through experiments analysis, adding members causes decreasing the inconsistency degree of a collective is not always helpful in making the quality of collective knowledge to be better. Instead, the quality of collective knowledge tends to be better if added members are closer to the real knowledge state.

Keywords: Collective knowledge · Inconsistency knowledge · Knowledge integration

1 Introduction

The problem of using multi-autonomous units such as agents or experts for solving the common real-life problems is being more and more popular. Although this phenomenon seems to be useful in giving a proper solution [1, 2], it also causes conflict because autonomous units can give different (or inconsistent) of knowledge states about a subject in the real world.

A collective is understood as a set of knowledge states given by autonomous units on the same subject. The knowledge states of a collective present autonomous units' knowledge state about the subject and each of them to some degree reflects the real knowledge state of the subject because of incompleteness and uncertainty. The collective knowledge of a collective is considered as the consensus of its knowledge states [3] and determined from the basic of the knowledge states of the collective. In addition, the real knowledge state exists but is not known by autonomous units when

© Springer-Verlag Berlin Heidelberg 2016
N.T. Nguyen et al. (Eds.): ACIIDS 2016, Part I, LNAI 9621, pp. 23–32, 2016.
DOI: 10.1007/978-3-662-49381-6_3

they are requested for giving knowledge states about the subject. It can be classified into objective or subjective case. Concretely, we consider the following table:

According to Table 1, the main difference between these cases is due to the existing of the real knowledge state. For subjective case, the real knowledge state exists dependently on the knowledge states of a collective such as in election committee or group decision making, and etc. However, for objective case, the existing of the real knowledge state is independently of the knowledge states of a collective such as the problem of weather forecast for a next day or the currency rate in a month and etc.

Table 1. Difference between Objective and Subjective cases

Subjective case	Objective case
• The real knowledge state is dependent on the knowledge states of the collective. • The collective knowledge and the real knowledge state are identical.	• The real knowledge state is independent of the knowledge states of the collective. • The collective knowledge reflects the real knowledge state to some degree.

The problem of determining the influence of the inconsistency degree of a collective on the quality of collective knowledge is a very important issue. On one hand, inconsistency plays an important role in giving a proper solution such as in [1, 2]. On the other hand, it can be undesired one such as in [4–7]. So far, the analysis of the quality of collective knowledge for objective case has been investigated in [8–10]. The collective knowledge is better than the worst knowledge state. In addition, if the distances from all knowledge states to the real state are identical, then the collective knowledge is the best one in compared with all knowledge states [9, 10] of the collective. In [8] the authors have investigated the influence of the number of collective members on the quality of collective knowledge. Through experiments analysis, which collective has more members will give better collective knowledge. Also in this work, if adding the collective knowledge of a collective to that collective, then the quality of collective knowledge is unchanged. However, the criteria for adding members and generating the real state are slightly simple. Besides, the authors have not investigated the relationship between the inconsistency degree of a collective and the quality of collective knowledge. From these limitations, in this work, we will investigate the influence of the inconsistency degree (in case of adding members to a collective with assumption the cost for inviting/implementing more autonomous units is acceptable) on the quality of collective knowledge. Concretely, adding members cause decreasing the inconsistency degree of a collective will cause the quality of collective knowledge to be better or worse? In other words, whether the collective knowledge of collectives with smaller inconsistency degree is better than those of collectives with higher inconsistency degree. The better collective knowledge is understood as the closer one to the real knowledge state. That is the main contribution of the paper.

The remaining part of the paper is organized as follows. Some basic notions related to consensus choice, collective of members' knowledge states, and collective knowledge determination are presented in Sect. 2. Section 3 presents methods for measuring the inconsistency degree of a collective and the quality of collective knowledge.

In Sect. 4 the method for the investigated problem is proposed. Section 5 presents the experimental results and theirs evaluation. Finally, Sect. 6 points out some conclusions and future works.

2 Background

2.1 Consensus Choice

Generally, a consensus choice is considered as a general agreement from the participations when they are not agreed on some matters [11]. A totally agreed situation as in [12] is not required because it's difficult to achieve in the real world. Consensus models have been proved to be useful in solving conflicts which arising in the knowledge integration process [11, 13, 14]. One of the most oldest consensus models was worked out for solving conflicts in which the conflicts content may be represented by two well-known problems: the Alternatives Ranking Problem, and Committee Election Problem [15]. Other models which have been investigated in [16–18] serve for solving conflicts with different structures such as: n-trees, semi-lattices, partitions or multi-attribute and multi-value. In this work, a consensus model which serves for solving conflicts in multi-attribute and multi-value structure is represented.

2.2 Collective of Knowledge States

In this work, we assume that U is a set of objects and each of them represents the potential of knowledge state referring to a subject in the real world. The members of U can be, for example, logic expressions, tuples, etc. By 2^U we denote the set of all subsets of U and $\prod_k(U)$ we denote the set of all k-member subsets with repetitions of set U (where k∈N, N is the set of natural numbers).

$$\prod(U) = \bigcup_{k=1}^{\infty} \prod_k(U)$$

Thus $\prod(U)$ is the set of all non-empty finite subsets with repetitions of set U. A set $X \in \prod(U)$ represents the knowledge states given by collective members on the same subject. Where each member $x \in X$ represents a knowledge state in a collective and X is also called a collective. The members in set U can be inconsistent with each other.

Based on the Euclidean space, a collective has the following form:

$$X = \{x_i = (x_{i1}, x_{i2}, \ldots x_{im}) : i = 1, 2, \ldots, n\}$$

where $x_{ik} \in R, k = 1, 2, \ldots, m$. It is a multi-dimensional vector.

2.3 Knowledge of a Collective

So far, a lot of algorithms which have been developed for collective knowledge determination with different knowledge representations such as: ordered partitions [19], relational structure [10, 20], hierarchical structure [21], ontology [10], and etc. As aforementioned, the collective knowledge of a collective is considered as the consensus of members' knowledge states and these knowledge states are taken into account for collective knowledge determination. There exist two consensus-based criteria for collective knowledge determination [10] such as: "*1-Optimality*" and "*2-Optimality*" or O_1 and O_2 for short. The collective knowledge of a collective is determined based on:

- criterion O_1 if:$d(x^*, X) = \min_{y \in U} d(y, X)$
- criterion O_2 if:$d^2(x^*, X) = \min_{y \in U} d^2(y, X)$

where x^* represents the collective knowledge of collective X. $d(x^*, X)$ represents the sum of distances from x^* to the knowledge states of collective X and $d^2(x^*, X)$ represents the sum of squared distances from x^* to the knowledge states of collective X. In case the collective knowledge (x^*) satisfies criterion O_2, then it has the following form:

$$x^* = \frac{1}{n}\left(\sum_{i=1}^{n} x_{i1}, \sum_{i=1}^{n} x_{i2}, \ldots, \sum_{i=1}^{n} x_{im}\right)$$

3   Inconsistency Degree and Quality of Collective Knowledge

3.1   Inconsistency Degree Measure

According to [10], the inconsistency degree presents the coherence and density levels of the knowledge states of collective members. Its measure is based on the distances between the knowledge states or from the collective knowledge to the knowledge states of collective members. The function for measuring the inconsistency degree of a collective is defined as follows:

Definition 1.   *The inconsistency degree of a collective is defined by function c as follows:*

$$c : \prod(U) \rightarrow [0, 1]$$

where each $X \in \prod(U)$ is a collective with repetition of members from universe U.

In [10] the authors have defined five functions serve for measuring the inconsistency degree of a collective. Some of them are considered as the representatives for the inconsistency degree of a collectives such as c_3, c_4, and c_5. Thus, we investigate determining the influence of these inconsistency functions on the quality of collective knowledge. With function c_3, the inconsistency degree of a collective takes into account the average of distances between the members of a collective. In this case, the smaller the average value, the better the inconsistency degree (higher consistency). Its definition is as follows:

$$c_3(X) = d_{mean}(X) = \begin{cases} \dfrac{2}{n(n-1)} \displaystyle\sum_{i=1}^{n-1}\sum_{j=i+1}^{n} d(x_i, x_j), & \text{for } n > 1 \\ 0, & \text{for } n = 1 \end{cases}$$

where $d(x_i, x_j)$ is the distance between x_i and x_j.

Function c_4 and function c_3 are mutually dependent. Its measure is taken into account the total average distance of all distances between the knowledge states of collective X. Its definition is as follows:

$$c_4(X) = d_{t\_mean}(X) = \frac{2}{n(n+1)} \sum_{i=1}^{n-1}\sum_{j=i+1}^{n} d(x_i, x_j) = \frac{(n-1)}{(n+1)} d_{mean}(X)$$

The third function is described by the minimal average of distances between a member of universe U and the knowledge states of collective X. This function reflects the coherence level of a collective. The member satisfying the minimal average of distances between a member of set U and the member of collective X is considered as the representative for that collective. Its definition is as follows:

$$c_5(X) = d_{min}(X) = \frac{1}{n} minD(X)$$

where $D(X) = \{d(x^*, X) : x^* \in U\}$ is the sum of distances from a member x^* to the knowledge states of collective X. That is:

$$d(x^*, X) = \sum_{i=1}^{n} d(x^*, x_i)$$

3.2 Quality of Collective Knowledge

As aforementioned, for objective case, there exists the real knowledge state of a subject in the real world. It is independently of the knowledge states of a collective and the collective knowledge reflects the real knowledge state of the subject to some degree. The measure of the quality is based on the difference between the collective knowledge and the real knowledge state [9, 10]. Thus, the best collective knowledge is understood as the closest one to the real knowledge state and the quality of collective knowledge is defined as follows:

$$Q_X = 1 - d(x^*, r)$$

where x^* represents the collective knowledge of collective X and r represents the real knowledge state.

4 The Proposed Method

In this section, a method for determining how the inconsistency degree influences the quality of collective knowledge for objective case by increasing the number of collective members is proposed. For this aim, we start from a collective with n members. Then the number of collective members is increased up to the maximal number of collective members (each step by 1 member). The criterion of an added member is to cause the inconsistency degree of the collective to be down. Meaning the inconsistency of a new collective (after adding 1 member) is smaller than that of the collective. The parameters of the investigated problem is presented in the following table (Table 2):

Table 2. The parameters of the investigated problem

Collective	Number of collective members	Knowledge of collective
X_1	n	x_1^*
X_2	n + 1	x_2^*
...	...	...
X_k	n + k	x_k^*

The knowledge states of a collective are randomly generated. In this work, the set U is all of points in the circle with center (0, 0) and radius 1.0. All of points represent autonomous unit's knowledge states about the same subject in real world. A set of n knowledge states belonging to the set U is called a collective. The procedure of the investigated problem is described follows:

The most important tasks in the procedure are generating a new member in step 2 and the real state in step 4. Because of randomly generating reason, adding a new member can cause the collective knowledge to be loser to or further from the real knowledge state and the inconsistency degree of the collective is also smaller or higher. Generally, we can't give any conclusion about the relationship between the inconsistency degree of the collective after adding a member and the quality of collective knowledge. Instead, in this work, the criterion for added members is decreasing the inconsistency degree of a collective. Meaning in step 2, every added member causes the inconsistency degree of the collective to be down. Concretely, we consider the following theorem about the relationship between an added member and the other members of a collective.

Theorem 1. For two collectives X, Y as follows:

$$X = \{x_1, x_2, \ldots, x_n\}, Y = \{X \dot{\cup} \{y\}\}$$

Let x^* be the knowledge of collective X, then we have the following statements:

(a) If $c_3(Y) \leq c_3(X)$, then $d(y, X) \leq \dfrac{\sum\limits_{i=1}^{n} d(x_i, X)}{(n-1)}$.

(b) If $c_5(Y) \leq c_5(X)$, then $d(x^*, y) \leq \frac{d(x^*, X)}{n}$.

Proof. This theorem is proved based on the Euclidean space and it is not included because of page limit.

Corollary 1. For two collectives X, Y as in Theorem 1. Let y^* be the knowledge of collective Y. Then the following statements are true:

(a) If $d(x^*, y) \leq \frac{d(x^*, X)}{n}$, then $d(x^*, y) \leq \frac{d(x^*, X)}{n} \leq \frac{d(y, X)}{n} \leq \frac{\sum\limits_{i=1}^{n} d(x_i, X)}{n(n-1)}$.

(b) If $d(x^*, y) \leq \frac{d(x^*, X)}{n}$, then $d(y^*, y) \leq d(x^*, y)$

Proof. This theorem is proved based on the Euclidean space and it is not included because of page limit.

In this work, the condition in Theorem 1(a) is used for generating added members. As aforementioned, the criterion for simulating the real state is also an important task because it exists independently of the set of knowledge states of a collective and it's not known at the moment when the autonomous units are invited for giving their knowledge states about a subject in the real world. Thus, the simulation of the real state is done when the collective reaches the maximal number of collective members. The criterion is based on the Theorem 1(b) and it is described as follows:

$$d(x^*, y) \leq \frac{d(x^*, X)}{n}$$

Meaning the real knowledge state is a member which is not too far from the collective knowledge of a collective. In our approach the criterion for the real knowledge state simulation is based on: the collective with n members; the collective with $(n + k)/2$ members; the collective with maximal $(n + k)$ members.

5   Experimental Results

From the procedure in Fig. 1 and the criteria for generating new members as well as simulating the real knowledge state, in this section some experimental results with collectives (from 3 to 150 members) are presented. In the following figures c_3 is the inconsistency degree of a collective. Symbol Q is the quality of collective knowledge. Firstly, Fig. 2 contains the results for experiments in case the real state is generated based on the first collective with 3 members.

Secondly, Fig. 3 contains the results in case the real state is generated based on the collective with $(n + k)/2$ members.

Lastly, Fig. 4 contains the results in case the real state is generated based on the collective with $(n + k)/2$ members.

From these figures, we can see that adding members to a collective such that causes decreasing the inconsistency degree of the collective is not helpful in improving

Step 1: Generate a collective with n members;

Step 2: Adding a new member to the collective such that decreasing the inconsistency degree of the collective.

Step 3: Repeating step 2 until the collective reaches the maximal member $(n + k)$.

Step 4: Generating the real sate.

Step 5: Determining collective knowledge for each collective.

Step 6: Calculating the inconsistency degree and the quality of each collective knowledge with the real state determined from step 4.

Step 7: Analyzing the relationship between the inconsistency degree and the quality of collective knowledge.

Fig. 1. The procedure of the proposed method

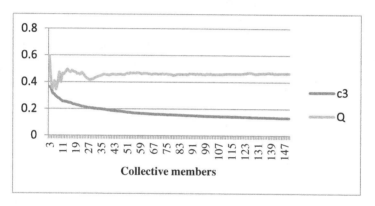

Fig. 2. The real state applied for collective with n members

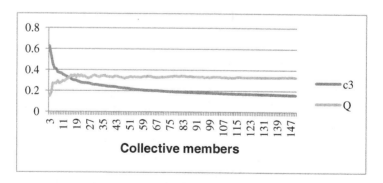

Fig. 3. The real state applied for collective with (n + k)/2 members

the quality of collective knowledge. However, it is only helpful in making the quality of collective knowledge to be stable. This phenomenon is due to the fact that the collective knowledge will be closer to the real knowledge state in case added members

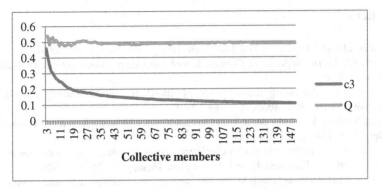

Fig. 4. The real state applied for collective with (n + k) members

which are closer to the real state (according to Theorem 2). Additionally, in some situations, collectives with higher inconsistency degree will give better collective knowledge than collectives with smaller inconsistency degree (collective with more consistency) [9, 10]. Concretely, we consider the following theorem about the influence of the adding a member to a collective on the quality of collective knowledge.

Theorem 2. For two collectives X, Y as follows:

$$X = \{x_1, x_2, \ldots, x_n\}, Y = \{X \dot\cup \{y\}\}$$

If $d(r, y) \leq d(r, x^*)$, then $d(r, y^*) \leq d(r, x^*)$.

Proof. The theorem is proved by contradiction.

6   Conclusions and Future Works

In this work the problem of the influence of the inconsistency degree on the quality of collective knowledge for objective case have been investigated by increasing the number of collective members. For this aim, we have proposed some theorems about the criteria for added members and the real knowledge state. Through experiments, adding members causes decreasing the inconsistency degree of a collective is not always helpful in improving the quality of collective. Besides, if the added members are closer to the real knowledge state, then the quality of the knowledge of a collective is better. The problem of analyzing the influence of the inconsistency degree on the quality of collective knowledge with other knowledge structures and determining the number of collective members should be enough for solving a given subject in the real world are considered as the future works and paraconsistent logics could be useful for these problems [22].

References

1. Shermer, M.: The Science of Good and Evil. Henry Holt, New York (2004)
2. Surowiecki, J.: The Wisdom of Crowds. Knopf Doubleday Publishing Group, New York (2005). Anchor
3. Nguyen, N.T.: Processing inconsistency of knowledge in determining knowledge of collective. Cybern. Syst. 40(8), 670–688 (2009)
4. Herrera-Viedma, E., et al.: Some issues on consistency of fuzzy preference relations. Eur. J. Oper. Res. 154(1), 98–109 (2004)
5. Francisco, C., et al.: Integration of a consistency control module within a consensus model. Int. J. Uncertainty Fuzziness Knowl. Based Syst. 16(supp01), 35–53 (2008)
6. Xu, Z.: An automatic approach to reaching consensus in multiple attribute group decision making. Comput. Ind. Eng. 56(4), 1369–1374 (2009)
7. Wu, Z., Xu, J.: A concise consensus support model for group decision making with reciprocal preference relations based on deviation measures. Fuzzy Sets Syst. 206, 58–73 (2012)
8. Nguyen, V.D., Nguyen, N.T.: A method for improving the quality of collective knowledge. In: Nguyen, N.T., Trawiński, B., Kosala, R. (eds.) ACIIDS 2015. LNCS, vol. 9011, pp. 75–84. Springer, Heidelberg (2015)
9. Nguyen, N.T.: Inconsistency of knowledge and collective intelligence. Cybern. Syst. 39(6), 542–562 (2008)
10. Nguyen, N.T.: Advanced Methods for Inconsistent Knowledge Management. Springer, New York (2008)
11. Day, W.H.E.: The consensus methods as tools for data analysis. In: Classifications and Related Methods of Data Analysis, IFC 1987. Springer, Heidelberg (1988)
12. Kline, J.A.: Orientation and group consensus. Cent. States Speech J. 23(1), 44–47 (1972)
13. Barthelemy, J.P., Guenoche, A., Hudry, O.: Median linear orders: Heuristics and a branch and bound algorithm. Eur. J. Oper. Res. 42(3), 313–325 (1989)
14. Nguyen, N.T.: Using consensus methods for determining the representation of expert information in distributed systems. In: Cerri, S.A., Dochev, D. (eds.) AIMSA 2000. LNCS (LNAI), vol. 1904, pp. 11–20. Springer, Heidelberg (2000)
15. Arrow, K.J.: Social Choice and Individual Values. Wiley, New York (1963)
16. Barthelemy, J.P., Janowitz, M.F.: A Formal Theory of Consensus. SIAM J. Discrete Math. 4(3), 17 (1991)
17. Barthelemy, J.P., Leclerc, B.: The median procedure for partitions. DIMACS Ser. Discrete Math. Theoret. Comput. Sci. 19, 3–33 (1995)
18. Day, W.H.E.: The complexity of computing metric distances between partitions. Math. Soc. Sci. 1(3), 269–287 (1981)
19. Danilowicz, C., Nguyen, N.T.: Consensus-based partitions in the space of ordered partitions. Pattern Recogn. 21, 269–273 (1988)
20. Nguyen, N.T.: Consensus system for solving conflicts in distributed systems. J. Inf. Sci. 147(1), 91–122 (2002)
21. Maleszka, M., Mianowska, B., Nguyen, N.T.: A method for collaborative recommendation using knowledge integration tools and hierarchical structure of user profiles. Knowl. Based Syst. 47, 1–13 (2013)
22. Nakamatsu, K., Abe, J.: The paraconsistent process order control method. Vietnam J. Comput. Sci. 1(1), 29–37 (2014)

Knowledge Base Refinement with Gamified Crowdsourcing

Daiki Kurita[1][✉], Boonsita Roengsamut[2],
Kazuhiro Kuwabara[1], and Hung-Hsuan Huang[1]

[1] College of Information Science and Engineering, Ritsumeikan University,
Kusatsu, Shiga 525-8577, Japan
is0165ee@ed.ritsumei.ac.jp
[2] Graduate School of Information Science and Engineering, Ritsumeikan University,
Kusatsu, Shiga 525-8577, Japan

Abstract. This paper discusses a gamification design for knowledge base refinement. The maintenance of a knowledge base involves human intervention and is one major application domain of human computation. Using the concept of crowdsourcing, refinement tasks can be delegated to many casual users over a network. In addition, gamification such as games with a purpose (GWAP) is a useful idea to motivate workers in crowdsourcing by making a task into a playful game. For effective gamification, designing the game rules is critical. In this paper, we present a model for simulating the gamified knowledge base refinement process to estimate the effects of different game rule designs beforehand.

Keywords: Crowdsourcing · Gamification · Knowledge base refinement · Linked data

1 Introduction

The Web has become a platform on which knowledge content is created. Although machine learning techniques are rapidly advancing in extracting meaningful data from a large volume of data, constructing a knowledge base of high quality still requires human effort, especially those of domain experts. However, since the cost of hiring domain experts remains rather high, it is important to harness the power of many casual users over networks. For this reason, the crowdsourcing concept is attracting much attention [6,8].

One issue in crowdsourcing with casual users is how to maintain their motivation, and the concept of gamification has been proposed to engage users [5]. Gamification refers to making a non-gaming context into a playful game so that a user (or player) can be engaged in a given task. In crowdsourcing, a task is divided into microtasks, each of which is assigned to a worker. In a typical crowdsourcing framework, an external reward is given to a worker, such as monetary one. When we make a microtask into a playful game, an intrinsic motivator to enjoy it is expected to be used to reward players instead of an external reward.

© Springer-Verlag Berlin Heidelberg 2016
N.T. Nguyen et al. (Eds.): ACIIDS 2016, Part I, LNAI 9621, pp. 33–42, 2016.
DOI: 10.1007/978-3-662-49381-6_4

The concept of games with a purpose (GWAP) [2] is a notable example of such an idea. In GWAP, the original task is indirectly executed as side effects of playing the game. Labeling an image is one typical example of GWAP, which is also used in the acquisition of knowledge. For example, in the OntoGame series [15], several games have been developed including SpotTheLink, which is targeted for ontology alignment [16]. Various types of games have also been developed to acquire semantic data [14]. In addition, gamification is applied in the maintenance of knowledge bases [10].

When the gamification approach is applied to knowledge base refinement, designing a game and its rules is an important issue to achieve high quality results. It is often difficult to estimate the effects of particular game rules before the game is played with human users. To tackle this problem, we make a simulation model of game execution. By simulating game execution, we expect to know the effectiveness of the game rule design beforehand and adjust the rules before letting human users play the game. As the first step toward this goal, we consider a model of executing a game that targets the refinement of knowledge bases. We focus on a knowledge base that is constructed as linked data [3], where the necessary information for updating the linked data is collected through game execution. Using this model, we estimate the effects of different game rule designs on the performance of knowledge refinement.

The rest of this paper is organized as follows. The next section describes related works. Section 3 presents an example of the knowledge base we used for this paper, and Sect. 4 explains our designed games. Section 5 presents a model that simulates the knowledge base refinement process and describes experimental results. The final section concludes this paper and describes future work.

2   Related Work

When we apply crowdsourcing to knowledge refinement, it is important to engage users (workers) in the task and guarantee the quality of the results. Gamification is an effective way to enhance user incentives. A model for appropriate incentive design was proposed [7] that investigated a combination of gamification and paid microtasks and created a predictive model for estimating a proper set of worker incentives.

Although gamification enhances the motivation of users, it does not necessarily lead to high quality output [11]. Games with a purpose (GWAP) rely on an ingenious game rule to obtain correct data [2]. The ESP game, one prominent example of GWAP, places an appropriate label on an image by two independent players who separately label a given image [1]. They only receive points when their labels match. Since no communication is assumed between the two players, the best strategy is to produce the correct label as much as possible. In this sense, correct data are expected to be obtained. However, such game rules are not necessarily easy to design for every domain.

To achieve high quality results, using qualified workers and engaging them in the task is also important. *Quizz* is a gamified crowdsourcing system for knowledge curating that asks users short quizzes [9]. This system estimates the competence of users using a calibration quiz, which is a special type of quiz whose answers are known beforehand. The system successfully collects and curates correct knowledge.

These works focus more on the detailed analysis of the behavior of the crowd-sourcing system with gamification. In contrast, in this paper we simulate game execution to estimate the effects of a particular game rule design beforehand.

3 Rental Apartment FAQ

3.1 Overview

We use the knowledge content of a rental apartment's Frequently Asked Question (FAQ) system [13] as a test-bed. This application was originally designed to support international students living in Japanese rental apartments. It contains some troubleshooting knowledge about various problems that might occur while living in the apartment, such as air conditioners that are not working. Its knowledge base is represented as linked data [3] or a Resource Description Framework (RDF), developed in the Semantic Web to facilitate sharing knowledge on the Web.

The knowledge base contains a simple domain ontology about the apartment including its typical floor plan and common fixtures or equipment used in it. Using the domain ontology, a link is derived between an FAQ entry and its most closely related floor plan section. We extended the original FAQ knowledge base to include the link's weight so that a relationship's strength can be represented. Figure 1 shows that a question statement of an FAQ entry, There's no hot water in my shower, is linked to both the kitchen and the bathroom with different weights.

There are some different ways to put weights to a link in the RDF model. For example, extending the RDF model was proposed to be a quadruplet instead of the original triple to include the weight of each link [4]. In this work, SPARQL, which is query language for RDF, is also extended to make queries for links with weight. In this paper, however, to use widely available RDF database software, we did not change the underlying data model and instead use the classical reification technique to store the link's weight.

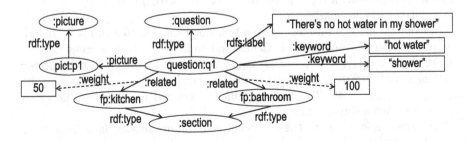

Fig. 1. Example link between question statement and floor plan section

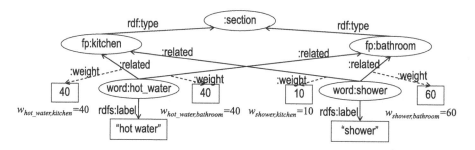

Fig. 2. Domain ontology with a link weight

3.2 Domain Ontology

The domain ontology mainly contains words corresponding to various equipment or fixtures often found in an apartment, such as an air conditioner or a shower. It contains words that describe a section of a floor plan such as living rooms or kitchens. The relationships between the fixtures and the floor plan sections are also described in the domain ontology. This knowledge is used to derive a link between a question statement and a relevant section of the floor plan and to calculate the link's weight.

3.3 Derivation of a Link

When a link is derived from a question statement to the section in the floor plan, a weight is put on the link. A question statement (an FAQ entry) might be associated with multiple places. In such cases, it is preferable to represent the strength of each association. Since the link between a question and a section is derived from the relationship between the fixtures and the section, the relationship between the fixtures and the section is also extended to have a weight.

The weight of the link between question statement (q) and floor plan section(s) $W_{q,s}$ is calculated as follows:

$$W_{q,s} = \sum_{k \in K(q)} e_k \cdot w_{k,s}, \tag{1}$$

where $K(q)$ represents a set of keywords associated with q and e_k represents the importance of the keyword k.

An example domain ontology with a link weight is depicted in Fig. 2. In this example, *shower* is connected to *kitchen* and *bathroom* with link weights of 10 and 60: $w_{shower,kitchen} = 10$, $w_{shower,bathroom} = 60$. *hot_water* is also connected to *kitchen* and *bathroom* but with different weights: $w_{hot\_water,kitchen} = 40$, $w_{hot\_water,bathroom} = 40$. The link's weight is intended to represent the closeness of the relationships.

The question statement, There's no hot water in my shower, has *shower* and *hot_water* keywords. Based on the weight of the links between these keywords and the floor plan section, the weights of the links between the question state-ment and floor plan section (*bathroom* and *kitchen*) are calculated as follows:

$$W_{q,bathroom} = w_{hot\_water,bathroom} + w_{shower,bathroom} = 40 + 60 = 100$$

$$W_{q,kitchen} = w_{hot\_water,kitchen} + w_{shower,kitchen} = 40 + 10 = 50.$$

Here, e_k is assumed to be 1 for all k.

3.4   Adding a New Question

Our rental apartment FAQ system contains a function to add a new question with a photo that further describes a question. A question statement can include hashtags. Hashtagged words are treated as keywords, allowing us to bypass the keyword extraction process. The place where the photo is taken (a section of the floor plan) is also specified using a hashtag.

For example, when a user inputs a question, such as There's no hot water in my shower, she may enter the following text: There's no #hot_water in my #shower. From this text, *hot_water* and *shower* are extracted as keywords. When a question is presented to other users, the hashtag itself may be omitted and the question will be displayed as There's no hot water in my shower.

If a photo is uploaded with a floor plan section, a correct link of the accom-panying question statement is considered to be given. In this case, only from the question statement text, the most related floor plan section is calculated using the domain ontology. If the result is identical as the place where the photo was taken, the domain ontology is considered acceptable. If not, an error correction procedure will be invoked. Since we manage the link with weights, the threshold of the weight with which an error is judged needs to be carefully set.

4   Game Design

Two types of games are discussed below. One is for collecting knowledge content, and the other is for fixing errors, more precisely, gathering information necessary to fix them. Both are extended from the Aparto Game developed for gamifying knowledge base construction of a rental apartment FAQ system [12].

4.1   Game for Knowledge Content Collection

This game, which gathers as many related keywords as possible with the floor plan section, is based on the *inversion-problem* game, which is one variation of games with a purpose (GWAP) [2]. The *inversion-problem* game has two players: A and B. Player A is shown a uploaded picture of the question and is instructed to produce words related to the picture's place (floor plan section). The produced words are shown to Player B, who is asked to answer where the original photo was taken. If Player B correctly answers the questions, she receives the points. If Players A and B do not have any means to communicate, Player A's best strategy is to produce correct words that are mostly related to the answer. The words produced by Player A can be added to the knowledge base as related words to the original photo's place.

Suppose that the shower's photo taken in the *bathroom* is uploaded with a question statement: There's no not water in my shower. Player A is shown this picture and produces words: *shower*, *bath*, and *hot_water*. Points are given if Player B's answer is *bathroom*.

4.2   Game for Knowledge Content Correction

When an error is found in the knowledge content, the data necessary to fix it are collected through a game. For example, suppose that the most related floor plan section of the question statement There's no hot water in my shower is *kitchen*. This is not intuitively correct. Since the weight of a link between a question statement and the floor plan section is calculated from the weights of a link between a keyword and a floor plan section, these weights might contain some errors.

To correct a link's weight, we created the following quiz. In the above example, the question statement contains two keywords: *hot_water* and *shower*. Since the weights of the links from keyword *shower* to the floor plan sections have different destinations, they are possible targets for updates. The quiz's choices can be created from the links from *shower* or a node that is an instance of an RDF class :section in the domain ontology.

A quiz is generated to ask a user about the most relevant floor plan section from a list of floor plan sections, and the weight of a link is updated according to the user's answer. The weight of the link the user answered is increased, and the other link weights are decreased. More specifically, the weight of link $w_{k,s}$ between floor plan section s and keyword k is updated by the following formula:

$$\begin{cases} w_{k,s} \leftarrow w_{k,s} + \alpha & (s = a) \\ w_{k,s} \leftarrow w_{k,s} - \frac{\alpha}{N-1} & (s \neq a), \end{cases} \tag{2}$$

where a represents the user's answer to the quiz, N refers to the number of choices in it, and α is a coefficient of the weight updates.

4.3   Assessing a User's Reliability

When knowledge content is being corrected, it is important to remove the effects of unreliable data from users. We assess user reliability by a *calibration* quiz whose correct answers are already known. This calibration quiz is mixed into a series of main quizzes. Based on its answer, user reliability R is evaluated. Using R, Formula 2 is rewritten as follows:

$$\begin{cases} w_{k,s} \leftarrow w_{k,s} + c(R) \cdot \alpha & (s = a) \\ w_{k,s} \leftarrow w_{k,s} - c(R) \cdot \frac{\alpha}{N-1} & (s \neq a), \end{cases} \tag{3}$$

where $c(R)$ is a function that returns the value of the coefficient of updating a link by user reliability R.

5   Experiment and Evaluation

5.1   Purpose

We conducted a simulation experiment to confirm that knowledge base refinement can be achieved by our proposed game. More specifically, this simulation focused on the changes in the weight of a link after the game presented in Sect. 4.2 was executed. In addition, we examined whether the effects of the changes in the game design can be simulated.

5.2   Methods

In the simulation experiment, we considered the situation mentioned in Sect. 4 and updated the weights of the links between *shower* and the floor plan section. Here we focus on *kitchen* and *bathroom* among the sections of the floor plan. The initial values of the link weight from *shower* to the *kitchen* and *bathroom* are set as follows: $w_{shower,kitchen} = 80$, $w_{shower,bathroom} = 30$.

The main quiz allows a user to select the most related floor plan section of *shower* from the list of *kitchen, bathroom, living_room*, and *toilet*. Since there are four choices on the list, N is set to 4. The calibration quiz has a list of four choices among them, and one choice is treated as a correct answer. There are two kinds of users: *reliable user* U_R and *non-reliable user* U_N. *Reliable user*'s reliability R is set to 1 and that of the *non-reliable user* is set to 0.

A *reliable user* is assumed to answer the calibration quiz correctly with a 0.92 probability and to select an answer for the main quiz from the choice list of *living_room, bathroom, kitchen*, and *toilet* with probabilities of 0.02, 0.92, 0.04, and 0.02. But a *non-reliable user* is assumed to answer the calibration quiz randomly, meaning that the probability of selecting a correct answer is 0.25 since there are four choices, and also assumed to answer the main quiz by selecting one from the four choices with an equal probability of 0.25.

We conducted a preliminary experiment. Based on its results, coefficient α in Formula 2 is set to 0.6 so that the most relevant floor plan section of the FAQ entry can be updated after a certain number of games are executed. In addition, $c(R)$ in Formula 3 is set to 2.0 for $R = 1$ (*reliable user*) and 0.5 for $R = 0$ (*non-reliable user*) so that the reliable user's answer has a greater effect on updating the link weight.

5.3   Results

We assume a huge number of users. Each game execution consists of randomly selecting one user (from the body of users) who plays the game. This game round is repeated 500 times. We examined the changes in the weights of a link of *shower* for each game round. We also considered three different sets of users with different *reliable user* percentages; the probabilities of *reliable user* $P(U_R)$ were set to 0.8, 0.5, and 0.2. The changes in the weight of a link from *shower* to *kitchen* and to *bathroom* in one trial run for three different user sets are

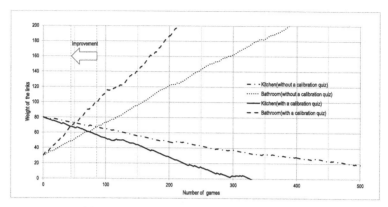

(a) Probability of reliable user: $P(U_R) = 0.8$

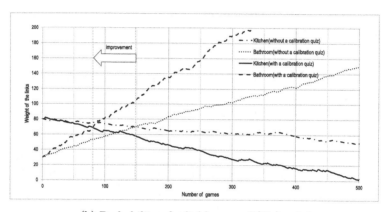

(b) Probability of reliable user: $P(U_R) = 0.5$

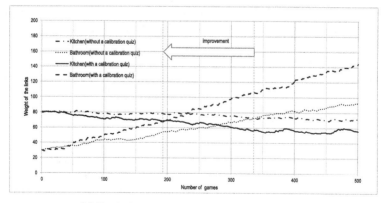

(c) Probability of reliable user: $P(U_R) = 0.2$

Fig. 3. Changes in link weight with and without a calibration quiz in one trial run

plotted in Fig. 3, where the x-axis represents the number of game rounds that were played. The plots include both cases with and without the calibration quiz.

5.4 Discussion

As shown in the graph in Fig. 3, the weight of the link between *shower* and *kitchen* decreases, but the weight of the link between *shower* and *bathroom* increases for all three cases of different probabilities of a reliable user. At some point, the weight for *bathroom* exceeds that of *kitchen*. This means that the most related floor plan section of the original question statement becomes *bathroom* after enough game rounds are executed. Thus, the error found in the knowledge base can be fixed through the game. In addition, when the calibration quiz is utilized, the intersection of the curves of the weight of two links comes earlier in the game rounds than when the calibration quiz is not used. Thus, the effects of the calibration quiz were confirmed.

We considered number of games G_{int} when two weight curves intersect and defined the *improvement rate* of the calibration quiz as the ratio of that number (G_{int}) with a calibration quiz to that number (G_{int}) without it. We ran the game 1,000 times and calculated the average of the *improvement rate* whose value was roughly equal for different user sets: 46.1 % for $P(U_R) = 0.8$, 45.5 % for $P(U_R) = 0.5$, and 42.9 % for $P(U_R) = 0.2$. This indicates that our calibration quiz is effective regardless of the percentage of reliable users.

In this experiment, the values of α in Formulas 2 and 3 were determined after the preliminary experiment so that the weight of the link can properly be updated after a certain number of game rounds. In other words, if each game round is played by a different user, we can determine the value of α based on the number of prospective players.

6 Conclusion and Future Work

This paper presented a gamification approach toward knowledge content creation and refinement. We presented two types of games: one for adding new knowledge content and another for revising it. Using a rental apartment FAQ as an example knowledge base, we created an execution model of the game for the latter knowledge base refinement to confirm our proposed gamification approach. We ran a simulation under a simple condition, and the simulation results indicate that playing games can update the weight of links in knowledge bases. The results also indicate that updates can be done more quickly by estimating user reliability.

Future work will extend a model to handle more complex situations and run simulations under various conditions with other methods of estimating user reliability. In addition, we plan to validate our simulation model by letting human users play the game under various conditions.

Acknowledgment. This work was partially supported by JSPS KAKENHI Grant Number 15K00324.

References

1. von Ahn, L., Dabbish, L.: Labeling images with a computer game. In: Proceedings of the SIGCHI Conference on Human Factors in Computing Systems, CHI 2004, pp. 319–326, April 2004

2. von Ahn, L., Dabbish, L.: Designing games with a purpose. Commun. ACM **51**(8), 58–67 (2008)

3. Bizer, C., Heath, T., Berners-Lee, T.: Linked data - the story so far. Int. J. Semantic Web Inf. Syst. **5**(3), 1–22 (2009)

4. Cedeño, J.P., Candan, K.S.: R2DF framework for ranked path queries over weighted RDF graphs. In: Proceedings of the International Conference on Web Intelligence, Mining and Semantics (WIMS 2011), pp. 40:1–40:12 (2011)

5. Deterding, S.: Gamification: designing for motivation. Interactions **19**(4), 14–17 (2012)

6. Doan, A., Ramakrishnan, R., Halevy, A.Y.: Crowdsourcing systems on the world-wide web. Commun. ACM **54**(4), 86–96 (2011)

7. Feyisetan, O., Simperl, E., Van Kleek, M., Shadbolt, N.: Improving paid microtasks through gamification and adaptive furtherance incentives. In: Proceedings of the 24th International Conference on World Wide Web (WWW 2015), pp. 333–343 (2015)

8. Howe, J.: Crowdsourcing: Why the Power of the Crowd Is Driving the Future of Business. Crown Business, New York (2009)

9. Ipeirotis, P.G., Gabrilovich, E.: Quizz: targeted crowdsourcing with a billion (potential) users. In: Proceedings of the 23rd International Conference on World Wide Web (WWW 2014), pp. 143–154 (2014)

10. Jovanovic, M.: Gamifying knowledge maintenance. https://www.essence-network.com/wp-content/uploads/2015/08/GamifyingKnowledgeMaintenance_Jovanovic.pdf. Accessed on 01 October 2015

11. Juźwin, M., Adamska, P., Rafalak, M., Balcerzak, B., Kąkol, M., Wierzbicki, A.: Threats of using gamification for motivating web page quality evaluation. In: Proceedings of the 2014 International Conference on Multimedia, Interaction, Design and Innovation (MIDI 2014), pp. 14:1–14:8 (2014)

12. Roengsamut, B., Kuwabara, K., Huang, H.H.: Toward gamification of knowledge base construction. In: Proceedings of the 2015 International Symposium on INnovations in Intelligent SysTems and Applications (INISTA 2015), pp. 20–26 (2015)

13. Saito, Y., Roengsamut, B., Kuwabara, K.: Incremental refinement of linked data: ontology-based approach. In: Nguyen, N.T., Attachoo, B., Trawiński, B., Somboonviwat, K. (eds.) ACIIDS 2014, Part I. LNCS, vol. 8397, pp. 133–142. Springer, Heidelberg (2014)

14. Šimko, J., Bieliková, M.: Semantic Acquisition Games - Harnessing Manpower for Creating Semantics. Springer International Publishing, Heidelberg (2014)

15. Siorpaes, K., Hepp, M.: Games with a purpose for the semantic web. IEEE Intell. Syst. **23**(3), 50–60 (2008)

16. Thaler, S., Simperl, E., Siorpaes, K.: SpotTheLink: playful alignment of ontologies. In: Proceedings of the 2011 ACM Symposium on Applied Computing (SAC 2011), pp. 1711–1712 (2011)

Argumentation Framework for Merging Stratified Belief Bases

Trong Hieu Tran[1]([✉]), Thi Hong Khanh Nguyen[2], Quang Thuy Ha[1], and Ngoc Trinh Vu[3]

[1] Vietnam National University, Hanoi, Vietnam
{hieutt,thuyhq}@vnu.edu.vn
[2] Electricity Power University of Vietnam, Hanoi, Vietnam
khanh@epu.edu.vn
[3] Vietnam Petroleum Institue, Hanoi, Vietnam
trinhvn@vpi.pvn.vn

Abstract. This paper introduces a new approach for belief merging by using argumentation technique. The key idea is to organize each belief merging process as a game in which participating agents use argumentation technique to debate on their own belief bases to achieve consensus i.e. a common belief base. To this end, we introduce a framework for merging belief by argumentation in which an argumentation-based belief merging protocol is proposed and a set of intuitive and rational postulates to characterize the merging results is introduced. Several logical properties of the family of argumentation-based belief merging operators are also pointed out and discussed.

Keywords: Argumentation · Belief merging

1 Introduction

Over the years Belief Merging has emerged as an active research field in Computer Science. It is close to the areas of data mining and big data [22], data fusion [7,19], information retrieval [2,23], multi-agent systems [15,20] and machine learning [14]. The main aim of Belief Merging is to achieve a common belief base which best represents for a given set of belief bases.

In literature, we adopted a large range of belief merging works which typically include the merging operators to deal with weighted belief bases [12], a family of the arbitration operators [21], the belief merging works with integrity constraints [11] or with stratified bases [17], and the belief merging works for description logics [13], possibilistic logic [3,16] and argumentation systems [6]. These works established the main stream of the researches on Belief Merging. The common characteristic of these works is that the belief bases which need to merge must be revealed explicitly and completely, and the agents, who possess these belief bases are ignored from the merging process. Therefore, it is impossible to apply them in the multi-agent systems where agents prefer consensus rather than imposition.

© Springer-Verlag Berlin Heidelberg 2016
N.T. Nguyen et al. (Eds.): ACIIDS 2016, Part I, LNAI 9621, pp. 43–53, 2016.
DOI: 10.1007/978-3-662-49381-6_5

Apart from the above belief merging works, another group of belief merging works is constructed based on the spirit of game theory. In each game, the participating agents use negotiation techniques or argumentation techniques together in order to reach the consensus. These works can overcome the above shortcomings. The works using negotiation techniques involve a two-stage belief merging process [4,5], a family of game-based merging operators [28], merging stratified belief bases as a game [18,24–27] and a solution for the bargaining game problem [28].

In the subgroup of works using argumentation technique, in order to resolve the conflicts of a committee, we organize a debate in which each agent uses its own belief base as arguments and manipulates argumentation skills to defend its own arguments and attacks the arguments of others. The typical works are the framework to merge possibilistic logic bases by Amgoud et al. [1] and the framework to merge weighted belief bases [9]. Unfortunately, both of these approaches are still affected by the well-known *drowning effect*[1].

In this paper, our contribution is twofold. Firstly, we introduce a new framework for belief merging by argumentation that can avoid the drowning effect. In which an argumentation protocol is proposed and a model is presented. This framework is more general and flexible than the above works because it allows to work with stratified belief bases instead of numbered belief bases. Secondly, we introduce a set of rational and intuitive postulates to characterize the properties of merging result and analyze the logical properties. Moreover, a representation theorem is introduced to present the connection between constructive and axiomatic directions.

The rest of this paper is organized as follows. In Sect. 2, we introduce some basic notations and concepts. Our framework for belief merging by argumentation including an argumentation protocol and an argumentation model is presented in Sect. 3. Section 4 introduces a set of postulates to characterize the family of argumentation-based merging operators and analyzes logical properties. Finally, the conclusion and future work are presented in Sect. 5.

2   Preliminaries

2.1   Stratified Knowledge Base

We consider a propositional language $\mathcal{L}$ built on a finite alphabet $\mathcal{P}$ and a set of logical connectives including $\neg, \wedge, \vee$ and $\rightarrow$. We also use symbols $\top$ and $\bot$ to present the truth values true and false, respectively. A *logic formula* (or *formula* for short) is constructed from $\mathcal{P}$ and these connectives. The consequence relation is denoted by the symbol $\vdash$, e.g. $\Phi \vdash \psi$ means ψ is a logical consequence of Φ where Φ is a set of formulas. A *belief base* is a finite set of formulas.

A *stratified belief base* $(K, \succeq)$ is a pair including a set of formulas K and a total pre-order relation $\succeq$ on K. Stratified belief base $(K, \succeq)$ is also presented as

---

[1] The drowning effect takes place when some beliefs are omitted because their preferences are lower than the preferences of conflict beliefs.

a sequence $(K, \succeq) = (L_1, \ldots, L_n)$, where each L_i is non-empty set of formulas and called a *stratum*. Given $\phi \in L_i$ and $\psi \in L_j$, $\phi \succeq \psi$ iff $i \leq j$, for all $i, j = 1, \ldots n$, we say that each formula in stratum L_i is preferred to any formula in stratum L_j. A *belief set* of (stratified) belief bases is a multi-set[2] of (stratified) belief bases.

We say that $(K, \succeq) = (L_1, \ldots, L_n)$ and $(K', \succeq') = (L'_1, \ldots, L'_m)$ are logically equivalent if $m = n$ and $L_i \equiv L'_i$ for all $i = 1, \ldots, n$ (it is written as $(K, \succeq) \equiv (K', \succeq')$). Further, belief sets $\mathcal{B} = \{(L_1, \succeq_1), \ldots, (L_n, \succeq_n)\}$ and $\mathcal{B}' = \{(L'_1, \succeq'_1), \ldots, (L'_n, \succeq'_n)\}$ are equivalent, denoted by $\mathcal{B} \equiv \mathcal{B}'$ iff $(L_i, \succeq_i) \equiv (L'_{\delta(i)}, \succeq'_{\delta(i)}), \forall i = 1, \ldots, n$ where δ is a permutation on $1, \ldots, n$.

2.2   Belief Merging

Belief Merging aims to achieve a common belief base from a given belief set which may include some jointly inconsistent belief bases. The belief merging approaches are widely investigated in a large range of works as we mentioned above. Hence, because of the lack of space in this paper, we do not discuss more about them. We only focus on the set of axioms to characterize the family of merging operators with integrity constraints in [11]. These axioms (also, called *(IC) axioms*) are stated as follows:

Definition 1 [11]. *Let $\mathcal{B}, \mathcal{B}_1, \mathcal{B}_2$ be belief sets, K_1, K_2 be consistent belief bases, and μ, μ_1, μ_2 be formulas from $\mathcal{L}$. Δ is an IC merging operator if and only if it satisfies the following axioms:*

(IC0) $\Delta_\mu(\mathcal{B}) \vdash \mu$.
(IC1) *if* $\mu \nvdash \perp$, *then* $\Delta_\mu(\mathcal{B}) \nvdash \perp$.
(IC2) *if* $\wedge \mathcal{B} \wedge \mu \nvdash \perp$, *then* $\Delta_\mu(\mathcal{B}) \equiv \wedge \mathcal{B} \wedge \mu$.
(IC3) *if* $\mathcal{B}_1 \equiv \mathcal{B}_2$ *and* $\mu_1 \equiv \mu_2$, *then* $\Delta_{\mu_1}(\mathcal{B}_1) \equiv \Delta_{\mu_2}(\mathcal{B}_2)$.
(IC4) *if* $K_1 \vdash \mu$ *and* $K_2 \vdash \mu$,*then* $\Delta_\mu(\{K_1, K_2\}) \wedge K_1 \nvdash \perp$ *iff* $\Delta_\mu(\{K_1, K_2\}) \wedge K_2 \nvdash \perp$.
(IC5) $\Delta_\mu(\mathcal{B}_1) \wedge \Delta_\mu(\mathcal{B}_2) \vdash \Delta_\mu(\mathcal{B}_1 \sqcup \mathcal{B}_2)$.
(IC6) *if* $\Delta_\mu(\mathcal{B}_1) \wedge \Delta_\mu(\mathcal{B}_2) \nvdash \perp$, *then* $\Delta_\mu(\mathcal{B}_1 \sqcup \mathcal{B}_2) \vdash \Delta_\mu(\mathcal{B}_1) \wedge \Delta_\mu(\mathcal{B}_2)$.
(IC7) $\Delta_{\mu_1}(\mathcal{B}) \wedge \mu_2 \vdash \Delta_{\mu_1 \wedge \mu_2}(\mathcal{B})$.
(IC8) *if* $\Delta_{\mu_1}(\mathcal{B}) \wedge \mu_2 \nvdash \perp$, *then* $\Delta_{\mu_1 \wedge \mu_2}(\mathcal{B}) \vdash \Delta_{\mu_1}(\mathcal{B}) \wedge \mu_2$.

(where Δ_μ is a belief merging operator under the integrity constraints μ.)

These axioms are deeply analyzed and discussed in many works (typically, [10,11,17]). In this paper, we refer to these axioms to investigate the properties of our family of argumentation-based belief merging operators. Moreover, we are working on stratified belief base, thus some of axioms need to be slightly modified. Particularly, axioms (IC2), (IC3) should be modified as follows [27]:

Given the belief sets $\mathcal{B} = \{(K_i, \succeq_i) | i = 1..n\}$, $\mathcal{B}' = \{(K'_i, \succeq'_i) | i = 1..n\}$ and formulas μ and μ'.

---

[2] The elements of a multi-set may be identical.

(IC2') Let $\wedge \mathcal{B} = \wedge_{i=1}^{n} K_i$, if $\wedge \mathcal{B} \wedge \mu \not\vdash \bot$, then $\Delta_\mu(\mathcal{B}) \equiv \wedge \mathcal{B} \wedge \mu$.
(IC3') If $\mu \equiv \mu'$ and $\exists \delta$ on $\{1, \dots, n\}$ s.t. $(K_i, \succcurlyeq_i) \equiv (K'_{\delta(i)}, \succcurlyeq'_{\delta(i)}), \forall i \in \{1, \dots, n\}$, then $\Delta_\mu(\mathcal{B}) \equiv \Delta_{\mu'}(\mathcal{B}')$.

2.3 Argumentation Framework

In this section, we recall Dung's general argumentation framework proposed in [8].

Definition 2 [8]. *An* argumentation framework *is defined as* $AF = \langle Arg, \mathcal{R} \rangle$ *where*

- *Arg is a set of arguments,*
- $\mathcal{R} \subseteq Arg \times Arg$ *is attack relationship between arguments,*

Definition 3 [8]. *Let* $X, Y \in Arg$, *we have*

- X *attacks* Y *iff* $\mathcal{R}(X, Y)$,
- *A set of arguments* $\mathcal{S}$ *defends* X *if* $\forall Y \in Arg, \mathcal{R}(Y, X)$, *then* $\exists Z \in \mathcal{S}, \mathcal{R}(Z, Y)$.

Definition 4 [8]. *Given a set of arguments* Arg *and* $\mathcal{S} \subseteq Arg$. S *is conflict-free iff* $\nexists X, Y \in S$ *s.t.* $\mathcal{R}(X, Y)$.

Definition 5 [8].

(1) *An argument* $X \in Arg$ *is acceptable w.r.t. a set* $\mathcal{S}$ *of arguments iff* $\forall Y \in Arg$ *if* $\mathcal{R}(Y, X)$, *then* $\exists Z \in \mathcal{S}$ *and* $\mathcal{R}(Z, Y)$.
(2) *A conflict-free set of arguments* S *is admissible iff* $\forall X \in S$, X *is acceptable w.r.t.* S.

Definition 6 [8]. *The characteristic function* $\mathcal{F}_{AF} : 2^{Arg} \to 2^{Arg}$ *is a map s.t.* $\forall \mathcal{S} \subseteq Arg, \mathcal{F}_{AF}(\mathcal{S}) = \{A | A \text{ is acceptable w.r.t } \mathcal{S}\}$.

It is easy to have the following remark:

Proposition 1 [8]. $\mathcal{F}_{AF}$ *is monotonic w.r.t set inclusion.*

The following is several common extensions:

Definition 7 [8]. *Given an argumentation framework* AF *and an admissible set of arguments* $\mathcal{S}$,

- $\mathcal{S}$ *is a* complete extension *of* AF *iff every acceptable argument w.r.t* $\mathcal{S}$ *belongs to* $\mathcal{S}$.
- $\mathcal{S}$ *is a* grounded extension *of* AF *iff it is the smallest element (w.r.t set inclusion) among the complete extensions of* AF.
- $\mathcal{S}$ *is a* preferred extension *of* AF *iff it is the largest element (w.r.t set inclusion) among the complete extensions of* AF.

Definition 8 [8]. *Given an argumentation framework AF be a conflict-free set of arguments S,*

$-$ *S is a stable extension of AF iff $\forall X \in Arg, X \notin S, \exists Y \in S$ s.t. $\mathcal{R}(Y, X)$.*

Remark that for each argumentation framework we have at most a ground extension and some stable extensions.

Example 1. For the sake of explanation, we use the situation in the example of [28] to illustrate our work. This example is briefly summarized that there are two parties to debate about the economical rescue plan of their country. Each party has several demands, ordered by the preference of this party and given as in Table 1. Clearly, there are several conflicts arisen here, e.g. {*banks*, *cars*, *(cars* $\wedge$ *banks)* $\rightarrow$ *deficit*, $\neg$*deficit*} or {$\neg$*(banks* $\wedge$ *mortgagers)*, *banks*, *mortgagers*}. In the following section, we will consider about a conflict resolution for this situation.

Table 1. The preferences of the parties *(the higher layer is, the more preferred belief.)*

Party 1	Party 2
(cars $\wedge$ *banks)* $\rightarrow$ *deficit*, $\neg$*deficit*	$\neg$*(banks* $\wedge$ *mortgagers)*
$\neg$*(banks* $\wedge$ *mortgagers)*	*(cars* $\wedge$ *banks)* $\rightarrow$ *deficit*
banks	*cars*
jobs	*mortgagers*

3   Framework for Belief Merging by Argumentation

3.1   Argumentation Protocol for Belief Merging

We consider a protocol, named *Simultaneous Submission protocol* for belief merging by argumentation as follows:

1. Argumentation process is organized in multiple rounds.
2. At each round, participants propose their arguments simultaneously.
   (a) If all the proposed arguments are jointly consistent with arguments collected in the previous rounds, they will be joined into an accepted set of arguments.
   (b) If the arguments of some agents jointly conflict to the accepted arguments, then the arguments of these agents are omitted and the remaining arguments will be joined into the accepted set of arguments.
   (c) If a participating agent proposes an argument and others can defeat it, this argument will be rejected by all.
3. Argumentation process will be terminated when no participating agent can propose any further argument.

Remark that because each agent is equipped a knowledge base with a finite set of formulas, i.e. a finite set of arguments, thus the argumentation process is always terminated. Moreover, because of (2b) agents have the incentive to stratify their own belief bases in fine-grained strata. This protocol also drives the participating agents to submit their more preferred arguments as soon as possible.

3.2   Argumentation-Based Model for Belief Merging

In this model, we consider a set of agents $\mathscr{A} = \{a_1, \ldots, a_n\}$, each agent a_i is equipped a *stratified belief base* $(K_i, \succcurlyeq_i)$.

Firstly, we consider two key concepts related to argument technique as follows:

Definition 9. *An* argument *is a non-empty consistent set of formulas.*

In general, an argument includes a support component and a conclusion component. However, in our model, it is suitable to simplify the support component as a tautology.

The attack is defined based on consistence as follows:

Definition 10. *Given two arguments Φ and Ψ, Φ attacks Ψ iff $\Phi \bigcup \Psi$ is inconsistent.*

Informally, two arguments attack each other if they are jointly inconsistent. Obviously, this attack relation is symmetric.

From multiagent viewpoint, in an argumentation game, agents manipulate their own beliefs to join a debate to achieve a (consistent) common knowledge under integrity constraints via some argumentation protocol. The argumentation game is defined as follows:

Definition 11. *An* argumentation game *is a double $\langle \{(K_i, \succcurlyeq_i)|a_i \in \mathscr{A}\}, \mu \rangle$, where $\{(K_i, \succcurlyeq_i)|a_i \in \mathscr{A}\}$ is the belief bases of participating agents and μ is integrity constraint.*
Given a set of agents $\mathscr{A}$, the set of all argumentation games from $\mathscr{A}$ is denoted by $g^{\mathscr{A}}$.

Each argumentation game involves a belief set of stratified knowledge bases equipped for the participating agents and integrity constraints as common knowledge.

We define argumentation solution as follows:

Definition 12. *An* argumentation solution f *is a function s.t. $\forall G = (\{(K_i, \succcurlyeq_i)|a_i \in \mathscr{A}\}, \mu) \in g^{\mathscr{A}}$, $f(G) = \{f_i(G)|a_i \in \mathscr{A}\} \in \mathscr{O}(G)$, where $f_i(G) \subseteq \mathcal{L}$.*

Now, we introduce the construction of belief merging by argumentation which complies Simultaneous Submission protocol. Argumentation process is organized in rounds. At each round, agents simultaneously submit their beliefs, depend on the current state of argumentation process, some arguments may be eliminated and others are added to the set of accepted beliefs. It causes a new argumentation state arise. The argumentation state is formally defined as follows:

Definition 13. *An argumentation state is a tuple* (G, O) *where* $G = (\{(K_i, \succcurlyeq_i) | a_i \in \mathscr{A}\}, \mu)$ *is an argumentation game and* $O = \{O_i | a_i \in \mathscr{A}\}$ *is a current possible outcome.*

We denote S^G the set of argumentation states of game G, and Λ the set of all finite sequences of argumentation states.

The set $X^* = \bigcup_{a_i \in \mathscr{A}} O_i \bigcup \{\mu\}$ is called the temporary accepted belief set.

In a round, if the argument proposed by some agent is jointly inconsistent with the accepted set of arguments (or other proposed arguments in this round), the argument should be marked to omit. We define the *under attack function* to indicate the list of agents who have the arguments to be omitted in a round as follows:

Definition 14. *The under attack function at round* i *is as follows:*
$$und : S^G \to 2^{\mathscr{A}}$$
where $und(S) \neq \emptyset$ *means there is a list of agents whose arguments are attacked.*

At a round, the agents under attack have to change their beliefs to get out of this list. The changing function will force these agents to make some concessions in their own beliefs to reach consensus at the current round of argumentation process, it will also change the current state to a new state.

Definition 15. *Changing function is a map:*
$$upd : 2^{\mathscr{A}} \times S^G \to S^G,$$
where $upd(A, S) = S'$ *means that at state* S, *the under attack set of agents* $A \subseteq \mathscr{A}$ *changes their belief to reach consensus, it causes the transformation of* S *into* S'.

Lastly, the following is the definition of argumentation solution:

Definition 16. *The solution to an argumentation state* S *for an argumentation game* $\langle (K_i, \succcurlyeq_i), C \rangle$ *is given by the function* $s^A : S^G \to \Lambda$, *defined as:*
$$s^A(S) = \lambda = (S_0, \dots, S_k), \ where$$

a. $S_0 = S$,
b. $S_k \neq S_{k-1}$ *and* $S_i = S_{i-1}, \forall i \in \{0, \dots, k\}$, *and*
c. $\forall j \in \{0, k\}$ *we have,* $\forall a_i \in \mathscr{A}$,

$$s_i^A(S_{j+1}) = \begin{cases} s_i^A(upd(und(S_j), S_j)) & if \ a_i \in und(S_j), \\ s_i^A(S_j) & otherwise. \end{cases}$$

Finally, the solution of belief merging by argumentation is the last argumentation state S^k and we call this solution *Argumentation Belief Merging solution*.

One instance of the Argumentation Belief Merging solution is an algorithm as follows:

Algorithm 1 - Input: *A game* $E = (\{(K_i, \succcurlyeq_i)|a_i \in \mathscr{A}\}, \mu)$.
  - Output: *Argumentation-based Belief Merging solution P.*

  Begin
    $E = (\{(K_i, \succcurlyeq_i)|a_i \in \mathscr{A}\}, \mu)$;
    $P = \{\emptyset|a_i \in \mathscr{A}\}$;
    $S_0 = (E, P)$;
    $i = 0$;
    While $und(S_i) \neq \emptyset$ *do*
      Begin
        $S_{i+1} = upd(und(S_i), S_i)$;
        $inc(i)$;
      End;
    $(E, P) = S_i$;
    return P;
  End;

Example 2. The argumentation process for *Example 1* is as follows:
-*Initial round*:
  $P = \{\emptyset, \emptyset\}$.
  $X^* = \{\emptyset\}$.
-*First round*:
  Party 1 proposes its most preferred argument:
    $(cars \wedge banks) \rightarrow deficit, \neg deficit$,
  Party 2 also proposes its most preferred argument:
    $\neg(banks \wedge mortgagers)$
  Hence,
  $P = \{ \{(cars \wedge banks) \rightarrow deficit, \neg deficit\},$
        $\{\neg(banks \wedge mortgagers)\}\}$.
  $X^* = \{(cars \wedge banks) \rightarrow deficit, \neg deficit, \neg(banks \wedge mortgagers)\}$.
......
Lastly, we have the result of the argumentation process as follows:
  $P = \{\{(cars \wedge banks) \rightarrow deficit, \neg deficit, \neg(banks \wedge mortgagers), jobs\},$
      $\{\neg(banks \wedge mortgagers), (cars \wedge banks) \rightarrow deficit, mortgagers\}\}$.
Thus,
  $X^* \equiv (cars \wedge banks) \rightarrow deficit \wedge \neg deficit \wedge \neg(banks \wedge mortgagers) \wedge jobs \wedge$
mortgagers.

4   Postulates and Logical Properties

In this section, we introduce a set of intuitive and rational postulates to characterize the family of belief merging operators based on argumentation technique.
Assume that we have an argumentation function f^A w.r.t argumentation game
$G = (\{(K_i, \succcurlyeq_i)|a_i \in \mathscr{A}\}, \mu)$, the postulates are as follows:
Firstly, the *Individual Rationality* postulate is stated as follows:

$(INDI)$. $f_i^A(G) \subseteq K_i$, $\forall a_i \in \mathscr{A}$

This postulate requires that the accepted beliefs of each agent after an argumentation process can not exceed its initial beliefs.

Secondly, the *Consistency* postulate is formulated as follows:

$(CONS)$. $\bigcup_{a_i \in \mathscr{A}} f_i^A(G) \not\vdash \neg\mu$.

Postulate $CONS$ require that the result of the argumentation process should be consistent with integrity constraints. Because each agent is equipped a finite belief base, this postulate also ensures that the argumentation process is always terminated.

Thirdly, the *Cooperativity* postulate is represented as follows:

$(COOP)$. $\bigcap_{H \in MAX\_CONS(\bigcup_{a_i \in \mathscr{A}} K_i, \mu)} H \subseteq \bigcup_{a_i \in \mathscr{A}} f_i^A(G) \bigcup\{\mu\}$.

Postulate $COOP$ states that if some belief does not appear in any conflict, it should be in the argumentation result. This postulate assures that the drowning effect is avoided.

Next, the *Pareto Optimality* is formulated as follows:

(PO). $\nexists f'^A$ s.t. $f_i^A(G) \subseteq f_i'^A(G) | \forall a_i \in \mathscr{A}$ and $f_j^A(G) \subset f_j'^A(G)$ for some $a_j \in \mathscr{A}$.

Similar to the idea of Pareto Efficiency largely used in Game theory, postulate PO requires that the result of an argumentation process should be Pareto efficiency i.e. all agents jointly preserve maximally their own beliefs.

Lastly, the *Symmetry* postulate is stated as follows:

$(SYMM)$. $f_i^A((\{(K_i, \succcurlyeq_i) | a_i \in \mathscr{A}\}, \mu)) \equiv f_{\delta(i)}^A((\{(X_{\delta(i)}, \succcurlyeq_{\delta(i)}) | a_i \in \mathscr{A}\}, \mu))$

where δ is a permutation function.

Postulate $SYMM$ requires that all participating agents in an argumentation process should be equality, i.e. each agent is independent of its identifier.

Theorem 1. *Argumentation Belief Merging solution f^A satisfies properties $INDI$, $CONS$, $COOP$, PO, and $SYMM$.*

In Social Choice theory, the *Unanimity* postulate is also widely used. This postulate is formulated as follows:

$(UNAN)$. if $\Phi \subseteq K_i | \forall a_i \in \mathscr{A}$, then $\Phi \subseteq f_i^A(G) | \forall a_i \in \mathscr{A}$

Postulate $UNAN$ states that if an argument is in the belief bases of all participating agents, then it should be in the solution of argumentation game. Although this postulate is intuitive and rational, unfortunately, it cannot meet the requirement in the case of merging stratified belief bases.

Let Δ^A be the argumentation-based merging operator defined from the above argumentation solution f^A as follows:

$$\Delta^A(G) = \bigcup_{a_i \in \mathscr{A}} f_i^A(G) \bigcup\{\mu\}. \tag{1}$$

$\Delta^A(G)$ is called the *Argumentation-based Belief Merging operator*. Based on the reference to the set of postulates on Subsect. 2.2, we have the following properties:

Proposition 2. *If a belief merging operator Δ is an Argumentation Belief Merging operator, it satisfies $(IC0)$, $(IC1)$, $(IC2')$, $(IC7)$, $(IC8)$ and, it does not satisfy $(IC3')$, $(IC4)$, $(IC5)$, $(IC6)$.*

5   Conclusion

In this paper, we have proposed a new argumentation-based framework for belief merging. The key idea of this work is from the behaviors of a group of participants when they want to reach agreement from a conflict situation. In this framework, an argumentation protocol is proposed and a model of belief merging by argumentation driven by the proposed protocol is presented. A set of intuitive and rational postulates to characterize the argumentation solutions is introduced, and the logical properties of family of argumentation-based belief merging operators are discussed and analyzed.

A lot of work still needs to be done. The proposed framework will be improved to become syntax irrelevance. The set of postulates will be deeply analyzed and computational complexities in this framework will also be evaluated.

Acknowledgments. This study was fully supported by Science and Technology Development Fund (B) from Vietnam National University, Hanoi under grant number QG.14.13 (2014–2015).

References

1. Amgoud, L., Kaci, S.: An argumentation framework for merging conflicting knowledge bases. Int. J. Approximate Reasoning **45**(2), 321–340 (2007). Eighth European Conference on Symbolic and Quantitative Approaches to Reasoning with Uncertainty (ECSQARU 2005)
2. Baeza-Yates, R.A., Ribeiro-Neto, B.A.: Modern Information Retrieval. ACM Press Addison-Wesley, Boston (1999)
3. Benferhat, S., Dubois, D., Kaci, S., Prade, H.: Possibilistic merging and distance-basedfusion of propositional information. Ann. Math. Artif. Intell. **34**, 217–252 (2002)
4. Booth, R.: A negotiation-style framework for non-prioritised revision. In: Proceedings of the 8th Conference on Theoretical Aspects of Rationality and Knowledge, TARK 2001, pp. 137–150. Morgan Kaufmann Publishers Inc. (2001)
5. Booth, R.: Social contraction and belief negotiation. Inf. Fusion **7**, 19–34 (2006)
6. Coste-Marquis, S., Devred, C., Konieczny, S., Lagasquie-Schiex, M., Marquis, P.: On the merging of dung's argumentation systems. Artif. Intell. **171**(10–15), 730–753 (2007)
7. de Amo, S., Carnielli, W.A., Marcos, J.: A logical framework for integrating inconsistent information in multiple databases. In: Eiter, T., Schewe, K.-D. (eds.) FoIKS 2002. LNCS, vol. 2284, pp. 67–84. Springer, Heidelberg (2002)
8. Dung, P.M.: On the acceptability of arguments and its fundamental role in non-monotonic reasoning, logic programming and n-person games. Artif.Intell. **77**(2), 321–357 (1995)

9. Gabbay, D., Rodrigues, O.: A numerical approach to the merging of argumentation networks. In: Fisher, M., van der Torre, L., Dastani, M., Governatori, G. (eds.) CLIMA XIII 2012. LNCS, vol. 7486, pp. 195–212. Springer, Heidelberg (2012)
10. Konieczny, S., Lang, J., Marquis, P.: Da2 merging operators. Artif. Intell. **157**, 49–79 (2004)
11. Konieczny, S., Pérez, R.P.: Merging information under constraints: a logical framework. J. Logic Comput. **12**(5), 773–808 (2002)
12. Lin, J.: Integration of weighted knowledge bases. Artif. Intell. **83**, 363–378 (1996)
13. Meyer, T., Lee, K., Booth, R.: Knowledge integration for description logics. In: Proceedings of the 20th National Conference on Artificial Intelligence, AAAI 2005, vol. 2, pp. 645–650. AAAI Press (2005)
14. Murray, K.S.: Learning as knowledge integration. Technical report, Austin, TX, USA (1995)
15. Olfati-Saber, R., Fax, J.A., Murray, R.M.: Consensus and cooperation in networked multi-agent systems. Proc. IEEE **95**(1), 215–233 (2007)
16. Qi, G., Du, J., Liu, W., Bell, D.A.: Merging knowledge bases in possibilistic logic by lexicographic aggregation. In: Grünwald, P., Spirtes, P. (eds.) UAI , Proceedings of the Twenty-Sixth Conference on Uncertainty in Artificial Intelligence, Catalina Island, CA, USA, 8–11 July 2010, pp. 458–465. AUAI Press (2010)
17. Qi, G., Liu, W., Bell, D.A.: Merging stratified knowledge bases under constraints. In: AAAI, pp. 281–286 (2006)
18. Qi, G., Liu, W., Bell, D.A.: Combining multiple prioritized knowledge bases by negotiation. Fuzzy Sets Syst. **158**(23), 2535–2551 (2007)
19. Reddy, M.P., Prasad, B.E., Reddy, P.G., Gupta, A.: A methodology for integration of heterogeneous databases. IEEE Trans. Knowl. Data Eng. **6**(6), 920–933 (1994)
20. Ren, W., Beard, R., Atkins, E.: A survey of consensus problems in multi-agent coordination. In: Proceedings of American Control Conference, vol. 3, pp. 1859–1864, June 2005
21. Revesz, P.Z.: On the semantics of arbitration. Int. J. Algebra Comput. **7**, 133–160 (1995)
22. Shaw, M.J., Subramaniam, C., Tan, G.W., Welge, M.E.: Knowledge management and data mining for marketing. Decis. Support Syst. Knowl. Manage. Support Decis. Making **31**(1), 127–137 (2001)
23. Sliwko, L., Nguyen, N.T.: Using multi-agent systems and consensus methods for information retrieval in internet. Int. J. Intell. Inf. Database Syst. (IJIIDS) **1**(2), 181–198 (2007)
24. Tran, T.H., Vo, Q.B.: An axiomatic model for merging stratified belief bases by negotiation. In: Nguyen, N.-T., Hoang, K., Jędrzejowicz, P. (eds.) ICCCI 2012, Part I. LNCS, vol. 7653, pp. 174–184. Springer, Heidelberg (2012)
25. Tran, T.H., Nguyen, N.T., Vo, Q.B.: Axiomatic characterization of belief merging by negotiation. Multimedia Tools and Applications, pp. 1–27, June 2012
26. Tran, T.H., Vo, Q.B., Kowalczyk, R.: Merging belief bases by negotiation. In: König, A., Dengel, A., Hinkelmann, K., Kise, K., Howlett, R.J., Jain, L.C. (eds.) KES 2011, Part I. LNCS, vol. 6881, pp. 200–209. Springer, Heidelberg (2011)
27. Tran, T.H., Vo, Q.B., Nguyen, T.H.K.: On the belief merging by negotiation. In: Jedrzejowicz, P., Jain, L.C., Howlett, R.J., Czarnowski, I. (eds.) 18th International Conference in Knowledge Based and Intelligent Information and Engineering Systems, KES 2014, Gdynia, Poland, 15–17 September 2014, vol. 35, Procedia Computer Science, pp. 147–155. Elsevier (2014)
28. Zhang, D.: A logic-based axiomatic model of bargaining. Artif. Intell. **174**, 1307–1322 (2010)

An Ontology-Based Knowledge Representation of MCDA Methods

Jarosław Wątróbski[1] and Jarosław Jankowski[1,2(✉)]

[1] West Pomeranian University of Technology,
Żołnierska 49, 71-210 Szczecin, Poland
jwatrobski@wi.zut.edu.pl,
jaroslaw.jankowski@pwr.edu.pl
[2] Department of Computational Intelligence,
Wrocław University of Technology,
Wybrzeże Wyspiańskiego 27, 50-370 Wrocław, Poland

Abstract. Multiple-criteria decision analysis methods are widely used as tools supporting a decision problem. The article presents the taxonomy of the methods, which takes into consideration the most essential characteristics. This taxonomy, in the conceptualization process, was written by means of description logic and then it was implemented in the OWL language in the form of ontology representing field knowledge in the scope of MCDA methods. The research also considers the ontology verification prepared with the use of competency questions.

Keywords: Ontology · Multiple-criteria decision analysis (MCDA) · Knowledge management

1 Introduction

Along with the development of operational research, an alternative approach evolution of MCDA methods has been observed. This alternative approach applies both in theoretical studies that result in the continuous development of existing methodologies and techniques, as well as the application layer covering new areas of application methods in business practice. The result of the aforementioned statement is the demand for the development of dedicated approaches adjusted to the specifics of the problem. This is supported by a detailed literature review, where research in various scientific disciplines is effectively conducted with the use of a number of multi-criteria methods [35, 39]. Combined with a variety of specific decision problems discussed by the authors of studies in this area, the natural direction of research can be an attempt to systematize the knowledge in this field [21]. Large heterogeneity of domain knowledge including available scientific publications and the existing decision support systems is an additional prerequisite for undertaking research in this field. In the literature, one can notice attempts to develop models of knowledge representation of MCDA problems and methods areas. For example, the paper [29] demonstrates an ontology designed to describe the structure of decision-making problems. In [31] an ontological representation of the AHP method and a set of inference rules was presented. Earlier studies of

N.T. Nguyen et al. (Eds.): ACIIDS 2016, Part I, LNAI 9621, pp. 54–64, 2016.
DOI: 10.1007/978-3-662-49381-6_6

systematized knowledge about various aspects of decision-making are shown in [32, 33]. Article [32] discusses the use of ontology knowledge model integrating knowledge about decision-making process (i.e. a set of alternatives, preferences). The proposed approach was later extended by additional ontology components [33]. Presented works deal with the problem of systematization of knowledge about the various MCDA methods only to a small extent. The knowledge about the characteristics of the different MCDA methods, their environmental context and use cases [30] is not included in characterized ontologies.

This article constitutes part of wider works which aim is to construct the ontology of MCDA methods which allows to choose a proper method depending on the characteristics of decision problems. This ontology, in its final form, should take into consideration aspects such as: characteristics of individual methods, information about the environmental context of their applications and concrete cases of application of the MCDA methods to solve specific decision problems. The possible construction of such a repository in the form of ontology allows formal specification and analysis of the various MCDA methods, as well as consequent sharing and reusing that domain knowledge [24]. The diagram depicting the construction of discussed ontology is presented in Fig. 1.

Fig. 1. Process of constructing ontology of MCDA methods and their applications

The aim of this paper is to develop the first stage of such solution, i.e. domain ontology containing knowledge model of MCDA methods. In order to construct such a solution, literature related to MCDA methods was reviewed and analyzed. This formed the basis of the development of a taxonomy and ontology. The study was divided into two parts: a discussion of the literature as well as the development of a taxonomy together with the practical verification of author's ontology using competency questions. The work constitutes a continuation of research described in the article [25]. The taxonomy, presented in the work, of a subset of MCDA methods was completed in this study. Furthermore, functional ontology for a broaden set of MCDA methods was worked out and verified. The engineer form of the ontology was presented with the use of the OWL standard and is available online [37].

2 Methods of Multi-Criteria Decision Analysis

The development of two main groups of methods and directions: approaches based on value/utility theory and outranking relations [19] is based on the research into the MCDA area. The utility theory-based approach derives from the American MCDA school. Two types of relationships between alternatives are determine: indifference (a_i I a_j) and preference (a_i P a_j) of one alternative over another. The methods in this group leave out non-comparability of the decision variants and assume transitivity and completeness of preference [19]. Methods based on outranking relations stem from the European MCDA school. Methods from this group frequently expand a set of basic preferential situations with the result that includes indifference of decision variants (a_i I a_j), weak preference one variant over another (a_i Q a_j), the strict preference of a variant of the decision-making relative to the other (a_i P a_j), and incomparability between data variations (a_i R a_j) [34]. Moreover, the preferential situations can be combined in "outranking" relation which contains the situations of indifference, strong and weak preference (a_i S a_j) [34]. In the literature two basic operational approaches can be distinguished to aggregate performance of variants: (1) aggregate to a single criterion (American school), (2) aggregation by using the outranking relationship (European MCDA school) [34]. Also, mixed (indirect) approaches, which combine elements of American and European decision-making schools, are applied. An example of this approach can be a group of PCCA (Pairwise Criterion Comparison Approach) methods [18]. A number of researchers acknowledge that the discussed groups of methods also differ in the occurrence of the criteria compensation effect. The compensation itself is that bad performance on one attribute can be compensated by good performance on other attributes [34]. While the difference between the two discussed groups of methods lies in the fact that in methods based on the value/utility theory there is a compensation, whereas methods employing the outranking relation by many researchers are considered non-compensatory [19]. Roy specifies that the difference refers to operational approaches in particular [34]. However, other researchers claim that the methods employing an exceeding relation are characterized by partial compensation [20]. Particularly Guitouni and Martel [21] state that there are no unanimous definitions or principles to characterize the degree of compensation. They distinguish three degrees of compensation (1) absolute compensation - a good performance on one criterion can easily counterbalance a poor one on another, e.g. weighted sum; (2) no compensation - some dimensions are important enough to refuse any kind of compensation or trade-offs, e.g. lexicographic method; (3) partial compensation - some kind of compensation is accepted between the different dimensions or criteria. They classify the majority of American and European Schools methods as the last group. MCDA methods also differ in nature and characteristics of data which are used in them [21]. The nature of date is closely related to the measurement scale. The data can be qualitative or quantitative and therefore can be expressed in the ordinal (qualitative) or cardinal (quantitative) scale. Moreover, the cardinal scale can be ratio or interval [22]. The characteristics of the data used refers to whether the data is certain or not [19]. Certain data, named also deterministic, is expressed in a crisp form, whereas uncertain data (non-deterministic) is represented by some kind of distribution (discrete or continuous) [21]. Furthermore, many new methods based on the fuzzy set theory enable to express uncertain data in a fuzzy form [34]. All the elements characterized above were taken into consideration in the prepared taxonomy and ontology. Based on an analysis of the literature, a complex set of available MCDA methods was identified. Part of the set was presented in the paper [25], whereas its development was included in this article.

In the future research one needs to take into consideration the decision-making issues considered by means of individual methods and their characteristics resulting from the uniqueness of a decision problem (e.g. the ability of a method to apply qualitative, quantitative or relative criteria weights, the ability to compare the productivity of variants, applying threshold values for the criteria comparisons of variants).

3   Constructing Ontology of Multi-Criteria Decision Analysis Methods

In the literature ontology is treated as the specification of conceptualization providing a description of the concepts and relationships that take place between them [27]. The application of ontologies as a solution supporting the choice of a given MCDA method is designed to assist the user in selecting the proper solution for a particular decision situation described using a specific criteria. Also, the ontology ought to provide detailed information about various MCDA methods. The first action in the construction of ontology is to develop a taxonomy of the MCDA methods. The identification and analysis of 20 MCDA methods was allowed to create a set of criteria and sub-criteria characterizing different solutions. A total set was formed comprising four main criteria (available binary relations, linear compensation effect, the type of aggregation and the type of preferential information) as well as 16 sub-criteria. This collection was the basis for the construction of taxonomies of analyzed solutions as well. Table 1 depicts the taxonomy of a subset of MCDA methods. The individual positons of Table 1 are characterized in Chapter 2. The subset extends the state of authors' research presented in [25].

Taxonomy presented in Table 1 should be converted to an ontological form and requires distinguishing the concept on the basis of criteria and sub-criteria and establishing their hierarchy [26]. In the ontology there are four types of taxonomic relations: the conclusion of the concepts, concepts separation, division, and total partition. Containment (subsumption) (Subclass-Of) concept C_1 in the concept of C means that C_1 is a subclass of (detailing of) C. This is due to concept C_1's inheritance of attributes of concept C. The subsumption of the concepts can be understood as the inclusion of the sets, as shown in Fig. 2(a). Severability (Disjoint-Decomposition) concepts C_1 and C_2 containing the concept of C means each occurrence (instance) of concept C affects the simultaneous occurrence of concept C_1 or C_2, but the occurrence of C_1 and C_2 cannot be at the same time. Furthermore, it may be the instance of C in the absence of the occurrence of concepts C_1 or C_2. Acceptable occurrences of concepts (instances I_1, I_2, I_3) while maintaining severability are shown in Fig. 2(b). The complete division (Exhaustive-Decomposition) concepts C_1 and C_2 containing a concept C is that each instance of C must be occurrence concept C_1, C_2 or both C_1 and C_2. In other words, the occurrence of concept C is also contained in the occurrence of total concepts C_1 and C_2. This situation is shown in Fig. 2(c) where a partition created with concepts C_1 and C_2 contained in the concept C is that each instance of concept C is also the occurrence concept C_1 or C_2. Partition concepts C_1 and C_2 can be understood as the sum of disjoint sets, as shown in Fig. 2(d). Figure 3 depicts a graphical diagram of a set of criteria and sub-criteria of constructed ontology. The authors decided to use concepts in the ontology, since the instances of concepts had been reserved in this case for reference

Table 1. Taxonomy of selected MCDA methods

Criterion	Available binary relations					Linear compensation effect			Type of aggregation			Type of preferential information					
Method name	I	P	Q	R	S	No	Total	Partial	Single criterion	Outranking	Mixed	Deterministic	Cardinal	Non-deterministic	Ordinal	Fuzzy	Reference
IDRA	Y	Y						Y			Y	Y	Y	Y	Y		[1]
MAPPAC	Y	Y	Y	Y				Y			Y	Y	Y	Y			[2]
PRAGMA	Y	Y	Y					Y			Y	Y	Y	Y			[3]
PACMAN	Y	Y	Y	Y				Y			Y	Y	Y	Y		Y	[4]
ARGUS	Y		Y	Y				Y		Y		Y			Y	Y	[5]
QUALIFLEX	Y		Y	Y				Y		Y		Y				Y	[6]
Lexicographic method	Y	Y				Y			Y			Y	Y			Y	[7]
TACTIC	Y	Y	Y					Y		Y		Y	Y	Y			[8]
MACBETH	Y	Y						Y	Y			Y	Y	Y	Y		[9]
Fuzzy AHP	Y	Y						Y	Y			Y	Y	Y		Y	[10]
ANP	Y	Y						Y	Y			Y	Y	Y			[11]
Fuzzy ANP	Y	Y						Y	Y			Y	Y	Y		Y	[12]
Fuzzy PROMETHEE I	Y	Y	Y					Y		Y		Y	Y	Y	Y	Y	[13]
Fuzzy PROMETHEE II	Y	Y						Y		Y		Y	Y	Y	Y	Y	[13]
NAIADE I				Y	Y			Y		Y		Y	Y	Y	Y	Y	[14]
NAIADE II					Y			Y		Y		Y	Y	Y	Y	Y	[14]
PAMSSEM I				Y	Y			Y		Y		Y	Y	Y	Y	Y	[15]
PAMSSEM II					Y			Y		Y		Y	Y	Y	Y	Y	[15]
Fuzzy TOPSIS	Y	Y					Y		Y			Y	Y	Y		Y	[16]
COMET	Y	Y					Y		Y			Y	Y	Y	Y	Y	[17]

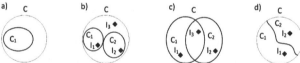

Fig. 2. Taxonomic relations between ontology concepts: (a) subsumption, (b) disjoint-decomposition, (c) exhaustive-decomposition, (d) partition

literature cases of applying individual methods in accordance with the structure of a decision problem. Such instances will be attached to the ontology in the future.

Part of ontology (set of criteria) written in the form of description logics [28] was concluded in expressions (1) – (10). Parts (1) – (4) of expressions indicate the criteria containing a (subsumption) in the concept of "Criterion". Expression (5) means the individual criteria are disjoint. Separation is used here because the individual criteria are independent of each other, but the taxonomy can be added to the new criteria. Records (6) – (8) describe the contents of the concept of "Linear compensation effect". The expressions (9) and (10) define a partition of concepts included in the criterion of "Linear compensation effect". It should be done due to the fact that the content of the concept of "Linear compensation effect" is complete and will not be added to it in the future. In addition, one method may meet only one of the sub-criteria (e.g. only supports the "Partial linear compensation effect"). In a similar manner a space is defined as the criterion "Type of aggregation".

$$\text{Linear compensation effect} \subseteq \text{Criterion} \tag{1}$$

$$\text{Available binary relations} \subseteq \text{Criterion} \tag{2}$$

$$\text{Type of aggregation} \subseteq \text{Criterion} \tag{3}$$

$$\text{Type of preferential information} \subseteq \text{Criterion} \tag{4}$$

$$\text{Linear compensation effect} \equiv \neg \text{Available binary relations} \ldots \\ \ldots \equiv \neg \text{Type of aggregation} \equiv \neg \text{Type of preferential information} \tag{5}$$

$$\text{No linear compensation effect} \subseteq \text{Linear compensation effect} \tag{6}$$

$$\text{Partial linear compensation effect} \subseteq \text{Linear compensation effect} \tag{7}$$

$$\text{Total linear compensation effect} \subseteq \text{Linear compensation effect} \tag{8}$$

$$\text{Linear compensation effect} \equiv \text{No linear compensation effect} \ldots \\ \ldots \cup \text{Partial linear compensation effect} \cup \text{Total linear compensation effect} \tag{9}$$

$$\text{No linear compensation effect} \equiv \neg \text{Partial linear compensation effect} \ldots \\ \ldots \equiv \neg \text{Total linear compensation effect} \tag{10}$$

A bit otherwise specified content criteria include "Type of preferential information" and "Available binary relations". In the case of the criterion "Type of preferential information" and its sub-criteria, complete division was applied, which describes the expressions (11) – (16). Complete division was used due to the fact that different methods of MCDA can simultaneously use different types of preferential information, but there is no other type of preferential information than those in the concept of "Type of preferential information" (the contents of this concept is complete).

$$\text{Cardinal} \subseteq \text{Type of preferential information} \tag{11}$$

$$\text{Fuzzy} \subseteq \text{Type of preferential information} \tag{12}$$

$$\text{Non-deterministic} \subseteq \text{Type of preferential information} \tag{13}$$

$$\text{Deterministic} \subseteq \text{Type of preferential information} \tag{14}$$

$$\text{Ordinal} \subseteq \text{Type of preferential information} \tag{15}$$

$$\text{Type of preferential information} \equiv \text{Cardinal} \cup \text{Fuzzy}\ldots$$
$$\ldots \cup \text{Non-deterministic} \cup \text{Deterministic} \cup \text{Ordinal} \tag{16}$$

In a similar way the space of criterion "Available binary relations" was defined. Inside criterion "Available binary relations" including relations R, S, I, P, Q complete division was applied. Exhaustive-decomposition was used due to the fact that the other type of relation between variants evaluated with the use MCDA methods does not exist. Meanwhile these relations can exist together in single method. The ontology offers a set of MCDA methods shown in Table 1, with a set of differentiating criteria and a network of taxonomic relationships between concepts (relations between different classes of instances). Using this ontology, it is possible to select methods based on selected criteria. This is the base for a simple reusable but structured domain knowledge area. Based on preset criteria a user can obtain detailed information about the satisfying method (methods) with its specific taxonomic characteristics. A sample set of results is depicted in Fig. 3a, illustrating a method (here Promethee II) which met the criteria for the query: binary relations P and I, the partial effect of linear compensation, aggregation using outranking relations, the type of preferential information – ordinal, cardinal and deterministic. To answer the ontology's competence question, the Protege editor's extension named "DL Query" was used. The tool allows formulating questions and asking the ontology the questions in accordance with the Manchester OWL (Web Ontology Language) [23] syntax and writing the question in the form of the ontology classes. The question had the form of: *"MCDA_Method and (hasCriterion some P and hasCriterion some I and hasCriterion some PartialLinearCompensationEffect and hasCriterion some OutrankingAggregation and hasCriterion some Ordinal and hasCriterion some Cardinal and hasCriterion some Deterministic)"*. A sample reasoning process [40] has the following course in this case: *PROMETHEE_II SubClassOf hasCriterion some I; isCriterion inverseOf hasCriterion; isCriterion Range MCDA_-Method; PROMETHEE_II SubClassOf hasCriterion some P; PROMETHEE_II SubClassOf hasCriterion some OutrankingAggregation; PROMETHEE_II SubClassOf hasCriterion some PartialLinearCompensationEffect; PROMETHEE_II SubClassOf hasCriterion some Ordinal; PROMETHEE_II SubClassOf hasCriterion some Cardinal; PROMETHEE_II SubClassOf hasCriterion some Deterministic*. The key stage in the presented reasoning is concluding a reverse relation isCriterion and hasCriterion as well as determining the scope of the relation isCriterion to the concept MCDA_-Method. On the basis a reasoning mechanism is able to conclude that the concept PROMETHEE_II is a subclass of the concept MCDA_Method. A further query to the ontology was created using the SPARQL language [36]. Inquiries to the knowledge base in SPARQL may relate only to the knowledge stored permanently and not that up to date by the inference. This allows a new structure of the knowledge base to be obtained that contains all the relationships between concepts and instances established through the mechanism of the applicant. Having deduced form prepared query ontology instances of multi-criteria methods, which use binary relations P and I and the aggregate results of the evaluation using outranking relations. The structure of competence query is listed below:

```
SELECT ?Method ?AvailableBinaryRelations ?TypeOfAggregation
WHERE {        ?Method rdf:type :MCDA_Method.
               ?Method rdf:type :MCDA_BinaryRelationP.
               ?Method rdf:type :MCDA_BinaryRelationI.
               ?Method rdf:type :MCDA_OutrankingAggregation.
               ?Method rdf:type ?AvailableBinaryRelations.
FILTER( REGEX(STR(?AvailableBinaryRelations),'BinaryRelation')).
               ?Method rdf:type ?TypeOfAggregation.
FILTER( REGEX(STR(?TypeOfAggregation),'Aggregation'))}
```

In the clause "SELECT" are defined variables that are to be displayed in the results, and in the clause "WHERE" are defined relationships that should exist between the variables. The relation of "rdf: type" specifies instances of a particular class. The competence query results are shown in Fig. 3b. The use of ontology as a tool to support the selection of the MCDA method allows a solution to be chosen that takes into account user-defined criteria on the basis of which only the MCDA methods or literature reference solutions that meet user-specified environmental determinants and decision-making are designated.

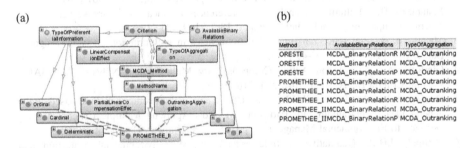

Fig. 3. Results of the Manchester OWL competence query (a), and results of the SPARQL language competence query (b)

The same ontology contains the complete set of domain knowledge about the MCDA methods. This ontology has been built using Protege 4, using the OWL [28]. The MCDA ontology is available in [37] and the effect of the reasoner is given in [38].

4   Conclusion

This article deals with the issue of the construction of ontology of MCDA methods. On the basis of the analysis of MCDA methods a taxonomy characterizing the different solutions was demonstrated. It constituted the basis for the construction of ontology of

MCDA methods. The findings confirmed the possibility of the conceptualization of knowledge in the area of MCDA methods. The application of the proposed ontology supports the decision-maker's correct choice of a multi-criteria method and allows for full domain knowledge about each one. It ought to be noted that the standard employed for the construction of the ontology ensures compliance with international semantic standards. This makes it possible to further use the developed solution as well as its connection to other ontologies in various fields within the growing trend of knowledge engineering. Additional research needs to be supplemented by ontology of reference cases of the application of each method in various areas (management, logistics, environment, medicine, etc.). For ontology, further criteria characterizing the various methods and the environmental context of their use can be attached. It allows for the greater use of the adequacy of the reasoner and asks for the use of various methods in decision problems using SWRL language rules.

Acknowledgments. The work was partially supported by European Union's Seventh Framework Programme for research, technological development and demonstration under grant agreement no 316097 [ENGINE] and by the National Science Centre, the decision no. DEC-2013/09/B/ST6/02317.

References

1. Greco, S.: A new PCCA method: IDRA. Eur. J. Oper. Res. **98**(3), 587–601 (1997)
2. Matarazzo, B.: Multicriterion analysis of preferences by means of pairwise actions and criterion comparisons. Appl. Math. Comput. **18**(2), 119–141 (1986)
3. Matarazzo, B.: Preference ranking global frequencies in multicriterion analysis (Pragma). Eur. J. Oper. Res. **36**(1), 36–49 (1988)
4. Giarlotta, A.: Passive and Active Compensability Multicriteria ANalysis (PACMAN). J. Multi-Criteria Decis. Anal. **7**(4), 204–216 (1998)
5. De Keyser, W.S.M., Peeters, P.H.M.: ARGUS – a new multiple criteria method based on the general idea of outranking. In: Paruccini, M. (ed.) Applying Multiple Criteria Aid for Decision to Environmental Management, pp. 263–278. Kluwer, Dordrecht (1994)
6. Paelinck, J.H.P.: Qualitative multiple criteria analysis, environmental protection and multiregional development. Pap. Reg. Sci. Assoc. **36**(1), 59–74 (1976)
7. Fishburn, P.C.: Exceptional paper-lexicographic orders, utilities and decision rules: a survey. Manage. Sci. **20**(11), 1442–1471 (1974)
8. Vansnick, J.C.: On the problem of weights in multiple criteria decision making (the noncompensatory approach). Eur. J. Oper. Res. **24**(2), 288–294 (1986)
9. Bana e Costa, C.A., Vansnick, J.C.: MACBETH — An interactive path towards the construction of cardinal value functions. Int. Trans. Oper. Res. **1**(4), 489–500 (1994)
10. Mikhailov, L., Tsvetinov, P.: Evaluation of services using a fuzzy analytic hierarchy process. Appl. Soft Comput. **5**(1), 23–33 (2004)
11. Saaty, T.L.: The Analytic Network Process. RWS Publications, Pittsburgh (2001)
12. Promentilla, M.A.B., Furuichi, T., Ishii, K., Tanikawa, N.: A fuzzy analytic network process for multi-criteria evaluation of contaminated site remedial countermeasures. J. Env. Manag. **88**(3), 479–495 (2008)

13. Wang, T.C., Chen, L.Y., Chen, Y.H.: applying fuzzy PROMETHEE method for evaluating IS outsourcing suppliers. In: Fifth International Conference on Fuzzy Systems and Knowledge Discovery, vol. 3, pp. 361 – 365 (2008)
14. Munda, G.: Multicriteria Evaluation in a Fuzzy Environment. Theory and Applications in Ecological Economics. Physica-Verlag, Heidelberg (1995)
15. Guitouni, A., Martel, J.M., Belanger, M., Hunter, C.: Managing a Decision Making Situation in the Context of the Canadian Airspace Protection. DOCUMENT DE TRAVAIL 1999-021 (1999)
16. Chen, C.T., Lin, C.T., Huang, S.F.: A fuzzy approach for supplier evaluation and selection in supply chain management. Int. J. Prod. Econ. **102**(2), 289–301 (2006)
17. Sałabun, W.: The characteristic objects method: a new distance-based approach to multicriteria decision-making problems. J. Multi-Criteria Decis. Anal. **22**(1–2), 37–50 (2015)
18. Martel, J.M., Matarazzo, B.: Other outranking approaches. In: Figueira, J., Greco, S., Ehrgott, M. (eds.) Multiple Criteria Decision Analysis: State of the Art Surveys, pp. 197–262. Springer, Boston (2005)
19. Bana e Costa, C.A., Vincke, P.: Multiple criteria decision aid: an overview. In: Bana e Costa, C.A. (ed.) Readings in Multiple Criteria Decision Aid, pp. 3–14. Springer, Berlin Heidelberg (1990)
20. Bagheri Moghaddam, N., Nasiri, M., Mousavi, S.M.: An appropriate multiple criteria decision making method for solving electricity planning problems, addressing sustainability issue. Int. J. Env. Sci. Technol. **8**(3), 605–620 (2011)
21. Guitouni, A., Martel, J.M.: Tentative guidelines to help choosing an appropriate MCDA method. Eur. J. Oper. Res. **109**(2), 501–521 (1998)
22. Roy, B.: Paradigms and challenges. In: Figueira, J., Greco, S., Ehrgott, M. (eds.) Multiple Criteria Decision Analysis: State of the Art Surveys, pp. 3–24. Springer, Boston (2005)
23. W3C Working Group, http://www.w3.org/TR/owl2-manchester-syntax/
24. Hepp, M.: Ontologies: state of the art, business potential, and grand challenges. In: Hepp, M., de Leenheer, P., de Moor, A., Sure, Y. (eds.) Ontology Management. Semantic Web, Semantic Web Services, and Business Applications, pp. 2–23. Springer, Heidelberg (2008)
25. Wątróbski, J., Jankowski, J.: Knowledge management in MCDA domain. In: Proceedings of the Federated Conference on Computer Science and Information Systems. Annals of Computer Science and Information Systems, vol. 5, pp. 1445–1450 (2015)
26. Ziemba, P., Jankowski, J., Watróbski, J., Becker, J.: Knowledge management in website quality evaluation domain. In: Núñez, M., Nguyen, N.T., Camacho, D., Trawinski, B. (eds.) ICCCI 2015. LNCS, vol. 9330, pp. 75–85. Springer, Heidelberg (2015). doi:10.1007/978-3-319-24306-1_8
27. Gruber, T.R.: A translation approach to portable ontology specifications. Knowl. Acquisition **5**(2), 199–220 (1993)
28. http://protege.stanford.edu/
29. Chai, J., Liu, J.N.K.: An ontology-driven framework for supporting complex decision process. In: World Automation Congress (WAC) (2010)
30. Wątróbski, J., Jankowski, J., Piotrowski, Z.: The selection of multicriteria method based on unstructured decision problem description. In: Hwang, D., Jung, J.J., Nguyen, N.-T. (eds.) ICCCI 2014. LNCS, vol. 8733, pp. 454–465. Springer, Heidelberg (2014)
31. Liao, X.Y., Rocha Loures, E., Canciglieri, O., Panetto, H.: A novel approach for ontological representation of analytic hierarchy process. Adv. Mater. Res. **988**, 675–682 (2014)
32. Kornyshova, E., Deneckere, R.: Using an ontology for modeling decision-making knowledge. Front. Artif. Intell. Appl. **243**, 1553–1562 (2012)

33. Kornyshova, E., Deneckère, R.: Decision-Making ontology for information system engineering. In: Parsons, J., Saeki, M., Shoval, P., Woo, C., Wand, Y. (eds.) ER 2010. LNCS, vol. 6412, pp. 104–117. Springer, Heidelberg (2010)
34. Roy, B.: Multicriteria Methodology for Decision Aiding. Springer, Dordrecht (1996)
35. Velasquez, M., Hester, P.T.: An analysis of multi-criteria decision making methods. Int. J. Oper. Res. **10**(2), 56–66 (2013)
36. Della Valle, E., Ceri, S.: Querying the semantic web: SPARQL. In: Domingue, J., Fensel, D., Hendler, J.A. (eds.) Handbook of Semantic Web Technologies, pp. 299–363. Springer, Berlin (2011)
37. http://tinyurl.com/ontoMCDAext
38. http://tinyurl.com/ontoMCDAext-inferred
39. Piegat, A., Sałabun, W.: Comparative analysis of MCDM methods for assessing the severity of chronic liver disease. In: Rutkowski, L., Korytkowski, M., Scherer, R., Tadeusiewicz, R., Zadeh, L.A., Zurada, J.M. (eds.). LNCS, vol. 9119, pp. 228–238Springer, Heidelberg (2015)
40. Ziemba, P., Jankowski, J., Wątróbski, J., Wolski, W., Becker, J.: Integration of domain ontologies in the repository of website evaluation methods. In: Proceedings of the Federated Conference on Computer Science and Information Systems. Annals of Computer Science and Information Systems, vol. 5, pp. 1585–1595 (2015)

Preliminary Evaluation of Multilevel Ontology Integration on the Concept Level

Adrianna Kozierkiewicz-Hetmańska[✉] and Marcin Pietranik

Faculty of Computer Science and Management, Wroclaw University of Technology,
Wybrzeze Wyspianskiego 27, 50-370 Wroclaw, Poland
{adrianna.kozierkiewicz,marcin.pietranik}@pwr.edu.pl

Abstract. In many real situations it is not possible to merge multiple knowledge bases into a single one using one-level integration. It could be caused, for example, by high complexity of the integration process or geographical distance between servers that host knowledge bases that expected to be integrated. The paralleling of integration process could solve this problem and in this paper we propose a multi-level ontology integration procedure. The analytical analysis pointed out that for presented algorithm the one- and multi-level integration processes give the same results (the same final ontology). However, the multi-level integration allows to save time of data processing. The experimental research demonstrated a significant difference between times required for the one- and multi-level integration procedure. The latter could be even 20 % faster than the former, which is important especially in the emerging context of Big Data. Due to the limited space we can only consider integration on the concept level.

1 Introduction

The processing of big sets of data is becoming essential problem in case of a company management. It could be stored in different, complex structures and reveal potential inconsistencies, therefore, its processing it is not an easy task. Especially, the integration of such datasets (combining a few separate data source into single one) can be both time- and cost-consuming due to the computational complexity of this process.

Let us imagine a situation in which some company needs to process a large amount of financial data coming from many different sources. Based on a final knowledge base obtained during such integration, the company's board would like to make some decisions about new investments. Too long time of processing could not be a problem in case of a longterm investments with distant deadlines. However, in many situations decisions such as selling or buying new assets, should be made quickly, even in real time and a potential delay could bring potential losses for the company. In other words, the dynamically changing environment requires easy and fast methods for data management and the time of data processing seems to be critical element for companies which need to make decisions based on a Big Data that constantly appear from different sources.

© Springer-Verlag Berlin Heidelberg 2016
N.T. Nguyen et al. (Eds.): ACIIDS 2016, Part I, LNAI 9621, pp. 65–74, 2016.
DOI: 10.1007/978-3-662-49381-6_7

Obviously ontologies, which are the main topic of the following article, shouldn't be treated only as raw data, but more complex knowledge representations. Nevertheless, the context of gathering large amount of data from different sources that can be further processed and eventually obtain intentional semantics require not only effective methods of aforementioned data processing, but also equally effective methods for dealing with large-scale knowledge bases.

In this paper we propose a procedure for ontology integration which can serve as such source. Due to the structure of these knowledge bases, their integration needs to be done on three levels: the concept's level, the instance's level and the relation's level. Due of the limited space available, authors concentrate only on a concept level, using an algorithm taken from [10]. The definition of the multi-level integration process is proposed and the results of one- and multi-level integrations are analysed analytically. However, as it was mentioned, in the case of Big Data, the critical issue is the time required for the integration. Therefore, we have used a set of example ontologies and alignments between them, in order to compare the times required by one- and two-level integration procedures to designate final results. To conduct described comparison, a dedicated experimental environment has been implemented using Python programming language and eventually used.

The remaining part of this paper is organized as follows. In the next section we give a brief summary of related works. Section 3 contains the introduction to ontologies and basic notions used throughout our research. In Sect. 4 the multi-level integration procedure is presented. Section 5 describes the results of analytical and experimental analysis of one- and multi-level ontology integrations. Section 6 concludes this paper.

2   Related Works

Since ontologies are becoming more and more popular, the problem of their integration (also referenced as merging) and their mapping are becoming very important. Cruz and Xiao in [4] discussed the role of ontologies in data integration. They considered two different settings depending on the system architecture: central and peer-to-peer data integration.

The problem of ontology integration were raised in many papers. In [15] authors describe activities that compose integration process like: identifying integration possibility, decomposition into modules of integration, initial assumptions and ontological commitments. In general, the process can be decomposed into choosing the right representation of knowledge in each module, selecting candidate ontologies, studying and analysing candidate ontologies, choosing source ontologies, applying integration operations and finally processing a resulting ontology. For each stage of a methodology it provides support and guidance to perform those activities.

In [2] authors presented the basic framework for ontology integration. They tried to answer how to specify the mapping between the global ontology and local ontologies and eventually have proposed a mechanisms based on queries. Noy and Musenl described a general approach to ontology merging and alignment called SMART [12] and PROMPT [13].

Li and his team in [8] described an agent-based framework of integration of similar ontologies coexisting in a distributed and heterogeneous environment. The basic remark which served as in initial inspiration was the fact that with the presence of ontology agents, newly generated ontologies can be reused many times. The proposed solution was tested in a prototype system implemented using Jade framework. Considered research pointed out that the proposed framework provides a flexible and effective modelling approach to tackle the integration over a variety of ontologies.

In [1] a hybrid approach for ontology integration is proposed. Authors distinguished two major approaches to integration of information: (i) the data warehouse (materialised approach) and (ii) a virtual approach (also referenced as mediator-based approach). They took advantages of both and proposed a hybrid framework.

In [5] authors have presented a set of methods facilitating the integration of independently developed ontologies using mappings.

In [10] author defined ontology and subsequently described some integration techniques. Due to the accepted ontology definition the integration process were decomposed into three levels: the concept's level, the relation's level and the instance's level. For each of these levels the suitable methods were proposed and analysed.

The integration on two or more levels is a new idea and so far it has not attracted much attention in literature. There are however some papers like [6,7] or [11] that address the one- and two-level consensuses and the problem of its' determination. Authors have developed a formal framework that can be easily used to designate the consensus in one- and two- steps for assumed macrostructure and microstructures. Next, for some criteria the analysis of obtained consensuses were made. The researches demonstrated that both one-level and two-level consensus in comparison to the optimal solution give acceptable results. Nevertheless, to the best of our knowledge the challenges of the multi-level ontology integration topic were not frequently addressed.

In our previous research we have also focused on the problem of ontology alignment [14] which can be treated as a pre-step to any other ontology integration process. In general, the task of designating an alignment an be described as a process of selecting elements of compared ontologies that refer to the same object taken from considered universe of discourse [16]. What distinguishes our work from other research is the fact that we have developed a framework built around four functions (namely $\lambda_A, \lambda_C, \lambda_R$ and λ_I) that are used to calculate the degree to which certain elements from some selected source ontology can be mapped to elements from a target ontology. What is worth emphasising, is the fact that these functions are not symmetrical. The reason behind this comes from straightforward remark - it is easier to align detailed representation of some object (no matter if it is an attribute, a concept, a relation etc.) into general representation that to map broad description into precise one without any loss of information. Therefore, our framework does not designate the closeness of two ontological elements, but the amount of knowledge that can be unequivocally transformed. Obviously, the above consideration does not entail formal asymmetry of concerned functions λ_A, λ_C, λ_R, λ_I.

3 Basic Notions

Lets assume that a pair (A, V) represents some real world in which A denotes a set of attributes that can describe objects from that world and V denotes a set of valid valuations of these attributes. In other words, we can say that $V = \bigcup_{a \in A} V_a$ where V_a is a domain of an attribute a.

On the simplest level we define ontologies as a following triple:

$$O = (C, R, I) \tag{1}$$

where C is a finite set of concepts, R is a finite set of relations between concepts $R = \{r_1, r_2, ..., r_n\}$, $n \in N$, $r_i \subset C \times C$ for $i \in [1, n]$ and I is a finite set of instances.

Elements of the set of concepts (also referenced as classes) C are defined as follows:

$$c = (Id^c, A^c, V^c) \tag{2}$$

where Id^c is a unique label, A^c is a set of attributes assigned to such concept and V^c is a set of domains of these attributes ($V^c = \bigcup_{a \in A^c} V_a$).

If the criteria $\forall_{c \in C} A^c \subseteq A$ and $\forall_{c \in C} V^c \subseteq V$ are met we can say that an ontology O is (A, V)-based.

Attributes from the set A by themselves do not carry any particular meaning. They obtain semantics by being included within particular concepts. In order to formally express it we need a set D_A of their atomic descriptions (e.g. *year_of_birth*) and in consequence a sublanguage of the sentence calculus constructed with members of D_A and elementary logic operators of conjunction, disjunction and negation. Eventually the semantics of attributes is given by a function:

$$S_A : A \times C \rightarrow L_s^A \tag{3}$$

The above equation allows to specify roles that variety of attributes obtain when they get included into different concepts. For example, an attribute *Address* means something different when used in the context of a concept *Home* and different when included in the concept *Website*. Furthermore, such approach to expressing attributes' semantics gave us a possibility to formally define *equivalency* (denoted as $\equiv$), *generalization* (denoted as $\uparrow$) and *contradiction* (denoted as $\downarrow$) between attributes [14].

We also accept the existence of a set D_R containing descriptions of relations.

By analogy, L_s^R denotes another sublanguage of the sentence calculus that is used to define a function that gives semantics of relations from the set R:

$$S_{R,O} : R \rightarrow L_s^R \tag{4}$$

Hence, we have provided a set of criteria for relationships between relations including equivalency, generalisation and contradiction.

An instance i (a member of the set I) of some concept is defined as a triple $i = (id, A_i, v_i)$, where id is its unique identificator, A_i stands for a set of assigned attributes and v_i denotes a function $v_i : A_i \to \bigcup_{a \in A_i} V_a$ which assigns values from the corresponding sets V_a to particular elements of the set A_i. We say that $i = (id, v_i, A_i)$ is an instance of a concept $c = (Id^c, A^c, V^c)$ only if $A^c \subseteq A_i$ and $\forall_{a \in A_i \cap A^c} v_i(a) \in V^c$. For convenience we will use the notation $Ins(O, c)$ to denote a set of instances of a concept c within ontology O.

4   Multi-level Integration

Out of many ways of defining the knowledge integration, we can describe it as a process of joining several, independent knowledge bases (in our case - ontologies) into a single one. In some cases it is impossible to do so during only one-level integration due to high complexity of required transformations or simply geographical distance between them that entails unacceptable latency due to too large data transfer.

A multi-level knowledge integration, i.e. simultaneously combining knowledge from a small number of sources for many subgroups and the eventual merging of the results into the one final knowledge base, might be applied to solve the described issue. The general idea for such approach is presented in Fig. 1.

The problem of ontology integration can be formulated as follows: *for given n ontologies $O_1, O_2, ..., O_n$ one should determine an ontology $O*$ which represents given ontologies in the best way.* As it was mentioned, the integration process can be conducted on one level or in special cases on two or more levels. The definitions of one level and multi level ontology integration process is presented below:

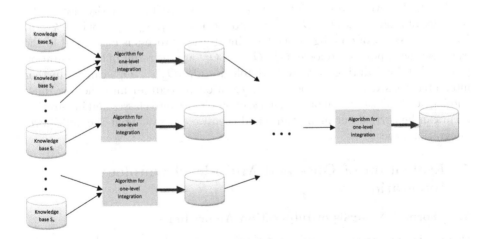

Fig. 1. The general idea for a multi-level integration process.

Definition 1. *The input of the one-level integration process is a sequence of* n *ontologies:* $O_1^1, O_2^1, ..., O_n^1$. *The output of the integration process is a single ontology* O^{1*}, *which is in multiple relationships with input ontologies, as defined by a group of criteria. Integration criteria* $K_1^1, K_2^1, ..., K_n^1$ *are the parameters of the integration task and tying* O^{1*} *with* $O_1^1, O_2^1, ..., O_n^1$ *each at least at a given level* $\alpha_1^1, \alpha_2^1, ..., \alpha_n^1$ $K_i^1(O^{1*}|O_1^1, O_2^1, ..., O_n^1) \geq \alpha_i^1$.

Based on the Definition 1 the multi-level integration is defined as follows:

Definition 2. *Let* $O_1^{m-1*}, O_2^{m-1*}, ..., O_n^{m-1*}$ *be ontologies obtained during* $m-1$ *level of the knowledge integration, where* $m \geq 2$. *The output of the* m-*level of integration is a single ontology* O^{m*}, *which is in multiple relationships with input structures, as defined by a group of criteria:* $K_1^m, K_2^m, ..., K_n^m$.

According to the literature [9], the following integration criteria are known: *completeness* (after the integration no data/elements are lost), *minimality* (the output of the integration is not much larger than its inputs), *precision* (the integration does not duplicate data), *optimality* (the output of the integration is the closest to inputs, in terms of some distance measure), *sub-tree agreement* (the output includes all the sub-trees from its inputs).

Due to the structure of an ontology which consists of three main elements: concepts, relations and instances, the problem of one-level ontology integration should be conducted in three steps: integration of concepts, integration of relations and integrations of instances. This problem has been solved in [10] where author decomposed problem of ontology integration into three phases and for each phase the appropriate algorithm were proposed. Integration on an instance level were solved using consensus methods, integration on a concept level required defining some additional postulates and an algorithm for relational level includes in the final set of relations only those relations which appear most often in the ontologies, and do not cause any contradiction.

The multi-level ontology integration task required to primarily divide the sequence of n ontologies $O_1, O_2, ..., O_n$ into k classes $X_1, X_2, ..., X_k$ where $k < n$. For each class X_i of ontologies one-level integration process is conducted in the way described above. Ontologies $O_1^{2*}, O_2^{2*}, ..., O_k^{2*}$ are the result of the integration process obtained during 2nd level. Ontologies $O_1^{2*}, O_2^{2*}, ..., O_k^{2*}$ can be further integrated (based on basic one-level integration procedure) into the one, final ontology O^{2*}. The division of a sequence of ontologies into classes and integrating them can be carried out many times and then we can say about the multi-level integration process.

5 Evaluation of One- and Multi-level Ontology Integration

5.1 Formal Analysis of Integration Algorithm

Due to the limited space available for this paper we have focused only on the evaluation of one- and multi-level concept integration. The base algorithm taken from [10] is conducted in the following steps:

Algorithm 1. Concept integration

Require: Concepts structures (A^i, V^i);

1: Set $A^* = \bigcup\limits_{i=1}^{n} A^i$;
2: **for all** $(a, b) \in A^* \times A^*$ **do**
3:     **if** $a \uparrow b$ **then**
4:         $A^* = A^* \setminus b$ if b does not occur in any relationship with any other attribute from A^*;
5:     **end if**
6: **end for**
7: **for all** $a \in A^*$ **do**
8:     determine its domain V_a as the sum of its domains from pairs (A_i, V_i)
9: **end for**

Theorem 1. *For an ontology integration on a concept level and for $m \geq 2$ the following condition is always satisfied: O^{m*} is equal to O^{1*}.*

Proof. In the first step we show that O^{m*} is equal to O^{1*} for $m = 2$. Due to the fact that we consider only concept integration we want to show that A^{m*} is equal A^{1*} and V^{m*} is equal V^{1*} where A^{m*}, A^{1*} are the results of attribute integration on multi- and one-level respectively and V^{m*}, V^{1*} are integrated values of attributes for multi- and one-level algorithm.

From Step 1 of Algorithm 1 it is obvious that $A^{1*} = \bigcup\limits_{i=1}^{n} A^i$. Two-level integration process is more complicated. Let us assume that $A_1, A_2 ..., A_n$ were divided into k classes. Therefore, $S_1 = \{i : A_i$ belongs to a class $1\}, S_2 = \{i : A_i$ belongs to a class $2\}, ..., S_k = \{i : A_i$ belongs to a class $k\}$. In the first step of the multi-level integration process we obtain $A_1^{1*} = \bigcup\limits_{i \in S_1} A^i, A_2^{1*} = \bigcup\limits_{i \in S_2} A^i, ..., A_k^{1*} = \bigcup\limits_{i \in S_k} A^i$. In the second step we get $A^{2*} = \bigcup\limits_{i \in S_1} A^i \cup \bigcup\limits_{i \in S_2} A^i \cup ... \cup \bigcup\limits_{i \in S_k} A^i$. Therefore, A^{2*} is A^{1*} equal because union of sets is associative. The same reasoning could be conducted for the set of attributes values. For $m \geq 2$ it is easy to show by using mathematical induction. $\square$

From Theorem 1 we know that the results of ontology integration for one- and multi-level give the same results. Therefore, in the next part of our paper we examine the influence that one- and multi-level integration processes have on the time required to determine the final ontology.

5.2 Experimental Evaluation

For experimental evaluation we have used ontologies taken from datasets provided by *Ontology Alignment Evaluation Initiative* (*OAEI*) for their annual evaluation campaigns. These campaigns are aimed at evaluation of plethora of ontology alignment frameworks which main goal is to designate a set of mappings that indicate equivalent elements taken from separate ontologies. The aforementioned evaluation methodology is based on a broad dataset containing pairs of ontologies (for convenience grouped into smaller subsets referred to as *tracks*) along with some gold standard - a reference mappings between them. During the actual evaluation of some selected alignment tool, its output is compared with such reference mappings and *Precision* and *Recall* values are calculated along with other quality metrics.

Due to the accessibility of the domain, for our particular experiment we have used four ontologies (namely *Sigkdd*, *Edas*, *ConfTool* and *Sofsem*) taken from the conference track of the latest 2015 evaluation campaign [17]. We have also used the provided reference alignments that have been designated between these ontologies in order to fulfil initial requirements: (i) selecting equivalent concepts that may be integrated into the final ontology and (ii) selecting equivalent attributes for the sake of Step 3 of Algorithm 1.

In our experiment we have tested the one- and two-level approach using a dedicated experimental environment written in Python programming language. The integration of all ontologies into the one, consistent version incorporating standard one-level approach took **0.0788 s**.

In Table 1 we present different times in seconds taken by the two-level integration approach. We have tested seven different selections of initial ontologies' classes X_1 and X_2. They can be understood as an initial division of used set of four ontologies into subsets containing respectively one, two or three of initial ontologies. Each of such division is represented as a row in Table 1.

Table 1. Experimental results for two-level integration process

Ontologies classes	1 level		2nd level	Total time
	X_1	X_2		
X_1={Sofsem, Sigkdd} X_2={Edas, ConfTool}	0.006 s	0.0072 s	0.06 s	**0.0672 s**
X_1={Sofsem, Edas} X_2={Sigkdd, ConfTool}	0.0026	0.0098 s	0.0564 s	**0.0662 s**
X_1={Sofsem, ConfTool} X_2={Sigkdd, Edas}	0.007 s	0.0071 s	0.0559 s	**0.0631 s**
X_1={Sofsem} X_2={Sigkdd, Edas, ConfTool}	0 s	0.0212 s	0.0628 s	**0.084 s**
X_1={Sigkdd} X_2={Sofsem, Edas, ConTool}	0 s	0.0323 s	0.0557 s	**0.088 s**
X_1={Edas} X_2={Sofsem, Sigkdd, Conftool}	0 s	0.0182 s	0.0573 s	**0.0756 s**
X_1={ConfTool} X_2={Sofsem, Sigkdd, Edas}	0 s	0.0348 s	0.0764 s	**0.1111 s**

Columns represent times taken by each level of performed integration for different classes and the time taken by the final step of the investigated method.

The presented values are obtained from 10 repeats of the same integration process and the arithmetic means of all of the times taken by partial iterations is provided. This allowed to rule out any potential distortions that may be caused by random technical issues such as memory access downtime etc.

From obtained results of our experiment we can draw a conclusion that the multi-level approach to the integration is significantly faster than the one-level procedure. As easily seen, from the last column of Table 1, the experimental verification pointed out that such integration process is shorter even by 20 % in comparison to the simpleone-level integration. In the context of Big Data [3] the shortest possible time required to obtain the expected results is a critical factor in providing reliable business solutions in due course.

6   Future Works and Summary

Because of the complexity of ontologies and their semantic expressiveness, managing them is a difficult task. Moreover, ontologies allow to easily store big sets of data (eventually enriching them with some intentional meanings), so methods for their convenient, quick, reliable and low-budget processing are critical.

In this work we have proposed the multi-level method of their integration. During this process, integrated ontologies are divided into some classes and for each of such class the one, consistent ontology is designated. Finally all of the partial results are merged into a final ontology. Such solution allows to decrease the time required for performing desired integration thanks to a parallelisation of the fragmentary calculations.

In our future work we would like to conduct more experiments using more ontologies and for more levels. Due to the limitations of this paper we were able only to examine four ontologies integrated only on two levels. Therefore, more sophisticated experiments could bring interesting conclusions. Additionally, we are planing to expand our framework with the integration of both instances and relations that are also important elements of ontologies.

Acknowledgment. This work was partially supported by the European Commission under the 7th Framework Programme, Coordination and Support Action, Grant Agreement Number 316097, ENGINE - European research centre of Network intelliGence for INnovation Enhancement (http://engine.pwr.wroc.pl/).

References

1. Alasoud, A., Haarslev, V., Shiri, N.: A hybrid approach for ontology integration. In: Proceedings of the 31st VLDB Conference, Trondheim, Norway (2005)
2. Calvanese, D., Giacomo, G., Lenzerini, M.: A framework for ontology integration. In: Proceedings of the 2001 International Semantic Web Working Symposium (SWWS 2001), pp. 303–316 (2010)

3. Chen, H., Chiang, R.H., Storey, V.C.: Business intelligence and analytics: from big data to big impact. MIS Q. **36**(4), 1165–1188 (2012). Society for Information Management and the Management Information Systems Research Center, Minneapolis, MN, USA

4. Cruz, I.F., Xiao, H.: The role of ontologies in data integration. J. Eng. Intell. Syst. **13**, 245–252 (2005)

5. Jiménez-Ruiz, E., Grau, B.C., Horrocks, I., Berlanga, R.: Ontology integration using mappings: towards getting the right logical consequences (2009). doi:10.1007/978-3-642-02121-3_16

6. Kozierkiewicz-Hetmańska, A.: Comparison of one-level and two-level consensuses satisfying the 2-optimality criterion (2012). doi:10.1007/978-3-642-34630-9_1

7. Kozierkiewicz-Hetmańska, A., Nguyen, N.T.: A comparison analysis of consensus determining using one and two-level methods, vol. 243. Advances in Knowledge-Based and Intelligent Information and Engineering Systems, pp. 159–168 (2012)

8. Li, L., Wu, B., Yang, Y.: Agent-based ontology integration for ontology-based applications. In: Meyer, T., Orgun, M. (eds.) Proceedings of the Australasian Ontology Workshop, vol. 58, Sydney, Australia (2005)

9. Maleszka, M., Nguyen, N.T.: A model for complex three integration tasks (2011). doi:10.1007/978-3-642-20039-7_4

10. Nguyen, N.T.: Advanced Methods for Inconsistent Knowledge Management. Springer, London (2008)

11. Nguyen, N.T.: Consensus Choice Methods and Their Application to Solving Conflicts in Distributed Systems. Wroclaw University of Technology Press, Wroclaw (2002). (in Polish)

12. Noy, N.F., Musen, M.A.: An algorithm for merging and aligning ontologies: automation and tool support. In: Proceedings of the Workshop on Ontology Management at the Sixteenth National Conference on Artificial Intelligence (AAAI 1999) (1999)

13. Noy, N.F., Musen, M.A.: PROMPT: algorithm and tool for automated ontology merging and alignment. In: AAAI/IAAI, pp. 450–455 (2000)

14. Pietranik, M., Nguyen, N.T.: A Multi-atrribute based framework for ontology aligning. Neurocomputing **146**, 276–290 (2014). doi:10.1016/j.neucom.2014.03.067

15. Pinto, M., Martins, J.P.: A methodology for ontology integration. In: Proceedings of K-CAP 2001, Victoria, British Columbia, Canada, 22–23 October 2001

16. Shvaiko, P., Euzenat, J.: Ontology matching: state of the art and future challenges. IEEE Trans. Knowl. Data Eng. **25**(1), 158–176 (2013)

17. http://oaei.ontologymatching.org/2015/

Temporal Ontology Representation
and Reasoning Using Ordinals and Sets
for Historical Events

Myung-Duk Hong, Kyeong-Jin Oh, Seung-Hyun Go,
and Geun-Sik Jo[✉]

Department of Information Engineering, Inha University, Incheon, Korea
{hmdgo, okjkillo}@eslab.inha.ac.kr, kosehy@gmail.com,
gsjo@inha.ac.kr

Abstract. In question-and-answer (QA) systems, various queries need to be processed. In particular, those queries such as temporal information require complex query generation processes. This paper proposes a temporal representation model that can support qualitative and quantitative temporal information on historical ontology by applying the concept of ordinals and sets and introduces operators that allow a QA system to easily handle complex temporal queries. To verify the effectiveness of the proposed model and operators, historical scenarios are presented to show that they can effectively handle complex temporal queries.

Keywords: Ontology · Temporal representation · Historical event · Question-Answering system

1 Introduction

The goal of a question-and-answer (QA) system is to provide correct natural language answers for given natural language questions consisting of natural languages to users. The most important task is to analyze given questions and the predefined text corpus. There are various techniques for extracting useful words and phrases [1]. However, it is difficult to present semantic relationships between keywords because keywords consist of words or sequences of words. If information on semantic relationships between keywords of a sentence can be used in the QA system, then the system can provide good performance. To implement the QA system, a knowledge base based on linked open data (LOD) such as DBPedia and Freebase can be constructed [2]. A knowledge base manages knowledge as a triple format, and the QA system uses knowledge to provide answers to given questions. However, if the QA system uses only an asserted triple in the knowledge base, then the system cannot support various questions from users. This is because some user queries require reasoning processing instead of simple queries for given queries.

This paper proposes a temporal representation model to support qualitative and quantitative temporal information on historical ontology through the application of the concept of ordinals and sets. In the proposed model, groups are built based on entities

© Springer-Verlag Berlin Heidelberg 2016
N.T. Nguyen et al. (Eds.): ACIIDS 2016, Part I, LNAI 9621, pp. 75–85, 2016.
DOI: 10.1007/978-3-662-49381-6_8

that frequently appear in user queries as questions, and operators are provided to represent qualitative and quantitative information on groups. Also, operators allow QA system to find out group information to extract facts for user queries. The proposed QA system performs forward reasoning process for the built groups without generating all new facts based on predefined inference rules. Later, the operators allow the proposed hybrid QA system to effectively respond to given complex questions which need a complex backward reasoning process.

The rest of this paper is organized as follows. Section 2 introduces related works. Section 3 presents a temporal representation scheme and a query operator for the proposed model. Section 4 describes a scenario and examples to verify the proposed model. Finally, Sect. 5 concludes with some suggestions for future research.

2 Background and Related Work

There are studies applying temporal information are as follows. Allen introduces 13 temporal relationships with distinct, exhaustive, and qualitative characteristics [3, 4]. As shown in Fig. 1, he describes 13 qualitative relationships between start and end instants. Temporal relationships also contain complement, composition, converse, intersection, and union operators for relationships. Complex temporal relationships are represented by defining the composition between basic interval relationships through the composition operator.

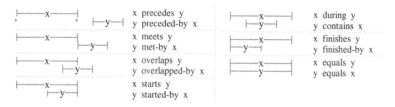

Fig. 1. Allen's basic temporal relationships

OWL-Time suggests a schema for temporal information in W3C [5, 6]. The schema also supports time properties including Allen's 13 relationships.

Numerous approaches have been proposed to represent temporal information additionally demanded under the triple structure. As shown in Fig. 2(a), reification is described as the statement class in RDF [7, 8]. In Fig. 2(b), the N-ary relationship permits only binary relationships for reification in both RDFS and OWL [9]. In Fig. 2(c), 4D fluents generate instances using the TimeSlice class in both subject and objects demanding a temporal information representation for representing temporal information to represent the timeSlice property [10, 11]. 4D fluents also represent time instances and intervals as instances of TimeInterval.

Perry proposes a support framework for temporal and spatial information [12]. In this approach, Perry designs an operator for an RDF-based model and queries for representing temporal and spatial information and proposes SPARQL-ST language to handle intersect and union functions in temporal information as shown in Fig. 3.

Fig. 2. A general representation technique

Fig. 3. Intersect and union operators for a temporal representation

Horrocks et al. propose the SWRL (semantic web rule language) [13]. The SWRL is written in human-readable grammar, and its format is a result of the following prerequisite:

$$Antecedent \Rightarrow consequent.$$

Prerequisites and results are the AND union between elements written as $a1 \wedge \ldots \wedge an$. Variables are represented by the basic rule that attaches the prefix as a question mark. A sentence "If x has y as a parent, then y and z are brothers, and x has z as an uncle" can be represented to result as the uncle's attribute based on the union of the parent's attribute and the brother's attribute, which can be defined as follows:

$$hasParent(?x, ?y) \wedge basBrother(?y, ?z) \Rightarrow hasUncle(?x, ?z)$$

where ?x, ?y, and ?z are variables, and hasParent, hasBrother, and hasUncle are attributes.

This paper represents temporal information with a historical model using ontology. Rules and operators are defined using the SWRL to verify the proposed temporal representation model. The scenario for the experiment is created using SPARQL.

3 Modeling for a Temporal Representation

3.1 Temporal Representation Schema

Assuming the expression of an instance containing time information to be presented in the knowledge base, the paper proposes a model that applies both ordinal numbers and set terms.

Figure 4 shows the basic temporal representation using the IntervalGroup class. The temporal representation generates instances for grouping meaningful sets of instances containing time information in the knowledge base. The condition of instances belonging to IntervalGroup is represented by generating instances using the GroupFilter class. GroupFilter is presented with the name and value of the given property. GroupFilter lists several conditions and generates several instances. Instances that meet the GroupFilter condition generate instances using the IntervalRef class to connect IntervalGroup. All properties have inverse properties. These concepts are represented by the following equations:

Given = { ?o = Object,   ?i = owl:Interval,   ?ir = InervalRef,
           ?ig = IntervalGroup,   ?gf = GroupFilter,   ?tf = TemporalProperty   }

TemporalOf(?i1, ?o1) ∧ ReferenceOf(?ir, ?i1) ∧ MemberOf(?ir, ?ig) ∧

ConditionOf(?gf1, ?ig) ∧ UseBeginDate(?ig, ?tf1) ∧ UseEndDate(?ig, ?tf2)

In Fig. 4, IntervalGroup, GroupFilter, UseBeginDate, UseEndDate, and IntervalRef are shown as dotted lines and are connected to generate a group using several interval instances.

Table 1. Relation-type property by Allen's rule

Interval ref.	Property	Relation type (Allen's rule)	Property	Interval ref.
?ir1	NextOf	precedes, meets, overlaps is finished by, contains	PreviousOf	?ir2
	EqualOf	starts, equals, is started by	EqualOf	
	PreviousOf	during, finishes, is overlapped by is met by, is preceded by	NextOf	

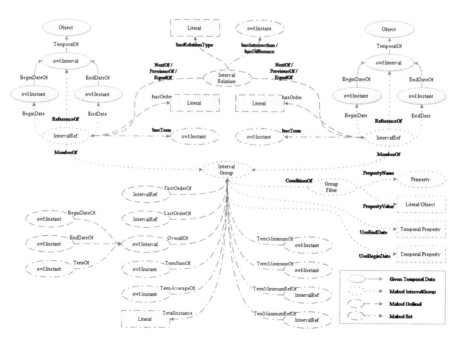

Fig. 4. A temporal ontology representation for historical events using ordinals and sets

Allen's rule is applied using BeginDate and EndDate between every interval connected to IntervalGroup to make conclusions on the subsequent ordinal relationship. Here temporal relationships between two intervals is concluded using Allen's rule. The conclusion and IntervalRef of each interval are connected by generating instances using the IntervalRelation class. To connect IntervalRelation and intervals between two IntervalRef classes, one of three properties (nextOf, previousOf, or equalOf) is selected and connected, as shown in Table 1.

The order between each instance is determined using nextOf, previousOf, and equalOf of Allen's rule. The calculated order is presented by Order property. IntervalRef for each first and last order is also presented by FirstOrder and LastOrder property, respectively. To calculate the Overall for every interval, a new interval is generated using FirstOrder's BeginDate and LastOrder's EndDate and presented as the Overall property.

Given = { ?ir = IntervalRef, ?il = IntervalRelation, ?ig = IntervalGroup,
        ?i = owl:Interval, ?d = owl:Instant }
((NextOf(?il1, ?ir1) ∧ Previous(?il1, ?ir2)) ∨(EqualOf(?il1, ?ir1) ∧ EqualOf (?il1, ?ir2)) ∨
(PreviousOf(?il1, ?ir1) ∧ NextOf (?il1, ?ir2))) ∧
(hasIntersection(?il1, ?d3) ∨ (hasDifference(?il1, ?d3)) ∧ hasOrder(?ir1, literal) ∧
hasOrder(?ir2, literal) ∧ MemberOf(?ir1, ?ig) ∧ FirstOrder(?ig, ?ir2) ∧
LastOrder(?ig, ?ir3) ∧ OverallOf(?i, ?ig) ∧ BeginDateOf(?d1, ?i) ∧EndDateOf(?d2, ?i)

In Fig. 4, IntervalRelation, Order, the intersection instance, the difference instance, FirstOrderOf's IntervalRef, LastOrderOf's IntervalRef, and OverallOf's IntervalRef are shown as dashed lines and represented as RelationType, Insection/Difference, and Order between each IntervalRef instance based on the application of Allen's rule between several instances. To conclude the Set relation, an instance in each IntervalRef property is set as UseBeginDate and UseEndDate, and then the time difference between BeginDate and EndDate is calculated to generate the owl:Instant class and connect IntervalRef to the Term property. In addition, the term of all IntervalRef Instances must be calculated, and each value should be compared. The lowest value should be presented as the TermMinimumOf property, but the value must be represented as the TermMinimumRefOf property to connect the real IntervalRef property. The highest value must be represented as the TermMaximumOf property, but the value must be represented as the TermMaximumRefOf property to connect IntervalRef. The TermSumOf property is represented by adding all term values. The TermAverageOf property is represented by calculating all average term values. The TotalInstance property is represented by calculating the number of every instance. All properties are represented as follows:

Given = { ?i = owl:Interval, ?ir = IntervalRef, ?d = owl:Instant, ?ig = IntervalGroup }
Term(?ir, ?d1) ∧ ReferenceOf(?ir1, ?i1) ∧ MemberOf(?ir1, ?ig) ∧
Overall(?i2, ?ig) ∧ Term(?2, ?i2) ∧ TermMinimumOf(?d2, ?ig) ∧
TermMaximumOf(?d3, ?ig) ∧ TermMinimumRefOf(?ir2, ?ig) ∧
TermMaximumRefOf(?ir3, ?ig) ∧ TermSumOf(?d4, ?ig) ∧
TermAverageOf(?d5, ?ig) ∧ TotalInstance(?ig, Literal)

All instances of TermSumOf, TermAverageOf, TermMinimumOf, TermMaxi-mumOf, TermMinimumRefOf's IntervalRef, TermMaximumRefOf's IntervalRef, and TotalInstance have the term instant depicted as alternative long-and-short dashed lines in Fig. 4. They are computed using each interval term.

3.2 Query Operator Design

An operator is designed using a temporal representation schema. This operator can be divided into the base operator and the extension operator. The extension operator cannot be used without the base operator. The base operator is distinguished between the construct operator for generating IntervalGourp and the select operator for query-ing. First, assume the following given knowledge:

Given: A temporal RDF graph G
    Let statement(e) = { (s, p, o) $\in$ TRIPLES(G) }, time_instant(t) = { t | t $\in$ T },
    start_time_instant(st) = { t | t $\in$ T, t $\in$ te }, end_time_instant(et) = { t | t $\in$ T, t $\in$ te },
    temporal(te) = { (e, st, et) | st <= et }

Suppose:
Let Interval_Group(ig) =
{ (gf1 ... gfn), (ir1 ... irn), fo, lo, ov, ts, ta, tmin, tmax, ti },
Group_Filter(GF) = { gf | gf $\in$ e },
Interal_Ref(IR) = { ir = (te, tm, od, pr, xt) | ir $\to$te },   Term(tm) = { t | et - st },
Order(od) = { od | order by (st of te) $\in$ IIR },
Previous(pr) = { ir, nil | if allens_rule(od-1, od) else nil },
Next(xt) = { ir, nil | if allens_rule(od, od+1) else nil },
First_Order(fo) = { ir | ir is first(st) $\in$ IIR) },   Last_Order(lo) = { ir | ir is last(st) $\in$ IIR) },
overall(ov) = { (st, et, t) | st is (st of fo), et is (et $\in$ lo), t is (et - st) },
TermSum(ts) = { ts | $\sum$tm $\in$ IIR },   TermAverage(ta) = { ta | ts / ti },
TermMinimum(tmin) = { tmin | $\lfloor tm \rfloor$ $\in$ IIR },   TermMaximum(tmax) = { tmax | $\lceil tm \rceil$ $\in$ IIR },
Total_Instance(ti) = { ti | count(tm) }

For a temporal representation based on given information, the above mentioned knowledge is generated: First, the base construct operator generates IntervalGroup and represents ordinals and sets if the group filter is given as MakeGroup (ub, ue, GF). The interval's BeginDate and EndDate are set as ub and ue, and a query is required. The operator is described as follows:

Operator: MakeGroup(ub, ue, GF)
Given: a ub is UserBeginDate,   a ue is UserEndDate,   GF are GroupFilters
Make: Temporal extension triples

Second, the operator queries IntervalGroup including each GF if the GF is given as SelectGroup (GF). The operator is described as follows:

Operator: SelectGroup(GF) $\to$IG
Given: GF are GroupFilters
Find: IG are IntervalGroups

Third, the base select operator queries to all IntervalRef belonging to each IntervalGroup's member if ig is given as SelectInterval (ig).

Operator: SelectInterval(ig) →IR
Given: a ig are IntervalGroup
Find: IR are IntervalRef

Finally, the base select operator retrieves to all IntervalRef belonging to each IntervalGroup's member if IG is given as SelectInterval (IG).

Operator: SelectInterval(IG) →IR
Given: IG are IntervalGroups
Find: IR are IntervalRef

Further, the proposed model provides an extension operator to support answers to complex questions with the base select operator.

The first extension operator is the condition and is expected to retrieve triples corresponding to the condition in the retrieved IntervalGroup by using either SelectGroup() or SelectInterval(). The operator is described as follows:

Operator: (SelectGroup(GF) ∨ SelectInterval(IG)) ∧ Condition →ir ∨ Literals ∨ Instances
Given: GF are GroupFilters,   IG are IntervalGroups,
Condition is first_order, last_order, overall, term_sum, term_average, erm_minimum, term_maximum or total instance
Find: ir ∨ Literals ∨ Instances is first_order, last_order, overall, term_sum, term_average, term_minimum, term_maximum or total instance

The second extension operator is Order() and expected to retrieve IntervalRef corresponding to order = ?x in the retrieved IntervalGroup by using either SelectGroup() or SelectInterval(). The operator is described as follows:

Operator: (SelectGroup(GF) ∨ SelectInterval(IG)) ∧ Order(?x) →ir
Given: GF are GroupFilters,   IG are IntervalGroups,
Order(?x) is Literal about Order
Find: a ir is IntervalRef

The third extension operator is tm() and expected to retrieve the relative temporal order of tm in the retrieved IntervalGroup's Overall by using either SelectGroup() or SelectInterval(). The operator is described as follows:

Operator: (SelectGroup(GF) ∨ SelectInterval(IG)) ∧ InstantRelation (tm(?x)) →triple
Given: GF are GroupFilters,   IG are IntervalGroups,
a tm(?x) is time_instant
Find: triple (s, p, o) is (IG :after ?x), (IG :before ?x) or (IG :contains ?x)

The fourth extension operator is twin tm() and expected to retrieve Allen's rule relationship by using Interval(?x), which includes two tm types for the retrieved

IntervalGroup's overall based on SelectGroup() as start time(?y) and end time(?z). The operator is described as follows:

Operator:	(SelectGroup()∨ SelectInterval()) ∧ IntervalRelation(Instance(?x) ∧
	tm(?y) ∧ tm(?z)) →triple
Given:	GF are GroupFilters, IG are IntervalGroups, a instance(?x) is instance,
	a tm(?y) is start_time_instant, a tm(?z) is end_time_instant
Find:	triple (s, p, o) is (s is IG, o is ?x, p is allen's 13 rule)

The final extension operator is ir() and expected to retrieve Allen's rule relationship for the typed IntervalRef for the retrieved IntervalGroup's Overall by using SelectGroup(). The operator is described as follows:

Operator:	(SelectGroup()∨ SelectInterval()) ∧ ir(?x) →triple
Given:	GF are GroupFilters, IG are IntervalGroups, a ir(?x) is IntervalRef
Find:	triple (s, p, o) is (s is IG, o is ?x, p is allen's 13 rule)

4   A Case Study: Query Verification

To verify the proposed model, temporal is used to express a history scenario to create the history of the Joseon monarchy in IntervalGroup. The proposed model is validated by executing queries.

Table 2 shows information on the Joseon monarchy from Wikipedia. This information is composed of the name of the monarch at the beginning and end of each reign, TempleName, and reignStartDate and presents the reignEndDate property.

Through operator MakeGroup, the following temporal representation is generated.

```
MakeGroup(
(UseBeginDate(reignStartDate) ∧ UseEndDate(reignEndDate)) ∧
hasJob("Monarch") ∧ nationalPlace("Joseon") )
→ConditionOf(?ig, ?gf1) ∧ ConditionOf(?ig, ?gf2) ∧
PropertyName(?gf1, "hasJob") ∧ PropertyValue(?gf1, "Monarch") ∧
PropertyName(?gf2, "nationalPlace") ∧ PropertyValue(?gf2, "Joseon") ∧
UseBeginDate(?ig, "reignStartDate") ∧ UseEndDate(?ig, "reignEndDate") ∧
∃MemberOf(?ir1Ö n, ?ig) ∧ ReferenceOf(?ir1Ö n, ?i1Ö n) ∧ ∃Term(?ir1Ö n, ?d1Ö n) ∧
∃(∃(NextOf(?i11Ö m, ?ir1Ö n) ∧ PreviousOf(?i11Ö m, ?ir1Ö n) ) ∨
∃(PreviousOf(?i11Ö m, ?ir1Ö n) ∧ NextOf(?i11Ö m, ?ir1Ö n)) ∨
∃(EqualOf(?i11Ö m, ?ir1Ö n) ∧ EqualOf(?i11Ö m, ?ir1Ö n ))) ∧
∃hasRelationType(?i11Ö m, RelationType) ∧ (∃hasIntersection(?i11Ö m, ?d28... 278) ∨
∃hasDifference(?i11Ö m, ?d28... 278)) ∧∃Order(?ir1Ö n, 1...27) ∧
FirstOrder(?irfo, ?ig) ∧ LastOrder(?irlo, ?ig) ∧ Overall(?io, ?ig) ∧
BeginDateOf(?ds, ?io) ∧ EndDateOf(?de, ?io) ∧ TermOf(?dt, ?io) ∧
TermSumOf(?dts, ?ig) ∧ TermAverageOf(?dta, ?ig) ∧ TotalInstance(?ig, 27) ∧
TermMinimumOf(?dtmin, ?ig) ∧ TermMaxmimumOf(?dtmax, ?ig) ∧
```

Table 2. A list of Joseon Monarchy

Temple name	Reign startdate (gYear)	Reign enddate (gYear)	Temple name	Reign startdate (gYear)	Reign enddate (gYear)	Temple name	Reign startdate (gYear)	Reign enddate (gYear)
Taejo	1392	1398	Yeonsangun	1494	1506	Sukjong	1674	1720
Jeongjong	1398	1400	Jungjong	1506	1544	Gyeongjong	1720	1724
Taejong	1400	1418	Injong	1544	1545	Yeongjo	1724	1776
Sejong	1418	1450	Myeongjong	1545	1567	Jeongjo	1776	1800
Munjong	1450	1452	Seonjo	1567	1608	Sunjo	1800	1834
Danjong	1452	1455	Gwanghaegun	1608	1623	Heonjong	1834	1849
Sejo	1455	1468	Injo	1623	1649	Cheoljong	1849	1863
Yejong	1468	1469	Hyojong	1649	1659	Gojong	1863	1907
Seongjong	1469	1494	Hyeonjong	1659	1674	Sunjong	1907	1910

The operator for the following SPARQL query using generated knowledge is as follows:

Q1. Who is the 25th monarch of the Joseon period? A1 : Cheoljong	

/* Example Q1 - Extension Sparql */
SELECT ?d
WHERE {
       ?a rdf:type   :IntervalGroup.
       ?b :ReferenceOf ?c
       ?c :TempleName   ?d.
}
EXTENSION_FILTER {
    SelectGroup(hasJob("Monarch"),
           nationalPlace("Joseon"), ?a).
    Order(?a, 25, ?b).
}
	/* Example Q1 - Common Sparql */
	SELECT ?d
	WHERE {
	      ?a :hasJob "Monarch".
	      ?a :nationalPlace "Joseon".
	      ?a :hasTemporal ?b.
	      ?b :hasBeginDate ?c.
	      ?a :TempleName ?d
	}
	ORDER BY ?c
	LIMIT 1
	OFFSET 25

For example 1, the query asks for the 25th monarch. At common SPARQL, the 25th monarch is identified by using LIMIT and OFFSET functions with corresponding properties. In the proposed extension, the query uses the SelectGroup operator to retrieve IntervalGroup first and extracts an instance in which the hasMember property value of the IntervalGroup class is 25 after IntervalRef instances connected by the hasMember property of IntervalGroup are evaluated using the order operator.

Q2: Who is the maximum term of Joseon's monarch? A2: Yeongjo	

/* Example Q2 - Extension Sparql */
SELECT ?e
WHERE{
       ?a rdf:type :IntervalGroup.
       ?a :TermMaximumRef ?b.
       ?b :ReferenceOf ?c.
       ?c :TemporalOf ?d
       ?d :TempleName ?e.
}
EXTENSION_FILTER {
    SelectGroup(hasJob("Monarch"),
           nationalPlace("Joseon"), ?a).
}
	/* Example Q2 - Common Sparql */
	SELECT ?f
	WHERE {
	      ?a :hasJob "Monarch".
	      ?a :nationalPlace "Joseon".
	      ?a :hasTemporal ?b.
	      ?b :hasBeginDate ?c.
	      ?b :hasEndDate ?d.
	      bind(year(?d)-year(?c) as ?e).
	      ?a :TempleName ?f
	}
	ORDER BY DESC(?e)
	LIMIT 1

In SPARQL of example 2, all instances of monarchs are extracted, and all distinctions between the hasBeginDate and hasEndDate properties are determined using the bind function. A solution is searched with calculated period. In the extension of SPARQL, the SelectGroup operator is used to extract IntervalGroup and instances connected to the TermMaximumRef property of IntervalGroup.

The aforementioned examples of queries show that the proposed temporal representation model and operators can make queries simple. Common SPARQL queries generally require additional costs to sort instances with given constraints.

5   Conclusions and Future Research

This paper proposes a temporal knowledge representation and operators for historical events by applying the concept of ordinals and sets. The model is evaluated using some historical events from Wikipedia. Extended temporal information is represented explicitly and formally for an easy understanding by ontology engineers and machines. Through the scenario, the proposed model can easily answer complex questions for historical events through queries using the proposed operators.

Future research should implement the modeled operator to be usable in SPARQL by extending it. In addition, future research should show how the proposed model can be used to represent spatiotemporal information applied to ordinals and sets and consider various types of temporal ontology beyond Korean historical ontology.

References

1. Kalyanpur, A., Murdock, J.W.: Unsupervised entity-relation analysis in IBM watson. In: Proceedings of the Third Annual Conference on Advances in Cognitive Systems ACS, p. 12 (2015)
2. Yahya, M., Berberich, K., Elbassuoni, S., Ramanath, M., Tresp, V., Weikum, G.: Natural language questions for the web of data. In: Proceedings of the 2012 Joint Conference on Empirical Methods in Natural Language Processing and Computational Natural Language Learning, Association for Computational Linguistics, pp. 379–390 (2012)
3. Allen, J.F.: Maintaining knowledge about temporal intervals. Commun. ACM **26**(11), 832–843 (1983)
4. Allen, J.F.: Temporal reasoning and planning. In: Reasoning About Plans, Morgan Kaufmann Publishers Inc., pp. 1–67 (1991)
5. Hobbs, J.R., Pan, F.: Time Ontology in OWL, W3C Working Draft 27 September 2006, W3C Working Draft27 (2006)
6. Pan, F., Hobbs, J.R.: Time in owl-s. In: Proceedings of the AAAI Spring Symposium on Semantic Web Services, pp. 29–36 (2004)
7. Beckett, D., McBride, B.: RDF/XML syntax specification (revised), W3C recommendation (2004)
8. Brickley, D., Guha, R.V.: RDF Schema 1.1., W3C Recommendation (2004)
9. Noy, N., Rector, A., Hayes, P., Welty, C.: Defining n-ary relations on the semantic web, W3C Working Group Note (2006)

10. Batsakis, S., Stravoskoufos, K., Petrakis, E.G.: Temporal reasoning for supporting temporal queries in OWL 2.0. In: König, A., Dengel, A., Hinkelmann, K., Kise, K., Howlett, R.J., Jain, L.C. (eds.) KES 2011, Part I. LNCS, vol. 6881, pp. 558–567. Springer, Heidelberg (2011)
11. Batsakis, S., Petrakis, E.G.: Representing temporal knowledge in the semantic web: the extended 4D fluents approach. In: Hatzilygeroudis, I., Prentzas, J. (eds.) Combinations of Intelligent Methods and Applications. SIST, vol. 8, pp. 55–69. Springer, Heidelberg (2011)
12. Perry, M.S.: A framework to support spatial, temporal and thematic analytics over semantic web data, Doctoral dissertation, Wright State University (2008)
13. Horrocks, I., Patel-Schneider, P.F., Boley, H., Tabet, S., Grosof, B., Dean, M.: SWRL: A semantic web rule language combining OWL and RuleML, W3C Member submission (2004)

Measuring Propagation Phenomena in Social Networks: Promising Directions and Open Issues

Dariusz Król[(✉)]

Department of Information Systems, Faculty of Computer Science and Management,
Wrocław University of Technology, Wrocław, Poland
Dariusz.Krol@pwr.edu.pl
http://www.ii.pwr.wroc.pl/~krol/eng_index.html

Abstract. The massive information is now spreading like wildfire in social media. As the usage of social data increased, the abuse of the media to spread distorted data also increased several times. To understand and predict the spread of information over a time period in online social networks researchers attempt to quantitatively model and measure the whole process. A number of different statistics aimed at measuring the spread were suggested. Many researchers have coupled these measures with various forgetting factor mechanisms to improve behavioural properties. Unfortunately, frequent unavailability of the full data record in social media prevents straightforward validation of such quantities. Moreover, since most known measures have global affects, they are rather inconvenient to evaluate for large networks. These difficulties lead us to contribute here a methodological identification of the propagation parameters to start afresh. The approach hinges on some recent results arising from the convergence between threshold models and cascade models. For example, three key concepts – distance, centrality and robustness – is successfully balanced by the proposed *scope–speed–failures* relationship. We conclude by identifying several open issues and possible directions for future research.

Keywords: Information diffusion · Spreading phenomenon · Cascading behaviour · Propagation speed · Behavioural metrics

1 Introduction

Over all 3 billion active users[1] and 5 billion web-facing devices enjoy instant access to approximately 876 million sites[2] on the Internet in May 2015. There are well over 1 million applications available, which have been downloaded more than 100 billion times. Access to information is easier than it was ever before. The digital universe is doubling in size every two years, and by 2020 the data – by creation and copying – will reach 44 ZB. YouTube has over a billion users

---

[1] http://www.internetworldstats.com/.
[2] http://news.netcraft.com/.

© Springer-Verlag Berlin Heidelberg 2016
N.T. Nguyen et al. (Eds.): ACIIDS 2016, Part I, LNAI 9621, pp. 86–94, 2016.
DOI: 10.1007/978-3-662-49381-6_9

and every day they upload 300 h of new videos per minute. Facebook currently stores over 260 billion photos, which translates to more than 300 million uploads every day. At the same time there are 83 million fake Facebook profiles. These numbers are both formidable and yet intimidating.

The massive information is now spreading like wildfire in social media. As the usage of social data increased, the abuse of the media to generate incomplete and distorted data also increased several times. To understand and predict the spread of information – and sometimes disinformation or misinformation – over a time period in online networks researchers attempt to quantitatively model and measure the whole process.

A number of different statistics aimed at measuring the spread were suggested over the years. The most prominent ones involve typical structural based network measures including PageRank and k-core [3]. Many researchers have coupled these measures with various forgetting factor mechanisms to improve behavioural properties [2]. Unfortunately, frequent unavailability of the full data record in social media prevents straightforward validation of such quantities. Moreover, since most known measures have global affects, they are rather inconvenient to evaluate for large networks. These difficulties lead us to contribute here a methodological approach to the identification of the propagation parameters [6].

We intend to contribute here a systematic approach to the identification of the parameters of a network that is based on our generic diffusion-based algorithm [8]. The approach hinges on some recent results arising from the convergence between threshold models and cascade models. We noticed that three key concepts – distance, centrality and robustness – is successfully balanced by the *scope–speed–failures* relationship.

Due to the lack of some data and severe privacy restrictions, it is important to develop and validate social network measures capable to cope with missing data and temporal aspects, respectively. To address this problem, an alternative set of statistics for spreading is proposed. We define measures, which capture the following details: scope, speed and failure of cascades on networks. In short, these can be comparable with well-known topological distance, centrality, efficiency and robustness.

This paper is organized as follows. Section 2 summarizes the features of propagation mechanism occurring in real world networks. In Sect. 3, an overview of applied methodologies for data propagation in different domains is provided. The short list of developed performance metrics are reported in Sect. 4. Finally, Sect. 5 concludes by identifying several open issues and possible directions for future research.

2   The Ambient Positive-Negative Propagation

Complex networks have been massively studied in many research fields such as biology, epidemiology, mathematics, physics, but also in social and behavioural sciences. Propagation and particularly error propagation are important properties of a complex network. By studying propagation phenomenon, we better understand the dissemination of information in a network to improve its efficiency and reliability. The propagation property of a network utilizes the underlying topological structure of a network to the utmost. In response to different kinds of disturbances propagation is extremely useful in protecting existing networks, e.g. power grids synchronization, large system behaviour prediction, resource discovery and monitoring, biological invasions finding and damages determining, virus propagation controlling and restraining, social and large scale infrastructure networks decomposition and immunizing. Recently, enormous interest has been devoted to modelling cascading failures in quest of designing 'better' networks, not only robust to the random loss of nodes like scale-free networks, but also less vulnerable to its fragility to the selective loss of the most connected ones. Many real networks have scale-free properties and therefore great effort is necessary in order to protect them from targeted cyber-attacks. As more and more real data become available, the interest in understanding the impact of positive-negative spread keeps growing.

Table 1. Real world examples of propagation phenomena

Networks	Nodes	Links	Spread mechanism
Internet	Routers, modems and other devices	Physical connections	Messages, services
WWW	The Web servers and applications	Hyperlinks	Documents and other web resources
Information	Documents, individuals	Citations, conversations, collaborations	Pieces of information, viruses
Semantic	Concepts	Subtype relationships, inheritance, causality	Logic values, messages
Power grid	Stations	Cables	Cascading failures
Phone and mobile	Stations and others	Wired and wireless transmissions	Voice, video and others
Biological	Species units, chemical compounds	Signals, biological contacts, protein interactions, gene associations	Noise and signalling pathways, gens mutations, cancers
Epidemiological	Individuals	Physical contacts	Diseases
Social	Individuals or organizations	Sets of dyadic ties, and other interactions	Trust, friendships, rumours, memes, messages

Real world networks, as the Internet, WWW, information and semantic networks, power grids, phone and cellular networks, biological, epidemiological and social networks have become permanently an important part of everyday life. When looking into these networks, one can find many instances of networks whose propagates due to the system's dynamics. Modelling, exploring and predicting propagation processes are now becoming well recognized as one of the most significant challenges in the complex network research. The different data domains in propagation analysis typically require dedicated techniques of different types. Some motivating examples are listed in Table 1.

Widely, the term 'propagation' is used to describe methods for automatically disseminating information. From epidemiology the propagation phenomenon is mimicked the spread of a contagious disease in fully decentralized manner. Just as infected individual organism transmits a virus to everyone who came into contact, the process in a system dispatches infected information it has received to selectively or randomly chosen neighbours in charge of forwarding it, and so on. In computer science, the propagation method involves transfer, reproduction or multiplication of data resulting in a repeatable action in the context of pursuing a defined goal.

On the other hand, the term 'error' entails different meanings in accordance with the specific background and their later usage. Generally, we understand an error as a difference between reality and the representation of reality. It includes not only deviations from accuracy, mistakes or faults, statistical variations, fallacies in reasoning, anomalies in software but also departures from norms as minor forms of social misbehaviour. Errors are defined in different ways. Three fundamental examples are as follows:

- Any result that has occurred and does not meet network (system) specification.
- Any state reflecting a fault during the fabrication process.
- Any difference between the current and the expected state of a network (system).

A couple of related terms are used to describe the 'error' concept, including failure, fault, defect, flaw, degradation, exception, bug, glitch, conflict and uncertainty, although 'error' is still the most popular term appearing in literature [1]. These terms are not exactly the same. For example, a fault is a violation of a system's underlying assumptions whereas a failure is an externally-visible deviation from specifications.

When we combine these term into one - 'error propagation' - we intend to narrow down the domain to network systems. Then the error from one system's component is propagated to the other. The error propagation in our representation is the series of connected error events. The error occurring in one element of the system can cause another error in the other part of the system. For example, the users are represented by their surname and initials of first and middle name, thus there is a source of error in distinguishing some of them. Trust or distrust may propagate through the relationship network [4]. Even a small piece of information about distrust can provide tangibly better judgments about how much one user should trust another. On social networks we are interested in spread of good-bad influence but also in the spread of obesity, sleep and drug use. Very disruptive effect is observed when incorrect event reports fast spread through the social media. It is quite essential to note, that the error propagates itself along with the communication flow.

We are interested in finding differences between the propagation mechanisms of different domains. For example, in general science, every scientific measurement is subjected to errors derived from various sources such as problem of definition, scale precision, intrinsic random uncertainty and systematic wrong values. In experiments, we need to understand how the uncertainties associated with the measurements of the certain quantities propagate in an uncertainty in the end value. This process is also known as propagation of uncertainty.

In software engineering, error propagation may occur at least twofold: globally between system components or modules, and locally between functions in a single program. Propagation of component errors, if it can be captured by the probability of occurrence, represents a meaningful extension to the reliability models. The error propagation in a program raises in the following intended and unintended ways: function calls, global variables updates, common extern data sharing and denial of resources. An error may be communicated in one of three ways: implicit, explicit and escaping. Both explicit and escaping errors have been represented in recent programming languages by the exception.

The propagation process in social networks is characterized by two aspects: its global structure, i.e. the graph neighbourhoods of users that defines who influenced whom, and its dynamics, i.e. the evolution of the propagation rate which is defined as the ratio of nodes that adopts the piece of information over time. The simplest way to interpret the process is to consider that a node can be either activated (i.e. has received the information and tries to propagate it) or not. We suppose that the dynamics of error propagation is much faster than the growth and change of the network itself.

The aim of this paper is to respond to many challenges of information propagation problem that the complex networks faces, and to present future directions which will shape the next generation of social systems. We have to take into account many realistic features such as coupling between networks, the dynamics of networks, interdependencies between structure, dynamics and functionality of networks, co-evolution of networks, and spatiotemporal properties of the networks.

3   Practical Methodologies for Social Propagation

Methodologies and techniques differ in how they consider information to propagate. There exist several perspectives that can be used to analyse them. The most frequent mentioned in most studies are: (1) the phase patterns that compose the methodology; (2) the strategies and techniques that are implemented in the methodology; (3) the measurements that are used in the methodology; (4) the types of data that are considered in the methodology; (5) the types of network topological structure in which all of the nodes are connected to transmit the information; (6) the organizations involved in the processes that create or update the data with their structure and norms.

Methodologies of social propagation adopt two general types of strategies, namely data-driven and user-driven. Data-driven strategies base directly on the values of data. For example, rumour propagation versus trust and viral dissemination. The objective in the latter case is a study of propagation of recommendations and opinions in social networks and an efficient selection of customers to focus of marketing efforts with the ultimate objective of receiving the highest reach of a favourable message. User-driven strategies includes an activity of the user that controls the whole process. One of such activity is microblogging called spread of word-of-mouth – communication over strong and weak ties in the network. It finds that weak ties are at least as important for information dissemination as the strong ties and interlinked blog posts are taking temporal information into account. it is also known that results obtained from a static representation of the network had little correspondence with results from a dynamic representation and therefore dynamics need to be taken into account. We observed a similar phenomenon as well [9]. Moreover, the focus of works to be done is on the local time-varying characteristics.

Network-level and local-level techniques are both a relevant perspective frequently considered, due to the effects of global-local activities. Existing network-level techniques for modelling propagation include e.g. independent and weighted influence cascades, the susceptible-infected and Reed-Frost models, and viral marketing approaches. Local-level techniques include tag-based cooperation, agent-based methods including learning and decision strategies, and probabilistic methods such as decision trees.

In the discussion until now, we have reviewed propagation methodologies and techniques. In the next section, we concentrate on different propagation measures which are applied in social networks. The definition of statistics to evaluate the process is a critical activity.

4   The Measurement of Social Propagation Attributes

In the field of social informatics, most authors either implicitly or explicitly distinguish two classes of performance measures: structural and behavioural. First class can be further divided into simple and aggregated version [7]. Due to computational complexity structural measures fit not well for social propagations with millions of active users. However, the second class is much more promising. For example, they can answer how wide propagation is, how fast it grows-decays, and how reliable it is.

Before we analyse the metrics, we should remember, that on a topological basis, most social networks are characterized with the small–world phenomenon. As observed in [8], the fraction of the initially infected nodes are more important in small–world networks than in scale–free networks. Consequently, the most critical parameters for the occurrence of propagation are as follows: (a) the probability of a process occurring, (b) the initial size and (c) speed of a process. Therefore, as expounded in [5], investigation of spreading activities, ordering them in time and analysing the *scope–speed–failure* relationship is very important. We propose *scope*, *speed* and failures which may be equivalents to distance, centrality, efficiency and robustness from classical network theory.

For the experiments we use the generic algorithm with the *pull* strategy on random graph with two parameters: the maximum influence between a node and infected neighbours needs to exceed the propagation threshold (θ), the latter applies to the *time-to-recovery*, which corresponds with the number of simulation steps. Figure 1 depicts the results of 100 simulation runs of the *pull* strategy

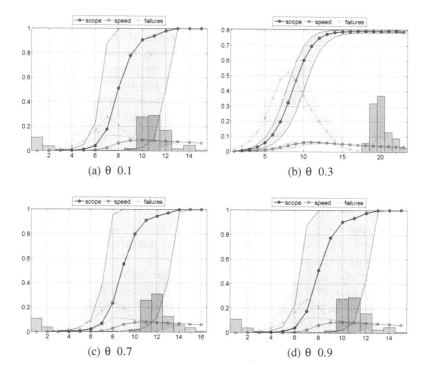

Fig. 1. The *scope–speed–failures* relationship. Scope (in black) as a function of time-steps compared with speed (in red) and failures (in green) with the pull strategy for different values of propagation threshold (Color figure online)

on Erdős-Rényi graph with $\theta = 0.1, 0.3, 0.7, 0.9$ and seeds randomly generated. The *scope* shows the number of successfully infected nodes, whereas *failures* depicts a number of unsuccessfully infected nodes. The last measure *speed* illustrates the rate at which the process covers a network. For each plots we print three different lines: the average, the 5th and 95th percentiles.

5   Summary

This paper has surveyed approaches, technologies and research issues related to propagation and error propagation mechanisms developed in large scale network systems. Particularly, it has reviewed social networks.

Better understanding of the different directions in which research has been done on propagation phenomenon was one of the goals for this short survey. Unfortunately, there is no universal procedure for efficiently utilizing these techniques within a system. Moreover, a set of new propagation problems are impending in future social systems, in which the line between reality and its representation will be blurred.

Following from the discussion of the methodologies, strategies and measures that concern propagation phenomenon in social networks, this section identifies directions in research activities and remaining open issues. These can be classified into seven categories based on different functional objectives of propagation phenomena. This classification is not meant to enumerate or describe all possible issues. However, due to the limit of space here, only the most interesting open issues from our point of view were selected and listed here.

- Maximizing robustness techniques including loosing trust and friends, fraudulent reputation building, false social annotations, emerging unwanted behaviour, combating crime, gossiping, spreading of rumours, memes, malwares, viruses, misinformation and disinformation.
- Enhancing scale and dynamic nature within and inter-domain specific issues including community detection, identifying influential spreaders, opinion maximization, uncertainty propagation, building disaster scenarios.
- Need for prompt network reliability estimation including deafness to social discord, locating and repairing faults like hegemony, dropping commitments, insincerity or dishonesty, and anonymity.
- The possibilities of reducing the rate of infection including new methods for protein synthesis and viral conductance, targeted cyber-attacks prevention.
- Comprehension the network of social networks including abnormal cascading identification failures, congestion and multi-layered profile analyses.
- Capturing the prediction time and the influence probabilities including effective centrality prediction and finding effectors.
- New structures in big social networks, precise graph sampling and crawling feasible to manage entire network.

The effectiveness of propagation measurements for a social network is still an open issue requiring further investigation. We are still far from a satisfactory adherence to reality and proper utilization of social propagation, but the recent advances outlined suggest a promising solution to these challenges. We hope that this paper presents a next step to exploit the propagation phenomenon in more effective way.

Acknowledgments. This research was partially supported by the statutory funds of the Wrocław University of Technology, Poland.

References

1. Albert, R., Jeong, H., Barabasi, A.L.: Error and attack tolerance of complex networks. Nature **406**(6794), 378–381 (2000)
2. Boccaletti, S., Latora, V., Moreno, Y., Chavez, M., Hwang, D.U.: Complex networks: structure and dynamics. Phys. Rep. **424**(4–5), 175–308 (2006). http://www.sciencedirect.com/science/article/pii/S037015730500462X
3. Bounova, G., de Weck, O.: Overview of metrics and their correlation patterns for multiple-metric topology analysis on heterogeneous graph ensembles. Phys. Rev. **85**, 016117 (2012). http://link.aps.org//10.1103/PhysRevE.85.016117

4. Hang, C.W., Wang, Y., Singh, M.P.: Operators for propagating trust and their evaluation in social networks. In: Proceedings of the 8th International Conference on Autonomous Agents and Multiagent Systems, AAMAS 2009, vol. 2, pp. 1025–1032. International Foundation for Autonomous Agents and Multiagent Systems, Richland (2009). http://dl.acm.org/citation.cfm?id=1558109.1558155
5. Król, D.: On modelling social propagation phenomenon. In: Nguyen, N.T., Attachoo, B., Trawiński, B., Somboonviwat, K. (eds.) ACIIDS 2014, Part II. LNCS, vol. 8398, pp. 227–236. Springer, Heidelberg (2014)
6. Król, D.: Propagation phenomenon in complex networks: theory and practice. New Gener. Comput. **32**(3–4), 187–192 (2014)
7. Król, D.: How to measure the information diffusion process in large social networks? In: Nguyen, N.T., Trawiński, B., Kosala, R. (eds.) ACIIDS 2015. LNCS, vol. 9011, pp. 66–74. Springer, Heidelberg (2015)
8. Król, D., Budka, M., Musiał, K.: Simulating the information diffusion process in complex networks using push and pull strategies. In: Proceedings of the European Network Intelligence Conference, ENIC 2014, USA, pp. 1–8. IEEE, New York (2014)
9. Król, D., Fay, D., Gabrys, B. (eds.): Propagation Phenomena in Real World Networks. Intelligent Systems Reference Library, vol. 85. Springer, Switzerland (2015). http://dx.org/10.1007/978-3-319-15916-4

Social Networks
and Recommender Systems

Visualizing Learning Activities
in Social Network

Thi Hoang Yen Ho, Thanh Tam Nguyen, and Insu Song[✉]

School of Business/IT, James Cook University Australia,
Singapore Campus, Singapore, Singapore
{yen.hothihoang, thanhtam.nguyen}@my.jcu.edu.au,
Insu.song@jcu.edu.au

Abstract. Proliferation of social networks and its use in education and higher learning have become an interesting topic. In order to accommodate diverse student cohorts in the age of massification of education market, this approach has become attractive. Recent advances in learning analytics and data visualization further proved to be useful in encouraging collaborative learning. On the other hand, information of social networks can be too complex to visualize often overloading users with too much possibly unwanted information. The question is what to show and what not to show when we visualize social relationships of users. In this paper, we propose a new visualization model called learning space and develop a method for visualizing learning activities on social network in a 3D virtual learning space. We evaluate the method using questionnaires to see whether visualization of social relations of learning will improve the learning in the following ways or not: make it more fun, make it easier, motivate more. The result shows that our method of visualization of learning activities in social network made learning more fun and easier. The result also shows that it helps student engage and motivate on the subjects.

Keywords: Engagement · Motivation · Encouragement · Collaborative learning

1 Introduction

Proliferation of social networks and its use in education and higher learning have become an interesting topic. In order to accommodate diverse student cohorts in the age of massification of education market, this approach has become attractive [1]. Recent advances in learning analytics and data visualization further proved to be useful in encouraging collaborative learning.

On the other hand, information of social networks can be too complex to visualize often overloading users with too much possibly unwanted information. The question is what to show and what not to show when we visualize social relationships of users. In particular, we are interested in whether visualization of social relations of learning will improve the learning in the following ways or not: make it more fun, easier, and motivational.

In this paper, we propose a method of visualizing learning activities in social networks using a 3D virtual world platform. First we review existing literature to identify

© Springer-Verlag Berlin Heidelberg 2016
N.T. Nguyen et al. (Eds.): ACIIDS 2016, Part I, LNAI 9621, pp. 97–105, 2016.
DOI: 10.1007/978-3-662-49381-6_10

gaps and opportunities in improving learning and teaching in educational social network. We then develop a new method for visualizing learning activities in an m-learning platform. We evaluate its effective in making learning more fun and easier as well as motivating students. Interviews and questionnaires are used for the evaluation. The result shows that our method of visualization of learning activities in social network made learning more fun and easier. The result also shows that it helps student engage and motivate on the subjects. We also observed changes in students learning behaviors. It also made easier for students to identify areas of focus for their studies. It also helped students avoid common mistakes by observing activities of other students.

2 Background

2.1 Data Visualization

One of the oldest document about data visualization was found in Turkey, 8000 years ago, it is the town map of Catal Hyuk in 6200 BC. At the time, people had learned to describe the information in graphical form so that we can understand it easier.

Visualization in general bring the huge advantage to our life in many fields. Burkhard, Meier [2] has presented a framework and model which can identify and related the key-aspects to help the communication. In this map, the target groups are presented as tubes. They found that the tube map visualization present the whole project and help the communication of a complex project with different target groups.

Jeffrey Heer has published a paper about the usage of visualization in online social network [3]. In this paper, he has presented a system which allow end user to navigate the large-scale online social networks. This model shows the connectivity of users and supports many visual feature functions for exploring the community structures. In result, author has provided the evidence of the usability of social network visualization on many purposes.

2.2 Use of Social Network for Education

Martınez et al. [4] have proposed a novel approach to evaluating the real experiences to encourage students' active and collaborative learning. This approach combines the traditional sources of data with computer logs with the social network analysis to evaluating. This proposal has provided an innovative techniques of studying. Fournier et al. [5] have presented a paper to illustrate the research methods used in exploring networked learning online. They proposed a hypothesis to analysis the online learning network and using visualization to improve the learning experiences.

2.3 Learning Analytics

Learning analytics has been examined by many researchers. Some researchers combined and analysed multiple sources of observations to provide a better understanding of learning and discovered new learning scenarios [6, 7]. Barre et al. [8] have set up experimentations of a collaborative e-learning system to argue how tracks arising from communication tools are analyzed and how useful it is in the reengineering purposes. Mazza, Milani [9] have published a paper of their research on the usage of learning systems and gaining insights from visualisations. Authors presented GISMO system to visualize data from courses collected in real settings, and to track on the status of the student to quickly discover which individuals need more attention.

2.4   Online Learning, E-Learning, M-Learning

Song and Vong [1] have introduced the Mobile Collaborative Experiential Learning (MCEL) to allow the students to interact with the system through mobile devices or any interactive devices which have internet connection. This system gives student ability to change their state of simulating learning lab at any time anywhere to archive their goal. In result, Song and Bhati [10] have applied MCEL on experiment. Students interacted with the system by low cost SMS message. This research has been evaluated by questionnaires in terms of User-friendliness, Technical feasibility, and Cost effectiveness.

On the research of online learning systems, MOODLE is known as one of the most popular open source course management system. With over than 65 million users, MOODLE was trusted by many large or small institutions and organizations as an effective online learning and courses management platform.

2.5   Use of Social Network for Education and Online Learning

Brady et al. [11] have argued that the trend of social network usage of students is increasing and the role of distance education in expanding of college or university level. In this paper, authors have figured the possibility of non-commercial, education-based SNSs such as Ning in the context of education. By doing the survey participating by the graduated students, authors found that education-based SNSs can be the most effective method in distance learning.

Santos et al. [12] have researched on an approach to promote the online learning. In this paper, they argue the fact that with the innovation of technology, teachers and students no longer solely interact directly face-to-face. This causes a difficulty to both sides on having an overview of the class and hard to discover the students' issues in-time. Therefore, the visualization of learning activities was proposed in order to solve this problem. Although they have evaluated the usability and user satisfaction, authors also provided a future plan for further evaluation.

2.6   Methodology

In order to identify gaps in literature and opportunities, we reviewed existing literature on online learning platforms, social network, their use in education and higher learning, and visualization of learning activities in social network.

Our literature review shows that not much research has been done in the visualization of social relations in learning. Therefore, we develop methods for visualizing education social network to investigate whether this would improve effectiveness of learning and motivational appeal, and encourage collaborative learning.

In this research, we create a survey for education social network users to find their needs, requirements and problems. An online questionnaire is created. The content covers these fields: ages, major, social network usage frequent for education purpose. With assumption of 3D data visualization for education activities, the questionnaire investigate user's experience to define the usability of 3D learning activities visualization in terms of engagement, motivation, encouragement and collaborative. Target of this survey will be JCU students and more focused on KOPO MES online mobile learning platform users (https://www.a.kopo.com/). After that, we analyzed the questionnaire data to justify the usability of visualizing social learning activities and identify the need of users.

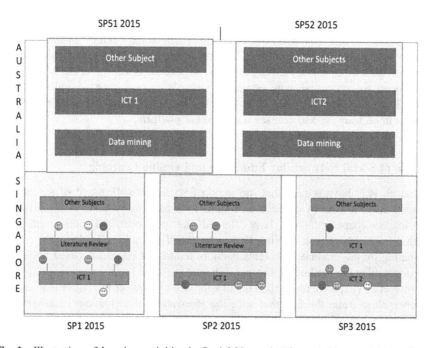

Fig. 1. Illustration of learning activities in Social Network. The activities are laid out in two dimensions: time and location. The horizontal axis represents time, such as semesters. The vertical axis represents locations. Each box represents a learning entity (a group of students in the same location and time).

Figure 1 illustrates the learning space of students in the world. Students participate in one or more learning activities in various locations and time. For instance, Fig. 1 shows two locations: JCU Singapore campus and JCU Australia campus over one year period. Students can move between campuses and same subjects offered in the two campuses but at different timing as the Singapore campus runs the tri-semester system whereas the Australia campus runs two-semester system.

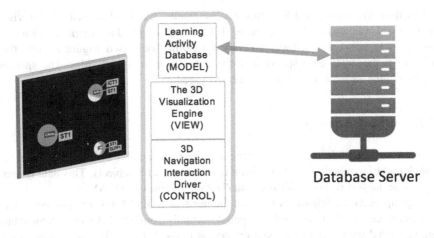

Fig. 2. The 3DLAV (Learning Activities Visualization) system of the social network is a HTML5 web application learning on any mobile devices. The system comprises of the learning activity database (model) which retrieves learning activity data from a remote server (https://www.a.kopo.com/), the 3D visualization engine (view), and 3D navigation interaction driver (control).

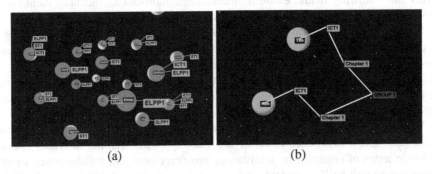

Fig. 3. Visualization of learning activities of students. Each circle represents a student and labels represent subjects and groups. Users use the standard touch interface on mobile devices and search tools to explore learning activities of other users in time and locaiton. (b) shows that two students Tim and Kim are studying ICT1 in a same group for chapter 1.

We often find that students are interested in what other students studying, where and when. Our system is to visualize these learning activities so that students can help each other forming close student communities.

Figure 2 illustrates the proposed system for visualizing these learning activities. Users work with the 3D Navigation Interaction Driver to interact with the system, the control component will send the request to the server to retrieve the data and filled into the learning activities database model component. The 3D visualization engine uses open-source ThreeJS library to visualize the learning activities on users' computer.

Based on the survey and literature review results, we develop methods for visualizing social relations of learning using HTML5 (css3 and JavaScript) to visualize learning social network. The three.js-JavaScript 3D library is used. Figure 3 shows that each user is presented as a sphere and connected to learning activities. The spheres which have the same activities are connected together.

3   Experiment Setup

In order to evaluate the new 3D learning activity visualization system (3DLAV), we created a set of synthetic data comprising of 27 users and 3 subjects. This data is used to generate the test database for the visualization system of 3DLAV.

A group of 31 participations who are the users of KOPO MES was invited to join the experiment, but there are only 23 people actually completed the test. Participants were presented with demo and given the access to the 3D visualization system to use the system and have an overview about the topic. The twelve of them were tested with a new module to see if they are motivated or whether the system gives them a better overview. The eleven participants were tested with the subject they are studying to see whether the system motivate them in their learning, and helping them to solve the problems. Finally, they were given a questionnaire to complete. The questionnaire covers the following fields: easiness, friendliness, motivation, encouragement, collaboration and the feedback on what need to be improved of the system.

In order to prove to usefulness of 3DLAV, we proposed an evaluation methodology of using questionnaire with 3DLAV users. The USE Questionnaire (Usefulness, Satisfaction, and Ease of Use Based) is used for this survey.

The content of the survey included 2 parts: Basic information of participants and the usefulness of 3DLAV with 5 Likert Scale (Strongly agree, agree, neutral, disagree, and strongly disagree).

With assumption of 30 data visualization for education activities, the questionnaire investigates user's experience to define the usability of 3D learning activities visualization in terms of engagement, motivation, encouragement and collaborative. Target of this survey will be JCU students and more focus on KOPO MES users. After that, we analyzed the questionnaire data to justify the usability of visualizing social learning activities and identify the need of users.

4   Results

There are 60.87 % of participants from 21 and 25 years old, 30.43 % of them are from 26 to 30 years old. Most of the participants are between 21 and 30 years old. Male participants took nearly 70 % of 23 people. The several of nationalities of participants helped to increase the accuracy of the result by getting data of different cultures and different methods of study. One thing in common is they all agreed with the hypothesis of the research.

More than 80 % of participants strongly agree or agree that they have problem with finding relevant information about courses, subjects and solutions online. This shows

Table 1. Questionnaire questions and survey result. 1 is strongly agree, 2 is agree, 3 is neutral, 4 is disagree, 5 is strongly disagree.

No	Question	Count	Av score
1.	3DLAV helps me be more effective in studying.	22	1.64
2.	3DLAV helps me be more productive in studying.	22	1.73
3.	3DLAV is useful in studying.	22	1.77
4.	3DLAV gives me a good overview of courses and students.	22	1.27
5.	3DLAV gives me more control over the activities in my study.	22	2.05
6.	3DLAV makes the things I want to accomplish easier to get done.	22	2.09
7.	3DLAV saves me time when I use it.	22	1.82
8.	3DLAV meets my needs.	22	1.95
9.	3DLAV does everything I would expect it to do.	22	2.00
10.	3DLAV is easy to use.	22	1.50
11.	3DLAV is simple to use.	22	1.41
12.	3DLAV is user-friendly.	22	1.64
13.	3DLAV requires the fewest steps possible to accomplish what I want to do with it.	22	1.91
14.	3DLAV is flexible.	21	1.90
15.	Using 3DLAV is effortless.	22	1.86
16.	I can use 3DLAV without written instructions.	22	1.91
17.	I don't notice any inconsistencies as I use 3DLAV.	22	1.95
18.	Both occasional and regular users would like 3DLAV.	22	2.05
19.	I can recover from mistakes quickly and easily.	21	2.00
20.	I can use 3DLAV successfully every time.	21	1.90
21.	I learned to use 3DLAV quickly.	22	1.50
22.	I easily remember how to use 3DLAV.	22	1.50
23.	3DLAV is easy to learn to use it.	22	1.68
24.	I quickly became skillful with 3DLAV.	22	1.73
25.	I am satisfied with 3DLAV.	22	1.86
26.	I would recommend 3DLAV to a friend.	22	2.09
27.	3DLAV is fun to use.	22	1.73
28.	3DLAV works the way I want it to work.	22	1.86
29.	3DLAV is wonderful.	22	2.00
30.	I feel I need to have 3DLAV.	22	2.09
31.	3DLAV is pleasant to use.	22	1.77
32.	I believe in the future, text lessons will be replaced by visualization.	22	1.73
33.	I think visualization of the social relations of other students' learning will motivate me in learning and improve my studying.	22	1.50

a strong need of 3DLAV and the potential benefits of the project. The fact is that many of the students do not know how to use social network effectively for their studies (Table 1).

The result has proved the hypothesis, the usefulness of 3DLAV in terms of improving students' experience, making study more fun and easier as well as motivating students more.

5 Conclusion

Our survey shows that students are interested or using social media during their study in order to find how other students studying, what they are studying, what subjects they have completed.

We proposed a new visualization model called learning space. We then developed a new way of visualizing learning activities on a 3D learning space. The results are promising that students find it intuitive to navigate learning activities of other student otherwise too complex or too time consuming to navigate manually. In the future, we believe visualization will play an important role in education field. Learning will be much easier and fun than ever when social media and easy to use learning activity explorer are combined.

References

1. Song, I., Vong, J.: Mobile collaborative experiential learning (MCEL): personalized formative assessment. In: 2013 International Conference on IT Convergence and Security (ICITCS), pp. 1–4. IEEE (2013)
2. Burkhard, R.A., Meier, M.: Tube map visualization: evaluation of a novel knowledge visualization application for the transfer of knowledge in long-term projects. J UCS **11**(4), 473–494 (2005)
3. Heer, J., Boyd, D.: Vizster: visualizing online social networks. In: 2005 IEEE Symposium on Information Visualization, INFOVIS 2005, pp. 32–39. IEEE (2005)
4. Martınez, A., Dimitriadis, Y., Rubia, B., Gómez, E., De La Fuente, P.: Combining qualitative evaluation and social network analysis for the study of classroom social interactions. Comput. Educ. **41**(4), 353–368 (2003)
5. Fournier, H., Kop, R., Sitlia, H.: The value of learning analytics to networked learning on a personal learning environment (2011)
6. Avouris, N., Fiotakis, G., Kahrimanis, G., Margaritis, M., Komis, V.: Beyond logging of fingertip actions: analysis of collaborative learning using multiple sources of data. J. Interact. Learn. Res. **18**, 231–250 (2007)
7. Marty, J.-C., Heraud, J.-M., Carron, T., France, L.: Matching the performed activity on an educational platform with a recommended pedagogical scenario: a multi-source approach. J. Interact. Learn. Res. **18**(2), 267–283 (2007)
8. Barre, V., El-Kechaï, H., Choquet, C.: Re-engineering of collaborative e-learning systems: evaluation of system, collaboration and acquired knowledge qualities. In: 12th Artificial Intelligence in Education AIED, pp. 9–16 (2005)
9. Mazza, R., Milani, C.: Gismo: a graphical interactive student monitoring tool for course management systems. In: International Conference on Technology Enhanced Learning, Milan, pp. 1–8 (2004)

10. Song, I., Bhati, A.S.: Automated tutoring system: mobile collaborative experiential learning (MCEL). In: 2014 IEEE 14th International Conference on Advanced Learning Technologies (ICALT), pp. 318–320. IEEE (2014)
11. Brady, K.P., Holcomb, L.B., Smith, B.V.: The use of alternative social networking sites in higher educational settings: a case study of the e-learning benefits of ning in education. J. Interact. Online Learn. 9(2), 151–170 (2010)
12. Santos, J.L., Govaerts, S., Verbert, K., Duval, E.: Goal-oriented visualizations of activity tracking: a case study with engineering students. In: Proceedings of the 2nd International Conference on Learning Analytics and Knowledge, pp. 143–152. ACM (2012)

A Mobility Prediction Model
for Location-Based Social Networks

Nguyen Thanh Hai, Huu-Hoa Nguyen, and Nguyen Thai-Nghe$^{(\boxtimes)}$

College of Information and Communication Technology,
Can Tho University, Can Tho, Vietnam
{nthai.cit, nhhoa, ntnghe}@ctu.edu.vn

Abstract. Mobility prediction plays important roles in many fields. For example, tourist companies would like to know the characteristics of their customer movements so that they could design appropriate advertising strategies; sociologists has made many research on migration to try to find general features in human mobility; polices also analyze human movement behaviors to seek criminals. Thus, for location-based social networks, mobility prediction is an important task. This study proposes a mobility prediction model, which can be used to predict the user (human) mobility. The proposed approach is conducted from three characteristics: (1) regular movement in human mobility, (2) the influence of relationships on social networks, (3) other features (in this work, we consider "hot regions" where attract more people coming to there). To validate the proposed approach, three datasets including over 500,000 check-ins which are collected from two location-based social networks, namely Brightkite and Gowalla, are used for the experiments. Results show that the proposed model significantly improves the prediction accuracy, thus, this approach could be promising for mobility prediction, especially for location-based social networks.

Keywords: Mobility prediction · Location-based social networks · Influence of relationships on location-based social networks · Human mobility

1 Introduction

Mobility prediction has important roles in many fields. It may enhance the handover performance in telecommunication. In addition, there are many advertising applications, which has applied mobility prediction. Tours companies would like to know the characteristics of human movement so that they could design advertising strategies. Taxi drivers, tour guides based on tour recommendations to direct tourists 1. Besides, polices also analyze human movement behaviors to seek criminals. In addition, sociologists have made many researches on migration to try to find general features in human mobility, etc.

In fact, there has been a dramatic influence of location-based social networks (LBSNs) on mobility prediction since information that the users shared on LBSNs may become recommendations to others. For example, a person (e.g., a potential user in Fig. 1) who has a LBSN account comes to an interesting place (e.g., a restaurant or a coffee shop,

© Springer-Verlag Berlin Heidelberg 2016
N.T. Nguyen et al. (Eds.): ACIIDS 2016, Part I, LNAI 9621, pp. 106–115, 2016.
DOI: 10.1007/978-3-662-49381-6_11

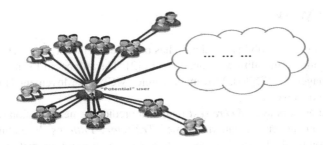

Fig. 1. Influence of a potential user *(picture source: adapted from mapexpo.com.au)*

etc.) and shares information on that location, e.g., he writes "What a Wonderful Place!" on his wall. When his friends see the feed, they may come to that place, thus, this is good for marketing/advertisement. Moreover, data on LBSNs are possibly used for analyzing human mobility. Data on human movement behaviors can be collected through LBSNs such as check-ins of users, information on locations. From collected data, we can use a mobility prediction model for capturing human movement in the future basing on their movement histories. In addition, social connections on LBSNs help to spread the information for influence users. Moreover, many users have been using LBSNs and the number of users of LBSNs may increase quickly so we can exploit the data on LBSNs for mobility prediction.

Although several researches have been studied in mobility prediction (to be presented in Sect. 2), some features such as relationships between human and hot regions have not been considered in detail. Therefore, in this work, we propose a model for mobility prediction **not only based on traditional prediction methods**, **but also based on relationships on LBSNs**, as illustrated in Fig. 2.

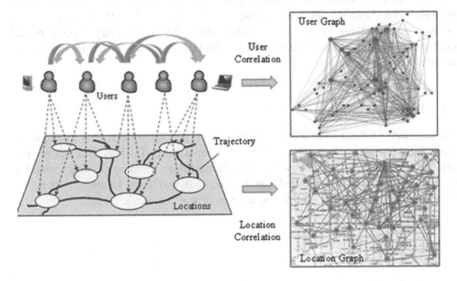

Fig. 2. Users' movement and relationships *(picture source: adapted from research.microsoft. com/apps/mobile/showpage.aspx?page =/en-us/projects/lbsn/)*

2 Related Work

Numerous models for mobility prediction has proposed in recent years. The difficulties of predicting human mobility were analyzed in [2]. The authors in [3] used Predictive Mobility Management (PMM), Movement Circle (MC), and Movement Track (MT) to predict the next movement.

In [4], the authors used *WhereNext* model to predict the next location of a moving object based on movement patterns named *Trajectory Patterns*. In addition, a considerable amount of studies has been attempted to propose prediction models in high-sampling-rate trajectories [5–7] . The authors in [8] proposed two places (home, place for work, etc.) namely "latent states" where people usually come to. Basing on the time, the proposed models in [8] could predict whether the user would be in one of latent states or not. In this study, social network relationships were also considered.

User preference, social influence and geographical influence were also considered in [9]. The authors used these characteristics to propose a model for recommendations.

Based on previous studies, we combine algorithms to predict whether a user comes to a given location or not. A similar model is mentioned in [10]. In this study, the model is described in detail with some improvements and experiments for evaluating the performance of the model.

Points of check-ins in Gowalla and Brightkite are discrete so they need to be clustered in OPTICS [11] to group the points of the datasets into a set of meaningful regions with Radius 50 m, 100 m, 150 m, 200 m and MinPts ranging from 2 to 5.

3 Proposed Method

In this section, we propose a model for predicting human mobility on LBSNs. First, algorithms for clustering are used to cluster points collected from the datasets because points from check-ins of users are discontinuous. Then, we analyze some crucial characteristics of human movement behaviors and suggest a model to capturing human movement.

3.1 Preliminaries

The definition of check-in action was mentioned in [10]. We collected the data of users' check-ins and relationships on location-based social networks. This data can be indicated as a graph $G = (U, L, W)$, where U indicates the set of nodes showing users, L is the link set representing the friendships between any two users. In the case, there is a link $l_{i,j}$ between two users user u_i and u_j, then user u_i and u_j are friends.

3.2 Predicting Mobility in the Near Future (PMNF)

We predict the next movements based on the estimate of covered space of movement of users. The geographical spaces where users are more likely to appear in need to be captured for prediction. The PMNF model (*in Algorithm 1*) which is proposed to solve this problem is based on 3 features: (1) the regular movement of users, (2) the movement of friends of users, (3) hot areas around the most visited areas of users and friends, and the most attractive regions for others.

George Liu et al. state that human movement includes random movement and regular movement [1]. We propose the PMNF to identify the movement patterns of users. For determining the regular movement, we consider two factors: (a) latent states where users are more likely to be in such as "home" and "work", and (b) circle movement patterns which means users have patterns in movement among some often visited locations. People have the tendency to repeat the movement or the sequence of movements. Therefore, we need to identify repetitive movements for predicting mobility accurately. For random movement, it is hard work because random movement behavior of individuals can change according to time. To do this, we determine hot regions where users are more likely to come and compute distance from friends of the users to areas where users usually visit.

We just focus on the influence of users on other users so we extract information about users, location, friendship that related to check-ins. This means we only consider users, friendships, locations that appeared in check-ins rows. Therefore, experiments proceed faster.

We have a strong observation that people have a trend in coming back to locations where they have already check-ins. The data set shows that more than 34 % of the users came back to locations where they had visited. Another interesting observation shows that 28 % of the check-ins of the users were made at regions where their friends had visited. Based on analyzing the history movement of users, we consider users' habits of movement. We compute the distance between locations where they had check-ins. We store previous movement of users into segments that recorded movement pattern of users.

If the given location is in the circle movement patterns of the user, it is likely that the probability of coming to the location of the user is rather high. These steps are explained in greater detail in the following sections:

In the beginning, an important small set of latent states (included 2 latent states that are considered as "home" and "work") must be identified by K-MEAN clustering method with K = 2. If the given location is in the determined set, we can make a guess that the user is more likely to come there. Otherwise, we move to the next step with considering the circle patterns. We store user's history movements as a sequence $M = <l_1, l_2, l_3, l_4, ..., l_n>$ in order to time with l_1, l_2... are locations. With the given location L, if there exists a match between $L1 = <l_n, l_x, l_{x+1}, l_{x+2}, L>$ and a part of L that means $L1 \subset M$, we can predict the user may come to L.

Algorithm 1: Predict the movement of user **(PMNF)**

Input: user_id, location_id

Output: the probability that the user will come to the given location

  1: prob=0;

  2: Determine the log of movement H_u of the user consisting of pattern movement.

  3: **if** location **in** H_u

  4:    prob= **Regular**+prob;

  5: **else** {

  6:    Consider 2 latent states of the given user. Estimate regions Regions_u around the states with average next movement distance considered as radius

  7:    **if** location_id **in** Regions_u

  8:      prob= **Regular**+prob;

  9: }

  10: Estimate a set Most_f including the most visited places of friends. Determine regions Regions_fri around Most_f with average next

---

For example, a user has a sequence of movement history such as $<$ ABCABDEC $>$, and the location needs to be predicted is D. We see that the last location where the user just visited is C while there exists a chain $<$ C<u>AB</u>D $>$ (4 locations in the chain) $\subset$ $<$ ABCABDEC $>$; therefore, we can predict that the user may come to D. The model supports the length of recognition chain is up to 5 locations.

If a region fits the features such as user preference, influence of friends and geographical influence, then a probability corresponding to the features will be added with values from variables **Regular, Influenced_by_friends, Hot** respectively. The values of 0.7, 0.1, 0.2 for **Regular, Influenced_by_friends, Hot** respectively, is chosen for the experiments. These values are conducted from . In, The authors stated that the factor of user preference accounts for at least 70 % of user decision while influence of friends and geographical influence occupy about 10 % and 20 % respectively.

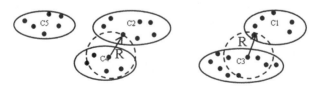

Fig. 3. An example shows points which are clustered by OPTICS for mobility prediction.

We consider the example from the Fig. 3. Points are clustered into 5 clusters: C1, C2, C3, C4, C5. Figure 3 indicates the bounded regions of 2 latent states with the points having the largest number of check-ins (in the cluster C3, C4) considered as centers. We do not consider points which are near the border of the cluster because of saving the run-time for considering so many points. The sequence of movement history of the user is given: $H = <C5C3C4C3C1C3>$. C2 belongs to the bounded regions of 2 latent states of the user with radius R considered as the average next movement distance of the user, although it does not present in user's movement log. Assuming that C1, C5 are attractive regions and C2 is the most visited place of a friend.

With the information from the Fig. 3 and the assumption, we calculates the probability that the user come to C1 is 0.9 (we have C is the last location where the user visited while there exists a chain $< C3C1 > \subset H$. Therefore, it fits regular movement of the user. In addition, C1 is a hot region) while the probability that the user travel to C2 is 0.8 (C2 is in the bounded regions of 2 latent states which belongs to the regular movement of the user. Besides, C2 is a place where the user may come by the influence of a friend). C5 is an attractive location, so the probability that the user will come C5 is 0.2.

4 Experimental Results

In this section, we used real datasets, online location-based social networks, to evaluate the performance of our algorithm. We use 3 baselines which are basic models for comparison.

4.1 Datasets and Settings

Three real datasets which collected from online location-based social networks (Brightkite and Gowalla) [8] are used for the experiments. The dataset which was extracted from Brightkite included 4,491,143 check-ins from 51,406 users. In this dataset, there were 772,967 check-in points where users have checked from March, 2008 to October, 2010. Besides, we have 2 kind of dataset from Gowalla. One dataset extracted from Gowalla, named Gowalla 1, includes 6,442,892 check-ins of 107,092 users at 1,280,969 points collecting from February, 2009 to October, 2010. The other dataset which was clustered by a grid-based clustering approach and extracted from the original Gowalla (called Gowalla 2), included 466,740 check-ins of 11,814 users at 5,789 regions.

We divide "Gowalla 2" into two parts (one for training, the other for the first testing data set) according to time of check-ins. The first section consists of 397,017 check-ins (85 %) collected from March, 2009 to June, 2010. The latter which is used for the first testing data set includes 69,723 check-ins in July, 2010 (15 %). In addition to testing tools, the second testing dataset (called "Dataset 4") included 100 rows contained locations users came or not is also introduced to evaluate fully performance of the model. This data set includes 3 columns: user_id, location_id, and the result of whether user came to the location (say "yes") or not ("no"). The data set evaluates the models exactly by reducing the location predicted inaccurately (by location labeled "no"). Similarly, Gowalla 2, Gowalla 1 is divided into 2 parts 80 %, 20 % for training and testing, respectively. The numbers of Brightkite are 75 % and 25 %, respectively.

To evaluate the performance, the 3 baselines are used to compare to the proposed prediction model.

Baseline 1: The Most Frequented Location Model (MFLM), we count the check-ins at locations, then we pick one location that had the largest number of check-ins and assume that the location is the location where the user may come. Assuming that user U has checked in 122 times at location L1. Location L2 has 345 check-ins from user U, and user U has 88 check-ins at location L3. Hence, location L2 (where the user has the largest number of check-ins) is predicted as the location the user may come in the future.

Baseline 2: RW has been proposed by [8].The model always predicts the next location where users checked in the last time. For instance, user A has 3 check-ins including 10 am July 1st, 2015 at location L1, 10 pm July 11th, 2015 at location L2, and 2 pm July 15th, 2015 at location L3. Then, location L3 is predicted as the location the user may come in the future.

Baseline 3: In addition to the baselines, a model namely "Full history", including full movement history of the users, also is introduced for the comparisons. The locations that appeared in the movement history of the user will be predicted as the location the user may come in the future.

Metrics: To evaluate the accuracy of the algorithm, the prediction accuracy is measured by

$$\frac{The\ number\ of\ rows\ is\ predicted\ exactly}{The\ number\ of\ rows\ is\ predicted}$$

3 baselines and the proposed model are trained by the training datasets, then algorithms predict users in the testing datasets whether they may come to the given location in the testing dataset or not.

4.2 Experimental Results

Figure 4 displays the performance of PMNF in comparison with the baselines. The Baseline 2 (RW) with prediction by the last known location of the user performs worst with the accuracy rates from 2–3 %, while the Baseline 1 (MFLM) has little better accuracy rates with the score from 3–5 %. "Full history" model scores a large improvement with over 32 %. Significantly, our model gives great improvements with over 2.5 times higher the "Full history" model, and over 15–20 times greater than Baseline 1 and Baseline 2, respectively.

Similarly, experiments with the data set including 100 rows (50 locations users came to, and the 50 others did not) also show similar results (Fig. 5). "Full history" presents an average performance with 71 %. The accuracy of PMNF model is up to 91 % with 48/50 locations predicted exactly. Similar to the first experiment, the accuracy rate is nearly 2.5 times than "Full history" model about the prediction of location users came to (48 compared to 21), while RW, MFLM show lower

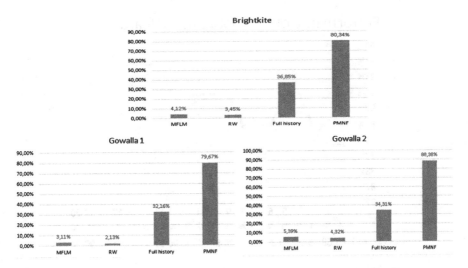

Fig. 4. Performance of PMNF compared to MFLM, RW, Full history model

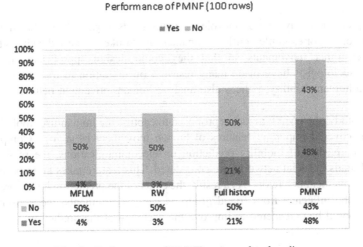

Fig. 5. Performance of PMNF compared to baselines

performances with the accuracy rate of only 53 %, 54 %, respectively. From these results, we can see that the proposed method is very promising, however, testing on other data sets will be continued in the future to consolidate the PMNF.

Another experiment also was employed to measure the influence of choice on the top places in the prediction. We observe in Fig. 6 that the accuracy increases to a threshold, and goes down. This explains that more locations to consider increases the probability of exact prediction for "yes" but also rises the number of locations predicted incorrect.

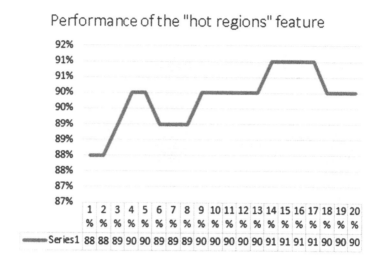

Fig. 6. Testing PMNF with different percentage of top hot regions

5 Conclusion

Mobility prediction plays important roles in many fields, especially for location-based social networks. For example, tourist companies would like to know the characteristics of their customer movements so that they could design appropriate advertising strategies; sociologists have made many research on migration to try to find general features in human mobility; polices also analyze human movement behaviors to seek criminals.

In this study, the PMNF model is proposed to capture the human/user mobility. The model has covered important features of human movement behaviors. Compared to different approaches of mobility prediction in telecommunications, the relationships between friends in social networks are considered. This feature takes an important role in human movement behaviors. As a result, the exact of the prediction has improved. In the future, the influence from friend group as well as comments, feeds from social networks should be considered to improve the accuracy of the prediction.

References

1. Yuan, J., Zheng, Y., Zhang, C., Xie, W., Xie, X., Sun, G., Huang, Y.: T-drive: driving directions based on taxi trajectories. In: Proceedings of the 18th SIGSPATIAL International Conference on Advances in Geographic Information Systems (2010)
2. Chan, J., Zhou, S., Seneviratne, A.A.: QoS adaptive mobility prediction scheme for wireless networks. In: Global Telecommunications Conference 1998, GLOBECOM 1998. The Bridge to Global Integration. IEEE, vol. 3, Sydney, NSW (1998)

3. Liu, G., Maguire Jr., G.: A class of mobile motion prediction algorithms for wireless mobile computing and communication. Mob. Netw. Appl. – Spec. Issue: Routing Mob. Commun. Netw. **1**(2), 113–121 (1996)
4. Monreale, A., Pinelli, F., Trasarti, R., Giannotti, F.: WhereNext: a location predictor on trajectory pattern mining. In: Proceedings of the 15th ACM SIGKDD International Conference on Knowledge Discovery and Data Mining, KDD 2009, New York, NY, USA (2009)
5. Tao, Y., Faloutsos, C., Papadias, D., Liu, B.: Prediction and indexing of moving objects with unknown motion patterns. In: Proceedings of the 2004 ACM SIGMOD International Conference on Management of Data, SIGMOD 2004, New York, NY, USA (2004)
6. Zhang, M., Chen, S., Jensen, C.S., Ooi, B.C., Zhang, Z.: Effectively indexing uncertain moving objects for predictive queries. Proc. VLDB Endow. **2**(1), 1198–1209 (2009)
7. Aggarwal, C.C., Agrawal, D.: On nearest neighbor indexing of nonlinear trajectories. In: Proceedings of the Twenty-Second ACM SIGMOD-SIGACT-SIGART Symposium on Principles of Database Systems, PODS 2003, New York, NY, USA (2003)
8. Cho, E., Myers, S.A., Leskovec, J.: Friendship and mobility: user movement in location-based social networks. In: Proceedings of the 17th ACM SIGKDD International Conference on Knowledge Discovery and Data Mining, KDD 2011, New York, NY, USA (2011)
9. Ye, M., Yin, P., Lee, W.-C., Lee, D.-L.: Exploiting geographical influence for collaborative point-of-interest recommendation. In: Proceedings of the 34th International ACM SIGIR Conference on Research and Development in Information Retrieval, SIGIR 2011, New York, NY, USA (2011)
10. Nguyen, T.H.: A novel approach for location promotion on location-based social networks. In: The 2015 IEEE RIVF International Conference on Computing & Communication Technologies - Research Innovation and Vision for Future (RIVF) (2015)
11. Ankerst, M., Breunig, M.M., Kriegel, H.-P., Sander, J.: OPTICS: ordering points to identify the clustering structure. In: Proceedings of the 1999 ACM SIGMOD International Conference on Management of Data, SIGMOD 1999, New York, NY, USA (1999)

Empirical Analysis of the Relationship Between Trust and Ratings in Recommender Systems

Kulwadee Somboonviwat[1(✉)] and Hisayuki Aoyama[2]

[1] Software Engineering Program, International College,
King Mongkut's Institute of Technology Ladkrabang, Bangkok, Thailand
kulwadee.so@kmitl.ac.th
[2] Department of Mechanical Engineering and Intelligent Systems,
The University of Electro-Communications, Tokyo, Japan
aoyama@mce.uec.ac.jp

Abstract. User-based collaborative filtering (CF) is a widely used recommendation method that suggests items to users based on ratings of other users in the system. The performance of user-based CF can be degraded due to its inherent weaknesses, such as data sparsity and cold start problems. To address these weaknesses, many researchers have proposed to incorporate trust information into user-based CF. However, as reported in many recent works on trust aware recommendation, effectively exploiting trust in recommendation is not straightforward due to insufficient understanding of the relationship between trust and ratings. This paper empirically analyses real-world ratings data and their associated trust networks. Specifically, we focus our analysis on comparative characteristics of cold users vs. non-cold users. Our results show that the characteristics of cold users and non-cold users are significantly different.

Keywords: Collaborative filtering · Trust network · Trust aware recommendation · Cold start problem · Experimentation

1 Introduction

In recent years, recommender systems have become a major component of the e-commerce ecosystem. They are designed and built to assist users in finding items (e.g. movies, music, books, etc.) of interest from a large collection of available choices. The recommender systems are being deployed at numerous websites to provide personalised recommendation and decision support services to end users. Examples of such websites include www.Amazon.com and www.NetFlix.com.

User-based collaborative filtering (aka. k-NN collaborative filtering [1]) is a well-known recommendation method that presents items to users based on

This work was supported by KMITL-UEC Global Alliance Lab (KMITL-UEC GAL). The KMITL-UEC GAL was established in 2014 under the collaboration between King Mongkut's Institute of Technology Ladkrabang, Bangkok, Thailand and the University of Electro-Communications, Tokyo, Japan.

© Springer-Verlag Berlin Heidelberg 2016
N.T. Nguyen et al. (Eds.): ACIIDS 2016, Part I, LNAI 9621, pp. 116–126, 2016.
DOI: 10.1007/978-3-662-49381-6_12

ratings of other users in the system. The underlying assumption of user-based CF is that if users agree about the quality of some items, then they will likely agree about other items as well – if Alice likes the same items as Jane, then Jane is likely to like the items Alice likes which she hasn't yet seen. Specifically, recommendations are generally made by computing the similarity between users in the system and recommending items that are liked by similar users.

In real world settings, the performance of user-based CF can be degraded due to its inherent weaknesses, such as data sparsity, cold start problem, and user profile based attack to name a few. In order to address these weaknesses, many researchers have proposed to improve recommendation performance by exploiting trust information derived from trust networks. This type of recommender systems is called *trust aware recommender systems*. Scores of trust aware recommendation algorithms have been proposed in the literature over the years (for example, [4–6,8,9,12]). Nevertheless, as already reported in some recent works ([10–12]), effectively exploiting trust in recommendation is not straightforward and there remains much work to be done towards effective trust aware recommendation.

One of the major obstacles to achieving effective trust aware recommendation is insufficient understanding of the relationship between trust and ratings. To alleviate this obstacle, in this paper, we aim to uncover the relationship between trust and ratings in the context of recommender systems by conducting empirical analyses of real-world ratings data and their associated trust networks. Specifically, we focus our study on the comparative characteristics of cold users versus non-cold users over trust networks. Main contributions of the paper are as follows.

- To the best of our knowledge, we are the first to consider the characteristics of cold/non-cold users separately.
- We empirically demonstrate that the characteristics of cold users and non-cold users over trust networks are significantly different.
- Based on our empirical results, we propose guidelines for designing an effective trust aware recommendation algorithm.

The rest of the paper is organised as follows. Section 2 introduces notation and terminology that will be used throughout the paper. Section 3 describes the characteristics of the datasets used in this study. Section 4 presents experimental analyses of the relationship between trust and ratings data. Finally, Sect. 5 concludes the paper with the future work.

2 Notation and Terminology

In this section, we introduce notation and terminology that are required for our empirical analyses of recommendation datasets in subsequent sections.

In a trust aware collaborative filtering (CF) system, *users* can express their preferences for various *items*. A preference of a user for an item is called a *rating*. Ratings are commonly represented with a triple (*UserId, ItemId, RatingValue*).

The *Rating Value* can be represented in several formats, for example, unary ratings (e.g. "has purchase"), binary ratings (e.g. "like/dislike"), integer-valued rating scales (e.g. "0–5 stars"), and real-valued rating scales (e.g. "0.0–10.0 score"). The set of all rating triples forms the ratings matrix (also known as the *U-I* matrix). In addition to specifying preferences or ratings, the users of a trust aware CF can implicitly or explicitly express trust statements toward other users in providing accurate and helpful ratings. A collection of trust statements can be organised as a trust network, where the users are represented as nodes and the trust statements are represented with edges or links between nodes.

In this paper, a set of users within a collaborative filtering system is denoted by U. A set of items is denoted by I. The set of items rated by user u is denoted by I_u. The set of users who have rated item i is denoted by U_i. The rating matrix is denoted by R. The rating value of user u for item i is denoted by $r_{u,i}$. $\bar{r}_u$ is the average rating of u. Typically, most users of recommender systems will provide ratings to only a small fraction of items in the system. This naturally results in highly sparse rating matrices. The sparsity of a rating matrix R is computed by:

$$Sparsity = (1 - \frac{\#Ratings}{\#Users \times \#Items}) \times 100\,\% \qquad (1)$$

In collaborative filtering systems literature, users with very small number of ratings are called *cold users*. It is well known that the generation of effective recommendation for the cold users is challenging (owing to the intrinsic properties of collaborative filtering systems). In this paper, we define a *cold user* as the user who has provided ratings to *less than five items* [8]. More formally, a cold user u_{cold} is defined as any user $u \in U$ such that $|I_u| < 5$, and a non-cold user $u_{noncold}$ is defined as any user $u \in U$ such that $|I_u| \geq 5$.

A trust network associated with a collaborative filtering system is represented as a directed graph $TN(V, E)$, where $V \subseteq U$ is a finite set of users $\{v_1, ..., v_n\}$, and E is a set of trust relations $\{e_1, ..., e_m\}$. A trust relation (or an edge) from user v_i to user v_j within the trust network TN indicates the existence of v_i's belief that ratings provided by v_j are trustworthy. The source node of the trust relation v_i is called a trustor, while the destination node v_j is called a trustee.

3   Datasets and Their Basic Characteristics

Four datasets are used in this work: Epinions, CiaoDVD.review, CiaoDVD.movie and FilmTrust. The Epinions dataset was collected by [7]. The CiaoDVD.review and the CiaoDVD.movie datasets were collected by [2]. The FilmTrust dataset was collected by [3]. All datasets consist of both user ratings data and the associated trust networks. We summarise characteristics of all datasets in Table 1.

The Epinions dataset [7] was crawled from Epinions.com during November to December 2003. Epinions.com is a website that allows users not only to write reviews for various products but also to read and rate the reviews written by other users using an integer-valued rating scale (from 1 to 5). Moreover, the users can explicitly express their trust toward the quality of other users' reviews

Table 1. Basic characteristics of the datasets.

Dataset	User ratings data				Trust networks	
	#Users	%ColdUsers	#Items	#Ratings	#Nodes	#Edges
Epinions	40,163	42 %	139,738	664,824	49,288	487,183
CiaoDVD.review	21,019	57 %	71,633	1,625,480	4,658	40,133
CiaoDVD.movie	17,615	85 %	16,121	72,665	4,658	40,133
FilmTrust	1,508	19 %	2,071	35,497	874	1,853

by adding the trusted users into their "Circle of Trust". According to Table 1, about 139 K items have received 664 K ratings from 40 K users (about 42 % of these users are cold users, i.e. they have provided less than 5 ratings). The rating matrix $R_{Epinions}$ is a highly sparse matrix, its sparsity is about 99.98 %. The trust network $TN_{Epinions}$ is a directed graph, consisting of 49 K users and 487 K trust relations. Note that, unlike other datasets, the number of users in the trust network ($|V_{TN_{Epinions}}| = 49,288$) is larger than the number of users in the user ratings data ($|U| = 40,163$). A plausible reason for this discrepancy is the limitation of the crawling method used to collect the data.

The CiaoDVD dataset [2] was crawled in December 2013 from the category DVD in the www.ciao.co.uk/. Like Epinions.com, users of www.ciao.co.uk/ can read/write reviews for various kinds of products that they purchased in the past; and other users can provide feedbacks on the helpfulness of these contributed reviews using an integer-valued rating scale (from 0 to 5). If a user finds any review writer to be trustworthy of valuable reviews, the user can add that review writer into his/her trust list. The trust lists of all users form the CiaoDVD's trust network. In this work, we have separated the CiaoDVD into two datasets: CiaoDVD.review and CiaoDVD.movie. The CiaoDVD.review dataset contains ratings data of review items; whereas the CiaoDVD.movie dataset contains ratings data of movie items. Both CiaoDVD.review and CiaoDVD.movie datasets share the same social trust network. As can be seen from Table 1, about 21 K users in the CiaoDVD.review dataset have contributed 1.6 M ratings for 71 K review items; around 57 % out of 21 K users are cold users. The rating matrix $R_{CiaoDVD.review}$ is slightly denser than $R_{Epinions}$, its sparsity is about 99.89 %. As for the CiaoDVD.movie dataset, 85 % of 17 K users are cold users; and there are 72 K ratings for 16 K items in the system. The sparsity of the rating matrix $R_{CiaoDVD.movie}$ is around 99.97 %. The trust network $TN_{CiaoDVD}$ is a directed graph, consisting of 4 K users and 40 K trust relations.

The FilmTrust dataset [3] was crawled from a movie sharing and rating website (http://trust.mindswap.org/FilmTrust) in June 2011. The FilmTrust users can rate and review movie using a real-valued rating scale (from 0.5 to 4.0 with step 0.5). The users can also explicitly specify trustworthiness of other users with a certain level of trust from 1 to 10. Unfortunately, due to the sharing policy, the FilmTrust dataset, that we adopted from [3], only provides the existence of trust statements between users with a binary values 0 and 1. As shown in Table 1, there are 1,508 users providing 35 K ratings to 2,071 items; and about 19 %

of these users are cold users. The sparsity of the rating matrix $R_{FilmTrust}$ is about 98.86%. The trust network $TN_{FilmTrust}$ is a directed graph; it consists of 874 nodes representing users and 1,853 edges representing trust relations.

It is worth noting that each dataset has different percentages of cold users and trust network sizes. The CiaoDVD.movie dataset represents the case where most users are cold users. The Epinions and the CiaoDVD.review datasets represent the case where a moderate number of users are cold users. The FilmTrust dataset represent the case where a small portion of users are cold users.

4 Empirical Analysis of Trust and Ratings Data

We conducted four types of analyses on the four datasets described earlier. The overall objective of our study is to understand the characteristics of the social trust networks associated with recommender systems in general and the comparative characteristics of those network nodes corresponding to cold versus non-cold users in particular.

4.1 Distribution of the Number of Trustees and Trustors

In the first experiment, we would like to study how many trustees and trustors that cold and non-cold users have. We counted the number of trustees (i.e. out-degree) and the number of trustors (i.e. in-degree) for both non-cold and cold users in the trust networks of all datasets, and created distribution plots for the number of trustees and the number of trustors.

Figure 1 shows the distributions of the number of trustees per users derived from trust networks associated with each dataset. Two observations deserve careful consideration here. First, the distributions of the number of trustees (for all but the FilmTrust dataset) follows the power-law. This means that a large portion of the users in the system has only a few number of trustees; and only a few users has a large number of trustees. Note that, although the distribution of the number of trustees in $TN_{FilmTrust}$ does not follow the power-law, a similar implication can be drawn regarding the proportion of users with a few number versus a large number of trustees. Second, from Fig. 1, one can see that the distributions of the number of trustees for *cold users* cases are significantly steeper than those for the *noncold users* case. This means that, compared to non-cold users, a far more sizeable portion of cold users have small number of trustees.

How can these two observations be exploited in trust-aware based CF where trust is used to replace user similarity computation? First, because most users (especially the cold users) have very few number of trustees, the trust-aware based CF algorithm should have some mechanism for selectively and effectively expanding the trusted neighbours (e.g. trust propagation and weight averaged with other similarity measures). On the other hand, for users with a large number of trustees, the trust-aware based CF algorithm should be able to intelligently filter out those trustees which might not contribute to the increase in recommendation performance (e.g. topic-based filtering technique, proposed by [12]).

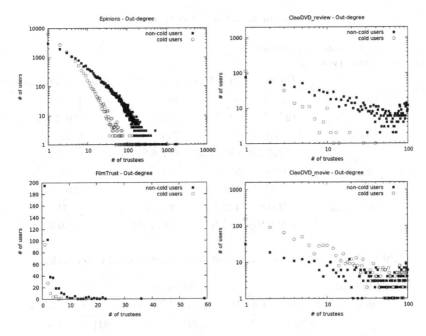

Fig. 1. Distribution of the number of trusted neighbors (out-degree)

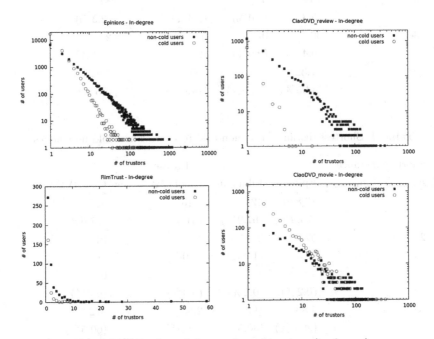

Fig. 2. Distribution of the number of trustors (in-degree)

Table 2. Ratio of link types

Source node type	Destination node type	Dataset			
		Epinions	FilmTrust	CiaoDVD.review	CiaoDVD.movie
u_{cold}	u_{cold}	15,128 (22.5 %)	61 (23.7 %)	107 (12.4 %)	12,533 (**56.3 %**)
	$u_{noncold}$	51,976 (**77.5 %**)	196 (**76.3 %**)	754 (**87.6 %**)	9733 (43.7 %)
$u_{noncold}$	u_{cold}	46,040 (11.0 %)	263 (16.5 %)	817 (2 %)	8,936 (**50.0 %**)
	$u_{noncold}$	374,039 (**89.0 %**)	1,333 (**83.5 %**)	38,455 (**98.0 %**)	8,931 (50.0 %)

Table 3. Shortest distance from cold to nearest non-cold user nodes

Shortest distance $u_{cold} \rightarrow u_{noncold}$	Dataset			
	Epinions	FilmTrust	CiaoDVD.review	CiaoDVD.movie
1	13,234 (50.8 %)	115 (46.0 %)	239 (25.6 %)	841 (23.7 %)
2	1,381 (5.3 %)	5 (2.0 %)	4 (0.4 %)	82 (2.3 %)
3-11	96 (0.4 %)	0 (0.0 %)	0 (0.0 %)	2 (0.1 %)
∞	11,325 (43.5 %)	130 (52.0 %)	690 (74.0 %)	2,623 (73.9 %)

Table 4. Shortest distance from non-cold to nearest non-cold user nodes

Shortest distance $u_{noncold} \rightarrow u_{noncold}$	Dataset			
	Epinions	FilmTrust	CiaoDVD.review	CiaoDVD.movie
1	17,880 (76.9 %)	424 (67.9 %)	1,168 (31.4 %)	417 (37.6 %)
2	528 (2.3 %)	9 (1.4 %)	1 (0.0 %)	21 (1.9 %)
3-11	40 (0.2 %)	1 (0.2 %)	0 (0.0 %)	0 (0.0 %)
∞	4,804 (20.6 %)	190 (30.5 %)	2,556 (68.6 %)	672 (60.5 %)

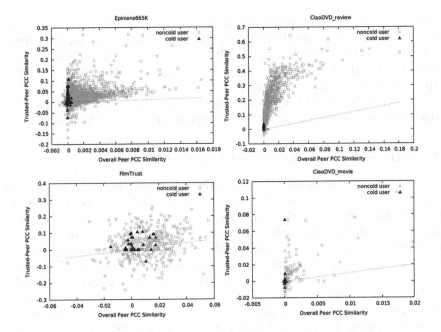

Fig. 3. Correlations of trust and interest similarities

The distributions of the number of trustors are shown in Fig. 2. Like the distributions in Fig. 1, the distributions of the number of trustors fit the power law (except for $TN_{FilmTrust}$ where the distributions is very close to exponential). The distribution plots for the *cold users* case are, like in Fig. 1, steeper than those for the *noncold users* case. Another interesting observation from Fig. 2 is that there are some *cold users* with a large number of trustors despite their limited number of ratings (<5). Notably, in $TN_{Epinions}$, some *cold users* even have more number of trustors than other *noncold users*.

Finally, we remark that the distribution plots of the CiaoDVD.movie dataset in Figs. 1 and 2 exhibit different phenomena than those of the other datasets: there are more cold users than non-cold users. This result is to be expected, because as we reported in Sect. 3, the $\%ColdUsers$ of CiaoDVD.movie dataset (85 %) is much larger than other datasets.

4.2 Ratio of Linkage Types

In this experiment, we look at the linkage patterns between pairs of nodes within the trust networks. In particular, we labelled each node in a trust network as either u_{cold} or $u_{noncold}$ according to the type of the user that the node represents. With two types of nodes within the trust network (i.e. u_{cold} and $u_{noncold}$), there would be four possible linkage patterns between a pair of nodes: $u_{cold} \rightarrow u_{cold}$, $u_{cold} \rightarrow u_{noncold}$, $u_{noncold} \rightarrow u_{cold}$, $u_{noncold} \rightarrow u_{noncold}$. We counted how many links in the trust network correspond to each linkage pattern; and computed the ratios of linkage types.

The number and the ratio of linkage types of the trust networks derived from our four datasets are reported in Table 2. It can be clearly seen that following a link in a trust network, in most cases, will lead to a $u_{noncold}$ node. However, the probability that the destination node will be a $u_{noncold}$ node will be higher if the source node is also a $u_{noncold}$ node. Note that, trust-aware based CF typically needs to search for trusted neighbours that correspond to $u_{noncold}$ nodes as these nodes could provide more item ratings information needed by the algorithms.

4.3 Shortest Distance to Nearest Non-cold Nodes

Most trust-aware recommendation algorithms traverse the trust network to acquire more trusted neighbours with a lot of ratings. Unfortunately, it is unclear how many hops away from the starting node should the algorithms attempt to seek for more trusted neighbours. In order to shade some light on this issue, we have conducted experiments to measure the shortest distance from any node to the nearest $u_{noncold}$ node in the trust networks of all four datasets. The results were summarised in Tables 3 and 4. From the Tables, we can conclude that the suitable number of hops for trust-aware recommendation algorithms that aim to seek for more $u_{noncold}$ nodes is just around 1–2 hops as it would be too expensive or even impossible (i.e. >12 hops is needed) to find more $u_{noncold}$ trusted neighbours afterwards.

It might be tempting to reason that the aggregate number of counts in Table 2 must be in agreement with the number of counts leading to non-cold user nodes in Tables 3 and 4. Nevertheless, that is not the case because in Table 2, we are counting different types of incident edges in the trust network; whereas, in Tables 3 and 4, we are counting shortest paths of different lengths.

4.4 A Closer Look at Correlations of Trust and Interest Similarities

In [13], it was found that there is a strong correlation between trust and interest similarity. This result indicated that trust information could be used to complement or to replace user similarity in CF based recommender systems. Unfortunately, to the best of our knowledge, there is no further work to investigate how the trust information should be exploited.

In this experiment, we compute the similarity between each user node u in a trust network TN and each of its adjacent node v using Pearson correlation:

$$PCC(u,v) = \frac{\sum_{i \in I_u \cap I_v}(r_{u,i} - \bar{r}_u)(r_{v,i} - \bar{r}_v)}{\sqrt{\sum_{i \in I_u \cap I_v}(r_{u,i} - \bar{r}_u)^2}\sqrt{\sum_{i \in I_u \cap I_v}(r_{v,i} - \bar{r}_v)^2}} \qquad (2)$$

The correlations of all-peers PCC similarity and trusted-peers PCC similarity for all four datasets are as shown in Fig. 3. Our main finding here is that, compared to trusted peers of $u_{noncold}$, trusted peers of u_{cold} nodes consistently gave better PCC scores. Consequently, the trust-aware CF based recommendation algorithms may safely choose to rely on a simple approach to exploiting trust information to generate recommendation for u_{cold} users. However, special care must be taken in order to effectively exploit trust information for $u_{noncold}$ users because, according to Fig. 3, there are a sizeable number of $u_{noncold}$ nodes whose trusted peer based PCC similarity scores are lower than those of the overall peer based PCC similarity scores.

5 Conclusion

We presented a detailed analyses of the comparative characteristics of *cold* versus *noncold* users over trust networks associated with user-based CF recommenders. Our results indicated that there are differences between the characteristics of *cold* and *noncold* users in several aspects. For the future work, we will apply these findings to the design of an effective trust aware recommendation algorithm.

References

1. Ekstrand, M.D., Riedl, J.T., Konstan, J.A.: Collaborative filtering recommender systems. Found. Trends Hum. Comput. Inter. 4(2), 81–173 (2011)
2. Guo, G., Zhang, J., Thalmann, D., Yorke-Smith, N.: ETAF: an extended trust antecedents framework for trust prediction. In: Proceedings of the 2014 International Conference on Advances in Social Networks Analysis and Mining, pp. 540–547 (2014)
3. Guo, G., Zhang, J., Yorke-Smith, N.: A novel bayesian similarity measure for recommender systems. In: Proceedings of the 23rd International Joint Conference on Artificial Intelligence, pp. 2619–2625 (2013)
4. Guo, G., Zhang, J., Thalmann, D.: A simple but effective method to incorporate trusted neighbors in recommender systems. In: Masthoff, J., Mobasher, B., Desmarais, M.C., Nkambou, R. (eds.) UMAP 2012. LNCS, vol. 7379, pp. 114–125. Springer, Heidelberg (2012)
5. Ma, H., King, I., Lyu, M.R.: Learning to recommend with social trust ensemble. In: Proceedings of the 32nd International ACM SIGIR Conference on Research and Development in Information Retrieval, pp. 203–210 (2009)
6. Ma, H., Lyu, M.R., King, I.: Learning to recommend with trust and distrust relationships. In: Proceedings of the 3rd ACM Conference on Recommender Systems, pp. 189–196 (2009)
7. Massa, P., Avesani, P.: Trust-aware bootstrapping of recommender systems. In: Proceedings of ECAI 2006 workshop on recommender systems, vol. 28, pp. 29 (2006)
8. Massa, P., Avesani, P.: Trust-aware recommender systems. In: Proceedings of the 1st ACM Conference on Recommender Systems, pp. 17–24 (2007)
9. O'Donovan, J., Smyth, B.: Trust in recommender systems. In: Proceedings of the 10th International Conference on Intelligent User Interfaces, pp. 167–174 (2005)
10. Shi, Y., Larson, M., Hanjalic, A.: Towards understanding the challenges facing effective trust-aware recommendation. Recommmender Systems and the Social Web 40 (2010)
11. Shi, Y., Larson, M., Hanjalic, A.: How far are we in trust-aware recommendation? In: Clough, P., Foley, C., Gurrin, C., Jones, G.J.F., Kraaij, W., Lee, H., Mudoch, V. (eds.) ECIR 2011. LNCS, vol. 6611, pp. 704–707. Springer, Heidelberg (2011)

12. Tang, J., Gao, H., Liu, H.: mTrust: discerning multi-faceted trust in a connected world. In: Proceedings of the 5th ACM Internaional Conference on Web Search and Data Mining, pp. 93–102 (2012)
13. Ziegler, C.N., Golbeck, J.: Investigating interactions of trust and interest similarity. Decis. Support Syst. **43**(2), 460–475 (2007)

Text Processing
and Information Retrieval

Integrated Feature Selection Methods Using Metaheuristic Algorithms for Sentiment Analysis

Alireza Yousefpour[1], Roliana Ibrahim[1(✉)],
Haza Nuzly Abdul Hamed[3], and Takeru Yokoi[2]

[1] Software Engineering Research Group, Faculty of Computing,
Universiti Teknologi Malaysia, 81310 UTM Skudai, Johor, Malaysia
aryousefpour@live.com, roliana@utm.my
[2] Tokyo Metropolitan College of Industrial Technology,
1-10-40 Higashi-oi, Shinagawa, Tokyo 140-0011, Japan
takeru@metro-cit.ac.jp
[3] Faculty of Computing, Universiti Teknologi Malaysia,
81310 UTM Skudai, Johor, Malaysia
haza@utm.my

Abstract. In text mining, the feature selection process can potentially improve classification accuracy by reducing the high-dimensional feature space to a low-dimensional feature space resulting in an optimal subset of available features. In this paper, a hybrid method and two meta-heuristic algorithms are employed to find an optimal feature subset. The feature selection task is performed in two steps: first, different feature subsets (called local-solutions) are obtained using a hybrid filter and wrapper approaches to reduce high-dimensional feature space; second, local-solutions are integrated using two meta-heuristic algorithms (namely, the harmony search algorithm and the genetic algorithm) in order to find an optimal feature subset. The results of a wide range of comparative experiments on three widely-used datasets in sentiment analysis show that the proposed method for feature selection outperforms other baseline methods in terms of accuracy.

Keywords: Feature selection · Integration feature selection methods · Harmony search · Genetic algorithm · Sentiment analysis

1 Introduction

With the exponential growth of internet usage, users can be easily posted their opinions and comments on the web. This posted-content is a valuable resources for companies, organizations, and individual as it can be used to inform better decision-making. This situation is creating growing interest in technologies for automatically analyzing and mining the opinions and sentiment from web documents. The main task of sentiment analysis can be carried out in two steps: the first step involves the selection and/or extraction of features from the textual opinions, and the second step involves the sentiment classification of the samples into multi-classes (e.g. positive, neutral, negative) [1].

© Springer-Verlag Berlin Heidelberg 2016
N.T. Nguyen et al. (Eds.): ACIIDS 2016, Part I, LNAI 9621, pp. 129–140, 2016.
DOI: 10.1007/978-3-662-49381-6_13

Saeys et al. [2] categorized feature selection techniques into three approaches: the filter, wrapper, and embedded approaches. The filter approach provides ways to choose an optimal subset of features by scaling features and eliminating low-scoring features. In the wrapper approach, the choice of an optimal feature subset is provided by generating and evaluating different subsets in the space of states. In the third approach, namely, the embedded approach, the search occurs through the combination of the model hypothesis and feature subset space into a classifier structure [1, 3].

In the case of generating the subsets in wrapper approach, heuristic search methods are used to solve the exponential time to find a feature subset; regarding the evaluation of subsets, the classifier evaluates the effect of selecting a feature subset on the performance of sentiment classification in order to find an optimal feature subset. For instance, two meta-heuristic algorithms including Harmony Search (HS) algorithm and Genetic Algorithm (GA) are efficient techniques for searching for the global-solution by overcoming the main problems of local-search techniques. Where, HS algorithm mimics the behavior of music improvisation process, as is the population-base and single-point search-base algorithm [4] and the GA belongs to the class of evolutionary algorithms which mimics the process of natural selection [5].

In this research, we propose a feature selection method to obtain a high-quality minimal feature subset from a real-world domain. In the method proposed in this research, the feature selection task is performed in two steps. In the first step, different middle feature subsets (MFSs) are obtained using a hybridization of filter and wrapper approaches to reduce the high-dimensional feature space. The MFSs are referred to as the local-solutions in feature space. In the second step, two different meta-heuristic algorithms are applied to integrate the different MFSs in order to find an optimal solution separately.

The remainder of this paper is organized as follows: Sect. 2 introduces the related works on sentiment analysis. Section 3 describes the proposed method for feature selection. The results of the comparative experiments and evaluation are presented in Sect. 4. Finally, Sect. 5 concludes the paper.

2 Related Works

Rogati and Yang [6] investigated filter techniques for textual feature selection. They scored features by five methods: document frequency (DF), information gain (IG), chi-square (CHI), mutual information, and term frequency. They removed 97 % of the low-scoring features and gained better classification results by using the CHI and IG methods. In some studies, feature selection methods such as the IG and CHI methods have been found to achieve better accuracy than other methods [7]. On the other hand, Yousefpour et al. [8] scored features based on the distribution of features by the standard deviation (SD) of the features. A hybrid approach has also been introduced for solving the problem of low-accuracy sentiment classification in the filter methods and the problem of the higher-computation burden in wrapper methods [9].

Feature selection using HS was proposed by Diao and Shen [4]. They found that their proposed method identified multiple solutions and overcome local-solutions with respect to the stochastic nature of the HS algorithm and control of its parameters.

They compared the HS with other meta-heuristic techniques such as the GA on the University of California Irvine benchmark datasets. They showed that the HS identified good-quality feature subsets for most of the test datasets. Wang et al. [10] used the HS algorithm for feature selection in email classification. Feature selection in credit risk assessment was proposed by Oreski and Oreski [5].

3   Methodology

In this research, we propose a feature selection method to obtain a high-quality minimal feature subset from a real-world domain which is described in detail in this section.

3.1   Feature Representation

The detection of features from a raw document is the main challenge in text mining; in other words, it is related to the question of whether a word or a sequence of words can be a feature. In the experiments conducted in this research, n-gram variable features were mainly used. The variable n-gram features were extracted using part of speech (POS) patterns to detect features with a maximum length of three. To identify the variable n-gram features from raw reviews, twenty-four pattern-based linguistic filters were developed in four categories. They are presented in Table 1.

Table 1. Twenty-four POS patterns used for detection of features

Class of patterns	POS-patterns
Noun	N, N N, N N N, N I, N N I, N I N, N V
Adjective	J, J N, J J N, J N N, J N I
Verb	V, V I, V N N, V N, V N I, V V, V V I, V J N
Adverb	R, R R, R R J, R R R

The POS patterns listed in Table 1 were categorized into four classes. In the first class, the patterns identify features that start with a noun (N). In the second class, the patterns capture features that begin with an adjective (J). The patterns in the third and fourth classes capture features that start with verbs (V) and adverbs (R). Letter (I) is a preposition or conjunction. For example, the feature of "excellent book" is identified by the pattern of "J N".

3.2   Feature Subset Selection Using Meta-Heuristic Algorithms

In this research, feature subset selection was performed in two steps. In the first step, different MFSs were created through a hybrid of filter and wrapper approaches in order to reduce the high-dimensional feature space to a low-dimensional feature space. In the second step, the MFSs were integrated using two meta-heuristic algorithms in order to find an optimal feature subset. Steps 1 and 2 are described in detail in this section:

Middle Feature Subsets. The feature subsets with the highest accuracy obtained in the hybrid approach were referred to as the MFSs. The left side of Fig. 1 shows the

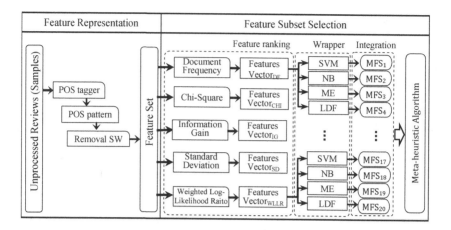

Fig. 1. Framework of proposed method showing different ways to create and integrate MFSs

preprocessing and feature representation based on POS. Three phases were used for the detection of features in the raw documents: POS tagging, linguistic POS patterns, and the removal of stop words (SW). To annotate documents with POS, we used the Stanford POS tagger.

To find a MFS, different feature subsets were generated based on the orders of feature ranks. After feature ranking, the features were sorted in descending order according to their weights in feature vector. Then, the different feature subsets were generated as follows:

$$\begin{cases} Feature\ Vector = (x_1, x_2, .., x_N), \forall_{i,j} i < j \rightarrow rank(x_i) \geq rank(x_j) \\ Feature\ subsets = \{\{x_1\}, \{x_1, x_2\}, \{x_1, x_2, x_3\}, \ldots, \{x_1, x_2, .., x_N\}\} \end{cases} \quad (1)$$

where x_i is a feature and N is the total number of features. In this representation, x_1 has the highest rank (or weight) and x_2 has the second highest rank in feature vector. Four well-known text classification algorithms, namely, Support Vector Machine (SVM), Naive Bayes (NB), Maximum Entropy (ME) and Linear Discriminant Function (LDF) evaluate different feature subsets to find the subset with the highest accuracy, which we called the MFS.

Integration of Different Middle Feature Subsets. The MFSs that were created in the first step were integrated using two meta-heuristic algorithms separately, namely, the HS algorithm and the GA.

HS Algorithm. As a meta-heuristic algorithm, the HS algorithm gains superb results in optimization problems. HS can be seen as a meta-heuristic algorithm that, seeks a global-optimal solution by a set of parameters including a musician (decision variable) who plays a note (value of the decision variable) to find the best harmony (global-optimal solution). The notes played by all musicians (called a harmony) are considered for optimization of an audience's aesthetic impression (objective function). The five key parameters of the HS algorithm are:

- Harmony memory size (HMS) - The HMS specifies the number of solution vectors in the harmony memory whereby a solution vector is the set of all decision variables, and a harmony memory is created by all the solution vectors.
- Harmony memory considering rate (HMCR) - The HMCR is the rate of selecting a value from the harmony memory.
- Pitch adjusting rate (PAR) - The PAR is the rate of choosing a neighboring value.
- Adjusting bandwidth (BND) - The BND is a fret width that specifies the amount of maximum change in pitch adjustment.
- Maximum iteration (maxIter) - The *maxIter* is the number of improvisations based on the stopping criteria.

Let $X = (x1, x2,..., xn)$ be a representation of a solution vector, where xi's are decision variables. The HS algorithm is shown as follows:

Step 1. Initialize the algorithm parameters
Step 2. Initialize the harmony memory with different MFSs
Step 3. Repeat:
    3.1. Improvise a new harmony from the harmony memory.
        -Memory considering
        -Pitch adjusting
        -Randomization
    3.2. Update the harmony memory.
Step 4. Repeat until the termination criterion is satisfied.

In Step 2 of the algorithm, the harmony memory is created by all the MFSs, whereby each row of the harmony memory represents one MFS. Each feature selector may choose one feature which has already been selected by another feature selector. In other words, all the feature selectors are allowed to choose the same feature. Figure 2 shows harmony memory encoded by the MFSs.

In step 3.1, a new solution (X^{new}) is improvised from the existing vectors in harmony memory as follow:

$$X^{new} = \left(x_1^{new}, x_2^{new}, \ldots, x_n^{new}\right) \tag{2}$$

$$\text{Where } x_i^{new} = \begin{cases} x_i(k) \in [valuerange] & w.p.(1 - HMCR) \\ x_i(k) \in \{x_i^1, x_i^2, .., x_i^{HMS}\} & w.p.HMCR \\ x_i(k \pm rand() * BND) & w.p.PAR \end{cases}$$

where x_i is a feature. In Step 3.2, the harmony memory is updated by a new solution vector if the evaluation of the new solution vector is better than the evaluation of the existing worst vector in the harmony search. The individual classifiers of SVM, NB,

Feature selector:	S_1	S_2	S_3	S_4	S_5	S_6	Feature subset
HM$_i$:	x_3	x_7	x_7	x_4	x_7	x_4	$\{x_3, x_4, x_7\} = MFS_i$
HM$_j$:	x_2	x_2	x_5	x_2	x_5	x_2	$\{x_2, x_5\} = MFS_j$
HM$_k$:	x_3	x_2	x_3	x_5	x_5	x_7	$\{x_2, x_3, x_5, x_7\} = MFS_k$

Fig. 2. Harmony memory encoded by MFSs

ME and LDF are used as objective functions for evaluation of the feature subsets and allocation of merit to each of the new feature subsets. In fact, the HS algorithm attempts to enhance existing accuracies in the harmony search.

Genetic Algorithm. The GA is inspired by Darwin's theory of evolution. It belongs to the class of evolutionary computing. The GA starts with a set of chromosomes (solutions) called a population. Solutions from one population are taken and used to form a new population. This is motivated by a hope that the new population will be better than the old one. Solutions which are selected to form new solutions (offspring) are selected according to their fitness the more suitable they are the more chances they have to reproduce. The GA is shown as follows:

Step 1. Initialize the algorithm parameters
Step 2. Initialize population (t) with different MFSs
Step 3. Determine fitness of population (t)
Step 4. Repeat:
    4.1. Select parents from population (t) according to their fitness
    4.2. Perform crossover on parents creating population (t + 1)
    4.3. Perform mutation of population (t + 1)
    4.4. Determine fitness of population (t + 1)
Step 5. Repeat until best individual is good enough.

Population size in the GA is equal to the number of MFSs, the size of the chromosome is set to the number of union MFSs, and the genes are binary in chromosome representation.

Figure 3 shows some chromosomes that are binary-encoded by the MFSs.

Features:	x_1	x_2	x_3	x_4	x_5	x_6	...	Feature subset	
Chrom$_i$:	0	1	0	0	1	1	...	$\{x_2, x_5, x_6,...\}$	$= MFS_i$
Chrom$_j$:	1	0	1	0	0	0	...	$\{x_1, x_3,...\}$	$= MFS_j$
Chrom$_k$:	1	1	0	1	1	0	...	$\{x_1, x_2, x_4, x_5,...\}$	$= MFS_k$

Fig. 3. Chromosomes that are binary-encoded by MFSs

4   Evaluation

Three document-level datasets that are widely used in sentiment analysis were employed in this research. The book, electronic, music review corpora were used to investigate the performance of the proposed method [11]. The review datasets comprised 1000 positive samples and 1000 negative samples. All the features were presented as POS-based features. All the feature weights in the term document matrix were set to the term frequency-inverse document frequency. All the results were obtained through 5-fold cross-validation with a random starting point in 3-repetitions on review datasets that were referred to as 5*3-FCV. Table 2 shows the average and SD of the

Table 2. Average and standard deviation of accuracy (ave ± σ) for classification algorithms on all POS-based features using 5*3-FCV

Classifier	Book review		Electronic review		Music review	
	# of features	Accuracy	# of features	Accuracy	# of features	Accuracy
SVM	53685	73.43 ± 2.12	29644	75.77 ± 2.34	41012	75.72 ± 1.70
NB		71.62 ± 1.72		75.37 ± 2.57		71.00 ± 2.51
ME		75.67 ± 2.12		77.92 ± 1.92		73.65 ± 2.39
LDF		74.08 ± 2.47		77.18 ± 2.05		72.13 ± 1.97

classification's accuracy on all training feature sets. The classifiers were trained using the training feature set. The classification accuracy was evaluated based on the testing feature set.

4.1 Performance Measures

Generally, the performance of sentiment classification is evaluated using four indexes, namely, accuracy, precision, recall and F1-score. The measures can be computed through the following equations:

$$Accuracy = \frac{TP + TN}{TP + TN + FP + FN} \tag{3}$$

$$F1 = \frac{2 \times Precision \times Recall}{Precision + Recall} \tag{4}$$

$$Precision(pos) = \frac{TP}{TP + FP} \tag{5}$$

$$Precision(neg) = \frac{TN}{TN + FN} \tag{6}$$

$$Recall(neg) = \frac{TN}{TN + FP} \tag{7}$$

$$Recall(pos) = \frac{TP}{TP + FN} \tag{8}$$

Where TP is the number of actual positive samples that were predicted to be positive, FN is the number of actual positive samples that were predicted to be negative, TN is the number of actual negative samples that were predicted to be negative, and FP is the number of actual negative samples that were predicted to be positive.

Table 3. Parameter settings of experimental environment

HS parameters		GA parameters	
Iteration	2000	Iteration	2000
HMS	20	Population size	20
HMCR	0.9	Crossover operator	Two point crossover
PMR	0.3	Crossover probability	0.9
BND	0.01	Mutation probability	0.3
Number of variables	Max size of MFSs	Chromosome size	Size of Union MFSs
Variables value	Feature	Gene	Feature
Evaluation function	SVM, NB, ME, LDF	Fitness function	SVM, NB, ME, LDF

4.2   Experimental Results and Discussion

The parameters of the HS algorithm and GA were adjusted based on the setting in Table 3 and were chosen empirically.

We obtained the performance results of the classifiers using two meta-heuristic algorithms on three review datasets. Table 4 presents the results when the HS algorithm was applied. Table 5 presents the results when the GA was used.

Based on a comparison of the performance of the HS algorithm and the GA, three main points can be made. First, the HS algorithm has some advantages over traditional optimization techniques such as the GA. For instance, the HS takes advantage of all existing quality vectors to obtain a new solution vector, whereas the GA only considers two parents in order to obtain a new offspring. Second, the HS algorithms has less time-complexity than the GA. Equations 9 and 10 compare the two algorithms based on the call number of evaluation function (CNEF).

$$CNEF_{HS} = N_{pop} + maxIter \qquad (9)$$

$$CNEF_{GA} = N_{pop} + N_{pop} * maxIter \qquad (10)$$

where N_{pop} is the size of the population and $maxIter$ is the number of iterations. In Eq. 9, CNEF is equal to the size of the initial population (N_{pop}) and only one new harmony is created in each iteration; however, in the GA, a new population or offspring is created in each iteration and the evaluation function is called once for each offspring.

Third, the HS algorithm outperforms the GA in terms of better accuracy in more time. Figure 4 shows the comparison between the two algorithms in terms of accuracy.

In addition, the proposed methods are compared with the average of the best MFSs (BMFS) as a baseline method. In order to assess whether there are any significant differences in terms of accuracy between the proposed methods and BMFS as a baseline method, a statistical test was conducted based on the accuracy results obtained from the 5*3-FCV experiments. The statistical test results showed significant differences between the proposed methods and the baseline method in all comparisons. We used a paired t-test to evaluate whether the differences between the two techniques were statistically significant. Table 6 shows the numerical results of the statistical test.

Table 4. Average and standard deviation (ave ± σ) of classification algorithm results and length of feature subset using the HS algorithm in 5*3-FCV

Dataset	Classifier	#features	Accuracy	Positive			Negative		
				Precision	Recall	F1	Precision	Recall	F1
Book	SVM	2413 ± 1368	86.93 ± 2.26	92.03 ± 3.16	83.80 ± 3.14	87.65 ± 2.10	81.83 ± 4.46	91.06 ± 2.71	86.10 ± 2.61
	NB	5347 ± 425	**94.75 ± 1.56**	**95.07 ± 1.15**	**94.52 ± 2.44**	**94.78 ± 1.51**	**94.43 ± 2.60**	**95.04 ± 1.13**	**94.72 ± 1.61**
	ME	2944 ± 1702	85.75 ± 1.57	86.30 ± 2.24	85.43 ± 2.03	85.84 ± 1.57	85.20 ± 2.41	86.15 ± 1.92	85.65 ± 1.62
	LDF	4634 ± 1342	86.70 ± 2.05	85.90 ± 3.62	87.35 ± 2.39	86.57 ± 2.21	87.50 ± 2.62	86.23 ± 3.00	86.81 ± 1.93
Elec.	SVM	3277 ± 2426	88.80 ± 2.40	90.33 ± 3.71	87.83 ± 3.16	89.00 ± 2.41	87.27 ± 3.88	90.07 ± 3.17	88.57 ± 2.48
	NB	3676 ± 353	**93.25 ± 1.18**	**93.73 ± 1.75**	**92.89 ± 2.07**	**93.29 ± 1.15**	**92.77 ± 2.34**	**93.71 ± 1.61**	**93.21 ± 1.23**
	ME	2031 ± 808	86.45 ± 1.84	84.00 ± 3.22	88.32 ± 2.19	86.07 ± 2.04	88.90 ± 2.40	84.87 ± 2.49	86.80 ± 1.72
	LDF	3244 ± 617	88.15 ± 1.52	87.53 ± 3.62	88.70 ± 1.87	88.05 ± 1.73	88.77 ± 2.34	87.81 ± 2.91	88.23 ± 1.38
Music	SVM	3962 ± 1434	86.28 ± 1.94	85.30 ± 3.65	87.07 ± 2.42	86.12 ± 2.07	87.27 ± 2.93	85.72 ± 3.05	86.43 ± 1.88
	NB	4280 ± 250	**92.53 ± 1.85**	**92.57 ± 2.06**	**92.55 ± 2.47**	**92.54 ± 1.84**	**92.50 ± 2.58**	**92.58 ± 1.96**	**92.52 ± 1.87**
	ME	1220 ± 512	83.83 ± 1.64	82.90 ± 3.08	84.55 ± 2.54	83.66 ± 1.72	84.77 ± 3.09	83.31 ± 2.23	83.98 ± 1.68
	LDF	4156 ± 740	86.10 ± 1.61	85.40 ± 3.07	86.68 ± 2.42	85.98 ± 1.70	86.80 ± 2.91	85.70 ± 2.40	86.20 ± 1.61

Table 5. Average and standard deviation (ave ± σ) of classification algorithms results and length of feature subset using the GA in 5*3-FCV

Dataset	Classifier	#features	Accuracy	Positive			Negative		
				Precision	Recall	F1	Precision	Recall	F1
Book	SVM	2432 ± 1098	86.25 ± 2.00	90.60 ± 3.02	83.50 ± 3.18	86.83 ± 1.80	81.90 ± 4.43	89.82 ± 2.57	85.58 ± 2.36
	NB	4584 ± 851	**94.98 ± 1.54**	**95.13 ± 1.39**	**94.88 ± 2.06**	**95.00 ± 1.52**	**94.83 ± 2.14**	**95.12 ± 1.38**	**94.97 ± 1.57**
	ME	3127 ± 1286	85.48 ± 1.88	85.87 ± 1.80	85.26 ± 2.46	85.55 ± 1.79	85.10 ± 2.79	85.76 ± 1.73	85.41 ± 1.98
	LDF	4446 ± 1124	88.05 ± 1.63	87.40 ± 2.09	88.57 ± 1.78	87.97 ± 1.66	88.70 ± 1.87	87.58 ± 1.92	88.13 ± 1.61
Elec.	SVM	2718 ± 1143	87.21 ± 1.26	89.29 ± 2.62	85.78 ± 2.19	87.46 ± 1.23	85.13 ± 2.86	88.94 ± 2.32	86.94 ± 1.36
	NB	3384 ± 725	**92.19 ± 1.31**	**91.58 ± 2.26**	**92.76 ± 1.80**	**92.14 ± 1.35**	**92.79 ± 1.90**	**91.71 ± 2.06**	**92.23 ± 1.28**
	ME	2484 ± 771	84.23 ± 1.12	82.38 ± 2.62	85.66 ± 1.69	83.95 ± 1.26	86.08 ± 2.02	83.00 ± 2.00	84.49 ± 1.04
	LDF	3016 ± 822	87.06 ± 1.33	87.29 ± 2.82	86.98 ± 2.32	87.08 ± 1.36	86.83 ± 2.83	87.32 ± 2.33	87.03 ± 1.38
Music	SVM	3283 ± 732	86.10 ± 1.68	84.00 ± 2.18	87.82 ± 3.13	85.81 ± 1.56	88.20 ± 3.62	84.68 ± 1.63	86.36 ± 1.86
	NB	3783 ± 438	**92.97 ± 1.59**	**93.17 ± 0.88**	**92.88 ± 2.87**	**93.00 ± 1.48**	**92.77 ± 3.14**	**93.14 ± 0.85**	**92.93 ± 1.71**
	ME	1859 ± 999	84.57 ± 1.97	84.17 ± 2.18	84.89 ± 2.51	84.51 ± 1.94	84.97 ± 2.82	84.31 ± 1.93	84.62 ± 2.03
	LDF	3413 ± 489	86.33 ± 2.16	87.27 ± 2.88	85.70 ± 2.41	86.45 ± 2.18	85.40 ± 2.71	87.07 ± 2.65	86.20 ± 2.18

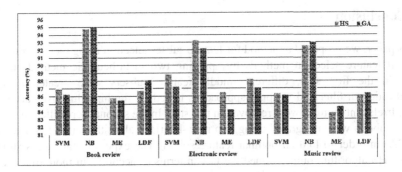

Fig. 4. Comparative average of accuracy between two meta-heuristic algorithms

Table 6. Comparison between two methods and BMFS as baseline method by T-test measure

Dataset	Classifier	BMFS	HS	T-test		BMFS	GA	T-test	
				P-value	H-value			P-value	H-value
Book	SVM	83.68	86.93	4.77e-04	1	84.05	86.25	6.05e-03	1
	NB	91.18	94.75	1.84e-05	1	91.63	94.98	4.69e-06	1
	ME	82.83	85.75	1.40e-05	1	82.92	85.48	2.15e-04	1
	LDF	83.72	86.70	1.67e-04	1	83.85	88.05	3.78e-07	1
Electronic	SVM	85.72	88.80	6.14e-04	1	85.88	87.21	1.07e-02	1
	NB	90.32	93.25	7.16e-08	1	90.19	92.19	5.62e-04	1
	ME	83.48	86.45	6.50e-05	1	83.08	84.23	1.84e-02	1
	LDF	85.43	88.15	1.83e-05	1	85.25	87.06	5.54e-04	1
Music	SVM	82.88	86.28	1.39e-04	1	82.40	86.10	3.28e-06	1
	NB	89.88	92.53	2.49e-04	1	91.55	92.97	1.92e-02	1
	ME	80.38	83.83	5.65e-05	1	80.65	84.57	2.99e-05	1
	LDF	83.25	86.10	2.40e-04	1	83.13	86.33	3.04e-04	1

The H-value indicates that the null hypothesis can be rejected at the 5 % significance level ("H = 0": not statistically significant; "H = 1": statistically significant).

On the other hand, the results in Table 6 show the extent of improvement in the two algorithms compared to the BMFS as the baseline method in terms of accuracy. In fact, 3.08 % was average of improvement using the HS algorithm and 2.51 % gained by GA. The BMFS is calculated by the average of the highest value of each FCV as follows:

$$BMFS = \frac{1}{M}\sum\nolimits_{i=1}^{M} HiMFS_i, \ \ HiMFS_i = \max_{1 \le j \le N} MFS_j \tag{11}$$

where M is the number of FCV, N is the number of MFSs and $HiMFS$ is the highest value in each FCV. For example, M is equal to the value of 15 and N is equal to the value of 20 in our study. In Table 6, the average value of the BMFS is shown and standard deviation value is left in short.

5 Conclusion

In this paper, we proposed a method for feature subset selection using the integration of different feature selection methods by applying two meta-heuristic methods in two steps. In the first step, different middle feature subsets were created using a hybridization of filter and wrapper approaches to reduce the high-dimensional feature space. Irrelevant features were removed in this step. In the second step, different middle feature subsets were integrated using two meta-heuristic algorithms (the HS algorithm and GA) in order to find an optimal feature subset. The experimental results showed that the HS algorithm outperformed the GA in terms of accuracy in more time. A hybrid harmony search can be used to integrate middle feature subsets in future work.

References

1. Ekbal, A., Saha, S.: Combining feature selection and classifier ensemble using a multiobjective simulated annealing approach: application to named entity recognition. Soft. Comput. **17**, 1–16 (2013)
2. Saeys, Y., Inza, I., Larrañaga, P.: A review of feature selection techniques in bioinformatics. Bioinformatics **23**, 2507–2517 (2007)
3. Yousefpour, A., Ibrahim, R., Abdull Hamed, H.N., Hajmohammadi, M.S.: A comparative study on sentiment analysis. Adv. Environ. Biol. **8**, 53–68 (2014)
4. Diao, R., Shen, Q.: Feature selection with harmony search. Syst., Man, Cybern., Part B: Cybern., IEEE Trans. **42**, 1509–1523 (2012)
5. Oreski, S., Oreski, G.: Genetic algorithm-based heuristic for feature selection in credit risk assessment. Expert Syst. Appl. **41**, 2052–2064 (2014)
6. Rogati, M., Yang, Y.: High-performing feature selection for text classification. In: Proceedings of the Eleventh International Conference on Information and knowledge management, pp. 659–661 (2002)
7. Uğuz, H.: A two-stage feature selection method for text categorization by using information gain, principal component analysis and genetic algorithm. Knowl.-Based Syst. **24**, 1024–1032 (2011)
8. Yousefpour, A., Ibrahim, R., Abdull Hamed, H.N., Hajmohammadi, M.S.: Feature reduction using standard deviation with different subsets selection in sentiment analysis. In: Nguyen, N.T., Attachoo, B., Trawiński, B., Somboonviwat, K. (eds.) ACIIDS 2014, Part II. LNCS, vol. 8398, pp. 33–41. Springer, Heidelberg (2014)
9. Warren Liao, T.: T.: Feature extraction and selection from acoustic emission signals with an application in grinding wheel condition monitoring. Eng. Appl. Artif. Intell. **23**, 74–84 (2010)
10. Wang, Y., Liu, Y., Feng, L., Zhu, X.: Novel feature selection method based on harmony search for email classification. Knowl.-Based Syst. **73**, 311–323 (2015)
11. Blitzer, J., Dredze, M., Pereira, F.: Biographies, bollywood, boom-boxes and blenders: Domain adaptation for sentiment classification. In: *ACL*, pp. 440–447 (2007)

Big Data in Contemporary Linguistic Research. In Search of Optimum Methods for Language Chronologization

Piotr Wierzchoń[✉]

Institute of Linguistics, Adam Mickiewicz University, Poznań, Poland
wierzch@amu.edu.pl

Abstract. The paper will concern the theoretical and practical problems of analysing the mass of linguistic data which has arisen in conjunction with the development of many fields of life. Moreover, the universe of texts is growing every day – both forwards and backwards. Forwards because every new article, book, blog, e-mail or text message expands the set of existing texts; and backwards because the same set is also expanded whenever a scan is made of another historical text. Our knowledge about past times is growing by leaps and bounds. We are therefore particularly interested in the analysis of historical texts that can be carried out in the second decade of the 21st century.

Keywords: Linguochronologisation · Photodocumentation · Polish language · 20th-century texts

1 Introduction

We consider here the theoretical and practical problem of making linguistic descriptions of the mass of data that has come into being with the development of many areas of human activity: academic expansion, involving for example the growth in the total number of researchers, diversification of linguistic disciplines, networking of laboratories, etc., as well as technical development, the development of civilization, and so on. We also wish to put forward the thesis that every historical epoch has its own types of accumulation, as described now by the term *big data*, and so the problem of making descriptions, including linguistic ones, of data of this type reappears according to some kind of cycle. Moreover, the accumulation of data, which is often unexpected – it may occur rapidly over a relatively short time, such as a few years or a couple of decades – leads to anomalies in existing theories. The multiplicity of linguistic examples deviating from those for which existing theses have been formulated is so large as to produce a crisis of existing concepts, and consequently an inability to formulate new theoretical ideas. Hence from time to time it occurs that a researcher is able to perceive the mass of new data, but for certain reasons is not able to create theoretical counter-proposals to the pre-existing model (cf. G. Lakoff, M. Rudnicki, etc.).

It can be stated as a general observation that the work of linguists in the 21st century is developing in the direction of experimental, corpus-based, quantitative studies. They are nearly always observing, and then calculating. While research prior to

© Springer-Verlag Berlin Heidelberg 2016
N.T. Nguyen et al. (Eds.): ACIIDS 2016, Part I, LNAI 9621, pp. 141–150, 2016.
DOI: 10.1007/978-3-662-49381-6_14

the big data era was dominated by the problem of identifying oppositions, at the present time more and more studies are devoted to the collection of numerical data relating to various linguistic objects.

To sum up – linguistics is becoming more and more quantitative. Consequently, due to the high costs of human work, it is becoming more and more computerized. The question then arises of how powerful a device needs to be used to perform a certain task – or perhaps more pertinently: how powerful a device is it not worth employing for the performance of a particular operation, especially due to economic reasons.

Most of the activities known to us concerning the automatic processing and searching of texts and extraction of information are carried out on private home devices. However, the linguistic challenges of the 21st century present significantly greater operational requirements [8]; that is, requirements that can only be met by employing supercomputers, clusters or grids. An important issue here is parallelization, the possibility of performing tasks simultaneously without collision and without holding up the queue. For example, the present author was once faced with the task of combining several tens of thousands of compounding forms with appropriate words (for the purpose of producing artificial vocabulary items and checking for their presence in real texts). This task required the combination of 60,000 prefixes with 4.7 million words, and searching for the resulting items in a list of 70 million words taken from 20th-century texts. A total of 282 billion objects were produced, and their presence was checked over 70 million lines. In home computing conditions this procedure required approximately one month of machine time, whereas a supercomputer with no definite location, distributed throughout the Polish academic infrastructure (http://www.plgrid.pl), completed it in just a few minutes.

We therefore aim to consider here the economics of the construction of big models, taking as an example the case of a derivational and lexical model constructed for the Polish language of the 19th and 20th centuries. In other words: how do we today conceive of the construction and effects of large models able to cope with any linguistic problem in a situation where it is known in advance that the quantity of new data affecting the linguistic picture is certain to grow? For example, a decade or more ago a researcher taking up a particular descriptive problem operated within a relatively predictable empirical base – the number of objects, such as texts, which he or she planned to study could be foreseen from the start. The present rate of growth in quantities of text – and text is always the object of study in the type of linguistic work described here – is an unprecedented phenomenon. The expansion of big data leads to the rapid ageing of linguistic information, particularly of the synchronic type (for Old and Middle Polish the problem has not yet arisen). In beginning work on a particular model, we can anticipate that a significant part of it will have become outdated by the time we finish. The acceptability of this phenomenon is one of the issues requiring consideration.

The concept of big data, in turn, relates to large, diverse and changing sets of data, which cannot be mastered by means of well-known and generally applied methods. The processing of large sets of data is a complex operational problem. Examples of data of this type include meteorological data, telephone billing items, DNA code, brain tomography images and so on. Today's CCTV surveillance systems, for instance, provide immeasurable portions of communicative information every day. The situation is similar with contemporary text – in other words, the information provided as input to linguistic analysis is multiplying day by day. Moreover, today's text, as a born-digital entity, is increasingly well described and structured. It comes with not only the content itself, but also a similarly large amount of meta-content, such as metrics and external descriptions. In every case we know the author, the date of creation of the text, place of publication, size, functional type of document, and so on. This information begins to accumulate, creating a new network of data which is not without interest to the linguistic model being created.

In contrast to investigations of an introspective nature, which require no more than a piece of paper and a pencil (and sometimes not even these), big data needs resources, in the form of premises, staff – generally speaking, economic resources. Researchers who are relatively well supplied with resources are able to carry out investigations on large sets of linguistic data. All of the large linguistic models known to the present author have been the result of projects funded by grants, which enabled and facilitated the researchers' work.

The concept of resources can be understood in many ways, for example in terms of researcher posts, computers, laboratories, cars, clothing, energy, size of monitors, temperature of rooms, comfortable chairs, and even the working atmosphere within the team. A particular category of resource, however, is *time*. Time is an issue which permeates the task of constructing large linguistic models. Time has to be conformed to; time is known in advance in the form of a deadline. It might even be suggested that deadlines are the category that distinguishes introspective research from research of an operational, decision-based nature.

In making a review of conscious strategic concepts, namely those fundamentally based on decision or choice, the first to be considered should be military programmes. Sun Zi's *The Art of War* is a Chinese military study from the 6th century BC. It consists of 13 chapters and contains operational principles expressed in condensed form, for example:

> *If one does not know the place of battle and the day of battle, then his left cannot aid his right, his right cannot aid his left, his front cannot aid his back, and his back cannot aid his front.*

A work of a similar nature is Carl von Clausewitz's *On War*, published in 1832, and entirely devoted to matters of strategy. With regard to the efficient organisation of work, mention should also be made of Karol Adamiecki (born 18 March 1866 in Dąbrowa Górnicza, died 16 May 1933 in Warsaw), originator of the study of organisation and management. These initial considerations lead on to the concept of **grammar of action**.

2 Grammar of Action

The term *grammar of action* was introduced to the literature on praxeology by Tadeusz Kotarbiński in the 1920 s and 1930 s. It referred to an explicit methodological programme which made it possible to distinguish certain significant components of an action [bolding – P.W.]:

> This is the field of enquiry that we have in mind when speaking of general methodology or praxeology. Generally, then, and briefly, by general methodology or praxeology we understand

the science of methods of doing anything, a science that considers work from the point of view of effectiveness, and in isolation both from the particular conditions of work in an exclusively defined specialty and from any evaluations of an emotional nature. That science has the right to be called methodology, since it considers methods, and it can be called general methodology since it considers the methods of all kinds of work [5].

These are the common concepts from the field of general methodology: **action, material, doer, goal, design, plan, work, product, tool, means, manner, method, organisation, ability to act, need to act, difficulty, obstacle, cooperation, counteraction, efficiency, effectiveness, performance, practicality** and others. The basic concept of action is so important here that the whole of the discipline being described might be called the theory of action, and so significant for it is the consideration of everything in terms of practicality that it would also be apt to apply the name **theory of practicality** [5].

Tadeusz Kotarbiński was not the only researcher to focus on issues concerning the components of action, including the design of operational processes. Others that might be listed include A. Espinas, B. Prus, G. Hostelet, N. Krupska, G. von Wright, and the founders of the Poznań "Complex of Operations" School. A more developed account of the components of action, particularly in the context of time- and task-based descriptions, can be found in the works of W. Gasparski, J. Węglarz, J. Błażewicz, R. Słowiński and others. Particularly in the work of Gasparski from the 1960 s and 1970 s, a description of a formalization of praxeology can be found – cf. *Kryterium i metoda wyboru rozwiązania technicznego w ujęciu prakseometrycznym* [The Criterion and Method of Selection of a Technical Solution from a Praxeometric Standpoint] [2], *Z zagadnień metodologii projektowania inżynierskiego* [From Problems of Methodology of Engineering Design] [3], and the more recent *Praxiology and the Philosophy of Technology* [1].

3 Digital Libraries

Digital libraries provide an ideal opportunity for the application of praxeological ideas in relation to operations in the field of linguistics. It might be suggested that the motivation for the creation of Polish digital libraries came from a series of thefts that took place in the late 1990 s. In 1998 a copy of Copernicus' *On the Revolutions of the Heavenly Spheres*, published in 1543, was stolen from the Library of the Polish Academy of Sciences. The greatest loss, however, was suffered in 1999 by the Jagiellonian Library, which discovered over a dozen items missing from its storeroom, including works by Galileo, Kepler and Bessarion.

As a result of these thefts, certain actions were taken that have consequences of interest to us. The path to the development of Polish digital libraries was opened by an Order of the Minister of Culture and Arts in 1998:

> **ORDER**
> **OF THE MINISTER OF CULTURE AND ARTS**
>
> of November 24, 1998
>
> on the determination of a list of libraries whose collections make up the national library resource, the organisation of that resource, and the principles and scope of its special protection.
>
> § 5. Special protection of the resource shall consist in:
>
> 1) the preparation of a plan for the protection of resources,
> 2) securing against damage in bunkers and protective structures within the organisational unit, and the organisation of appropriate protection,
> 3) restriction of the making available of the resources solely for academic and display purposes to other libraries or institutions providing appropriate conditions for their protection,
> 4) **recording of the resources on other media** [bolding – P.W.]

It was partly because of the provision contained in item (4) that an extensive digitization campaign began in Poland. In 2002 the country's first digital library, the Digital Library of Wielkopolska, was established, using *dLibra* software. *dLibra* is a system for the storage, editing and sharing of digital publications, developed at the Poznań Supercomputing and Networking Centre. The publications (magazines, books, photographs, films, etc.) were made available – and are planned to continue to be – to all users in an identical manner, in most cases permanently and entirely free of charge. At the end of the first decade of the 21st century the total number of publications was approximately 300,000:

> The collection of approximately 300,000 digital objects currently available in the PIONIER network is being constantly expanded thanks to the cooperation of several hundred institutions of culture in Poland. The resource is the result of many years of effort and cooperation by institutions of science and culture, effort oriented towards enriching the services and opportunities offered to the information society. Thanks to the PIONIER network these valuable resources can be used widely in academic circles and in education, as well as by individuals [6].

The digital libraries, which have already been the subject of a significant number of reports in the subject literature (cf. e.g. [7, 9]), store various types of collections (printed publications, press cuttings, audiovisual materials); the most popularly used library items are old printed materials, press publications and manuscripts, particularly from the 20th century, such as those shown here: (Fig. 1)

The historical material accumulated in the digital libraries has provided an ideal basis for the development of a linguistic theory which enables the identification of historical trends in language, particularly during the 19th and 20th centuries (1801–2000). Moreover, the material contained in the digital libraries was ideally suited to being investigated under a *big data* model. This results from the fact that every publication in the *dLibra* system is described according to the Dublin Core Schema, e.g. oai:www.wbc.poznan.pl:359665:

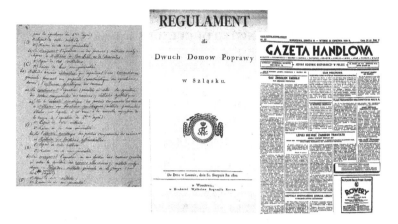

Fig. 1. Selected categories of materials stored by *dLibra*.

dc:title (pl): Amtsblatt der Königlichen Regierung zu Posen. 1902.05.06 Nro.18
dc:subject (pl): Grand Duchy of Poznań
dc:subject (pl): statutes
dc:subject (pl): Poznań in the 19th cent.
dc:subject (pl): official announcements
dc:subject (pl): Poznań Province
dc:subject (pl): official orders
dc:subject (pl): Wielkopolska people of the 19th cent.
dc:subject (pl): regulations
dc:description (pl): Gothic font
dc:description (pl): From 1821 alternative title "Dziennik Urzędowy Regencji w Poznaniu".
dc:description (pl): co-published supplement: Oeffentlicher Anzeiger http://www.wbc.poznan.pl/dlibra/publication/408392
dc:type (pl): magazines
dc:type (pl): official publications
dc:format (pl): image/x.djvu
dc:identifier: http://www.wbc.poznan.pl/Content/359665
dc:identifier: oai:www.wbc.poznan.pl:359665
dc:language (pl): ger
dc:rights (pl): Biblioteka Poznańskiego Towarzystwa Przyjaciół Nauk
dc:rights (pl): for all without restriction

Access (http://fbc.pionier.net.pl/owoc/oai-hosts) to more than two million publications with descriptions of this type provides linguists with previously unimaginable analytical opportunities, because each of the publications is accompanied by bibliographic parameters (*title, subject, type*, etc.). This wealth of resources gave rise to the theory of linguochronologization, which is oriented towards generating hypotheses concerning the chronology of the language of the 19th and 20th centuries.

4 The Theory of Linguochronologization (*TLCH*)

The theory of linguochronologization (*TLCH*), also called *chronologizational grammar* [9], describes those properties of language units which are relevant to chronologization, namely the assignment of dates (such as specific years) to language units. The date assigned to a language unit is equal to the date of the text from which it is extracted.

The purpose of the theory of linguochronologization, as formulated by the present author, is to enable the acceptance or rejection of hypotheses concerning linguistic chronology, particularly for vocabulary from the Modern Polish era (1750–2000). Such hypotheses are formulated using a class of observational sentences obtained in the process of excerption of empirical material. These sentences consist of assertions concerning a language unit (a word, for example) and its dating. To ensure that examples of language units appear in a form that is as faithful as possible to the original, it is proposed that they be presented in the form of photodocumentation.

4.1 Methods of Chronologization

The fundamental task faced by a language historian is to determine at what time a given word or language phenomenon first appeared. We therefore consider criteria for the evaluation of novelty or neonymy (the property of being a neonym). There exist several criteria for determining the neonymy of a given lexical unit with respect to a given date limit, such as the year 1945. Every neonymic hypothesis, namely one which states that a given unit came into being after a specified date, derives from the criteria adopted for determining neonymy. Among the main criteria considered are those based on: (a) consultation; (b) lexicographic listing; (c) appearance in dictionary source material; (d) appearance in texts; (e) introspection; (f) derivational analysis.

In the case of the **consultation criterion** the chronologization of a given unit is decided by appropriate consultation with experts. Questions are addressed to a group of experts, who may be subject matter specialists, specialists in the history of the discipline in question, etc. The **lexicographic criterion** is based on the use of a list of the headwords of a particular dictionary – usually one of general type, a so-called national dictionary, reference dictionary, etc. It is checked whether the item in question appears in the list of headwords. If it is found not to appear, it is considered to be a neonym with respect to that dictionary. The **dictionary source material criterion** is a variant of the lexicographic criterion. Its application consists in investigating whether a unit appears in the dictionary's source material prior to the expected date. In other words, under this criterion the date is assigned based on the absence of attestation in the quotations contained in the dictionary source material. The **textual criterion** involves analysis of a certain set of texts optimally chosen based on pre-selective analysis, and determination of whether a given unit appears within that set in texts with a given chronologization. The **introspective criterion** involves the dating of a given unit based on the researcher's own intuition regarding its chronology. Under the **derivational**

criterion it is determined that a derivative form cannot have an earlier chronologization than its basic form, whose chronologization is determined by another criterion.

TLCH, which serves to generate chronological hypotheses for particular words, morphemes, etc., is a theory of applied linguistics, and as such must contain certain directives. Directives are formulated for the purpose of achieving the best possible excerption results, that is, the largest possible set of examples. Directives are divided into: (a) requirements; (b) prohibitions; (c) recommendations.

Requirements and prohibitions are directives which must be unconditionally adhered to in ***TLCH***, whereas recommendations are directives whose application is suggested, but which need not be realized unconditionally. For example, in constructing a chronologizational model of the vocabulary of the 19th and 20th centuries, it is necessary to determine:

(a) what empirical basis should be taken for the applied theory leading to the obtaining of a model for the vocabulary of the 19th and 20th centuries, that is, what set of sources (documents) should be contained in such an empirical base;
(b) whether that set ought to be subject to additional classification;
(c) how to present given excerpts;
(d) how to order excerption tasks, that is, what tasks should be given priority as regards excerption for the 19th and 20th centuries;
(e) what time frame to adopt for a task designed in this way;
(f) a team of how many people, having what resources at their disposal, would be able to develop a chronologizational model of a particular language, of what size and with what excerption features;
(g) etc.

The above questions might be answered by stating that:

(a) for the creation of a chronologizational model of 19th- and 20th-century Polish the optimum textual resource is the ***dLibra*** system, created in 2002;
(b) from *dLibra* it is necessary to select those documents for which the best technical form of digital photograph is available (**higher scan resolution, preserved scan colour, paper source for scan**, etc.);
(c) the optimum form of presentation of excerpts from 19th- and 20th-century texts is **photodocumentation**;
(d) the optimum ordering of excerption tasks for 19th- and 20th-century vocabulary gives priority to the excerption of **affixational** derivatives and **compounds**;
(e) the optimum time frame for the excerption of 19th- and 20th-century vocabulary from the material contained in *dLibra* is up to **several years**;
(f) the task of excerption and chronologization for 19th- and 20th-century vocabulary ought to be carried out by a team consisting of between **one** and **over a dozen** people;
(g) etc.

5 Results

The creation of a diachronic resource in the form of headwords from 19th- and 20th-century vocabulary will provide a powerful tool for researchers studying the language in general or the language of press reports, of smaller textual forms such as advertisements and press announcements, etc. An example model of such a resource is presented at www.nfjp.pl (the *National Photocorpus of the Polish Language*). Such a collection also provides a tool that can be used by historians and cultural anthropologists. Researchers studying 20th-century topics might use it to make interpretations from many angles. A possible question to be addressed is that of identity, determining the consciousness of language users in the light of the systems of concepts which they used during that century. The general interpretation will aim to determine by what mechanism a resource came into being, spread, vanished, or changed in function. This task, in view of its explanatory component, is one that will never be complete. Undoubtedly several generations of linguists, historians and anthropologists could make use of this resource and make a variety of interpretations [4]. Already now it is possible to identify certain patterns governing the vocabulary of the 19th and 20th centuries; cf.: (Figs. 2 and 3)

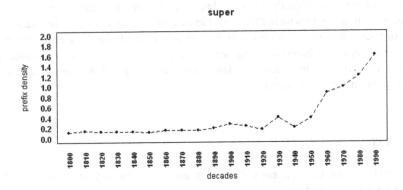

Fig. 2. Prevalence of the prefix *super-* in 1800–2000

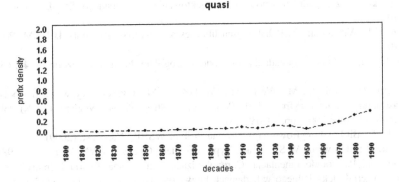

Fig. 3. Prevalence of the prefix *quasi-* in 1800–2000

It should be emphasised that magazines in particular are an especially valuable resource for linguistic research. For this reason, free access to large quantities of texts in the OAI-PMH (Open Archives Initiative Protocol for Metadata Harvesting) system is particularly important. Polish digital libraries fulfil these conditions. We should remember that newspaper texts constitute a register of the language created day by day in many different subject areas, and moreover by a variety of authors. By reading newspapers we can discover interesting words, sentence constructions, prefixes, suffixes, etc. Newspapers also have the significant advantages that (a) there are many of them, and (b) they are relatively precisely dated. This means that a sufficiently exact date (a specific year) for a word of interest can be determined easily and reliably. In this way it is possible to establish, for example, the earliest known date of use of a given word in print – examples might include *telewizor* ("television"), *aspiryna* ("aspirin"), *komputer* ("computer"), etc. Of course, this will not always be the earliest date at which the word in question appeared. Researchers will very rarely be certain that they have searched all of the printed material in which the word might conceivably have appeared. The situation is somewhat reminiscent of the work of archaeologists – a discovered object, such as an axe, may be the oldest known item of its kind in a given area, but there is no certainty that there did not exist older objects which have not survived or which have not yet been uncovered. Naturally both a linguist and an archaeologist may, based on their knowledge, suitably interpret their discovery and put forward hypotheses as to whether the object in question might in fact be the oldest. The power of such a hypothesis depends on the accumulation of textual research material. In this sense, research based on a big data model represents an extremely promising path to a discovery of the secrets of language – certainly not the final one, but an extremely extensive and exhaustive one.

References

1. Gasparski, W. Airaksin, T. (eds.): Praxiology and the Philosophy of Technology. New Bruswick – London (2007)
2. Gasparski, W.: Kryterium i metoda wyboru rozwiązania technicznego w ujęciu prakseometrycznym. PWN, Warszawa (1970)
3. Gasparski, W.: Z zagadnień metodologii projektowania inżynierskiego. SDOBN, Warszawa (1970)
4. Górny, M., Wierzchoń, P.: Polish digital libraries as a philologists' tools. IJ UAM, Poznań (2010)
5. Kotarbiński, T.: O istocie i zadaniach metodologii ogólnej. Przegląd Filozoficzny **41**, 68–75 (1938)
6. Mazurek, C., Stroiński, M., Werla, M., Węglarz, J.: Możliwości i wyzwania dla polskiej infrastruktury bibliotek cyfrowych. In: Mazurek, C., Stroiński, M., Węglarz, J. (eds) Polskie biblioteki cyfrowe 2009, pp. 11–19 (2010)
7. Wałek, A.: Biblioteki cyfrowe na platformie dLibra. WSBP, Warszawa (2009)
8. Węglarz, J.: Sterowanie W Systemach Typu Kompleks Operacji. PWN, Warszawa (1981)
9. Wierzchoń, P.: Fotodokumentacja. Chronologizacja. Emendacja. Teoria i praktyka weryfikacji materiału leksykalnego w badaniach lingwistycznych. IJ UAM, Poznań (2008)

Improving Twitter Aspect-Based Sentiment Analysis Using Hybrid Approach

Nurulhuda Zainuddin[(⊠)], Ali Selamat, and Roliana Ibrahim

Faculty of Computing, Universiti Teknologi Malaysia,
81310 Johor Bahru, Johor, Malaysia
alhuda710@gmail.com, {aselamat,roliana}@utm.my

Abstract. Twitter sentiment analysis has emerged and become interesting in many field that involves social networks. Previous researches have assumed the problem as a tweet-level classification task where it only determines the general sentiment of a tweet. This paper proposed hybrid approach to analyze aspect-based sentiments for tweets. We conducted several experiments to identify explicit and implicit aspects which is crucial for aspect-based sentiment analysis. The hybrid approach between association rule mining, dependency parsing and Sentiwordnet is applied to solve this aspect-based sentiment analysis problem. The performance is evaluated using hate crime domain and other benchmark dataset in order to evaluate the results and the finding can be used to improve the accuracy for the aspect-based sentiment classification.

Keywords: Twitter · Aspect-based sentiment analysis · Aspect extraction · Aspect classification

1 Introduction

Major studies in twitter research area apply sentiment analysis to classify opinions of user's expressions and thoughts from their tweets. The major area being highlighted in tweets is included to develop variety of techniques to monitor twitter in real time for the major events such as political events, business (stock markets movements, movies) and also society.

Using social network sites and micro-blogging sites as a source of data still need deeper analysis [1]. The major problem in supervised learning technique is the data problem such as data sparsity issues [2]. The availability of training dataset, the language used in twitter, and also determining the structure of learned function is might be one of the issues. In certain situation, unsupervised learning techniques can be applied because it does not require any training data but are not mathematically well-defined [3].

Previous researches considered twitter sentiment analysis problem as a tweet-level sentiment classification task which is similar to document-level sentiment classification. Tweet-level sentiment classification will decide the tweet sentiment in general. However, it is more crucial to determine what exactly the opinions of

© Springer-Verlag Berlin Heidelberg 2016
N.T. Nguyen et al. (Eds.): ACIIDS 2016, Part I, LNAI 9621, pp. 151–160, 2016.
DOI: 10.1007/978-3-662-49381-6_15

tweets rather than getting an overall positive or negative sentiment which might not be useful to the organizations. It would appear tweets can still contain various sentiments about a service or organization even though Twitter has limitation of characters [4].

In this research, we discuss an aspect-based sentiment analysis problem on twitter. As a result, we proposed a hybrid approach for the aspect-based sentiment analysis and we also presented our own Hate Crime Twitter Sentiment (HCTS) dataset in order to give the solution for this problem. Additionally, we benchmark with existing Stanford Twitter Sentiment (STS) dataset for this purpose.

The rest of this paper is organized as follows; Sect. 2 describes related work for twitter aspect-based sentiment classification task. Section 3, describes the proposed framework. Then, Sect. 4 describes the results and discussion obtained from the experiments. Finally, the last section presents the conclusion and suggested future work.

2   Related Work

The main focus is to conduct a research on the following twitter aspect-based sentiment analysis. Aspect-based sentiment analysis is the research problems that focus on the identification of all sentiment expressions within a given document and also to determine specifically the aspects sentiment of the target [5]. Similarly, according to [6], aspect-based sentiment analysis is focus to identify all sentiment expressions within a document including the specific aspects of the target. The aspect-based sentiment analysis can be divided into two main task which are aspect-based feature extraction and aspect-based sentiment classification [7].

The first task of aspect-based sentiment analysis is aspect-based feature extraction which is to identify aspects of the entity or also known as information extraction task. To perform aspect-based feature extraction, we need to find the major aspects of entities in a specific domain. There are mainly four types of methods for extracting feature for aspect-based sentiment analysis which are extraction by frequent nouns and noun phrases that considered as aspect candidates, extraction by opinions and aspect relation, extraction by supervised methods, and extraction by topic modelling [8].

Meanwhile, aspect sentiment classification will determines whether the opinions on different aspects are either subjective or objectives. The main task in aspect-based sentiment classification is to extract the opinion words and also find the polarity of opinion words towards its context. One of the methods being used by researcher is to extract all the adjectives, adverbs, verbs from the tagged sentences [9].

3   The Proposed Framework

The proposed framework consists of data collection and preparation, twitter preprocessing, and two main aspect-based sentiment analysis phases which is aspect-based feature extraction and aspect-based sentiment classification.

3.1 Data Collection and Preparation

There are only a few free resources available for sentiment analysis in microblogging. The corpus that will be used for sentiment analysis must be labelled to know which documents express positive opinion and also negative opinion. We consider this work as a target-dependent and automatic aspect sentiment classification by using opinion words extraction.

Here, we proposed our own Hate Crime Twitter Sentiment (HCTS) Dataset. We use Twitter Search API v1.1 for our data collection though in order to produce manual labelled corpus is hard and time consuming [10]. The most regular and reliable method for gathering data is to request a paged set of data based on a given query. The query selected for this research is the types of hate crime which is "anti-muslim", "anti-black", "anti-white", "anti-jews" and "anti-feminist". We also considers twitter hashtags to make sure the tweet is about the context such as hashtags #racist and #MuslimCyberHateCrime. The dataset produced a study corpus of 622 tweets. This dataset will be published once it is verified by the experts. However, this is a relatively small corpus for a machine learning classification task but in order to validate the results it was essential to derive a gold standard dataset that was coded according to the specific target (query) it contained. It is common to use a smaller dataset where this is required [11].

The method to classify tweets is based on work by [12] where emoticons are stripped off to avoid affecting the classifier accuracy. The twitter search API allows retrieving the most recent tweets based on a search query. However, it returns only 100 tweets per page and up to 15 such pages. The API also limits the number of requests per hour from an Internet Protocol address. Considering the hourly limits and allowing enough time for our requests to yield unique tweets, our system sent out a search request for each word every 30 min.

This corpus was domain-oriented and not very large because the annotation procedure was very complex and manual. In order to build the training data, we use Sentiwordnet from [13] to identify whether the tweets contain sentiment words. Then we decide the sentiment orientation of each tweets according to context it appear. From this experiment, we separate factual (neutral) from opinionated text by removing all neutral tweets from the database to avoid unnecessary process.

We also used Stanford Twitter Sentiment (STS) Dataset which is publicly available to evaluate the methods. The Stanford Twitter Sentiment (STS) corpus (http://help.sentiment140.com/) was introduced by [12]. It consists of a training and testing set. But we only considers testing dataset because we want the tweets is about the given target. The testing dataset consists of 173 negative, 180 positive and 139 neutral tweets which are manually classified. The targets such as "malcolm gladwel", "nike", "kindle2", "night at the museum", "time warner", "Bobby Flay", "google" and "obama".

3.2   Twitter Preprocessing

Twitter language has some unique attributes that may not provide relevant information in order to reduce the feature space [12]. The unique attributes that will be eliminated is included username. Username is included in the tweets in order to direct the messages to another one. It can be recognized with the symbol @ followed by the usernames. We also remove links and hashtags. Link is used to include web direction in tweets. In this process, links, hashtags and retweets will be removed because it will not be analyzed. The problem arise because retweets have similar content and for that reason it will be removed to avoid redundancies.

Another problem with tweet is the language used in tweets is more casual and informal. So it is important to do some filtering on the raw tweets before feeding to classifier. The filtering process involves remove new lines and opposite emoticons where some of the tweets contain two parts which are positive and negative. This tweet has to be removed to avoid ambiguity. Furthermore, remove emoticons with no clear sentiment and also repeated letters, laugh and punctuation marks.

The datasets will go through the preprocessing task of tweets such as tokenization, stop word removal, lowercase conversion and stemming. Tokenization is the process of splitting a text into words, phrases, or other meaningful parts, namely tokens. Next, remove the stop words from tweets that are commonly encountered in texts such as conjunctions, prepositions, etc. Then all uppercase characters are usually converted to their lowercase forms before the classification stages. Finally, the stemming process was performed to obtained root and stem of the derived words.

3.3   Twitter Aspect-Based Feature Extraction

The task is performed to find the explicit aspects and implicit aspects. For the explicit aspects, we proposed association rule mining for this task.In this experiment, we focus on finding features that appear explicitly as nouns and noun phrases in the tweets by using part-of-speech (POS) tagging. Then, to find the important feature/aspects of the hate crime based on people commented on their twitter, and ranking the hate crime features according to the frequencies they appear on twitter.

Association Rule Mining for Explicit Aspect Extraction. Association rule mining is used to find those important aspects for the given target [14]. In this case, we define aspects as important if it appears in more than 1 % (minimum support) of the tweets.

Association rule mining is stated as follows:

Let $I = i_1, ..., i_n$ be a set of items, and D be a set of transactions (the dataset). Each transaction consists of a subset of items in I. An *association rule* is an implication of the form $X \rightarrow Y$, where $X \subset I$, $Y \subset I$, and $X \cap Y = \emptyset$. The rule $X \rightarrow Y$ holds in D with confidence c if c%of transactions in D that support

X also support Y. The rule has support s in D if $s\%$ of transactions in D contain $X \cap Y$. The problem of mining association rules is to generate all associations rules in D that have support and confidence greater than the user-specified minimum support and minimum confidence.

Association rule mining is based on Apriori algorithm. We use Apriori algorithm to find the $frequent(important)aspects$ from a set of transactions that satisfy a user-specified $minimumsupport$.

Dependency Parsing for Implicit Aspect Extraction. We also proposed Stanford Dependency Parser [15] in order to capture grammatical relations occurring in twitter. In our experiment, we use this dependency parser to find out the implicit aspects from tweets. We have obtained the list of sentiment tweets from preprocessing steps, then we fed this tweets to dependency parser to find certain relations which is found to be useful than the others. Here we considered relation nsubj(nominal subject), amod(adjectival modifier), (advmod(adverbial modifier), xcomp(clausal component with external subject) and negation modifier. The reason behind using this grammatical relations will not be explained further in this paper because it is beyond the scope.

For instance, we take from this tweet: *"tumblr user melepeta for being anti-feminist and now she's claiming reverse racism is real and that people are overreacting"*

The dependency parser yields the following results: *root (ROOT-0, claiming-12), amod (melepeta-4, unfollowed-1), compound (melepeta-4, tumblr-2), compound (melepeta-4, user-3), nsubj (claiming-12, melepeta-4), mark (anti-feminist-7, for-5), cop (anti-feminist-7, being-6), acl (melepeta-4, anti-feminist-7), cc (anti-feminist-7, and-8), advmod (she-10, now-9), conj (anti-feminist-7, she-10), aux (claiming-12, 's-11), amod (racism-14, reverse-13), nsubj (real-16, racism-14), cop (real-16, is-15), ccomp (claiming-12, real-16),cc (real-16, and-17), mark (overreacting-21, that-18), nsubj (overreacting-21, people-19), aux (overreacting-21, are-20), conj (real-16, overreacting-21)*

From this example, there are two aspects identified from twitter which is 'anti-feminist' as an explicit aspects and 'reverse racism'. 'Reverse racism' is an implicit aspect that actually referring to anti-white sentiment. From the dependency parser, we also get opinion words for the aspect 'anti-feminist' and 'reverse racism'. This grammatical relation is important to identify the sentiment of the aspects. Besides, when negation words is found, the opinion word will be changed to opposite sentiment.

3.4   Opinion Word Scoring Using Sentiwordnet

In this experiment, we use features which are useful for opinion words extraction to determine whether the tweets contain the sentiment. It is asserted opinion words is the word that people use to express a positive or negative opinion [14]. For that reason, opinion words which are located around the hate crime feature/aspects in the sentence can be extracted using all the remaining frequent features especially when people express their views. From these words it is possible to infer the sentiment present in tweets.

One reason is research has shown that adjectives and adverbs are good indicators of subjectivity and opinions [16]. A nearby adjective refers to the adjacent adjective that modifies the noun or noun phrase that is a frequent feature for the aspect-based extraction. Consequently, it is considered an opinion word when such an adjective is found.

The Sentiwordnet [17] is used to assign the polarity scores for each opinion word appear on tweets. For example in this tweet: *"you are speculating, based on racial anti-white propaganda. Don't regurgitate, think for yourself. Study the facts/evidence"*

The polarity score of the word 'speculate' based on Sentiwordnet is 0.75 because it is definitely a negative opinion when the user tweets based on review in an idle or casual way and with an element of doubt. But it can become positive opinion when speculate referring to believe especially on uncertain or tentative grounds. So it will get other polarity scores. Since our research is to focus on aspect-based sentiment analysis, so different polarity scores will be assigned depending on the context it appear. This process we addressed nominally before we move to aspect-based classification problem to improve the accuracy of the tweet sentiment.

3.5   Evaluation Measures

We use accuracy measures to evaluate the whole classification performance with binary classes (positive and negative). For positive and negative sentiments on entities, we employ the standard evaluation measures of precision and recall.

Four effective measures used in this study are based on confusion matrix output, which are True Positive (TP), False Positive (FP), True Negative (TN), and False Negative (FN) [18].

- Precision(P) = TP/(TP+FP)
- Recall(R) = TP/(TP+FN)
- Accuracy(A) = (TP+TN)/(TP + TN + FP + FN)

4   Results and Discussion

4.1   Twitter Aspect-Based Feature Extraction

In twitter aspect-based sentiment analysis, the first phase is aspect-based feature extraction. Here we focus on finding aspects that appear explicitly as nouns or noun phrases in the tweets by using part-of-speech (POS) tagging. Association rule mining is used to find the important aspects from the given target. Association rules are created by analyzing data for frequent if/then patterns and using the criteria support and confidence to identify the most important relationships. In other words, support is an indication of how frequently the items appear in the database whereas confidence indicates the number of times the if/then statements have been found to be true. In this experiment, we applied different values for minimum support and minimum confidence before we get suitable value for minimum support which is 0.1 and minimum confidence value to be 0.8.

The process began by taking each tweet for analysis. Then the Part of Speech (POS) tags of the tweets were identified and nouns and noun phrases were extracted. This process is repeated for all sentences. The next process would be, counting the frequency of extracted words and take most frequent one. Each tweet is annotated with a list of aspects that is relevant to both dataset. It identifies all main aspects which represent company, event, location, misc, movie, person, and product which are relevant to the known aspects listed from Stanford Twitter Sentiment (STS) dataset. Besides, there are four explicit aspects from Hate Crime Twitter Sentiment (HCTS) dataset including "anti", "muslim", "feminist" and "jewish" which are relevant according to our targets. The results also show aspects words which represent all the main aspects such as "racist", "girl", "muslim", "feminist", "jewish", "women", "israel". The generated frequent explicit aspects are stored to the feature set for further processing.

Another task is to find the implicit aspects. We proposed Stanford Dependency Parser to find the implicit aspects based on grammatical relation between words in the sentence. The dependency parsing would obtain list of implicit aspects from tweets such as 'nigga', 'nigger', 'anti-black', 'neo-nazism', 'anti-oppresion', 'anti indigeneity', 'anti-semitic', 'reverse racism' and 'anti equality'. The process continues with opinion word scoring using Sentiwordnet [17].

4.2   Twitter Aspect-Based Sentiment Classification

In this section, Support Vector Machine (SVM) is used to get the sentiment classification accuracy on the Stanford Twitter Sentiment(STS) dataset and Hate Crime Twitter Sentiment (HCTS) dataset. We divided our Twitter Aspect-Based Sentiment Classification into two level of classification which are tweet-level classification and tweet aspect-based classification. We conducted several experiments and presented that our aspect-based classifier can improve the performance by giving detail sentiment of tweets compared to conventional tweet-level classifier.

Twitter Aspect-Based Level Classification. This experimental result for twitter aspect-based sentiment analysis involves the polarity orientations for each aspect from the tweets. Figure 1 shows the polarity scores according to Sentiwordnet for HCTS Dataset. The hybrid approach successfully performed classification according to the aspects. For racial aspects from hate crime, the approach successfully classified 28 % as neutral, 53 % as hate tweets and 19 % as free tweets. Besides, it also classified 24 % as neutral tweets, 58 % as hate tweets and 18 % as free tweets towards sexual aspects. Finally, for religion aspects, the hybrid approach classified 32 % as neutral, 48 % as hate tweets and 20 % as free tweets.

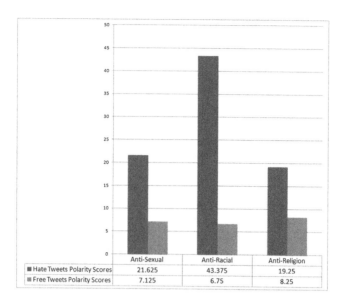

Fig. 1. Aspect-based sentiment classification results for different aspects in HCTS dataset

Twitter Tweet-Level Classification. Here we compared with tweet-level classification where we used precision, recall, and accuracy evaluation measures. We obtained classification results with Hate Crime Twitter Sentiment (HCTS) Dataset and Stanford Twitter Sentiment (STS) dataset. Table 1 shows the performance of tweet-level classifier on both dataset. Besides that, it shows the different classification results when using different features such as ngram and Part-of-Speech (POS) Tags. It can be seen from Table 1 that the accuracy results with Support Vector Machine (SVM) classifier is achieved 75.11 % with HCTS dataset. Besides, we obtained different results with STS dataset where the highest accuracy achieved is 81.58 %.

Table 1. HCTS and STS dataset tweet-level classification results

		Accuracy	Precision	Recall
HCTS	Unigrams	62.70	62.91	62.04
	Bigrams	54.05	53.56	62.15
	Trigrams	50.82	50.33	60.07
	POS Tags	75.11	84.12	65.34
STS	Unigrams	81.58	83.08	78.64
	Bigrams	66.57	84.56	39.32
	Trigrams	50.71	49.58	33.93
	POS Tags	80.73	82.32	77.48

5   Conclusion

Most of the work in Twitter sentiment classification has performed tweet-level sentiment classification. In this paper, we proposed a combination of association rule mining, dependency parsing and Sentiwordnet to perform aspect-based sentiment classification for Twitter. The novelty lies in the use of hybrid approach for twitter aspect-based sentiment analysis. Our approach can be applied and tested in subjectivity detection and polarity classification. The pre-processing steps are also important because twitter contains informal language and restrictions of 140 characters which are also help to improve classification performance.

Acknowledgments. The Universiti Teknologi Malaysia (UTM) under Research University funding vot number 02G31 and Ministry of Higher Education (MOHE) Malaysia under vot number 4F550 are hereby sincerely acknowledged for providing the research fundings to complete this research.

References

1. Medhat, W., Hassan, A., Korashy, H.: Sentiment analysis algorithms and applications: a survey. Ain Shams Eng. J. **5**(4), 1093–1113 (2014)
2. Bhuta, S., Doshi, A., Doshi, U., Narvekar, M.: A review of techniques for sentiment analysis of twitter data. In: 2014 International Conference on Issues and Challenges in Intelligent Computing Techniques (ICICT), pp. 583–591, February 2014
3. Khan, F.H., Bashir, S., Qamar, U.: Tom: twitter opinion mining framework using hybrid classification scheme. Decis. Support Syst. **57**, 245–257 (2014)
4. Lek, H.H., Poo, D.: Aspect-based twitter sentiment classification. In: 2013 IEEE 25th International Conference on Tools with Artificial Intelligence (ICTAI), pp. 366–373, November 2013
5. Feldman, R.: Techniques and applications for sentiment analysis. Commun. ACM **56**, 82–89 (2013)
6. Brychcin, T., Konkol, M., Steinberger, J.: Uwb: machine learning approach to aspect-based sentiment analysis. SemEval **2014**, 817 (2014)
7. Liu, K.-L., Li, W.-J., Guo, M.: Emoticon smoothed language models for twitter sentiment analysis. In: AAAI, vol. 2, pp. 1678–1684 (2012). cited By (since 1996)1
8. Li, S., Zhou, L., Li, Y.: Improving aspect extraction by augmenting a frequency-based method with web-based similarity measures. Inf. Process. Manage. **51**(1), 58–67 (2015)
9. Kansal, H., Toshniwal, D.: Aspect based summarization of context dependent opinion words. In: Procedia Computer Science, vol. 35, pp. 166–175, Knowledge-Based and Intelligent Information & Engineering Systems 18th Annual Conference, KES-2014 Gdynia, Poland, Proceedings, September 2014
10. Camara, E.M., Martin-Valdivia, M.T., Lopez, L.A.U., Montejo-Raez, A.: Sentiment analysis in twitter. Nat. Lang. Eng. **20**(1), 1–28 (2014). cited By (since 1996)3
11. Pak, A., Paroubek, P.: Twitter as a corpus for sentiment analysis and opinion mining. In: Proceedings of the Seventh Conference on International Language Resources and Evaluation (LREC 2010), (Valletta, Malta), European Language Resources Association (ELRA), May 2010

12. Go, A., Bhayani, R., Huang, L.: Twitter sentiment classification using distant supervision. CS224N Project Report, Stanford, pp. 1–12 (2009)
13. Baccianella, S., Esuli, A., Sebastiani, F.: Sentiwordnet 3.0: an enhanced lexical resource for sentiment analysis and opinion mining. In: Calzolari, N., Choukri, K., Maegaard, B., Mariani, J., Odijk, J., Piperidis, S., Rosner, M., Tapias, D., (eds.) LREC, European Language Resources Association (2010)
14. Hu, M., Liu, B.: Mining opinion features in customer reviews. In: Proceedings of the 19th National Conference on Artifical Intelligence, AAAI 2004, pp. 755–760. AAAI Press (2004)
15. De Marneffe, M.-C., Manning, C.D.: The stanford typed dependencies representation. In: Coling 2008: Proceedings of the Workshop on Cross-Framework and Cross-Domain Parser Evaluation, pp. 1–8, Association for Computational Linguistics (2008)
16. Liu, B.: Sentiment analysis and subjectivity. In: Handbook of Natural Language Processing, 2nd edn. Taylor and Francis Group, Boca Raton (2010)
17. Esuli, A., Sebastiani, F.: Sentiwordnet: a publicly available lexical resource for opinion mining. In: Proceedings of LREC, vol. 6, pp. 417–422 (2006)
18. Zainudin, N., Selamat, A.: Sentiment analysis using support vector machine. In: International Conference on Computer, Communications, and Control Technology (I4CT), pp. 333–337, September 2014

Design of a Yoruba Language Speech Corpus for the Purposes of Text-to-Speech (TTS) Synthesis

Théophile K. Dagba[1]([⊠]), John O.R. Aoga[2], and Codjo C. Fanou[3]

[1] School of Applied Economics and Management,
University of Abomey-Calavi, Cotonou, Benin
theophile.dagba@eneam.uac.bj
[2] Polytechnic School of Abomey-Calavi,
University of Abomey-Calavi, Cotonou, Benin
johnaoga@gmail.com
[3] School of Administration,
University of Abomey-Calavi, Cotonou, Benin
chcodjo@yahoo.fr

Abstract. This paper deals with the design of a speech corpus for a corpus-based Text-To-Speech (TTS) synthesis approach. The purposes are first to provide enough speech to develop Yoruba corpus-based TTS system and second, to provide a simple methodology for other languages corpus design. The paper focuses on text analysis, selection of the reliable sentences, selection of the reader, and sentences recording. The analysis is performed to ensure a good balance of the corpus. Then, 2,415 sentences are gathered (essentially affirmative sentences). Those sentences have been read by a Yoruba language journalist who is a native speaker of the language. There is one speaker for the whole corpus.

Keywords: Yoruba language · Language corpus · TTS · Unit selection

1 Introduction

One of the crucial problems that have to be solved when speech recognition or a speech synthesis system is developed is the availability of a proper speech corpus for the system training and testing. The problem is usually solved in the following way: first, a set of suitable sentences are selected from a database of phonetically transcribed sentences; next the set of selected sentences are read by a group of speakers and, as the last step, the utterances are used to form the training and the test datasets [19]. Several works have been realized in the field of speech technology in general and TTS in particular. Among them we can mention those on Spanish [17], French [10], Czech [13], etc. The main obstacle to African languages in speech applications is the lack of sufficient speech material for the study of speech events and for training, development, and testing of algorithms and systems [10]. A 2013 review on prosody realization in Text-to-Speech applications showed that Yoruba is under-researched in the area of prosody and speech synthesis in general [5]. So far, Yoruba language has not

© Springer-Verlag Berlin Heidelberg 2016
N.T. Nguyen et al. (Eds.): ACIIDS 2016, Part I, LNAI 9621, pp. 161–169, 2016.
DOI: 10.1007/978-3-662-49381-6_16

experienced enough studies in speech corpus design for TTS synthesis. This paper provides an overview of ongoing research on the development of a Yoruba corpus. The goal of our work is to provide base material for the development and evaluation of TTS synthesis systems. We focus on the analysis, selection of text material and reader.

The paper is organized as follows. Sections 2 and 3 presents the Yoruba sound system and related works respectively. In Sect. 4 the methodology for designing Yoruba text corpus is presented. Section 5 deals with the conditions for recording a quality speech corpus. Section 6 is dedicated to our results and discussion. Finally, Sect. 7 contains the conclusion and outlines our future work in this field.

2 Yoruba Sound System

Yoruba is an African language of the family of Niger-Congo languages. It is natively spoken in southwestern Nigeria (the second largest ethnic group in number), Benin and Togo by over 30 million people [11]. There are three sets of sounds which make up Yoruba words: these are vowels, consonants and tones [4]. In Yoruba there are 12 vowels which are classified into 2 types, oral and nasalized vowels. Oral vowels are produced entirely through the mouth and nasalized ones are produced through both the mouth and the nose. Orthographically, nasalized vowels are written with an 'n' following an oral vowel. Yoruba has 18 consonants. It is a tonal language. It has three surface tones of different pitch levels. Tones are marked on vowels and syllabic nasals. The tones and their orthographic representations are as in Table 1 with the corresponding musical note[1].

Table 1. Yoruba language tones

Tone	Mark	Corresponding musical note
High	´	mi
Mid	Unmarked	re
Low	`	do

Indeed, a word may have different lexical meanings depending on whether it is said with a high, a mid or a low pitch. This shows the extent to which tones are important in Yoruba. The wrong pronunciation of a word could involve a wrong comprehension as illustrated in Table 2. Then, the tonal information removes the ambiguity in the pronunciation of well written and properly accented Standard Yoruba texts [15].

Table 2. Illustration of tone use on the vowel *o*

Word	Tone on the vowel	Meaning of the word
Kọ́	High	To build
Kọ	Mid	To sing
Kọ̀	Low	To refuse

---

[1] http://www.africa.uga.edu/Yoruba/phonology.html.

3 Related Work and Motivation

In the past few years, Yoruba TTS study has drawn a wide attention. TTS is then the area of speech technology that attracts more research effort. In 2004, Odéjobi et al. [16] have presented the design and analysis of an intonation model for Text-To-Speech synthesis applications using a combination of Relational Tree and Fuzzy Logic technologies. The model was demonstrated using Standard Yoruba language. In the proposed intonation model, phonological information extracted from text is converted into Relational Tree. Mean opinion Scores of 9.5 and 6.8, on a scale 1–10, was obtained for intelligibility and naturalness respectively. In 2011, a text markup system for text intended as input to standard Yoruba speech synthesis was presented by Odéjobi [15]. In 2012, van Niekerk and Barnard [23] have investigated the acoustic realization of tone in short continuous utterances in Yoruba. Fundamental frequency (F0) contours were extracted for automatically aligned syllables from a speech corpus collected for speech recognition development. Extracted contours were processed and analyzed statistically to describe acoustic properties in different tonal contexts. In 2013, Afolabi and Wahab [2] have focused their research work on the use of E-learning Text-To-Speech to teach Yoruba language online. A database was created for the recorded syllables in the tree tones of Yoruba language. In 2014, Akinadé and Odéjobi [3] examined the process underlying the Yoruba numeral system and described a computational system that is capable of converting cardinal numbers to their equivalent Standard Yoruba number name. In 2015, Adeyemo and Idowu [1] considered the development of TTS in Yoruba to assist Yoruba language speaking people especially the visually impaired users. They therefore created inventory of syllable pronounceable in Yoruba and recorded all of them. In other hand, Dagba et al. [7] investigated the integration of Yoruba into eSpeak[2] system for the purposes of mobile phone applications. They have defined 54 phonemes of Yoruba language by using existing phoneme tables such as Base table, English table and French table. They have built also rules which indicate how to pronounce certain groups of words. They finally have a dictionary file with 70 rules.

As shown by the above review, apart from the work of van Niekerk and Barnard [23] these studies are not corpus oriented or they relied on relatively small samples based on carefully designed corpora.

4 Design of Yoruba Text Corpus

4.1 Background of Corpus Building

The whole corpus building process is diagrammed as shown in Fig. 1 [21]. The criteria in designing speech corpus are size, coverage, domain and quality.

The recently developed corpus-based speech synthesizers tend to rely on large scale database, ranging from a few hours to more than 10 h of speech corpora, to provide

---

[2] http://espeak.sourceforge.net.

sufficiently natural output speech [12]. But an increase of the corpus size will affect the performance of the used method and slacken the synthesis process.

For unit selection synthesizer, the quality of speech is highly dependent on unit coverage of speech corpus. The corpus database must be phonetically rich. In other words, it must involve as many phonetic combinations as possible, including intra-syllabic and inter-syllabic structures, in a corpus of acceptable size [6]. The corpus words should also include at least one instance of all units [14].

As argued in [18], a system with a good selection module and a high quality speech corpus may yield output speech of extremely high quality, even if the signal processing module is rather simple.

The domain or focus application of a corpus-based Text-To-Speech is very important since a limited domain can reduce the corpus size and yet preserve the quality of synthetic speech. Several projects have been developed in restricted domains such as in weather forecasts and talking clock contexts [9, 14].

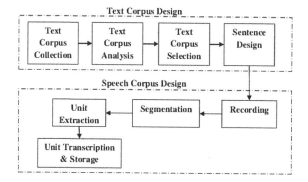

Fig. 1. Corpus building process

Our speech corpus construction requires the collection of texts written in Yoruba with well-spelled words. The reading of each sentence constitutes the audio corpus. The reader must respect the rules of pronunciation, tones and punctuation while adopting a consistent pace in a sound proof environment. Ideally, a recording studio is a suitable environment for this kind of recording. The speech corpus is a set of audio and text corpora, in the same folder, with a link between the text and the corresponding record.

4.2   Text Collection and Preprocessing

The collection of textual data was done from a Yoruba version of the Holy Bible[3]. The first 50 chapters of Genesis were taken into account in the construction of the corpus. However, after processing, some sentences were made entirely of personal names.

---

[3] http://www.jw.org/yo.

Those sentences have been deleted. Also some sentences of genesis 7, 13 are too long and have been deleted too. Paragraphs are extracted to expect overall consistency of the corpus. Because the meaning of sentences matters here, it is important to have correct sentences for easier reading and to allow the reader to be in a real and coherent context, and to help to enrich the speech corpus in emotions. Illustrations and tables are deleted as well as characters that are not taken into account such as +,−, *, %, etc. It is decided to use sentences as units of the corpus.

4.3   Text Analysis

This step allows the analysis of the text corpus at sentences level. In the first step, the statistics on the size of the corpus of sentences and words (number of words, number of distinct words, the average number of words in sentences, etc.) are produced. Then, the proportion of co-occurrence P(u,v) (see Eq. 1) is computed with f(u) (frequency of word u), f(v) (frequency of word v) and f(u,v) (frequency of word u and v occurring in the same sentence).

$$P(u, v) \;=\; f(u, v)/(f(u) + f(v) - f(u, v)) \tag{1}$$

After that, the existence of different phonemes of Yoruba in every sentence and in the corpus is assessed. It is then ensured that there is no excessive difference between the frequencies of occurrence of different phonemes. In addition, most common contexts of use are represented. It is this tradeoff (between frequencies of phonemes and contexts) which defines the sound balance in the corpus. Finally, a K-means classification is performed based on the frequency of occurrence of words and phonemes to better appreciate the different lexical categories in the corpus. All the above analysis is repeated till acceptable tradeoff between frequencies of phonemes and utilization contexts is achieved.

5   Recording of the Sentences

After the design of text corpus, focus is placed on speech corpus design as illustrated in Fig. 1.

5.1   Choice of the Reader

To find the suitable reader, some criteria are used. First of all, the reader should be someone that practises the language in his/her daily life. A Yoruba language radio journalist who is also a native speaker has been selected. Thus, it is sure to have a voice respecting the rules of pronunciation and tones, but also prosodic parameters such as rhythm, intonation and emphasis. We took into account the playback speed because it is an important factor affecting the proper articulation of words. The recording is done in a recording studio preferably late at night.

5.2   Reading

This step is the recording of the text corpus. It is based on the use of Redstart of MaryTTS [20]. This tool allows us to calibrate the system on the recording settings prior to any playback/recording. Speech audio and timing parameters are given as shown in Table 3.

Table 3. Speech audio and timing parameters

Parameter	Value
Frequency	44,100 Hz
Number of bit	16
Input type	Mono
Timing before reading	2,000 ms
Timing after reading	2,000 ms
Pause	0 ms

The Redstart tool also allows listening to the sound, and re-recording or viewing the signal spectrum, pitch and energy diagram. A meticulous handling of the records is conducted. We listened to each sentence looking at the written version and the spectrum of the signal to verify that the sound is not clipped (Fig. 2) or misread. If one and/or another of the above cases of mistakes occur, the recording is repeated.

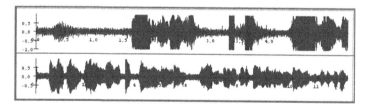

Fig. 2. Spectrum of a clipped signal (top) and a normal spectrum (bottom)

The audio tool convertor of MaryTTS [20] is used for the normalization of recordings in wave format to meet the conditions of the voice synthesis system. This tool also permits to address the overall amplitude and power of the recording sentence by sentence, to filter the noise frequencies below 50 Hz and remove start and end silence of wave sounds.

6   Results and Discussion

6.1   Results

The text corpus collected on the Internet, after analysis and balancing contains 2,415 sentences. Most of these sentences are affirmative sentences (88.65 %) with only 6.05 % of interrogative sentences and 5.30 % of exclamation sentences (see Table 4).

This corpus contains 46,117 words (2,275 distinct words). The average occurrence of words is 20.27 with a standard deviation of 93.22. This standard deviation reflects the unequal distribution of words in the corpus. This state is justified by the fact that words such as pronouns and prepositions appear more than 1,000 times in the corpus while common nouns, verbs, adjectives, and adverbs appear 100 times. Proper nouns and cardinal numbers appear less than 10 times. K-means classification confirmed the three categories of words that we had previously identified and the balance of phonemes.

Table 4. Features of the corpus

Item	Value
Sentences	2,415
Affirmative sentences	88.65 %
Interrogative sentences	6.05 %
Exclamation sentences	5.30 %
Average of word per sentence	11.38
Phonemes	148,823
Frequency of phonemes	2,705.87
Phonemes of high tone	24.48 %
Phonemes of mid tone	16.02 %
Phonemes of low tone	18.60 %
Phonemes of consonant	18.60 %
Size of the corpus	234 mn

6.2   Evaluation of the Corpus

To have an idea about the quality of this corpus, an experimental TTS corpus-based system using "unit selection algorithm" for Yoruba language is built by applying MaryTTS. The Mean Opinion Score (MOS) was used to evaluate the general output of the system. We have got the result as presented in Table 5. In subjective testing, a MOS is the arithmetic mean of all of the individual opinion scores resulting from a single test [8, 24]. Then, 10 native Yoruba speakers were selected. They were between 11 and 30 years old. Each person had listened to 10 synthesis sentences and had given a mark between 0 and 5. The MOS is equal to 2.9. This score is equivalent to a good perception of the voice in the system output. At this stage, we have integrated Yoruba localization into MaryTTS which is available on Gitub branch[4] of this tool. The next version of the tool will merge it with the master branch.

6.3   Discussion

We can first notice that the detailed methodology used to design our speech corpus can be used in similar work for other languages. Second, this corpus can be used in

---

[4] https://github.com/johnaoga/marytts

Table 5. Result of MOS evaluation

Appreciation	Mark	Average score in %
Excellent	5	14
Very good	4	30
Good	3	20
Fair	2	14
Poor	1	12
Poor	0	10

synthesis systems. It can also be noticed that the contribution of linguistic analysis to ensure sound balance proved to be very useful. It has helped to know how to present all the phonemes in the corpus. This has also allowed us to better understand the constitution and the features of our corpus. Indeed, the texts of the corpus must be recorded by the same person who must be qualified to do this work. Studies on the possibility of combining heterogeneous voice sources may allow the use of a great mass of heterogeneous data. Those issues were previously mentioned in the literature [14, 22].

7 Conclusion

This paper deals with the design of a speech corpus for corpus-based Text-To-Speech synthesis approach. First, texts have been collected and analyzed. After that, we have proceeded to the recording of the sentences. We have obtained a speech corpus which contains 2,415 sentences with 148,823 phonemes. The corpus has been tested in an experimental TTS system with a good result. Our future work will increase the size of the corpus, and take into account more interrogative and exclamation sentences.

Acknowledgments. The authors acknowledge the contribution of Vincent AWE, radio journalist of Yoruba language at the Office of Radio and Television of Benin (ORTB), for the recording of the corpus.

References

1. Adeyemo, O.O., Idowu, A.: Development and integration of text to speech usability interface for visually impaired users in Yoruba language. Afr. J. Comput. ICT **8**(1), 87–94 (2015)
2. Afolabi, A.O., Wahab, A.S.: Implementation of Yoruba text-to-speech e-learning system. Int. J. Eng. Res. Technol. **2**(11), 1055–1064 (2013)
3. Akinadé, O.O., Ọdẹ́jọbí, O.A.: Computational modelling of Yorùbá numerals in a number-to-text conversion system. J. Lang. Model. **2**(1), 167–211 (2014)
4. Akinlabi, A.: Yorùbá sound system. In: Understanding Yoruba life and culture. Africa world press Inc. pp. 453–468. (2004)
5. Akinwonmi, A.E.: A prosodic text-to-speech system for yorùbá language. In: 8th IEEE International Conference for Internet Technology and Secured Transactions (ICITST), pp. 630–635, London (2013)

6. Chou, F.-C., Tseng, C.-Y., Lee, L.-S.: A set of corpus-based text-to-speech synthesis technologies for mandarin Chinese. IEEE Trans. Speech Audio Process. **10**, 481–494 (2002)
7. Dagba, T.K., Aoga, O.R., Fanou, C.C.: eSpeak support of Yoruba language for the purposes of mobile phone applications. In: 3rd IEEE Pan African Conference on Science Computing and Telecommunication (PACT'2015), pp. 137–141, Kampala (2015)
8. Dagba, T.K., Boco, C.: A text to speech system for Fon language using multisyn algorithm. Procedia Comput. Sci. **35**, 447–455 (2014). Elsevier
9. Fék, M., Pesti, P., Németh, G., Zainkó, C., Olaszy, G.: Corpus-based unit selection TTS for Hungarian. In: Sojka, P., Kopeček, I., Pala, K. (eds.) TSD 2006. LNCS (LNAI), vol. 4188, pp. 367–373. Springer, Heidelberg (2006)
10. Gauvain, J.-L., Lamel, L., Eskenazi, M.: Design considerations and text selection for BREF, a large french read-speech corpus. In: 1st International Conference on Speech and Language Processing, vol. 2, pp. 1097–2000 (1990)
11. Igue, A.M.: Grammaire Yorùbá de base abrégée. Center for Advanced Studies of African Society (CASAS), monograph 238 (2009)
12. Kawai, H, Tsuzaki, M.: Study on time-dependent voice quality variation in a large-scale single speaker speech corpus used for speech synthesis. In: Proceeding of the IEEE Workshop on Speech Synthesis, pp. 15–18 (2002)
13. Matousek, J., Psutka, J., Kruta, J. : Design of speech corpus for text-to-speech synthesis. In: Interspeech (Eurospeech), pp. 2047–2050 (2001)
14. Nagy, A., Pesti, P., Németh, G., Böhm, T.: Design issues of a corpus-based speech synthesizer. Hung. J. Commun. **6**, 18–24 (2005)
15. Odéjobí, O.A.: Design of a text markup system for Yorùbá text-to-speech synthesis applications. In: Conference on Human Language Technology for Development, pp. 74–80, Alexandria, Egypt (2011)
16. Odéjobí, O.A., Beaumont, A.J., Wong, S.H.S.: A computational model of intonation for Yorùbá text-to-speech synthesis: design and analysis. In: Sojka, P., Kopeček, I., Pala, K. (eds.) TSD 2004. LNCS (LNAI), vol. 3206, pp. 409–416. Springer, Heidelberg (2004)
17. Ortega-Garcia, J., Gonzalez-Rodriguez, J., Marrero-Aguiar, V.: AHUMADA: a large speech corpus in Spanish for speaker characterization and identification. Speech Commun. **31**(2), 255–264 (2000)
18. Piits, L., Mihkla M., Nurk T., Kiissel, I.: Designing a speech corpus for Estonian unit selection synthesis. In: Proceedings of the 16th Nordic Conference of Computational Linguistics NODALIDA-2007, pp. 367–371 (2007)
19. Radová, V., Vopálka, P.: Methods of Sentences Selection for Read-Speech Corpus Design. In: Matoušek, V., Mautner, P., Ocelíková, J., Sojka, P. (eds.) TSD 1999. LNCS (LNAI), vol. 1692, pp. 165–170. Springer, Heidelberg (1999)
20. Schröder, M., Trouvain, J.: The German text-to- speech synthesis system MARY: a tool for research, development and teaching. Int. J. Speech Technol. **6**, 365–377 (2003)
21. Tan, T.-S., Hussain, S.: Scorpus design for Malay corpus-based speech synthesis system. Am. J. Appl. Sci. **6**(4), 696–702 (2009)
22. Taylor, P.: Text-to-Speech Synthesis. Cambridge University Press, Cambridge (2009)
23. Van Niekerk, D.R., Barnard, E.: Tone realisation in a Yoruba speech recognition corpus. In: 2012 Workshop on Spoken Language Technologies for Under-Resourced Languages (SLTU), Cape Town, South Africa. http://www.mica.edu.vn/sltu2012/files/proceedings/11.pdf
24. Viswanathan, M., Viswanathan, M.: Measuring speech quality for text-to-speech systems: development and assessment of a modified mean opinion score (MOS) scale. Comput. Speech Lang. **19**(1), 55–83 (2005)

Named Entity Recognition for Vietnamese Spoken Texts and Its Application in Smart Mobile Voice Interaction

Phuong-Nam Tran[✉], Van-Duc Ta, Quoc-Tuan Truong, Quang-Vu Duong,
Thac-Thong Nguyen, and Xuan-Hieu Phan

University of Engineering and Technology,
Vietnam National University (VNU), Hanoi, Vietnam
{namtp.mi13,ductv_53,tuantq_57,vudq_57,thongnt_57,hieupx}@vnu.edu.vn

Abstract. Named entity recognition (NER) for written documents has been studied intensively during the past decades. However, NER for spoken texts is still at its early stage. There are several challenges behind this: spoken texts are usually less grammatical, all in lowercase, and even have no punctuation marks; continuous text chunks like email, hyperlinks are interpreted as discrete tokens; and numeric texts are sometimes interpreted as alphabetic forms. These characteristics are real obstacles for spoken text understanding. In this paper, we propose a lightweight machine learning model to NER for Vietnamese spoken texts that aims to overcome those problems. We incorporated into the model a variety of rich features including sophisticated regular expressions and various look-up dictionaries to make it robust. Unlike previous work on NER, our model does not need to rely on word boundary and part-of-speech information – that are expensive and time-consuming to prepare. We conducted a careful evaluation on a medium-sized dataset about mobile voice interaction and achieved an average F_1 of 92.06. This is a significant result for such a difficult task. In addition, we kept our model compact and fast to integrate it into a mobile virtual assistant for Vietnamese.

Keywords: Natural language understanding · Named entity recognition · Vietnamese spoken text processing · Mobile virtual assistant

1 Introduction

With the recent important advances [6,9,10] in speech recognition technology, voice-based communication and interaction are becoming increasingly popular. We can easily see this shift via the use of speech-to-speech translation[1] [1], call center automation, in-car voice control, and the emergence of mobile virtual assistants like Apple Siri, Google Now, and Microsoft Cortana [17]. These applications serve different purposes but they all have two main phases: automatic speech recognition (ASR) and spoken text understanding [19]. The former has

---

[1] Microsoft Skype Translator and AT&T Speech-to-Speech Translation.

© Springer-Verlag Berlin Heidelberg 2016
N.T. Nguyen et al. (Eds.): ACIIDS 2016, Part I, LNAI 9621, pp. 170–180, 2016.
DOI: 10.1007/978-3-662-49381-6_17

been being studied for decades and now achieved significant results while the latter is still at its early stage. Among spoken text processing tasks, NER is one of the fundamental problems that is essential for further language understanding.

NER, as defined in MUC conference series [5,7], is the process of extracting text chunks that are names of person, organization, location, or time mentioned in written documents. Starting two decades ago, NER, however, is still far from being solved in terms of its scope and recognition accuracy. Named entity, in a broader sense, is any type of interest like a book, song, movie title or the name of a bacteria, symptom, or medical condition. This diversity and the ambiguity nature of human language make this task difficult and still an open problem.

NER for spoken texts, however, is even more challenging than working with written documents. Firstly, spoken sentences are commonly shorter and less grammatical. Sentence constituents like subject or object are sometimes omitted. Also, there are no punctuation marks in the texts. It, therefore, is non-trivial to segment and parse spoken sentences correctly. Secondly, the ASR's output texts are all in lowercase. This makes it extremely difficult to recognize proper nouns or personal names because capitalization, the most reliable feature for NER, is now not available. The third difficulty is the alphabetic and numeric ambiguity, that is, numbers are sometimes produced by ASR in alphabetic texts. For example, spoken time expressions like *8:25'* or *8h25'* may be interpreted as *8 giờ hai mươi lăm phút* (*8 o'clock twenty five minutes*), a much longer text chunk and much harder to handle. Lastly, continuous text patterns like email addresses and hyperlinks may be interpreted as discrete tokens. For instance, a URL like *dantri.com.vn* will be produced as *dân trí chấm com chấm vn* (*dan tri dot com dot vn*), a discrete, longer, and not a well-formed text pattern.

The above obstacles, however, are the main motivations for our work. In this paper, we will present how we put our effort to solve NER for Vietnamese spoken texts (NER4VST for short). In our work, we aimed to build a NER4VST model that is robust, fast, and compact. It is capable of recognizing named entities in Vietnamese spoken texts about the interaction contents between mobile users and their smart devices. The NER4VST model was also integrated into a virtual assistant for Vietnamese (VAV[2]). VAV is capable of interpreting users' spoken natural language commands to perform a variety of tasks on their smartphones such as locating an address, finding direction from location A to location B on the map, playing a song, making a call to a particular contact name, asking for weather information, arranging a meeting in their calendar, and many more. In VAV, the NER4VST model helps to identify the command arguments (i.e., the named entities) in the users' inputs like a contact name or number, a hyperlink, an address or a location, a datetime expression, or an application name. Figure 1 shows the general design of the VAV and how the NER4VST model is integrated into this application. In order to make the model accurate, fast, compact, and applicable, we had to deal with a lot of problems, including the above challenges and other engineering issues. All in all, the following points make our work different from the previous studies and are also our main contributions:

---

[2] VAV: https://play.google.com/store/apps/details?id=com.mdnteam.vav.

- To the best of our knowledge, this work is the first study conducting NER for Vietnamese spoken texts. We also built a medium-sized set of data with named entities annotated for evaluation as well as application in the VAV.
- The NER4VST model can deal with the challenges stated above like less grammatical issue, all-lowercase texts, no punctuation, numeric–alphabetic ambiguity, and discrete interpretation of continuous patterns (e.g., email, URLs).
- Our model does not need to rely on the word boundary and part-of-speech information as almost all previous NER systems. We incorporated into the model a variety of rich features that help overcome the lack of these types of information.
- The model is lightweight and compact in order that NER can be performed right on mobile devices. This is obviously an advantage because normal NER models are usually large and need powerful machines.

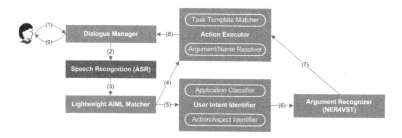

Fig. 1. NER4VST is a core component of the mobile virtual assistant for Vietnamese

2   Related Work

NER for written texts has a long history of development. Most previous papers focused on NER tasks defined in the MUC conferences [5,7] and the automatic content extraction conference (ACE). There have been a lot of studies about NER for written documents [3,4]. We, however, only describe here most related researches working on NER for Vietnamese written or spoken texts.

Nguyen et al. 2007 [14] used conditional random fields (CRFs) to perform NER for Vietnamese texts and achieved an average F_1 of 85.51. In order to improve the question answering performance, Molla et al. 2007 [13] performed NER for speech data. However, they describe that they used various kinds of features including capitalization, that is not available in our spoken texts. Pan et al. 2005 [16] proposed two approaches to NER for Mandrin Chinese spoken documents: the first is based on the observation that NER is easier if they jointly consider the multiple occurrences of a name in different contexts of the whole document; the second is based on external knowledge, e.g., the Web, to expand the context information of each named entity. And Hatmi et al. 2013 [8] proposed a multi-level methodology based on CRFs to NER for speech transcripts of more than 40 h of data recorded from different French speaking television and radio stations. Most studies above worked on speech transcripts or stories – that are much different from the output texts of ASR.

3   NER for Vietnamese Spoken Texts (NER4VST)

3.1   NER4VST as a Segmentation Problem for Sequence Data

NER4VST is the task of identifying named entities in Vietnamese spoken text sentences. In this study, named entities are limited to the list of eight types $L = \{location, datetime, url, cnumber, cname, aname, email, number\}$ as shown in Table 1. Let $\mathbf{x} = (x_1, x_2, \ldots, x_n)$ be an input Vietnamese spoken text sentence, performing NER4VST for $\mathbf{x}$ is to recognize a list of named entities $E_{\mathbf{x}} = \{e_i(l_i, a_i, b_i)\}$ mentioned in $\mathbf{x}$. For each $e_i(l_i, a_i, b_i) \in E_{\mathbf{x}}$, l_i ($\in L$) is its named entity type, a_i and b_i are the indexes of the first and the last tokens of e_i in $\mathbf{x}$ ($1 \leq a_i \leq b_i \leq n$). Named entities are non-overlapping, that is, given two entities $e_i, e_j \in E_{\mathbf{x}}$, the first and the last indexes of the two entities must satisfy: $b_i < a_j$ or $b_j < a_i$. Table 2 gives samples of Vietnamese spoken sentences with named entities marked. In the 7th sentence $\mathbf{x} = $ "*thời tiết vũng tàu ngày kia*", there are two named entities: $E_{\mathbf{x}} = \{e_1(location, 3, 4), e_1(datetime, 5, 6)\}$.

Table 1. Named entity types (i.e., command argument types) defined in VAV

NE types	Description
location	Location name, e.g., a province, city, river, ... or a specific address
datetime	Time or date expression, any form of time, date, or time and date
url	Uniform resource locator or hyperlink: e.g., a web page
cnumber	Contact number (phone/fax), any form of phone/fax number
cname	Contact name, e.g., names saved in the contact list
aname	Application name, e.g., dropbox, facebook, skype, evernote, ...
email	Email address
number	Number, e.g., integer, real, or percentage

NER can be seen as a segmentation problem for sequence data. In most previous studies, a machine learning model is trained to recognize named entities as chunks of texts in sentences [3,4]. In addition to text content, these models used information about part-of-speech (POS) tags of words. For languages like Vietnamese, word boundary must first be identified. Hence, these models actually use three kinds of information: word tokens, word boundary, and POS tags of segmented words. These add richer features to the models and, therefore, help to achieve better NER performance. We, on the other hand, did not use information about word boundary and POS tags for NER4VST. There are reasons for this.

Table 2. Sample natural (spoken) language commands with labeled named entities

[ngã tư sở] ***ở đâu*** *(**where is** [so intersection])*
đánh thức lúc *[7 giờ kém 15 phút sáng]* *(**wake up at** [quarter to 7 o'clock am])*
gọi số *[0903206714]* *(**call number** [0903206714])*
vào trang *[dân trí chấm com chấm vn]* *(**open page** [dan tri dot com dot vn])*
đặt lịch họp với ibm lúc *[9 giờ 15 sáng thứ tư tuần sau]*
*(**arrange a meeting with ibm at** [quarter pass 9 wednesday morning next week])*
tìm đường từ *[ga hà nội]* ***đến*** *[88 láng hạ]*
*(**find direction from** [hanoi station] **to** [88 lang ha])*
thời tiết *[vũng tàu]* *[ngày kia]*
*(**weather in** [vung tau] [the day after tomorrow])*
âm lịch *[hôm nay]* ***ngày bao nhiêu*** *(**what lunar date is** [today])*
gửi email cho *[anh nam]* *(**send email to** [anh nam])*
mở *[skype]* *(**open** [skype])*
gửi số điện thoại của *[yến]* ***cho*** *[dương vũ]*
*(**send the phone number of** [yen] **to** [duong vu])*

Firstly, performing word segmentation and POS tagging are also challenging for spoken texts. Secondly, if we attempt to do that, we have to annotate a lot of labeled data, this is expensive and time-consuming. Lastly, we want to keep the NER4VST model compact to fit and execute it right on mobile devices.

3.2 Machine Learning Approaches for NER4VST

For segmentation problems, linear-chained graphical models like conditional random fields (CRFs) [11] have been proven really effective. One of the advantages is that CRFs can encode the sequential dependencies between consecutive positions. However, in our work, we decided to use maximum entropy (MaxEnt) classification [2] because of several reasons. First, MaxEnt is suitable for sparse data like natural language [3,4,15,18]. Second, like CRFs, MaxEnt can also encode various rich and overlapping features at different levels of granularity. Third, we observed that sequential dependencies in Vietnamese spoken texts are not as strong as in written texts, therefore we do not really need sequential models like CRFs. Lastly, MaxEnt is very fast in training and inference in comparison with CRFs. Also, the MaxEnt model is really compact to run on mobile devices.

The MaxEnt principle is to build a classification model based on what have been known from data and assume nothing else about what are not known. This means MaxEnt model is the model having the highest entropy while satisfying constraints observed from empirical data. Berger et al. (1996) [2] showed that MaxEnt model has the following mathematical form:

$$p_\theta(y|x) = \frac{1}{Z_\theta(x)} \exp \sum_{i=1}^{n} \lambda_i f_i(x, y) \qquad (1)$$

where x is the data object that needs to be classified, y is the output class label. $\theta = (\lambda_1, \lambda_2, \ldots, \lambda_n)$ is the vector of weights associated with the feature vector $F = (f_1, f_2, \ldots, f_n)$, and $Z_\theta(x) = \sum_{y \in \mathcal{L}} \exp \sum_i \lambda_i f_i(x, y)$ is the normalizing factor to ensure that $p_\theta(y|x)$ is a probabilistic distribution. Feature in MaxEnt is defined as a two-argument function: $f_{<cp, \, l>}(x, y) \equiv [cp(x)][y = l]$, where $[e]$ returns 1 if the logical expression e is *true* and returns 0 otherwise. Intuitively feature $f_{<cp, \, l>}(x, y)$ indicates correlation between a useful property, called *context predicate* (cp), of the data object x and an output class label $l \in \mathcal{L}$.

To build a MaxEnt model for NER4VST, we defined the set of class labels $\mathcal{L} = \{$ *b-location, i-location, b-datetime, i-datetime, b-url, i-url, b-cnumber, i-cnumber, b-cname, i-cname, b-aname, i-aname, b-email, i-email, b-number, i-number, o* $\}$. The *b-<ne type>* indicates the first token of a named entity and *i-<ne type>* is the next or the last token of that named entity. The label *o* indicates outside of named entities. This is the IOB2 format, a common label representation for sequential labeling and segmentation problems.

Training or estimating parameters for MaxEnt model is to search the optimal weight vector $\theta^* = (\lambda_1^*, \lambda_2^*, \ldots, \lambda_n^*)$ that maximizes the conditional entropy $H(p_\theta)$ or maximizes the log-likelihood function $L(p_\theta, \mathcal{D})$ with respect to a training data set $\mathcal{D}$. Because the log-likelihood function is convex, the search for the global optimum is guaranteed. Recent studies have shown that quasi-Newton methods like L-BFGS [12] are more efficient than the others. Once trained, the MaxEnt model will be used to predict class labels for new data. Given a new object x, the predicted label is $y^* = \text{argmax}_{y \in \mathcal{L}} \, p_{\theta^*}(y|x)$.

3.3   Feature Templates for Training NER4VST Segmentation Model

Features are important part of NER4VST model. We attempted to incorporate in the model a variety of highly discriminative features like: n-grams, regular expressions, dictionaries, and combinations of them as listed in Table 3.

The first feature type is n-gram. The top of Table 3 shows the context predicate templates for generating 1-grams, 2-grams, and 3-grams. We used a window of size 5 sliding along text sentence. The leftmost index is -2, the next token is -1, the centered (the current) token is 0, the next one is 1, and the rightmost is 2. We took 1-grams, 2-grams (combinations of two consecutive tokens) and similarly 3-grams. The second feature type is regular expressions. We used certain regular expression patterns like *number, day, date, week, full–date, phone–number, street–name, url*, etc. For these patterns, we used sliding windows of different sizes (5, 7, 9, 11, 13) and in each window we combined 2, 3, 4, 5, 6, and 7 consecutive tokens for matching.

We also used a variety of dictionaries for looking-up features: **url** (popular URLs in Vietnam like *vnexpress.net* and *24 giờ chấm com* (*24h.com*)); **part–of–url** (a list of top and second-level domains: *.vn, .com*, and *edu.vn*); **address-word** (a list of words indicating address text like *đường* (*street*) and *ngã tư* (*intersection*)); **time–word** (a list of words about time expressions like *nửa đêm* (*midnight*) and *rưỡi* (*half*)); **day–word** (words about day expressions

Table 3. Feature templates for training the maxent NER4VST model

N-grams	Context predicate templates
1-grams	$[w_{-2}]$, $[w_{-1}]$, $[w_0]$, $[w_1]$, $[w_2]$
2-grams	$[w_{-2}w_{-1}]$, $[w_{-1}w_0]$, $[w_0w_1]$, $[w_1w_2]$
3-grams	$[w_{-2}w_{-1}w_0]$, $[w_{-1}w_0w_1]$, $[w_0w_1w_2]$

Reg. expressions	Text templates for matching regular expressions
is–number, is–day,	1–tokens: $-2\ldots2 \rightarrow [w_{-2}]$, $[w_{-1}]$, $[w_0]$, $[w_1]$, $[w_2]$
is–date, is–week,	2–tokens: $-2\ldots2 \rightarrow [w_{-2}\ w_{-1}]$, $[w_{-1}\ w_0]$, $[w_0\ w_1]$, $[w_1\ w_2]$
is–month, is–year,	3–tokens: $-2\ldots2 \rightarrow [w_{-2}\ w_{-1}\ w_0]$, $[w_{-1}\ w_0\ w_1]$, $[w_0\ w_1\ w_2]$
is–full–date	4–tokens: $-3\ldots3 \rightarrow [w_{-3}\ w_{-2}\ w_{-1}\ w_0]$ $\ldots$ $[w_0\ w_1\ w_2\ w_3]$
is–phone–number,	5–tokens: $-4\ldots4 \rightarrow [w_{-4}\ w_{-3}\ w_{-2}\ w_{-1}\ w_0]$ $\ldots$
is–street–number,	$[w_0\ w_1\ w_2\ w_3\ w_4]$
is–email, is–url,	6–tokens: $-5\ldots5 \rightarrow [w_{-5}\ w_{-4}\ w_{-3}\ w_{-2}\ w_{-1}\ w_0]$ $\ldots$
	$[w_0\ w_1\ w_2\ w_3\ w_4\ w_5]$
	7–tokens: $-6\ldots6 \rightarrow [w_{-6}\ w_{-5}\ w_{-4}\ w_{-3}\ w_{-2}\ w_{-1}\ w_0]$ $\ldots$
	$[w_0\ w_1\ w_2\ w_3\ w_4\ w_5\ w_6]$

Dictionaries	Text templates for matching dictionaries
url, part–of–url,	1–tokens: $-2\ldots2 \rightarrow [w_{-2}]$, $[w_{-1}]$, $[w_0]$, $[w_1]$, $[w_2]$
address-word,	2–tokens: $-2\ldots2 \rightarrow [w_{-2}\ w_{-1}]$, $[w_{-1}\ w_0]$, $[w_0\ w_1]$, $[w_1\ w_2]$
day–word, time–word,	3–tokens: $-2\ldots2 \rightarrow [w_{-2}\ w_{-1}\ w_0]$, $[w_{-1}\ w_0\ w_1]$, $[w_0\ w_1\ w_2]$
period–word,	4–tokens: $-3\ldots3 \rightarrow [w_{-3}\ w_{-2}\ w_{-1}\ w_0]$ $\ldots$ $[w_0\ w_1\ w_2\ w_3]$
app–name,	5–tokens: $-4\ldots4 \rightarrow [w_{-4}\ w_{-3}\ w_{-2}\ w_{-1}\ w_0]$ $\ldots$
street–name,	$[w_0\ w_1\ w_2\ w_3\ w_4]$
organization–word,	6–tokens: $-5\ldots5 \rightarrow [w_{-5}\ w_{-4}\ w_{-3}\ w_{-2}\ w_{-1}\ w_0]$ $\ldots$
contact–word,	$[w_0\ w_1\ w_2\ w_3\ w_4\ w_5]$
location–word,	7–tokens: $-6\ldots6 \rightarrow [w_{-6}\ w_{-5}\ w_{-4}\ w_{-3}\ w_{-2}\ w_{-1}\ w_0]$ $\ldots$
province–name	$[w_0\ w_1\ w_2\ w_3\ w_4\ w_5\ w_6]$

Additional feature templates (regular expressions and dictionaries)
street number (reg. expression) + street name (dict.)
address word (dict.) + street name (dict.)
street number (reg. expression) + address word (dict.) + street name (dict.)

like *hôm nay* (*today*) and *ngày tới* (*the next day*)); **period–word** (words about time periods like *hàng ngày* (*everyday*) and *thứ tư hàng tuần* (*weekly on wednesday*)); **location–word** (words or phrases indicating location or direction like *định vị* (*locate*), *từ* (*from*), *đến* (*to*)); **app-name** (popular applications like *facebook*, *dropbox*, and *my talking tom*); **street–name** (street names in Vietnam: *trần hưng đạo* and *võ văn tần*); **province–name** (names of provinces/cities in Vietnam: *hà nội*, *vũng tàu*); **contact–word** (words related to contact names/numbers: *gọi cho* (*call to*) and *nhắn tin tới* (*sms to*)); **organization–word** (words that are prefixes of organizations, such as, *sân bay* (*airport*) or *ngân hàng* (*bank*)). We also combined regular expressions and dictionaries in order to capture long addresses: street number + street name, address word + street name, and street number + address word + street name.

4   Evaluation

4.1   Experimental Data and Settings

In order to evaluate the NER4VST model, we created a medium-sized dataset including 3,029 Vietnamese spoken text sentences. In this dataset all named entities of eight types described in Table 1 were annotated in XML format. A group of ten students were asked to use Google Now to speak natural language questions or commands (sample sentences are shown in Table 2) and obtain the output text sentences from this ASR service. The resulting sentences are then corrected to obtain the final Vietnamese spoken text dataset. We used the MaxEnt implementation – that is one component of the Java-based Text Processing Toolkit (JTextPro[3]) – for training and testing. We trained four MaxEnt models for 4-fold cross-validation tests. For each fold, we train a model with 200 L-BFGS [12] iterations and we also performed inference on the test data for evaluation. The experimental results will be reported in the next subsection.

4.2   Experimental Results and Analysis

Table 4 shows the experimental results of the 4th fold. *Human* is the number of manually annotated named entities in the test set. *Model* is the number of named entities predicted by the trained MaxEnt model. *Match* is the number of named entities correctly recognized by the MaxEnt model, i.e., the true-positive. The last three columns are the precision, recall, and F_1-measure calculated based on *Human*, *Model*, and *Match* values. We achieved the macro-averaged F_1-measure of 94.62 and the micro-averaged F_1-measure of 93.38. This is a significantly high performance because we did not use any information about punctuation, capitalization, word boundary, and part-of-speech tags. We carefully performed 4-fold cross validation evaluation and the results are reported in Table 5. The lowest and highest micro-averaged F_1 values are 91.06 and 93.38, respectively. The average value of the micro-averaged F_1-score of the four folds is 92.06.

Figure 2 gives us an insight into the performance of each named entity type. We calculated the average precision, recall, and F_1-measure of each type over the four folds. The NER4VST models achieved the highest recognition accuracy for the three classes cnumber, **email**, and **number** with the average F_1-score values of 98.04, 98.69, and 97.01, respectively. This is because these types have well-formed formats that can be easily captured using regular expressions. We also achieved a significant F_1 value (95.85) for application names (aname) thanks to look-up features based on a dictionary of popular mobile

---

[3] JTextPro: http://jtextpro.sourceforge.net.

Table 4. The precision, recall, and F_1-score of NE types of the best fold

NE types	Human	Model	Match	Precision	Recall	F_1-score
aname	296	302	289	95.70	97.64	96.66
datetime	159	159	152	95.60	95.60	95.60
cname	80	79	72	91.14	90.00	90.57
cnumber	31	32	31	96.88	100.00	98.41
location	145	141	119	84.40	82.07	83.22
url	46	49	45	91.84	97.83	94.74
email	6	6	6	100.00	100.00	100.00
number	21	20	20	100.00	95.24	97.56
Average$_{macro}$				**94.44**	**94.80**	**94.62**
Average$_{micro}$	784	788	734	**93.15**	**93.62**	**93.38**

Table 5. The results of the 4-fold cross validation evaluation

Folds	Human	Model	Match	Pre.$_{micro}$	Rec.$_{micro}$	F_1-score$_{micro}$
Fold 1	778	777	708	91.12	91.00	91.06
Fold 2	782	782	724	92.58	92.58	92.58
Fold 3	788	797	723	90.72	91.75	91.23
Fold 4	784	788	734	93.15	93.62	93.38
Average				**91.89**	**92.24**	**92.06**

native applications like *skype, calendar, candy crush saga*. However, the models are sometimes confused to distinguish between the name of a native application and the hyperlink of the web-based version of that application. For example, a user command like "*mở vnexpress*" can be interpreted as "open the native app named *vnexpress*" or "open the web page *vnexpress.net*". As stated earlier, ASR interprets URLs as long, discrete, and irregular text patterns (e.g., *thesaigon-times.vn* as "*the saigon times chấm vn*"). This diversity and irregularity lead to an average F_1-score of 92.78 for the url class. Date and time (datetime) is another hard class (average F_1 of 91.60). Datetime expressions in spoken texts are very diverse and long such as "*8 giờ 30 phút sáng thứ tư hàng tuần*" (*8:30 on wednesday morning every week*). There is no general regular expression to capture these datetime patterns. We used various rich local features to infer the whole datetime chunks.

Contact names (cname) in contact lists are even more ambiguous. Mobile users tend to save contacts in different ways. Sometimes, names are associated with the relationship, title, profession, or even the nickname of a person like "*chị thảo*" (*sister thảo*), "*thầy nguyễn văn thành*" (*teacher nguyen van thanh*), "*ngọc anh microsoft*", or "*nam doremon*". Contact name recognition from spoken texts, therefore, is much harder than personal name recognition in written

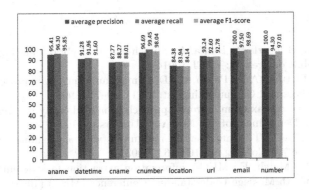

Fig. 2. The average precision, recall, and F_1-score of NE types over 4-fold CV tests

documents. For this reason the NER4VST models achieved an average F_1 value of 88.01 for this type. Among the eight named entity types, **location** is probably the most difficult-to-recognize class. In our work, a location is any form of a postal address or a specific location ranging from a bridge, an intersection, a street, a building, a university, a shopping mall ... to a city or a country. Some location is very long like *"ngã tư nguyễn trãi khuất duy tiến"* (*the intersection between nguyễn trãi and khuất duy tiến streets*). Also, there is no regular expressions or dictionary can cover the diversity of location names. That is why we achieved the lowest F_1-score (84.14) for this class. Fortunately, in practice we can perform well in most cases and that is why NER4VST models can be applied in VAV. We also conducted experiments with Conditional Random Fields (CRFs) [11] and the results are about 2 % higher than those of MaxEnt.

5 Discussion

In this paper, we have presented a lightweight machine learning approach to the problem of NER4VST. To overcome the challenges derived from the nature of spoken language, we incorporated into the NER4VST models a wide spectrum of features ranging n-grams, regular expressions to look-up dictionaries. We decided to use MaxEnt classification to build the NER4VST models rather than using complex graphical models like CRFs to keep the models fast and compact. Also, we did not rely on word boundary and POS tags because these kinds of information are expensive and time-consuming to acquire. The experimental results show that with limited labeled training data and limited information, we can obtain an average F_1 value of 92.06, a significantly high performance for a difficult task like NER4VST. The trained NER4VST models were also integrated into VAV, a virtual assistant for Vietnamese that allows mobile users to interact with their smartphones using natural spoken questions and commands.

Acknowledgment. This work was supported by the project QG.15.29 from Vietnam National University, Hanoi (VNU).

References

1. Angelov, K., Bringert, B., Ranta, A.: Speech-enabled hybrid multilingual translation for mobile devices. In: EACL (2014)
2. Berger, A., Pietra, S.A.D., Pietra, V.J.D.: A maximum entropy approach to natural language processing. Comput. Linguist. **22**(1), 39–71 (1996)
3. Borthwick, A.: A maximum entropy approach to named entity recognition. Ph.D. dissertation, Department of CS, New York University (1999)
4. Chieu, H.L., Ng, H.T.: Named entity recognition with a maximum entropy approach. In: The 7th CoNLL, pp. 160–163 (2003)
5. Chinchor, N., Marsh, E.: MUC-7 information extraction task definition (version 5.1). In: The 7th Message Understanding Conference (MUC) (1998)
6. Graves, A., Jaitly, N.: Towards end-to-end speech recognition with recurrent neural networks. In: ICML (2014)
7. Grishman, R., Sundheim, B.: Message understanding conference 6: a brief history. In: The 6th Message Understanding Conference (MUC-6) (1995)
8. Hatmi, M., Jacquin, C., Morin, E., Meignier, S.: Named entity recognition in speech transcripts following an extended taxonomy. In: SLAM (2013)
9. Hannun, A., Case, C., Casper, J., Catanzaro, B., Diamos, G., Elsen, E., Prenger, R., Satheesh, S., Sengupta, S., Coates, A., Ng, A.Y.: Deep speech: scaling up end-to-end speech recognition (2014). arXiv:1412.5567v2
10. Hinton, G., Deng, L., Yu, D., Dahl, G., Mohamed, A., Jaitly, N., Senior, A., Vanhoucke, V., Nguyen, P., Sainath, T., Kingsbury, B.: Deep neural networks for acoustic modeling in speech recognition. IEEE Signal Process. Mag. **29**, 82–97 (2012)
11. Lafferty, J.D., McCallum, A., Pereira, F.: Conditional random fields: probabilistic models for segmenting and labeling sequence data. In: ICML, pp. 282–289 (2001)
12. Liu, D., Nocedal, J.: On the limited memory BFGS method for large-scale optimization. Math. Program. **45**, 503–528 (1989)
13. Molla, D., Zaanen, M., Cassidy, S.: Named entity recognition in question answering of speech data. In: The Australasian Language Technology Workshop (2007)
14. Nguyen, C.T., Tran, T.O., Phan, X.H., Thuy, H.Q.: Named entity recognition in Vietnamese free–text and web documents using CRFs. In: ADD (2007)
15. Nigam, K., Lafferty, J., McCallum, A.: Using maximum entropy for text classification. In: IJCAI Workshop on Machine Learning for Information Filtering, pp. 61–67 (1999)
16. Pan, Y.C., Liu, Y.Y., Lee, L.S.: Named entity recognition from spoken documents using global evidences and external knowledge sources with applications on mandarin chinese. In: IEEE Automatic Speech Recognition and Understanding (2005)
17. Popkin, J.: Google, Apple Siri and IBM Watson: the future of natural-language question answering in your enterprise. Gartner Technical Professional Advice (2013)
18. Ratnaparkhi, A.: A maximum entropy model for part-of-speech tagging. In: The Empirical Methods in Natural Language Processing Conference (1996)
19. Tur, G., Mori, R.D.: Spoken Language Understanding: Systems for Extracting Semantic Information from Speech. Wiley, New York (2011)

Explorations of Prosody in Vietnamese Language

Tang Ho Lê[1(✉)] and Anh-Viêt Nguyên[2]

[1] Université de Moncton, Moncton, Canada
letangho@yahoo.ca
[2] Essence Network Informatic J. S. C., Ho Chi Minh City, Vietnam

Abstract. In this paper, we attempt to analyze the intonation of Vietnamese prosody in order to produce a natural Vietnamese synthesizer. We continue to go into the advanced research of the intonation (intensity, altitude and duration). We study the pitch change between words in a phrase, the "swallowing sound" phenomenon at the junction of two words, the simulations of the duration of words in the Vietnamese TTS and the echo problem in processing the change of word's duration.

Keywords: Vietnamese language · Text to speech (TTS) · Prosody · Intonation · Intensity · Altitude · Duration

1 Introduction

In a previous study [1], we have aborded the impact and role of intonation in Vietnamese. Accordingly, we have split the meaning of "Prosody" from the sense of Western into 2 parts: word-prosody (effect on the individual words - local effect) and sentence-prosody (effect on the whole statement - global effect). We went into the advanced research of the prosody and suppose that in Vietnamese language, the most obvious factor which effects to Vietnamese prosody is the duration. This is different from polyphonic languages (English, French, etc.) in which the intonation and intensity of prosody are extremely important in making natural voices.

In this report, we only study the altitude and the intensity factors of tone, and the effects of them in the phrase to make a more natural voice. Precisely, we have not yet considered the sentence-prosody in a full sentence, because in the polyphonic language where questions pronounced by a rising tone at the end, that raises up the pitch of that word. In Vietnamese, we can't do that because it will change the accent of the word making a different meaning or have no meaning at all. We also do not study the context sensitive voice because it must be done with the semantic analysis to understand the pronounced words, such as "to speak with a low tone", "to whisper", "to talk confidentially" or "to speak in a tone of command", etc. Moreover, to do so, instead of changing the elements of sound (intensity, altitude and duration) we also need the special PUs (phonetic units), and they must be recorded separately.

© Springer-Verlag Berlin Heidelberg 2016
N.T. Nguyen et al. (Eds.): ACIIDS 2016, Part I, LNAI 9621, pp. 181–189, 2016.
DOI: 10.1007/978-3-662-49381-6_18

2 Comparing the Concatenation Method Between the Vietnamese and the Polyphonic Languages

In our recent research [1], when we synthesize the Vietnamese by the concatenating method, we record only 1984 basic phonetic units (called PU or diphones) and concatenate 2 PUs to generate about 12,000 Vietnamese simple words (e.g.: là + àm = làm). While in a polyphonic language, a word can be concatenated from many PU (e.g.: lo-co-mo-ti-ve = locomotive).

In concatenating a pair of Vietnamese PU together, we have cut the first part of the first PU to merge it with the second part of the second PU. In doing so we have avoided a difficult problem, that is the two factors (volume and pitch) at the juncture of two PUs are already equal, or easily be adjusted so that they are equal. We just decided the duration (by percentage) of each PU when splicing and merging them. While in polyphonic languages, this adjustment is very complex (one must decide the stress, the duration and intensity of each PU).

Now, considering the Vietnamese prosody, we encounter similar problems as with the polyphonic languages. That is, after the decision on the appropriate duration for each word in the phrase, we have to adjust the pitch of each word. However, this problem only occurs in the joint between two words in a phrase or in a compound words, and the change is relatively small for a very short time (only about 5-10 ms), but its effect is very clear (to make more natural voice rather than disjointed and jerky voice). This is the reason that we must consider this juncture: if two words were spoken continuously (they are spoken stuck together), then the pitch of the following word will be affected by the tail of the previous word, sometime this may cause the echo making it very difficult to hear.

3 The Pitch Change Between Words in a Phrase

In Vietnamese, we have up to 6 accents (6 levels of intonation). Therefore, it depends on the accent of the two words that we must adjust their pitches.

3.1 Research

The Table 1 below shows all cases of adjusting the pitch after many experimentations:

If we do not adjust these parameters, the voice is audible but not natural, because the difference between the tail of the previous word and the head of the next word is almost equal to zero (since we recorded each PU separately). To adjust the pitch, we have to know about the pitch of each word which is difficult to hear. Unfortunately, the program tool for the speech spectrum only shows the pitch by the density of the sound waves, which is difficult to recognize. Therefore, we need to use an appropriate tool such as Xitona [3] software (Fig. 1).

Table 1. Adjusting the pitch for the first part of the second word (that is the current word) depending on the accent of the previous word.

Pre.	Cur. (Adjust pitch for the first part of word)					
	Không	Sắc	Huyền	Hỏi	Ngã	Nặng
Không		Inc..1	Dec..1	Inc..1	Inc..1	Inc..1
Sắc	Dec..1		Dec..2	Inc..1	Inc..1	Inc..1
Huyền	Inc..1	Inc..2		Dec..1	Inc..1	Inc..1
Hỏi	Dec..1	Inc..1	Dec..2			Inc..1
Ngã	Dec..1	Inc..1	Dec..2			Inc..1
Nặng	Inc..1	Inc..1	Dec..1	Inc..1	Inc..1	

Inc..2 = Increment 15 % of pitch

Inc..1 = Increment 9 % of pitch

Blank = Not change

Dec..1 = Decrement 9 % of pitch

Dec..2 = Decrement 15 % of pitch

Pre.: Previous word Cur.: Current word Next: Next word

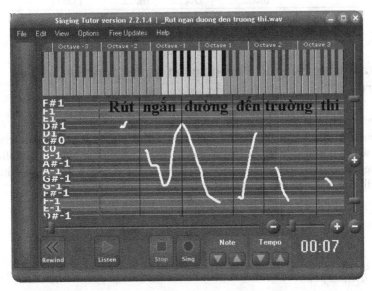

Fig. 1. In this graph, the white line illustrates the pitch of the phrase "Rút ngắn đường đến trường thi" (*Reduce the path to the school*).

3.2 Technique of Changing Pitch at the Junction Between Two Words

According to the Tables 1 and 2 above, we conduct an experiment with the phrase "Rút ngắn đường đến trường thi" to change the pitch for the **tail** of the previous words:

"Rút	ngắn	đường	đến	trường	thi"
	Dec..2	Inc..2	Dec..2	Inc..1	

Table 2. Adjusting the pitch for the last part of the current word depending on the accent of the next word.

Không	Sắc	Huyền	Hỏi	Ngã	Nặng	Next
	Inc..1	Dec..1				Không
Dec..1		Dec..2			Dec..1	Sắc
Inc..1	Inc..2		Inc..1	Inc..1	Inc..1	Huyền
Dec..1		Dec..1				Hỏi
Dec..1		Dec..1			Dec..1	Ngã
Dec..1	Inc..1	Dec..1				Nặng

(table header: Cur. (Adjust pitch for the last part of word))

To change the head of the following words:

"Rút	ngắn	đường	đến	trường	thi"
		Inc..2	Dec..2	Inc..2	Dec..1

4 "Swallowing Sound" Phenomenon at the Junction of Two Words

While we investigate about the junction between two words, we found a special phenomenon which is called "word swallowing". This phenomenon occurs when the previous words is ended with a strong sonant [2] (such as a - o - e - ê) and the following word is begun with a sensitive consonant [2] (such as b, c, ch, d, k, kh, p, t, tr), then this sonant will cause the following word's consonants to disappear (Table 3).

Table 3. Example of three phrases

Phrase	Sonant	Sensitive consonant
tạo **ra-từ** (*created from*)	a	t
sinh vật nơi đây thật phong phú và **đa-dạng** (*really riche and multiforms*)	a	d
cả-trên biển và núi (*both on the sea and on the mountain*)	ả	tr

In order to avoid this problem, we just insert a silence about 5 ms so that all words could be heard clearly. For examples (Fig. 2):

With the phrase "Rút ngắn đường đến trường thi" (*Reduce the path to the school*), the Figs. 4 and 5 illustrate a sound spectrum **before** and after changing the pitch at the junction between two words, analyzed by Praat [4] software (Fig. 3):

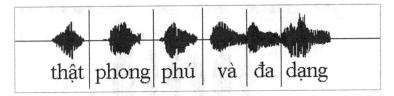

Fig. 2. The sound spectrum of the phrase " thật phong phú và đa dạng "

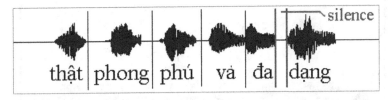

Fig. 3. The sound spectrum of that one after the silence's inserting

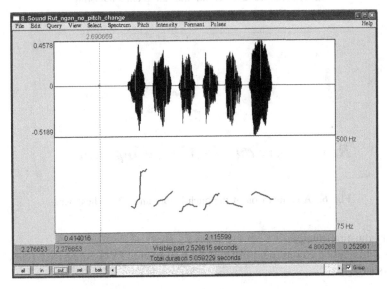

Fig. 4. The sound spectrum before the pitch's adjusting.

Audio files of a paragraph both before and after pitch adjustment may refer to the URL (Fig. 6):

– No adjustment pitch:
  https://drive.google.com/file/d/0Byzy9ZhFKHPuYnpNWjRHS0RBSW8/view?
  usp=sharing

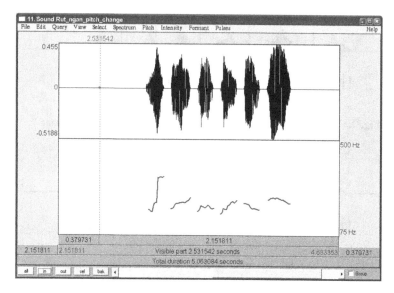

Fig. 5. The sound spectrum after the pitch's adjusting.

Fig. 6. A comparison of the pitch before and after adjustment.

– After adjusting pitch:
  https://drive.google.com/file/d/0Byzy9ZhFKHPuSzQ2YVQxSmpMNEk/view?
  usp=sharing

5 Normalizing of the Word's Duration

When we synthesize Vietnamese by using concatenating method, in order to make
synthesized voice more natural, we have to normalize the word's duration. According
to [2, 6], the Vietnamese words have different durations depending on their components. Especially, some words containing sonants will have the largest lengths.

We calculate and test the duration of each word by using the below formula, with L being the length of the element in milliseconds (ms):

$$L_W = L_{FC} + L_{MV} + L_{LC} \qquad (1)$$

W : Word   FC : First Consonant (L1)
MV : Middle Vowel (L2)   LC : Last Consonant (L3)

For words which does not contain all of the 3 components such as "ba" (without LC), "an" (without FC) or "oa" (without both FC and LC), we use this formula instead (by adding the average length for L1 and L3) (Table 4):

Table 4. A formular to normalize the word's length in all cases.

Category	Word	L1	L2	L3
Has **3 parts**	nam	L_{FC}	L_{MV}	L_{LC}
Without **LC**	na_	L_{FC}	L_{MV}	36
Without **FC**	_am	20	L_{MV}	L_{LC}
Without both **FC and LC**	_a_	36	L_{MV}	45

The Table 5 shows some vowels and consonants's average length.

Table 5. Length of some consonants and vowels.

Consonant	Length	Consonant	Length	Vowel	Length
Ch	97	B	62	Iêu	62
Gi	73	C	62	Yêu	62
Gh	70	D	51	Oai	51
Kh	86	Đ	49	Oao	49
Ng	86	G	48	Ai	48
Nh	86	H	47	Ao	47
Ph	73	J	47	Au	47
Qu	62	K	47	A	47
Tr	85	L	65	Ă	65
Th	79	M	47	Â	47

This is a statistical comparison of some words' duration spoken by a broadcaster (PTV) and their durations are simulated by our sound synthesis (TTS):

The table below is a statistical comparison of some words' duration (in ms) spoken and recorded by a broadcaster and their durations are used to normalize the Phonetic Unit in our voice synthesizer (TTS) (Table 6):

Original wave files read by a Broadcaster can be listened or downloaded from this URL:

https://drive.google.com/open?id=0Byzy9ZhFKHPuNlowTUg2OHUtMVU

Table 6. Duration of words spoken by a broadcaster and our TTS:

Word	Cù	lao	Chàm	nằm	trong	địa	bàn	Hành	chính	của	xã	Tân	Hiệp
PTV (ms)	92	182	313	209	174	90	249	167	184	107	193	130	179
TTS (ms)	114	149	253	218	144	136	226	162	171	125	196	124	136
Compare (%)	23.9	−18.1	−19.2	4.3	−17.2	51.1	−9.2	−3.0	−7.1	16.8	1.6	−4.6	−24.0

TTS wave files which we made before normalizing can be listened to or down-loaded from the URL:
https://drive.google.com/file/d/0Byzy9ZhFKHPuQTNLU2ZfN3dtbVE/view?usp=sharing
TTS wave files that we perform after normalizing can be listened or downloaded from the URL:
https://drive.google.com/file/d/0Byzy9ZhFKHPuZGdRQzdjT0NiX2s/view?usp=sharing.

6 Reducing the Echo in Processing the Word's Duration

Nowadays, there are several methods which can be used to stretch the duration of the sound such as PSOLA [5], WSOLA [7]. However, the echo is going to appear if the stretching (prolonging or shortening) is great. To overcome this obstacle, we partition each of the words from 3 to 5 segments depending on the properties of words. After stretching each segment, we merge all of them into one word to minimize the echo while synthesizing Vietnamese (Fig. 7).

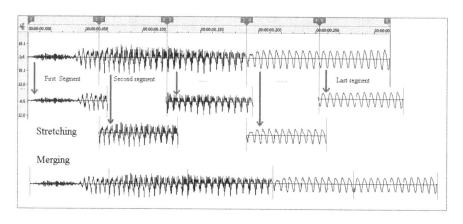

Fig. 7. How to minimize the echo while synthesizing Vietnamese.

The TTS wave files with no splitting segment can be heard or downloaded from the URL:
https://drive.google.com/file/d/0Byzy9ZhFKHPucklSSF8yN3FtMDg/view?usp=sharing
The TTS wave files in which each word is splitting into segments to reduce the echo can be heard or downloaded from the following URL:
https://drive.google.com/file/d/0Byzy9ZhFKHPuMk5DT0U5SjVuYjA/view?usp=sharing.

7 Future Research

In this paper, we only analyze and make the change of the pitch in some general cases listed in Tables 1 and 2 (based on the previous and the following word's accent). In the future, we will continue to statistically analyze for more details and try to formulate some rules so that the change of the pitch of the previous word's accent and the following word could be more appropriate.

References

1. Lê, T.-H., Nguyen, A.-V., Truong, H.V., Van Bui, H., Lê, D.: A Study on Vietnamese Prosody. In: Nguyen, N.T., Trawiński, B., Jung, J.J. (eds.) New Challenges for Intelligent Information and Database Systems. SCI, vol. 351, pp. 63–73. Springer, Heidelberg (2011). doi:10.1007/978-3-642-19953-0_7. http://www.researchgate.net/publication/220963896_A_Study_on_Vietnamese_Prosody
2. Lê, T.-H.: A Research on Vietnamese Phonetic - Khảo cứu ngữ âm Việt Nam. http://vietsciences.free.fr/vietnam/tiengviet/daytiengviet01.htm
3. Xitona. "Singing Tutor Duet" Software– URL:http://www.xitona.com/
4. Praat. "Doing phonetics by computer" – URL:http://www.fon.hum.uva.nl/praat/
5. Douglas O'Shaughnessey, Speech Communications – Human and Machine, Second Edition, IEEE Press, 2000
6. Lê Hồng Minh, et al.: Nghiên cứu và xây dựng phần mềm "Tự động đọc văn bản tiếng Việt" bằng phương pháp tổng hợp Formant. http://117.3.71.125:8080/dspace/bitstream/DHKTDN/1876/1/4911.pdf
7. Verhelst, W., Roelands, M.: An overlap-add technique based on waveform similarity (WSOLA) for high quality time-scale modification of speech. In: ICASSP (1993)

Identifying User Intents in Vietnamese Spoken Language Commands and Its Application in Smart Mobile Voice Interaction

Thi-Lan Ngo[1,2(✉)], Van-Hop Nguyen[2], Thi-Hai-Yen Vuong[2],
Thac-Thong Nguyen[2], Thi-Thua Nguyen[2], Bao-Son Pham[2],
and Xuan-Hieu Phan[2]

[1] University of Information and Communication Technology, Thai Nguyen University,
Thai Nguyen, Vietnam
[2] University of Engineering and Technology, Vietnam National University,
Hanoi, Vietnam
{ntlan,hopnv_57,yenvth_57,thongnt_57,thuant.mi13,sonpb,hieupx}@vnu.edu.vn

Abstract. This paper presents a lightweight machine learning model
and a fast conjunction matching method to the problem of identifying user intents behind their spoken text commands. These model and
method were integrated into a mobile virtual assistant for Vietnamese
(VAV) to understand what mobile users mean to carry out on their smartphones via their commands. User intent, in the scope of our work, is an
action associated with a particular mobile *application*. Given an input
spoken command, its *application* will be identified by an accurate classifier while the *action* will be determined by a flexible conjunction matching algorithm. Our classifier and conjunction matcher are very compact
in order that we can store and execute them right on mobile devices.
To evaluate the classifier and the matcher, we annotated a medium-
sized data set, conducting various experiments with different settings,
and achieving impressive accuracy for both the application and action
identification.

Keywords: Natural spoken language understanding · User intent identification · Vietnamese spoken text processing · Mobile virtual assistant

1 Introduction

With the recent major advances in speech recognition technology [8–10], voice
interaction has become an alternative way for mobile users to interact with their
smart devices. We can easily see this trend via the application of speech-2-speech
translation[1] [1], the growing adoption of in-car voice control, call-center automation, and the popularity of mobile virtual assistants like Apple Siri, Google Now,
and Microsoft Cortana [12]. These applications deal with different issues but all

---

[1] Microsoft Skype Translator and AT&T Speech-to-Speech Translation.

© Springer-Verlag Berlin Heidelberg 2016
N.T. Nguyen et al. (Eds.): ACIIDS 2016, Part I, LNAI 9621, pp. 190–201, 2016.
DOI: 10.1007/978-3-662-49381-6_19

of them have to solve two major problems: automatic speech recognition (ASR) and spoken language understanding [15]. While the former has a long history of research and development, the latter is still at its early stage.

Unfortunately, parsing and understanding spoken texts are much more challenging than working with written documents. The output texts of ASR services are commonly short, less grammatical, all in lowercase, and even have no punctuation marks. In addition, ASR commonly interprets numbers as alphabetic texts and recognizes continuous text chunks (e.g., email addresses, hyperlinks) as discrete tokens. These inherent characteristics of spoken texts are obviously big obstacles for deep spoken text analysis and understanding.

In this work, we will present our effort to understand spoken text commands from mobile users, that is, we will propose a two-step method to identify the intent behind their spoken commands. According to Bratman, intent or intention, in the general sense, is a mental state that represents a commitment to carrying out an action or actions in the future [7]. In scope of this study, we consider and represent a user intent of a spoken command as a pair of an *application* and an *action*. For example, a spoken command like *"đặt lịch họp với công ty microsoft vào 8 giờ 30 sáng thứ 4 tuần sau"* (*arrange a meeting with microsoft at eight thirty on next wednesday morning*) means that we need to insert or add (the action) an entry about the meeting into the calendar (the application). Our method takes the command sentence as input and classifies it into one of a pre-defined set of applications (see Table 4 for the 23 applications) and then matching the command together with its application with a list of pre-defined conjunctions to determine the action (see Table 5 for a list of 53 actions).

There have been several studies working on transforming spoken language sentences or commands into structured and actionable forms so far. Wit.ai[2] is a platform that converts natural language sentences into structured forms that can be applied for robot control, wearable devices, home automation, and internet-of-thing applications. Bastianelli et al. (2014) proposed a method to understand natural language for human-robot interaction [2]. Branavan et al. (2009, 2010) presented a reinforcement learning method for mapping natural language instructions to actions [5,6]. Tellex et al. (2011) also proposed a natural language understanding technique for robotic navigation and mobile manipulation [14].

Our work is different from those previous studies in several points and these are also our main contributions. First, we propose a definition of user intent for human-mobile voice interaction commands. Second, to the best of our knowledge, this is the first attempt to identify user intents in Vietnamese spoken texts. Third, we propose a lightweight approach with an accurate classifier and a flexible conjunction matcher to perform application and action identification. To deal with the obstacles in spoken text processing, we came up with a variety of rich and sophisticated features to make the classifier really robust. Also, the classifier and the matcher are fast and compact to be executed right on mobile devices. We also built a medium-sized labeled data set for evaluation and analysis. The classifier built on this data set and the conjunction matcher were applied

---

[2] Wit.ai: https://wit.ai.

in our mobile virtual assistant for Vietnamese (VAV[3]). VAV allows mobile users to do a lot of tasks via spoken natural language commands like setting an alarm, arranging a meeting, finding a direction from point A to point B on the map, searching a contact, asking weather information, finding and playing a song and many more. The experimental results show that our solution can meet important requirements: compact, fast, and highly accurate for real-world applications.

2 User Intents in Spoken Texts: Definition and Application

As stated earlier, Bratman (1987) defined intent or intention as a mental state that represents a commitment to carrying out an action or actions in the future [7]. In scope of this work, we focused on the user-mobile interaction domain, that is, capturing the user intents behind their spoken text commands.

In our study, a user intent of a spoken text command **c**, denoted **I(c)**, is defined as a pair of an application a and an action f associated with a.

$$\mathbf{I}(\mathbf{c}) = \langle a, f \rangle \tag{1}$$

Table 1 shows several examples of user spoken commands and their intents. The command "*vào trang dân trí chấm com chấm vn*" (*open page dantri dot com dot vn*) means using the web browser to open the web page "*dân trí chấm com chấm vn*". Thus, the intent should be $\langle browser, open \rangle$ or *browser::open*. The user

Table 1. Sample Vietnamese spoken language commands and their user intents

Vietnamese spoken language command	Intent ($a::f$)
ngã tư tây sơn chùa bộc ở đâu (*where is tay son chua boc intersection*)	map::locate
tìm đường từ hà đông đến 88 láng hạ (*find direction from ha dong to 88 lang ha*)	map::find–direction
gọi số 0903206714 (*call number 0903206714*)	phone::call
đặt lịch họp với ibm 9 giờ 15 sáng thứ tư tuần sau (*arrange a meeting with ibm at quarter pass 9 next wednesday morning*)	calendar::set
vào trang dân trí chấm com chấm vn (*open page dan tri dot com dot vn*)	browser::open
mở bài hạ trắng (*play song ha trang*)	music::open
thời tiết vũng tàu ngày kia (*the weather in vung tau the day after tomorrow*)	weather::query
hôm nay ngày bao nhiêu âm lịch (*what lunar date is today*)	calendar::query
gửi số điện thoại của yến cho dương vũ (*send phone number of yen to duong vu*)	contact::share
mở twitter (*open twitter*)	other–app::open

---

[3] VAV: https://play.google.com/store/apps/details?id=com.mdnteam.vav

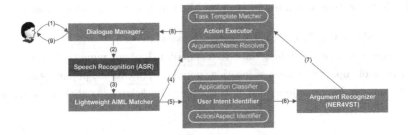

Fig. 1. The overall architecture of the mobile virtual assistant for Vietnamese (VAV). The **user intent identifier** module is a core component in VAV.

input *"ngã tư tây sơn chùa bộc ở đâu"* (*where is the tay son chua boc intersection*) is actually a question. However, the intent behind this question is to locate the *tay son chua boc intersection* on the map (*map::locate*).

In our work, we defined a set of 23 mobile applications or features like *phone*, *sms*, *calendar*, *email*, etc. (see Table 4 for a complete list of applications). We have a special class called *other–app* that refers to any other applications installed on the user device, such as *facebook*, *dropbox*, *candy crush saga*, etc. We also defined a list of actions associated with each application, such as *open*, *set*, *query*, *add*, *turn–on*, ... (see Table 5 for a complete list of actions). We combined applications and actions to obtain a list of totally 53 intents as shown in Table 5.

As mentioned earlier, user intent identification is one of the core components of the VAV. The general architecture of the VAV application is shown in Fig. 1. In order to understand and execute a user spoken command, VAV first records the user voice command and converts to a spoken text sentence using an ASR service. The spoken text command is then analyzed to capture the intent that the user meant to do as well as to extract any necessary parameters for execution. Given a command *"tìm đường từ hà đông đến 88 láng hạ"* (*find direction from ha dong to 88 lang ha*), VAV needs to determine its proper intent as map::find–direction. In addition, finding a direction requires a source and a destination as arguments. Therefore, VAV will ask the **argument recognizer** to recognize the source *"hà đông"* and destination *"88 láng hạ"* in the command. The resulting intent and arguments are finally sent to the **action executor** for execution. There are totally six core components in the VAV. However, we will only present the technical details of the **user intent identifier** module. As depicted in Fig. 1, this module consists of two sub-modules: *Application Classifier* for application identification and *Action/Aspect Identifier* for action recognition.

3   Identifying User Intent from Vietnamese Spoken Texts

Understanding a user intent $\mathbf{I}(\mathbf{c})$ behind a spoken text command $\mathbf{c}$ is to identify the pair $\langle a, f \rangle$ as defined in Eq. 1. In this paper, we proposed a two-step solution

Table 2. Feature templates for training MaxEnt-based application classifiers

1-gram (word tokens), 2-gram (two consecutive word tokens)
regular expressions: for matching date–time, email, url, phone number, location ...
dictionary look-up: dictionaries of location, popular applications, popular songs ...
conjunction matching: conjunctions of words that are skeletons of spoken commands

to determine the application a and the action f. The subsequent sections will describe how we built the application classifier and the conjunction matcher.

3.1   Application Classification with Maximum Entropy

Basically, we can use any classification method for building a classifier. However, we decided to use maximum entropy (MaxEnt) for several reasons. First, MaxEnt is suitable for sparse data like natural language [3,4,13]. Second, MaxEnt can encode a variety of rich and overlapping features at different levels of granularity for better classification. Lastly, MaxEnt is very fast in training and inference, and its performance is competitive with most advanced statistical learning methods. The principle behind MaxEnt is to build a classifier having the highest entropy while satisfying all constraints (features) observed from the empirical data. MaxEnt model has the following mathematical form [3]:

$$p_\theta(y|x) = \frac{1}{Z_\theta(x)} \exp \sum_{i=1}^{n} \lambda_i f_i(x, y) \qquad (2)$$

where x is input object, y is the classified label. $\theta = (\lambda_1, \lambda_2, \dots, \lambda_n)$ is the vector of weights associated with the feature vector $F = (f_1, f_2, \dots, f_n)$. $Z_\theta(x) = \sum_{y \in \mathcal{L}} \exp \sum_i \lambda_i f_i(x, y)$ is the normalizing factor. Training MaxEnt model is to search for the optimal $\theta^* = (\lambda_1^*, \lambda_2^*, \dots, \lambda_n^*)$ that maximizes the entropy function $H(p_\theta)$ or maximizes its convex log-likelihood function $L(p_\theta, \mathcal{D})$ for training set $\mathcal{D}$. Recent studies have shown quasi-Newton methods like L-BFGS [11] are fast and efficient. Once trained, the MaxEnt model will be used to predict class labels for new data. Given a new object x, the predicted label is $y^* = \text{argmax}_{y \in \mathcal{L}} \, p_{\theta^*}(y|x)$.

In order to train the MaxEnt models, we used different of feature templates from the data. Table 2 shows four types of features used for our classifiers.

3.2   Action Identification with Conjunction Matching

After performing application classification using the *Application Classifier* (see Fig. 1) for a given command **c**, we will obtain its application $a_\mathbf{c}$. Both **c** and $a_\mathbf{c}$ will be sent to the *Action/Aspect Identifier* to determine the action $f_\mathbf{c}$. In order to identify $f_\mathbf{c}$, we proposed a fast and efficient conjunction matching method. The main idea is that for each user intent $a{::}f$, we manually created a list of conjunctions associated with it. Each conjunction consists of a set of matching

Table 3. Sample conjunctions for identifying the action/function of user intents

Conjunction string: <raw conjunction string> $\mathbf{r} = \langle a_{\mathbf{r}}(\text{app.}), f_{\mathbf{r}}(\text{action}), p_{\mathbf{r}}(\text{priority}), b_{\mathbf{r}}(\text{begin}), e_{\mathbf{r}}(\text{end}), \mathbf{s_r}(\text{segments}), \mathbf{t_r^u}(\text{terms}) \rangle$
Conjunction string: "calendar, set, 1–^ *đặt+lịch*" $\mathbf{r}_1 = \langle \text{calendar}, \text{set}, 1, \textit{true}, \textit{false}, (\textit{đặt lịch}), \{(\textit{đặt}, 1), (\textit{lịch}, 1)\} \rangle$
Conjunction string: "weather, query, 1–^ *thời+tiết ở*" $\mathbf{r}_2 = \langle \text{weather}, \text{query}, 1, \textit{true}, \textit{false}, (\textit{thời tiết}, \textit{ở}), \{(\textit{thời}, 1), (\textit{tiết}, 1), (\textit{ở}, 1)\} \rangle$
Conjunction string: "calendar, query, 1–*âm+lịch bao+nhiêu*$" $\mathbf{r}_3 = \langle \text{calendar}, \text{query}, 1, \textit{false}, \textit{true}, (\textit{âm lịch}, \textit{bao nhiêu}), \{(\textit{âm}, 1), (\textit{lịch}, 1),$ $(\textit{bao}, 1), (\textit{nhiêu}, 1)\} \rangle$
Conjunction string: "map, find–direction, 1–^ *tìm+đường từ đến*" $\mathbf{r}_4 = \langle \text{map}, \text{find–direction}, 1, \textit{true}, \textit{false}, (\textit{tìm đường}, \textit{từ}, \textit{đến}), \{(\textit{tìm}, 1), (\textit{đường}, 1),$ $(\textit{từ}, 1), (\textit{đến}, 1)\} \rangle$

Algorithm 1. Conjunction indexing

1: **procedure** INDEXCONJUNCTIONS($\mathbf{R}$)
2:    create a map: $\langle t \Rightarrow \langle \mathbf{r} \Rightarrow x_{\mathbf{r}t} \rangle \rangle : \langle Term \Rightarrow \langle Conjunction \Rightarrow Frequency \rangle \rangle$
3:    **for each** $\mathbf{r} = \langle y_{\mathbf{r}}, p_{\mathbf{r}}, b_{\mathbf{r}}, e_{\mathbf{r}}, \mathbf{s_r}, \mathbf{t_r^u} \rangle \in \mathbf{R}$ **do**
4:       **for each** $(t, g) \in \mathbf{t_r^u} = \{(t_1^u, g_1), (t_2^u, g_2), \dots, (t_l^u, g_l)\}$ **do**
5:          $x_{\mathbf{r}t} \leftarrow g$
6:       **end for**
7:    **end for**
8:    **return** the map $\langle t \Rightarrow \langle \mathbf{r} \Rightarrow x_{\mathbf{r}t} \rangle \rangle$
9: **end procedure**

---

conditions. Given a command, the matching algorithms will find a (small) subset of matched conjunctions. In this subset, the conjunction with the highest priority will be selected and its action will be chosen for the command.

Mathematically, let $\mathbf{R} = \{\mathbf{r}_1, \mathbf{r}_2, \dots, \mathbf{r}_n\}$ be the set of all conjunctions associated with all intent types. Each conjunction $\mathbf{r} \in \mathbf{R}$ is defined as a 7-tuple: $\mathbf{r} = \langle a_{\mathbf{r}}, f_{\mathbf{r}}, p_{\mathbf{r}}, b_{\mathbf{r}}, e_{\mathbf{r}}, \mathbf{s_r}, \mathbf{t_r^u} \rangle$, in which $a_{\mathbf{r}}$ and $f_{\mathbf{r}}$ are the application and action associated with the conjunction $\mathbf{r}$; $p_{\mathbf{r}}$ is the conjunction priority (1 is the highest); $b_{\mathbf{r}}$ is *true* if the first term of this conjunction must be the first word in a matched command and is *false* otherwise; similarly $e_{\mathbf{r}}$ is *true* if the last term of $\mathbf{r}$ must be the last word of a matched command; $\mathbf{s_r} = (s_1, s_2, \dots, s_q)$ is the list of all text segments of $\mathbf{r}$ (a text segment can be a single word/term or a chunk of consecutive words/terms); $\mathbf{t_r^u} = \{(t_1^u, g_1), (t_2^u, g_2), \dots, (t_l^u, g_l)\}$ is the set of all unique words/terms together with their frequency (count) in $\mathbf{r}$.

Table 3 shows several sample conjunctions. A conjunction can simply be represented as a text string like "map, find–direction, 1–^ *tìm+đường từ đến*" ("map, find–direction, 1–^ *find+direction from to*"). The '+' character is used to connect consecutive terms in a segment. The character '^' indicates the segment "*tìm đường*" must appear at the beginning of a matched command ($b_{\mathbf{r}} = true$). 1 indicates that this conjunction is at the highest priority. Therefore, this conjunction

---

Algorithm 2. Get relevant conjunctions of a user input spoken text command

---

1: **procedure** GETRELEVANTCONJUNCTIONS(**c**, **R**, $\langle t \to \langle \mathbf{r} \to x_{\mathbf{r}t} \rangle \rangle$)
2:     create an empty set of conjunctions: $\mathbf{R}' = \emptyset$
3:     create a temporary map: $\langle \mathbf{r} \Rightarrow z_{\mathbf{r}} \rangle$: $\langle Conjunction \Rightarrow Count \rangle$
4:     **for** each $(w, h) \in \mathbf{w_c^u} = \{(w_1^u, h_1), (w_2^u, h_2), \ldots, (w_k^u, h_k)\}$ **do**
5:         get the sub–map $\langle \mathbf{r} \Rightarrow x_{\mathbf{r}w} \rangle$ from the map $\langle t \Rightarrow \langle \mathbf{r} \Rightarrow x_{\mathbf{r}t} \rangle \rangle$ with $t = w$
6:         **for** each map entry $(\mathbf{r} \Rightarrow x_{\mathbf{r}w}) \in \langle \mathbf{r} \Rightarrow x_{\mathbf{r}w} \rangle$ **do**
7:             **if** $h \geq x_{\mathbf{r}w}$ **then**
8:                 **if** $\mathbf{r}$ is a key in the map $\langle \mathbf{r} \Rightarrow z_{\mathbf{r}} \rangle$ **then**
9:                     $z_{\mathbf{r}} \leftarrow z_{\mathbf{r}} + 1$
10:                **else**
11:                    add the entry $(\mathbf{r} \Rightarrow 1)$ to the map $\langle \mathbf{r} \Rightarrow z_{\mathbf{r}} \rangle$
12:                **end if**
13:            **end if**
14:        **end for**
15:    **end for**
16:    **for** each map entry $(\mathbf{r} \Rightarrow z_{\mathbf{r}}) \in \langle \mathbf{r} \Rightarrow z_{\mathbf{r}} \rangle$ **do**
17:        **if** the number of unique words in conjunction $\mathbf{r}$ (i.e., l) $= z_{\mathbf{r}}$ **then**
18:            $\mathbf{R}' \leftarrow \mathbf{R}' \cup \{\mathbf{r}\}$
19:        **end if**
20:    **end for**
21:    **return** the set of relevant conjunctions: $\mathbf{R}'$
22: **end procedure**

---

can be converted into mathematical form as: $\mathbf{r_4} = \langle \text{map}, \text{find–direction}, 1, true, false, (\text{tìm đường, từ, đến}), \{(\text{tìm}, 1), (\text{đường}, 1), (\text{từ}, 1), (\text{đến}, 1)\} \rangle$.

User input command, denoted **c**, is defined as a pair: $\mathbf{c} = \langle \mathbf{w_c}, \mathbf{w_c^u} \rangle$, where $\mathbf{w_c} = (w_1, w_2, \ldots, w_m)$ is the list of all words/terms and $\mathbf{w_c^u} = \{(w_1^u, h_1), (w_2^u, h_2), \ldots, (w_k^u, h_k)\}$ is the list of unique words/terms together with their frequency in the command. For example, given a user command like *"thời tiết ngày mai ở xuân mai thế nào"* (*"what is the weather like in xuan mai tomorrow"*), **c** will be converted to $\langle (\text{thời, tiết, ngày, mai, ở, xuân, mai, thế, nào}), \{(\text{thời}, 1), (\text{tiết}, 1), (\text{ngày}, 1), (\text{mai}, 2), (\text{ở}, 1), (\text{xuân}, 1), (\text{thế}, 1), (\text{nào}, 1)\} \rangle$.

To perform conjunction matching, all the conjunctions must first be indexed using the IndexConjunctions described in Algorithm 1. The key data structure is a map from terms to their conjunctions together with their frequency in each conjunction: $\langle t \Rightarrow \langle \mathbf{r} \Rightarrow x_{\mathbf{r}t} \rangle \rangle$. Given an input command **c**, the procedure GetRelevantConjunctions in Algorithm 2 will find a smaller subset of conjunctions ($\mathbf{R}'$) that are relevant to **c**. The main idea is that a conjunction $\mathbf{r}$ is relevant to **c** if **c** contains all unique words/terms in $\mathbf{r}$ and the frequency of each unique word/term in **c** is always greater or equal to that of the corresponding word/term in $\mathbf{r}$. Algorithm 2 performs inverse look-up and checking the frequency condition very fast and efficiently. The number of relevant conjunctions of a command is much smaller than the total number of conjunctions, i.e., $|\mathbf{R}'| \ll |\mathbf{R}|$. All relevant conjunctions are then checked for the conditions $b_{\mathbf{r}}$ and $e_{\mathbf{r}}$ as well as the list of segments $s_r = (s_1, s_2, \ldots, s_q)$ of r appear in c with the same order. Conjunctions satisfying these conditions will be added to the set of matched conjunctions R^*. From R^*, the highest-priority conjunction satisfying $a_r \equiv a_c$ will be selected. Finally, the action of the command is f_r of the selected conjunction.

Table 4. The precision, recall, and F_1-score of application types in the best fold (features: n-gram, regular expression, dictionary look-up, conjunction matching)

Application Type	Human	Model	Match	Precision	Recall	F_1-score
setting–3g	17	21	17	80.95	100.0	89.47
music	20	20	20	100.0	100.0	100.0
alarm	41	41	40	97.56	97.56	97.56
setting	8	8	8	100.0	100.0	100.0
other–app	27	23	23	100.0	85.19	92.00
browser	52	54	52	96.30	100.0	98.11
photo	22	22	22	100.0	100.0	100.0
setting–orientation	17	17	17	100.0	100.0	100.0
sms	22	22	22	100.0	100.0	100.0
setting–wifi	17	16	16	100.0	94.12	96.97
phone	31	29	29	100.0	93.55	96.67
setting–bluetooth	16	16	16	100.0	100.0	100.0
email	12	11	11	100.0	91.67	95.65
map	26	25	25	100.0	96.15	98.04
reminder	47	47	47	100.0	100.0	100.0
setting–volume	28	28	28	100.0	100.0	100.0
calendar	29	28	27	96.43	93.10	94.74
web–search	26	27	26	96.30	100.0	98.11
weather	24	25	24	96.00	100.0	97.96
contact	19	21	19	90.48	100.0	95.00
note	24	24	24	100.0	100.0	100.0
camera	19	19	19	100.0	100.0	100.0
setting–brightness	11	11	11	100.0	100.0	100.0
Macro average				**98.00**	**97.88**	**97.94**
Micro average	555	555	543	**97.84**	**97.84**	**97.84**

4 Evaluation

4.1 Experimental Data and Settings

To evaluate the proposed method, we asked a group of students to speak natural language commands to an ASR service (Google Voice) then took the output spoken texts. We obtained a medium-sized data set consisting of 2368 Vietnamese spoken text sentences as shown in Table 1. The output spoken text sentences are all in lowercase and have no punctuation marks. We then randomly divided the data set into four parts to perform 4-fold cross-validation evaluation.

4.2 Experimental Results and Analysis

Table 4 shows the performance of application classification of the fold giving the highest accuracy. The feature templates include 1,2-gram, regular expressions, dictionary look-up, and conjunction matching. The macro-averaged F_1-score of 97.94 means that the balance of performance among classes. The micro-averaged F_1 of 97.84 means that with this set of feature templates, we can achieve a very high accuracy level. We also conducted the experiments for different feature choices and reported the results in Fig. 2. On average (over the four folds), we can achieve an micro-averaged F_1 of 93.11 with 1,2-gram features only. We added regular expressions and the result is 94.59. Dictionary look-up features

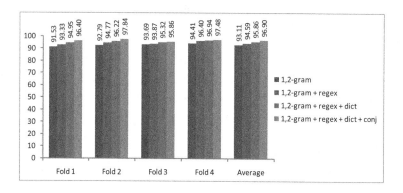

Fig. 2. The MaxEnt accuracy of the 4 folds with different feature templates

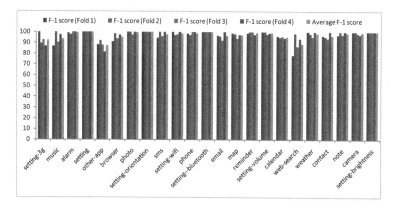

Fig. 3. The F_1 scores of each class in the 4 folds and their average value (features: n-gram, regular expression, dictionary look-up, conjunction matching)

can improve the performance and gave the micro-averaged F_1 of 95.86. And finally, with all feature types (including conjunctions), we achieved the result of 96.90. This is a significantly high accuracy.

The results reported in Fig. 3 show the performance of each class over the four folds. Most of the classes are very stable, achieving very high accuracy in every fold. For some others like *setting-3g*, *other-app*, and *web-search*, the accuracy values are less stable and lower than the others. This is understandable because commands for web-search are highly ambiguous. The classifier needs a search action is about *calendar::query*, *contact::query*, *map::locate*, *phone::query*, or *web-search::query*. For example, a command like "*tìm phố trần hưng đạo*" will

Table 5. The accuracy of action recognition using conjunction matching

Application	Action	Accuracy	Application	Action	Accuracy
alarm			phone	call	100.0
	turn–off	100.0		query	100.0
	set	99.06		open	90.0
	delete	100.0	reminder	turn–off	100.0
	open	96.78		set	100.0
browser	open	88.27		delete	97.96
calendar	set	92.30		open	90.91
	query	84.91	sms	send	96.15
	delete	100.0		open	100.0
	open	88.20	weather	query	100.0
camera	take–photo	98.77	web–search	query	72.97
	record–video	100.0		open	50.00
contact	add	100.0	setting–wifi	turn–off	100.0
	query	96.00		turn–on	100.0
	share	71.43	setting	open	100.0
	open	100.0	setting–3g	turn–off	93.94
email	query	75.00		turn–on	95.92
	send	100.0	setting–volume	turn–down	95.24
	open	100.0		set	80.95
map	find–direction	87.80		turn–up	95.56
	locate	83.95	setting–orientation	turn–off	100.0
	open	88.89		turn–on	100.0
music	open	100.0	setting–brightness	turn–down	93.33
note	add	97.56		set	92.86
	open	96.67		turn–up	93.75
other–app	open	94.74	setting–bluetooth	turn–off	100.0
photo	open	100.0		turn–on	100.0

ask the VAV to locate *Trần Hưng Đạo* street on the map rather than doing a web search for the information about *Trần Hưng Đạo*, a legendary supreme commander in the history of Vietnam. This is very ambiguous because most of the features indicate that this is a web search command. The class *other–app* is also ambiguous because it includes all commands related to other applications installed on the user device. Table 5 shows the action recognition accuracy using conjunction matching. As we can see, manually created conjunctions can work well for most actions. There are only some that gave lower accuracy than the others like *contact::share* (71.43), *email::query* (75.00), *web–search::query* (72.97), and *web–search::open* (50.00). As explained above, *web–search* is a highly ambiguous class and their actions are also hard to recognize. The two actions *email::query* and *web–search::query* are also difficult to distinguish.

5 Conclusions

In this paper, we have presented a lightweight approach to identify the user intents behind their spoken text commands. We proposed to use a MaxEnt classifier to classify applications and a conjunction matching to identify actions. We evaluated our method on a medium-sized data set consisting of 23 application classes and 53 intents and achieved significant results. Both the application classifier and the conjunction matcher are highly accurate, very fast and compact so that they can be run right on mobile devices. Also, these classifier and matcher were successfully integrated into our mobile virtual assistant for Vietnamese.

Acknowledgment. This work was supported by the project QG.15.29 from Vietnam National University, Hanoi (VNU).

References

1. Angelov, K., Bringert, B., Ranta, A.: Speech-enabled hybrid multilingual translation for mobile devices. In: EACL (2014)
2. Bastianelli, E., Castellucci, G., Croce, D., Basili, R., Nardi, D.: Effective and robust NLU for human-robot interaction. In: ECAI, vol. 263, pp. 57–62 (2014)
3. Berger, A.L., Pietra, V.J.D., Pietra, S.A.D.: A maximum entropy approach to natural language processing. Comput. Linguist. **22**(1), 39–71 (1996)
4. Borthwick, A.: A maximum entropy approach to named entity recognition. Ph.D. dissertation, Deptartment of CS, New York University (1999)
5. Branavan, S.R.K., Chen, H., Zettlemoyer, L.S., Barzilay, R.: Reinforcement learning for mapping instructions to actions. In: ACL/IJCNLP, pp. 82–90 (2009)
6. Branavan, S.R.K., Zettlemoyer, L.S., Barzilay, R.: Reading between the lines: learning to map high-level instructions to commands. In: ACL, pp. 1268–1277 (2010)
7. Bratman, M.: Intention, Plans, and Practical Reason. Harvard University Press, Cambridge (1987)
8. Graves, A., Jaitly, N.: Towards end-to-end speech recognition with recurrent neural networks. In: ICML (2014)

9. Hannun, A., Case, C., Casper, J., Catanzaro, B., Diamos, G., Elsen, E., Prenger, R., Satheesh, S., Sengupta, S., Coates, A., Ng, A.Y.: Deep Speech: scaling up end-to-end speech recognition (2014). arxiv.org/abs/1412.5567v2

10. Hinton, G., Deng, L., Yu, D., Dahl, G., Mohamed, A., Jaitly, N., Senior, A., Vanhoucke, V., Nguyen, P., Sainath, T., Kingsbury, B.: Deep neural networks for acoustic modeling in speech recognition. IEEE Signal Process. Mag. **29**, 82–97 (2012)

11. Liu, D., Nocedal, J.: On the limited memory BFGS method for large-scale optimization. Math. Program. **45**, 503–528 (1989)

12. Popkin, J.: Google, apple siri and IBM watson: the future of natural-language question answering in your enterprise. Gartner Technical Professional Advice (2013)

13. Ratnaparkhi, A.: A maximum entropy model for part-of-speech tagging. In: EMNLP, vol.1, pp. 133–142 (1996)

14. Tellex, S., Kollar, T., Dickerson, S., Walter, M.R., Banerjee, A.G., Teller, S., Roy, N.: Understanding natural language commands for robotic navigation and mobile manipulation. In: AAAI (2011)

15. Tur, G., Mori, R.D.: Spoken Language Understanding: Systems for Extracting Semantic Information from Speech. Wiley, New York (2011)

A Method for Determining Representative of Ontology-Based User Profile in Personalized Document Retrieval Systems

Bernadetta Maleszka[(✉)]

Faculty of Computer Science and Management, Wroclaw University of Technology,
Wybrzeze Wyspianskiego 27, 50-370 Wroclaw, Poland
Bernadetta.Maleszka@pwr.edu.pl

Abstract. Information overload is one of the most important problems in context of personalized document retrieval systems. In this paper we propose to use ontology-based user profile. Ontological structures are appropriate to represent relations between concepts in user profile. We present a method for determining representative profile of users' group. Two users are in the same group when their interests (profiles) are similar. If a new user is classify to a group, a system can recommend him a representative profile to avoid ,,cold-start problem". Results obtained in experimental evaluation are promising. Method presented in this paper is a crucial part of developed personalized document retrieval system.

Keywords: Ontology-based user profile · User preference · Representative profile · Personalization

1 Introduction

Nowadays personalization systems are more and more popular due to the problem with information overload. In the Internet one can find a lot of information but the problem is to find interesting documents that really correspond with users needs. In this context personalization system can be treated as a filter that selects only relevant documents. Personalized document retrieval system stores information about the user: his queries and selected results. Based on them it builds user profile which is used during the recommendation process.

Our general research area is connected with developing a personalized document retrieval system. In the system we can distinguish three modules. The first module is associated with user profiles and methods for the clustering procedure. In the second one we consider problems with determining representative profile of the users group. The last module is associated with a new user that comes to the system – he is classified into a proper group based on his demographic data and a representative profile of this group is recommended for him. Despite of the lack of history information about user activities we can avoid "cold start" problem which is important problem in many personalization systems.

N.T. Nguyen et al. (Eds.): ACIIDS 2016, Part I, LNAI 9621, pp. 202–211, 2016.
DOI: 10.1007/978-3-662-49381-6_20

In this paper we consider methods for the second module. We assume that we have a group of users with similar interests (similar profiles) and the main aim is to determine a representative profile of this group. A quality condition is also proposed to check if the obtained profile represents users' interests for the whole group. We have developed a series of simulations to show dependencies between systems' parameters.

The rest of the paper is organized as follows. In Sect. 2 we present a survey of methods connected with ontology-based user profile building and integrating information from many profiles. The model of documents set and user profile are presented in Sect. 3. The proposition of methods for building user profile based on his activities and for determining representative profile of users group are described. Section 4 presents the idea of evaluation methodology and analysis of the results. In the last Sect. 5 we gather the main conclusions and future works.

2   Related Works

In this paper we consider methods for building user profile based on his behaviour and determining representative profile of a group. In this section we present the most popular approaches connected with personalized document retrieval systems. Multiple methods in those areas were proposed.

Recommendation problem can be defined as a function that measures the gain or usefulness of item i_n to user u_m; $i_n \in \{i_1, i_2, \ldots, i_N\}$ and $u_m \in \{u_1, u_2, \ldots, u_M\}$, N is a number of items and M is a number of users [1]. The authors of [1] distinguish kinds of recommender systems:

- content-based system – the user is recommended items similar to those he preferred in the past;
- collaborative filtering systems – the user is recommended items that people with similar tastes and preferences liked in the past;
- hybrid recommender systems – uses and combines methods from both previous systems.

Due to the fact that either content-based or collaborative filtering systems have many constraints [4], the most popular are hybrid ones. In a recommender system information about a user is stored in a profile. The structure of profile has changed over time – from "bag of words", through vector [9,16] or hierarchical structure [10] to ontology-based structure [5,6].

Authors of [6] present an approach based on ontologies to represent the interaction process between user profile and its context for collaborative learning. They also analyzed rules-based methods connected with role assignments, permissions, restrictions and context of location to obtain appropriate profile of the user. The profile can be then used to improve query results (or make search results more relevant) [5].

In this paper we consider ontology-based user profile. Such structure has been extensively used in data integration systems. Ontology provides an explicit and machine-understandable conceptualization of a domain [3].

The first problem in recommender system is to build a user profile. Cantador et al. [1] have proposed an ontological representation of the domain of discourse where user interests are defined. User profiles are initially described as weighted lists measuring the users' interests for the concepts from the reference ontology. Relation among users are extracted from links between users and concepts. Users that share interests of a specific groups of concepts are assigned to the same group.

The second aspect of recommendation is to cluster users into groups of similar information needs. In this research area one can find a wide spectrum of approaches, especially methods for integration of ontology-based profiles.

Pinto et al. [15] identify three meanings of ontology "integration":

- to build a new ontology by reusing (by assembling, extending, specializing or adapting) other ontologies already available;
- to build an ontology by merging several ontologies into a single one that unifies all of them;
- to build an application by using one or more ontologies.

Depending on aim of integration and conditions that are required during integration process, the result ontology could contain eg. common parts of all input ontologies or all concepts from input ontologies. Obtained ontology can be used instead of initial ontologies or only as an intermediary between initial systems (based on initial ontologies).

One can differentiate the following levels of ontology integration: alignment, partial compatibility and unification. Alignment is the weakest form – it requires minimal change, but it can only support limited kinds of interoperability.

To perform ontology integration, it is necessary to consider the following aspects [14]:

- possibility of integration – one should define a set of possible actions;
- modules in ontology – identify the modules into which the ontology can be divided;
- assumption and requirements – identify the assumptions and ontological commitments that each module should comply to;
- knowledge – identify what knowledge should be represented in each module;
- candidates – identify candidate ontologies that could be used as modules;
- get candidate ontologies in an adequate form;
- study and analize candidate ontologies to choose the most adequate source ontologies to be reused;
- integrate knowledge;
- analyze resulting ontology.

It is a difficult task to integrate ontologies while there can appear inconsistencies [11]. Due to this fact many systems offer semi-automatic approach to ontology merging and alignment. Then the result ontology should be verified and improved. An exemplary system is presented in [2], where authors propose a method for the unification of concepts with ontologies by grounding concepts to a shared representation in the form of Wordnet and Wikipedia.

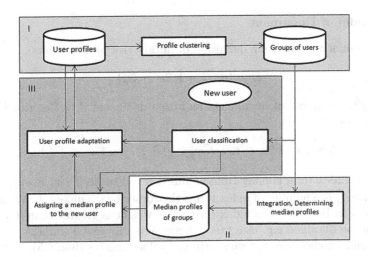

Fig. 1. Schema of personalized document retrieval system

In this paper we investigate the problem of integration ontology in the context of user profiles. We present a method for building user profile which integrates instances (documents) into user profile and a method for determining representative profile of a users group – we integrate ontology-based user profiles into one ontology.

3 Model of Personalized Document Retrieval System

A schema of the considered collaborative recommendation system is depicted in Fig. 1 ([7]). The architecture of the system consists of three main modules: (I) User Profile Module, (II) Representative Profile Module and (III) Module of a New User.

User Profile Module consists of database with user profiles, method for profile clustering to determine groups of users. Groups of users are stored in a database and can change when user profiles change with time.

The second module refers to determining representative profile of each group. In each group we have users whose interests and information needs are similar. The representative profile can be obtained using knowledge integration methods and tools.

The last module is connected with a new user that has register to the system. To avoid ,,cold-start" problem, we propose to ask a user for some demographic data. Based on this data a user is classified into one of existing group. A representative profile of this group is assigned to the user. Then users' activity is observed and his profile is updated according to his current information needs.

In our previous work [8] we have presented method for clustering users based on theirs profiles. We were comparing ontology-based profiles on levels of concepts and instances. In this paper we consider methods in the second modules connected with determining representative profile of a users group.

3.1   Model of Document

Let us consider the following definition of library (set of documents).

$$D = \{d_i : i = 1, 2, \ldots, n_d\} \tag{1}$$

where n_d is a number of documents and each document d_i is described in the following way:

$$d_i = \{(t^i_j, w^i_j) : t^i_j \in T \wedge w^i_j \in [0.5, 1), j = 1, 2, \ldots, n^i_d\} \tag{2}$$

where t^i_j is index term coming from assumed set of terms T, w^i_j is appropriate weight and n^i_d is a number of index terms that describe document d_i. We assume that term can be treated as index term if its weight is greater than 0.5 (index terms can be obtained e.g. using term frequency – inverse document frequency method and we choose only terms with the values greater than 0.5).

3.2   Model of User Profile

We assume that user profile has ontological structure. Definition of ontology was provided in [13].

Ontology-based user profile is defined as a triple:

$$O = (C, R, I) \tag{3}$$

in which by C we denote a finite set of concepts, by R a finite set of relations between concepts $R = \{r_1, r_2, \ldots, r_n\}$, $n \in N$ and $r_i \subset C \times C$ for $i \in \{1, n\}$ and by I a finite set of instances.

Assuming the existence of a finite set A of all possible attributes and a finite set V of their valid valuations such that $V = \bigcup_{a \in A} V_a$, where V_a is a domain of an attribute a by real world we will call a pair (A, V).

The structure of a concept c is defined as a triple:

$$c = (Id^c, A^c, V^c) \tag{4}$$

where Id^c is its unique identificator, A^c is a set of attributes assigned to c and V^c is a set of domains of attributes from A^c defined as $V^c = \bigcup_{a \in A^c} V_a$.

An instance i taken from the set I is defined as:

$$i = (id, A_i, v_i) \tag{5}$$

where id is its identificator, A_i is a set of attributes describing the instance i and v_i is a function with a signature $v_i : A_i \to \bigcup_{a \in A_i} V_a$, which assigns to attributes from the set A_i specific values taken from their domains.

In our model of ontology-based profile we consider index term and its synonyms as a concept, relations between concepts are relations between the terms (eg. relation "is-a-part" or generalization – specification relation) and instance is a set of documents that are described using particular keyword.

Algorithm 1. Algorithm for determining user profile.

Input: a set of k documents that are relevant for user information needs
Output: user profile
for $c_j \in C$ **do**
$\quad$ $n \leftarrow 0$;
$\quad$ $s \leftarrow 0$;
$\quad$ **for** $d_i \in D$ **do**
$\quad\quad$ **if** $w_j^i > \eta$ **then**
$\quad\quad\quad$ $s = s + w_j^i$;
$\quad\quad\quad$ $n = n + 1$;
$\quad\quad$ **end**
$\quad$ **end**
$\quad$ $w_{avg} = s/n$;
$\quad$ Assign calculated w_{avg}^j to weight of concept c^j in user profile.
end

3.3 Method of Determining the User Profile

In this section we consider the problem of determining a profile for a user based on his activity. We observe user queries and documents that he selected as relevant to find out what are his interests.

Let us assume that user has k documents that are relevant for his information needs (and a level of usefulness). Each document is an instance of reference ontology (thesaurus). We propose the following idea of determining user profile: first system calculates average value of each concepts' weights from all relevant documents. The concept will be added to user profile if its average weight is greater than assumed threshold η.

3.4 Determining Representative Profile for a Group of Users

The main aim of this paper is to present a method for determining representative profile of users group. Each user profile is based on reference ontology which means that the set of concepts in user profile is a subset of all concepts in reference ontology. Such approach is certainly helpful because we need matches to refer to the same upper ontology or to conform to the same reference ontology [12].

Let us assume that in a group we have m users. Each profile $p_j, j \in \{1, 2, \ldots, m\}$ is based on reference ontology. The representative profile p_r will be also based on reference ontology. To calculate weights of each concept ($t_i, i \in \{1, 2, \ldots, \#T\}$) in representative profile we propose the following procedure (Algorithm 2).

4 Experimental Evaluation

In this section we present an idea of experiments. The objective of the evaluation is to check the quality of developed method of determining representative

---

Algorithm 2. Algorithm for determining representative profile of group of users.

Input: a set of m users' profiles
Output: representative profile
for $j \in \{1, 2, \ldots, m\}$ do
   | $n \leftarrow 0$;
   | $s \leftarrow 0$;
   | for $t_i \in p_r$ do
   | | if $w_j^i > \gamma$ then
   | | | $s = s + w_j^i$;
   | | | $n = n + 1$;
   | | end
   | end
   | $w_{pr}^j = s/n$;
   | Assign calculated w_{pr}^j to weight of concept c^j in representative profile.
end

---

profile. We propose a series of simulations to obtain the results while performing experiments with real users are time- and cost-consuming.

4.1   Quality Condition

Let us assume that profile P_{base} is a profile built using Algorithm 1 based on all instances from users in this group and P_{repr} is a representative profile of group of users. The main difference between determining baseline profile and representative profile is as follows: profile P_{base} can be treated as a mean profile from all relevant documents and profile P_{repr} is determined based on profiles of users from the group. The P_{repr} will be acceptable if it satisfy quality condition.

We propose the following *quality condition*:

Distance between representative profile P_{repr} and baseline profile P_{base} should be less than ϵ.

$$d(P_{repr}, P_{base}) < \epsilon$$

4.2   Plan of the Experiments

To check the quality of proposed methods we perform experiments with the following plan:

1. Generate a set of N documents.
2. Generate a set of M users.
3. For each user u_i, $i \in \{1, 2, \ldots, M\}$
   (a) Determine a set of relevant documents based on his preference.
   (b) Determine user profile using Algorithm 1.
4. Cluster profiles into group of users with similar interests (methods for this point are presented in [8]).

5. Determine representative profile P_{repr} of each group of users.
6. Determine profile P_{base}.
7. Calculate distance between profiles: $d(P_{repr}, P_{base})$ and check if Quality Condition is satisfied.

4.3 Analysis of the Results

In the experiments we performed, we have checked if there is any dependencies between number of users in a group and number of relevant documents for each user and number of concepts in each user profile. We have explored the system for different numbers of users: starting from $M = 10$ to $M = 190$. We have considered smaller and greater set of documents ($N \in \{100, 200, 500\}$) and different cardinality of ontology concepts (10, 20 or 100).

Results of performed simulation are presented in Figs. 2 and 3.

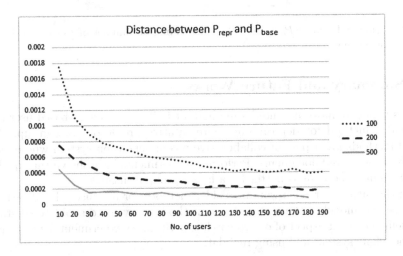

Fig. 2. Distance between P_{repr} and P_{base} depending on number of users for different number of documents.

In Fig. 2 we present the distance between P_{repr} and P_{base} depending on number of users for different number of documents. The obtained results have shown that greater number of users implies the smaller distance ($\epsilon < 0.002$). It is worth to note that when we assume the value of ϵ we can estimate how many users we should have in the group.

When we analyze results for different number of documents we can see that if users have more relevant documents, the distance between representative and baseline profile is smaller. It means that it is better to gather more information about the users before determining representative profile.

Comparing results for different numbers of concepts in the ontology we can notice that there is no significant difference between obtained distances. It has shown that presented method is working not only for 10 or 20 concepts but only for greater ontologies.

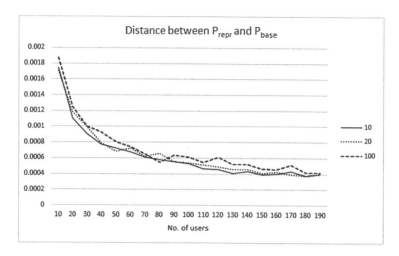

Fig. 3. Distance between P_{repr} and P_{base} depending on number of users for different number of concepts.

5 Summary and Future Works

In this paper we have presented a method for building ontology-based user profile and a method for determining representative profile of users group. Proposed methods are a part of collaborative recommendation system where users obtain personalized documents. Performed experiments have shown that quality of determined methods is better for a larger group of users.

In the future works we would consider methods for new user classification – in presented model of the system these methods allow us to avoid "cold start" problem. The next aspect of our research is performing experimental evaluations for the whole system and using real data.

Acknowledgments. This research was partially supported by Polish Ministry of Science and Higher Education.

References

1. Cantador, I., Bellogin, A., Castells, P.: A multilayer ontology-based hybrid recommendation model. J. AI Commun. - Recomm. Syst. **21**(2–3), 203–210 (2008)
2. Cantador, I., Szomszor, M., Alani, H., Fernandez, M., Castells, P.: Enriching ontological user profiles with tagging history for multi-domain recommendations. In: Proceedings of 1st International Workshop on Collective Semantics: Collective Intelligence & the Semantic Web (2008)
3. Cruz, I.F., Xiao, H.: The role of ontologies in data integration. J. Eng. Intell. Syst. **13**(5), 245–252 (2005)
4. Etaati, L., Sundaram, D.: Adaptive tourist recommendation system: conceptual frameworks and implementation. Vietnam J. Comput. Sci. **2**, 95–107 (2015). doi:10.1007/s40595-014-0034-5

5. Limbu, D.K., Connor, A.M., MacDonell, S.G.: A framework for contextual information retrieval from the WWW. arXiv preprint (2014). arXiv:1407.6100

6. Luna, V., Quintero, R., Torres, M., Moreno-Ibarra, M., Guzman, G., Escamilla, I.: An ontology-based approach for representing the interaction process between user profile and its context for collaborative learning environments. Comput. Hum. Behav. **51**, 1387–1394 (2015)

7. Maleszka, M., Mianowska, B., Nguyen, N.T.: A method for collaborative recommendation using knowledge integration tools and hierarchical structure of user profiles. Knowl.-Based Syst. **47**, 1–13 (2013)

8. Maleszka, B.: A method for profile clustering using ontology alignment in personalized document retrieval systems. In: Núñez, M., et al. (eds.) ICCCI 2015. LNCS, vol. 9329, pp. 410–420. Springer, Heidelberg (2015). doi:10.1007/978-3-319-24069-5_39

9. Mianowska, B., Nguyen, N.T.: Tuning user profiles based on analyzing dynamic preference in document retrieval systems. Multimedia Tools Appl. **65**, 93–118 (2012). doi:10.1007/s11042-012-1145-6

10. Montaner, M., Lopez, B., Rosa, J.: A taxonomy of recommender agents on the internet. Artif. Intell. Rev. **19**, 258–330 (2003)

11. Noy, N.F., Musen, M.A.: An algorithm for merging and aligning ontologies: automation and tool support. In: Proceedings of the Workshop on Ontology Management at the Sixteenth National Conference on Artificial Intelligence (1999)

12. Noy, N.F.: Semantic integration: a survey of ontology-based approaches. SIGMOD Rec. **33**, 65–70 (2004)

13. Pietranik, M., Nguyen, N.T.: A multi-attribute based framework for ontology aligning. Neurocomput. **146**, 276–290 (2014)

14. Pinto, H.S., Martns, J.P.: Ontology integration: how to perform the process. In: Proceedings of the IJCAI-01 Workshop on Ontologies and Information Sharing (2001)

15. Pinto, H.S., Gomez-Perez, A., Martins, J.P.: Some issues on ontology integration. In: Proceedings of the IJCAI-99 workshop on Ontologies and Problem-Solving Methods (1999)

16. Zhu, Y., Xiong, L., Verdery, C.: Anonymizing user profiles for personalized web search. In: Proceedings of the 19th International Conference on World Wide Web, WWW 2010 (2010)

Database Systems
and Software Engineering

Data Quality Scores for Pricing on Data Marketplaces

Florian Stahl[1]([⊠]) and Gottfried Vossen[1,2]

[1] ERCIS, Leonardo-Campus 3, 48149 Münster, Germany
{florian.stahl,gottfried.vossen}@ercis.de
[2] University of Waikato Management School, Private Bag 3105,
Hamilton 3240, New Zealand

Abstract. Data and data-related services are increasingly being traded on data marketplaces. However, value attribution of data is still not well-understood, in particular when two competing offers are to be compared. This paper discusses the role data quality can play in this context and suggests a weighted quality score that allows for 'quality for money' comparisons of different offerings.

1 Introduction

As information has become an important production factor, we have reached a point at which data – the basic unit in which information is exchanged – is increasingly being traded on data marketplaces [6,10]. Basically, a data marketplace is a platform which allows providers and consumers of data and data-related services (e.g., data mining algorithms) to interact with each other. This is particularly beneficial for small and medium-sized enterprises, as they would otherwise not be able to access and analyse such data. A core problem here is the *pricing* of data; in this paper we suggest to relate data pricing to quantifiable data quality criteria through which both parties involved can satisfy their interests.

While data *quality* has been researched for quite some time already (see NAUMANN [7] for an excellent account), the topic of data *pricing* has only recently been put on the database research agenda by authors such as BALAZINSKA et al. [2,3] or TANG et al. [11]. The authors of [2] cast the problem in a relational setting, argue that relational *views* can be interpreted as versions of the 'information good' data – an assumption that will also be made here – and identify three open problems: (1) pricing of data updates; (2) pricing of integrated data for complex value chains; and (3) pricing of competing data sources that provide essentially the same data but in different quality.

The first challenge can be addressed by calculating the difference between the full price of the new and the old product, which is similar to the approach suggested by TANG et al. [11] for buying samples of XML data. The second problem may be addressed by introducing an intermediary pricing for all providers refining the raw data. This means the raw data vendor operates using established

© Springer-Verlag Berlin Heidelberg 2016
N.T. Nguyen et al. (Eds.): ACIIDS 2016, Part I, LNAI 9621, pp. 215–224, 2016.
DOI: 10.1007/978-3-662-49381-6_21

means; all vendors following in the value chain have to deal with the output price of the lower level vendor as cost and build their prices accordingly, which can be achieved by an appropriate data marketplace infrastructure. The last question has been addressed in [9] on which this paper builds by presenting a quality-centric price comparison model.

The remainder of this paper is structured as follows. Firstly, Sect. 2 reviews relevant quality criteria. The actual scoring model is developed and illustrated by an example in Sect. 3. Finally, this paper is concluded in Sect. 4.

2   A Review of Data Quality Criteria

BALAZINSKA et al. [2] have argued that often multiple vendors offer similar data products which mainly differ in quality. Concurring with this observation, we here develop a quality scoring model that allows for the comparison of two offers from different providers. To this end, this section will first give an overview of applicable data quality criteria and then develop a quality score appropriate for data marketplaces in Sect. 3.

To simplify the discussion, we focus on data quality in a Web context and start from the seminal work by NAUMANN [7], who came up with four categories of quality criteria; it aggregates several earlier studies on data quality, most notably [5,8,13,14]. In the following, we will review these categories and elaborate on one score that can be meaningfully computed in each; others can be found in [9]. In general, the setting we consider is that of *relational* data like in [2], where the subject of a market transaction is a *view*; in particular, a view to be sold (or bought) will be denoted u with attribute set X_u; an element of u will be denoted as μ.

The four categories of quality criteria are (a) *content-related*, i. e., directly rooted in the data; (b) *technical*, i. e., related to the organization and delivery of the data; (c) *intellectual*, i. e., related to the knowledge of eventual users; and (d) *instantiation-related* i. e., related to the presentation of the data. A brief description of each measure is given next, along with an elaboration of how it is relevant in the context of data marketplaces, i. e., whether it (a) is automatically calculable, (b) can be used for an automated intra-marketplace comparison, and (c) can be used for an automated inter-marketplace comparison.

The following *content-related* criteria are specified in [7]:

- *Accuracy,* the percentage of correct values in the data set;
- *Completeness,* the percentage of non-null values in the data set;
- *Customer Support,* the amount and usefulness of available human help;
- *Documentation,* the extent of available meta data regarding the data sets;
- *Interpretability,* the match between a user's understanding and the data;
- *Relevancy,* the degree to which the data satisfies a user's information needs;
- *Value-added,* the value which usage of the data provides to its users.

Clearly the *value-added* criterion of data is highly customer-dependent. Given that it is the aim of this work to approximate this criterion through the other quality dimensions, it will not be further analyzed. Most of the other criteria can at most be calculated partially, as they require knowledge that goes beyond the actual data. Indeed, the existence and the extent of *customer support* and *documentation* can be evaluated, but it does not say anything about the actual quality. Next, *accuracy* can be compared to verified accurate data, which can, however, be very difficult, so that for a comparison between different offerings, *accuracy* has only limited applicability. Moreover, *interpretability* and *relevancy* cannot be automatically examined, as both require an in-depth understanding of potential users, which is difficult to achieve in an automated fashion. As a consequence, these measures have only limited applicability to intra-marketplace and inter-marketplace comparison.

The exception is *completeness* discussed next. To this end, the closed world and open world assumptions (CWA and OWA, respectively) can be distinguished [4]. In a relational setting, *CWA* means that only values actually present in a relation represent true facts about the real world, i. e., if a value is missing, the data is false or at least incomplete. Under the *OWA*, for a null value one cannot state if the corresponding value is really missing (i.e., currently unknown) or does not exist [4]. For the purpose of this work, it is not particularly relevant why a value is missing; the fact is, it cannot be delivered to the customer. Hence, the *CWA* will be used here as general assumption for all quality criteria. Thus, we assume that all information necessary for pricing is available on the data marketplace(s) under consideration. As a consequence, *completeness* can be interpreted as *null-freeness* and can be evaluated by measuring the number of table entries not containing a null-value ($\perp$) in the relation to be sold, compared to the maximum amount of data possible:

$$c(u) = \frac{|\{\mu[A]\,|\,\mu \in u, A \in X_u, \mu[A] \neq \perp\}|}{|u| \times |X_u|} \tag{1}$$

NAUMANN [7] mentions the following *technical* criteria:

- *Availability,* the probability that a query is answered within a given time;
- *Latency,* the time between issuing a query and receiving its first response;
- *Price,* the amount of money a user has to pay for the data;
- *Quality of Service,* the error rate when transmitting the data;
- *Response Time,* the time needed for receiving the full query response;
- *Security,* the degree of protection through encryption and anonymization;
- *Timeliness,* the freshness of the data.

Following the argument of *value-added* above, *price* will be excluded. Here, an automated calculation is possible for most of these measures. The exemption is *quality of service* for which it has to be specified what quality specifically means; thus, it is generally not very precise and hence excluded from further consideration. *Availability* requires multiple measurements over time in order to be properly evaluated. In conclusion, all measures discussed in this group have a limited applicability in an intra-marketplace comparison – as data is supplied through the same infrastructure – and all are highly important in an inter-marketplace comparison.

As an example for a measure in this category, we consider *timeliness*. According to [4], *timeliness*, i.e., the freshness of data, depends on a number of characteristics, including (a) delivery time, i.e., the time at which the datum is being delivered; (b) input time, i.e., the time at which the datum was entered into the system; (c) age, i.e., the age of the datum when entered into the system; and (d) volatility, i.e., the typical time period a datum keeps its validity. In this paper, we abstract from age, as it is supposed that time-sensitive data is entered into the system immediately. Furthermore, in most cases it is only relevant when a datum was last updated and how long it remains valid. Adopting the definition of [4], the *timeliness* of a record or tuple t_μ is a function of delivery time (DT), input time (IT), and volatility:

$$t_\mu(DT, IT, Volatility) = max\left\{0, 1 - \frac{DT - IT}{Volatility}\right\} \qquad (2)$$

In order to make *timeliness* measurable, we assume that a *LastUpdated* attribute and a volatility constant v exist for each view u; also, the intended delivery time must be known. Then, the overall timeliness score can be calculated as average timeliness for all tuples in u:

$$tim(u) = \frac{\sum_{\mu \in u} t_\mu(\mu[DT, LastUpdated], v)}{|u|} \qquad (3)$$

The *intellectual* criteria mentioned in [7] are:

- *Believability,* the expected accuracy;
- *Objectivity,* the degree to which the data is free of any bias;
- *Reputation,* the degree of high standing of the source perceived by customers.

All of them but *objectivity* cannot be assessed automatically without any further user input, since they are inherently dependent on the users. *Objectivity,* could be partially calculated automatically if there are technical means to verify the data (supposing verifiable data is also objective). *Reputation* and *objectivity* may both be used for intra-marketplace comparison to some degree where the first requires some infrastructure on the data marketplace to measure the reputation such as a rating system for customers on the platform. *Objectivity* could be used as a measurement if the requirements for an automated assessment are fulfilled. In this case it could even be used for an inter-marketplace comparison, which seems very unlikely for the other two because it is hard to measure *reputation* objectively in an automatic fashion across platforms and for *believability* the argument is that it is inherently unmeasurable given its subjectivity.

Finally, the group of *instantiation-related* criteria is comprised of:

- *Amount of Data,* the number of bytes returned as a query result;
- *Representational Conciseness,* how well the representation matches the data;
- *Representational Consistency,* how well the representation matches previous representations of the same data;
- *Understandability,* the degree to which the data can be understood by users;
- *Verifiability,* the degree to which a data set can be checked and verified.

While the *amount of data* and the *representational consistency* can be assessed automatically, the other three cannot. For *representational conciseness* and *understandability* the reason is that only humans can judge whether the data format matches the data or whether they understand the data. *Verifiability* very much depends on the actual use-case and is hard to generalize; thus it has been categorised as not automatically assessable. Consequently, *amount of data* and *representational consistency* can be used for intra- and inter-marketplace comparison. Moreover, *representational consistency* is suitable for automated assessment, but requires multiple measurements, similar to *availability*. The representational conciseness cannot be automatically assessed and is, thus, neither suitable for either comparison mode.

As an example of this category, *amount of data* is chosen. Given that measuring the size of the data in bytes can be difficult when comparing two offerings, as providers might store the very same data at different compression rates, here the *amount of data* will be measured by calculating the proportion of selected rows and columns compared to the maximum available between providers. As auxiliary means, let $Y \subseteq X_u$ and let $A_i \Theta_i a_i$ be a selection condition with $A_i \in X_u$, $a_i \in dom(A_i)$, $\Theta_i \in \{<, \leq, >, \geq, =, \neq\}$, where *dom(A) denotes the domain of the attribute* A. Then, several selection criteria can be combined into a condition C by using disjunction or conjunction. Whereas most of the other scores consider each provider individually, *amount of data* needs to consider more than one as the score is more expressive, if it is compared to the *amount of data* of other providers. Therefore, it is supposed that several providers are to be compared. Regarding *amount of data*, the offering that contains the largest response to the query is denoted u^*. To allow for comparability, it is further supposed the views under consideration have the same schema. Now, the score for *amount of data AoD* can be defined as:

$$AoD(u) := \frac{|\pi_Y(\sigma_C(u))|}{|\pi_Y(\sigma_C(u^*))|} \tag{4}$$

To conclude this section, an overview of all the criteria, their automated computability, and their applicability for intra- and inter-marketplace evaluation is presented in Table 1, where criteria that are applicable without restrictions are annotated with ✔, those that have limited applicability are annotated with ✔/✘, and those not applicable at all are annotated with ✘.

3 A Data Quality Score for Pricing

The aforementioned quality criteria can be categorized by whether they can be assessed automatically A, manually M, or whether they are hybrids H, which results in the following three sets:

$A = \{$*Amount of Data, Availability, Completeness, Latency, Representational Consistency, Response Time, Security, Timeliness*$\}$

$M = \{$*Believability, Interpretability, Relevancy, Representational Conciseness, Reputation, Understandability, Verifiability*$\}$

$H = \{$*Accuracy, Customer Support, Documentation, Objectivity*$\}$

Inspired by BAETGE et al. [1], who suggest using a scoring model to evaluate professional football players for which it is difficult to find an objective value like in the case of data, we propose to approach data pricing utilizing a scoring model for data quality. Note that a similar approach was proposed by NAUMANN [7] in the context of quality-driven query planning.

Given sets A, M, and H of criteria, let a, m, and h denote scoring values associated with their members, resp. These values, as evident from the previous

Table 1. Overview of quality criteria

Category	IQ criterion	Automated	Intra DM	Inter DM
Content-related	Accuracy	✔/✘	✔/✘	✔/✘
	Completeness	✔	✔	✔
	Customer support	✔/✘	✔/✘	✔/✘
	Documentation	✔/✘	✔/✘	✔/✘
	Interpretability	✘	✘	✘
	Relevancy	✘	✘	✘
Technical	Availability	✔	✔/✘	✔
	Latency	✔	✔/✘	✔
	Response time	✔	✔/✘	✔
	Security	✔	✔/✘	✔
	Timeliness	✔	✔/✘	✔
Intellectual	Believability	✘	✘	✘
	Objectivity	✔/✘	✔/✘	✔/✘
	Reputation	✘	✔/✘	✘
Instantiation-related	Amount of data	✔	✔	✔
	Representational conciseness	✘	✘	✘
	Representational consistency	✔	✔	✔
	Understandability	✘	✘	✘
	Verifiability	✘	✘	✘

examples, are scaled in the interval $[0, 1]$, with 0 being the worst and 1 being the highest score, i.e., let a, m, and h be in $[0, 1]$. Then, a simple overall score can be defined as follows:

$$QS_{\text{simple}} = \sum_{a \in A} \frac{a}{|A|} + \sum_{m \in M} \frac{m}{|M|} + \sum_{h \in H} \frac{h}{|H|} \tag{5}$$

As it is very likely that users have different preferences for quality criteria, it is sensible to allow users to provide a weighting vector $W = (w_1, \ldots, w_n), w_i \in \mathbb{R}_0^+$ to express their preferences, where n denotes the overall number of quality criteria. As in [7], it is requested that the sum of all weights must equal 1.

Several methods exist to ask users for their preferences. NAUMANN [7] mentions direct specification, pair wise comparison, and an eigenvector model. Here, direct specification is suggested, technically realized by means of a slider-based *GUI* which allows for visual feedback. Based on this, the overall, more sophisticated score QS can be defined as follows:

$$QS = \sum_{i=1}^{|A|} w_{a_i} a_i + \sum_{j=1}^{|M|} w_{m_j} m_j + \sum_{k=1}^{|H|} w_{h_k} h_k \tag{6}$$

At a less abstract level, QS determines the overall quality of a data set, and it should be intuitively clear that on a data marketplace infrastructure mostly the automated and hybrid criteria will be relevant. While QS is not a price, it can be seen as a relative value when comparing offers of different vendors. Moreover, given a price P for a data set, quality score QS can be used to give customers guidance regarding the *quality for money* (QM) they receive, by calculating:

$$QM = \frac{QS}{P} \tag{7}$$

We now consider an example: Providers A and B both provide past, current, and forecast weather data, i.e., they constantly fill their database with new data as well as update forecast data, resp. It is further assumed that the data is not complete, as some entries get lost due to sensor malfunctions. Provider A uses very reliable sensors but fewer (measuring *AirPresure*, *Temperature (Temp)*, and *Precipitation*), which results in more complete but less extensive data. Provider B collects more data (in addition to provider A *Cloudage*), using less reliable sensors. The data sets of both providers also include the date and station for which the weather is forecast, as well as when the data were last updated.

Now suppose that an airline wants to buy weather forecast data at 5 pm (17:00) on 7th May 2017 for the next three days from three different airports (FRA,LHR,AMS). We assume that the volatility of weather forecast data is 24 h. After the airline has submitted its query, the data marketplace calculates possible result sets for providers A (Table 2) and B (Table 3), resp., and applies the quality score calculation with user supplied weights (i.e., preferences). For simplicity, only the measures previously described will be demonstrated; clearly, other measures – as described in [9] or newly created ones – could be applied in the same way.

Table 2. Relation u_A for provider A

Station	AirPressure	Temp	Precipitation	Date	LastUpdated
FRA	$\perp$	17	0	2017-05-08	14:00
FRA	1020	19	0	2017-05-09	15:00
FRA	1005	15	41	2017-05-10	16:00
LHR	1025	16	17	2017-05-08	14:00
LHR	1008	14	85	2017-05-09	15:00
LHR	1003	12	70	2017-05-10	16:00

Table 3. Relation u_B for provider B

Station	AirPressure	Temp	Precipitation	Cloudage	Date	Last Update
FRA	1022	18	0	70	2017/05/08	14:00
FRA	$\perp$	20	$\perp$	25	2017/05/09	14:00
LHR	1015	$\perp$	19	$\perp$	2017/05/08	13:00
LHR	1004	13	$\perp$	93	2017/05/09	13:00
AMS	$\perp$	13	16	$\perp$	2017/05/08	12:00
AMS	1002	12	23	97	2017/05/09	12:00

Supposing that all weights are equal, i.e., $w_a = w_m = w_h = \frac{1}{3}$, the overall quality scores per provider can be calculated as:

$$QS = \frac{1}{3}AoD(u) + \frac{1}{3}c(u) + \frac{1}{3}tim(u) \tag{8}$$

Pluging in the respective formulae results in:

$$QS = \frac{1}{3}\frac{|\pi_Y(\sigma_C(u))|}{|\pi_Y(\sigma_C(u^*))|} + \frac{1}{3}\frac{|\{\mu[A], \mu \in u, A \in X_u | \mu[A] \neq \perp\}|}{|u| * |X_u|}$$
$$+ \frac{1}{3}\frac{\sum_{\mu \in u} t(\mu[LastUpdated^*], v^*)}{|u|} \tag{9}$$

The actual values for each score as well as the overall score for $w_i = \frac{1}{3}$ and an alternative weighting ($w_1 = \frac{3}{20}, w_2 = \frac{1}{2}, w_3 = \frac{7}{20}$) are presented in Table 4.

From these calculations it is evident that, using equal weights, provider A has the higher quality data. However, if the airline has a strong interest in the amount of data, it may be better off buying from provider B. Transferring the quality score to pricing, suppose provider A offers their data for \$ 1,200.00 and provider B for \$ 1,100.00. Then, a customer with an equal appreciation for all quality criteria can calculate the respective quality-for-money ratio as quality score point per \$ 1,000.00; from this it follows that provider B offers the better quality for money as evident from Eqs. 10 and 11.

$$QM_A = \frac{0.90\bar{5}}{1.2} \qquad\qquad \approx 0.7546 \qquad\qquad (10)$$

$$QM_B = \frac{0.8796}{1.1} \qquad\qquad \approx 0,7997 \qquad\qquad (11)$$

An issue that becomes evident in this example is the fact that the amount of data is very hard to judge without further domain knowledge: Indeed, considering only the number of records and attributes one might miss that provider A offers no data for the station at AMS or that provider B does only two days of forecasting. This is left to the customer, who has to adjust their queries to ensure they receive all the data they want.

Table 4. Quality score results for providers A and B

	Provider A	Provider B
Amount of data	0.83	1.00
Completeness	0.97	0.81
Timeliness	0.92	0.83
Weighted overall score ($w_i = \frac{1}{3}$)	0.91	0.88
Weighted overall score ($w_1 = \frac{1}{2}, w_2 = \frac{3}{20}, w_3 = \frac{7}{20}$)	0.88	0.91

4   Conclusions

In this paper we have studied the problem of pricing data on a data marketplace, and we have established a relationship between quality criteria and pricing. As demonstrated, quality scoring can help customers to choose a data provider. The approach presented allows for a comparison of different providers on a data marketplace based on the quality of different offerings, while considering personal preferences of customers. Additionally, it supports customers in choosing the product which fits their needs best. If a consensus can be reached on what quality framework to use, the approach even allows for an inter-marketplace comparison, probably improving competition and, as a result thereof, also the products on offer.

Moreover, providers may also benefit as it makes evident how their offer performs compared to the competition's. Also, the scoring model can serve to determine relative prices based on the relative quality of two offerings. This can be particularly helpful if data is to be traded for data. Furthermore, providers can learn which criteria are demanded by customers and, thus, they should emphasize more often. As a result, they may adapt prices or quality to improve their competitiveness. With additional knowledge gathered on a data marketplace it would then be possible to use this score as price indication, i.e., by learning from customer choices which data with what quality is sold at what price. This can further improve the market for data over time.

The usage of quality criteria has vast potential with regard to versioning, i.e., creating different product versions of one relational data product [2], a concept known as second-order price discrimination. The idea behind the creation of different versions is to discriminate customers by their willingness to pay, e. g., offering low quality weather data to people interested in a rough forecast with a small budget and a high-quality product to customers using it for in-depth analysis willing to spend a higher amount. This concept has been applied, for instance, in [11,12] for individual quality criteria (completeness and accuracy). On a broader level, based on all quality criteria presented herein and modelled as multiple-choice knapsack, this idea has been explored further in [9].

We acknowledge the fact that besides data quality, data novelty (i. e., the degree to which data is new to a customer) is important. This we want to explore in future research. Furthermore, a practical case study, evaluating the possibilities to implement such a scoring model on data market places, is an important future research topic.

References

1. Baetge, J., Klönne, H., Weber, C.: Möglichkeiten und Grenzen einer objektivierten Spielerbewertung im Profifußball. In: KoR 12.06, pp. 310–319 (2013)
2. Balazinska, M., Howe, B., Koutris, P., Suciu, D., Upadhyaya, P.: A discussion on pricing relational data. In: Tannen, V., Wong, L., Libkin, L., Fan, W., Tan, W.-C., Fourman, M. (eds.) Buneman festschrift 2013. LNCS, vol. 8000, pp. 167–173. Springer, Heidelberg (2013)
3. Balazinska, M., Howe, B., Suciu, D.: Data markets in the cloud: an opportunity for the database community. PVLDB 4(12), 1482–1485 (2011)
4. Batini, C., Scannapieca, M.: Data Quality: Concepts, Methodologies and Techniques. Data-Centric Systems and Applications. Springer, Heidelberg (2006)
5. Chen, Y., Zhu, Q., Wang, N.: Query processing with quality control in the World Wide Web. World Wide Web 1(4), 241–255 (1998)
6. Muschalle, A., Stahl, F., Löser, A., Vossen, G.: Pricing approaches for data markets. In: Proceedings of the WorkshopBusiness Intelligence for the Real Time Enterprise, Istanbul,Turkey (2012)
7. Naumann, F.: Quality-Driven Query Answering for Integrated Information Systems. LNCS, vol. 2261, p. 3. Springer, Heidelberg (2002)
8. Redman, T.: Data Quality for the Information Age. Artech House Telecommunications Library, Artech House, Boston (1996)
9. Stahl, F.: High-Quality web information provisioning and quality-based data pricing. Ph.D. thesis. University of Münster (2015)
10. Stahl, F., Löser, A., Vossen, G.: Preismodelle für Datenmarktplätze". In: Informatik-Spektrum 37.1 (2014)
11. Tang, R., Amarilli, A., Senellart, P., Bressan, S.: Get a sample for a discount. In: Decker, H., Lhotská, L., Link, S., Spies, M., Wagner, R.R. (eds.) DEXA 2014, Part I. LNCS, vol. 8644, pp. 20–34. Springer, Heidelberg (2014)
12. Tang, R., Shao, D., Bressan, S., Valduriez, P.: What you pay for is what you get. In: Decker, H., Lhotská, L., Link, S., Basl, J., Tjoa, A.M. (eds.) DEXA 2013, Part II. LNCS, vol. 8056, pp. 395–409. Springer, Heidelberg (2013)
13. Wang, R.Y., Strong, D.M.: Beyond accuracy: what data quality means to data consumers. J. Manage. Inf. Syst. 12(4), 5–33 (1996)
14. Weikum, G.: Towards guaranteed quality, dependability of information services. In: Buchmann, A.P. (ed.) Datenbanksysteme in Büro, Technik und Wissenschaft, pp. 379–409. Springer Verlag, Heidelberg (1999)

Extraction of Structural Business Rules from C#

Bogumila Hnatkowska[⊠] and Marcin Ważeliński

Wroclaw University of Technology,
Wyb. Wyspiańskiego 27, 50-370 Wroclaw, Poland
Bogumila.Hnatkowska@pwr.edu.pl, mwazelinski@gmail.com

Abstract. Business rules are very important assets of any enterprise. Very often they are directly coded in existing software systems. As business rules evolve during a time, the software itself becomes the only valuable source of the rules applied. The aim of the paper is to present an approach to automatic business rules extraction from existing system written in C#. Considerations are limited to structural business rules. The proposed approach was implemented in a tool which usefulness was confirmed by examples. In comparison with existing solutions for reverse-engineering it gives better results, characterized by high correctness, and accuracy.

Keywords: Business rules · Structural business rules · C# · ORM · Extraction · Reverse engineering

1 Introduction

Business rules belong to the main assets of any enterprise. Very often they are directly coded in existing software systems. Typically such types of systems process a large amount of data, involve thousands of business rules of different types, and are maintained because business rules change over the time [1]. As business rules evolve it is very likely that documentation of the software systems becomes (if it is available at all) out-of-date. In the case of system re-engineering or necessity of system migration to a new technology, the system itself could be the only up-to-date source of knowledge of the rules applied [2]. So, there is a strong need of business rules extraction from existing systems.

The list of potential benefits resulting from extraction process is not limited to system re-documentation. It also includes [3, 4]: (a) definition of mappings between business-rules and source code what supports understanding of the system, (b) support in system re-engineering and migration, and general in system maintenance, (c) support for validation of the system against its requirements, (d) support in further business rules evolution.

The paper presents an approach to automatic extraction of business rules from existing systems written in c#. According to Tiobe index [5] c# belongs to the most popular programming languages (5.6 % share in the market, August 2015), and the trend is growing up. Additional assumption is that the system uses one of the

© Springer-Verlag Berlin Heidelberg 2016
N.T. Nguyen et al. (Eds.): ACIIDS 2016, Part I, LNAI 9621, pp. 225–234, 2016.
DOI: 10.1007/978-3-662-49381-6_22

Object-Relationship Mapping (ORM) frameworks. At that moment two of them are considered: NHibernate [6], and EntityFramework [7]. Extracted rules are represented in Unified Modeling Language (UML) [19] and Object Constraint Language (OCL) [20]. Both languages are often selected for this purpose, e.g. [8, 9].

The proposed approach is program-centric, automatic, and provides repeatable results. Similarly to the approach presented in [3], it uses a set of deterministic mappings among different business rule types and their representation in the source code to knowledge retrieving, and it addresses extraction of structural business rules. While authors of [3] extract business rules from a legacy database, we extract them from the source code written in c#. In comparison, other attempts could lead to different results as they often rely on manual activities, e.g. identification of variables representing domain entities [2, 10, 11] used in combination with program slicing. Sometimes extracted business rules are not clearly named against existing classifications (mainly calculation or derivation rules are discovered) or even clearly presented, e.g. in [2] only the names of variables, their location and synonyms are presented, in [10] the extraction methods are shortly presented not their results, in [4] only examples of derived variables are presented.

The rest of the paper is structured as follows. Section 2 delivers a short description of business rules and used classification. Section 3 contains the description of business rules extraction method. Section 4 provides a simple but representative example of the extraction method application and compares its result with the alternative solutions. Last Sect. 5 concludes the paper.

2   Business Rules

"Business Rule is a statement that defines or constrains some aspect of the business. It is intended to assert business structure or to control or influence the behavior of the business" [12].

Within this paper we put attention to the former of the above mentioned roles which business rules can play, what limits the scope of our interest to the structural (definitional) rules according to SBVR classification [13].

SBVR (Semantics of Business Vocabulary and Business Rules) is an OMG standard in which a meta-model for describing business vocabularies and business rules is given. This meta-model could be instantiated in different ways (we can use different languages), e.g. the specification itself uses SBVR Structured English to define the SBVR vocabularies and rules.

According to SBVR the other kind of rules are operative business rules. They influence the behavior of the business either by: definition of obligation ("It is obligatory that …"), or definition of prohibition (e.g. "It is prohibited that …"), or definition of restricted permission (e.g. "It is permitted that … only if … ").

Both, structural and operative business rules remove some degree of freedom. "The degree removed by a rule might concern the behavior of people (in the case of operative business rules), or their understanding of concepts (in the case of structural rules)" [13].

According to [13] structural rule statement should be expressed "purely in terms of noun concepts and verb concepts, as well as certain logical/mathematical operators, quantifiers, etc." Those terms of noun concepts and verb concepts introduce both: the main elements in the domain of discourse (noun concepts), and relationships among them (verb concepts). So, it should be noted, that definition of business rules needs the domain vocabulary be defined first.

Structural business rules define assertions true about all instances of a concept or propose necessary characteristics of it, meaning that the characteristic is always true of each instance of the concept. Some structural rules can impose multiplicity constraints for binary verb concepts. Computations and derivations also belong to that group of rules.

Below you find some examples of structural business rules written in SBVR Structured English or using RuleSpeak templates [14]. All examples are taken from [15]:

- It is necessary that each rental *has* exactly one requested car group (SBVR).
- Each rental always *has* exactly one requested car group (RuleSpeak).
- It is permitted that a rental *is open* only if each driver *of* the rental *is* not a barred driver (SBVR).
- A rental may *be open* only if each driver *of* the rental *is* not a barred driver (RuleSpeak).

There are many other but similar business rules classifications, that emerged before SBVR, e.g. defined by Odell [16] (derivation rules, constraints), Bubenko et al. [17] (derivation rules, event-action rules, constraints), Business Rules Group (BRG) [12] (structural assertions, action assertions), Traveter and Wagner [18] (static and dynamic constraints, derivation rules, event rules, production rules, authorization rules). We are mainly interested in (static) constraints, and derivations (calculations) as well as structural assertions. It should be noted that BRG contributed to SBVR standard.

3 Business Rules Extraction

3.1 Mappings Between Source Code and Business Elements

As was mentioned in Sect. 2, business rules strongly depend on business vocabulary (noun concepts and verb concepts). In consequence, business vocabulary must be extracted first. Next, we are able to extract other business elements, in our case structural constraints.

The mappings between an object-oriented application implemented in c#, java or c++ and business vocabulary are commonly known and implemented in many existing reverse engineering tools, e.g. Visual Paradigm or Visual Studio 2013. The rules implemented in our tool are very similar to those defined in [3]. The only exceptions results from the assumption that the domain classes are stored in a database the application communicates with by the use of NHibernate or Entity Framework. In consequence, for example, from all the classes we filter only those being a part of domain model. Dependently on the ORM used, we need either to find out classes in

XML configuration file for NHibernate or - for Entity Framework - to find out classes used as concrete parameters in the following contexts:

1. Properties of *DbContext* class of type *DbSet<T>*,
2. Classes which inherit from *EntityTypeConfiguration<T>*,
3. Calling the method *DbModelBuilder.Entity<T>* in *OnModelCreating* method of *DbContext* class

where *T* is a class we are looking for.

The newly proposed element is the extraction of structured business rules from the source code. We take into consideration the role of properties of classes as well as constraints represented by properties' attributes (features specific to c# language).

Below selected mappings between c# specific elements and business rules (constraints, derivations, and computations) are shortly described.

Source code element: attribute [StringLength(…)] defined for property *P* in class *C*
Source code example:

```
class Department {
  [StringLength(50, MinimumLength = 3)]
  public string Name { get; set; }
}
```

UML/OCL: OCL constraint; property *P* always has to be … characters long
UML/OCL example:

```
context Department inv:
    self.Name.size() >= 3 and self.Name.size() <= 50
```

Each *Name* always <u>has</u> to be at least 3 and at most 50 characters long
Comment: property *P* must be of string type
Source code element: attribute [Required] defined for property *P* in class *C*
Source code example:

```
class Person {
  [Required]
  public string LastName { get; set; }
}
```

UML/OCL: OCL constraint; Each *C* always has *P*
UML/OCL example:

```
context Person inv:
    not self.LastName.oclIsUndefined()
```

Each *Person* always <u>has</u> *LastName*
Comment: none
Source code element: Property *P* in class *C* for which only a getter is defined
Source code example:

```
    class Person {
    public string FullName {
       get{ return LastName + ", " + FirstMidName; }
    }}
```

UML/OCL: UML derived attribute *P* (proceeded with '/') in class *C*

UML/OCL example: *FullName* derived attribute in *Person* class
Comment: that rule could also be treated as computation rule

3.2   Applied Technologies

The application supporting extraction process was implemented in c# using Visual Studio 2013 Professional. It is based on .NET Framework 4.5, and Windows Presentation Foundation. .NET Compiler Platform "Roslyn" is used to read and manipulate existing c# solutions, and project files. GraphViz [21], and GraphViz4Net library are used to create UML class diagrams presenting extraction results.

3.3   Extraction Process

The extraction process consists of 4 stages, presented in Fig. 1.

Fig. 1. Extraction procedure – main stages.

The aim of the first stage is to load a solution or a project representing an application written in c# from which we would like to extract business rules. That is done with Roslyn. Solution/project files are organized in a tree structure, and next displayed in Solution Explorer Window (see Fig. 2).

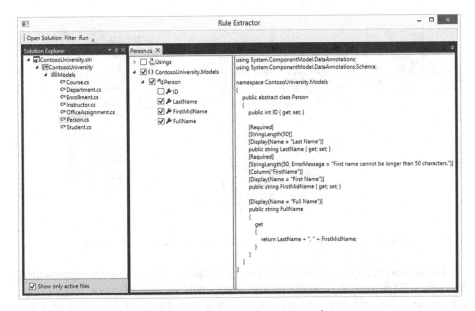

Fig. 2. Solution extraction stage results.

The second stage aims in selecting elements for which the extraction procedure is to be run. The elements could be filtered automatically, dependently on the ORM, or manually (a user can select both classes and their features).

A filter implemented for Entity Framework first looks for classes that inherit from *EntityTypeConfiguration<T>*. Class *T* is very likely a member of the application domain model. Next the filter looks for properties of *DbSet<T>* type in *DbContext*. In the last step, the filter visits all method calls trying to find *DbModel-Builder.Entity<T>*. In all cases class *T* is considered to be a part of the domain model. Having the classes the filter is ready to identify associations between them. That is done by looking for methods *HasRequired*, *HasOptional*, *HasMany*, and *WithMany* which are used to define dependencies between tables. Class property which include 'ID' is treated as artificial key, and it is automatically excluded from results of filtering process. Artificial keys typically do not represent domain values. Similarly, the filter also excludes class properties labeled with one of the .NET attributes: [Timestamp], [Key], [ForeignKey].

A filter implemented for NHibernate first analyses files with *.hbm.xml extension looking for such that has as a root of xml document <class> node. Attribute 'name' contains information about class names representing tables in our database (we consider these classes as being the elements of domain model). Nodes <property>, <many-to-one>, <bag> help in identification of class attributes as well as associations between classes.

The third stage aims in building an internal representation of extracted domain model. The model is represented as an instance of the meta-model presented in Fig. 3.

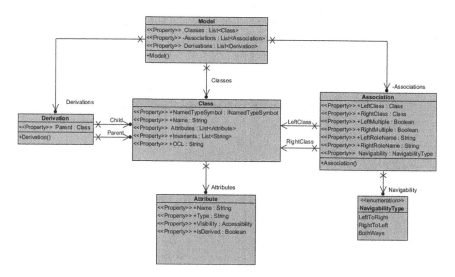

Fig. 3. Meta-model of domain model and structural business rules.

The process itself consists of five steps: (a) extraction of noun concepts represented by classes (b) extraction of verb concepts represented by generalization relationship, (c) extraction of verb concepts represented by associations, (d) extraction of noun concepts represented by class attributes, (e) extraction of structural business rules represented by constraints or derived attributes.

The fourth stage aims in translating the internal representation of the business model into graphical representation with the use of GraphViz and GraphViz4Net.

4 Example

4.1 Test Project Description

To demonstrate the results of proposed extraction method we use a tutorial project from Microsoft. "The Contoso University sample web application demonstrates how to create ASP.NET MVC 5 applications using the Entity Framework 6 and Visual Studio 2013 (...) It includes functionality such as student admission, course creation, and instructor assignments" [22].

The domain model used here is quite simple. It consists of only 7 domain classes, e.g. *Course, Person, Student, Instructor.* Correctness of input data is validated with the use of attributes from System.ComponentModel.DataAnnotations namespace.

4.2 Business Rules Implemented Within the Test Project

We have identified and counted structural business rules implemented within the test project. The project contains definition of:

- 20 general noun concepts (e.g. Course, Title, Credits, Department, Name, Budget, Start Date, Student, Person, Instructor, First Name, Last Name);
- 20 fact types (e.g. Student specializes Person, Instructor specializes Person, Course has Title, Course has Credits, Student enrolls for Course);
- 13 constraints (e.g. Each Person always has exactly one First Name; Each Person always has exactly one Last Name; Each Course always has to be tought by exactly one Department; Each Name always has to be at least 3 and at most 50 characters long).

4.3 Extraction Results

The result of extraction method is represented by a class diagram with a set of OCL constraints (see Fig. 4).

4.4 Comparison with Alternative Tools

We decided to compare results of structural business rules extraction with alternative tools existing on the market, i.e. Visual Paradigm, and Visual Studio 2013. Both mentioned tools are able first of all to extract business vocabulary (classes, attributes, generalization relations, associations).

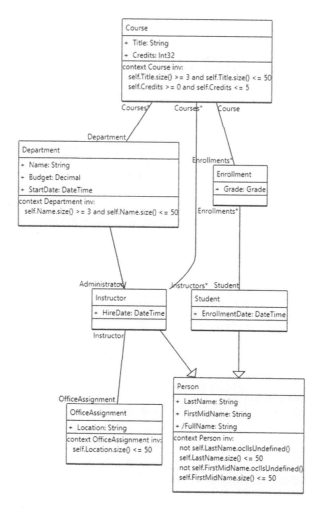

Fig. 4. Example extraction results - UML class diagram with the set of OCL constraints.

It should be mentioned that our tool offers two main functionalities, not available in the competitors:

- It is able to limit presented results to domain classes automatically; e.g. in the case of Visual Paradigm and Visual Studio a user has to manually check out classes belonging to the domain, otherwise also other classes like views or controllers will be extracted; moreover our tool hides irrelevant data, e.g. operations, artificial keys, timestamps etc. That is possible as domain classes are extracted from underlying ORM.
- It is able to extract not only the vocabulary but also structural constraints connected to class properties. That is possible as the tool is able to interpret annotations.

Table 1 presents comparison results in terms of number of extracted elements. The first column shows a real number of business rules belonging to particular category (counted manually). The following columns show numbers of business rules extracted by (a) implemented tool, (b) Visual Paradigm, (c) Visual Studio 2013. It can be noticed that proposed approach bits the other in the number of extracted constraints. The other tools are only able to extract multiplicity constraints, when one association end is limited to 1. The implemented tool does not fully support enumeration types. It is why it extracted 12 from 13 constraints (enumeration Grade type is not represented as a separate class). On the other side, the tool as the only one is able to extract derivation/computation rules.

Table 1. Effectiveness of implemented tool in comparison with alternative tools

Element type	Test project	Implemented tool	Visual Paradigm	Visual Studio
General noun concepts	20	20	20	20
Fact types	20	20	20	19
Constraints	13	12	1	4
Derivations/Computations	1	1	0	0
Total	54/100 %	53/98 %	41/76 %	43/80 %

5 Summary

The paper presents a method of automatic extraction of structural business rules from application source code written in c#. The proposed method is similar to that presented in [3], but the main difference is that the source of knowledge in [3] is a legacy database while in our case application vocabulary (noun and verb concepts) is extracted from an ORM tool. At that moment the implemented solution supports two such tools: Entity Framework, and NHibernate. Some heuristics that help in hiding irrelevant data are implemented, however a user still has a possibility to extract these elements if he or she is interested in them.

The way in which application vocabulary is retrieved can be met also in the existing tools supporting reverse engineering. However, the other tools are unable to extract most of structural business rules, e.g. constraints. What more, the other tools need a user to select folders containing the classes representing domain knowledge. Otherwise, the resulting model will be huge and difficult to understand. We obtained 302 elements: classes, attributes, operations, associations, etc. in Visual Paradigm for the test project.

In the future we are going to validate the method with larger applications, containing more implemented business rules. We hope that will help to discover and implement more filtering rules. Also a support for enumeration type is planned. The other direction of the research is to extend it with operational rules.

References

1. Putrycz, E., Kark, A.W.: Recovering business rules from legacy source code for system modernization. In: Paschke, A., Biletskiy, Y. (eds.) RuleML 2007. LNCS, vol. 4824, pp. 107–118. Springer, Heidelberg (2007)
2. Wang, X., Sun, J. Yang, X., He, Z.: Business rules extraction from large legacy systems. In: CSMR 2004, pp. 249–258 (2004)
3. Chaparro, O., Aponte, J., Ortega, F., Marcus, A.: Toward the automatic extraction of structural business rules from legacy databases. In: WCRE 2012, pp. 479–488 (2012)
4. Shao, J., Pound, C.J.: Extracting business rules from information systems. BT Technol. J. **17**(4), 179–186 (1999)
5. Tiobe index for August. http://www.tiobe.com/index.php/-content/-paperinfo/-tpci/index. html
6. http://nhibernate.info/. Accessed 28 November 2015
7. http://www.entityframeworktutorial.net/what-is-entityframework.aspx
8. Demuth, B., Hussmann, H., Loecher, S.: OCL as a specification language for business rules in database applications. In: Gogolla, M., Kobryn, C. (eds.) UML 2001. LNCS, vol. 2185, pp. 104–117. Springer, Heidelberg (2001)
9. Bajwa, I.S., Lee, M.G.: Transformation rules for translating business rules to OCL constraints. In: France, R.B., Kuester, J.M., Bordbar, B., Paige, R.F. (eds.) ECMFA 2011. LNCS, vol. 6698, pp. 132–143. Springer, Heidelberg (2011)
10. Jain, A., Soner, S., Rathore, A.S., Tripathi, A.: An approach for extracting business rules from legacy C++ code. In: ICECT 2011, pp. 90–93 (2011)
11. Cosentino, V., Cabot, J., Albert, P., Bauquel, P., Perronnet, J.: A model driven reverse engineering framework for extracting business rules out of a Java application. In: Bikakis, A., Giurca, A. (eds.) RuleML 2012. LNCS, vol. 7438, pp. 17–31. Springer, Heidelberg (2012)
12. Hay, D., Healy, K.A.: Defining business rules – what are they really? http://www.businessrulesgroup.org/first_paper/br01c0.htm. Accessed 28 November 2015
13. OMG Semantics of Business Vocabulary and Business Rules (SBVR), Version 1.2. (2013)
14. RuleSpeak®. http://www.rulespeak.com/en/. Accessed 28 November 2015
15. OMG Semantics of Business Vocabulary and Business Rules (SBVR), Version 1.2., Annex H – The RuleSpeak® Business Rule Notation (2013)
16. Odell, J.J.: Business Rules, Advanced Object-oriented Analysis and Design using UML, pp. 99–107. Cambridge University Press, Cambridge (1998)
17. Bubenko, J., Persson, A., Stirna, J.: D3: user guide of the knowledge management approach using enterprise knowledge patterns. Technical report, Stockholm, Sweden (2001). ftp.dsv.su.se/users/js/d3_km_using_ekp.pdf. Accessed 28 November 2015
18. Taveter, K., Wagner, G.: Agent-oriented enterprise modeling based on business rules. In: Kunii, H.S., Jajodia, S., Sølvberg, A. (eds.) ER 2001. LNCS, vol. 2224, pp. 527–540. Springer, Heidelberg (2001)
19. OMG Unified Modeling Language (UML), Version 2.5 (2013)
20. OMG Object Constraint Language (OCL), Version 1.3.1 (2012)
21. GraphViz – Graph Vizualization Software. http://www.graphviz.org
22. Dykstra, T.: Getting Started with Entity Framework 6 Code First using MVC 5. http://www.asp.net/mvc/overview/getting-started/getting-started-with-ef-using-mvc/creating-an-entity-framework-data-model-for-an-asp-net-mvc-application

Higher Order Mutation Testing to Drive Development of New Test Cases: An Empirical Comparison of Three Strategies

Quang Vu Nguyen[(⊠)] and Lech Madeyski

Faculty of Computer Science and Management,
Wroclaw University of Technology,
Wybrzeze Wyspianskiego 27, 50-370 Wroclaw, Poland
{Quang.vu.nguyen,Lech.Madeyski}@pwr.edu.pl

Abstract. Mutation testing, which includes first order mutation (FOM) testing and higher order mutation (HOM) testing, appeared as a powerful and effective technique to evaluate the quality of test suites. The live mutants, which cannot be killed by the given test suite, make up a significant part of generated mutants and may drive the development of new test cases. Generating live higher order mutants (HOMs) able to drive development of new test cases is considered in this paper. We apply multi-objective optimization algorithms based on our proposed objectives and fitness functions to generate higher order mutants using three strategies: HOMT1 (HOMs generated from all first order mutants), HOMT2 (HOMs generated from killed first order mutants) and HOMT3 (HOMs generated from not-easy-to-kill first order mutants). We then use mutation score indicator to evaluate, which of the three approaches is better suited to drive development of new test cases and, as a result, to improve the software quality.

Keywords: Mutation testing · Higher order mutation testing · Live mutants · Equivalent mutants · Multi-objective optimization algorithm

1 Introduction

In 1970 s, a fault-based technique was introduced by DeMillo et al. [1] and Hamlet [2] as a way to measure the effectiveness of test suites, called mutation testing (first order mutation testing). Mutants are the different versions of an original program generated by inserting, via a mutation operator, only one semantic change (or fault) into the original program. Mutation operators depend on programming languages, but there are some traditionally used mutation operators, e.g., deletion of a statement, replacement of Boolean expressions, replacement of arithmetic, replacement of a variable. Given set of test cases (TCs) is executed on the original program and all its mutants. If output result of mutant is different than the output result of original program, with any test case (TC), we say that the mutant is killed. In other words, the test case kills mutant. If a mutant was killed by all of given TCs, it is named "Easy to kill".

Conversely, if output results of mutant and original program are the same with all test cases, the mutant is called "live" or "not killed". In this case, none of the test cases in the given set of test cases can kill the mutant. This could be for two reasons: (1) The

N.T. Nguyen et al. (Eds.): ACIIDS 2016, Part I, LNAI 9621, pp. 235–244, 2016.
DOI: 10.1007/978-3-662-49381-6_23

given set of test cases is "not good enough" to detect the difference between the original program and its mutants; it drives developers to create new test cases able to kill live mutants. (2) The mutant is an equivalent mutant; it means that the mutant has the same semantic meaning as the original program and there is no test case able to kill the mutant.

The equivalent mutant problem (EMP) is one of the crucial problems in mutation testing [4, 7, 12]. This is one of the reasons why mutation testing is not yet widely adopted in practice. A lot of approaches have been proposed for overcoming the EMP (and the mutation testing's problems in general) including second order mutation testing [7–11] or higher order mutation testing [3–6] in general. Higher order mutation testing is an idea presented by Jia and Harman [3] and in a manifesto by Harman et al. [4]. This promising idea offers solutions to overcome the limitations of traditional mutation testing. Mutants can be classified into two types: First Order Mutants (FOMs) and Higher Order Mutants (HOMs). The first are used in traditional mutation testing and generated by applying mutation operators only once in each mutant. The second are used in higher order mutation testing and constructed by inserting two or more changes per mutant.

Mutation score (or mutation adequacy) was defined as the ratio of the number of killed mutants to the number of non-equivalent mutants [13]. The number or non-equivalent mutants is a difference between total number of generated mutants and number of equivalent mutants.

Mutation score indicator (MSI) is another quantitative measure of the quality of test cases. Different from MS, MSI was defined as the ratio of killed mutants to all generated mutants [7, 14–17]. MSI lies between 0 and 1. If MSI is 0, all generated mutants are live mutants. If MSI is 1, all mutants are killed. Ignoring equivalent mutants means that we accept the lower bound on mutation score. In addition, in fact many mutation operators can produce equivalent mutants of the same behaviour as the original program, while detection of equivalent mutants often involves additional human effort.

Live mutant problem includes equivalent mutants and non-equivalent mutants, which could be killed by adding high quality TCs. So, existing live mutants can drive development of new high quality TCs. Development of new high quality TCs decreases the number of live mutants due to new TCs able to kill some non-equivalent live mutants. New high quality TCs have a positive impact on software quality. Our goal is to investigate which strategy to generate HOMs gives more opportunities to drive development of high quality TCs. Three considered strategies are: (1) HOMT1 - HOMs generated from all first order mutants, (2) HOMT2 - HOMs generated from killed first order mutants and (3) HOMT3 - HOMs generated from not-easy-to-kill first order mutants.

In this paper, we apply three multi-objective optimization algorithms – NSGAII, NSGAIII and eMOEA (Epsilon-MOEA) – to generate HOMs based on our objectives and fitness functions. We use mutation score indicator (MSI) as the indicator of usefulness of higher order mutation in driving development of TCs. Furthermore, HOMs simulate faults, which require more than one change to correct them. This kind of faults, represented by HOMs, is even more realistic than faults represented by FOMs. For example, Purushothaman and Perry [23] found that there is less than 4 % probability that a one-line change will introduce a fault in the code. Hence, HOMs complement FOMs and enhance realism of mutation testing giving opportunity to simulate more realistic faults and create test cases able to spot these kind of faults. Also the number of equivalent mutants in each of the strategies (FOMT, HOMT1, HOMT2 and HOMT3) would be different. Hence, our dependent variable (MSI) is indeed an approximate measure of how useful each of the strategies can be in driving TCs development.

The rest of the paper is organized as follows. Section 2 includes our objectives and fitness function, which are applied to multi-objective optimization algorithms. Section 3 presents the experimental procedure, the proposed multi-objective optimization algorithms and real-world projects under test. Section 4 shows results of the empirical evaluation. Section 5 discusses threats to validity, while the last section presents conclusions and proposition of future works.

2 Objectives and Fitness Functions

Based on the idea "number of test cases (TCs) which can kill HOMs is as small as possible", we have proposed objectives and fitness functions [29], which we will apply in the different multi-objective optimization algorithms to generate HOMs. Some notations are explained below (See Fig. 1):

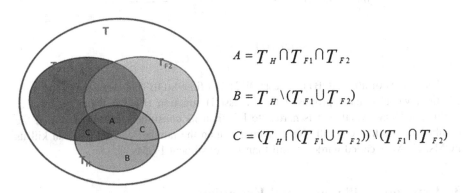

$$A = T_H \cap T_{F1} \cap T_{F2}$$

$$B = T_H \setminus (T_{F1} \cup T_{F2})$$

$$C = (T_H \cap (T_{F1} \cup T_{F2})) \setminus (T_{F1} \cap T_{F2})$$

Fig. 1. The combination of sets of TCs

H: a HOM, constructed from FOMs: F_1 and F_2
T: The given set of test cases
$T_{F1} \subset T$: Set of test cases that kill FOM1
$T_{F2} \subset T$: Set of test cases that kill FOM2
$T_H \subset T$: Set of test cases that kill H
$A \subset T_H$: Set of test cases that can kill H and all its constituent FOMs.
$B \subset T_H$: Set of test cases that kill H but cannot kill any its constituent FOMs.
$C \subset T_H$: Set of test cases that kill H and can kill FOM1 or FOM2.

The Fig. 1 showed that the set of TCs that kills a HOM can be divided into 3 subsets:

- The first is the subset that can kill HOM and all its constituent FOMs (subset A)
- The second is the subset that kills HOM and can kill FOM1 or FOM2 (subset C)
- The third is the subset that only kills HOM and cannot kill any FOMs (subset B)

From that, we have proposed objectives and their fitness functions [29] (see Equations below) to apply multi-objective optimization algorithms to construct HOMs as follows:

Objective 1: Minimize the number of TCs that kill HOM and also kill all its constituent FOMs (The fitness function is fitness(OB1) in Eq. 1).

Objective 2: Minimize the number of TCs that kill HOM but cannot kill any their constituent FOMs (The fitness function is fitness(OB2) in Eq. 2).

Objective 3: Minimize the number of TCs that kill HOM and can kill FOM1 or FOM2 (The fitness function is fitness(OB3) in Eq. 3).

$$fitness(OB1) = \frac{\#(T_H \cap T_{F1} \cap T_{F2})}{\#T_H} \tag{1}$$

$$fitness(OB2) = \frac{\#(T_H \setminus (T_{F1} \cup T_{F2}))}{\#T_H} \tag{2}$$

$$fitness(OB3) = \frac{\#((T_H \cap (T_{F1} \cup T_{F2})) \setminus (T_{F1} \cap T_{F2}))}{\#T_H} \tag{3}$$

$$fitness(H) = \frac{\#T_H}{\#(T_{F1} \cup T_{F2})} \tag{4}$$

The values of fitness(OB1), fitness(OB2) and fitness(OB3) lie between 0 and 1. In addition, we also have proposed the fitness(H) function (Eq. 4) which is used to evaluate a HOM whether it is harder to kill than its constituent FOMs or not. If the number of TCs that can kill HOM is smaller than the number of TCs that can kill its FOMs, HOM is called harder to kill than its constituent FOMs.

3   Experiment Planning and Execution

The aim of our experiment was to answer the research question: How to combine FOMs to create hard to kill HOMs (well suited to evaluate the quality of test cases and drive their development)?

3.1   Supporting Tool

We use Judy tool [7, 17] to conduct the empirical studies. Judy (http://www.mutationtesting.org/) is a mutation testing tool for Java programs. It supports large set of mutation operators, as well as HOM generation, HOM execution and mutation analysis.

3.2   Multi-Objective Optimization Algorithms

NSGA-II is the second version of the Non-dominated Sorting Genetic Algorithm that was proposed by Deb et al. [18] for solving non-convex and non-smooth single and multi-objective optimization problems. Its main features are: it uses an elitist principle; it emphasizes non-dominated solutions; and it uses an explicit diversity preserving mechanism. NSGA-III is the extension of NSGA-II which is based on the supply of a set of reference points and demonstrated its working in 3 to 15-objective optimization problems [19]. The εMOEA (eMOEA) is a steady state multi-objective evolutionary algorithm that co-evolves both an evolutionary algorithm population and an archive population by randomly mating individuals from the population and the archive to generate new solutions [20, 21].

3.3   Projects Under Test (PUT)

In our empirical study we use five real-world, open source projects (see Table 1) which were downloaded from the SourceForge website (http://sourceforge.net). Table 1 shows the projects selected for the experiment along with their number of classes (NOC), lines of code (LOC) and number of given test cases (#TCs).

Table 1.   Projects under test

Project	NOC	LOC	#TCs
BeanBin	72	5925	68
Barbecue	57	23996	190
JWBF	51	13572	305
CommonsChain 1.2	103	13410	17
CommonsValidiator 1.4.1	144	25422	66

3.4   Experimental Procedure

For each project under test, we ran the process, which was described in following experimental procedure, 5 times. HOMs were generated in three ways. Firstly, HOMs were created by combining FOMs from the set of all generated FOMs. And second one, delete first live FOMs from set of generated FOMs, then create HOMs by combining FOMs from the set of killed FOMs. And the last, first delete all of easy to kill FOMs, which were killed by all of given TCs, from set of generated FOMs, then create HOMs by combining FOMs from the set of not-easy-to-kill FOMs Then we calculated the average value of each program for each algorithm. We set out the experimental procedure as follows:

```
for each software under test do
Generate all possible FOMs by applying the set of Judy
mutation operators
Count and save MSI of first order mutation testing
Set objectives and fitness functions
   for each multi-objective optimization algorithm do
      - set populationSize =100
      - set maxMutationOrder =15
      - from set of all FOMs, generate and evaluate HOMs,
guided by objectives and fitness functions
      - count and save MSI of higher order mutation testing
      - delete the live FOMs from set of all generated FOMs
      - from set of remaining-FOMs, generate and evaluate
HOMs, guided by objectives and fitness functions
      - count and save MSI of higher order mutation testing
      - delete the easy-to-kill FOMs from set of all
generated FOMs
      - from set of remaining-FOMs, generate and evaluate
HOMs, guided by objectives and fitness functions
      - count and save MSI of higher order mutation testing

   end

end
```

4 Results and Analysis

Results were shown in Table 2. FOMT is implementation of first order mutation testing. HOMT1 is the implementation of higher order mutation testing where HOMs are generated on a basis of all FOMs. HOMT2 is the implementation of higher order mutation testing where HOMs are generated on a basis of killed FOMs. HOMT3 is the implementation of higher order mutation testing where HOMs are generated on a basis of not-easy-to-kill FOMs.

The results presented in Table 2 indicate that the given sets of TCs of PUTs have lower MSI in first order mutation testing. It means that there are many live FOMs and the given sets of TCs are not good enough to detect the difference between original program and their mutants and, therefore, need to be improved following the results of mutation analysis based on the FOMT strategy. The numbers of live FOMs makes up from 52 % to 87 % of generated mutants. Only a small number of FOMs were killed by the given sets of TCs. In the case of live FOMs, we have to check whether the live FOMs are equivalent mutants or not, but it often involves additional human effort. If mutants are not equivalent, developers or testers create new TCs and check whether they are able to kill live FOMs. If live FOMs are equivalent mutants, TCs, which can kill them, do not exist.

Table 2. The mean value of MSI for each project under test (%)

Project Under Test (PUT) / Strategy		Barbecue	BeanBin	Commons Chain	Commons Validator	JWBF
FOMT		15.79	15.11	42.65	47.10	12.96
HOMT 1	NSGAII	70.59	36.32	89.92	92.41	94.54
	NSGAIII	69.67	43.04	84.38	92.31	91.30
	eMOEA	69.19	41.04	87.55	93.15	91.20
HOMT 2	NSGAII	100	100	100	100	100
	NSGAIII	100	100	100	100	100
	eMOEA	96.15	100	100	99.31	100
HOMT 3	NSGAII	64.01	62.11	40.12	83.93	77.78
	NSGAIII	54.23	59.23	50.28	86.05	80.00
	eMOEA	73.59	69.42	49.67	87.76	90.91

The most striking result is that the HOMT2 strategy appeared to be useless as it gives a false impression that TCs are of high quality (MSI is equal or close to 100 %) and the usefulness of HOMT2 is strongly limited, i.e., opportunities of test case improvement guided by results of HOMT2 mutation analysis are rare if any. Almost all of higher degree mutants, which were constructed by combining the killed FOMs, are also killed. This indicates that, combining first order killed mutants to create higher degree mutants is not a good way to evaluate and improve the quality of given set of test cases because the generated HOMs are easy to kill.

The HOMT1 and HOMT3 strategies seem to be better and offer more opportunities to improve the quality of given set of test cases, as MSI (and the number of killed mutants) decreased in comparison to HOMT2.

The experimental results indicated that, we should not use first order live mutants to create difficult (but possible) to kill higher order mutants. And using not-easy-to-kill mutants to generate higher order mutants is a promising method, which could be applied to the area of higher order mutation testing to evaluate and improve the quality of given set of TCs.

5 Threats to Validity

Equivalent mutants constitute a threat to validity because the ratio of equivalent mutants in each strategy is unknown, while the problem of detecting equivalence between two mutants is an undecidable problem [7]. Furthermore, using five selected projects under test (PUTs) may not be representative of all Java programs in general and therefore, the results of the study may not be generalizable to all Java programs. Additionally the number of evaluated strategies is limited and, we think that further investigations would allow proposing new strategies for generating difficult (but possible) to kill higher order mutants. Applying other multi-objective optimization algorithms as well as the large PUTs is also needed to improve the obtained results.

6 Conclusions and Future Work

We applied our objectives and fitness functions to multi-objective optimization algorithms for constructing HOMs from the set of generated FOMs in three ways. In the first one, we used all of the FOMs, in the second one, we used the FOMs, which were killed by at least one TC, while in the third one, we used not-easy-to-kill FOMs. The results indicated that applying multi-objective optimization in the area of higher order mutation testing to generate HOMs could be an interesting complementary approach to FOMT, but the strategy of selecting FOMs to build HOMs is of great importance. The strategy one should absolutely avoid is to build HOMs on a basis of killed FOMs (i.e., HOMT2). The alternative strategies HOMT1 and HOMT3, where HOMs are built on a basis of all FOMs give better results. The obtained results suggest the direction of further investigation, which could be a strategy where HOMs are build, for example, on a basis of live FOMs and/or HOMs of lower degree.

Applying multi-objective optimization algorithms to generate higher order mutants is a promising way for overcoming the limitations of mutation testing [5, 22]. In our previous work [22], the results of our experiment indicated that our approach is able to reduce the generated HOMs compared with FOMs as well as is useful in constructing higher order mutants. In this paper, we shed additional light on usefulness of higher order mutation strategies to drive development of new, high quality test cases.

In further research, we will investigate how process metrics [24, 25], based on development of test cases, combined with product metrics [26], based on mutation testing, can improve software defect prediction models we build in collaboration with our industrial partners [27, 28].

References

1. DeMillo, R.A., Lipton, R.J., Sayward, F.G.: Hints on test data selection: help for the practicing programmer. IEEE Comput. **11**(4), 34–41 (1978)
2. Hamlet, R.G.: Testing programs with the aid of a compiler. IEEE Trans. Softw. Eng. **SE-3**(4), 279–290 (1977)
3. Jia, Y., Harman, M.: Higher order mutation testing. Inf. Softw. Technol. **51**, 1379–1393 (2009)
4. Harman, M., Jia, Y., Langdon, W.B.: A manifesto for higher order mutation testing. In: Third International Conference on Software Testing, Verification, and Validation Workshops, (2010)
5. Langdon, W.B., Harman, M., Jia, Y.: Efficient multi-objective higher order mutation testing with genetic programming. J. Syst. Softw. **83**, 2416–2430 (2010)
6. Jia, Y., Harman, M.: Constructing subtle faults using higher order mutation testing. In: Proceedings of the Eighth International Working Conference Source Code Analysis and Manipulation (2008)

7. Madeyski, L., Orzeszyna, W., Torkar, R., Józala, M.: Overcoming the equivalent mutant problem: a systematic literature review and a comparative experiment of second order mutation. IEEE Trans. Softw. Eng. **40**(1), 23–42 (2014). doi:10.1109/TSE.2013.44

8. Mresa, E.S., Bottaci, L.: Efficiency of mutation operators and selective mutation strategies: an empirical study. Softw. Test. Verification Reliab. **9**(4), 205–232 (1999)

9. Papadakis, M., Malevris, N.: An empirical evaluation of the first and second order mutation testing strategies. In: Proceedings of the 2010 Third International Conference on Software Testing, Verification, and Validation Workshops, ser. ICSTW 2010, IEEE Computer Society, pp. 90–99 (2010)

10. Vincenzi, A.M.R., Nakagawa, E.Y., Maldonado, J.C., Delamaro, M.E., Romero, R.A.F.: Bayesian-learning based guidelines to determine equivalent mutants. Int. J. Softw. Eng. Knowl. Eng. **12**(6), 675–690 (2002)

11. Polo, M., Piattini, M., Garcia-Rodriguez, I.: Decreasing the cost of mutation testing with second-order mutants. Softw. Test. Verification Reliab. **19**(2), 111–131 (2008)

12. Nguyen, Q.V., Madeyski, L.: Problems of mutation testing and higher order mutation testing. In: van Do, T., Thi, H.A.L., Nguyen, N.T. (eds.) Advanced Computational Methods for Knowledge Engineering. AISC, vol. 282, pp. 157–172. Springer, Heidelberg (2014). doi:10.1007/978-3-319-06569-4_12

13. Zhu, H., Hall, P.A.V., May, J.H.R.: Software Unit Test Coverage and Adequacy. ACM Comput. Surv. **29**(4), 366–427 (1997)

14. Madeyski, L.: On the effects of pair programming on thoroughness and fault-finding effectiveness of unit tests. In: Münch, J., Abrahamsson, P. (eds.) PROFES 2007. LNCS, vol. 4589, pp. 207–221. Springer, Heidelberg (2007). doi:10.1007/978-3-540-73460-4_20

15. Madeyski, L.: The impact of pair programming on thoroughness and fault detection effectiveness of unit tests suites. Softw. Process: Improv. Pract. **13**(3), 281–295 (2008). doi:10.1002/spip.382

16. Madeyski, L.: The impact of test-first programming on branch coverage and mutation score indicator of unit tests: An experiment. Inf. Softw. Technol. **52**(2), 169–184 (2010). doi:10.1016/j.infsof.2009.08.007

17. Madeyski, L., Radyk, N.: Judy - a mutation testing tool for Java. IET Softw. **4**(1), 32–42 (2010). doi:10.1049/iet-sen.2008.0038

18. Deb, K., Pratap, A., Agarwal, S., Meyarivan, T.: A fast and elitist multi objective genetic algorithm: NSGA-II. IEEE Trans. Evol. Comput. **6**(2), 182–197 (2002)

19. Deb, K., Jain, H.: An evolutionary many-objective optimization algorithm using reference-point-based nondominated sorting approach, Part I: solving problems with box constraints. IEEE Trans. Evol. Comput. **18**(4), 577–601 (2014)

20. Kollat, J.B., Reed, P.M., The value of online adaptive search: a performance comparison of NSGAII, ε-NSGAII and ε-MOEA. In: Coello Coello, C.A., Aguirre, A.H., Zitzler, E. (Eds.), Evolutionary Multi-Criterion Optimization, Third International Conference, EMO 2005 Guanajuato, Mexico, March 9–11 (2005)

21. Deb, K., Mohan, M., Mishra, S.: A fast multi-objective evolutionary algorithm for finding well-spread pareto-optimal solutions. KenGAL, Report No. 2003002. Indian Institute of Technology, Kanpur, India (2003)

22. Nguyen, Q.V., Madeyski, L.: Searching for strongly subsuming higher order mutants by applying multi-objective optimization algorithm. In: Le Thi, H.A., Nguyen, N.T., Do, T.V. (eds.) Advanced Computational Methods for Knowledge Engineering. AISC, vol. 358, pp. 391–402. Springer, Heidelberg (2015). doi:10.1007/978-3-319-17996-4_35

23. Purushothaman, R., Perry, D.E.: Toward Understanding the Rhetoric of small source code changes. IEEE Trans. Softw. Eng. **31**(6), 511–526 (2005)

24. Madeyski, L., Jureczko, M.: Which process metrics can significantly improve defect prediction models? An empirical study. Softw. Qual. J. **23**(3), 393–422 (2015). doi:10.1007/s11219-014-9241-7

25. Jureczko, M., Madeyski, L.: A review of process metrics in defect prediction studies. Metody Informatyki Stosowanej **30**(5), 133–145 (2011). http://madeyski.e-informatyka.pl/download/Madeyski11.pdf

26. Jureczko, M., Madeyski, L.: Towards identifying software project clusters with regard to defect prediction. In: Proceedings of the 6th International Conference on Predictive Models in Software Engineering (PROMISE 2010). ACM, New York, NY, USA, Article 9, 9:1–9:10. doi: 10.1145/1868328.1868342

27. Madeyski, L., Majchrzak, M.: Software measurement and defect prediction with DePress extensible framework. Found. Comput. Decis. Sci. **39**(4), 249–270 (2014). doi:10.2478/fcds-2014-0014

28. Hryszko, J., Madeyski, L.: Bottlenecks in software defect prediction implementation in industrial projects. Found. Comput. Decis. Sci. **40**(1), 17–33 (2015). doi:10.1515/fcds-2015-0002

29. Nguyen, Q.V., Madeyski, L.: Empirical evaluation of multi-objective optimization algorithms searching for higher order mutants. Cybern. Syst. Int. J. (Accepted 2016). DOI: 10.1080/01969722.2016.1128763 URL: http://madeyski.e-informatyka.pl/download/NguyenMadeyski16CS.pdf

On the Relationship Between the Order of Mutation Testing and the Properties of Generated Higher Order Mutants

Quang Vu Nguyen[(✉)] and Lech Madeyski

Faculty of Computer Science and Management,
Wroclaw University of Technology, Wroclaw, Poland
quang.vu.nguyen@pwr.edu.pl

Abstract. The goal of higher order mutation testing is to improve mutation testing effectiveness in particular and test effectiveness in general. There are different approaches which have been proposed in the area of second order mutation testing and higher order mutation testing with mutants order ranging from 2 to 70. Unfortunately, the empirical evidence on the relationship between the order of mutation testing and the desired properties of generated mutants is scarce except the conviction that the number of generated mutants could grow exponentially with the order of mutation testing. In this paper, we present the study of finding the relationships between the order of mutation testing and the properties of mutants in terms of number of generated high quality and reasonable mutants as well as generated live mutants. Our approach includes higher order mutants classification, objective functions and fitness functions to classify and identify generated higher order mutants. We use four multi-objective optimization algorithms for constructing higher order mutants. Obtained empirical results indicate that 5 is a relevant highest order in higher order mutation testing.

Keywords: Mutation testing · Higher order mutation · Higher order mutants · Multi-objective optimization algorithm

1 Introduction

Mutation testing has been considered as one of the most effective techniques for evaluating the quality of given sets of test data which are used in software testing. The technique is applied to assess the quality of given sets of test cases (TCs) based on their ability of detecting the differences between program under test (PUT) and its mutants [1,2]. The mutants are different PUT versions, which are produced by syntactically altering the source code of the PUT. The syntactic changes are called mutation operators. After executing the given set of test cases on the original program (PUT) and each of its mutants, mutation testing evaluates the quality of test cases by mutation score (MS) or mutation score indicator (MSI). MS is defined as the ratio of killed mutants to the differences

© Springer-Verlag Berlin Heidelberg 2016
N.T. Nguyen et al. (Eds.): ACIIDS 2016, Part I, LNAI 9621, pp. 245–254, 2016.
DOI: 10.1007/978-3-662-49381-6_24

of all generated mutants and equivalent mutants [1,2]. While MSI is the ratio of killed mutants to all generated mutants [9–12].

Many approaches (e.g., selective mutation, sampling mutation and weak mutation) have been proposed to overcome limitations of mutation testing [6,14]. Higher order mutation testing, an idea of Jia and Harman in 2009 [3,5], is one of the most promising solutions. Instead of using only one simple change as the traditional mutation testing [1,2], higher order mutation testing uses more complex changes to generate mutants by applying two or more mutation operators. An n-order mutant is created by n mutation operators, for example, it can be generated by combining n first order mutants. Hence, higher order mutants reflect more realistic complex faults and can be harder to kill than first order mutants [3–5,8]. Strongly subsuming higher order mutants (HOMs) [3,5] can be used to replace all of n constituent first order mutants (FOMs). This is not only without loss of test effectiveness but potentially can reduce the cost of mutation testing execution by reducing the number of generated mutants.

Equivalent mutant is the one that has the same semantic meaning as the original program and thus cannot be detected any test suite [1,2]. Higher order mutation testing can also be helpful to overcome equivalent mutants problem (EMP) [12] which is a serious, long-standing problem of mutation testing.

In this paper, our research focuses on finding the "relevant highest order" of higher order mutants based on the relationships between the order of mutation testing and the properties of mutants. We apply multi-objective optimization algorithms to search for valuable HOMs and investigate the relationships between the order of mutation testing and the properties of mutants in term of ability for constructing high quality and reasonable HOMs, as well as generating live HOMs. High quality and reasonable HOM, one of 11 HOM types that were classified by us [15,16], is a HOM which is harder to kill than any constituent FOMs and is only killed by the subset of the intersection of set of test cases that kill each constituent FOM. This definition is the same as the definition of Strongly subsuming and coupled HOM in the classification by Harman et al. [3,5] which we extended. Live mutants are the mutants which cannot be killed by the given test suite but could be killed by new quality TCs [17]. In this case, we have to create new TCs to improve the fault detection effectiveness of the existing set of TCs.

The rest of the paper is organized as follows. Section 2 is the overview of the proposed approaches in higher order mutation testing. Section 3 presents the experiment goals, the multi-objective optimization algorithms and implementation details. Section 4 includes the results to answer the posed research questions. The last section includes the conclusions and further work.

2   Related Work

Second order mutation testing is a specific case of higher order mutation testing. According to results of works on second order mutation testing, not only at least 50 % of mutants were reduced [12,18,19] without loss of effectiveness of testing, but also the number of equivalent mutants can be reduced (i.e. the reduction in the mean percentage of equivalent mutants passes from about 18.66 % of total of FOMs to about 5 % of total of HOMs [19] or the mean reduction of equivalent HOMs is about 50 % compared with FOMs [12]) and generated second order mutants can be harder to kill than first order mutants [7,12,18,19].

In 2009, Jia and Harman [3,5] defined a new paradigm for mutation testing—higher order mutation testing—and the rules to classify HOMs. They then used search-based optimization algorithms to find and evaluate the proportion of subsuming, as well as strongly subsuming HOMs to all generated HOMs. Their experiments showed approximately 15 % of all found Subsuming HOMs are strongly subsuming HOMs and they also indicated that finding such HOMs may not be too difficult. The highest order of their experiment is 9 and they summarized that "the highest order mutants may find application in attempts to reduce mutation effort because they subsume the largest number of FOMs" [3].

Langdon et al. [8] suggested inserting "semantically close" faults instead of inserting "syntactically close" faults to the original program under test in order to produce better mutants. Their opinions are based on the claim of Purushothaman and Perry [20], who indicated that a modification made to fix one real fault needs several source code changes. From analysis of the relationship between syntax and semantics, Langdon et al. [8] introduced two objectives, small semantic distance and minimum syntactic changes, which were applied using NSGA-II multi-objective optimization algorithm. Their goal is to find higher order mutants that represent more realistic complex faults and are harder to kill. The highest order of their experiment is 70 and the number of mutants grows exponentially with order.

Omar et al. [17] presented the approach using search-based algorithms for finding subtle HOMs with a new objective function to identify subtle HOMs. They defined subtle HOMs as HOMs that are not killed by a given set of test cases but can be killed by other new test cases. They set up different maximum orders for each algorithm. For example, 25 is the maximum HOM degree for Random Search Algorithm and 15 is the maximum HOM degree for Genetic Algorithm. Their results indicate that the ability in finding subtle HOMs of lower degrees or higher degrees belongs to different algorithms [17].

In our previous work [15,16], with 15 being the highest considered degree of mutants, we used multi-objective optimization algorithms for finding valuable high quality and reasonable HOMs (strongly subsuming and coupled HOMs) based not only on a new classification of HOMs but also new objective and fitness functions. The results indicated that our approach can be useful in searching for available high quality and reasonable HOMs, and among them, eNSGA-II is one of the best algorithms. In this paper we use the classification of HOMs, as well as objective and fitness functions proposed by us [15,16]. There are eleven categories of HOMs (see Table 1) and they are identified on a basis of the values of 4 fitness functions [15,16].

H1 category describes live (and potentially equivalent) mutants, which cannot be killed by the given set of test cases.

Table 1. Eleven categories of HOMs (see [16] for details)

Name	HOM is
H1	Live (potentially equivalent) Mutant (T_H is null)
H2	Non-Quality, Un-Reasonable and With New TCs
H3	Non-Quality, Un-Reasonable and With Mixed TCs
H4	Non-Quality, Reasonable and With New TCs
H5	Non-Quality, Reasonable and With Mixed TCs
H6	Non-Quality, Reasonable and With Old TCs
H7	**High quality and Reasonable**
H8	Quality, Reasonable and With Mixed TCs
H9	Quality, Reasonable and With Old TCs
H10	Quality, Un-Reasonable and With Old TCs
H11	Quality, Un-Reasonable and With Mixed TCs

3   Experiment Goals and Set up

3.1   Goals

The study will answer the research questions posed as follows:

RQ1. What are the ratios of number of HOMs in the identified mutant categories (H1-H11) to all generated HOMs for different orders?

By means of this question, we want to obtain the number of generated HOMs in the identified mutant categories. A number of generated HOMs will be collected and classified according to the kind of HOMs and according to the order of HOMs.

RQ2. What are the ratios of high quality and reasonable HOMs (H7) to all generated HOMs for different orders?

This question is, in fact, a part of the previous research question focused on a kind of mutants being of special interest. The aim is to obtain the frequency of generating high quality and reasonable HOMs (H7). Such mutants not only reflect harder to kill, realistic, complex faults but also could be used to replace all of its constituent FOMs.

RQ3. What are the ratios of "live (potentially equivalent) mutants" (H1) to all generated HOMs for different orders?

Answering this question may shed some light while trying to find the relevant highest order of mutation testing.

3.2   Experimental Units and Material

In this study we use the same mutation testing tool for Java called Judy, including also multi-objective optimization algorithms for searching HOMs, and three different projects under test for this study as in our previous works [15,16].

Judy[1] [12,13] mutation testing tool for Java not only provide the large set of mutation operators but also has build-in support for HOMs generation, higher order mutation testing execution and mutation analysis.

Four multi-objective optimization algorithms (NSGA-II, eNSGA-II, NSGA-III and eMOEA) are implemented to produce and evaluate HOMs based on our objective and fitness functions [16].

In our empirical study, we use three projects under test (PUT) [16], which are real-world software projects. Table 2 shows lines of code (LOC), number of classes (NOC) and number of given test cases (NOT) of the three selected open source projects.

Table 2. Software projects under test

Project under test (PUT)	LOC	NOC	NOT
BeanBin[a]	5925	72	68
Barbecue[b]	23996	57	190
JWBF (Java Wiki Bot Framework)[c]	13572	51	305

[a]http://beanbin.sourceforge.net
[b]http://barbecue.sourceforge.net
[c]http://jwbf.sourceforge.net

3.3   Approach

We set out the experimental procedure as follows (for each software, we run each algorithm 3 times, after then we calculate the average numbers to evaluate):

```
for each software under test do
    for each algorithm do
        loop 3 times
            .generate all possible FOMs by applying
            the set of Judy mutation operators
            .set objective and fitness functions
            for each multi-objective optimization algorithm do
                .set populationSize =100
                .set maxMutationOrder =15
                .from set of FOMs, generate and evaluate HOMs,
                guided by objectives and fitness functions
                .calculate the numbers to answer RQs
            end for
        end loop
        .calculate the mean values
    end for
end for
```

---

[1] http://www.mutationtesting.org/.

4   Results and Analysis

The maximum mutation order in our experiment is 15. Differences in the source code of the three SUTs resulted in some variations in mutation operators finally used to generate HOMs [15, 16].

Table 3. The ratios of number of HOMs in the identified mutant categories to all generated HOMs [%]

HOMs	eMOEA	NSGAII	eNSGAII	NSGAIII
H1	35.58	34.16	37.61	34.26
H2	0.06	0.06	0	0.14
H3	5.11	6.72	4.18	6.28
H4	4.32	3.66	6.89	4.81
H5	0.33	0.59	0	0.2
H6	46.45	46.11	38.46	45.41
H7	3.68	4.12	6.18	3.62
H8	0.06	0	0	0
H9	0.26	0.45	0.71	0.4
H10	3.54	3.53	4.75	4.16
H11	0.59	0.59	1.18	0.73

Table 4. The ratios of number of HOMs of a particular order (2–15) to all generated HOMs [%]

Order	eMOEA	NSGAII	eNSGAII	NSGAIII
2	11.65	12.57	9.34	12.18
3	15.67	14.70	14.12	14.82
4	14.69	16.11	14.01	14.94
5	13.57	15.08	14.73	12.56
6	11.07	8.93	7.44	9.39
7	7.81	6.93	7.08	8.53
8	5.96	5.57	3.34	5.22
9	4.13	4.47	5.28	4.31
10	3.60	4.08	3.13	3.51
11	2.10	2.52	5.14	3.37
12	3.01	3.55	5.03	2.84
13	2.81	1.49	4.67	3.31
14	1.63	2.19	3.47	2.38
15	2.30	1.81	3.22	2.64

Answer to RQ1

To answer this question, we calculate the ratios of number of HOMs in each of the identified mutant categories (H1-H11) to all generated HOMs (see Table 3), as well as the ratios of number of HOMs of a particular order (2–15) to all generated HOMs (see Table 4).

The ratios of number of HOMs in the identified mutant categories to all generated HOMs are similar for four algorithms, see also Fig. 1.

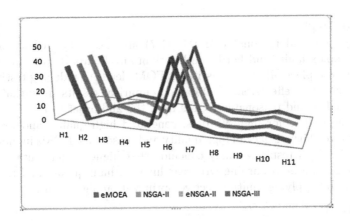

Fig. 1. The ratios of number of (H1-H11) to all generated HOMs [%]

The mean ratio of H1 mutants to all HOMs is around 35 % and the ratio of high quality and reasonable HOMs to all HOM is from 3.68 % to 6.42 %. H1 mutants in this case are live mutants, which cannot be killed by the given test suite but could be killed by some other new quality TCs [17]. We need further investigation to evaluate whether the HOMs are really equivalent mutants or not. It includes creating the new quality TCs to detect the difference between PUT and its non-equivalent mutants which belong to the set of live mutants.

A high number of H6 (see Table 3 and Fig. 1) shows that there are many generated HOMs which are more difficult to be killed than FOMs and only be killed by TCs belonging to the union of sets of TCs that can kill their constituent FOMs, except the TCs that can kill simultaneously all their constituents. The ratio of total of reasonable HOMs (H4-H9) to all of generated HOMs is fairly high, over 55 % of total generated HOMs. This indicates that we can find the mutants that are harder to kill and more realistic (reflecting real, complex faults) than FOMs by applying multi-objectives optimization algorithm. Approximately 9 % of reasonable HOMs (H4-H9) are classified as high quality and reasonable HOMs (H7). This number is high because the ratio of all reasonable HOMs to all generated HOMs is quite a large.

Table 4 describes the ratios of generated HOMs of a particular order to all of generated HOMs. The results indicated that generally for lower orders the number of generated HOMs is larger than for higher orders, for all of our four search-based algorithms.

Table 5. The mean ratios of H7 to all generated HOMs per order [%]

Order	2	3	4	5	6	7	8	9	10	11	12	13	14	15
eMOEA	11.80	6.27	5.76	1.88	0.53	0.00	0.99	0.00	1.64	0.00	0.00	0.00	0.00	0.00
NSGAII	14.37	7.53	3.98	2.19	0.65	0.84	0.00	1.30	0.00	0.00	0.00	0.00	0.00	0.00
eNSGAII	20.38	11.96	8.46	4.88	0.00	0.00	0.00	0.00	0.00	0.00	2.14	0.00	0.00	0.00
NSGAIII	13.05	6.69	3.98	1.11	0.00	0.70	0.00	1.38	0.00	0.00	0.00	0.00	0.00	0.00

Answer to RQ2

High quality and reasonable HOMs (H7) are the HOMs which are more realistic complex faults and harder to kill than any FOMs [15,16]. In addition, using them to replace all of its constituent FOMs leads to reducing testing costs without loss of test effectiveness. Obtained empirical results show that we can find high quality and reasonable HOMs from the 2nd-order to the 5th. For the 6th-order, as well as for higher orders, generated high quality and reasonable HOMs are rare. There is lack of high quality and reasonable HOMs in many cases (see Table 5). As a result, we may conclude that higher order mutation up to the 5-th order can be rewarding wrt. searching for high quality and reasonable HOMs (H7) by applying multi-objective optimization algorithms.

Answer to RQ3

Table 6 shows the mean ratios of live (and potentially equivalent) mutants (H1) to all produced HOMs according to orders. The number of H1 mutants is quite large–22 to 55 % of the generated HOMs. Live mutants include non-equivalent mutants and equivalent mutants. Non-equivalent mutants can be killed by some new quality TCs. Equivalent mutants are really same-semantic-meaning versions of the original program under test and cannot be killed by any test suite. In this case, we need a further investigation to evaluate whether live mutants are equivalent or not (a thorough review of the possible approaches and their classification is presented by Madeyski et al. [12]). This leads to creating new high quality TCs to improve the fault detection effectiveness of the existing set of test cases.

Table 6. The mean ratios of H1 to all generated HOMs per order [%]

Order	2	3	4	5	6	7	8	9	10	11	12	13	14	15
eMOEA	59.53	41.41	33.07	22.75	25.40	35.26	32.01	36.67	27.32	37.38	30.72	37.06	36.14	40.17
NSGAII	50.54	43.59	32.53	26.13	28.91	29.97	38.33	27.39	33.33	30.77	27.32	35.06	26.55	32.26
eNSGAII	58.85	50.13	26.41	22.68	22.71	30.46	35.48	40.82	42.53	53.85	42.86	46.15	41.24	36.67
NSGAIII	48.45	42.24	28.82	29.54	23.89	33.26	26.62	27.65	41.24	37.06	48.95	22.16	35.83	30.08

FINDING: *5 is a relevant highest order in higher order mutation testing as the ratio of high quality and reasonable HOMs (H7) to total number of HOMs (generated using multi-objective optimization algorithms) is high for orders between 2 and 5. This ratio is low, close to zero, for orders higher than 5, while the ratio* *of live (and potentially equivalent) mutants to total number of HOMs is large for every order* (see Fig. 2).

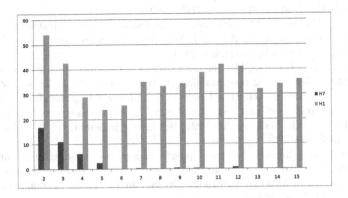

Fig. 2. The ratios of H1 and H7 to all generated HOMs per order [%]

5   Conclusions and Future Work

In this paper, we have investigated the relationships between the order of mutation testing and the properties of generated higher order mutants. We evaluated the results on a basis of generated different kinds of HOMs (H1-H11), especially the number of generated high-quality-reasonable HOMs (H7) and the number of generated live HOMs (H1). The empirical results indicated that 5 can be a relevant highest order in higher order mutation testing.

Using only three selected projects under test (PUTs) may not be a representative sample of all Java programs in general and therefore, the results of the study may not be generalizable to all Java programs as well as other programming language. Hence further research is recommended. Nevertheless, we believe that this study is a step towards unveiling the relationship between the order of mutation testing and the properties of generated mutants represented by our classification of higher order mutants.

References

1. DeMillo, R.A., Lipton, R.J., Sayward, F.G.: Hints on test data selection: help for the practicing programmer. IEEE Comput. **11**(4), 34–41 (1978)
2. Hamlet, R.G.: Testing programs with the aid of a compiler. IEEE Trans. Software Eng. **3**(4), 279–290 (1977)
3. Harman, M., Jia, Y., Langdon, W.B.: A manifesto for higher order mutation testing. In: Proceedings of the 2010 Third International Conference on Software Testing, Verification, and Validation Workshops. ICSTW 2010, pp. 80–89. IEEE Computer Society, Washington, DC (2010)
4. Jia, Y., Harman, M.: Constructing subtle faults using higher order mutation testing. In: Source Code Analysis and Manipulation. pp. 249–258 (2008)

5. Jia, Y., Harman, M.: Higher order mutation testing. Inf. Softw. Technol. **51**(10), 1379–1393 (2009)
6. Jia, Y., Harman, M.: An analysis and survey of the development of mutation testing. IEEE Trans. Softw. Eng. **37**(5), 649–678 (2011)
7. Kintis, M., Papadakis, M., Malevris, N.: Evaluating mutation testing alternatives: a collateral experiment. In: 2010 17th Asia Pacific Software Engineering Conference (APSEC), pp. 300–309, November 2010
8. Langdon, W.B., Harman, M., Jia, Y.: Efficient multi-objective higher order mutation testing with genetic programming. J. Syst. Softw. **83**(12), 2416–2430 (2010)
9. Madeyski, L.: On the effects of pair programming on thoroughness and fault-finding effectiveness of unit tests. In: Münch, J., Abrahamsson, P. (eds.) PROFES 2007. LNCS, vol. 4589, pp. 207–221. Springer, Heidelberg (2007). doi:10.1007/978-3-540-73460-4_20
10. Madeyski, L.: Impact of pair programming on thoroughness and fault detection effectiveness of unit test suites. Softw. Process Improv. Pract. **13**(3), 281–295 (2008). doi:10.1002/spip.382. http://madeyski.e-informatyka.pl/download/Madeyski08.pdf
11. Madeyski, L.: The impact of test-first programming on branch coverage and mutation score indicator of unit tests: an experiment. Inf. Softw. Technol. **52**(2), 169–184 (2010). doi:10.1016/j.infsof.2009.08.007
12. Madeyski, L., Orzeszyna, W., Torkar, R., Józala, M.: Overcoming the equivalent mutant problem: a systematic literature review and a comparative experiment of second order mutation. IEEE Trans. Softw. Eng. **40**(1), 23–42 (2014). doi:10.1109/TSE.2013.44
13. Madeyski, L., Radyk, N.: Judy - a mutation testing tool for java. IET Softw. **4**(1), 32–42 (2010). doi:10.1049/iet-sen.2008.0038
14. Nguyen, Q.V., Madeyski, L.: Problems of mutation testing and higher order mutation testing. In: van Do, T., Thi, H.A.L., Nguyen, N.T. (eds.) Advanced Computational Methods for Knowledge Engineering. AISC, vol. 282, pp. 157–172. Springer, Heidelberg (2014). doi:10.1007/978-3-319-06569-4_12
15. Nguyen, Q.V., Madeyski, L.: Searching for strongly subsuming higher order mutants by applying multi-objective optimization algorithm. In: Le Thi, H.A., Nguyen, N.T., Do, T.V. (eds.) Advanced Computational Methods for Knowledge Engineering. AISC, vol. 358, pp. 391–402. Springer, Heidelberg (2015). doi:10.1007/978-3-319-17996-4_35
16. Nguyen, Q.V., Madeyski, L.: Empirical evaluation of multi-objective optimization algorithms searching for higher order mutants. An International Journal on Cybernetics andSystems (2016). doi:10.1080/01969722.2016.1128763, http://madeyski.e-informatyka.pl/download/NguyenMadeyski16CS.pdf
17. Omar, E., Ghosh, S., Whitley, D.: Constructing subtle higher order mutants for Java and aspectJ programs. In: 2013 IEEE 24th International Symposium on Software Reliability Engineering (ISSRE), pp. 340–349, November 2013
18. Papadakis, M., Malevris, N.: An empirical evaluation of the first and second order mutation testing strategies. In: 2010 Third International Conference on Software Testing, Verification, and Validation Workshops (ICSTW), pp. 90–99, April 2010
19. Polo, M., Piattini, M., García-Rodríguez, I.: Decreasing the cost of mutation testing with second-order mutants. Softw. Test. Verification Reliab. **19**(2), 111–131 (2009). doi:10.1002/stvr.392
20. Purushothaman, R., Perry, D.E.: Toward understanding the rhetoric of small source code changes. IEEE Trans. Softw. Eng. **31**(6), 511–526 (2005)

Intelligent Information Systems

Responsive Web Design: Testing Usability of Mobile Web Applications

Jarosław Bernacki, Ida Błażejczyk, Agnieszka Indyka-Piasecka,
Marek Kopel, Elżbieta Kukla, and Bogdan Trawiński[⊠]

Department of Information Systems, Wrocław University of Technology,
Wybrzeże Wyspiańskiego 27, 50-370 Wrocław, Poland
i.blazejczyk@hotmail.com,
{jaroslaw.bernacki,
agnieszka.indyka-piasecka,marek.kopel,
elzbieta.kukla,bogdan.trawinski}@pwr.edu.pl

Abstract. Responsive web design (RWD) allows applications for adapting dynamically to diverse screen sizes, proportions, and orientations. RWD is an approach to the problem of designing for the great number of devices ranging from small smartphones to large desktop monitors. The goal of the paper was to test the usability of an application for management of a scientific conference developed using responsive design paradigms. Two versions of the responsive applications were implemented using different design patterns. Various techniques of usability were employed including tests with prospective users, experts' inspections as well automated tools. The obtained results were thoroughly analysed and recommendations on the utilization of individual design patterns in developing mobile web applications were formulated.

Keywords: Responsive web design · Usability testing · Web applications · Mobile applications

1 Introduction

The rapidly growing market for mobile devices along with technological progress caused a considerable growth of the number of mobile websites and applications. Usability models, methods, and metrics have been an intensively developed area of research for many years. During the last decade the area has been extended to include usability problems of mobile systems and applications. ISO 9241-11 presents usability model consisting of three attributes: effectiveness, efficiency and satisfaction [1]. Nielsen's model of usability is composed of five following attributes: efficiency, satisfaction, learnability, memorability, and errors [2]. In turn Harrison et al. [3] devised the PACMAD usability model for mobile applications which mentions seven attributes effectiveness, efficiency, satisfaction, learnability, memorability, errors and cognitive load. Extensive overview of challenges and problems in usability testing of mobile applications can be found in Zhang and Adipat [4]. Shitkova et al. [5] worked out and present a set of 39 usability guidelines for mobile applications and websites. Gomez et al. [6] complied heuristic evaluation checklists and adapted them to mobile

© Springer-Verlag Berlin Heidelberg 2016
N.T. Nguyen et al. (Eds.): ACIIDS 2016, Part I, LNAI 9621, pp. 257–269, 2016.
DOI: 10.1007/978-3-662-49381-6_25

interfaces. A series of works reporting usability evaluation of mobile applications using different approaches have been published recently [7–10].

In this paper we present a new website worked out to support the preparation of the 8th Asian Conference on Intelligent Information and Database Systems to be held during 14-16 March 2016 in Da Nang, Vietnam. ACIIDS 2016 is co-organized by Vietnam-Korea Friendship Information Technology College (Vietnam) and Wrocław University of Technology (Poland). The application was developed using responsive design paradigms [11] to address the technological progress and market trends. The goal our research was to conduct usability evaluation of two versions of the application: the first one designed for conservative users using traditional laptops and the second one intended for users utilizing smartphones. Both versions were compared in terms of effectiveness, efficiency, and satisfaction. Moreover, the ACIIDS 2016 website's usage statistics produced by Google Analytics [12], one of the most commonly used web analytic tools, were analysed.

2 Web System for ACIIDS 2016 Conference

A new web system was developed and exploited to aid in preparing the ACIIDS 2016 conference. The main goal of the system was to provide the scientific community with all organizational information about the conference including instructions how to submit papers and participate in the event. Therefore, the majority of information was placed in the textual form on the website. The proper information architecture and navigation was necessary to enable the users to find quickly information they needed. The overall design of the website adhered to a standard structure of internet conference services to be recognizable and easy to use to any researcher who was interested in ACIIDS 2016. The system did not support the processes of paper submission and reviewing which were accomplished by another system, namely the EasyChair conference management system. The only function requiring input of data was the Registration which enabled the authors of accepted papers to register for the conference and provide data for invoice receipt.

We assumed that our service should be accessible and usable to researchers possessing different devices ranging from small smartphones through laptops and tablets to big desktop monitors. Therefore we applied responsive web design (RWD) approach to develop our conference service (see Fig. 1). Responsive design represents significant challenges for developers because it depends on shuffling elements around the page, which makes the design and development work closely together to provide a usable experience across various screen resolutions, sizes, and orientations. RWD requires conducting suitable usability tests.

Google Analytics, one of the most commonly used web analytic tools, was employed to gather ACIIDS 2016 website's usage statistics. Selected metrics provided by Google Analytics based on data collected during the second half of 2015 are presented in the rest of this section. They illustrate how intensively the website was used and how important role it played in preparing the conference. The definition of the metrics can be easily found in the Internet, e.g. on the official website of the tool [13],

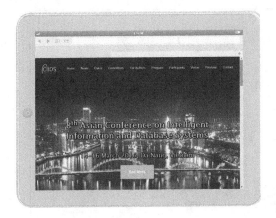

Fig. 1. Illustration of responsive design of ACIIDS 2016 website.

and in one of the numerous books on web analytics, e.g. in the Clifton's book [12]. Basic metrics produced by Google Analytics are presented in Table 1. The Map Overlay report in Fig. 2 illustrates that 18,349 visits came from 102 countries and 6 continents and Vietnam and Poland are the most frequent locations. In turn, ten top countries and cities where the most visits came from are placed in Table 2. They are located mainly in South-East Asia, Japan, Korea, and Europe, what is depicted in Fig. 3.

Table 1. Basic metrics provided by Google Analytics

Metrics	Value	Metrics	Value
No. of visits (sessions)	18,349	No. of visitors	7,529
No. of page views	71,906	New visitors	41.4 %
Avg. visit duration [min]	3:39	Returning visitors	58.6 %
Pages/Visit	3.92	Bounce rate	37.0 %

The daily distribution of ACIIDS 2016 site visits is presented in Fig. 4, where peaks of the curve reflect the increased number of visits caused by some controlled events. These events occurred when either mailing actions were carried out (M) or deadlines for paper submission, notification of acceptance, registration, camera-ready papers, and payment expired (D).

The pie graphs in Fig. 5 indicate that the vast majority of potential conference participants still use laptops and desktop computers. Nevertheless, the number of visits with mobile devices, such as smartphones and tablets is 1572 and 351, respectively. In turn, 820 new users utilized smartphones and 151 new visitors accessed the website with tablets. These figures justify our decision to develop the website using the RWD approach.

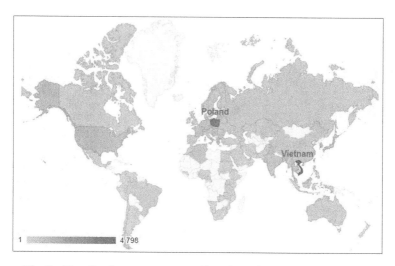

Fig. 2. Map Overlay report on countries where site's visitors came from

Table 2. Ten top countries and cities where the most visits came from

	Country	Visits	City	Cnty	Visits
1	Vietnam	4,798	Ho Chi Minh City	VN	1,970
2	Poland	3,739	Wroclaw	PL	1,753
3	United States	714	Hanoi	VN	1,449
4	Japan	672	Da Nang	VN	691
5	Taiwan	625	Gliwice	PL	488
6	South Korea	550	Warsaw	PL	469
7	Thailand	528	Bangkok	TH	385
8	India	492	Singapore	SG	238
9	Malaysia	449	Can Tho	VN	218
10	Czech Republic	434	Seoul	KR	193

3 Experimental Evaluation Setup

The main goal of evaluation experiments was to identify the usability problems of two versions of the ACIIDS 2016 website. The first version was designed for conservative researchers using laptops and desktop computers whereas the second one was devoted to those who utilize small touchscreen devices like mobile phones. The next goal was to compare both versions in terms of effectiveness, efficiency, and satisfaction using two different sorts of commonly used devices, namely standard laptops and smartphones. The third goal was to work out the recommendations on how the ACIIDS 2016 website might be improved.

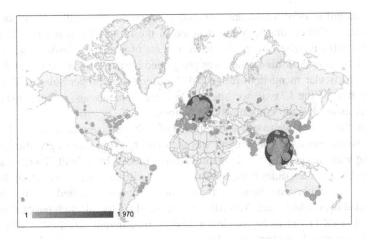

Fig. 3. Map Overlay report on cities where site's visitors came from

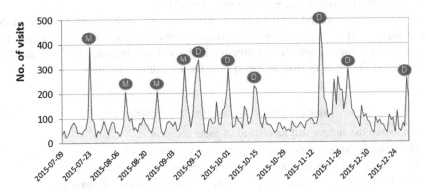

Fig. 4. Daily distribution of ACIIDS 2016 website visits

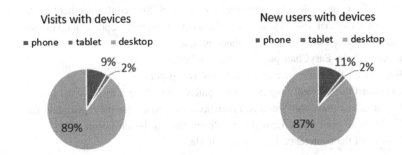

Fig. 5. Distribution of device types over visits and new users

Two experimental versions of the ACIIDS 2016 web applications were developed using the RWD approach to carry out usability tests across platforms. Both versions encompassed all functions of the official conference website being actively exploited but they differed in information and navigation structures. We called them classic and vertical stack websites, and they are denoted in the rest of the paper as **Classic** and **Vstack**, respectively. The former consisted of a set of separate subsites with common graphic layout. Menu of this version was positioned horizontally at the top of the page. Clicking a particular menu item redirected the user to the proper subsite. The layout and navigation of the Classic website met the requirements and expectations of the researchers using desktop computers and laptops.

In turn, the **Vstack** website was composed of only one page comprising the whole content divided into tabs. The content was then laid out in a scrolling vertical column Navigating was provided by a side menu that was hidden by default. To see the menu the user had to click a special button positioned in a visible place at the top of the page. When the user clicked an item on the menu, the page "scrolled" to the tab that corresponded the clicked item. According to Tidwell [14] mobile web pages should use the vertical stack layout to be usable on devices of different sizes, especially if they contained text-based content and forms. Both versions were implemented with HTML5, CSS3 and Bootstrap framework [15] allowing for RWD. Dynamic elements of the website, like forms, were implemented in PHP.

15 tasks were prepared for the users to complete (see Table 3). They reflected the most common actions performed by the visitors of the conference website. All tasks but two consisted in finding specific information in the website. In turn, tasks T11 and T13 encompassed the input of data to the system through online forms.

Table 3. 15 tasks to complete by the participants

T1. Find when the authors will be notified of paper acceptance.
T2. Find what is the registration fee for authors of accepted papers.
T3. Find what is the page limit for a paper submitted to the conference.
T4. Check whether John Smith from USA is the member of Program Committee.
T5. Find the topic of the plenary lecture delivered by Professor John Smith from USA.
T6. Find where the conference proceedings will be published.
T7. Display the photo gallery from the previous ACIIDS 2014 conference held in Bangkok.
T8. Find to what special session you could submit a paper on traffic optimization in a smart city.
T9. Find the date and time of the conference banquet.
T10. Proceed to the EasyChair paper submission system.
T11. Register for the conference as an author of two papers.
T12. Download the detailed program of the conference in the pdf file.
T13. Log in as the user John Smith and download the invoice for the registration fee.
T14. Find how to book hotel room at the Pullman Danang Beach Resort 5*.
T15. Download the conference flyer in the pdf file.

Effectiveness, efficiency, and satisfaction measures recommended by the ISO Standard 9241-11 [1] were applied to usability evaluation experiments.

Participants of usability testing were volunteers recruited from among the researchers of Wroclaw University of Technology who were engaged in research into intelligent information and database systems. They fully corresponded the target users of the ACIIDS 2016 website. In total, 32 users took part in research and 47 percent of them participated in previous editions of the ACIIDS conference. The participants were divided into four separate groups. Each group had to complete the tasks using the *Classic* or *Vstack* versions of the applications run on either laptops (L) or smartphones (S). Thus, we obtained four variants of application/device combinations which we called *Classic_L*, *Classic_S*, *Vstack_L*, and *Vstack_S*, respectively. The basic characteristics of the participants are given in Fig. 6.

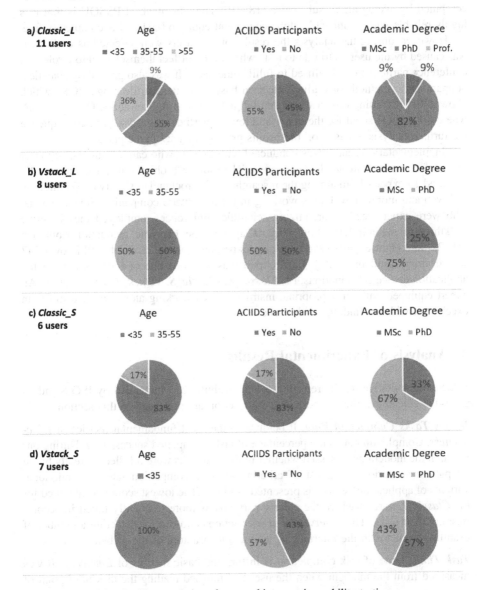

Fig. 6. Characteristics of users taking part in usability testing

Usability testing was carried out by individual users separately under the supervision of a moderator. Each session of a participant using laptop was recorded by the Morae tool [16], in turn the user's activity during testing with a smartphone was captured by the Lookback tool [17]. Having completed a session the moderators filled the forms with results using their own notes and Morae or Lookback recordings.

The Post-study System Usability Questionnaire (PSSUQ) was applied to measure users' perceived satisfaction with the conference application. PSSUQ is one of the most popular standardized questionnaires which encompasses 16 items grouped in four sub-scales: overall, system, information, and interface quality. The final score is computed by averaging all sub-scales [18, 19]. The advantage of PSSQU is that it is highly reliable with Cronbach's alpha coefficient equal to 0.94 and is completely free.

However, during the analysis the results of five users were rejected as outliers. It was caused by the users' attitude to tests who could not feel themselves into a role of a conference participant and refused to fulfil some tasks. It was also proved by statistical nonparametric method for outlier detection based on interquartile range (IQR) which allowed for discarding observations that fell below Q1–3 IQR or above Q3 + 1.5 IQR, where Q1 and Q3 stand for the upper and lower quartiles respectively. In consequence the further analysis is based on the results produced by 27 participants.

Supplementary research was conducted with experts who carried out inspection of the conference website to discover potential usability problems in an interface. We recruited 19 experts from among the researchers of Wroclaw University of Technology and web and mobile developers working in local software companies. Seven experts, who were experienced in organizing scientific conferences, employed the cognitive walkthrough method [20]. Four experts applied the heuristic evaluation approach [21, 22] and eight experts used control lists prepared on the basis of well-known 247 web usability guidelines [23]. Each expert was assigned one of four variants of the application/device combinations, i.e. *Classic_L, Classic_S, Vstack_L*, or *Vstack_S*. An expert equipped with an appropriate instruction was working alone and had to fill an excel form with his findings within one week.

4 Analysis of Experimental Results

Selected quantitative results referring to the attributes recommended by ISO Standard 9241-11: efficiency, effectiveness, and satisfaction are presented in this section.

Binary Tasks Completion Rate. It belongs to the most fundamental metrics of *Effectiveness*. Completion rate is the percentage of tasks completed successfully. During our research the moderators assessed whether individual tasks were fulfilled correctly based on pattern scenarios and a list of criteria. Average completion rates for individual variants of application/devices is presented in Fig. 7. The lowest score was obtained for the *Classic_L* variant where the elderly participants more frequently failed in accomplishing their tasks. The observation of user actions allowed for detecting a number of usability problems in the interface as well as the structure of the website.

Task Time. Time of task completion is in turn the basic metrics of *Efficiency*. It was measured from the moment when the user had finished reading the task until point of

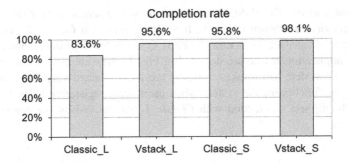

Fig. 7. Average binary tasks completion rate for individual variants of application/devices

time when the user had found the answer or declared verbally that had finished the task. The average completion time for individual tasks is depicted in Fig. 8. It can be clearly seen that time needed to accomplish task 11 was the longest because it consisted in filling the online registration form. This activity took more time on smartphones. Having excluded tasks 11 and 13, which required filling online forms, we made nonparametric Wilcoxon tests to compare the task time between each pair of application/device. The significant differences were observed only between variants *Classic_L* and *Classic_S*. The younger participants using *Classic_S* were completing tasks faster than older users employing *Classic_L*.

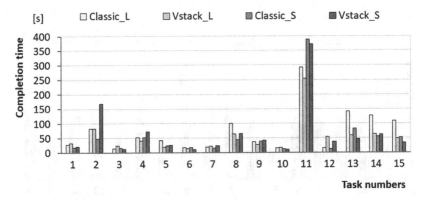

Fig. 8. Average completion time for individual tasks

Satisfaction Ratings. Each participant filled the Post-study System Usability Questionnaire (PSSUQ) at the end of the usability session. 34 percent of PSSUQ were filled completely whereas 66 percent were incomplete. According to Sauro and Lewis [19] PSSUQ items produce four scores including one overall and three subscales. The System Quality (SysQual) subscale is computed as average of items 1 through 6, the Information Quality (InfoQual) subscale is obtained as average of items 7 through 12, the Interface Quality (IntQual) subscale is produced by averaging items 13 through 15, and finally the Overall score is achieved by averaging all the items, i.e. 1 through 16. The final scores may have values between 1 and 7, where the lower score the higher degree of satisfaction. The PSSUQ scores for individual scales and for all variants of application/device are shown in Fig. 9. It is clearly seen that *Classic_L* provided the highest degree of satisfaction. In turn, the PSSUQ scores for individual items and for all variants of application/device are depicted in Fig. 10. The items 7–9 concerning the reaction of the system to mistakes made by the users attained very bad scores. This reflects the malfunction of the system when users were registering using the Firefox browser. The highest satisfaction with *Classic_L* was proved by the non-parametric Wilcoxon test.

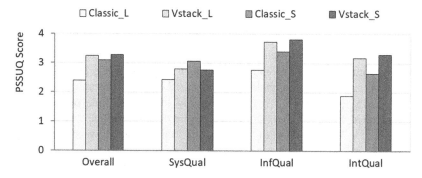

Fig. 9. PSSUQ scores for individual scales.

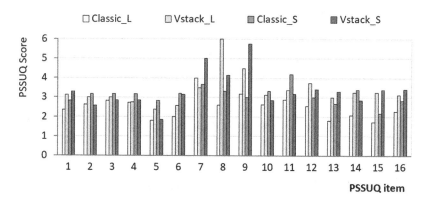

Fig. 10. PSSUQ scores for individual items.

Final Recommendations. 111 recommendations on how to improve the application and the structure of website were formulated based on the results of usability testing and expert inspections. 32 problems were rejected by the moderators as less important. The summary of problems identified for individual versions and devices and combinations of them is shown in Table 4. 13 problems out of 79 were identified by users only, 35 by experts only and 31 by both groups. The percentage of problems detected by users only, experts only and by both groups is depicted in Fig. 11.

Table 4. Summary of usability problems identified during research

Problems applied to	No.
Each version and each device	59
Only *Classic* version	5
Only *Vstack* version	9
Only smartphones	1
Only laptops	1
Only *Classic_L*	0
Only *Classic_S*	2
Only *Vstack_L*	0
Only *Vstack_S*	2
Sum for *Classic_L*	64
Sum for *Classic_S*	66
Sum for Vstack_L	68
Sum for Vstack_S	70
Total	79

Problems identified by

■ only users   ■ only experts   ■ both groups

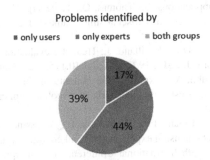

Fig. 11. Percentage of usability problems identified by individual groups of researchers

5 Conclusions and Future Work

The results of usability testing of the conference website developed according to responsive web design paradigm are presented in the paper. The experiments conducted with 32 participants and 19 experts provided an extensive list of recommendations on how to improve the application and information structure of the website.

The highest degree of satisfaction was perceived by the participants using the *Classic_L* application/device variant. Usability testing with this variant was carried out with the oldest group of users who got used to standard applications run on popular devices such as laptops. This group of participants failed in completing tasks more frequently than other groups. However, they blamed themselves for that rather than the interface deficiency of the application they tested. The younger participants accomplished their tasks with any application/device variants easily. Nevertheless, they noticed many interface elements which should be improved, especially in regard to the information structure and online forms.

It is planned to work out an improved version of the ACIIDS website based on the recommendations formulated in this study. The second round of usability testing will be carried out with the improved version of the conference website.

Acknowledgments. This paper was partially supported by the statutory funds of the Wrocław University of Technology, Poland.

References

1. ISO 9241-11:1998 Ergonomic requirements for office work with visual display terminals (VDTs) - Part 11: Guidance on usability
2. Nielsen, J., Budiu, R.: Mobile Usability. New Riders Press, Berkeley (2012)
3. Harrison, R., Flood, D., Duce, D.: Usability of mobile applications: literature review and rationale for a new usability model. J. Interact. Sci. **1**, 1 (2013). doi:10.1186/2194-0827-1-1
4. Zhang, D., Adipat, B.: Challenges, methodologies, and issues in the usability testing of mobile applications. Int. J. Hum. Comput. Interact. **18**(3), 293–308 (2005)
5. Shitkova, M., Holler, J., Heide, T., Clever, N., Becker, J.: Towards usability guidelines for mobile websites and applications. In: Thomas, O., Teuteberg, F. (Eds.) Proceedings of the 12th International Conference on Wirtschaftsinformatik (WI 2015), Osnabrück, Germany, pp. 1603–1617 (2015)
6. Gómez, R.Y., Caballero, D.C., Sevillano, J.: Heuristic evaluation on mobile interfaces: a new checklist. Sci. World J. **2014**, 19 (2014). doi:10.1155/2014/434326. Article ID 434326
7. Biel, B., Grill, T., Gruhn, V.: Exploring the benefits of the combination of a software architecture analysis and a usability evaluation of a mobile application. J. Syst. Softw. **83**(11), 2031–2044 (2010)
8. Plechawska-Wojcik, M., Lujan Mora, S., Wojcik, L.: Assessment of user experience with responsive web applications using expert method and cognitive walkthrough - a case study. In: Proceedings of the 15th International Conference on Enterprise Information Systems, ICEIS-2013, pp. 111–118 (2013)
9. Gatsou, C., Politis, A., Zevgolis, D.: An exploration to user experience of a mobile tablet application through prototyping. Int. J. Comput. Sci. Appl. **11**(1), 56–74 (2014)
10. Lestari, D.M., Hardianto, D., Hidayanto, A.N.: Analysis of user experience quality on responsive web design from its informative perspective. Int. J. Softw. Eng. Appl. **8**(5), 53–62 (2014)
11. Marcotte, M.: Responsive Web Design. A Book Apart, New York (2011)
12. Clifton, B.: Advanced Web Metrics with Google Analytics, 3rd edn. Wiley, New York (2012)
13. Google Analytics website. http://www.google.com/analytics/. Accessed 1st December 2015
14. Tidwell, J.: Designing Interfaces, 2nd edn. O'Reilly Media, Sebastopol (2011)
15. Bootstrap website. http://getbootstrap.com/. Accessed 1st December 2015
16. Morae website. https://www.techsmith.com/morae.html. Accessed 1st December 2015
17. Lookback website. https://lookback.io/. Accessed 1st December 2015
18. Lewis, J.R.: Psychometric evaluation of the PSSUQ using data from five years of usability studies. Int. J. Hum. Comput. Interact. **14**(3&4), 463–488 (2002)
19. Sauro, J., Lewis, J.R.: Quantifying the User Experience: Practical Statistics for User Research. Morgan Kaufmann, Waltham (2012)

20. Mahatody, T., Mouldi Sagar, M., Kolski, C.: State of the art on the cognitive walkthrough method, its variants and evolutions. Int. J. Hum. Comput. Interact. **26**(8), 741–785 (2010)
21. Nielsen, J., Molich, R.: Heuristic evaluation of user interfaces. In: Proceedings of the SIGCHI Conference on Human Factors in Computing Systems: Empowering People (CHI 1990), pp. 249–256. ACM, New York, USA (1990)
22. Weinschenk, S., Barker, D.T.: Designing Effective Speech Interfaces. Wiley, New York (2000)
23. 247 Web usability guidelines. http://www.userfocus.co.uk/resources/guidelines.html. Accessed 1st December 2015

Person Name Disambiguation for Building University Knowledge Base

Piotr Andruszkiewicz[✉] and Szymon Szepietowski

Institute of Computer Science, Warsaw University of Technology, Warsaw, Poland
P.Andruszkiewicz@ii.pw.edu.pl, S.Szepietowski@stud.elka.pw.edu.pl

Abstract. In this paper we propose a new algorithm for person name disambiguation within authors of scientific publications. The algorithm is effective, elastic, and tailored to a scientific knowledge base. Besides the common properties of publication; namely, title, venue, author and co-authors names, it also exploits references. One of the reasons is that we decided to enrich the University Knowledge Base with connections between publications, not only references represented by a reference (i.e. author's name, title, etc.). Our algorithm utilises the unsupervised approach which does not require creating a training set, which is time and resources consuming. However, we want to leverage additional information available from crowd sourcing or authorised users which confirms authorship and citation relations between papers. By utilising this information default parameters of the unsupervised algorithm can be optimised for a given case by means of a genetic algorithm in order to increase the accuracy. The proposed method can be applied for three tasks: assigning a publication to a specific researcher, indicating that a new author is yet unknown to the database and clustering a set of publications into clusters that contain papers of one researcher. Validation results confirm high accuracy of the new algorithm and its usefulness in the process of populating a scientific knowledge base.

Keywords: Person name disambiguation · Unsupervised approach · Genetic algorithm · Scientific knowledge base

1 Introduction

In the process of building a digital library it is crucial to identify a person who wrote certain article and to find all her/his publications. Usually information provided by a publication is not enough to precisely indicate an author of that piece of work. Many solutions were proposed to tackle this problem, nevertheless it still remains unsolved.

A knowledge base that is useful for researchers should consist of publications and their authors. Furthermore, in order to increase its usefulness, references

Research has been supported by the National Centre for Research and Development under grant No SP/I/1/77065/10 and the Institute of Computer Science, Warsaw University of Technology under Grant No. II/2015/DS/1.

© Springer-Verlag Berlin Heidelberg 2016
N.T. Nguyen et al. (Eds.): ACIIDS 2016, Part I, LNAI 9621, pp. 270–279, 2016.
DOI: 10.1007/978-3-662-49381-6_26

used in publications could be extracted and connected with corresponding publications creating a citation graph. Researcher's profiles are also useful, e.g., topics that are of interest for a given researcher. All the above allow for providing additional functionality for a knowledge base, for instance, expert finding in a given field, evaluating researcher's contribution by calculating indices, e.g., h-index. In order to extend functionality of our University Knowledge Base [1], we decided to include references and researcher's profiles in the knowledge base.

While extracting information about researchers, their activities and publications, one of the crucial problems is the name disambiguation. Usually, we have a set of publications with a given text representing author's name and a set of researches (also with text strings representing their names and sometimes with additional attributes forming a researcher's profile) and we would like to assign each publication to one of the researchers. The problem is complicated because usually there are various forms of researcher's names, e.g., a name can be written with the first and last name of a researcher (with or without middle name), or with initials for the first name, with or without initials for the middle name. It often happens that the authors' first and last names are from the dictionary of first names, in this case family names are mistakenly assumed to be first names (and then are replaced by initials). Last but not least, for popular names it happens very often that two different persons hold the same pair (first name, family name).

While building the University Knowledge Base [1] we also encountered the issue of person name disambiguation. Hence, we designed an approach that is effective and tailored to our knowledge base. This paper describes in more detail the approach sketched in [1].

The algorithm, besides the regular attributes; that is, co-author name, article title, and publication venue, exploits the citation relation between publications because our knowledge base is collecting this information.

We used unsupervised approach in order to avoid creating training sets for each researcher. However, a training set can be used in our solution to update default parameters of the algorithm and increase effectiveness. This functionality is motivated by the fact that data in the University Knowledge Base is verified by crowd sourcing and authorised users. Especially, information confirmed by authorised users is of high quality and could be used to tune the algorithm parameters. The update of the parameters is performed by a genetic algorithm.

The proposed algorithm can be applied for person name disambiguation in two ways, by clustering a set of publications into clusters of publications authored by one researcher and by assigning a publication to a particular researcher. Furthermore, it is able to detect that a given publication is written by a new author who has not been listed in the knowledge base yet.

In this paper we focus on the person name disambiguation task for a scientific knowledge base. In particular we present a new approach to disambiguation of authors. Moreover, we show how the proposed solution can be applied in our University Knowledge Base and how it influences its functionality and the quality of data.

The remainder of this paper is organized as follows: Sect. 2 presents related work. In Sect. 3 the proposed approach to person name disambiguation is described. The experimental results are highlighted in Sect. 4. Finally, Sect. 5 summarises the conclusions of the study and outlines future avenues to explore.

2   Related Work

Various methods have been employed for Person Name Disambiguation. A review of this field of study can be found in [2,3]. We can distinguish two main approaches: (1) supervised [4–6] and (2) unsupervised [7–10] regarding the requirements for a training set.

In supervised methods the idea is to build a predictive model, which is used to decide if a pair of publications belong to the same author or not. The drawback of supervised methods is that they need a training set, which in turn requires human labour to be annotated. Moreover, models are usually trained for each name separately; it means that huge amount of data should be manually annotated. Furthermore, a training set should be up-to-date regarding the subject of papers published by researchers with the same name. The consequence of that is the constant need of updating a training set with new domains that will appear in considered publications. A training set usually consists of pairs of publications labeled with positive category, they belong to the same author, or negative category if they are of different authors.

In [4] two supervised solutions that use Naive Bayes and SVM classifiers were proposed. Regarding the training set construction the most important difference between these two approaches is that the former uses only positive cases whereas the latter both positive and negative.

[11] introduced a different method called DISTINCT. It uses SVM to learn a model for weighing different types of linkages and applies agglomerative hierarchical clustering that connects the most similar pairs of clusters.

Unsupervised methods do not need a manually annotated training set, which is the advantage over supervised methods. In order to solve name disambiguation problem, unsupervised methods use clustering to partition publications. Each cluster is a set of publications written by a given researcher. These methods can be divided into three steps: building data representation model, calculating similarity between publications, and clustering.

In [12] citation-term matrix and tf-idf were used to represent data. A different approach, where relations between attributes of publications (e.g., whether two publications have the same venue, the same co-authors, or cite each other) were used, was presented in [8].

In step 2 the similarity measure is calculated. In [12] cosine similarity was used and a categorical set similarity measure was proposed in [9] for that purpose.

As the last step, clustering is performed. Many classical clustering algorithms were used, e.g., agglomerative clustering [9], K-way spectral clustering [12], and Hidden Markov Random Fields (HMRF) [8].

Considering a different criterion, person name disambiguation algorithms can be divided into two groups, depending on the approach they utilise. One approach is to group publications into clusters written by the same author [12,13]; and another approach is to assign a given publication to a specific author [4,8].

Some person name disambiguation methods utilise external information, e.g., researcher's web pages on the Internet, search engines. [14] uses web search engine in order to find single author documents and extract citations. [15] proposes a measure of Web correlation, which calculates a number of co-occurrence in a web page. User feedback helps in [16] to obtain better disambiguation.

One of the reasons why our method can employ a more convenient unsupervised approach is that default values of all the parameters are easy to define at the beginning, which makes the training phase unnecessary. However, if there is a training set available, the parameters can be anytime updated and tuned by a genetic algorithm. The proposed method is suitable for two approaches: for publications' grouping and a publication assignment to a person. Moreover, it considers citation relation and is easily extensible to other information, e.g., topics of interests of researchers that form researcher's profiles. Those properties are not common which makes our algorithm particularly interesting and new. Its detailed description can be found in Sect. 3.

3 Person Name Disambiguation for Researchers

Our algorithm includes two steps. The initial clustering that forms sets of publications probably authored by one researcher. Those sets of publications are clustered again in the second step.

3.1 Initial Clustering

The initial clustering is based on matching the first and last names; we skip middle names. Having a set of publications written by researchers with the same last name, we try to match first names and create clusters of publications whose authors' first and last names match. We reduce first names to initials if at least one first name is given by an initial. In this way we obtain basic name disambiguation: small clusters containing publications which supposed to be written by the same researcher. Apart from this it is almost certain that publications from different clusters do not belong to the same author.

The relations discovered in this step are used in the further steps of the name disambiguation algorithm.

3.2 Final Clustering

In the next step we perform clustering for each initial cluster separately because initial clustering cannot separate publications written by authors with the same first, middle and last name. The process of final clustering is described in detail in the next sections.

Similarity of Publications. The aim of a method is to cluster publications based on relations between them. As a starting point we use the relations proposed in [7]; namely, co-authorship, citations, extended co-authorship, and user's restrictions. We calculate them as shown in [7]. As an additional relation we add the similarity of two publications' titles because we observed that a scientist usually publishes within one domain and his/her publications often have the same words in titles. To calculate titles relation we used Apache Lucene[1] similarity measure. In the calculation of similarity between publications we assign a weight to each relation. The similarity of two publications is defined as follows.

$$P = \sum_{i=1}^{5} p_i w_i, \tag{1}$$

where:

- p_i – is value of the i-th relation,
- w_i – is the weight for i-th relation,
- P – is the similarity of publications.

Similarity between a publication and a group of publications is defined as an average of similarities between the publication and all publications from the group.

In order to choose the relation weights we propose a genetic algorithm, which is a new approach. Moreover, we propose a clustering method for assigning publications to groups iteratively. This process is described in the next section.

Iterative Clustering. The pseudocode for the clustering method is presented in Algorithm 1. The algorithm iteratively scans publications and searches for the nearest cluster of publications. If the similarity between a cluster and the publication is high enough, i.e., greater then *similarityThreshold*, the publication is assigned to the most similar cluster. Otherwise, a new cluster with only one publication is brought into being.

The output of the algorithm are groups of publications authored by one researcher.

Genetic Algorithm. In order to assess the similarity between publications, the measure defined in Eq. 1 needs xto be calculated. This requires optimisation of the weights for all considered relations and a *similarityThreshold*. To avoid the manual process of choosing the weights and the *similarityThreshold* parameter values, we propose to use a genetic algorithm.

A candidate solution, a creature, in our algorithm contains 6 values, 5 for weights of relations (co-authorship, citations, extended co-authorship, user's restrictions, and the similarity of titles) and one for a *similarityThreshold*. The sample creature is shown in Table 1.

---

[1] https://lucene.apache.org/.

Algorithm 1. The iterative clustering algorithm

Data: $\mathcal{P}$ // a set of publications
Data: *similarityThreshold* // a minimal threshold for a publication to be assigned to an existing cluster
Result: $\mathcal{C}$ // a set of clusters

begin
   $\mathcal{C} \leftarrow \emptyset$
   foreach $p \in \mathcal{P}$ **do**
      slist $\leftarrow \emptyset$
      foreach $c \in \mathcal{C}$ **do**
         $s \leftarrow similarity(p, c)$ // calculate the similarity between a publication and a cluster
         slist.add(s) // store the similarity
      end
      smax $\leftarrow$ *slist.getMax()* // find the highest similarity
      if *smax.isNull()* $||$ *smax* $<=$ *similarityThreshold* **then** // if there are no clusters or there is no cluster similar enough
         $\mathcal{C}.addNewCluster().add(p)$ // add new cluster with the given publication
      end
      else
         maxSimCluster $\leftarrow$ *smax.getCluster()* // get the cluster; it is the most similar cluster for the publication
         maxSimCluster.add(p) // add the given publication to the most similar cluster
      end
   end
end

Where:

- $\mathcal{P}$ – a set of publications,
- $\mathcal{C}$ – clustering results,
- *similarityThreshold* – a parameter that reflects minimal similarity threshold for a publication to be assumed that it belongs to a cluster.
- *similarity(p, g)* – calculates the similarity between a publication and a cluster of publications.

Table 1. The sample candidate solution.

r_1	r_2	r_3	r_4	r_5	b
0.23	0.85	0.3	0.44	0.91	1.35

In our algorithm an initial population is randomly generated. The population has 32 creatures. We use the truncation selection and take the best half of the creatures. The crossover operator calculates the average of all recalibrated weights and the threshold. The mutation operator adds a random value drawn from $< -1; 1 >$ range to the threshold parameter and the weights are modified according to the equation $w_{new} = 1 - w_{old}$.

The next generation consists of half of the best parents and half of the generated children. As a fitness function we use the measure defined in Eq. 2.

4   Experimental Evaluation

Data set consists of 2305307 publications and their authors, collected from the following digital libraries: Association for Computing Machinery Digital Library (ACM), DBLP Computer Science Bibliography, Microsoft Academic Search, and CiteSeer Digital Library. Unfortunately, only 142832 of publications had information about references to other publications (links between publications). The rest only provided a title and author(s) of the referenced publication. In the experiments we used only the 142832 publications with reference information.

In our experiments, the clustering accuracy measure was defined as follows.

$$S = \frac{|G_t \cap G_a|}{|G_t|}, \tag{2}$$

where:

- G_t – is cluster assignments based on a test set,
- G_a – is the result of the clustering algorithm.

Thus, the accuracy of clustering is the number of clusters that are present both in the test set (the ground truth clusters) xand in the clustering results provided by the algorithm divided by the number of clusters from the test set.

The first experiment was conducted for the last name Johnson. For this last name we prepared user's restrictions that led to perfect clustering. Using this first experiment we adjusted the algorithm parameters; namely the weights of the relations and the threshold parameter. With the adjusted values of weights and parameter we performed the experiments on other last names, and the results are presented in Fig. 1. The accuracy for clustering is around 0.80 on our data set. Only the results for Li and Wilson significantly deviate from the average with the accuracy of 0.70 and 0.96, respectively.

We also verified whether we could find better parameters' values for each last name by applying the genetic algorithm separately for each last name. Table 2 shows the results of disambiguation with fixed weights and threshold values, that were used in the previous experiment, and with the weights tailored for each last name independently.

Tailoring the parameters for each name separately did not improve the output to a high extent, only for 5 out of 8 last names we were able to find better set

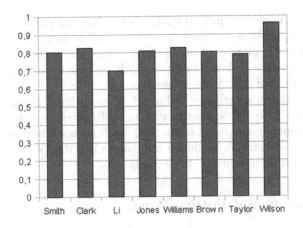

Fig. 1. Clustering accuracy measure in person name disambiguation.

of parameters. This suggests that the parameters can successfully be optimised based on a small fragment of the data set, which solves the problem of creating a big training set with labeled data. The optimal parameters for a set of names can be estimated for example by taking the average of the weights and threshold values for several last names.

Table 2. Clustering accuracy with fixed weights and threshold values, parameters chosen separately for each last name for all researchers and those with at least two publications (denoted as 2r).

Lastname	Smith	Clark	Li	Jones	Williams	Brown	Taylor	Wilson
Fixed params accuracy	0.8046	0.8276	0.6967	0.8108	0.8276	0.8056	0.7917	0.9630
Chosen params accuracy	0.8390	0.8620	0.8279	0.8378	0.8620	0.8056	0.7917	0.9630
Chosen params acc. 2r	0.6250	0.6670	0.5670	0.5550	0.7000	0.7330	0.6250	0.8000

The high accuracy of the algorithm obtained in the already described experiment suggests that name disambiguation is performed properly; that is, the algorithm distinguishes different authors with the same last name. However, these very promising results may derive from the fact that there are many researchers who have only one publication in our data set. Therefore we repeated the experiment using only researchers that have at least two publications. The results are presented in Table 2 (the row denoted as 'Chosen params acc. 2r'). We can observe that the results dropped significantly. The reason could be that there is not enough links between publications, especially not all publications of one author have linked references, and our algorithm cannot find the relations. This is to some extent probably caused by the lack of references linking publications in our collected set (over 2.3 million publications) that was reduced by removal of publications without linked references. Publications removal resulted in almost 94% reduction of the number of publications. Thus our citation graph was very sparse compared to the real one. In future work, we plan to investigate this issue and enrich our test set with more references.

To sum up, the proposed method apart from authors' names additionally takes into account also title similarity which enables the system to assign a publication to a proper researcher, i.e., cluster, in an unsupervised manner. Moreover, additional information (included in a supervised data set) can be used to update the algorithm parameters at anytime.

5   Conclusions and Future Work

In this publication the problem of person name disambiguation for building a scientific knowledge base has been investigated. We propose an algorithm which is tailored to our University Knowledge Base, however, it can successfully be applied to other knowledge bases, provided they contain objects connected by some relations.

The algorithm utilises unsupervised approach and lets the user improve its default parameters by adding supervised data such as information provided by crowd sourcing and authorised users. Not only can it be applied for clustering a set of publications in order to obtain clusters of publications authored by one researcher but it also allows one publication to be assigned to a specific researcher who is known by the knowledge base. Moreover, it detects publications written by new authors that are not represented in the knowledge base yet.

The results of the experiments suggest that our algorithm can be used for person name disambiguation in the University Knowledge Base, increase its accuracy and enrich its functionality.

In future work, we plan to investigate the possibility of a training set self generation for the initial setup of parameters application. Furthermore, we plan to enhance our algorithm by using additional information from researcher's profiles, investigate the issue of many authors with only one publication and enrich our data set with more references.

References

1. Koperwas, J., Skonieczny, L., Kozłowski, M., Andruszkiewicz, P., Rybiński, H., Struk, W.: AI platform for building university research knowledge base. In: Andreasen, T., Christiansen, H., Cubero, J.-C., Raś, Z.W. (eds.) ISMIS 2014. LNCS, vol. 8502, pp. 405–414. Springer, Heidelberg (2014)
2. Smalheiser, N.R., Torvik, V.I.: Author name disambiguation. ARIST **43**(1), 1–43 (2009)
3. Ferreira, A.A., Gonçalves, M.A., Laender, A.H.F.: A brief survey of automatic methods for author name disambiguation. SIGMOD Rec. **41**(2), 15–26 (2012)

4. Han, H., Giles, C.L., Zha, H., Li, C., Tsioutsiouliklis, K.: Two supervised learning approaches for name disambiguation in author citations. In: Chen, H., Wactlar, H.D., Chen, C., Lim, E., Christel, M.G. (eds.) Proceedings of ACM/IEEE Joint Conference on Digital Libraries, JCDL 2004, Tucson, AZ, USA, 7–11 June 2004, pp. 296–305. ACM (2004)

5. Ferreira, A.A., Veloso, A., Gonçalves, M.A., Laender, A.H.: Effective self-training author name disambiguation in scholarly digital libraries. In: Proceedings of the 10th Annual Joint Conference on Digital Libraries, pp. 39–48. ACM (2010)

6. Veloso, A., Ferreira, A.A., Gonçalves, M.A., Laender, A.H.F., Meira, Jr., W.: Cost-effective on-demand associative author name disambiguation. Inf. Process. Manage. vol. 48(4), pp. 680–967 (2012)

7. Tang, J., Yao, L., Zhang, D., Zhang, J.: A combination approach to web user profiling. ACM Trans. Knowl. Discov. Data 5(1), 2: 1–2: 44 (2010)

8. Tang, J., Fong, A.C.M., Wang, B., Zhang, J.: A unified probabilistic framework for name disambiguation in digital library. IEEE Trans. Knowl. Data Eng. 24(6), 975–987 (2012)

9. Li, S., Cong, G., Miao, C.: Author name disambiguation using a new categorical distribution similarity. In: Flach, P.A., De Bie, T., Cristianini, N. (eds.) ECML PKDD 2012, Part I. LNCS, vol. 7523, pp. 569–584. Springer, Heidelberg (2012)

10. Liu, Y., Li, W., Huang, Z., Fang, Q.: A fast method based on multiple clustering for name disambiguation in bibliographic citations. JASIST 66(3), 634–644 (2015)

11. Yin, X., Han, J., Yu, P.S.: Object distinction: Distinguishing objects with identical names. In: Chirkova, R., Dogac, A., Özsu, M.T., Sellis, T.K., (eds.) Proceedings of the 23rd International Conference on Data Engineering, ICDE 2007, The Marmara Hotel, Istanbul, Turkey, 15–20 April 2007, pp. 1242–1246. IEEE (2007)

12. Han, H., Zha, H., Giles, C.L.: Name disambiguation in author citations using a k-way spectral clustering method. In: Marlino, M., Sumner, T., III, F.M.S., (eds.) Proceedings of ACM/IEEE Joint Conference on Digital Libraries, JCDL 2005, Denver, CO, USA, 7–11 June 2005, pp. 334–343. ACM (2005)

13. Cota, R.G., Ferreira, A.A., Nascimento, C., Gonçalves, M.A., Laender, A.H.F.: An unsupervised heuristic-based hierarchical method for name disambiguation in bibliographic citations. JASIST 61(9), 1853–1870 (2010)

14. Pereira, D.A., Ribeiro-Neto, B.A., Ziviani, N., Laender, A.H.F., Gonçalves, M.A., Ferreira, A.A.: Using web information for author name disambiguation. In: Heath, F., Rice-Lively, M.L., Furuta, R., (eds.) Proceedings of the 2009 Joint International Conference on Digital Libraries, JCDL 2009, Austin, TX, USA, 15–19 June 2009, pp. 49–58. ACM (2009)

15. Peng, H., Lu, C., Hsu, W., Ho, J.: Disambiguating authors in citations on the web and authorship correlations. Expert Syst. Appl. 39(12), 10521–10532 (2012)

16. de Souza, E.A., Ferreira, A.A., Gonçalves, M.A.: Combining classifiers and user feedback for disambiguating author names. In: II, P.L.B., Allard, S., Mercer, H., Beck, M., Cunningham, S.J., Goh, D.H., Henry, G., (eds.) Proceedings of the 15th ACM/IEEE-CE on Joint Conference on Digital Libraries, Knoxville, TN, USA, 21–25 June 2015, pp. 259–260. ACM (2015)

Improving Behavior Prediction Accuracy by Using Machine Learning for Agent-Based Simulation

Shinji Hayashi[✉], Niken Prasasti, Katsutoshi Kanamori, and Hayato Ohwada

Department of Industrial Administration, Faculty of Science and Technology, Tokyo University of Science, Tokyo, Japan
shintetu84@gmail.com, niken.prasasti@sbm-itb.ac.id, {katsu,ohwada}@rs.tus.ac.jp

Abstract. This study models an integration between agent-based simulation and machine learning in order to achieve comprehensive behavior prediction. The model is applied to the case of customer churning in a subscription-based business. Providing a good model for behavior prediction requires dynamic simulation based on social structure. In this study, we first executed an agent-based simulation to capture the dynamic structure of human behavior. Next, we conducted machine learning to classify human behavior using a classification algorithm. Finally, we verified the agent-based simulation and machine learning results by comparing the accuracy of both models. Based on the agent-based simulation results, we provide some recommendations to improve the accuracy of agent-based simulation based on the classification results from machine-learning procedures.

Keywords: Agent-based simulation · Machine learning · C4.5 · Action prediction · Behavior prediction

1 Introduction

A large amount of information is now available due to the rapid spread of the Internet. While analyzing a large amount of data, there is a growing momentum to use it in business and in people's personal lives. Until now, many companies have analyzed a large amount of data and have used it in business. Rapid progress is being made in further information terminals such as smart phones, and various services such as electronic money and SNS are spreading. People now perform various actions using the Internet.

Machine learning can predict behavior. However, it is difficult to use the results of machine learning to influence policies and measures after the behavior prediction [1]. For instance, the machine-learning result may only lead to a decision that the company needs a new action in marketing strategy (e.g., promotion, cross-selling, or up-selling) to decrease the number of customer defections, without knowing when the ideal time is to implement such marketing strategies. There is a need for a dynamic perspective

© Springer-Verlag Berlin Heidelberg 2016
N.T. Nguyen et al. (Eds.): ACIIDS 2016, Part I, LNAI 9621, pp. 280–289, 2016.
DOI: 10.1007/978-3-662-49381-6_27

that can look for the detailed structure inside the customer churning process to complement improving the prediction techniques. One method commonly used for modeling the detailed social structure is agent-based modeling and simulation. Many earlier works on customer behavior and customer behavior prediction have executed agent-based simulations to obtain a better result in modeling the behavior of the customer as an agent. This is supported by the fact that each customer has different preferences in making decisions, and by using agent-based simulation we can see certain dynamics through this behavior. The prediction accuracy of human behavior can be improved by combining machine learning with agent-based simulations. However, the setting values used in the agent-based simulations are often determined by human judgment, and it is not easy to determine the optimal parameters.

In this study, in performing action predictions, we propose a method for determining the appropriate parameters of the agent-based simulation by using the simulation results of machine learning.

2 Related Work

With the capability for behavior prediction using machine learning, Jourge et al. predicted customer behavior using a dataset provided by a Brazilian mobile company [2]. They applied four machine-learning techniques: neural networks, decision trees, a genetic algorithm, and neuro-fuzzy simulations. These four methods have been verified through demonstrations and experiments to any degree of accuracy the cancellation action of the user. Their study uses 37 kinds of data, including information on billing data and monthly fees, communication history data, age, gender, demographic data, the residential city limits, and contract data such as discount plans. As a result, the neural network and the decision tree were able to predict the action of the customer with relatively high accuracy. In contrast, the genetic algorithm and neuro-fuzzy method did not provide good predictions.

Scott et al. built an analytical prediction model using methods such as Bayesian filters, decision trees, neural networks, discriminant analysis, and clustering on the data for 100,000 customers, provided by Tera Data Corporation [3]. Their study uses 171 attributes, including the talking time, monthly fee, functions, contract period, the corresponding number of calls to the customer center, and certain demographic data such as age and sex. As a result, a relatively high degree of accuracy is obtained by a decision tree and logistic regression. However, high accuracy could not be obtained using the other approach. Integrating the use of machine-learning techniques and agent-based simulation, Rand et al. performed an experiment by using agent-based simulation as well as machine learning for action or behavior prediction [4]. Their work explored the use of the machine-learning cycle as a model refinement engine for agent-based modeling and simulation. It focused on agent-based modeling, sending data to the machine-learning cycle to handle the model refinement.

3   Proposed Methodology

3.1   Problem Formulation

In this study, we propose a method to improve the prediction accuracy by changing the parameters of the simulation using the result of machine learning.

In agent simulation used to analyze human behavior, we simulate of the dynamic structure. The key to agent-based simulation is how the agents perform with other agents or within the environment.

In the model, G is a set of agents representing each customer.

$$G = \{g_1, g_2, \ldots \ldots g_n\} \tag{1}$$

A is set of actions representing customer behavior.

$$A = \{\text{Cancelation, Renew}\} \tag{2}$$

S is set of states representing each human's state. We set that the probability that the state s for each agent g takes action a. is parameters that affect the simulation.

$$\Theta = \{\theta_1, \theta_2, \ldots \ldots, \theta_n\} \tag{3}$$

The behavior of each agent may be determined by its state and actions regardless of the agent. We propose a method for determining the appropriate parameters and improving the accuracy of the agent simulation by modifying certain parameters characteristic of machine learning.

3.2   Methodology and Framework

Machine learning and agent-based modeling and simulation can be combined in many ways, as some research has examined [4]. We performed machine learning and agent-based simulation in each experiment using the same data in order to compare and verify their results. Machine learning and agent-based simulation both utilize an algorithm to control their flow of operation. In this study, we determine the parameters Θ seeking threshold in machine learning and agent-based simulation flow are generalized as follows: initialization of the agent, creation of action and state, agent action, state update of agents, and finally we obtain the result Fig. 1.

4   Experiment

4.1   Research Object

This study analyzes the behavior of the customer in customer churning, in which customers are willing to terminate the services they are subscribed for. Specifically, we focus on predicting the behavior of a customer regarding an auto-renewal service.

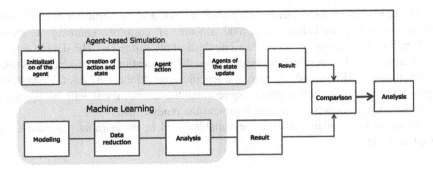

Fig. 1. Framework of proposed model

An auto-renewal service is updated automatically just before the expiration period of the products subscribed to by the customer. The auto-renewal service is intended to recover customer from their renewal fee. The auto-renewal update procedure is performed every year. However, the customer can cancel their subscription at any time during the subscription period. The customer can "opt-in" or "opt-out"; if the customer chooses to opt-in, it indicates that they would like to continue their subscription and they agree to the auto-renewal procedures. Choosing opt-out indicates that they would not like to continue their subscription.

4.2 Dataset

We obtained a dataset from a security-software subscription-based company's e-commerce site that contains historical customer purchasing data. Figure 2 indicates the termination time after the auto-renewal update on customers' subscriptions. It presents the number of cancellations during the time period when customers are enrolled in their subscriptions (Fig. 2a) and during the time just before their subscriptions expire (Fig. 2b).

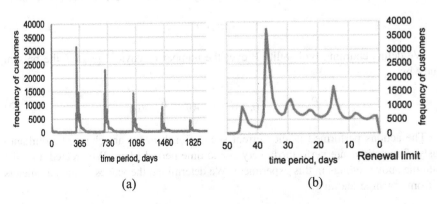

Fig. 2. Detection of customer churning (a) over a long term and (b) over a short term.

The horizontal axis represents the number of days, and the vertical axis represents the number of cancellations. The total amount of customer churning is about 1.82 million, accounting for more than one-third of all customers. The date of termination in the Fig. 2 corresponds to the day when the notification of auto-renewal is sent to the customer. This e-mail was sent at a specific time: 45 days, 37 days, 15 days, or eight days before the subscription expires and the auto-renewal will be updated. Therefore, a peak occurs on the graph around the renewal date.

Based on the available dataset, the variables used in the learning procedures are listed in Table 1.

Table 1. Selected variables used in machine learning procedures

Variables	Description
UPDATE_COUNT	Total count of renewals and repurchases (first purchase is excluded)
PRODUCT_PRICE	Recently purchased product price
ACTIVATION_FLAG	Whether it has registered the product
ORG_FLAG	Type of customer, whether individual or company
VSSA_FLAG	Whether they are using the optional (additional) service
MAIL_STATUS	Delivery status of e-mail notification
RIHAN_FLAG	Whether the subscription is cancelled

4.3 Churn Prediction by Agent-Based Simulation

We propose a method to predict customers' churning behavior using agent-based simulation by Hayashi et al. [5]. To generate a customer churning prediction, we model the customer churning behavior and simulate the dynamic structure. In this experiment, state s shall depend on the purchase price and available days for the customer. We adopt the definition of the hazard function (failure rate) of a personal computer (PC) as the failure rate in the number of days elapsed. t_{ar} is churn rate regarding the e-mail received in the number of days elapsed. t_{pri} is churn rate by the price of products purchased. t_{pc}, t_{ar} and t_{pri} are no difference in each customer, changing pri as parameters in this experiment. t_{pc}, t_{ar} and t_{pri} change by the state S. θ consists of pri, λ and p.

$$\theta = (\text{pri}, \lambda, p) \tag{4}$$

We define churning probability P over the number of days elapsed in the present study by

$$P_{s,a}(\theta) = F_{(t_{pc}(s))} + F_{(t_{ar}(s))} + F_{(t_{pri}(s))} \tag{5}$$

The actions performed in the current state of each agent are changed simultaneously. Simulations are performed every single time period (day). We created a model with the above settings in this experiment. We determined the values of the parameters Θ from machine learning.

Churn Rate. We assume that the failure probability of a PC will be higher for longer periods. We calculate the failure probability according to the period of use of the PC in the hazard function. In a former work, Junxiang et al. analyzed churning predictions using the hazard function [6].

$$F(t) = 1 - \exp(-(\lambda t)^p) \tag{6}$$

Figure 3 illustrates the failure rate of a PC according to the hazard function.

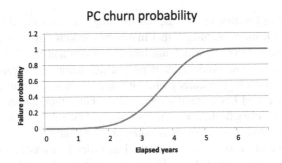

Fig. 3. PC churn probability by hazard function

We assume that customers tend to churn when the ending contract period from the reception of the e-mail has a certain number of days left. Furthermore, we assume that the product purchase price also impacts churning. Thus, the churn rate is calculated by considering the average price of the purchased product as well.

Result. We performed several experiments by changing the churning rate parameters. The results of an experiment conducted assuming that 1 % of users will cancel or churn on the date the notification mail is received had a correlation coefficient of 0.61. There was a large difference between the actual churning transition and the simulation results. Thus, in the next experiment, we assumed that 5 % of the users will churn. The correlation coefficient increased to 0.79, so the prediction was improved.

Higher values were obtained in both the correlation and mean-square error using the 5 % churn rate in the conducted experiments Table 2.

4.4 Churning Prediction by Machine Learning

The purpose of churn prediction in this study is to use historical information on customers who continued and customers who previously churned and to predict the continuation and churning patterns. However, it is not possible to create a longer learning period and highly original precision of predictor. Therefore, it is necessary to examine the algorithm as well as the features and attributes to be used for learning.

Table 2. Predicted results of simulation

	result
Correlation coefficient	0.7908
Mean square error	10327.39

When the customer is churning, Fig. 2 shows that the width is biased toward a particular timing with respect to the number of days until the renewal date. Therefore, we focus on whether the customer will continue the next update to perform learning and prediction.

Result of Churning Prediction. In machine learning, we use a technique called C4.5. C4.5 is a machine-learning technique that makes predictions based on a classification tree [7]. It is possible to check which attributes are important in the classification by analyzing the tree. To make predictions using machine learning, we can also use other techniques that provide high accuracy but with more complex learning. However, in this study, the data used has a small number of attributes that mainly have binary or discrete values. The classification tree in C4.5 has a relatively simple learning mechanism as a learning technique, so it is considered appropriate.

We use a 10-fold cross-validation method in order to measure the accuracy of prediction. Customers are divided into 10 groups, creating a learned classification tree using a machine-learning technique. Each customer in the group that was not used for learning is used to predict whether they will continue or churn. In this case, the group that was predicted to cancel is not included in the learning. By doing this for all 10 groups, it is possible to make a prediction of churning for all customers. It is possible to measure the prediction accuracy of the machine-learning techniques from the prediction result. Figure 4 depicts part of a decision tree obtained in the experiments using the C4.5 algorithm.

Table 3 represents the churning predicted by machine learning.

Table 3 indicates a high prediction accuracy, at around 84.8 %. However, the number that could not be correctly predicted was significant. Therefore, to perform machine learning using a non-variable, it is necessary to improve the prediction accuracy.

5   Integrating Machine Learning and Agent-Based Simulation to Improve Prediction Accuracy

In order to improve the agent-based simulation accuracy based on the results of the machine learning, we needed to change some parameter in the agent-based simulation based on the machine learning classification result. To change the parameters of the simulation, we looked for a common point of the simulation and decision tree. It can be seen that one feature that could be affecting the customer decision is the price. Thus, an analysis was performed that changed the criteria for churn by changing the *pri*value.

```
UPDATE_COUNT <= 3↓
│  CC_PRODUCT_PRICE <= 4700↓
│  │  VSSA_FLAG = FALSE: TRUE (37729.0/14872.0)↓
│  │  VSSA_FLAG = TRUE: FALSE (2.0)↓
│  CC_PRODUCT_PRICE > 4700↓
│  │  CC_PRODUCT_PRICE <= 5648: FALSE (6311.0/2667.0)↓
│  │  CC_PRODUCT_PRICE > 5648↓
│  │  │  CC_PRODUCT_PRICE <= 8600↓
│  │  │  │  CC_PRODUCT_PRICE <= 7667↓
│  │  │  │  │  CC_PRODUCT_PRICE <= 6800: TRUE (9.0)↓
│  │  │  │  │  CC_PRODUCT_PRICE > 6800↓
│  │  │  │  │  │  ORGFLAG = FALSE: FALSE (4.0)↓
│  │  │  │  │  │  ORGFLAG = TRUE: TRUE (3.0/1.0)↓
│  │  │  │  CC_PRODUCT_PRICE > 7667: TRUE (143.0/1.0)↓
│  │  │  CC_PRODUCT_PRICE > 8600↓
│  │  │  │  CC_PRODUCT_PRICE <= 13315↓
│  │  │  │  │  CC_PRODUCT_PRICE <= 11315: FALSE (9905.0/3431.0)↓
│  │  │  │  │  CC_PRODUCT_PRICE > 11315↓
│  │  │  │  │  │  CC_PRODUCT_PRICE <= 13000: TRUE (24.0)↓
│  │  │  │  │  │  CC_PRODUCT_PRICE > 13000: FALSE (103.0/37.0)↓
│  │  │  │  CC_PRODUCT_PRICE > 13315↓
│  │  │  │  │  CC_PRODUCT_PRICE <= 18500: FALSE (162.0/30.0)↓
│  │  │  │  │  CC_PRODUCT_PRICE > 18500: TRUE (15.0)↓
UPDATE_COUNT > 3↓
│  UPDATE_COUNT <= 5↓
│  │  UPDATE_COUNT <= 4↓
│  │  │  CC_PRODUCT_PRICE <= 4500: FALSE (50946.0/13088.0)↓
│  │  │  CC_PRODUCT_PRICE > 4500↓
│  │  │  │  CC_PRODUCT_PRICE <= 4743: FALSE (7543.0/2370.0)↓
│  │  │  │  CC_PRODUCT_PRICE > 4743↓
│  │  │  │  │  CC_PRODUCT_PRICE <= 5648: TRUE (54.0)↓
│  │  │  │  │  CC_PRODUCT_PRICE > 5648: FALSE (9606.0/1995.0)↓
│  │  UPDATE_COUNT > 4↓
│  │  │  CC_PRODUCT_PRICE <= 8600↓
│  │  │  │  CC_PRODUCT_PRICE <= 4700: FALSE (18713.0/6045.0)↓
│  │  │  │  CC_PRODUCT_PRICE > 4700↓
│  │  │  │  │  CC_PRODUCT_PRICE <= 5648↓
│  │  │  │  │  │  VSSA_FLAG = FALSE↓
│  │  │  │  │  │  │  MAIL_STATUS = TRUE: FALSE (731.0/342.0)↓
│  │  │  │  │  │  │  MAIL_STATUS = FALSE: TRUE (189.0/66.0)↓
│  │  │  │  │  │  VSSA_FLAG = TRUE: FALSE (56.0/11.0)↓
│  │  │  │  │  CC_PRODUCT_PRICE > 5648: FALSE (214.0/42.0)↓
│  │  │  CC_PRODUCT_PRICE > 8600↓
│  │  │  │  CC_PRODUCT_PRICE <= 8800: TRUE (340.0/1.0)↓
│  │  │  │  CC_PRODUCT_PRICE > 8800↓
│  │  │  │  │  CC_PRODUCT_PRICE <= 9200: TRUE (14.0/4.0)↓
│  │  │  │  │  CC_PRODUCT_PRICE > 9200: FALSE (163.0/41.0)↓
│  UPDATE_COUNT > 5: TRUE (805.0/36.0)↓
```

Fig. 4. Part of decision tree for predicting customer churning

Table 3. Results predicted by machine learning

	Product
Correctly churn number predicted	2442878
Correctly churn number that could not be predicted as churn	30851
Correctly non-churn number that could be predicted non-churn	88052
Correctly non-churn number that could not be predicted non-churn	28358
Accuracy	84.8 %

5.1   Experiment Result

Comparing the simulation results and the above results, a correlation coefficient exceeding the above was obtained by changing the parameters. The correlation coefficient increased from 0.79 (highest correlation coefficient) to 0.84. Next, towards the result of changing the parameters and indicates a low value, it is shown that the mean square error is brought closer to the actual churning transition by combining the simulation information with machine learning Table 4.

In Fig. 5, the simulation results and the actual churning transitions compared to the number of days elapsed since the start time of a subscription is similar, but a deviation occurs in the graph from the first year through the second year. It is considered necessary to extend our experiments by including other factors and adjusting the churning rate by mail notification.

Table 4. Comparison of correlation coefficient and mean square error

	Result of chapter 3.4	Proposed method
Correlation coefficient	0.7908	0.8417
Mean square error	10327.39	8326.74

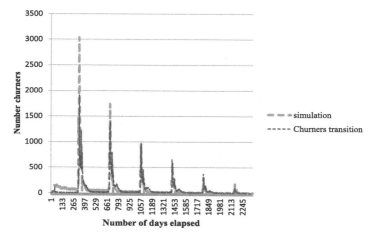

Fig. 5. Simulation results and churning transitions (Color figure online)

6   Conclusion

In this study, we conducted experiments to improve the prediction accuracy by combining the results of machine learning and agent-based simulation for customer churning prediction. In performing the experiments, prediction with classification trees formed using C4.5 utilized its relatively simple learning mechanism. Furthermore, we examined the agent-based simulation assuming the elements of several churning possibilities. Based on the results obtained from integrating the machine-learning results and the agent-based simulation, we can improve the accuracy of the simulation result. It has been shown that it is possible to realize an improvement in prediction accuracy by applying the features obtained by the classification tree in the simulation. This study successfully shows how the results of machine learning can be useful for parameter determination in the modeling process of an agent-based simulation.

Future work will seek to increase accuracy by conducting experiments using a number of variables. Based on information obtained from the machine-learning result, we can explore what changes are needed to change the parameters in the simulation. Moreover, in future work it will be necessary to improve the accuracy of the machine learning by using a variable from the agent-based simulation.

References

1. Niken, P., Masato, O., Katsutoshi, K., Hayato, O.: Customer lifetime value and defection possibility prediction model using machine learning: an application on cloud-based software company. In: Nguyen, N.T., Attachoo, B., Trawiński, B., Somboonviwat, K. (eds.) ACIIDS 2014, Part II. LNCS(LNAI), vol. 8398, pp. 595–604. Springer, Heidelberg (2014)
2. Jorge, B.F., Marley, V., Marco, A.P., Carlos, H.B.: Data mining techniques on the evaluation of wireless churn. In: ESANN 2004 Proceeding, European Symposium on Artificial Neural Networks Bruges, pp. 483–488 (2004)
3. Scott, A.N., Sunil, G., Wagner, K., Junxiang, L., Charlotte, H.M.: Defection detection: measuring and understanding the predictive accuracy of customer churn models. J. Mark. Res. XL(3), 204–211 (2006)
4. Rand, W.: Machine learning meets agent based modeling: when not to go to bar. In: Proceedings (2006)
5. Shinji, H., Ohwada, H., Kanamori, K.: Customer churn prediction using the agent model simulation and machine learning. Journal of Information Processing (2015)
6. Junxiang, L.: Predicting customer churn in the telecommunications industry-an application of survival analysis modeling using SAS. In: SAS Conference Proceeding SAS Users Group International vol. 27,14–17 April 2002, Orland, Florida (2002)
7. Quinlan, J.R.: C4.5 Programs for Machine Learning. Morgan Kaufmann Publishers, San Francisco (1993)

A Model for Analysis and Design of Information Systems Based on a Document Centric Approach

Bálint Molnár[✉], András Benczúr, and András Béleczki

Information Systems Department, Eötvös Loránd University of Budapest, Pázmány Péter sétány
1/C, Budapest 1117, Hungary
{molnarba,abenczur}@inf.elte.hu, beleczki.andras@outlook.com

Abstract. The recent tendency in analysis and design of Information Systems is that the emphasis is placed on the documents that are ubiquitous around information systems and organizations. The proliferation of computer literacy led to the general use of electronic documents. To understand the anticipated behaviour of Information Systems and the actual operation of a particular organization, the analysis of documents play increasingly important role. The behaviour of Information Systems can be interpreted in a framework of Enterprise Architecture and its models that are contained in it. Certain parts and entirety of various types of documents connected to business processes, tasks, roles and actors within organization. The tracking of life cycle of documents and representing the complex relationships is essential both at analysis and operation time. We propose a theoretical framework that makes use previous results of modelling and well-founded mathematical techniques.

Keywords: Information System · Document modeling · Information System architecture · Zachman framework · Hypergraphs

1 Introduction

The Information Systems become more and more complex for several reasons. The use of various electronic document types is commonplace in organizations. The interactive forms, Web pages, the structured and semi structured documents - on the one hand- are the stimulus to start the chain of activities within organization - on the other hand - they are the end of business processes. The Business and System Analysts typically encounter documents during the process of analysis, the conceptualization of requirements by users. The documents emerge in the disguise as requirements, subject of actual business processes and tasks, essential component of data processing internal to the Information Systems.

There are several modeling and descriptive approaches for modeling Information Systems. Some of them make use of formalism grounded in mathematics. It seems to be beneficiary and feasible if we elaborate a method that focus on documents and their life cycle and places them into the context of Information Architecture of Information Systems. The adequate mathematical formalism that is capable to represent the complex relationships is the generalized *hypergraph* [5].

© Springer-Verlag Berlin Heidelberg 2016
N.T. Nguyen et al. (Eds.): ACIIDS 2016, Part I, LNAI 9621, pp. 290–299, 2016.
DOI: 10.1007/978-3-662-49381-6_28

In Sect. 2, we present the previous researches reported in the literature, in Sect. 3 we describe the required mathematical background, in Sects. 4 and 5 we outline our method making use of the previous approaches in a document centric approach, and Sect. 6 provides a summary and conclusions.

2   Literature Review

The Web based and Web Information Systems are the typical examples that make extensive use of the various document formats. The emphasis on Web technologies slowly diminished as the application of Web technology, definitely at user interfaces, became commonplace. A systematic design approach to construct web-based applications is discussed in [10]. The shown method makes use of semi-structured and interactive documents represented by XML. Another paper presents an approach for a well-founded, concepts-based modeling process for a Web site. For designing of Web Information Systems, Rossi presents a design procedure [19]. To grasp the complex behavior of Information Systems the notion of the enterprise architecture give a helping hand, namely the Blokdijk's perception of Information Systems, Zachman ontology and TOGAF, all of them was created for information systems [4, 16, 17, 21].

An Information System supports business processes (Business Process Modeling, BPM) and is usually tightly coupled to other IS. A fairly standard way to model business processes is the application of Business Process Modeling (BPM) methods. The Information Systems can also be perceived as a structure with underlying databases for structured, semi-structured as well as unstructured documents (XML-based). The documents play important role at the interface, interaction level and at core activities of data processing. The integration level and the degree of reconciliation between Business Processes and organization can be analyzed on the base of ontologies and semantic approaches [7]. A model driven and formal comprehensive approach for Information Systems modelling can be found in Luković et al.'s paper [11]. There are various input data format for communication to services: (1) HTML pages, (2) SOAP messages, (3) semi-structured and unstructured documents (XML-based) [3, 6, 15].

There were some previous papers that tried to put the before-mentioned approaches into a unified framework by essentially semi-formal way [12–14]. The Enterprise Architecture framework as Zachman and TOGAF provides a supporting environment, [16, 17, 21]. Blokdijk's collection of Information System Models yields a structuring guideline [4], moreover, the axiomatic design approach employed for the Information System environment offers an opportunity for a method that is interesting not only for theoretic modeling point of view but provides a chance to support practical design methods [20].

3   Formal Mathematical Background

Hypergraphs. As we have outlined previously, the problem to be solved can be described as a set of complex, heterogeneous relationships. The basic components that appear as constituents participate sometimes in hierarchical, sometimes rather network-like relationships. These two kind of relations are different from each other. The hypergraphs as mathematical structure seems to be apt to representing the inter-relationships among the models, views, viewpoints, perspectives, and the over-arching documents and business processes [21].

We start with the basic definitions of hypergraphs in order to employ for depicting the before-mentioned complex relationships.

Definition 1. A *hypergraph H* is a pair *(V, E)* of a finite set $V = \{v_1, ..., v_n\}$ and a set *E* of nonempty subsets of *V*. The elements of *V* are called vertices (nodes), the elements of *E* are called edges [5].

Definition 2. *Generalized* or extended hypergraph. The notion of hypergraph may be extended so that the hyperedges can be represented – in certain cases – as vertices, i.e. a hyperedge *e* may consist of both vertices and hyperedges as well. The hyperedges that are contained within the hyperedge *e* should be different from *e* [5].

Considering a document model, a particular document type hierarchy can be perceived as a "hierarchy" of hyperedges. The free variables or placeholders to be filled-in may occur as ultimate vertices within hyperedges that represents the instance of extension of particular document type. In a document subpart hierarchy, a specific subpart of document may be denoted by a vertex within a particular hyperedge that describes this document that contains the subpart, although that subpart as a vertex may include a document type hierarchy that can be depicted by a hyperedge.

Definition 3. A *directed hypergraph* is an ordered pair

$$\vec{H} = \left(V; \vec{E} = \{\vec{e_i} : i \in I\} \right) \tag{1}$$

Where *V* is a finite *set of vertices* and $\vec{E}$ is a set of *hyperarcs* with finite index set *I*. Every *hyperarc* $\vec{e_i}$ can be perceived as an ordered pair

$$\vec{e_i} = \left(\vec{e_i^+} = (e_i^+, i) ; \vec{e_i^-} = (i, e_i^-) \right) \tag{2}$$

Where $e_i^+ \subseteq V$ is the set of vertices of $\vec{e_i^+}$ and $e_i^- \subseteq V$ is the set of vertices $\vec{e_i^-}$. The elements of $\vec{e_i^+}$ (hyperedges and/or vertices) are called *tail* of $\vec{e_i}$, while elements of $\vec{e_i^-}$ are called *head* [5]. We may use as shorthand notation for ordered pairs, e.g. a vertex and a directed hyperedge as ordered pair $<v_i, e_j>$.

The underlying graph representation is based on the hypergraphs and directed hyper-graphs. The potential implementations of hypergraphs in a hypergraph database allows for linking attributes to vertices, even more to hyperedges. The target domain, namely documents and model of Information Systems within organizations, contains complex n-ary relationships. The hypergraph provides the opportunity to depict recursive construction, to describe logical relations, to store compound structures along with their values [1, 8, 9].

As an illustration of the basic concepts of directed hypergraph, an example can be seen in Fig. 1 that makes sense of the representation for the domain by hypergraph. The essential characteristics is that vertices contain composite constituents that are themselves may be graphs; generalized hyperedge may contain other hyperedges but not itself and nodes. Detailed description about the *Architecture Describing Hypergraph* can be found in [22].

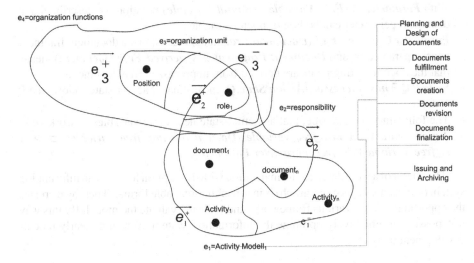

Fig. 1. Example for directed hypergraph representing a sample of essential relationships

Description Logics. One of the most common approaches of formalization is the use of some mathematical-logical language. The Description Logics belongs to the theories of mathematical logics, and their purpose is to create a formal knowledge representation [2, 18]. Compared to propositional calculus (or propositional logic), the expressiveness of description logic is higher, and it has a more effective algorithm for the decision problem than the first-order predicate logic. On the other hand, the network like knowledge representation - where the elements of the network are nodes and links are relationships as e.g. the semantic net-work - can be related to the theory of hypergraphs. In both case, nodes can be used to define concepts, and links can be used to characterize the relationships among them. On bearing this in mind, it is obvious to apply description logic on a system based on the mathematical background of hypergraphs.

The knowledge representation systems based on Description Logics contains two main components: the TBox, and the ABox. The TBox introduces the terminology, i.e., the basic concepts, which denote sets of individuals (atomic and complex), and roles, which define binary relations between individuals. These are forming the vocabulary of an application domain. The ABox contains assertions among named individuals and the vocabulary.

There are many variations of the Description Logics (based on the description languages varieties) and there is an informal convention, where their name indicates which operators are allowed. For example, a basic logical language is the Attributive

Language – AL, which allows: atomic negation, concept intersection, limited existential quantification and universal restriction. This can be extended with other operators, as e.g. concept union (U), full existential qualification (E), cardinality restriction (N), or complex concept negation (C). The description language lays the ground for the description logic. To illustrate the use of the Description Logic in a document centric environment, we give some examples below:

- With *Parameter* $\sqsubseteq$ (*Free Variable* $\sqcup$ *Bound Variable*) notation we describe that a document parameter can be free or bound variable.
- *Parameter* $\sqsubseteq \exists$ *is_part_of.*(*Document Fragment*) means, that a document fragment consist parameters, and *Document Fragment.*$\exists$ *is_derived*(*Free Document*) means that the document fragments are derived from unprocessed free documents.
- *State.P* $\sqsubseteq \exists$ *has_successor.*(*Action State.Q*) means that Q action-state follows the P state.
- The following line describe, that an action-state needs free variables to work with: *Action State* $\sqsubseteq \exists$ *has_free_variables*(*Document*); *has_free_variables* $\equiv \geq 1$ *is_free_variable* $\sqcap$ *is_free_parameter.Parameter.*

The output of a well-designed formalization with description logics of an information system (represented by a hypergraph) is in a machine-readable format. Thereby it creates the opportunity to use various frameworks and tools to evaluate the model. By this way it is possible to effectively optimize the information system even in the early model-development phase.

4   Formalized Document Centric Approach

In the case of a particular organization, we can imagine there is a comprehensive document that is a representation - in conceptual sense – of all potential documents. This overarching document is composed of generic document types. Generic document types are hierarchical structures that can be described by configuration hyperedges that reflect the composition of documents. There are hierarchical relations among the members of a generic document. The hierarchical relationships can be described by configuration hyperedges; the instances of a generic document member can be perceived as extensions and can be represented by extensional hyperedges.

A generic document type *GDT* is a hierarchy of document types *DTH*. The elements of *DTH* can belong to a configuration hyperedge e_{Ci} as vertices. The generalized hypergraphs allow that the vertices may appear as complex structures, as hyperedges. Therefore, a node can be a hyperedge that itself a configuration hyperedge that contains a hierarchy of document types. Thereby, the representation makes possible for a recursive definition of document types and gathering them into a generic document type.

The direction of the hyperarc shows whether a document plays the *input* or *output* role in a particular context. The definition is given above (see formula (2)) permits the differentiation between the *information* represented by the *head* and *tail* of a hyperarc, and the *information* that are represented in the form of nodes that are contained within the heads and the tails [5].

Definition 4. The *Document Subhypergraph* consists of:

- A finite set of documents that are represented by vertices $DOC = \{doc_1, ..., doc_n\}$;
  - The documents contain variables, the variables belong to attribute types $Attr = \{T_1, ..., T_n\}$;
  - The finite set of domains is $DOMSET = \{D_1, ...D_k\}$ that contains the domain of each single type, T_i;
- The relationship between a generic document type GDT hierarchy and its constituents document types belonging to a DTH can be described by hyperarcs representing *is-a* relationships; the hierarchy is a mapping of supertype-subtype relationships between document types. The relationships can be deduced from the variables, their attributes and the types of attributes.
- The relationship between a document doc_i and a document type DT can be described by a hyperarc representing the *instance-of* relationship.

The concept of generic document type offers possibilities for derivation of new document types from other document types that can be regarded as templates. The derivation rules can be formalized by logical statement that may create either a slightly different document type according to the structure of documents and then an instance of it or operate during the lifecycle of an instance of the document types. A document type may contain business rules in the form of predicates, data retrieving and calculation rules.

5 Information Architecture and Documents

Beside the essential documents, Information Systems can be described by various models that are ordered into a reasonable structure by Enterprise Architecture approach.

The models' descriptions appear usually in semi-structured document forms as XML and/or JSON that offers a chance for uniform treatment of documents and models of Information Systems. As structuring principal for models of Information Systems, we can use Zachman ontology and/or TOGAF [17, 21]. The set of relations among models and the internal structure of models plays essential role.

The models can be arranged into three meta-groups namely *organization*, *documents* and *activities* related models. For modeling, the relationships and interactions among these three meta-groups and the underlying collections of data are significant. The models, documents and concepts of Information Systems and a node representing the external environment compose a hypergraph that embraces all important parts of the application domain that may be called as *System Hypergraph*.

In the case of documents, a directed hyperedge can express the input and output roles of documents that they may play within activities of business processes. The document may be attached to organization units and actors through a *responsibility hyperedge* (labeled directed hyperedge). The variables of documents may be connected to data nodes of D that is organized into reasonable partitions that are represented by nodes contained in hyperedges that can be mutually mapped to specific data collections. These sub-hypergraphs may be called *Sub-system Hypergraph*s. Between the models, a refinement relation can be identified within an architectural perspective and represented by a directed hyperarc *is-a-refinement*. The documents and their structures can be described by documents model.

Definition 5. *Models* of Information Systems represented in the Architecture Describing Hypergraph are:

- The set of vertices is divided up into two basic subsets V_{Doc} and V_{Model};
- $V_{Doc} \supseteq \{OGDT\}$ where *OGDT* signifies the *overarching generic document*, that is the supertype of all other document types and their instances;
- $V_{Model} \supseteq \{EA, \{external\_evironment\}\}$, where *EA* designates the overall *Enterprise Architecture* consisting of models, the *external_evironment* refers to the outside world that is typically the source of *stimulus* that is generated by either humans or any other systems;
- $V_{Configuration} = \cup\, h_i$ where $h_i \in \boldsymbol{E}_C$, and $\cap\, h_i = \varnothing$ where $h_i \in \boldsymbol{E}_C$.

The expressions articulate the fact that the configuration hyperedges represents the structure of artifacts of models and documents in the form of structural constituents as nodes.

- The set of arcs (directed edges of graphs) A is partitioned into subsets A_{Doc_Target}, A_{Model_Target}, $A_{Interaction}$, where $A_{Doc\_Target} \subseteq V_{Configuration} \times V_{Doc}$, $A_{Model\_Target} \subseteq V_{Configuration} \times V_{Model}$.
  The directed edges, the arcs map a complex structure, a configuration of elements (nodes) to a node that represents either a document or a model.
- $HA_{Interaction} \subseteq V_{Model} \times V_{Doc}$, $HA_{Interaction} \subseteq E_D$;
  The interaction between certain models and specific documents can be expressed by a hyperedge $h \in HA_{Interaction}$.
- E_C can be partitioned into two subsets $E_{Configuration\_Document}$ and $E_{Configuration\_Model}$.

The hyperedges $h_{i,\,cd} \in E_{Configuration\_Document}$, $h_{j,\,cd} \in E_{Configuration\_Model}$ represent an inheritance structure. The inheritance structure conforms to the object-oriented paradigm, i.e. the configuration of documents and models inherit the attributes of super-classes, and may have extra attributes as well. Each attribute of a certain configuration can be represented by a vertex of the hyperedge. An attribute linked to a node either in V_{Model} or in V_{Doc}, its value represented by a link to a $d \in D$ when it is valuated. If the attribute is multi-valued then the attribute is connected to hyperedge $h \in Power(D)$ (the power set of D).

- The set of extensional hyperedges E_E is split into two subsets $E_{Superclass}$ and $E_{Extension}$
  - The hyperarc $h \in E_{Superclass}$, if $h \in E_E$, (h set of nodes)
    - Either $h \in V_{Doc}$ and $OGDT \in h$
    - or $h \subset V_{Model}$ and $EA \in h$.
    - Given a node $v_i \in h$ and $h' \in E_{Superclass}$, then either valid that $<v_i, h'> \in E_{Super\_doc}$, then $h' \subseteq h$
    - Or $<v_i, h'> \in E_{Super\_model}$. then $h' \subseteq h$

- Notation: $E_{Super\_doc} = V_{Doc} \setminus \{OGDT\}) \times E_{Superclass} \subset E_D$;
- Notation: $E_{Super\_model} = ((V_{Model} \setminus \{EA, \{external\_evironment\}\}) \times E_{Super\_class} \subseteq E_D)$;

The hyperedges $h \in E_{Superclass}$ provide the association between a class of objects (models or documents) and its super-classes in compliance to the object-oriented paradigm. For the reason of our modeling approach, we make distinction between the two top super-classes, namely *OGDT*, the *overarching generic document*, *EA* the overall *Enterprise Architecture*. The conditions above specify the transitivity of *is-a* relationship for the relation between class and its super-classes.

- The instances of models can be represented by $E_{Instance\_model} \subset V_{Model} \times E_E$ (extensional);
- The instances of documents can be represented by $E_{Instance\_doc} \subset V_{Doc} \times E_E$;
- $h \in E_E$, (h set of nodes) is $h \in E_{Attribute\_Set}$ if $h \subset D$. The following statement is valid as well: $\cup\, h_i = D$, $h_i \in E_{Attribute\_Set}$. The hyperarcs $h \in E_{Attribute\_Set}$ are used to represent the attributes domains, and the associated values.
- The hyperarc $h \in E_{Extension}$, if $h \in E_E$, (h set of nodes) and
  - Given a node $v_i \in h \subset V_{Doc}$ and $h \in E_{Extension}$, then $<v_i, h> \in E_{Instance\_doc}$, $<v_i, h'> \in E_{Super\_doc}$, then for each $n \in h$ and each $d_t \in h'$ ∃ $ha \in E_E$ (hyperarc) where $<d_t, ha> \in E_{Instance\_doc}$;
  - Or
  - Given a node $v_i \in h \subset V_{Model}$ and $h \in E_{Extension}$, then $<v_i, h> \in E_{Instance\_model}$, $<v_i, h'> \in E_{Super\_model}$, then for each $n \in h$ and each $d_t \in h'$ ∃ $ha \in E_E$ (hyperarc) where $<d_t, ha> \in E_{Instance\_model}$;

  A hyperedge $h \in E_{Extension}$ represents an extension for models and documents respectively as well. The above described statement formalizes the transitivity of *instance-of* relationship.

- The intensional hyperarc $h \in E_I$, $<d, h> \in E_{Intension}$ if $E_{Intension} \subset V_{Doc} \times E_I$, $d \in V_{Doc}$, $h \in E_{Configuration\_Document}$, $h \subset V_{Doc}$; the intensional hyperarc defines the hierarchical relationship between templates, rule-based document types and extensional document types that are instantiated.
- The set of hyperedges in E_D (hyperarcs) can be arranged into several subsets according to the notion of Enterprise Architecture:
  - The hyperarc $h \in E_{View} \subseteq E_G$, $h \subseteq V_{Model}$, represents a stakeholder's view that puts together models that describe the specific viewpoint of a role within organization.
  - The hyperarc $h \in E_{Perspective} \subseteq E_G$, $h \subseteq Powerset\,(V_{Model})$, embodies a hierarchy of models according to a refinement hierarchy;
  - The hyperarc $h \in E_{Doc\_Life\_cycle} \subseteq E_G$, $<d, h> \in E_{Instance\_doc} \times E_{Instance\_model}$, $d \in V_{Doc}$, that depicts the life cycle of document through the interactions with models.

6 Conclusion

In this paper we proposed an Architecture Describing Hypergraph as representation for Enterprise Architectures and related Documents. The suggested descriptive method takes advantages from the basic properties of generalized hypergraphs, i.e. unequivocal representation of complex relationships; moreover, there are some distinguished features

– Uniform treatment of both intensional and extensional aspects of documents and models within Enterprise Architecture;
– Direct depiction of hierarchical relationships through instance-of, sub-class-of, super-class-of relationships;

The outlined approach can also be considered as a formal background to analyze and design Information Systems. The documents play important roles in Information Systems in the time of analysis, design, specification and operation with strong coupling to roles of organizations. The unified framework provides an opportunity for uniform handling of models and documents on a formal foundation.

The hypergraph-based approach offers the chance to apply further mathematical tools for assistance in the design, verification and validation to maintain the integrity and consistency of Information Systems.

References

1. Ausiello, G., Franciosa, P., Frigioni, D.: Directed hypergraphs: problems, algorithmic results, and a novel decremental approach. In: Restivo, A., Ronchi Della Rocca, S., Roversi, L. (eds.) ICTCS 2001. LNCS, vol. 2202, pp. 312–328. Springer, Heidelberg (2001)
2. Baader, F.: The Description Logic Handbook: Theory, Implementation, and Applications. Cambridge university press (2004)
3. Bernauer, M., Schrefl, M.: Self-maintaining web pages: from theory to practice. Data Knowl. Eng. **48**, 39–73 (2004)
4. Blokdijk, A., Blokdijk, P.: Planning and Design of Information Systems. Academic Press, London (1987)
5. Bretto, A.: Hypergraph Theory: An Introduction. Springer, Heidelberg (2013)
6. Chiua, C.-M., Bieber, M.: A dynamically mapped open hypermedia system framework for integrating information systems. Inf. Softw. Technol. **43**, 75–86 (2001)
7. Gábor, A., Kő, A., Szabó, I., Ternai, K., Varga, K.: Compliance check in semantic business process management. In: Demey, Y.T., Panetto, H. (eds.) OTM 2013 Workshops 2013. LNCS, vol. 8186, pp. 353–362. Springer, Heidelberg (2013)
8. Gallo, G., Longo, G., Pallottino, S., Nguyen, S.: Directed hypergraphs and applications. Discrete Appl. Math. **42**(2), 177–201 (1993)
9. Iordanov, B.: HyperGraphDB: a generalized graph database. In: Shen, H.T., Pei, J., Özsu, M., Zou, L., Lu, J., Ling, T.-W., Yu, G., Zhuang, Y., et al. (eds.) WAIM 2010. LNCS, vol. 6185, pp. 25–36. Springer, Heidelberg (2010)
10. Köppen, E., Neumann, G.: Active hypertext for distributed web applications. In: Proceedings of 8th IEEE International Workshops on Enabling Technologies: Infrastructure for Collaborative Enterprises (WET-ICE 1999), pp. 297–302 (1999)

11. Luković, I, Ivančević, V, Čeliković, M, Aleksić, S.: DSLs in action with model based approaches to information system development. In: Mernik, M. (ed). Formal and Practical Aspects of Domain-specific Languages: Recent Developments, pp. 502–532. IGI Global, Hershey (2013). ISBN 978-1-4666-2092-6, doi:10.4018/978-1-4666-2092-6

12. Molnár, B.: Applications of hypergraphs in informatics: a survey and opportunities for research. Ann. Univ. Sci. Budapest. Sect. Comput. **42**, 261–282 (2014)

13. Molnár, B., Tarcsi, A.: Architecture and system design issues of contemporary web-based information systems. In: Proceedings of 5th International Conference on Software, Knowledge Information, Industrial Management and Applications (SKIMA 2011), Benevento, Italy, 8–11 September 2011

14. Molnár, B., Benczúr, A.: Facet of modeling web information systems from a document-centric view. Int. J. Web Portals (IJWP) **5**(4), 57–70 (2013). IGI Global

15. Nama, C.-K., Jang, G.-S., Ba, J.-H.: An XML-based active document for intelligent web applications. Expert Syst. Appl. **25**, 165–176 (2003)

16. OASIS: A reference model for service-oriented architecture. White Paper, Service-Oriented Architecture Reference Model Technical Committee, Organization for the Advancement of Structured Information Standards, Billerica, MA, February 2006

17. Open Group: TOGAF: The Open Group Architecture Framework, TOGAF® Version 9 (2010). http://www.opengroup.org/togaf/

18. Rudolph, S.: Foundations of description logics. In: Reasoning Web. Semantic Technologies for the Web of Data, pp. 76-136. Springer Berlin Heidelberg (2011)

19. Rossi, G., Schwabe, D., Lyardet, F.: Web application models are more than conceptual models. In: Chen, P.P., Embley, D.W., Kouloumdjian, J., Liddle, S.W., Roddick, J.F. (eds.) ER 1999. LNCS, vol. 1727, pp. 239–252. Springer, Heidelberg (1999)

20. Suh, N.P.: Axiomatic Design: Advantages and Applications. Oxford University Press, New York (2001)

21. Zachman, J.A.: A framework for information systems architecture. IBM Syst. J. **26**(3), 276–292 (1987)

22. Molnár, B., Benczúr, A., Béleczki, A.: Formal approach to modelling of modern information systems. Int. J. Inf. Syst. Proj. Manag. (2016, to be published)

MobiCough: Real-Time Cough Detection and Monitoring Using Low-Cost Mobile Devices

Cuong Pham[✉]

Computer Science Department and Machine Learning and Applications Lab,
Posts and Telecommunications Institute of Technology, Km 10, Hadong, Hanoi, Vietnam
cuongpv@ptit.edu.vn

Abstract. In this paper we present MobiCough, a method and system for cough detection and monitoring on low-cost mobile devices in real-time. MobiCough utilizes the acoustic data stream captured from a wirelessly low-cost microphone worn on user's collar and connected to the mobile device via Bluetooth. Mobi-Cough detects the cough in four steps: sound pre-processing, segmentation, feature & event extraction, and cough prediction. In addition, we propose the use of a simple yet effective robust to noise predictive model that combines Gaussian Mixture model and Universal Background model (GMM-UBM) for predicting cough sounds. The proposed method is rigorously evaluated through a dataset consisting of more than 1000 cough events and a significant number of noises. The results demonstrate that cough can be detected with the precision and recall of more than 91 % with individually trained models and over 81 % for subject independent training. These results are really potential for health-care applications acquiring cough detection and monitoring using low-cost mobile devices.

Keywords: Cough detection · Monitoring · Machine learning · Healthcare · Ubiquitous computing · Mobile devices · Acoustic sensors · Universal background model · Gaussian mixture model · Health monitoring

1 Introduction

Cough is a prevalent symptom of many related respiratory diseases from minor ailment to severely lung tuberculosis or chronic cough. According to a report [1] the prevalence of cough-related diseases is around 29 % of hospital consultant episodes. One of the most serious cough-related diseases is pneumonia killing nearly 6.6 million children under five every year [2], and it is the leading disease causing of death in children. Especially, the situation is even worse in Vietnam due to climate characteristics of hot and humid leading to the high rate of pharyngitis and pneumonia infection. Every year, up to 4,500 children died and 2.9 million under five years old infectious pneumonia, adding about 40 children are hospitalized to be treated for respiratory disease everyday [2]. This high rate indicates the lack of local healthcare and an effective monitoring pneumonia method for patients and people who are highly potentially infectious and living in remote areas.

© Springer-Verlag Berlin Heidelberg 2016
N.T. Nguyen et al. (Eds.): ACIIDS 2016, Part I, LNAI 9621, pp. 300–309, 2016.
DOI: 10.1007/978-3-662-49381-6_29

Monitoring cough is a task that is able to provide cough frequency information to users and doctors. Such information is useful for diagnose and treatment of the cough-related diseases. Traditionally, cough is monitored by manually diarizing from nurses and patients themselves. This would possibly report imprecise cough and whooping frequency as manually counting cough for every time and everywhere is infeasible. Therefore, an alternative is to automatically monitor cough using electronic devices. Recent technologies such as wearable computing can easily allow users to capture cough sounds which are analysed for detecting and counting the cough. Recently, although research on cough detection has made significantly progress such as detection of cough using mobile phones [5, 14] while preserving privacy [5], or analysis sounds from an audio recorder [6, 9, 13]. These works achieved relatively high accuracies of cough detection rate. However, real-time cough detection and monitoring is omitted from these works. As cough relates to the diversity of respiratory diseases, it is understandable that the detection of cough in real-time is crucial for opportune interventions, diagnosis, cure, treatment and even emergency aids made by the doctors. Therefore, in this work we propose a method and system that can detect and monitor cough by analysing sounds from a low-cost, wireless microphone in real-time. In addition, to improve the detection accuracy, we propose the Gaussian Mixture model (GMM) combined with the Universal Background models (UBM), an adaptive version of Gaussian Mixture models with maximum a posteriori scheme, for discriminating cough and noise sounds (i.e. from human's speaking, environment etc.). The proposed method is evaluated on a cough dataset consisting of more than 1,000 cough events, and a significant number of background noise events. The results of this research can be a complementary tool for real-time monitoring pneumonia as well as other respiratory related diseases. In addition, we have implemented the pre-trained GMM-UBM models (the GMM-UBM models after trained with offline data) on the smart phones for real-time cough detection and the monitoring module on the smart phones.

2   Related Work

Automatic cough detection has attracted by researchers, medical experts, and doctors for long years as cough is the most frequent symptom appearing on the people when asking medical advice [5]. Approaches to cough detection can be the use of array of audio sensors [6, 8, 10, 13], a single microphone worn on user's body [4] or mobile phones [3, 14]. Previously, arrays of sensors installed in the environment surroundings are proven effectively for context recognition and situated services [11], while wearable sensors commonly utilized for human activity recognition and fall detection [12]. While the use of either multiple sensors installed in the environment surroundings can possibly be limited by the range of sensing signals (i.e. a room or a house), the mobile phones or wearable sensors, in contrast, can allow users to detect and to monitor coughs at everywhere and at every time.

Multiple acoustic sensors are widely used for cough detection as the cough detector can achieve performance accuracies as high as over 95 %. Work by Drugman, T. et al. [6, 8] for example, proposed a cough detection method based on Artificial Neural

Networks which were trained feature vectors comprising of 222 feature elements, While [6] investigated how the acoustic sensors worn on different positions of user's body can impact on the performance accuracy of the cough detector, work by Vizel E. et al. [7] analyzed acoustic data captured from both microphones worn on user's chest and ambient sensors (installed in the environment surroundings). Similarly, Zheng, S. et al. [12] proposed CoughLoc which analyzes acoustic signals from a non-intrusively wireless sensor network. CoughLoc exploits the location of cough occurrence to enhance the detection accuracy. [8] uses various sensors including Siemens EMT 25 C accelerometer (Siemens); PPG 201 accelerometer (PPG); Sony ECM-T150 electret condenser microphone with air coupler etc. connected to an electret condenser microphone for comparisons of the effectiveness of lung sound transducers. In contrast, cough detection using a single microphone analyses only one audio stream from a microphone worn on chest of the user. For instance, [4] proposed Leicester Cough Monitor (LCM) using an audio recorder on patient's chest. LCM was rigorously evaluated and achieved sensitivity and specificity over 91 % on their (offline) dataset of 15 patients with chronic cough and 8 healthy subjects. Another study [11] proposed hidden Markov models trained with more than 800 min of ambulatory recordings and achieved the detection accuracy rate of 82 % with the false alarm rate is as low as 7 events per hour.

With the incorporation of multi-processors, cache memory, and many sensors such as accelerometer, GPS, gyroscope, digital camera, microphone etc., mobile phones become powerful platforms for health-care applications. A cough detector on mobile phones utilizes audio stream captured by the microphone embedded inside the mobile phone. For example, [3] proposed a cough detection method based on mean decibel energy, component weight features extracted from Fast Fourier transform coefficients from acoustic raw data captured from the mobile phone carried in the participant's shirt pocket or using a neck strap (the phone's microphone facing up in the direction of the mouth). [3] achieved the true positive rate is as high as 92 % while preserving privacy for the users, using a neck strap might be inconvenient for users. Although multiple acoustic sensors might enhance the cough detection accuracy, it might lead to several limitations such as high cost and being limited within the range of the sensing signals, and being uncomfortable and invasive for the users. Mobile device based cough detection, in contrast, is low-cost while cough (and many other diseases) might be monitored every time and everywhere. Therefore, in this investigation we propose the cough detection and monitoring method using mobile devices that is convenient for the users while being cheap. Our proposed method is distinct from multiple sensor based cough detection as we use only one single microphone for capturing the sound from the user while cough detection system runs on a mobile platform. Our proposed method is more convenient for the users than [3, 13, 16] as the users are not be required to wear the phone on the user's chest area while the detection performance is not affected by phone's positions.

3   Real-Time Cough Detection & Monitoring

As depicted in the Fig. 1, cough is detected in 4 steps: *sound pre-processing*, *audio segmentation*, *feature & event extraction*, and *cough prediction*.

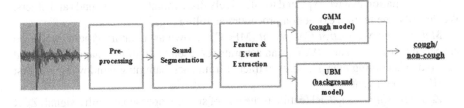

Fig. 1. Cough detection system

3.1 Hardware

The hardware used in this study includes a tiny wifi microphone and a mobile phone platform. The microphone is a low-cost, mini wireless Bluetooth earphone speaker stereo with microphone provided by the OEM (cost at $5.00). The microphone can communicate to the mobile phone via Bluetooth wireless technology. The battery is completely embedded inside the case, and it has a long talking time of 6 h without any external power. With the size of $33 \times 12 \times 7$ mm, the microphone is very easy to be worn on the user's collar for capturing sounds from the user while the user's mobile phone can be flexible to be positioned at hand or pocket.

3.2 Sound Pre-Processing

The sampling rate is down-sampled to 8 kHz. It is noticed that the sampling rate of 8 kHz is good enough for practical applications. After that sound signals are filtered using a low-pass filtering procedure to eliminate background noise and silent. Also, a high-pass filtering with a transfer function $H(x) = 1-\alpha x^{-1}$ where $\alpha = 0.95$ for emphasizing on higher frequency bands of the cough signals.

3.3 Audio Segmentation

Continuously audio stream is segmented into 2-second sliding windows with 50 % overlap between two consecutive sliding windows. A sliding window can contain a silent, a cough, or an unknown event. Silent sliding windows are discarded on the ground using a simple energy feature based threshold. Each sliding window is segmented into 25 ms hamming windows. Therefore, a sliding window comprises of 80 hamming windows, each contains 200 samples. The reason for using small segments such as hamming windows is the acoustic signals are constantly changing, but a short time scale of 25 ms is assumed statistically stationary.

3.4 Feature & Event Extraction

For each hamming window, the features *Mel-frequency Cepstral Coefficient* (MFCC), *Zero-crossing Rate* (ZCR) and *Entropy* are extracted. These features also contain rich sound information that we expected to effectively discriminate cough sounds and others. We describe the feature extraction procedure as followings.

Mel-frequency Cepstral Coefficient (MFCC): known as a feature type widely used in speech recognition, an MFCC is an accurate representation of the shape of the vocal tract (sounds coming out). Features extracted from each hamming window are the first 12 MFCC coefficients.

Zero-crossing rate (ZCR) feature is the rate of sign changes along with a signal, ZCR is computed from a hamming window as:

$$\text{ZCR} = \frac{1}{199} \sum_{i=0}^{199} I\{x_i * x_{i-1} < 0\} \tag{1}$$

x_i is the sample i^{th} of the hamming window, $I\{A\}$ is an indicator function.

$$I\{A\} = \begin{cases} 1 & \text{if A is true} \\ 0 & \text{otherwise} \end{cases} \tag{2}$$

The entropy feature over a hamming window, the measure of the amount of information that is missing reception, can be calculated as:

$$\text{Entropy}(x) = - \sum_{i=0}^{199} p(x_i) \log(p(x_i)) \tag{3}$$

Where x_i is a sample value; $p(x_i)$, a probability distribution of x_i within the hamming window, can be estimated as the number of x_i in the hamming window divided by 200; and the probability $0*\log(0)$ is assumed to be 0.

12 MFCC, ZCR, and Entropy features are combined into one feature vector sized of 14. Features are normalized to ensure all feature values in the range of [0, 1]. With feature size of 14, we can avoid unnecessary delay (good for real-time implementation) while we can achieve reasonable detection accuracy.

As an effective event extraction and processing can significantly reduce the computation resources, we develop a simple threshold-based event extraction algorithm for pruning the audio stream, and searching (highly potential) cough event candidates over the audio stream. The thresholds are estimated by using 4-fold cross validation procedure on the subset of the dataset. To select appropriate features for searching cough events, we test feature by feature over the dataset under 4-fold cross validation protocol, and then manually select 4 out of 14 features that perform best with high true positive rate and lowest false positive rate. The candidate cough events are afterward used for predicting cough.

3.5 Cough Prediction

Two models Gaussian Mixture model (GMM) and Universal Background model (UBM) are proposed for the prediction of cough. In this work, cough sound is modeled using

GMM while UBM is used for modeling background sounds which are any sounds out of cough, including noise, speech, etc.

In brief, the Gaussian Mixture model for modeling the cough G is a triple of 3 parameters: $\vec{\mu}_i$, C_i, $\vec{w}_i$ extracted from the training data:

In which $\vec{\mu}_i$ is the mean vector; C denotes the covariance matrix; and $\vec{w}_i$ represents the prior probabilities (w_i is the prior probability of ith mixture component). Given the feature vector $\vec{f}$ computed from a hamming window, the likelihood for the cough is computed as follows.

$$p(\vec{f}|G) = \sum_{i=1}^{M} w_i N(\vec{f}\vec{\mu}_i, C_i) \tag{4}$$

Where N denotes the Normal probability distribution with the mean vector $\vec{\mu}_i$ and covariance matrix C:

$$N(\vec{f}|\vec{\mu}_i, C_i) = \frac{1}{|\sqrt{2\pi C}|} e^{-\frac{1}{2}(\vec{f}-\vec{\mu}_i)^T C^{-1}(\vec{f}-\vec{\mu}_i)} \tag{5}$$

A sliding window w, comprising of 80 hamming windows, the likelihood is approximated based on the independent assumption over hamming windows:

$$p(w|G) = \prod_{i=1}^{80} p\left(\vec{f}_i| G\right) \tag{6}$$

$\vec{f}_i$ is the feature vector computed from the hamming window i^{th}

Mixture models for (known) cough are trained using feature vectors computed from audio data labeled with cough. The training process is straight forward by using k-Means clustering and Maximum Likelihood (ML) optimization. Model parameters $\vec{\mu}_i$, C_i, $\vec{w}_i$ are estimated on class-specific training data. The number of Gaussian mixtures M = 7 is estimated by a 4-fold cross validation procedure on the training dataset.

Training a Universal Background model (UBM) with background noise data is different from does the cough model (GMM). Similar to a GMM, a UBM is also a triple of $U = \{\vec{\mu}_i, C_i, \vec{w}_i\}_{i=1}^{M}$ where M is the number Gaussian mixtures. As UBM is trained to capture general characteristics of all background noises, M is significantly larger than that is used in GMM for modeling cough. In our study, to estimate M, we vary M to 10, 20, 30, 50, 100 on a 4-fold cross validation procedure on a subset of the training data. M = 50 is selected as it performs best (highest true positive rate). This leads to the covariance matrices are large, to facilitate fast computation, only diagonal form of the covariance matrices are considered.

The UBM U is trained using Expectation-Maximization (EM) algorithm. The model parameters are initialized as.

$w_i = \frac{1}{50}$; $C_i = I$ (unit matrix); and $\vec{\mu}_i$ is randomly selected from training data.

The EM algorithm performs iterations until parameters are stable. After training process, we have a pre-trained GMM model and a pre-trained UBM model. These models are combined into the prediction stage that we can deploy on the mobile devices for cough detection.

A cough in a sliding window is detected if p(w|G) ≥ p(w|U); that means:

$$\prod_{i=1}^{80} p(\vec{f_i} |G) \geq \prod_{i=1}^{80} p(\vec{f_i} |U) \tag{7}$$

Otherwise, a non-cough is detected.

3.6 Cough Monitoring

Once a cough is detected in real-time, its information including the time (precise in millisecond) and place (GPS data) will be written into a log file. In addition, we also remark the fits of coughing or whoops when there are more than 3 coughs are detected within 3 s. The Fig. 2 (right) is an example of a fit of coughing.

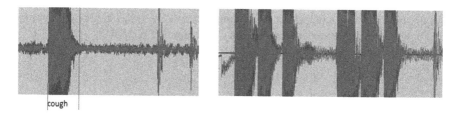

Fig. 2. A cough sound (*left*) and a fit of coughing (whoops) (*right*).

As cough information would need to be available for accessing by the doctor anytime, we have implemented some functions (as shown in the Fig. 3) that can easily be accessed and retrieved by both users and doctors.

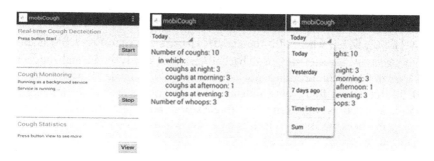

Fig. 3. MobiCough GUI and Cough monitoring.

4   Experimental Evaluation

This section presents an empirical experiment for verifying our proposed method. The section is divided into two parts: data collection & annotation which describe the procedure of data collection from 10 pharyngitis or pneumonia patients; and followed by the evaluation.

4.1   Data Collection & Annotation

As so far no publicity of the cough dataset is available, so we ourselves need to collect cough data. We develop a simple audio logging program on Android phones for data collection. The logging program is deployed on 10 Android-based cell phones including Samsung, LG, FPT phones etc. 10 participants having with pharyngitis or pneumonia infection are willing to involve the study. The subjects were asked to use the cell phones installed audio logging program and to wear the mini wireless Bluetooth earphone speaker stereo on his/her collar for 3–6 h at different time of the day. The collected data is annotated by experimenters using Praat tool [15] with two labels: *cough* and *non-cough*. After annotation step, collected audio data is labeled with 1,117 cough events and more than 12,677 background events (any events out of the cough and silent in the audio logging files).

4.2   Evaluation

The cough event extraction can detect correctly 1,091 cough events out of 1,117 cough events of the ground truth. This results that our cough event extraction algorithm can achieve as high as 97.6 % of true positive rate. However, it misses out about approximately 3 % cough events while it incorrectly extracts 2,661 non-cough events. Details are shown in the Table 1.

Table 1. The test results for the event extraction algorithm

TP	FP	TN	FN
1,091	2,661	8,898	27

As the cough event extraction finds out 3,752 cough "candidate" events which are used for detecting cough from GMM-UBM models proposed in Sect. 3. An event can contain from a few to several hundreds of sliding windows. We do two evaluations: subject-dependent and subject independent protocols.

Under the subject dependent protocol, we use training and testing data from the same subject. For each subject, we divide the data into two equal portions; one is used for training, and the other is for testing; and the two portions are permuted; after that the results are averaged. We repeat the process for all subjects and the results are aggregated. It is noticed that the subject dependent implementation is particularly useful for cough detection systems that need to be adaptive to the users.

Under the subject independent, we use data from 9 subjects for training, and left out another subject to test. Then we repeat the process for all subjects and the results are aggregated. It is noticed that the testing data is not included into the training data and comes from other subject. The subject independent protocol is useful for cough detection systems acquiring pre-trained models (the cough model is trained from other person's coughs).

The evaluation results are represented in the Fig. 4. Subject-dependent result (91 %) is significantly higher than subject-independent (81 %). This is reasonable as test and train data are the same person for subject-dependent evaluation. This is highly recommended that models trained for cough detection and monitoring systems on personal devices coming from the owner would improve the detection accuracy. With the precisions and recalls are as high as 81 % and 91 % for subject-independent and subject-dependent respectively, MobiCough demonstrates that the detection of coughs using low-cost mobile devices is feasible. These results are very potential for heal-care applications that acquire cough information for diagnose and treatment cough-related diseases. The evaluation results are comparable to other works [3, 4, 6, 7, 9, 16] while we mainly focus on the detection of cough in real-time which is very crucial for opportune interventions, diagnosis, cure, treatment and even emergency aids made by the doctors and remarkably more convenient use.

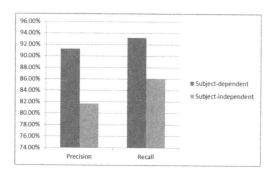

Fig. 4. Precision and Recall of cough detection evaluations

5   Conclusion

We present MobiCough, a method and system for cough detection and monitoring on low-cost mobile devices in real-time. Our proposed method combines predictive GMM and UBM models to enhance the detection performance. The proposed method is rigorously evaluated over a dataset consisting of more than 1000 cough events and a significant number of noises. With the detection accuracies are more than 91 % precision and recall for subject dependent training, and over 81 % precision and recall for subject independent training, the results are very comparable to other works while being feasible for real-time processing, low-cost and pretty convenient use. MobiCough is very promising for heal-care applications that acquire cough information for diagnose and treatment cough-related diseases.

References

1. WHO's pneumonia fact sheet. http://www.who.int/mediacentre/factsheets/fs331/en/
2. KidsHealth. http://kidshealth.org/parent/infections/lung/pneumonia.html
3. Larson, E.C., et. al.: Accurate and privacy preserving cough sensing using a low cost microphone. In: Proceedings of UbiComp, pp. 375–384. Beijing (2011)
4. Birring, S.S., et al.: The leicester cough monitor: preliminary validation of an automated cough detection system in chronic cough. Eur. Respir. J. **31**(5), 1013–1018 (2008)
5. Schappert, S., Burt, C.: Ambulatory care visits to physician offices, hospital outpatient and emergence. Vital Health Stat. **13**, 1–66 (2006)
6. Drugman, T., et al.: Audio and contact microphone for cough detection. In: Proceedings of INTERSPEECH, pp. 1303–1306. IEEE Press, Portland (2012)
7. Vizel, E., et al.: Validation of an ambulatory cough detection and counting application using voluntary cough under different conditions. Cough **6**(3), 1–8 (2008)
8. Kraman, S.S., et al.: Comparisons of lung sound transducers using a bioacoustics transducer testing system. J. Appl. Physiol. **101**(2), 169–176 (2006)
9. Masto, S., et al.: Detection of cough signals in continuous audio recordings using hidden Markov models. IEEE Biomed. Eng. **53**(6), 1078–1083 (2006)
10. Zheng, S., et al.: CoughLoc: location-aware indoor acoustic sensing for non-intrusive cough detection. In: International Workshop on MobiSys (2011)
11. Pham, C., et al.: The ambient kitchen: a pervasive sensing environment for situated services. In: Proceedings of ACM Conference on Designing Interactive Systems, Newcastle, UK (2012)
12. Pham, C., et al.: A wearable sensor based approach to real-time fall detection and fine-grained activity recognition. J. Mobile Multimedia **9**, 15–26 (2013)
13. Drugman, T., et al.: Assessment of audio features for automatic cough detection. In: Proceedings of 19th European Signal Processing Conference, pp. 1289–1293 (2011)
14. Mark, S., Hyekyun, H., Mark, B.: Automated cough assessment on a mobile platform. J. Med. Eng. **2014**, 1–9 (2014)
15. Praat. http://www.fon.hum.uva.nl/praat/
16. Shin, S.H., et al.: Automatic detection system for cough sounds as a symptom of abnormal health condition. Trans. Inf. Tech. Bio. **13**(4), 486–493 (2008)

Database of Peptides Susceptible to Aggregation as a Tool for Studying Mechanisms of Diseases of Civilization

Pawel P. Wozniak[1], Jean-Christophe Nebel[2], and Malgorzata Kotulska[1]([✉])

[1] Faculty of Fundamental Problems of Technology,
Department of Biomedical Engineering,
Wrocław University of Technology, Wrocław, Poland
`malgorzata.kotulska@pwr.edu.pl`
[2] Faculty of Science, Engineering and Computing,
School of Computing and Information Systems, Kingston University,
Kingston upon Thames, UK

Abstract. We introduce a database containing peptides related to diseases arising from protein aggregation. The general database AmyLoad includes all experimentally studied protein fragments that could be involved in erroneous protein folding, leading to amyloid formation. The database has been extended since its first release with regard to new instances of peptides or their fragments. Moreover, information of related diseases has been added to all entries, whenever available. Currently the database includes all available peptides tested for their potential amyloid properties, obtained from diverse resources, creating the largest dataset available at one place. This enables comparison between properties of amyloid and non-amyloid peptides. We could also select candidates for the most pathogenic peptides, involved in several diseases related to protein aggregation. We also discuss a need for sub-databases of different structures, such as related to $\beta\gamma$-crystallins - a protein family occurring in the eye lens. Misfolding of these proteins may lead to various forms of cataract. Those freely available internet services can facilitate finding the link between a protein sequence, its propensity to aggregation and the resulting disease, as well as support research on their pharmacological treatment and prevention.

1   Introduction

Many diseases, especially neurodegenerative, result from protein fragments forming aggregates. This occurs when a cell environment fosters the partial unfolding of protein chains or their fragmentation, in a way that the parts prone to joining with other protein fragments are exposed. For the majority of proteins, considerable conformational rearrangement must have occurred to initiate the aggregation process. Such changes cannot take place in the typical tightly packed native protein conformation, due to the constraints of the tertiary structure. Thus, formation of a non-native partially unfolded conformation is required, presumably

N.T. Nguyen et al. (Eds.): ACIIDS 2016, Part I, LNAI 9621, pp. 310–319, 2016.
DOI: 10.1007/978-3-662-49381-6_30

enabling specific intermolecular interactions, including electrostatic attraction, hydrogen bonding and hydrophobic contacts. This partial unfolding can be influenced by various factors, such as protein high concentration, high temperature, low pH, binding metals, or exposition to UV light.

Initially, the resulting molecules form clusters consisting of a few elements, which are called oligomers. Next, they grow into larger aggregates. Aggregation of proteins or their fragments may lead to amorphous (unstructured) clusters or amyloid (highly ordered) unbranched fibrils. Independently of the protein sequence and its original structure, aggregates always display a common cross-β structure. The distinctive structure of the steric zipper enables the selective detection of amyloids from amorphous aggregates using either a variety of microscopic techniques or fluorescence of probes with which they form compounds.

Amyloid fibrils have been observed in the brains of people suffering from Alzheimer's disease. They are also associated with Parkinson's disease, amyotrophic lateral sclerosis and Huntington's disease, as well as many other conditions, even non-neurodegenerative diseases such as type 2 diabetes and some types of cataract. Cells in tissues containing these fibrils exhibit very high mortality. However, the reasons for this cytotoxicity have not been resolved. In recent years the occurrence rate of diseases characterized by accumulation of protein deposits has increased significantly. These disorders are sometimes called diseases of civilization since they are more prevalent in developed countries where life expectancy is higher. Unfortunately, their mechanisms are still poorly understood. Although studies indicate that these diseases to some extent have a genetic basis, the influence of lifestyle cannot be excluded. Unfortunately dissolution of peptide aggregates is very difficult, especially for amyloids which are resistant to activity of proteolytic enzymes and chemical compounds due to the specific and highly ordered structure of their steric zipper.

Cataract is among the diseases associated with protein aggregation. Age-related cataract is a major burden on public health: this is the most common cause of blindness worldwide, affecting tens of millions of people. The lens fibre cells are essentially composed of crystallin proteins, which are among the most highly concentrated intracellular proteins in the human body. $\beta\gamma$-crystallins define a superfamily of crystallins sharing similar sequence and structure. Despite a large set of literature concerning the family of $\beta\gamma$-crystallins, there is no dedicated service containing all available genotypic and phenotypic data, as well as tools for resolving molecular mechanisms of the disease and supporting the development of potential pharmacological treatments.

Currently, it is believed that short peptide sequences of amyloidogenic properties (called hot-spots) can be responsible for aggregation of amyloid proteins. These 4-10 residue long fragments (typically hexapeptides) have a high propensity for strong interactions that lead to protein aggregation. Previous studies have suggested that amyloidogenic fragments may have regular characteristics, not only with regard to averaged physicochemical properties of their amino acids, but also the order of amino acids in the sequence. There have been attempts to predict the sequence of such peptides by computational modelling. Physics and chemistry based models have been used, including FoldAmyloid [1]. This method is based on the density of the protein contact sites. Other methods perform threading a peptide on an amyloid fiber backbone, followed by determination of its energy and stability [2–4]. Statistical approaches include production of frequency profiles, such as the WALTZ method [5] and machine learning methods, which have been used by our team [6,7]. Some other approaches, mostly biophysical-based, enable classification of hot spots for non-amyloid aggregation. Recently, AGGRESCAN3D has been proposed to estimate more accurately aggregation propensity by performing 3D structure based analysis [8]. All of these methods, although promising, have faced difficulties due to limited amount of experimental data available for their construction and validation.

Knowledge in the field of diseases related to protein aggregation is still patchy, no global view of the problem is available and the link between the molecular level and the phenotype is still generally missing. In addition, although a large amount of information is available, its dispersion in separate publications, data sets and web services hampers research development. To fill in this gap, we proposed a new web service, AmyLoad [9], which is devoted to protein aggregation and can facilitate global research in the field. The service is focused on general amyloidogenic peptides. However studies on specific cataract related aggregation have led us to the idea of separate sub-database services, including more specific and detailed information, even if some is only predicted by software tools.

2 Results

AmyLoad is a website which gathers information about known, experimentally-derived, amyloidogenic and non-amyloidogenic amino acid protein fragments [9]. The data comes from literature studies and various data sources, which are WALTZ-DB [5], AmylFrag, AmylHex [10], and datasets used to validate such methods like TANGO [11] and AGGRESCAN [12]. Although these data sources contain protein fragments according to different specific features, the fact that one fragment can belong to more than one dataset makes these databases difficult to work with. In addition, data filtration is usually unavailable. In this category of datasets WALTZ-DB and AmylHex contain only protein fragments which are composed of six amino acids. AmylFrag possesses 45 literature-derived fragments which are longer than six residues. TANGO and AGGRESCAN are well known amyloidogenicity prediction methods, which are based on different sequence analysis algorithms. They were also validated using original experimental data. TANGO uses statistical mechanics algorithm based on the physicochemical principles of beta-sheet formation to predict protein aggregation. It was tested on a set of 179 peptides obtained from the literature and 71 new peptides derived from human disease-related proteins [11]. Alternatively, AGGRESCAN takes advantage of the aggregation-propensity scale for amino acids. It was trained on a database of 57 experimentally known amyloidogenic proteins [12]. These training data have been included into our database.

The website was built using the Django web framework and a MySQL database. Figure 1 presents the AmyLoad database tables and relations between them.

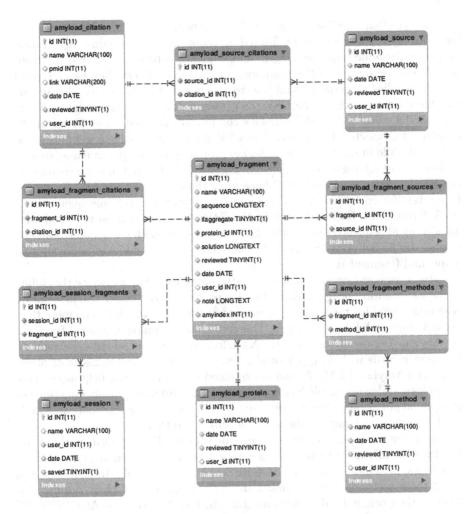

Fig. 1. AmyLoad database structure

The most important table contains detailed information about each fragment such as name, residue sequence, aggregation ability, experimental conditions, and data about its first occurrence in the database: user and date. Each protein fragment is related to only one protein record stored in another table. While there can be more than one protein fragment related to a single protein record, the residue sequence and protein record id attributes are unique for each record of the protein fragment. Other tables store information about references related to the record of the fragment, datasets of origin, experimental methods used for their discovery, and their saves in private users sessions. Records of these tables are in many-to-many relation with protein fragment record. Information about registered users and administrators are stored securely taking advantage of the common user authentication system provided by Django.

The principal function of the AmyLoad website is data browsing and filtering. Currently, the data can be filtered according to protein name, aggregation propensity, residue sequence length, and sequence substring. Selected fragment records can be moved to the temporary session field where they can be saved for later studies. The data about selected fragments can be downloaded in several file formats, such as the CSV, SSV (Semicolon Separated Values), XML, and FASTA. The SSV and XML files contain all information gathered in the database for selected fragments. The CSV file includes only that information which is visible in the browsing table, i.e. protein name, fragment name, residue sequence, and aggregation propensity. The FASTA file contains information in the FASTA format, which is well known for those who work in bioinformatics. It is a text-based format which represents nucleotide or protein sequences as a single-letter code. Each sequence in the FASTA format can be preceded with an additional sequence information line which begins with the greater-than ($>$) symbol. In the AmyLoad website, that line contains the AmyLoad index, protein name, and fragment name.

New fragments can be submitted into the database in two ways by either filling in the website form or uploading a file in a proper XML format. The first option is recommended for the submission of a single fragment which has no more than one reference and dataset of origin. This is the most common manner of fragment submission. Taking advantage of the XML format, users can add at once multiple fragment records including potentially several references and datasets of origin each. The AmyLoad XML format is explained in details on the help page where examples and downloadable files are provided. Submitted fragments are visible to the general public only after reviewer approval. Until then, they are only accessible on the personal web page of the user who submitted them.

Following submission of protein sequences in the FASTA format, the AmyLoad website allows for their analysis with several implemented amyloidogenicity predictors, i.e. FoldAmyloid [1], AGGRESCAN, and FISH [7]. Implementations of all these methods were validated through comparison of their results with those of their original online implementations. FoldAmyloid and AGGRESCAN are based on sliding window algorithms. FoldAmyloid analyzes experimentally-derived expected probability of hydrogen bonds formation and expected packing density of amino acids in the sequence of interest [1]. AGGRESCAN, on the other hand, calculates the aggregation-propensity scale for residues derived from in vivo experiments [12]. Finally, FISH is a machine learning method which was created in our group, based on the site-specific co-localization of aminoacid pairs. The results of the analysis for any of the chosen methods are sent by email to users using a binary format, where 1 means that the associated residue in the submitted sequence belongs to the amyloidogenic fragment. Together with the implemented methods, AmyLoad allows its users to search its database for sequence fragments which occur in submitted FASTA sequences. The advantage of using the AmyLoad implementations instead of the original modelling tools is that users can run calculations for several submitted fragments at once. In addition, results of different methods are presented in the email in a comparable format which makes AmyLoad a simple consensus amyloidogenicity predictor. On the analysis website, references and links to the original implementations of FoldAmyloid, AGGRESCAN, FISH, and other 13 well known amyloidogenicity predictors are also provided.

There are three types of users interacting with the AmyLoad website. The first one is the common non-registered user who can use the entire analysis website, and browse and filter the data about sequence fragments gathered in the database. The second one is a registered user who can create temporary sessions and save them for later studies. Furthermore, only the registered user is able to submit new fragments into the database. Finally, there are the database administrators who have all the privileges of non-registered and registered users, together with the ability to review the submitted fragments. Only after the administrator-reviewer approval, a fragment becomes visible to the general public on the browsing web page.

Currently, there are 1477 unique entries in the AmyLoad database, which come from over 150 different proteins. Figure 2 shows the distribution of sequence lengths within the deposited fragments. The website contains also information about almost 100 references related to the amyloidogenicity topic. According to the literature, peptide fragments deposited in the AmyLoad were analysed by almost 20 different experimental methods such as electron microscopy or thioflavin dyeing.

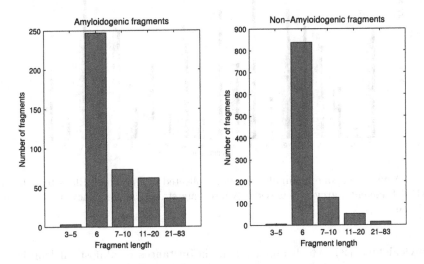

Fig. 2. Lengths of fragments deposited in AmyLoad

Collecting all sequences diagnosed for amyloidogenicity enabled finding a pattern in the contents of amyloid and non-amyloid fragments. We tested several sets of peptides within different length ranges. Exemplary results are shown in Fig. 3 (hexapeptides which are the best studied with regard to their aggregation propensity) and in Fig. 4 (fragments of all lengths). The study showed that amyloid fragments are rich in valine (V), isoleucine (I), leucine (L) and

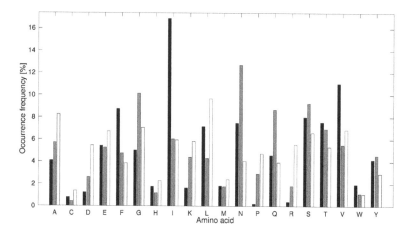

Fig. 3. Amino acid contents in hexapeptides collected in AmyLoad. Black bars denote amyloid fragments, grey non-amyloid, white bars statistical frequency as in Uniprot database

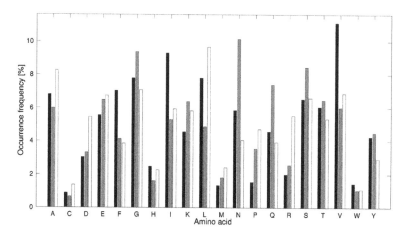

Fig. 4. Amino acid contents in all fragments collected in AmyLoad. Black bars denote amyloid fragments, grey non-amyloid, white bars statistical frequency as in Uniprot database

phenyloalanine (F), which appeared in the instances of almost all lengths. It proves an excess of non-polar neutral aminoacids with high propensity to form beta strands. Interestingly, non-amyloidogenic peptides are characterized with increased contents of non-polar charged aminoacids, especially positively charged lysine (K) and arginine (R). Enrichement of peptides with charged aminoacids impacts on their non-amyloidogenic properties, which may be considered for applications towards changing the amyloidogenic properties of peptides.

The database has been extended by diseases related to amyloid fragments. It enabled to observe that majority of the fragments (44) are related to the

Alzheimer's disease, then to prion diseases (26 fragments), type II diabetes (20 fragments), dialysis-related amyloidosis (16 fragments), and Amyotrophic Lateral Sclerosis (11 fragments). Interesingly, some of the amyloidogenic fragments have been shown as involved in up to three diseases. These include certain fragments of alpha synuclein - 8 fragments of this protein can lead either to Alzheimer's disease, or Parkinson's disease, or dementia with Lewy bodies. Another protein involved in 3 different diseases is τ-protein - with 6 fragments that can lead to Alzheimer's disease, Pick's disease, and progressive supranuclear palsy. However, a number of entries has not been associated with any disease, yet.

A general database, such as AmyLoad cannot contain sequences that have not been experimentally confirmed with regard to their misfolding properties, even though they may contribute to understanding of a disease. On the other hand, several protein features could be predicted with modelling tools, which would be very helpful in studying molecular mechanisms of the diseases of interest. Such information is produced by specific modelling tools and it is associated with some uncertainty, especially when predictors produce contradicting results. As a consequence it needs to be very carefully examined and tagged in a database. Although addition of modelling results could be considered for a general database such as AmyLoad, this would require an immense amount of work of the database curators to add these results with regard to every protein potentially aggregating, taking into account all available modelling tools. We believe that general databases should include less information, i.e. fewer fields, but from more reliable sources. Therefore, we decided to include only experimental data into AmyLoad and extend it into related subdatabases, dedicated to specific protein families that may be of research interest with regard to certain diseases. The first such database will be devoted to the $\beta\gamma$-crystallins and their involvement in cataract development. One of the specificities of these proteins is their aggregation pathway: many crystallins that are involved in the development of cataract can form amorphous aggregates rather than amyloids, others follow an amyloid aggregation path, whereas some may be involved in both paths. The sub-database of crystallins will separate and extend the number of their current entries in AmyLoad, while increasing the amount of available information about each of them.

3   Conclusions

We reported an internet database system dedicated to aggregating peptides, which may underlie several diseases of civilization, such as neurodegenerative diseases, diabetes type 2, and cataract. The general peptide database AmyLoad contains all currently known sequences of aminoacids whose propensity to amyloid aggregation has been published, based on experimental results. The entries also include some more specific information regarding each record.

Analysis of the amyloidogenic peptides, collected in the database showed a strong excess of neutral non-polar aminoacids with high propensity to for beta strands, such as valine, isoleucine, leucine and phenyloalanine. It appeared in the instances of all lengths. Interestingly, non-amyloidogenic peptides are characterized with increased contents of aminoacids with a positive charge and to lesser extent of negative charge. Since the enrichement of peptides with charged aminoacids impacts on non-amyloidogenic properties of peptides, it may be considered for applications towards changing the amyloidogenic properties of peptides.

AmyLoad may be too general for a specific family of proteins related to diseases in which aggregation may assume different forms, or modelling results would be crucial for further studies. Hence, more specialized sub-databases of different fields and including modelling results are required, such as the one containing $\beta\gamma$-crystallin proteins - underlying various forms of cataract. The sub-database should allow more detailed information than AmyLoad, focused more on different aggregate structures, leading to different disease phenotypes and different potential treatments for which pharmacophores could be designed based on available data.

Databases of aggregating proteins are needed to support further research in related diseases. We believe that our freely available internet service will facilitate the identification of the link between a protein sequence, its propensity to aggregation and the resulting disease, and discovering molecular events behind development of diseases related to protein aggregation, as well as their pharmacological treatment and prevention.

The AmyLoad database, as well as other tools, could be found at Comprec server: http://comprec-lin.iiar.pwr.edu.pl/amyload/

Acknowledgements. This work was in part supported by the grant N N519 643540 from National Science Centre of Poland.

References

1. Garbuzynskiy, S.O., Lobanov, M.Y., Galzitskaya, O.V.: FoldAmyloid: a method of prediction of amyloidogenic regions from protein sequence. Bioinformatics **26**(3), 326–332 (2010)
2. Goldschmidt, L., Tenga, P.K., Riek, R., Eisenberg, D.: Identifying the amylome, proteins capable of forming amyloid-like fibrils. PNAS **107**, 3487–3492 (2010)
3. Bryan Jr., A.W., O'Donnell, C.W., Menke, M., Cowen, L.J., Lindquist, S., Berger, B.: STITCHER: dynamic assembly of likely amyloid and prion -structures from secondary structure predictions. Proteins, vol. 80, pp. 410–420 (2011)
4. O'Donnell, C.W., Waldispühl, J., Lis, M., Halfmann, R., Devadas, S., Lindquist, S., Berger, B.: A method for probing the mutational landscape of amyloid structure. Bioinformatics **27**, i34–i42 (2011)
5. Beerten, J., Van Durme, J., Gallardo, R., Capriotti, E., Serpell, L., Rousseau, F., Schymkowitz, J.: WALTZ-DB: a benchmark database of amyloidogenic hexapeptides. Bioinformatics **31**(10), 1698–1700 (2015)
6. Stanislawski, J., Kotulska, M., Unold, O.: Machine learning methods can replace 3D profile method in classification of amyloidogenic hexapeptides. BMC Bioinformatics **14**, 21 (2013)

7. Gasior, P., Kotulska, M.: FISH Amyloid - a new method for finding amyloido-genic segments in proteins based on site specific co-occurrence of aminoacids. BMC Bioinformatics **15**, 54 (2014)
8. Zambrano, R., Jamroz, M., Szczasiuk, A., Pujols, J., Kmiecik, S., Ventura, S.: AGGRESCAN3D (A3D): server for prediction of aggregation properties of protein structures. Nucleic Acids Res. (2015). doi:10.1093/nar/gkv359
9. Wozniak, P.P., Kotulska, M.: AmyLoad - website dedicated to amyloidogenic pro-tein fragments. Bioinformatics (2015). doi:10.1093/bioinformatics/btv375, http://comprec-lin.iiar.pwr.edu.pl/amyload/
10. Thompson, M.J., Sievers, S.A., Karanicolas, J., Ivanova, M.I., Baker, D., Eisenberg, D.: The 3D profile method for identifying fibril-forming segments of proteins. Proc. Natl. Acad. Sci. U.S.A. **103**(11), 4074–4078 (2006)
11. Fernandez-Escamilla, A.M., Rousseau, F., Schymkowitz, J., Serrano, L.: Prediction of sequence-dependent and mutational effects on the aggregation of peptides and proteins. Nat. Biotechnol. **22**(10), 1302–1306 (2004)
12. Conchillo-Sol, O., de Groot, N.S., Avils, F.X., Vendrell, J., Daura, X., Ventura, S.: AGGRESCAN: a server for the prediction and evaluation of "hot spots" of aggregation in polypeptides. BMC Bioinformatics **8**, 65 (2007)

Using a Cloud Computing Telemetry Service to Assess PaaS Setups

Francisco Anderson Freire Pereira[1], Jackson Soares[2],
Adrianne Paula Vieira Andrade[1], Gilson Gomes Silva[1],
and João Paulo Souza Medeiros[1]([✉])

[1] Department of Computing and Technology (DCT),
Universidade Federal do Rio Grande do Norte (UFRN), Caicó, RN, Brazil
[2] Universidade Federal do Rio Grande do Norte (UFRN), Caicó, RN, Brazil
joaomedeiros@ufrnet.br

Abstract. Cloud Computing (CC) is a new paradigm in which capabilities and resources related to Information Technology (IT) are provided as services. This provision could be done via the Internet and on-demand, and is accessible without requiring detailed knowledge of the underlying technology. In this paper we assess different service platforms using cloud computing telemetry services. Using an OpenStack telemetry service, namely Ceilometer, we design experiments to assess the performance of different Platform as a Service (PaaS) setups for basic purposes (e.g. database and Web server). The assessment could be use to decide between common used platforms, comparing requisites of storage, processing time and processor load. Given the costs of each metric at a CC platform and the performance of each PaaS setup, the IT manager could choose the most advantage one.

Keywords: Could computing · Telemetry · Electronic systems

1 Introduction

Cloud Computing (CC) represents a new way of enhancing and simplifying the use of Information Technology (IT) resources. It emerges as a new paradigm of computing, a new proposal for managing the computing infrastructure for a better use of resources in a given environment. These resources are offered on demand, with centralized management in order to avoid idle equipments use.

The efficient use of CC infrastructure and resources can reduce the high cost of hardware, software, IT maintenance and improve management control in institutional environment [14].

One of the incentives for companies to adopt CC is predominantly from a costs perspective, as organizations increasingly discover that their substantial capital investments in IT are often grossly underutilized. Furthermore, equally pertinent are the maintenance and service costs that have proved to be a drain on the possible scarce corporate resources [9].

N.T. Nguyen et al. (Eds.): ACIIDS 2016, Part I, LNAI 9621, pp. 320–329, 2016.
DOI: 10.1007/978-3-662-49381-6_31

There are many definitions regarding CC, however the definition commonly accepted is that of the National Institute of Standards and Technology (NIST), where cloud computing is defined as a model for enabling ubiquitous, convenient, on-demand network access to a shared pool of configurable computing resources (e.g., networks, servers, storage, applications, and services) that can be rapidly provisioned and released with minimal management effort or service provider interaction [10].

Depending on the type of provided capability, there are three distinct scenarios where CC is used, they are: Infrastructure as a Service (IaaS), Platform as a Service (PaaS) and Software as a Service (SaaS). In IaaS, it is possible to manage a large set of computing resources, such as storing and processing capacity. Through virtualization, they are able to split, assign and dynamically resize these resources to build ad-hoc systems as demanded by customers. In PaaS, cloud systems can offer an additional abstraction level, instead of supplying a virtualized infrastructure, they can provide the software platform where systems run on. In the SaaS scenario, various services hosted on cloud systems are offered for users. An example of this is the online alternatives of typical office applications such as word processors [15].

Cloud computing has recently received attention in computer science and information systems disciplines [11]. Previous studies have investigated CC from different perspectives such as the final users adoption [11,12] and companies adoption [7]. Other studies present prototypes of architectures and testing techniques for the IT infrastructure in computing environment [2,3,6,8,13]. However, few studies have investigated the use of PaaS in cloud associated with open source solutions, which is the focus of this study. The objective of this research is to compare the performance of three different platforms in an open source cloud environment using telemetry services. According to Rossigneux et al. [13], the consumptions monitoring of all equipments in cloud is required for exploring further improvement in energy efficiency and evaluate the impact of system-wide policies.

This work is organized as follows. In Sect. 2, we present related works. In Sect. 3, we describe the proposed testbed architecture and present the experiments results. We analyze the results with applications in e-learning, IT management and Green IT in Sect. 4. We present some concluding remarks and future work in Sect. 5.

2   Related Work

When conducting a literature search, we identified some studies related to this work. They are presented below in chronological order.

Malhotra and Jain (2013) [8] emphasizes the importance of cloud tests to obtain success in the cloud implantation. Cloud testing is a form of testing in which Web applications uses cloud computing environment and infrastructure to simulate real world user traffic. The authors present techniques available for cloud testing which are functional and non-functional. The functional ones include: system testing, integration testing and user acceptance testing. The non-functional ones include: business requirement testing, cloud security testing, cloud scalability and performance testing; and ability testing techniques, that included compatibility and interoperability testing, disaster recovery testing and multi-tenancy testing.

Brinkmann et al. (2013) [2] developed a monitoring architecture and highlighted the multiples demands on CC monitoring systems, such as regular checks of the Service Level Agreement (SLA) and the precise billing of the resource usage. They mention some CC platforms like OpenStack, which provide resources for the implementation and cloud resource management methods. They propose an approach that the monitoring data is organized in a distributed and easily scalable tree structure and is based on the Device Management Specification of the Open Mobile Alliance (OMA) and the Device Management Tree (DMT) Admin Specification of the Open Services Gateway initiative (OSGi).

Xiaojiang and Yanlei (2013) [16] proposed a cloud computing service platform based on OpenStack that supports service management, auto-scaling, security control and high availability. They implemented a resource monitoring subsystem, which can monitor performance metrics like Central Processing Unit (CPU) Utilization, memory usage and network Input/Output (I/O) of physical and virtual resources.

Rossigneux et al. (2014) [13] reported their experiences on monitoring large-scale systems and introduce an energy monitoring software framework called KiloWatt API (KWAPI) that interfaces with OpenStacks Ceilometer to provide power consumption information collected from multiple heterogeneous probes. Experimental results demonstrate that the overhead posed by the monitoring framework is small.

In our study, we also use the OpenStack Ceilometer [13]. However, our work has a different perspective that is the evaluation of the database and Web servers.

3   Experiments

We used an OpenStack based testbed to assess different PaaS setups. The proposed architecture is depicted in Fig. 1.

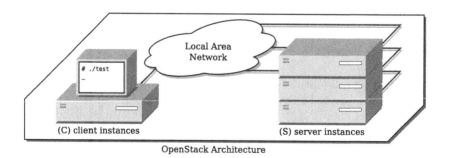

OpenStack Architecture

Fig. 1. Testbed architecture.

For the application layer we analyzed Apache and MariaDB. The different PaaS setups are described in Table 1.

Table 1. The different PaaS setups.

#	Operating system	Server
1	FreeBSD 10.2	Apache 2.4.16
2	Windows Server 2012	Apache 2.4.16
3	Debian 8.0 (Linux 3.16)	Apache 2.4.16
4	FreeBSD 10.2	MariaDB 10.0
5	Windows Server 2012	MariaDB 10.0
6	Debian 8.0 (Linux 3.16)	MariaDB 10.0

The PaaS setups were based on the same virtual hardware: 5 VCPUs and 6 GB of VRAM memory. For the operating system layer we analyzed one represent for Windows, BSD and Linux. The Ceilometer metrics used are listed in Table 2.

Table 2. The Ceilometer metrics used in the experiments.

Metric	Description
cpu	CPU time used
cpu_util	Average CPU utilization
disk.read.bytes	Volume of reads
disk.read.bytes.rate	Average rate of read requests
disk.write.bytes	Volume of writes
disk.write.bytes.rate	Average rate of write requests

The choice of the metrics was based in the characteristics of each metric for the processing performed by machines in database and Web server applications.

3.1 Web Server

In the first experiment, the Ceilometer was configured to perform the metrics publication every second[1]. After that, requests were sent from a client machine to the servers. The Apache Web server was used. In order to identify the measurements related to the evaluation procedure, we stored the time at the beginning and the end of the process. We retrieved metrics samples collected in the stored interval. In Figs. 2 and 3, we present each Web server PaaS CPU utilization.

Based on Fig. 2, we could affirm that Windows had the worst performance, with the peak of their use through the 140 % (more than one CPU), similar to FreeBSD. Debian had the best performance. The CPU usage time in nanoseconds

---

[1] Instructions to reproduce the experiments in https://github.com/labepi/aciids-2016.

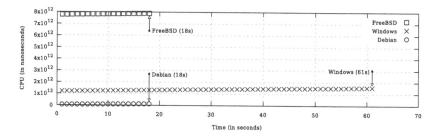

Fig. 2. CPU usage for the Web server PaaS experiment.

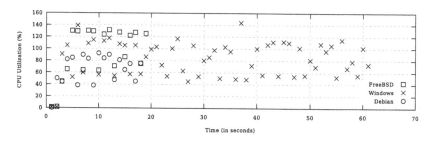

Fig. 3. CPU load for the Web server PaaS experiment.

shows FreeBSD consuming more than Debian. The Windows consumed approximately seven times more than others. The results suggest that the amount of bytes read from disk was higher than Windows that have the worst performance, while Debian got better performance.

In Fig. 4 and Table 3 we present the results of Apache Bench, a tool for benchmarking HyperText Transfer Protocol (HTTP) servers.

Regarding the time spent processing the requests sent by clients, the average Debian response was 21ms and FreeBSD 25ms, almost the same value. Although the Debian has the best average responsiveness, FreeBSD is recommended for real-time applications, since it responsiveness is more stable. Windows has the worst performance with an average time of 92ms. In this experiment we do not assess the disk metrics because the tests does not affect them significantly.

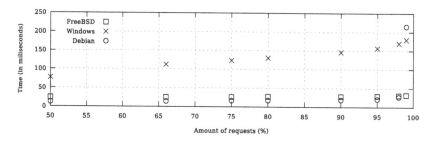

Fig. 4. The percentage of requests response times (in milliseconds).

Table 3. The time performance for the different Web server PaaS setups.

Amount	Time (≤ milliseconds)		
	Debian 8.0 (Linux 3.16)	FreeBSD 10.2	Windows Server 2012
50 %	12	24	77
66 %	14	25	112
75 %	16	26	123
80 %	17	27	130
90 %	19	28	146
95 %	21	29	157
98 %	27	31	170
99 %	215	32	180
100 %	3015	47	3179
Min.	3	15	3
Mean	21	25	92
Max.	3015	47	3179

3.2 Database

In the second experiment, the Ceilometer was configured to perform the metrics publication every 60 s. To stress the server, eight clients were used simultaneously. Clients sent requests for insertion, search and deletion of data on servers. In Figs. 5 and 6, we present the measurements for CPU utilization.

Debian performed the tasks faster (32 mins.) and FreeBSD was the slowest (52 mins.). The CPU load demonstrates that Windows had the highest peak of usage (30 %). FreeBSD and Debian consumed less CPU percentage. The CPU usage time presents FreeBSD consuming bulk 7 times more than others.

For database systems it is convenient to assess the disk usage. In Figs. 7, 8, 9 and 10, we present the measurements for I/O operations on disk extracted from the disk metrics presented in Table 2, namely: volume of reads, average rate of read requests, volume of writes, and average rate of write requests.

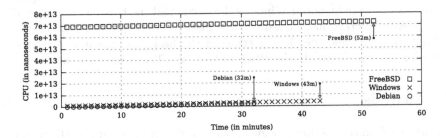

Fig. 5. CPU usage for the database PaaS experiment.

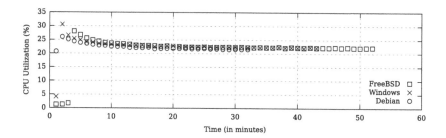

Fig. 6. CPU load for the database PaaS experiment.

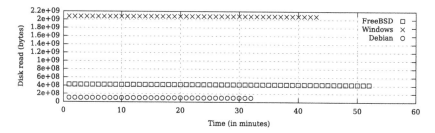

Fig. 7. Disk read activity.

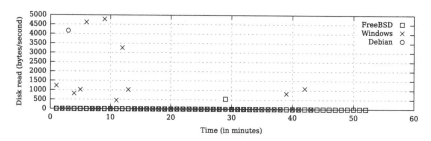

Fig. 8. Disk read activity rate.

The amount of bytes read from disk were higher in Windows, which had the worst performance, while Debian obtained the best. The amount of requests readings rates were higher in Windows. FreeBSD had the best result among all.

However, FreeBSD had the worst performance regarding the amount of bytes written to disk, while the Debian achieved the best. The amount of requests written rates were higher in FreeBSD. In this case, Windows outperforms.

4   Analysis

The use of CC infrastructure provide important benefits in e-learning [4] systems, IT management [5] and Green IT [1]. According Dong et al. [4], cloud computing allows an e-learning ecosystem with the infrastructure which is reliable, flexible, cost-efficient, self-regulated, and QoS-guaranteed.

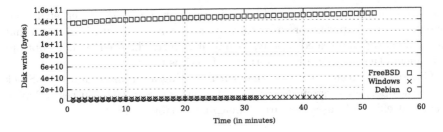

Fig. 9. Disk write activity.

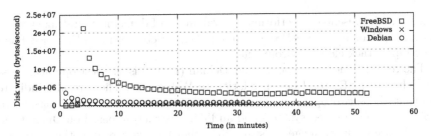

Fig. 10. Disk write activity rate.

The e-learning cloud architecture is composed of three layers: Infrastructure layer, Content layer and Application layer. The Infrastructure layer includes the hardware and software virtualization technologies that are used to ensure the stability and reliability of the infrastructure. The Content layer consists of e-learning contents, such as Web file systems, database systems and Web services. The Application layer consists of e-learning services, systems, tools, and so on [4]. E-learning systems using CC infrastructure can better use available resources through their instances, especially security and database intensive.

The results of this research can be used by the Chief Information Officer (CIO) as a parameter in the choice of e-learning systems. For example, when analyzing the response time and processing operations they can choose the operating system that will host your database and your Web server in open source cloud, namely OpenStack. The results showed that Debian operating system has a better performance when using a database in an open source platform.

Cloud Computing can help the CIOs to reduce budgets and increase service's levels of their applications. In this regard, the cloud computing solution allows in a SaaS scenario to be defined metrics that can help in the solution response time and then include these measures within the SLA. For PaaS and IaaS solutions, the IT manager initially will esteem the capacity's level of the solution and develop the resources for the solution that works satisfactorily [5].

Cloud Computing can be considered as an alternative to the consumption's reduction, optimization and efficient use of computing resources in organizations. The concept of IT efficiency and Green IT involves not only the computing resources used more efficiently [9], but further, the computers can be physically located in geographical areas that have access to cheap electricity while their computing power can be accessed long distances away over the Internet. In this context, cloud computing provides that companies use computational tools that can be deployed and scaled rapidly, even as it reduces the need for huge upfront investments that characterize enterprise IT setups today.

Our research results indicate that when hosting a database in a OpenStack cloud, the operating system that generates a better use of the resources is Debian because it efficiently uses the virtual machine capabilities.

5   Conclusion

This research presented a performance evaluation of different operating systems (Windows, Debian and FreeBSB) used to provide Web server and Database PaaS on an OpenStack cloud environment.

The results generally support that Debian (Linux) is the best operating system. The FreeBSD stood out in relation to Web server. These results indicate also that these two operating systems seem to be the most advantageous to implement e-learning ecosystems and an IT infrastructure based on OpenStack.

As future work, we suggest testing with other Web servers and other database systems, especially the non-relational. Furthermore, its suggested the use of other analytical metrics such as network load analysis and also testing the operation of simultaneous servers in a CC environment.

Acknowledgement. The authors would like to thanks the staff of the Elements of Information Processing Laboratory (LabEPI) for the technical and theoretical support.

References

1. Baliga, J., Ayre, R.W.A., Hinton, K., Tucker, R.S.: Green cloud computing: balancing energy in processing, storage, and transport. Proc. IEEE **99**(1), 149–167 (2011)
2. Brinkmann, A., Fiehe, C., Litvina, A., Luck, I., Nagel, L., Narayanan, K., Ostermair, F., Thronicke, W.: Scalable monitoring system for clouds. In: Proceedings of the IEEE/ACM 6th International Conference on Utility and Cloud Computing (UCC), pp. 351–356, December 2013
3. Buyya, R., Yeo, C.S., Venugopal, S.: Market-oriented cloud computing: vision, hype, and reality for delivering IT services as computing utilities. In: Proceedings of the 10th IEEE International Conference on High Performance Computing and Communications (HPCC), pp. 5–13, September 2008
4. Dong, B., Zheng, Q., Yang, J., Li, H., Qiao, M.: An e-learning ecosystem based on cloud computing infrastructure. In: Proceedings of the Ninth IEEE International Conference on Advanced Learning Technologies (ICALT), pp.125–127, July 2009
5. Jamsa, K.: Cloud Computing: SaaS, PaaS, IaaS, Virtualization, Business Models, Mobile Security and More. Jones & Bartlett Learning, Burlington (2012)
6. Kumar, S., Murthy, O.: Cloud computing for universities: a prototype suggestion and use of cloud computing in academic institutions. Int. J. Comput. Appl. **70**(14), 1–6 (2013)

7. Lin, A., Chen, N.C.: Cloud computing as an innovation: Percepetion, attitude, and adoption. Int. J. Inf. Manage. **32**(6), 533–540 (2012)
8. Malhotra, R., Jain, P.: Testing techniques and its challenges in a cloud computing environment. SIJ Trans. Comput. Sci. Eng. Appl. (CSEA) **1**(3), 88–93 (2013)
9. Marston, S., Li, Z., Bandyopadhyay, S., Zhang, J., Ghalsasi, A.: Cloud computing: the business perspective. Decis. Support Syst. **51**(1), 176–189 (2011)
10. Mell, P.M., Grance, T.: SP 800-145. The NIST Definition of CloudComputing. Technical report, Gaithersburg, MD, USA (2011)
11. Park, S.C., Ryoo, S.Y.: An empirical investigation of end-users switching toward cloud computing: A two factor theory perspective. Comput. Hum. Behav. **29**(1), 160–170 (2013)
12. Ratten, V.: Cloud computing: A social cognitive perspective of ethics, entrepreneurship, technology marketing, computer self-efficacy and outcome expectancy on behavioural intentions. Australas. Mark. J. (AMJ) **21**(3), 137–146 (2013)
13. Rossigneux, F., Lefevre, L., Gelas, J.P., De Assuncao, M.: A generic and extensible framework for monitoring energy consumption of openstack clouds. In: Proceedings of the IEEE Fourth International Conference on Big Data and Cloud Computing (BdCloud), pp. 696–702, December 2014
14. Tumbas, P., Matkovic, P., Sakal, M., Tumbas, S.: Exploring the potentials of cloud computing in higher education. In: Proceedings of the 8th International Technology, Education and Development Conference, pp. 2624–2631, March 2014
15. Vaquero, L.M., Rodero-Merino, L., Caceres, J., Lindner, M.: A break in the clouds: towards a cloud definition. SIGCOMM Comput. Commun. Rev. **39**(1), 50–55 (2008)
16. Xiaojiang, L., Yanlei, S.: The design and implementation of resource monitoring for cloud computing service platform. In: Proceedings of the 3rd International Conference on Computer Science and Network Technology (ICCSNT), pp. 239–243, October 2013

Towards the Tradeoff Between Online Marketing Resources Exploitation and the User Experience with the Use of Eye Tracking

Jarosław Jankowski[1,3]($\boxtimes$), Paweł Ziemba[2], Jarosław Wątróbski[3], and Przemysław Kazienko[1]

[1] Department of Computational Intelligence,
Wrocław University of Technology,
Wybrzeże Wyspiańskiego 27, 50-370 Wrocław, Poland
{jaroslaw.jankowski,przemyslaw.kazienko}@pwr.edu.pl
[2] The Jacob of Paradyż University of Applied Sciences in Gorzów Wielkopolski,
Teatralna 25, 66-400 Gorzów Wielkopolski, Poland
pziemba@pwsz.pl
[3] West Pomeranian University of Technology,
Szczecin, Żołnierska 49, 71-210 Szczecin, Poland
jwatrobski@wi.zut.edu.pl

Abstract. Online systems are often overloaded with marketing content and as a result, perceived intrusiveness negatively affects the user experience and the evaluation of the website. Intentional and unintentional avoidance of the commercial content creates the need for compromise solutions from both the perspective of user experience and business goals. The presented research shows a unique approach to search for tradeoffs between the editorial content and the intensity of marketing components with the use of eye tracking and the multiple-criteria decision analysis methods.

Keywords: User experience · Online marketing · Marketing exploitation · Eye tracking · HCI

1 Introduction

In many organizations, Internet systems have become a key component of business models and they play an important role in their commercial activities. Various revenue models used within online services are based on subscription services, electronic commerce, or support from advertisers or sponsors [20]. Despite the development of various concepts, the profit from selling the advertising space still plays an important role for many business models and excessive online marketing resources exploitation takes place. This applies to social media platforms, news portals and entertainment services [19]. While the increase of advertising expositions within websites is directly related to profits, the negative side effects can also be observed. Users perceive advertising clutter with a high share of advertising components among editorial content [1]. There is an observed drop of user satisfaction when more and more intrusive

© Springer-Verlag Berlin Heidelberg 2016
N.T. Nguyen et al. (Eds.): ACIIDS 2016, Part I, LNAI 9621, pp. 330–343, 2016.
DOI: 10.1007/978-3-662-49381-6_32

advertising techniques are used to attract potential customers' attention [2, 18]. Earlier research in this field was addressed especially to the effect of intrusiveness on brand awareness and memory [13]. Dedicated measures were introduced to evaluate the level of intrusiveness based on scales defined by Li et al. [5] and were later utilized in various studies [4, 18]. Most of the earlier methods focused on measuring intrusiveness or improving the performance of online marketing [6]. Research presented in this paper is based on tradeoff solutions and demonstrates a multi-criteria approach based on the integration of MCDA methods with eye-tracking feedback. During the designed experiment, the editorial content was used with embedded advertisement and the focus level on either editorial or commercial content was measured. The results based on these multi-criteria methods revealed that the method of selecting advertising content may be efficiently used within the website, if taking into account criteria related to the user experience and web portal profits. The paper is organized as follows. Section 2 consists of the literature review. In Sect. 3, the conceptual framework and assumptions for empirical research are presented. In the Sect. 4, empirical results from research are presented, followed by a multi-criteria analysis and results in Sect. 5. Section 6 includes conclusions and directions for future work.

2 Literature Review

Design of online systems and their integration with marketing content requires several decisions at all stages of the design process and takes into account factors related to all engaged parties. First of all, user experience should be considered towards better functionality and the creation of solutions that better address the needs of web users [18]. This approach can be conflicting with the perspective of portal owner, who have a high focus on profits. Decisions aimed at improving economic performance are made by online ventures and often result in negative feedback from web users. The situation is complicated even more when the point of view of external advertisers, with a focus on performance, is taken into account by the web portal operator. Questions arise as to which level the intrusiveness of marketing content can be increased to attract user attention and keep profits at an acceptable level without invading user experience what was discussed in our earlier research in the relation to the recommending interfaces [14], web conversions [15] and repeated contacts with the marketing content [16]. Other studies in this field addressed several aspects related to the intrusiveness of online content. Intrusiveness in relation to online advertisements is defined as "a perception or psychological consequence that occurs when an audience's cognitive processes are interrupted" [5]. Ha and Litman defined intrusiveness as "[...] the degree to which advertisements in a media vehicle interrupt the flow of an editorial unit" [8]. To measure intrusiveness, Li et al. [5] introduced an approach based on a seven-point scale defining content as distracting, disturbing, forced, interfering, intrusive, invasive and obtrusive, with seven levels ranging from "strongly agree" to "strongly disagree". The scale, which was reduced from an initial eleven items, is based on psychological mechanisms and was later used for different experiments by other authors [4, 18]. The excessive usage of video, audio and animations within online content causes an overload problem of commercial content and leads to side effects, followed by negatively affecting user experience [9]. These effects where analyzed in the earlier studies. For example, the field experiments performed by Moe were based on the timing of popup-messages with on across-page delay and within-page delay and their effect on click-throughs and site exit behaviors were studied [6]. Goldstein et al., in turn, emphasized the conflict within companies operating web portals where advertising content is a source of income, but intensive exploitation of advertising space leads users to abandoning the website [7]. Earlier research reported several usability problems related to online advertising with misleading information and difficult to find options to remove advertising content [3]. In a situation when most web portals are trying to attract the attention of web users with commercial content, online users are overloaded by the marketing content and only part of it receives attention due to the limited ability to process information [17]. Experienced web users are focused on completing their tasks, while they ignore irrelevant content [13]. Unfortunately, this problem can be only partially solved by adaptive personalization of advertisements [36]. Some undesirable side effects during web-based tasks are observed based on the unintentional avoidance. They were identified as "banner blindness" by Benway [11] and extended by other researchers [10]. Apart from cognitive avoidance and banner blindness, the physical avoidance of marketing content takes place [12]. While earlier research focused mainly on the user experience or efficacy of advertising content separately, the research presented in this paper showed an integrated approach with the main goal of supporting decisions and multi-criteria evaluation from the perspective of web users and the efficiency of commercial content. The conceptual framework is presented in the next stage, followed by results from the experiment based on searching for tradeoffs between content intrusiveness and user experience with the use of eye tracking and multi-criteria methods.

3   Methodological Background

The presented method is based on the scaling parameters of advertising content related to its ability to attract user attention. Within advertising content, the factors related to the intensity of influence on a user were identified. The first group of factors used in this research is related to external characteristics of the content, such as animations or flashing effects intended to attract user attention. All of those factors can affect the level of intensity of advertising content with the vector of intensity, $I = [I_1, I_2,..., I_m]$ with distinguished m levels of intensity, such as $I_i > I_j$ for all $i > j$. The level of intensity can be evaluated by a designer or obtained as a result of perceptual experiments or questionnaires. The second dimension of marketing content related to the intrusiveness measured by its size $S = [S_1, S_2,..., S_n]$ with n sizes and ordered elements, such as $S_i > S_j$ for each $i > j$. The third distinguishing factor is related to the location of advertising content within a webpage. A higher location is having usually a greater impact on users. Location is represented by k available, distinct placements, $L = [L_1, L_2,..., L_k]$ such as $L_i > L_j$ if $i > j$. The next considered factor is the cost of emission of advertisement within the portal, which is directly related to website operator profits. Evaluation of the costs of advertisements is based on the location, size, format used and techniques. More intrusive formats usually deliver a higher number of interactions but their emission cost is higher as well. Following these specifics, the profits are a function of parameters related to advertisements $P = f(I,S,L)$. For the measurement of the performance of advertisements within experimental content, the used factors are based on eye-tracking measurements and are related to the total time that the web user has looked at the marketing content. Used factors include the number of viewers to the marketing content (MV), percentage of total time spent on the website with the focus on advertisement (MTP), focus time on the marketing content (MT), time to the first view of the advertising content after the website is fully loaded (MFV), the number of repeated visits to the advertising content (MRV) and the total number of returning visitors (MRVN). The equivalent factors based on this eye-tracking can be assigned to the editorial content in a form of total time spent on editorial content (ET), the percentage representation of it (ETP), and the number of viewers (EV), time to the first viewing of the editorial content (EFV), the number of revisits (ERV), and the number of returning visitors (ERVN). While users are usually goal-oriented with a focus on reading editorial content, time spent on advertisements can be treated as negatively affecting user experience. Depending on the web portal approach to the user experience, advertising content can be selected to maximize profits and minimize time spent with the focus on the advertisement $Sel(A) \rightarrow [\min(MT), \max(P)]$. The final evaluation should deliver a set of advertisements with the potential for profits and impact on users based on the requirements from advertisers. Due to the relation of profits P to factors assigned to the advertisement, the final result is characterized by the level of intensity I, size S and location L. For the analysis of the tradeoff solutions, a multi-criteria approach should be used with proper methods selection based on characteristics of decision problem and criteria selections [39, 40]. The decision-making process can be identified in four main stages [25]. In the first stage, the object of the decision and a definition of the set of potential decision variants A and the determination of the reference problematic to A is formulated. In the second stage, analyzing of consequences and developing the consistent set of criteria C is performed. The third stage includes modelling comprehensive preferences and operationally aggregating performances. The last stage consists of investigating and developing the recommendation, based on the results of Stage Three and the problematic defined in Stage One. In the first stage, Roy [26] distinguished four problematics, including α - selection, β - sorting, γ - ranking, and δ - description. Various methods and operational approaches can be used for aggregating performances. The first approach is based on the use of a single synthesizing criterion without incomparabilities, including the relations of indifference (I), weak preference (Q) and strict preference (P). The second approach is based on the synthesis of outranking with incomparabilities, including the relations of: S - outranking, R - incomparability. The third approach is based on the interactive local judgments with trial-and-error iterations. Used in this research, the Promethee method is based on the outranking relation and pairwise comparisons. Promethee I/II solves the sorting problem, delivers the ranking of variants, and indicates the best of them (in terms of Pareto evaluation) [21]. This method uses the outgoing and incoming flows of preferences, specifying how much greater one alternative is than the other, and how much it is surpassed by other variants. Using the Promethee methods, a decision maker can choose between six preference functions: usual criterion, quasi-criterion, criterion with linear preference, level-criterion, criterion with linear preference and indifference area and Gaussian criterion [22]. Other approaches based on the characteristic objects method with a new distance-based approach can be applied [37, 38].

4 Empirical Results

For the purpose of the experiment, a design having three levels of intensity $I = [1,2,3]$, three levels of size $S = [1,2,3]$ and three levels of localization $L = [1,2,3]$ was used. As a result, the experimental space $[3 \times 3 \times 3]$ with its 27 combinations of websites with integrated marketing content was used in a form of design variant. Three components of each website presented editorial content with the headline content concerning current e-commerce news. As a fourth element, advertisement was used built on the basis of three parameters: intensity of marketing object, size, and location on the page. Acceptable values of each of the characteristics are shown in Table 1.

Table 1. Parameters of advertising content and it's variants

Value	Intensity (P)	Localization (L)	Size (S)
1	Low	Down	Small
2	Medium	Middle	Medium
3	High	Top	Big

In stage two, a set of criterions was used based on the variables assigned to the collected data. The experiment includes 16 participants who have agreed to take part, and average age did not exceed 30 years. None of the subjects showed any visual defects that could distort the results. In addition, each of the respondents confirmed that they regularly use web portals and have contact with online advertisements. The content used in the experiment was designed to target their needs and none of the subjects had seen them before the experiment. During the eye-tracking study, the experimental page was presented for 15 s and eye-tracking patterns were saved with 60 Hz frequency. Both visual and statistical analysis of results was performed. In the first stage heat maps and scan paths were generated for each design variant. Example results are showed in Figs. 1 and 2.

The data collected during the study was used for the decision criteria. The set of criteria was extended with the cost, which is the sum of the values shown in Table 3. For example, the cost of a small ad with a high-intensity presence, located at the bottom of the page is 1 (Size) + 3 (Intensity) + 1 (Localization) = 5. The values of decision-making criteria for design variants are shown in Table 2.

5 Searching for Tradeoff Solutions with MCDA Method

Proper selection of the advertising components to be used within a website provides the ability to generate the desired interactions and performance. Design solutions can deliver different results in terms of user experience and marketing performance. To solve the problem of design variants, the Promethee II was used, which uses a synthesis

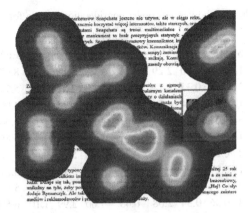

Fig. 1. Heat map for the design variant with middle ad location and medium size with high intensity

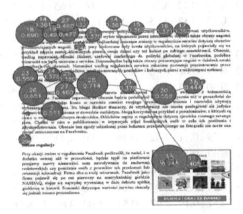

Fig. 2. Scan path for the example design variant with down ad localization and big size with medium intensity

of criteria based on outranking relation [21]. This choice is related to its wide versatility, resulting from, among others, the use of six different preference functions [22]. In an initial stage, the directions of preference for each criterion, defined weighting of the criteria and their groups, selected functions preferences, and preferences with set threshold values were established. The proper selection of both functions, as well as the thresholds, has a substantial influence on the sequence variants in the rankings [30, 31]. In addition, the type of the preference function depends on the type of criteria. For quantitative criteria functions with linear preference, linear preference and indifference area or Gaussian preference should be used [27]. The models based on the linear preference and Gaussian preference create two separate scenarios for decision making. It features preference use the thresholds like criterion with linear preference p and s for Gaussian criterion. The s threshold can be interpreted as the standard deviation values

Table 2. The values of the variants under consideration criteria for decision-making

Variant				Advertisement				Editorial content				Cost
No	P	L	S	MV	MT	MFV	MRV	EV	ET	EFV	ERV	
1	1	3	1	0.36	0.90	2.29	0.21	1.00	5.11	0.47	8.07	5.00
2	1	2	2	0.64	0.33	1.28	1.00	1.00	4.57	0.64	5.71	5.00
3	2	2	1	0.64	0.24	4.77	0.00	1.00	3.75	0.68	4.00	5.00
4	2	1	3	0.57	0.28	0.91	0.07	1.00	4.51	1.10	6.64	6.00
5	3	1	1	0.57	0.19	3.77	0.07	1.00	4.01	1.54	5.86	5.00
6	3	3	2	0.57	0.38	2.69	0.71	1.00	5.36	0.65	7.36	8.00
7	1	2	1	0.29	0.07	1.58	0.14	1.00	6.11	0.45	5.57	4.00
8	2	2	2	0.64	0.23	6.33	0.43	1.00	3.55	0.69	5.57	6.00
9	1	2	3	0.36	0.34	2.08	0.43	1.00	3.99	1.10	6.50	6.00
10	1	3	2	0.50	0.20	2.60	0.71	1.00	5.43	0.99	8.64	6.00
11	2	3	2	0.50	0.18	2.41	0.50	1.00	4.37	1.12	6.93	7.00
12	3	3	3	0.64	0.55	2.03	0.93	1.00	2.64	0.50	4.50	9.00
13	3	2	2	0.57	0.22	4.76	0.14	1.00	4.65	1.08	6.07	7.00
14	1	1	1	0.43	0.13	3.95	0.00	1.00	3.80	0.90	6.50	3.00
15	3	2	3	0.50	0.12	1.80	0.14	1.00	3.83	0.84	8.29	8.00
16	1	1	3	0.36	0.14	1.21	0.00	1.00	4.12	0.71	7.86	5.00
17	1	1	2	0.29	0.06	0.96	0.21	1.00	3.42	1.51	4.64	4.00
18	2	2	3	0.50	0.15	3.78	0.50	1.00	3.86	0.71	4.07	7.00
19	2	1	2	0.50	0.20	0.16	0.50	1.00	3.52	0.78	6.36	5.00
20	1	3	3	0.36	0.04	2.93	0.07	1.00	4.91	0.62	5.71	7.00
21	3	2	1	0.29	0.11	2.42	0.29	1.00	4.33	1.09	6.93	6.00
22	2	1	1	0.43	0.15	0.26	0.57	1.00	2.80	1.18	6.29	4.00
23	3	1	2	0.43	0.15	1.70	0.21	1.00	3.50	0.92	7.29	6.00
24	2	3	3	0.79	0.30	2.87	1.14	1.00	4.32	0.85	5.71	8.00
25	2	3	1	0.57	0.19	2.97	0.50	1.00	3.63	1.68	6.29	6.00
26	3	1	3	0.57	0.14	3.43	0.79	1.00	3.47	1.37	6.07	7.00
27	3	3	1	0.85	0.34	4.60	0.62	1.00	3.10	1.64	7.46	7.00

of all options for a given decision criterion [29]. On the other hand, the p threshold can be established between credible *min* and *max* values that can take a given criterion [28]. In the developed decision model, the value of p threshold was assumed to be twice the standard deviation. Preference model, taking into account weighting of the criteria and thresholds for individual preference function within example decision process with the use of external expert are shown in Table 3.

The last stage relied on the preparation of recommendation decisions on the basis of the multi-criteria procedure. These recommendations were based on the assessment carried out using the two preference functions mentioned earlier. In addition, a verification of the results with the GAIA analysis (Geometrical Analysis for Interactive Assistance) was used [33] together with sensitivity analysis [34]. As a result, preference flows were determined with the use of two preference functions. Values of the preference flows are given in Table 4. The analysis of the rankings indicates that they are

Table 3. Preferences for criterions in proposed model

Criterion and group	Direction	Weight	Weight [%]	Preference function	
				Linear (p)	Gaussian (s)
Advertisement		**4**	**40.00**		
Ad Viewers	Max	5	16.70	0.28	0.14
Ad Viewed Times (s)	Max	4	13.30	0.34	0.17
Ad 1st view (s)	Min	2	6.70	2.90	1.45
Ad Revisits	Max	1	3.30	0.64	0.32
Editorial content		**4**	**40.00**		
Txt Viewers	Max	5	16.70	1.00	0.00
Txt Viewed Times (s)	Max	4	13.30	1.60	0.80
Txt 1st view (s)	Min	2	6.70	0.70	0.35
Txt Revisits	Max	1	3.30	2.38	1.19
Economic factor		**2**	**20.00**		
Cost	Min	6	2000	2.82	1.41

very similar. These similarities are well illustrated by a co-occurring factor correlation between rankings, which amounted to 0.9945.

The analysis of results can be divided into two main groups of decision variants. The worst variant in each rank are: A15, A17, A18, A20, A21, A26. In turn, the best options are A1 and A2. Other variants fall between these groups. The next step was to verification of the results through the use of GAIA analysis for the rankings. The GAIA methodology is concerning the k-criteria of the decision problem in the k-dimensional space that is projected onto a plane [35]. Alternatives are represented by points and criteria preferences are symbolized by vectors. If these vectors are oriented in the same direction, it means that they are represented by the criteria in a similar way, which affects the global assessment of variants. The length of the vector indicates the strength of a given criterion to assess the variant. The closer a vector is to the particular alternative, the more the vector supports this alternative [21]. The results of GAIA analysis are shown in Figs. 3 and 4.

In this analysis, to increase its readability, the different groups of criteria, not individual criteria, were taken into account. In analyzing the GAIA charts, their similarity should be noted. When seeing the GAIA plane from the perspective of groups of criteria, it can be observed that the criteria related to advertising and expense have a slightly greater impact on the final solution than the criteria related to the text presented on the site. For each of the graphs, it is visible that the criteria related to advertising the most appropriate variants include variants A27, A25, A4, A8 and A2. On the other hand, criteria related to text support variants A20, A21 and the criterion cost rewards variants A22, A14 and A17. Furthermore, the vector of cost criterion is set almost orthogonal to a vector of advertising and text criteria, which means that they are to some extent independent of cost. In contrast, vector sets of criteria related to advertising and text are positioned opposite each other, which means that these criteria are mutually conflicting. Vector Л, representing a compromise, shows A2 as the best solution, which is consistent with those previously obtained rankings. The last step in

Table 4. Net preferences flow values

Ranking	Linear preference				Gaussian preference			
	ϕ^+	ϕ^-	ϕ_{netto}	Variant	ϕ^+	ϕ^-	ϕ_{netto}	Variant
1	0.362	0.061	0.301	A2	0.301	0.047	0.254	A2
2	0.389	0.129	0.260	A1	0.349	0.107	0.242	A1
3	0.26	0.133	0.127	A10	0.295	0.175	0.120	A7
4	0.301	0.180	0.121	A6	0.217	0.101	0.116	A10
5	0.329	0.209	0.120	A7	0.255	0.153	0.102	A6
6	0.237	0.132	0.105	A4	0.187	0.103	0.084	A4
7	0.286	0.202	0.084	A24	0.246	0.169	0.077	A24
8	0.224	0.142	0.083	A19	0.182	0.109	0.073	A19
9	0.234	0.176	0.058	A3	0.188	0.148	0.040	A3
10	0.235	0.207	0.028	A14	0.202	0.166	0.036	A14
11	0.199	0.189	0.010	A16	0.160	0.153	0.008	A16
12	0.227	0.230	−0.003	A22	0.192	0.193	0.000	A22
13	0.185	0.194	−0.009	A5	0.228	0.237	−0.009	A27
14	0.192	0.203	−0.011	A8	0.144	0.160	−0.016	A5
15	0.260	0.273	−0.013	A27	0.151	0.171	−0.020	A8
16	0.286	0.320	−0.035	A12	0.250	0.291	−0.041	A12
17	0.167	0.211	−0.044	A9	0.128	0.173	−0.045	A9
18	0.164	0.223	−0.06	A13	0.109	0.164	−0.056	A11
19	0.143	0.206	−0.064	A11	0.126	0.185	−0.058	A13
20	0.146	0.216	−0.071	A25	0.111	0.177	−0.067	A25
21	0.127	0.218	−0.091	A23	0.095	0.173	−0.078	A23
22	0.109	0.254	−0.145	A18	0.153	0.276	−0.123	A17
23	0.178	0.323	−0.145	A17	0.082	0.206	−0.124	A18
24	0.122	0.268	−0.146	A21	0.093	0.220	−0.127	A21
25	0.122	0.274	−0.152	A15	0.090	0.220	−0.130	A26
26	0.117	0.269	−0.152	A26	0.095	0.224	−0.130	A15
27	0.136	0.295	−0.159	A20	0.112	0.244	−0.132	A20

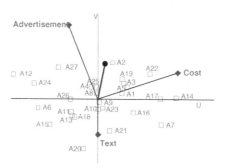

Fig. 3. The GAIA plane sets of criteria with the use of linear preference

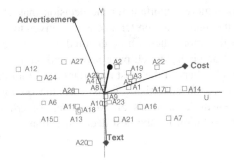

Fig. 4. The GAIA plane sets of criteria with the use of Gaussian preference

preparing the recommendation was to conduct the rankings sensitivity analysis to changes in weights of the individual groups of criteria. Charts showing the analysis for linear function of a preference for selected top options are shown in Fig. 5. Due to the fact that a sensitivity analysis for Gaussian preference function gave very similar results. it was not presented here.

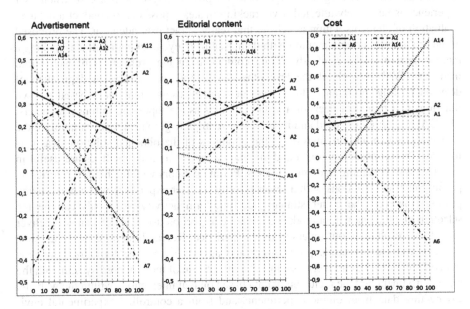

Fig. 5. Sensitivity analysis for top variants

The sensitivity analysis on weight sets of criteria indicates a high stability of the resulting solution for the best decision variants. Variants A1 and A2 are the best in terms of weight: 20 % to 83 % for the criteria relating to advertising; 0 % to 89 % for the criteria relating to the text; and 3 % to 48 % for the criterion of cost. It can be assumed that these two solutions offer a good compromise between text content and advertising pages. If it is more important for the decision maker that users pay more attention to the editorial content, then preference will be given to variant A1. However, if it is more important that more attention be devoted to available advertising content then the preference will be given to web page layout and ad attributes represented by the A2 variant. On the other hand if the most important aspect for the decision maker is advertising outcome based on advertising costs, then the optimal choice may be a variant of A14. The sensitivity analysis on weight sets of criteria indicates a high stability of the resulting solution for the best decision variants.

6   Summary

The design of websites and the integration of marketing content requires the consideration of some factors related to website usability and user experience. However, parameters related to the business indicators are also important, as it is difficult to maintain the websites without any advertising content. The presented research showed the possible tradeoff between online marketing resources exploitation and user experience based on multicriteria selection of marketing content and integrating it with the editorial part of a webpages. A high focus on profits may lead to a case when user experience is negatively affected. Eye tracking makes it possible to conduct a detailed evaluation of elements within a website that really attract user attention. The proposed usage of the multi-criteria method with data used from eye tracking facilitates making a more objective selection of advertising content using the preferences of web portal managers. With the use of the Promethee II method for the presented problem of decision making, preference flows were determined using preferences based on two functions that represent decision makers attitudes as the input and output parameters. The GAIA method was used and it delivered a complete graphical representation of the decision problem and an analysis of the goodness of the obtained solutions was performed. The proposed approach and the integration of the used methods are useful in the process of obtaining tradeoff solutions. Portal operators with the highest priority on profits can select solutions with a lower priority for user experience while maximizing profits. However, the better strategy for long-term activity and goals can be the selection of elements with a moderate negative impact on users, which can still deliver acceptable results without invading user experience. Research showed how the game between website operators, portal users and advertisers can be formalized towards tradeoff solutions. Future work assumes the integration of the proposed system with the real environment and the inclusion of data based on real interactions within the website. Connecting data from online experiments and from a controlled experimental environment can deliver tools for wider use for marketing purposes.

Acknowledgments. The work was partially supported by European Union's Seventh Framework Programme for research, technological development and demonstration under grant agreement no. 316097 [ENGINE] and by the National Science Centre, the decision no. DEC-2013/09/B/ST6/02317.

References

1. Ha, L., McCann, K.: An integrated model of advertising clutter in offline and online media. Int. J. Advertising **27**(4), 569–592 (2008)
2. Brajnik, G., Gabrielli, S.: A review of online advertising effects on the user experience. Int. J. Hum. Comput. Interact. **26**(10), 971–997 (2010)
3. Gibbs, W.: Examining users on news provider web sites: a review of methodogy. J. Usability Stud. **3**, 129–148 (2008)
4. McCoy, S., Everard, A., Polak, P., Galletta, D.F.: The effects of online advertising. Commun. ACM **50**(3), 84–88 (2007)
5. Hairong, E., Edwards, S.M., Joo-Huyn, L.: Measuring the intrusiveness of advertisements: scale development and validation. J. Advertising **31**(2), 37–47 (2002)
6. Moe, W.W.: A field experiment to assess the interruption effect of pop-up promotions. J. Interact. Mark. **20**(1), 34–44 (2006)
7. Goldstein, D.G., McAfee, R.P., Suri, S.: The cost of annoying ads. In: Proceedings of the 22nd International Conference on World Wide Web (WWW 2013), International World Wide Web Conferences Steering Committee, pp. 459–470. Republic and Canton of Geneva, Switzerland (2013)
8. Ha, L., Litman, B.R.: Does advertising clutter have diminishing and negative returns? J. Advertising **26**(1), 31–42 (1997)
9. Rosenkrans, G.: The creativeness and efectiveness of online interactive rich media advertising. J. Interact. Advertising **9**(2), 18–31 (2009)
10. Burke, M., Hornof, A., Nilsen, E., Gorman, N.: High cost banner blindness: ads increase perceived workload hinder visual search and are forgotten. ACM Trans. Comput. Hum. Interact. **12**(4), 423–445 (2005)
11. Benway, J.P., Lane, D.M.: Banner blindness: web searchers often miss "obvious" links (1998). http://www.internettg.org/newsletter/dec98/banner_blindness.html
12. Krammer, V.: An effective defense against intrusive web advertising. In: Proceedings of the 2008 Sixth Annual Conference on Privacy, Security and Trust (PST 2008), pp. 3–14. IEEE Computer Society, Washington, USA (2008)
13. Chatterjee, P.: Are unclicked ads wasted? Enduring effects of banner and pop-up ad exposures on brand memory and attitudes. J. Electron. Commer. Res. **9**(10), 51–61 (2008)
14. Jankowski, J.: Modeling the structure of recommending interfaces with adjustable influence on users. In: Selamat, A., Nguyen, N.T., Haron, H. (eds.) ACIIDS 2013, Part II. LNCS, vol. 7803, pp. 429–438. Springer, Heidelberg (2013)
15. Jankowski, J.: Balanced approach to the design of conversion oriented websites with limited negative impact on the users. In: Bădică, C., Nguyen, N.T., Brezovan, M. (eds.) ICCCI 2013. LNCS, vol. 7803, pp. 527–536. Springer, Heidelberg (2013)
16. Jankowski, J.: Increasing website conversions using content repetitions with different levels of persuasion. In: Selamat, A., Nguyen, N.T., Haron, H. (eds.) ACIIDS 2013, Part II. LNCS, vol. 7803, pp. 439–448. Springer, Heidelberg (2013)
17. Lang, A.: The limited capacity model of mediated message processing. J. Commun. **50**(1), 46–70 (2000)
18. Zha, W., Wu, H.D.: The impact of online disruptive ads on users' comprehension evaluation of site credibility and sentiment of intrusiveness. Am. Commun. J. **16**(2), 15–28 (2014)

19. Andrew, Y.T., Kim, D.J.: A comparative analysis of online social networking sites and their business models. E-marketing: Concepts, Methodologies, Tools and Applications, pp. 803–813. IGI Global Publications, Hershey (2012). doi:10.4018/978-1-4666-1598-4.ch048
20. Lai, V.S., Wong, B.K.: Business types e-strategies and performance. Commun. ACM **48**(5), 80–85 (2005)
21. Brans, J.P., Mareschal, B.: Promethee methods. In: Figueira, J., Greco, S., Ehrgott, M. (eds.) Multiple Criteria Decision Analysis, pp. 163–195. Springer, Boston (2005)
22. Brans, J.P., Vincke, P.: A preference ranking organisation method: (the Promethee method for multiple criteria decision-making). Manag. Sci. **31**, 647–656 (1985)
23. Behzadian, M., Kazemzadeh, R.B., Albadvi, A., Aghdasi, M.: PROMETHEE: a comprehensive literature review on methodologies and applications. Eur. J. Oper. Res. **200**, 198–215 (2010)
24. Ghafghazi, S., Sowlati, T., Sokhansanj, S., Melin, S.: A multicriteria approach to evaluate district heating system options. Appl. Energy **87**, 1134–1140 (2010)
25. Roy, B.: Multicriteria Methodology for Decision Aiding. Springer, Dordrecht (1996)
26. Roy, B.: Paradigms and challenges. In: Figueira, J., Greco, S., Ehrgott, M. (eds.) Multiple Criteria Decision Analysis: State of the Art Surveys, pp. 3–24. Springer, Boston (2005)
27. Deshmukh, S.C.: Preference ranking organization method of enrichment evaluation (Promethee). Int. J. Eng. Sci. Invention **2**(11), 28–34 (2013)
28. Roy, B.: The outranking approach and the foundations of ELECTRE methods. In: Bana e Costa, C.A. (ed.) Readings in Multiple Criteria Decision Aid, pp. 155–183. Springer, Heidelberg (1990)
29. Amponsah, S.K., Darkwah, K.F., Inusah, A.: Logistic preference function for preference ranking organization method for enrichment evaluation (PROMETHEE) decision analysis. Afr. J. Math. Comput. Sci. Res. **5**(6), 112–119 (2012)
30. Podvezko, V., Podviezko, A.: Dependence of multi-criteria evaluation result on choice of preference functions and their parameters. Technol. Econ. Dev. Econ. **16**(1), 143–158 (2010)
31. Podvezko, V., Podviezko, A.: Use and choice of preference functions for evaluation of characteristics of socio-economical processes. In: 6th International Scientific Conference on Business and Management, Vilnius, pp. 1066–1071 (2010)
32. Macharis, C., Brans, J.P., Mareschal, B.: The GDSS PROMETHEE procedure. J. Decis. Syst. **7**, 283–307 (1998)
33. Mareschal, B., Brans, J.P.: Geometrical representations for MCDA. Eur. J. Oper. Res. **34**, 69–77 (1988)
34. Saltelli, A., Tarantola, S., Chan, K.: A role for sensitivity analysis in presenting the results from MCDA studies to decision makers. J. Multi-Criteria Decis. Anal. **8**, 139–145 (1999)
35. Janssens, G.K., Pangilinan, J.M.: Multiple criteria performance analysis of non-dominated sets obtained by multi-objective evolutionary algorithms for optimisation. In: Papadopoulos, H., Andreou, A.S., Bramer, M. (eds.) AIAI 2010. IFIP AICT, vol. 339, pp. 94–103. Springer, Heidelberg (2010)
36. Kazienko, P., Adamski, M.: AdROSA - adaptive personalization of web advertising. Inf. Sci. **177**(11), 2269–2295 (2007)
37. Sałabun, W.: The characteristic objects method: a new distance-based approach to multicriteria decision-making problems. J. Multi-Criteria Decis. Anal. **22**(1–2), 37–50 (2015)
38. Piegat, A., Sałabun, W.: Identification of a multicriteria decision-making model using the characteristic objects method. Appl. Comput. Intell. Soft Comput. (2014)

39. Wątróbski, J., Jankowski, J., Piotrowski, Z.: The selection of multicriteria method based on unstructured decision problem description. In: Hwang, D., Jung, J.J., Nguyen, N.-T. (eds.) ICCCI 2014. LNCS, vol. 8733, pp. 454–465. Springer, Heidelberg (2014)

40. Ziemba, P., Piwowarski, M., Jankowski, J., Wątróbski, J.: Method of criteria selection and weights calculation in the process of web projects evaluation. In: Hwang, D., Jung, J.J., Nguyen, N.-T. (eds.) ICCCI 2014. LNCS, vol. 8733, pp. 684–693. Springer, Heidelberg (2014)

Using Cognitive Agents for Unstructured Knowledge Management in a Business Organization's Integrated Information System

Marcin Hernes[(✉)]

Wrocław University of Economics ul,
Komandorska 118/120, 53-345 Wrocław, Poland
marcin.hernes@ue.wroc.pl

Abstract. Management of unstructured knowledge in business organizations, mainly by using integrated information system, is a very important process. This type of knowledge allows supporting decision-making process to a high degree. The aim of this paper is to present using a cognitive agent's architecture for knowledge management in integrated information system running in business organization. Analysis of the existing works in considered field is presented in the first part of paper; next an unstructured knowledge management process using The Learning Intelligent Distribution Agent architecture has been described. The last part of paper presents the research experiment performed in order to verification developed solution.

Keywords: Knowledge management · Unstructured knowledge · Integrated management information systems · Cognitive agents

1 Introduction

Integrated management information systems process mainly structured knowledge [3]. This type of knowledge is stored by using different kind of structures (e.g. tables, sets, tuples, and trees). Management of unstructured knowledge in business organizations, mainly by using integrated information system, is a very important process. This type of knowledge allows supporting decision-making process to a high degree. Therefore, an unstructured knowledge must be processed in parallel to a structured knowledge. An unstructured knowledge is related mainly to various types of texts stored in natural language. Websites, social media or users' opinions about products, experts' opinions about forecast securities or currency exchange rates may serve as examples. Taking into consideration business organizations, processing of this type of knowledge is related to documents describing the organization's environment. A knowledge contained in these documents can be very useful. For example, on the basis of users' opinions about products it is possible to predict a sales volume or on the basis of currency exchange rates trading on financial markets can be performed. Unstructured knowledge processing, however, is more difficult than structured knowledge processing.

© Springer-Verlag Berlin Heidelberg 2016
N.T. Nguyen et al. (Eds.): ACIIDS 2016, Part I, LNAI 9621, pp. 344–353, 2016.
DOI: 10.1007/978-3-662-49381-6_33

The aim of this paper is to present using a cognitive agent's architecture for knowledge management in integrated information system running in business organization and verification of this agent's performance.

This paper is organized as follows: analysis of the existing works in considered field is presented in the first part of paper; next an unstructured knowledge management process using The Learning Intelligent Distribution Agent architecture has been described. The last part of paper presents the research experiment performed in order to verification developed solution.

2 Related Works

Knowledge management in business organization is related to structured and unstructured knowledge processing performed mainly by integrated information systems. It is understood in the literature of subject in different ways, inter alia as [2, 6, 13]:

- knowledge creation process and use it to improve the efficiency of the organization,
- organization's information, knowledge and experience management,
- encouraging employees to share knowledge.

Knowledge management can be divided into the following sub-processes [2, 4, 5]:

- identification,
- acquiring,
- creation,
- organization,
- utilization,
- storing.
- evaluation,
- transfer,
- integration,
- triggering creativity in employees.

Processing of unstructured knowledge is performed by using different methods; often two groups of a hybrid approaches are used:

1. The unstructured knowledge is structuralized and processed in symbolic way (this type of processing can by performed, for example, by expert systems or genetic algorithms)
2. The unstructured knowledge is converted into numerical representation and processed in emergent way (this type of processing can by performed, for example, by neural networks or fuzzy logic systems).

There are two categories of analysis of text documents [2, 8]:

- Deep Text Processing, involving a linguistic analysis of grammatical relationships and context (meaning) determining occurring in text documents. This type of processing can be very difficult and complex.

- Shallow Text Processing, relies on identification of particular parts of sentence (e.g. subject, predicate) and parts of speech (e.g. noun, adjective, verb) without analysis more complex relationships.

There are different techniques for text document analysis, including:

- machine learning [7, 12] (Bayesian classifiers, support vector machines, used, for example, to recognizing contextual polarity [14]).
- rule based systems [10, 11] (used, for example, to the extraction of spatial relationship, identify the requirements for the projects, expressed on Internet forums [15] or extraction of information from real estate ads [14]).

More often the cognitive agents' architectures are used in order to unstructured knowledge processing. The literature of subject presents many solutions of such architectures and, according [16], their main features include:

- memory organization
- learning methods.

Cognitive agents' architectures are divided as follows [16]:

- **symbolic architectures** processing an unstructured knowledge in symbolic way.
- **emergent architectures** processing an unstructured knowledge in an emergent way.
- **hybrid architectures** processing an unstructured knowledge parallel in symbolic and emergent way.

3 Knowledge Management Using LIDA Cognitive Agent

In this research The Learning Intelligent Distribution Agent (LIDA), developed by Cognitive Computing Research Group [1], has been used in order to unstructured knowledge management. This is a hybrid architecture and as it's practical implementation can serve automatic searching job opportunity by sailors serving in the US Navy (sailors are employed within several months contracts, and the role of cognitive agent program was based on an analysis of the body of the e-mail message sent by the sailors, who ended a contract, and search for these contracts in accordance with the requirements set out in the body of these messages).

The LIDA agent is architecture with tools to grounding the meaning of symbols. It allows processing the unstructured knowledge. Basic operations in the frame of this architecture are performed by the codelets - specialized, mobile programs processing information in the model of global workspace. The LIDA consists of the following modules [1]:

- Sensory Memory,
- Perceptual-Associative Memory,
- Workspace,
- Transient Episodic Memory,
- Declarative Memory,

- Attentional Codelets,
- Global Workspace,
- Procedural Memory,
- Action Selection,
- Sensory-Motor Memory.

It is running in the cognitive cycle [9] divided into following phases [1]:

- perception,
- memory retrieval,
- conscious broadcast,
- action.

Cognitive cycle is asynchronous and it is repeated at a frequency of 5 - 10 times per second, depending on the number of tasks that must be executed in a given cycle. Also learning processes are performed in the frame of particular phases. The LIDA architecture is implemented in Java and available as framework software, therefore it can be use by a wider range of scientists or developers. A framework includes classes of objects performing the operations in the architecture (definitions and methods for all types of memory, communication protocols, methods for performing the agent's operations on the real world objects - for example, a searching and recognition of objects, identifying the characteristics of the objects, identifying associations between objects). All of tasks' parameters can be configured by using the "xml" file. The programmers' job is to complete the tools provided by the framework LIDA on aspects related to the specific area of the problem - for example, economics, management (writing the desired program code).

A cognitive agent of the above-discussed structure may be used in supporting the process of knowledge management in business organizations. This agent can be implemented in integrated information system. As an example, one can have a look at a situation when an agent analyses opinions of experts about forecast securities or currency exchange rates (Fig. 1). An agent is permanently searching for documents containing opinions of experts (in sources such as entries posted on subject-related portals, social networking sites, as well as on brokerage houses websites) and it are storing them in the sensory memory. Next, opinions are interpreted in the perceptual memory (e.g. it is determined which currency/security a given opinion refers to, and whether an expert predicts an increase or decrease in the exchange marks of a given currency/security). Using events stored in the episodic memory (e.g. "in the previous period a decrease/fall in the exchange marks was noted") and rules stored in the declarative memory (e.g. "if opinions of experts forecast a decrease in the exchange marks, a given currency/securities shall be sold") a current situational model is created in the form of objects (e.g. description of currencies/securities), events (e.g. investors' activities) and connections existing between them (e.g. "investors are panicking so prices of currencies/securities have started falling"). In the next step, the current situational model is sent to the global workspace, and specific patterns of action are selected from the procedural memory (e.g. "Buy", "Sell"). A selected action is sent to the sensory-motor memory in which algorithms of action are started (e.g. sending an order for opening/closing a given item on the currency market).

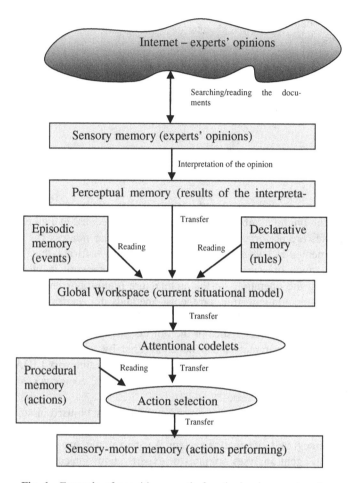

Fig. 1. Example of cognitive agent's functioning in an enterprise.

The LIDA cognitive agents can be used for performing all of the sub-processes of knowledge management (Fig. 2).

Sub-processes of identification and acquisition are mainly related to data and information– they consist of recording essential phenomena taking place around an organization in the sensory memory of an agent. The sub-process of knowledge creation starts the moment the perceptive memory is activated, and it consists of recognizing phenomena, distinguishing their traits (features, attributes) and determining relations (dependencies) between the phenomena or attributes, and rules describing the dependencies. In this way the sub-processes of creating, organizing and storing are realized. While the sub-process of storing continues, the sub-process of use starts the moment an agent has generated the current situational model of the environment, recorded in the global working memory, and in the module of current consciousness, or an agent's awareness and understanding of current phenomena taking place in the environment of a company. Knowledge may be used by selecting possible patterns

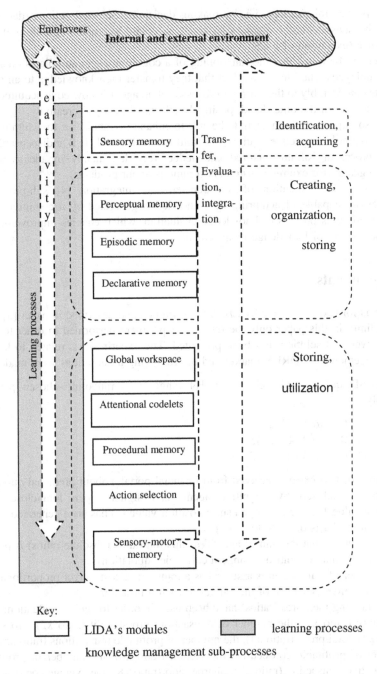

Fig. 2. Knowledge management process performed by LIDA agent.

from the procedural memory, i.e. by making decisions. Selected actions (decisions) activate the sensory-motor memory in order to create a proper algorithm of their performance (execution of a decision).

It needs to be noted that the functioning of a cognitive agent stimulates creativity among employees who, due to the fact that they transfer their knowledge to an agent, contribute considerably to the learning processes of an agent. Knowledge acquired by a cognitive agent, on the other hand, positively affects employees' creativity.

It is also important for an agent to be able to enhance symbols, i.e. to assign natural language symbols respective equivalents from the real world. It is indispensable for accurate processing of unstructured knowledge, recorded mainly by means of the natural language (for example customers' opinions about products).

Thanks to the execution of the transfer and integration [4] sub-processes, employees are capable of acquiring new knowledge generated by a cognitive agent along a continuous process of knowledge evaluation, which enables improvement of the whole process of knowledge management.

4   Experiments

The method for text processing by LIDA has been detailed described in paper [3]. Due to pages limit, in this paper only the research experiment performed in order to verification developed solution, has been presented. The experiment is related to knowledge management on FOREX market and the following assumptions were made:

1. EUR/USD quotes were selected in H4 range from randomly chosen periods, notably:

   - 01-09-2015 to 12-09-2015,
   - 13-09-2015 to 25-09-2015,
   - 26-09-2015 to 09-10-2015.

2. Opinions of two expert (received from financial portals) about predicted quotation has been analyzed by cognitive agent, and signals open long/close short position-value 1, close long/open short position-value -1) has been generated by the agent on the basis of considered opinions.
3. It was assumed that the unit of agents' evaluation ratios (absolute ratios) is pips (a pip is a minimal amount of change a currencies quotation).
4. The number of transaction is assumed as a transaction cost (direct proportionality).
5. The trader invests 100 % of the capital in each transaction.
6. The following measures (ratios) have been used in order to agents evaluation: Rate of Return (ratio x1), the number of transaction, gross profit (ratio x2), gross loss (ratio x3), total profit (ratio x4), the number of profitable transactions (ratio x5), the number of profitable consecutive transaction (ratio x6), the number of profitable consecutive transaction (ratio x7), Sharpe ratio (ratio x8), the average coefficient of volatility (ratio x9).
7. In order to compare achieved results, the following evaluation function has been used [17]:

Table 1. Experimental results.

Ratio	Expert 1			Expert 2			B&H		
	Period 1	Period 2	Period 3	Period 1	Period 2	Period 3	Period 1	Period 2	Period 3
Rate of return [Pips]	−650	19200	2320	220	−9240	3100	−220	7300	2950
The number of transactions	10	18	20	5	5	7	1	1	1
Gross profit [Pips]	2090	8840	2390	1580	1450	3580	0	7300	2950
Gross loss [Pips]	1800	−2030	−1090	1400	−1750	−1200	−220	0	0
Total profit [Pips]	1860	24850	4870	3200	1000	4850	0	0	0
The number of profit. trans.	5	15	14	3	2	6	0	1	1
The number of profitable consecutive transaction	2	6	6	2	1	5	0	1	1
The number of unprofitable consecutive transaction	2	1	3	2	1	1	1	0	0
Sharpe ratio	−2,32	2,75	13,25	15,69	−1,02	3,14	0,00	0,00	0,00
The average coefficient of volatility [%]	2,80	69,73	1,75	0,14	90,70	9,87	0,00	0,00	0,00
The average rate of return per transaction	−65,00	1066,67	116,00	44,00	−1848,0	442,86	−220,00	7300,00	2950,00
Value of evaluation function (y)	0,32	0,69	0,33	0,12	0,31	0,41	0,11	0,34	0,38

$$y = (a_1x_1 + a_2x_2 + a_3(1 - x_3) + a_4x_4 + a_5x_5 + a_6x_6 + + a_7(1 - x_7) + a_8x_8 \\ + a_9(1 - x_9) + a_{10}x_{10})$$

where x_i denotes the normalized values of evaluation ratios from x_1 to x_{10}. In this experiment the values of coefficients a_1 to a_{10} equal 1/10.

The domain of the function is in the range [0..1], and agents' range is directly proportional to the function value.

The results of agents' decisions were compared with the decisions generated by the Buy-and-Hold benchmark (the trader make buy decision on the beginning of considered period and sell decision on the end of considered period). The results of the research experiment have been presented in Table 1.

The values of evaluation ratios are differ in each period. Taking into consideration particular agents' decisions, the values of such ratios as Gross profit and the number of profitable consecutive transaction are approximative, however the values of rate of return, gross loss, Sharpe ratio and the average rate of return per transaction are characterized by high volatility. It may also be noticed that in case of the Expert 1 and Expert 2, the values of ratios have shown variability in particular periods. Taking into consideration value of evaluation function, a ranking of Experts' differs in particular periods. In the first period the Expert 1 was the best and the Expert 2 was ranked higher than the Buy-and-Hold. In the second period the Expert 1 was ranked higher and Expert 2 was ranked lower than B&H. The Expert 2 was ranked higher and Expert 1 was ranked lower than Buy-and-Hold in the third period.

Considering value of evaluation function in relation to all the periods, the Expert 1 was ranked highest most often (2 out of 3 periods) despite the fact, that it's rate of return was not always the highest. Often the risk level is very important for traders. On the basis of results of research experiment it can be state, that low level of risk has been achieved by Expert 1 and Expert 2.

The results of research experiment allow also to drawing conclusion that although different values of evaluation function, the cognitive agents can be used for support decision making process.

5   Conclusions

The cognitive agents can be used for unstructured knowledge management in a business organization's integrated information system. Thereby activities related to the process can proceed in a manner similar to human behavior. Cognitive agents allow conducting depth analysis of information, drawing conclusions and taking certain actions. These properties enable the organization to achieve competitive advantage through efficient management of their knowledge and reduce the costs of the company by making fast and accurate decisions also at the strategic level, and also save the time required for all activities, which previously have been operated by human.

Further research shall focus, among other things, on enlarging set of experts involved in the experiments and using longer time periods. Also verification cognitive agents performing in relation to unstructured knowledge management in other areas, for example manufacturing, logistics or controlling, is planned.

Acknowledgement. This research was financially supported by the National Science Center (decision No. DEC-2013/11/D/HS4/04096).

References

1. Franklin, S., Patterson, F.G.: The LIDA architecture: adding new modes of learning to an intelligent, autonomous, software agent. In: Proceedings of the International Conference on Integrated Design and Process Technology, Society for Design and Process Science, San Diego, CA (2006)
2. Hecker, A.: Knowledge beyond the individual? making sense of a notion of collective knowledge in organization theory. Organ. Stud. **33**(3), 423–445 (2012)
3. Hernes, M.: Performance evaluation of the customer relationship management agent's in a cognitive integrated management support system. In: Nguyen, N.T. (ed.) Transactions on CCI XVIII. LNCS, vol. 9240, pp. 86–104. Springer, Heidelberg (2015)
4. Maleszka, M., Nguyen, N.T.: Integration computing and collective intelligence. Expert Syst. Appl. **42**(1), 358–378 (2015)
5. Nguyen, N.T.: Inconsistency of knowledge and collective intelligence. Cybern. Syst. **39**(6), 542–562 (2008)
6. Pham, L.V., Pham, S.B.: Information extraction for Vietnamese real estate advertisements. In: Fourth International Conference on Knowledge and Systems Engineering (KSE), Danang (2012)
7. Sebastiani, F.: Machine learning in automated text categorization. ACM Comput. Surv. (CSUR) **34**(1), 1–47 (2002)
8. Tomassen, S.L.: Semi-automatic generation of ontologies for knowledge-intensive CBR. Norwegian University of Science and Technology (2002)
9. Tran, C.: Cognitive information processing. Vietnam J. Comput. Sci. **1**, 207–218 (2014)
10. Trandabăţ, D. Using semantic roles to improve summaries. In: Proceedings of the 13th European Workshop on Natural Language Generation (ENLG 2011), Association for Computational Linguistics, Stroudsburg, PA, USA, pp. 164–169 (2011)
11. Vlas, R.E., Robinson, W.N.: Two rule-based natural language strategies for requirements discovery and classification in open source software development projects. J. Manage. Inf. Syst. **28**(4), 11–38 (2012)
12. Wawer, A.: Mining opinion attributes from texts using multiple kernel learning. In: IEEE 11th International Conference on Data Mining Workshops (2011)
13. Owoc, M., Weichbroth, P.: Validation model for discovered web user navigation patterns. In: Mercier-Laurent, E., Boulanger, D. (eds.) AI4KM 2012. IFIP AICT, vol. 422, pp. 38–52. Springer, Heidelberg (2014)
14. Wilson, T., Wiebe, J., Hoffmann, P.: Recognizing contextual polarity: An exploration of features for phrase-level sentiment analysis. Comput. linguist. **35**(3), 399–433 (2009)
15. Zhang, C., Zhang, X., Jiang, W., Shen, Q., Zhang, S.: Rule-based extraction of spatial relations in natural language text. In: International Conference on Computational Intelligence and Software Engineering (2009)
16. Duch, W., Oentaryo, R.J, Pasquier, M.: Cognitive architectures: where do we go from here? In: Wang, P., Goertzel, P., Franklin, S. (eds.) Frontiers in Artificial Intelligence and Applications, Vol. 171, pp. 122–136. IOS Press (2008)
17. Korczak, J., Hernes, M., Bac, M.: Risk avoiding strategy in multi-agent trading system. In: Proceedings of Federated Conference Computer Science and Information Systems (FedCSIS), pp. 1119–1126 (2013)

A Norm Assimilation Approach
for Multi-agent Systems in Heterogeneous
Communities

Moamin A. Mahmoud[✉], Mohd Sharifuddin Ahmad,
and Mohd Zaliman M. Yusoff

Center for Agent Technology, College of Information Technology,
Universiti Tenaga Nasional, Kajang, Selangor, Malaysia
{moamin,sharif,zaliman}@uniten.edu.my

Abstract. In a heterogeneous community, which constitutes a number of social groups that adopt different social norms, norm assimilation is considered as the main problem for a new member to join a desired social group. Studies in norm assimilation seem to be lacking in concept and theory within this research domain. Consequently, this paper proposes a norm assimilation approach, in which a new agent attempts to join a social group via assimilating with the social group's norms. Several cases are considered for an agent's decision which are, can assimilate, could assimilate, and cannot assimilate. We develop the norm assimilation approach based on the agent's internal belief about its ability and its external belief about the assimilation cost of a number of social groups. From its beliefs about its ability and assimilation cost, it is able to decide whether to proceed or decline the assimilation with a specific social group or join another group.

Keywords: Social norm · Normative agent systems · Norm assimilation · Heterogeneous community

1 Introduction

While empirical research on norms have been the subject of interest for many researchers [1–4], norm assimilation has not been formally discussed. Crudely put, norm assimilation is the process of joining and abiding by the rules and norms of a social group [5, 6].

The problems of norm assimilation are attributed by the ability and capacity of an agent to assimilate in a heterogeneous community, which entails a number of social groups that adopt different social norms (in compliance and violation) and the motivation required for the agent to assimilate with a better-off group [5]. Thus, agents in heterogeneous communities do not join other social groups randomly, but their decisions are built upon their ability to assimilate the norms of a desired social group. Agents should be able to check whether they can cover the cost of joining a desired social group.

This work aims to answer the question, how is norm assimilation practiced by agents in heterogeneous communities? The goal of norm assimilation can be achieved

© Springer-Verlag Berlin Heidelberg 2016
N.T. Nguyen et al. (Eds.): ACIIDS 2016, Part I, LNAI 9621, pp. 354–363, 2016.
DOI: 10.1007/978-3-662-49381-6_34

based on the social theory of assimilation that has been developed by Eguia [5], in which the decision to assimilate is influenced by two main elements which are the cost of assimilation and the ability of agents.

2   Related Work

The literature in normative agent domain proposes several phases for norm life cycle which are norm creation, norm emergence, norm detection or identification, norm internalization, and norm removal [1, 2, 7]. We argue that the norm assimilation phase should be included as a major phase and located between norm detection and norm internalization. When an agent detects the norms of a particular social group, it should decide whether to proceed with or abstain from assimilating with that group before internalizing the detected norms.

We distinguish the difference between norm internalization and norm assimilation as follows. Conte et al. [8] defined norm internalization as a mental process, in other words, it is an internal process inside the agent's mind. We suggest that norm assimilation is an external process between an agent and a social group. The related research in this area are few and mainly limited to those proposed by social studies' researchers such as Eguia [5] and Konya [9].

Eguia [5] proposed a theory in norm assimilation, in which there are two types of agents; advantaged agents and disadvantaged agents and there are also two types of groups; better-off group and worse-off group. Any disadvantaged agents can choose to join the worse-off group without cost, or it can learn to enhance its ability to be able to assimilate with the better-off group but the enhancement is costly. He found that advantaged agents screen those who want to assimilate by imposing a difficult assimilation process such that the agents who assimilate are those whose abilities are sufficiently high so that they generate a positive externality of the group. Members of the relatively worse-off group face an incentive to adopt the social norms of the better-off group and assimilate with it. The cost of assimilation is chosen by the better-off group to screen those who wish to assimilate.

Studies on norm assimilation are not conducted within the domains of normative multi-agent systems. Instead, they are developed based on social science's perspective. Consequently, this work attempts to exploit the related work to build an assimilation approach within the framework of normative multi-agent systems and propose it as a new major normative multi-agent process.

3   The Assimilation Approach

To present the development of an assimilation approach for multi agent systems, we begin with two pertinent topics, which are the assimilation model, and the influential elements on assimilation.

We develop an assimilation model based on an agent's internal belief about its ability and its external belief about the cost of assimilation with a specific social group. As shown in Fig. 1, while the agent has its internal belief about its ability,

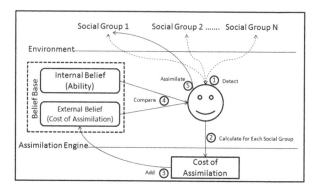

Fig. 1. The assimilation model

it detects the various social groups' norms and computes the cost of assimilation for each group. Based on its ability and the cost, it then decides which social groups it should assimilate with.

From Fig. 1, there is an agent and a number of social groups. (1) The agent first detects the norms of these groups. Then, (2) it calculates the cost of assimilation with each social group based on their enacted norms. (3) The agent then adds the various costs of these groups to its external belief. (4) From its internal belief that contains its ability and its external belief that contains the cost of assimilation with these social groups, (5) it decides which group to assimilate with, and in Fig. 1, we assume that the agent chooses the Social Group 1. However, norm detection is beyond the scope of this paper, we therefore assume that agents are able to detect the social groups' norms.

According to Eguia [5], the decision to assimilate is influenced by two elements, one belongs to the agent's internal belief, which is the ability, and the other belongs to its external belief, which is the assimilation cost of a specific social group. The assimilation cost consists of the maximum and threshold costs [5]. We define these elements as follows:

Definition 1: The Agent's Ability, ABL, is the qualification and competence of an agent to assimilate with the norms of a social group.

Definition 2: The Assimilation Cost, C, is the total effort and expenses incurred by an agent to assimilate with a social group. It consists of two types: the maximum and the threshold assimilation costs. In fact, the assimilation cost reflects how hard or easy for an agent to join a particular social group.

Definition 3: The Maximum Assimilation Cost, C_m, is the highest cost imposed by a social group for assimilation. Any agent which is able to meet this cost is considered as optimal because it has the competence to practice all the required norms of a group.

Definition 4: The Threshold Assimilation Cost, C_t, is the minimum acceptable cost to assimilate with a group. The threshold cost differs from one group to another.

Based on these elements, agents or social groups can decide to accept or reject any assimilation. There are three cases to consider, two of which are favorable to an agent for assimilation.

Case 1: The first case is when the value of an agent's ability is greater than the value of a social group's, δ, threshold assimilation cost. We say that the agent, α, can assimilate with the social group, δ.

$$\text{ABL}(\alpha) = C_m(\delta) \vee \text{ABL}(\alpha) > C_t(\delta) \Rightarrow \text{can(assimilate}\,(\alpha, \delta))$$

Case 2: The second case is when the value of the agent's ability is equal to the value of threshold assimilation cost. We say that the agent, α, could assimilate with δ,

$$\text{ABL}(\alpha) = C_t(\delta) \Rightarrow \text{could(assimilate}\,(\alpha, \delta))$$

Case 3: The third case is when the value of the threshold assimilation cost is greater than the agent's ability. In this case, we say that the agent, α, cannot assimilate with δ,

$$\text{ABL}(\alpha) < C_t(\delta) \Rightarrow \text{cannot(assimilate}\,(\alpha, \delta))$$

4 Computing Ability, Maximum Cost, and Threshold Cost

After we defined and explained the possible decision cases based on the elements of ability, ABL; maximum cost, C_m; and threshold cost, C_t; we show here our methods in computing the values of ABL, C_m, and C_t.

Since the elements of ability, maximum cost, and threshold cost deal with norms, the methods begin from exploiting the types of norms. There are three types of norms that regulate an agent's behaviour which are convention, obligation, and prohibition [10–13]. Thus, a set of norms for a social group is categorized into these types.

Next, we discuss further each norm type (convention, obligation, prohibition norms) and its property as well as its effect on ABL, C_m, and C_t. But initially, we define some terms that are pertinent to our discussion on norms types.

Definition 5: A Positive Effect Norm (Reward or Non-Penalty Type), n^+, is a type of norm that brings reward or avoids penalty if an agent practises it such as obligation-type norms.

Definition 6: A Negative Effect Norm (Penalty Type), n^-, is a type of norm that brings penalty if an agent practises it such as prohibition type of norms.

Definition 7: A Weight Parameter, WT, relies on the norm's type and norm's strength, ST, of a social group. The Norm's Strength refers to the number of agents that practises a specific norm in a social group. The weight parameter can have values of 1, 0, or -1 depending on the norm's type, n^+ or n^-, and the norm's strength within the range $0 \leq ST \leq 1$. According to Konya [9], the majority is represented by ≥ 0.5 and minority by < 0.5 of population.

$$WT = \begin{cases} 1 & ST(n \in n^+) \geq 0.5 \\ 0 & ST(n \in n^+) < 0.5 \end{cases} \tag{1}$$

This means that when the norm, n, belongs to n^+ (a reward or non-penalty type) and its strength, ST, is greater than or equal to 0.5 of population, then its Weight equals 1. But when its strength is less than 0.5 of population, then its Weight equals 0. Similarly, we define the weight of penalty type norms as follows:

$$WT = \begin{cases} 0 & ST(n \in n^-) \leq 0.5 \\ -1 & ST(n \in n^-) > 0.5 \end{cases} \tag{2}$$

which means that when the norm, n, belongs to n^- (a penalty type) and its strength is less than or equal 0.5 then its Weight equals 0. But when its strength is more than 0.5, then its Weight equals −1.

Definition 8: The Ability Parameter, ABL, depends on the ability of an agent, α. The parameter can have values of 1, 0, or -1 based on whether the agent adopts the positive effect norms, or avoids negative effect norms.

$$ABL = \begin{cases} 1 & \text{adopt } (\alpha, (n \in n^+)) \\ 0 & \neg \text{adopt } (\alpha, (n \in n^+)) \end{cases} \tag{3}$$

This means that when the norm, n, belongs to n^+ and an agent, α, adopts it, then its Ability on assimilating n equals 1. But, when the norm, n, belongs to n^+ and an agent, α, does not adopt it, then its Ability on assimilating n equals 0. Similarly, for negative effect type norms:

$$ABL = \begin{cases} 0 & \neg \text{adopt } (\alpha, (n \in n^-)) \\ -1 & \text{adopt } (\alpha, (n \in n^-)) \end{cases} \tag{4}$$

which means that when the norm, n, belongs to n^- and an agent, α, does not adopt it, then its Ability on assimilating n equals 0. But, when the norm, n, belongs to n^- and an agent, α, adopts it, then its Ability on assimilating n equals -1.

Definition 9: Conventions, N^{cnv}, are the type of norms that are adopted by every member of a social group. Any agent joining a social group must be able to assimilate this type of norm. Consequently, conventions belong to the positive norm's type, and it is always adopted by the majority of population, thus, having the norm strength of > 0.5 implying that its weight parameter value equals 1.

Definition 10: Obligation Norms, N^{obl}, are the type of norms that avoids penalty when an agent adopts them. A social group's adoption of this norm type is based on the extent of applying penalty on violators. If penalties are often applied, very high population of group members adopt these norms. Since an unassimilated obligation norm causes a penalty, it therefore belongs to the positive norms. The norm's Weight either equals 1 when it is adopted by a majority or equals 0 otherwise.

Definition 11: Prohibition Norms, N^{phb}, are the type of norms that causes penalty when an agent adopts them. A social group's adoption of this norm type is based on the extent of applying penalty on violators. If penalties are often applied, very low population of group members adopt these norms. Since an assimilated prohibition norm causes a penalty, it therefore belongs to the negative norms. The norm's Weight either equals -1 when it is adopted by a majority or equals 0 otherwise.

From the above arguments, we evaluate the different types of norms that have different weight parameter values based on their norm's strength.

Computing the Maximum Assimilation Cost, Cm: Here we define a formula for computing the maximum assimilation cost, C_m. Since C_m deals with the optimal assimilation, the formula of C_m must include the positive type norms (convention, obligation) and exclude the negative type (prohibition) norms. If N is a set of norms that is adopted by a society, δ, then,

$$N = \{N^{cnv} \cup N^{obl} \cup N^{phb}\} \tag{5}$$

This means that the set of norms, N, equals the union of all norms types sets (convention, obligation, prohibition).

Since the maximum cost deals with convention and obligation norms and exclude prohibition norms, then the maximum cost,

$$C_m = N \backslash \{N^{phb}\} \tag{6}$$

This means the maximum cost equals the set of convention and obligation norms only.

Let N^{cnv} be a set of convention norms, n_k^{cnv}; N^{obl} be a set of obligation norms, n_k^{obl}; and N^{phb} be a set of prohibition norms, n_k^{phb}, where $k = 1, 2, \ldots n$. Accordingly, we redefine N and C_m,

$$N = \bigcup n_k^{cnv}, n_k^{obl}, n_k^{phb} \tag{7}$$

$$C_m = \bigcup n_k^{cnv}, n_k^{obl} \tag{8}$$

Computing the Threshold Cost, Ct: Since the threshold cost, C_t represents the minimum assimilation cost, we include the weight parameter value within the formula of the threshold cost. The weight for obligation norms equals 1 when $ST \geq 0.5$, or 0 when $ST < 0.5$. Therefore, the norm's strength, ST, is the prominent parameter which has a direct effect on the weight. For example, smoking is not a norm in many social groups but due to a large number of members who smoke (i.e. high norm's strength), an agent who smokes and would like to join the group is able to assimilate quite easily because the assimilation cost is low.

Norm assimilation entails a social group's commitment with its norms. If the members are very committed, the assimilation cost is high, only new high quality members join them [1]. But if they are not committed and often violate their norms, they are bound to accept new low quality members. Consequently, the threshold cost is based on the norms' weights of a particular social group. In other words, the threshold cost equals the set of social group norms excluding the norms of weight equal 0 because their strengths are less than 0.5.

$$C_t = N \setminus \{WT(n) \in N : WT = 0\} \tag{9}$$

From Eq. 7,

$$C_t = \bigcup n_k^{cnv}, n_k^{obl}, n_k^{phb} \setminus \{WT(n_k) \in N : WT = 0\} \tag{10}$$

The equation means that the threshold cost equals the union of all norms types but excluding the norms whose weight parameter, WT, equals 0.

Computing the Agent's Ability, ABL: The ability, ABL, computation is based on the formula of threshold assimilation cost and maximum assimilation cost. After an agent detects a social group's norms and calculates the maximum and threshold costs, it then calculates its ability based on what convention and obligation norms to assimilate and what prohibition norms to avoid. Consequently,

$$ABL = N \setminus \{ABL(n) \in N : ABL = 0\} \tag{11}$$

From Eq. 7,

$$ABL = \bigcup n_k^{cnv}, n_k^{obl}, n_k^{phb} \setminus \{ABL(n_k) \in N : ABL = 0\} \tag{12}$$

5   An Example to of Norm Assimilation

In this example, we assume that there are three cafés which represent three social groups $(\delta_1, \delta_2, \delta_3)$ of similar norms, but are different in the range of compliance with their norms. Let us assume that the social norms (sit, order, smoke, litter, pay, tipping) are classified as follows:

Conventions:	{sit, order, pay}
Obligation Norms:	{tipping}
Prohibition Norms:	{smoke}

We also assume that the norms' strengths, ST, for the norms of each of the social group are as follows (note that a norm's strength of 0.7 means that 70 % of the population enact the norm),

From Eqs. 1 and 2, the weight of each norm can be identified.

Table 1. The assumed values of ST for δ_1, δ_2, δ_3

Norm	δ_1	δ_2	δ_3
	ST	ST	ST
sit	0.95	0.97	0.96
order	0.97	0.96	0.977
pay	0.99	0.98	0.98
tipping	0.85	0.20	0.02
smoke	0.03	0.08	0.80
litter	0.05	0.10	0.10

Table 2. WT of each norm for each of δ_1, δ_2, δ_3

Norm	δ_1	δ_2	δ_3
	WT	WT	WT
sit	1	1	1
order	1	1	1
pay	1	1	1
tipping	1	0	0
smoke	0	0	-1
litter	0	0	0

We assume that there are three agents α_1, α_2, α_3, calculating the cost of each social group to eventually assimilate with the appropriate one. However, all of them desire the more advantaged group. Based on Eqs. 8 and 10, the agents compute the maximum cost, C_m, and the threshold cost, C_t, for all social groups (δ_1, δ_2, δ_3). The maximum cost of all social groups is the same because they are practicing similar norms. From Eq. 8 and Table 1,

$$C_m = \cup n_k^{cnv}, n_k^{obl} \Rightarrow C_m(\delta_1, \delta_2, \delta_3) = \{sit, \ order, \ pay, \ tipping\}$$

But the threshold cost, C_t, is different from one social group to another. From Eq. 10 and Tables 1, 2, 3;

$$C_t = \bigcup n_k^{cnv}, \ n_k^{obl}, \ n_k^{phb} \setminus \{WT(n_k) \in N : WT = 0\}$$

$C_t(\delta_1) = \{sit, \ order, \ pay, \ tipping\};$ $C_t(\delta_2) = \{sit, \ order, \ pay\};$ $C_t(\delta_3) = \{sit, \ order, \ pay, \ smoke\}$

Let us assume that the ability parameter values for the three agents are as follows: According to Eq. 12 and Table 3, the ability of α_1, α_2, α_3;

$$ABL = \bigcup n_k^{cnv}, \ n_k^{obl}, \ n_k^{phb} \setminus \{ABL(n_k) \in N : ABL = 0\}$$

$ABL(\alpha_1) = \{sit, \ order, \ pay, \ tipping\};$ $ABL(\alpha_2) = \{sit, \ order, \ pay\};$ $ABL(\alpha_3) = \{sit, \ order, \ pay, \ smoke\}$

Table 3. The assumed values of ABL for α_1, α_2, α_3,

Norm	α_1	α_2	α_3
	ABL	ABL	ABL
sit	1	1	1
order	1	1	1
pay	1	1	1
tipping	1	0	1
smoke	0	0	-1
litter	0	0	0

Table 4 exemplifies the agents ability and the assimilation costs of the social groups. It also presents the results, reasons, comments and decision of each agent on each social group.

Table 4. The decision on assimilation of each agent towards the three social groups

Agent	ABL	Society	Cost		Result	Reason	Comments	Decision
			Maximum Cost, C_m	Threshold cost, C_t				
α_1	{sit, order, pay, tipping}	δ_1	{sit, order, pay, tipping}	{sit, order, pay, tipping}	case 1:can assimilate	$ABL(\alpha_1) = C_m(\delta_1)$	α_1 ability equals to the maximum assimilation cost of all societies	assimilates (α_1, δ_1) As δ_1 is more advantaged than δ_2, and δ_3.
		δ_2	{sit, order, pay, tipping}	{sit, order, pay}	case 1:can assimilate	$ABL(\alpha_1) = C_m(\delta_2)$		
		δ_3	{sit, order, pay, tipping}	{sit, order, pay, smoke}	case 1:can assimilate	$ABL(\alpha_1) = C_m(\delta_3)$		
α_2	{sit, order, pay}	δ_1	{sit, order, pay, tipping}	{sit, order, pay, tipping}	case 3: cannot assimilate	$ABL(\alpha_2) < C_t(\delta_1)$	No tipping within the ability of α_2	assimilates (α_2, δ_2) As δ_2 is more advantaged than δ_1.
		δ_2	{sit, order, pay, tipping}	{sit, order, pay}	case 2: could assimilate	$ABL(\alpha_2) = C_t(\delta_2)$	The agent ability set equals to the threshold assimilation cost of δ_2	
		δ_3	{sit, order, pay, tipping}	{sit, order, pay, smoke}	case 1:can assimilate	$ABL(\alpha_2) > C_t(\delta_3)$	can avoid prohibited norm (smoke)	
α_3	{sit, order, pay, smoke}	δ_1	{sit, order, pay, tipping}	{sit, order, pay, tipping}	case 3: cannot assimilate	$ABL(\alpha_3) < C_t(\delta_1)$	No tipping norm and the ability involves the prohibited norm (smoke)	assimilates (α_3, δ_3)
		δ_2	{sit, order, pay, tipping}	{sit, order, pay}	case 3: cannot assimilate	$ABL(\alpha_3) < C_t(\delta_2)$	the ability involves the prohibited norm (smoke)	
		δ_3	{sit, order, pay, tipping}	{sit, order, pay, smoke}	case 2: could assimilate	$ABL(\alpha_3) = C_t(\delta_3)$	The agent ability equals to the threshold assimilation cost of δ_3	

6   Conclusion and Further Work

In this paper, we present a novel theory on norm assimilation in a heterogeneous community where there are a number of social groups practicing different norms. Any agent, which would like to join a social group, has to be able to assimilate with its norms. The suggested assimilation approach is based on the agent's internal and external beliefs. The internal belief represents the agent's ability and the external belief represents the assimilation cost of a social group.

For our future work, we shall study the issue of norm assimilation based on morality of norms and an agent emotional state towards a particular group. In this work, we compute the cost of assimilation based on the ability of the agent to determine if it can or cannot assimilate.

References

1. Hollander, C., Wu, A.: The current state of normative agent-based systems. J. Artif. Soc. Soc. Simul. **14**(2), 6 (2011)
2. Savarimuthu, B.T.R.: Mechanisms for norm emergence and norm identification in multi-agent societies (Thesis, Doctor of Philosophy). University of Otago (2011)
3. Mahmoud, M.A., Mustapha, A., Ahmad, M.S., Ahmad, A., Yusoff, M.Z.M., Hamid, N.H.A.: Potential norms detection in social agent societies. In: Omatu, S., Neves, J., Rodriguez, J.M.C., Santana, J.F.P., Gonzalez, S.R. (eds.) Distributed Computing and Artificial Intelligence. AISC, vol. 217, pp. 419–428. Springer, Heidelberg (2013)
4. Mahmoud, M.A., Ahmad, M.S., Ahmad, A., Mustapha, A., Yusoff, M.Z.M., Hamid, N.H.A.: Building norms-adaptable agents from potential norms detection technique (PNDT). Int. J. Intell. Inf. Technol. (IJIIT) **9**(3), 38–60 (2013)
5. Eguia, J.X.: A theory of discrimination and assimilation. New York University (2011)
6. Mahmoud, M.A., Ahmad, M.S., Ahmad, A., Yusoff, M.Z.M., Mustapha, A.: Norms detection and assimilation in multi-agent systems: a conceptual approach. In: Lukose, D., Ahmad, A.R., Suliman, A. (eds.) KTW 2011. CCIS, vol. 295, pp. 226–233. Springer, Heidelberg (2012)
7. Mahmoud, M.A., Ahmad, M.S., Yusoff, M.Z.M., Mustapha, A.: A review of norms and normative multiagent systems. Sci. World J. **2014**, 23 (2014). doi:10.1155/2014/684587. Article ID 684587
8. Conte, R., Andrighetto, G., Campenni, M.: Internalizing norms: a cognitive model of (social) norms' internalization. Int. J. Agent Technol. Syst. (IJATS) **2**(1), 63–73 (2010)
9. Konya, I.: A dynamic model of cultural assimilation. Boston College Working Papers in Economics 546 (2002)
10. Mahmoud, M.A., Ahmad, M.S., Yusoff, M.Z.M., Mustapha, A.: Norms assimilation in heterogeneous agent community. In: Dam, H.K., Pitt, J., Xu, Y., Governatori, G., Ito, T. (eds.) PRIMA 2014. LNCS, vol. 8861, pp. 311–318. Springer, Heidelberg (2014)
11. Boella, G., Torre, L.V.D.: Substantive and procedural norms in normative multiagent systems. J. Appl. Log. **6**(2), 152–171 (2008)
12. Mahmoud, M.A., Ahmad, M.S., Ahmad, A., Yusoff, M.Z.M., Mustapha, A., Hamid, N.H.A.: Obligation and prohibition norms mining algorithm for normative multi-agent systems. In: KES-AMSTA, pp. 115–124, May 2013
13. Mahmoud, M.A., Ahmad, M.S., Ahmad, A., Mohd Yusoff, M. Z., Mustapha, A.: A norms mining approach to norms detection in multi-agent systems. In: 2012 International Conference on Computer & Information Science (ICCIS), vol. 1, pp. 458–463. IEEE, June 2012

Knowledge in Asynchronous Social Group Communication

Marcin Maleszka[✉]

Faculty of Computer Science and Management, Wroclaw University of Technology,
Wyb. Wyspianskiego 27, 50-370 Wroclaw, Poland
Marcin.Maleszka@pwr.edu.pl

Abstract. Multi-agent systems are one of many modern distributed approaches to decision, optimization and other problem solving. Among others, multi-agent systems have been often used for prediction, but those approaches require a supervisor agent for integrating the knowledge of other agents. In this paper we discuss the shortcomings of such approach and propose a switch to decentralized groups of agents with asynchronous communications. We show that this approach may obtain similar results, while avoiding the pitfalls of centralized architecture.

Keywords: Knowledge integration · Multiagent system · Asynchronous communication

1 Introduction

Knowledge integration has been discussed in many theoretical works and used in some practical applications. It is often treated as an instantaneous calculation that occurs in some place in a larger system. In many situations this may be the case, but the motivation for this paper is the question whether this is a valid assumption.

Many multi-agent systems use a special supervisor agent, which only exists to integrate knowledge from other agents. This single agent becomes the most critical in the entire system, as without it, no result can be determined. The agent may be taken offline by some external factor, but also by simple communication overload. This may occur e.g. when all other agents try to communicate with it at the same time. Theoretical papers do not approach this subject, as the experiments are conducted only on small number of agents. Practical applications deal with the problem by optimizing the communication protocols and upgrading the hardware, but this may still lead to limited scalability options.

At the same time social network analysis in many research papers has lead to many interesting results, due in much part to analysis on large number of data. Most of those are decentralized structures, that at most have some additional stronger ties between select nodes (*friend* relationship). With many real social groups working in an decentralized manner, the centralized approach of multi-agent system is somewhat inadequate.

© Springer-Verlag Berlin Heidelberg 2016
N.T. Nguyen et al. (Eds.): ACIIDS 2016, Part I, LNAI 9621, pp. 364–373, 2016.
DOI: 10.1007/978-3-662-49381-6_35

In our research we focused on developing a fully decentralized multi-agent system for knowledge integration and dissemination. This required the adoption of asynchronous communication - at no points a single agent knows if his communicates will be received and taken into account. We developed a simulation environment to compare such system with a centralized one and conducted some basic experiments in it. We also adapted one of our previous centralized multi-agent systems to a new approach and observed its behaviour in a practical application.

This paper is organized as follows: in Sect. 2 we discuss a selection of other research done on centralized and decentralized multiagent systems and social group communications; in Sect. 3 we present the basics of our approach to decentralized multi-agent system; in Sect. 4 we simulate a large scale multi-agent system and in Sect. 5 we apply our observation to an existing application; we conclude this paper with some final remarks, possible applications and future work aspects in the last Sect. 6.

2   Related Works

Multiple previous research papers considered both centralized and decentralized agent systems, but the influence of the agent relations structure on the integration of their knowledge was not taken into account.

Among others [5] presents a multi-agent decision-support system with a supervisor agent. This agent integrated investment knowledge from other agents and presents it to the user to increase the speed of decision by reducing the items required for human analysis. Other papers name the main agent a facilitator. In [13] the system deals with power system restoration. Agents may make decisions based on local information and even on information from neighbours, but only the facilitator can act as a manager in a decision process. Centralized systems are also often used in traffic control system [7,14]. Basic agents are usually controllers for some subsystems focused on a single functionality or a single geographical area, while the supervisor manages the overall system. Another situation is the use of special verification agents, where the work of a fully decentralized network of agents is observed and corrected. Such approach has been used e.g. for clustering [1].

The authors of [9] work with fully decentralized systems of either uniform or diverse agents to decide on moves in the game of Go. They prove that the group of random diverse agents using majority vote gives better results than a group of uniform agents. In [16] a decentralized system of agents is used for surveillance. Each agent combines its own observations with knowledge received from other agents. The authors discuss the limits of optimization possible by using multi-agent approach. The authors of [18] develop an approach to move from centralized to decentralized multi-agent system, but the applicability of this method to knowledge integration is negligible. The centralized and decentralized approaches are mixed in the hybrid system used in [4]. In this paper authors consider delays and asynchronous communication, but do not take into account time required for processing data. The experiment is run on a group of eight physical agents, where the communication delay plays a role, but the calculation time and correctness of results does not.

Related area of research considered social group communications. Both centralized (hierarchical employee structures) and decentralized (friend and social networks) situations were analyzed. The authors of [12] consider a group of students working on some research problems. They show that the underlying social network is not important to the final (integrated) result, unless strong ties may be determined between participants. In that case the group result is visibly improved. This shows that more structures than centralized and decentralized one may need to be considered. In [17] the authors present the model of disease outbreak based on observations of real world human interaction. They use data gathered by observing public transit and create a mathematical model of disease dissemination. In [3] a model of Twitter social network is used to predict disease outbreaks, where each user may treat his contacts as sensors.

Similar model may be used to represent knowledge dissemination. In fact, authors of [10] propose building such models and using them to improve teaching processes and in *mouth-to-mouth* marketing campaigns. One such model is presented in [2] with the focus on the importance of non-linguistic behaviors, the relative time consistency of the social roles played by a given person during the course of a meeting, and the interplays and mutual constraints among the roles enacted by the different participants in a social encounter.

Overall, it may be determined that most multi-agent systems are either centralized, or the decentralization occurs only on a small scale. The research on agents also does not take into account the important results from the area of social group communication. This constitutes parts of motivation for our research.

3   Asynchronous Group Communication

In this paper we focus on asynchronous communication in groups of social agents (people or computer programs). We differentiate this type of communication with the following assumptions:

- Each social agent starts communication in irregular intervals.
- Social agent communication is unidirectional - after agent sends a message, he does not expect a reply.
- Receiver agent may use the knowledge in the message as he sees fit. He may also discard it.
- Each agent may have a group of receivers that he communicates with more often (this assumption is for our future work).

In comparison we use the centralized organization of computer agents with a supervisor agent. The assumptions for this type of group are as follows:

- There is a single supervisor agent.
- Each agent periodically sends messages to the supervisor.
- Supervisor processes knowledge and periodically sends the integrated knowledge back to other agents.

It may be discerned that in the centralized approach the processing takes place only in one place and only after each agent has communicated with the supervisor. On the other hand in the asynchronized approach the processing may occur after each communication and the overall knowledge of the group changes constantly. In both cases the collective may achieve the same result, but while the centralized approach has one critical node, the decentralized one does not.

We will use the following notation to represent this situation:

- $A = \{a_i : i \in \{0, 1, \ldots, n\}\}$ is the set of n observation agents and, if it is necessary, one supervisor agent (denoted a_0).
- $T_{i,j}$ is the average time between communication from a_i to a_j.

In the centralized system the time before the knowledge is shared with the supervisor and received back is:

$$T_C = max_{i \in \{1, \ldots, n\}} (T_{j,0}) + max_{i \in \{1, \ldots, n\}} (T_{0,j})$$

In a fully random case of asynchronous decentralized system, the time to achieve consensus is:

$$T_D = max_{i,j \in \{1, \ldots, n\}} \left(min\{T_{i,j}, T_{i,x_1} + T_{x_1,j}, \ldots, T_{i,x_1} + \sum_{z=1}^{n-1} T_{x_z,x_z+1} + T_{x_n,j}\} \right)$$

which is the time needed for each agent to send its knowledge state to all other agents.

The worst case of the decentralized system may in fact be denoted similar to ring topology networks. Agent a_1 can only communicate to a_2 and receive from a_n, agent a_2 only to a_3 and receive from a_2, etc. In that case the time may be denoted as:

$$T_R = 2 \sum_{i=1}^{n-1} T_{i,i+1} + T_{n-1,n} + T_{n,1}$$

The information is send from some agent to the next one in the ring, which adds its own information. Once it reaches the last agent in the ring, it may conduct the integration and then send the information again to all the other agents using the ring. Thus, the ring must be navigated twice. In this case any agent is critical to achieving final consensus spread to all agents. As such this situation should be avoided.

In further parts of this paper we will focus on the centralized and fully random cases as applied to selected problems of multi-agent knowledge integration.

4   Integration in Simulated Environment

In order to study the changes in the collective knowledge during asynchronous group communication we have developed a simulation environment and agent systems working on real-world data. This section focuses on experiments in the simulated environment, while the next one focuses on using asynchronous communication in practical applications.

The simulation environment was developed in Java using JADE Agent Framework [8]. Two types of agents exist, *SocialAgents* that start with some partial knowledge and *SupervisorAgent* used in the centralized approach. *SocialAgents* have two modes of communication: to other agents of this type (asynchronous case) and to/from the supervisor agent (centralized case). Each agent has I different elements of knowledge (issues), each being an integer value between 0 and K. The number of agents may be fully randomized up to 5000, due to memory limitations of used hardware. In the asynchronous case, every t milliseconds each *SocialAgent* may initiate communication to some randomly selected other agent with probability p. During communication, the agent sends value of between 1 and I issues to the receiver. The receiver may use different strategies to deal with this new knowledge (from disregarding it, to discarding own knowledge and adopting the received one). In the centralized case the *SocialAgents* send all their issues to the supervisor, which integrates them and sends back the results. The *SocialAgents* adopt the result while discarding own knowledge states. Every few cycles of communication (once in centralized case) the states of each agent are observed for the experiment.

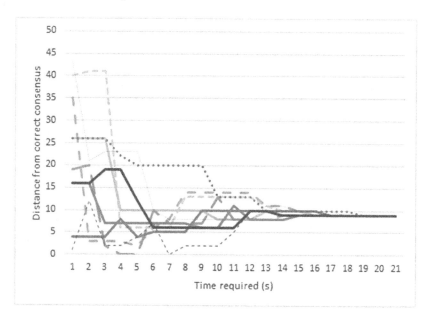

Fig. 1. Evolution of *SocialAgent* knowledge. Ten agents, single issue in range $(1-100)$, observation of agent knowledge state every second.

An example of a small simulation run for the decentralized case is shown in Fig. 1. In our experiment we wanted to observe how the centralized and decentralized systems behave for different number of agents. We conducted test runs using groups of between 200 and 5000 agents, using the same hardware and parameters. The simulation environment had insufficient hardware resources for any larger groups to be run consistently.

We observed time needed to achieve the consensus and spread it to all agents (for default value of one observation cycle = 1s) and error rate of each approach. For decentralized approach the error was the difference from the real consensus (determined in centralized case). For the centralized case the error was the number of runs, where the consensus was not calculated or not spread to all agents, due to communication problems (overload).

The results, presented in Table 1 are averaged over 20 runs. The time component is also presented in Fig. 2. As may be observed, the centralized approach starts having problems with around 1000 agents and stop working at 2000 agents. With the same communication protocols and setup, the decentralized approach continues working up to 5000 agents. Meanwhile the error rate of the asynchronous case slowly drops to 1 % or less. In both cases, the calculation time rises with the number of agents (in centralized case, due to communication queue), but the centralized approach is generally much faster.

Table 1. Time required and errors in centralized and decentralized case, for various number of agents.

No. Agents	Cycles - Centr	Cycles - Decentr	Errors - Centr	Errors - Decentr
200	1	22	0 %	6 %
400	1	27	0 %	6 %
600	2	35	0 %	4 %
800	3	46	0 %	3 %
1000	4	54	5 %	2 %
1500	9	72	5 %	1 %
2000	∞	91	100 %	1 %
2500	∞	130	100 %	1 %
3000	∞	188	100 %	1 %
4000	∞	301	100 %	1 %
5000	∞	417	100 %	1 %

5   Integration of Forecasts

Multiple multi-agent systems for forecasting by knowledge integration exist. Among others, in our previous work [11] we created a multi-agent system for weather prediction. The system operated on three levels:

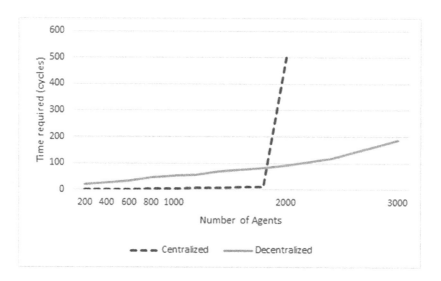

Fig. 2. Time required for integration in centralized and decentralized case, depending on the number of agents. For 2000 agents and more the centralized approach does not work (time is infinite).

- Source level - where a group of agents downloaded weather forecasts from Internet websites or created own forecasts based on historical data (using regression and other mathematical methods).
- Integration level - where a single supervisor agent integrated forecasts from the source agents using consensus theory and created a single forecast.
- Interaction level - where the single forecast was displayed to the user.

The system was robust in the sense that any source agents may drop out of the system until only one is left and the supervisor agent will still work correctly and display the weather forecast. Our experiment using real data in Wroclaw, Poland in April-May 2015 has shown that the system has not only smaller error than the worst random forecast from the sources (as predicted by consensus theory [15]), but also smaller than the average error of those sources, when using dominant value (choice theory [9]).

In the current research we redesigned this system to work without the supervisor agent. We applied the same communication protocols as in the simulated experiment, that is:

- A source agent A_1 determines its own prediction by downloading or calculating it.
- Agent A_1 at random intervals sends this prediction to another random agent A_2.
- Agent A_2 receives the data and integrates it with its own knowledge state.
- Agent A_2 at random intervals sends the corrected prediction to another agent A_3 (where A_3 may be the same as A_1).

This communication cycle continues for some time. In our test system we determined that an agent stops sending out communicates after it didn't change its own state after the previous received message. The communication starts again after the next forecasts are created (in our test system, once a day).

In the centralized system we used various approaches to data integration, based on consensus and choice theory. The only one applicable to decentralized approach is the average. We performed the experimental run, comparing the system results with real weather, for the period of several weeks in October 2015 in Wroclaw, Poland. The results are presented in Table 2. For comparison, we added results from the previous run in April-May 2015. Note that we present values for using average for integration, where dominant value gave better results.

Table 2. Centralized and decentralized weather forecast system. Prediction of a random agent was used as the overall prediction of the decentralized system.

System-run	MAE	Comp. with best source	Comp. with worst source	Comp. with avg. source
Centralized (April-May)	2,018	82%	7% better	95%
Centralized (October)	2,132	75%	2% better	90%
Decentralized (October)	1,984	83%	9% better	97%

In general the multi-agent weather prediction system generated worse results in October than in April-May. Surprisingly, the decentralized system was marginally better than the centralized one. This may be due to the forecasts in Internet sources being biased towards high or low predictions.

This short run shows that the decentralized approach may be successfully used in practical applications that were previously using a central supervisor agent.

6   Conclusions

In this paper we show the feasibility of a decentralized multi-agent system with asynchronous communication used for knowledge integration, both in a simulated environment and in a practical application. The large scale experiment in the simulated environment shows that the decentralized approach is not much worse than the centralized one and gets better with larger number of agents. The decentralized approach may also work with a much larger number of agents with identical hardware and software. The results of using a decentralized system in a practical application shows that the difference from the centralized consensus may as well be towards the real state, that the system was required to predict.

We will be continuing the research on decentralized knowledge processing multi-agent systems, focusing on practical applications of such systems. Among others, the results of [12] has shown that even in a decentralized system, some underlying *friend* relationship between agents may be used to improve the integration results. We will also address this type of more complex social groups in our future research.

Additionally, during this research we have observed that integration of large amount of knowledge takes enough time that it should be taken into consideration. Many works (e.g. [6,15]) assume that integration is instantaneous. In our future work we will analyze situation where this is not the case.

Acknowledgment. This research was co-financed by Polish Ministry of Science and Higher Education grant.

References

1. Chaimontree, S., Atkinson, K., Coenen, F.: A multi-agent based approach to clustering: harnessing the power of agents. In: Cao, L., Bazzan, A.L.C., Symeonidis, A.L., Gorodetsky, V.I., Weiss, G., Yu, P.S. (eds.) ADMI 2011. LNCS, vol. 7103, pp. 16–29. Springer, Heidelberg (2012)
2. Dong, W., Lepri, B., Pianesi, F., Pentland, A.: Modeling functional roles dynamics in small group interactions. IEEE Trans. Multimed. **15**(1), 83–95 (2013)
3. Garcia-Herranz, M., Moro, E., Cebrian, M., Christakis, N.A., Fowler, J.H.: Using friends as sensors to detect global-scale contagious outbreaks. PloS one **9**(4), e92413 (2014)
4. Hale, M.T., Nedic, A., Egerstedt, M.: Cloud-based centralized/decentralized multi-agent optimization with communication delays. arXiv preprint. (2015). arxiv:1508.06230
5. Hernes, M., Sobieska-Karpiska, J.: Application of the consensus method in a multiagent financial decision support system. Inf. Syst. e-Bus. Manage., Springer, Heidelberg (2015). doi:10.1007/s10257-015-0280-9
6. Korczak, J., Hernes, M., Bac, M.: Performance evaluation of decision-making agents in the multi-agent system. In: Proceedings of Federated Conference Computer Science and Information Systems (FedCSIS), Warszawa, pp. 1177–1184 (2014)
7. Iscaro, G., Nakamiti, G.: A supervisor agent for urban traffic monitoring. In: IEEE International Multi-Disciplinary Conference on Cognitive Methods in Situation Awareness and Decision Support (CogSIMA), pp. 167–170. IEEE (2013)
8. JADE, Java Agent Development Framework. http://jade.tilab.com/
9. Jiang, A., Marcolino, L.S., Procaccia, A.D., Sandholm, T., Shah, N., Tambe, M.: Diverse randomized agents vote to win. In: Advances in Neural Information Processing Systems, pp. 2573–2581 (2014)
10. Maleszka, M., Nguyen, N.T., Urbanek, A., Wawrzak-Chodaczek, M.: Building educational and marketing models of diffusion in knowledge and opinion transmission. In: Hwang, D., Jung, J.J., Nguyen, N.-T. (eds.) ICCCI 2014. LNCS, vol. 8733, pp. 164–174. Springer, Heidelberg (2014)
11. Mercik, J., Tolkacz, O., Wojciechowska, J., Maleszka, M.: Wykorzystanie integracji wiedzy do zwiekszenia efektywnosci prognozowania w warunkach niepewnosci. In: Porebska-Miac T. (Ed.) Systemy Wspomagania Organizacji , Wydawnictwo Uniwersytetu Ekonomicznego w Katowicach, Katowice 2015 (2015)

12. De Montjoye, Y.-A., Stopczynski, A., Shmueli, E., Pentland, A., Lehmann, S.: The Strength of the Strongest Ties in Collaborative Problem Solving. Scientific reports 4, Nature Publishing Group (2014)
13. Nagata, T., Sasaki, H.: A multi-agent approach to power system restoration. IEEE Trans. Power Syst. **17**(2), 457–462 (2002)
14. Nakamiti, G., da Silva, V.E., Ventura, J.H., da Silva, S.A.: Urban traffic control and monitoring – an approach for the brazilian intelligent cities project. In: Wang, Y., Li, T. (eds.) Practical Applications of Intelligent Systems. AISC, vol. 124, pp. 543–551. Springer, Heidelberg (2011)
15. Nguyen, N.T.: Advanced Methods for Inconsistent Knowledge Management. Advanced Information and Knowledge Processing. Springer, London (2007)
16. Peterson, C.K., Newman, A.J., Spall, J.C.: Simulation-based examination of the limits of performance for decentralized multi-agent surveillance and tracking of undersea targets. In: SPIE Defense+ Security, pp. 90910F–90910F. International Society for Optics and Photonics (2014)
17. Sun, L., Axhausen, K.W., Lee, D.H., Cebrian, M.: Efficient detection of contagious outbreaks in massive metropolitan encounter networks. Scientific reports, 4, Nature Publishing Group (2014)
18. Xuan, P., Lesser, V.: Multi-agent policies: from centralized ones to decentralized ones. In: Proceedings of the First International Joint Conference on Autonomous Agents and Multiagent Systems: part 3, pp. 1098–1105. ACM (2002)

Decision Support and Control Systems

Interpreted Petri Nets in DES Control Synthesis

František Čapkovič[(⊠)]

Institute of Informatics, Slovak Academy of Sciences, Bratislava, Slovakia
Frantisek.Capkovic@savba.sk

Abstract. Discrete event systems (DES) control based on interpreted Petri nets (IPN) is presented in this paper. While place/transition Petri nets (P/T PN) are usually used for modelling and control in case of controllable transitions and measurable places, the IPN-based models yield the possibility for the control synthesis also in case when P/T PN models contain some uncontrollable transitions and unmeasurable places. The creation of the IPN model from such a P/T PN model is introduced and the control synthesis is performed. The illustrative examples as well as the case study on a robotized assembly cell are introduced.

Keywords: Control · Discrete event systems · Modelling · Place/ transition Petri nets · Interpreted Petri nets

1 Introduction

Petri nets (PN) [1,2,6–8] are a special kind of bipartite directed graphs. Namely, they have two kinds of nodes (places and transitions) and two kinds of edges (from places to transitions and conversely). The PN structure [1] is a triple $N = \langle P, T, B \rangle$, where P is a finite set of n places and T is a finite set of m transitions. B represents the PN edges. Thus,

$$P \cup T \neq \emptyset; \quad P \cap T = \emptyset \tag{1}$$

$$B \subseteq (P \times T) \cup (T \times P) \tag{2}$$

where $B = F \cup G$ with $F \subseteq (P \times T)$ and $G \subseteq (T \times P)$.

Except the structure PN have also their *dynamics*. Thus, the complete PN definition is $PN = (N, M_0)$ with M_0 being the initial marking (represented below in (3) by the initial state vector $\mathbf{x}_0$).

A transition $t \in T$ may be fired at a marking M if it is enabled - i.e. if all of its input places have at least one token. The firing of t removes one token from each of its input places and adds one token to each of its output places. After firing t a new PN marking M' is reached. This process is expressed below by the state Eq. (3). The set $R(M_0)$ expresses all markings reachable from the initial marking M_0. The PN marking development can be understood to be PN *dynamics*, formally expressed [1] by the quadruplet $\langle X, U, \delta, \mathbf{x}_0 \rangle$. Here,

F. Čapkovič—Partially supported by the grant VEGA 2/0039/13.

N.T. Nguyen et al. (Eds.): ACIIDS 2016, Part I, LNAI 9621, pp. 377–387, 2016.
DOI: 10.1007/978-3-662-49381-6_36

$\delta : X \times U \to X$, where X is a set of state vectors $\mathbf{x}_k$ and U is a set of control vectors $\mathbf{u}_k$. While the entries of state vectors $\mathbf{x}_k$ express states of elementary places $\sigma_{p_i}^k$, $i = 1, \ldots, n$ (i.e. the number of tokens), in different steps $k = 0, 1, \ldots, K$, the entries of control vectors $\mathbf{u}_k$ represent the states of elementary transitions $\gamma_{t_j}^k$, $j = 1, \ldots, m$ (i.e. their disabling or enabling). The sets B, F, G can be replaced by the incidence matrices $\mathbf{B}$, $\mathbf{F}$, $\mathbf{G}$ of PN edges. Here, $\mathbf{B} = \mathbf{G}^T - \mathbf{F}$. Consequently, PN dynamics can be described by the restricted linear discrete integer system (all of its parameters and variables are non-negative integers) as follows

$$\mathbf{x}_{k+1} = \mathbf{x}_k + \mathbf{B}.\mathbf{u}_k; \quad k = 0, 1, \ldots, K \tag{3}$$

$$\mathbf{F}.\mathbf{u}_k \leq \mathbf{x}_k \tag{4}$$

Here, $\mathbf{x}_k = (\sigma_{p_1}^k, \ldots, \sigma_{p_n}^k)^T$ with $\sigma_{p_i} \in \{0, 1, \ldots, \infty\}$, $i = 1, \ldots, n$, and $\mathbf{u}_k = (\gamma_{t_1}^k, \ldots, \gamma_{t_m}^k)^T$ with $\gamma_{t_j} \in \{0, 1\}$, $j = 1, \ldots, m$.

The mathematical model (3) - (4) describes the PN marking development. It can be successfully used for modelling DES when all places are measurable and all transitions are controllable. However, in PN models of real systems unmeasurable places and uncontrollable transitions can occur. In such a case Interpreted PN (IPN), being an extension of PN, are applied. IPN were defined e.g. in [3–5,9,10] and in other sources.

1.1   Interpreted Petri Nets

In the sense of the definition introduced in [10] the IPN is the following sextuplet

$$Q = \langle (N, \mathbf{x}_0), \Sigma, \Phi, \lambda, \Psi, \varphi \rangle \tag{5}$$

where,

$PN = (N, \mathbf{x}_0)$ is the original PN;

$\Sigma = \{\alpha_1, \alpha_2, \ldots, \alpha_r\}$ is the input alphabet with α_i, $i = 1, \ldots, r$, being the input symbols;

$\Phi = \{\delta_1, \delta_2, \ldots, \delta_s\}$ is the output alphabet with δ_i, $i = 1, \ldots, s$, being the output symbols;

$\lambda : T \to \Sigma \cup \{\varepsilon\}$ labels the transitions. Either an input symbol $\alpha_i \in \Sigma$ or the internal event ε is assigned to each PN transition by this function. Thus, two sets of transitions arise - the set T_c of controllable transitions and the set T_u of uncontrollable transitions. Of course, $T = T_c \cup T_u$.

$\Psi : P \to \Phi \cup \{\varepsilon\}$ labels the places. Either an output symbol $\delta_i \in \Phi$ or the null event ε is assigned to each PN place by this function. Thus, two sets of places arise - the set P_m of the measurable places and the set P_{nm} of unmeasurable places. $P = P_m \cup P_{nm}$.

φ is the output function assigning the output vector $\mathbf{y}_k = \varphi.\mathbf{x}_k$ to the PN state vector $\mathbf{x}_k$. The entries of the output vector $\mathbf{y}_k$ represent the states of measurable places.

More details about the functions are introduced in [10].

2   IPN in Modelling and Control

To indicate the importance of IPN models for DES control, let us introduce two illustrative examples concerning (i) the principle of creating the IPN model of DES at existence of uncontrollable transitions and unmeasurable places; (ii) the principle of its control.

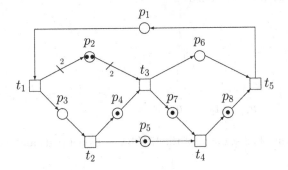

Fig. 1. An example of the PN-based model

To show how the IPN model can be created consider the simple PN-model given in Fig. 1. Suppose that the measurable places are $P_m = \{p_1, p_5, p_6\}$ and the unmeasurable ones are $P_{nm} = P \backslash P_m = \{p_2, p_3, p_4, p_7, p_8\}$. Consider that the controllable transitions are $T_c = \{t_1, t_5\}$, while the uncontrollable ones are $T_u = T \backslash T_c = \{t_2, t_3, t_4\}$. Thus, for such IPN the input alphabet $\Sigma = \{\alpha_1, \alpha_5\}$ and the output alphabet $\Phi = \{\delta_1, \delta_2, \delta_3\}$. Consequently, $\lambda(t_k)_{k=1,\ldots,5} = \{\alpha_1, \varepsilon, \varepsilon, \varepsilon, \alpha_5\}$, $\Psi(p_i)_{i=1,\ldots,8} = \{\delta_1, \varepsilon, \varepsilon, \varepsilon, \delta_2, \delta_3, \varepsilon, \varepsilon\}$. The output equation is as follows

$$\mathbf{y}_k = \varphi.\mathbf{x}_k \qquad \text{where} \qquad \varphi = \begin{pmatrix} 1\,0\,0\,0\,0\,0\,0\,0 \\ 0\,0\,0\,0\,1\,0\,0\,0 \\ 0\,0\,0\,0\,0\,1\,0\,0 \end{pmatrix} \qquad (6)$$

Thus for the initial state $\mathbf{x}_0 = (0, 2, 0, 1, 1, 0, 1, 1)^T$ the output vector is $\mathbf{y}_0 = (0, 1, 0)^T$.

In order to explain how the IPN model is controlled, consider a segment shown in Fig. 2 left. While the *upper line* containing p_4 and t_3 represents the fragment of the model of the control system PN_{cs} (containing the control specifications), the *lower line* represents the fragment of the IPN model of the controlled plant PN_{pl}. The RG of the model is given in Fig. 2 right, where $\mathbf{x}_0 = (1, 0, 0, 1)^T$, $\mathbf{x}_1 = (0, 1, 0, 1)^T$, $\mathbf{x}_2 = (0, 0, 1, 1)^T$ and $\mathbf{x}_3 = (0, 0, 1, 0)^T$.

Here, the controllable transition t_1 is enabled because it is required to reach the stated output while p_4 represents the state of a sensor. The self-loop between them represents the relation between the place of the control specification and the plant controllable discrete event. The transition t_3 represents enabling the event expressing the situation when the plant and control specification have the same output and p_3 represents the state of the sensor. The self-loop between them expresses the relation between the plant measured place and the control specification. Note that the fragment of RG accordant with the segment of the controlled plant is straight-lined. Such fragments occur also in more complicated structures. Of course, RG of more complicated structures of the plant models will not be entirely straight-lined as in Fig. 2. Some branchings occur in such cases as well - it is apparent in Fig. 4 as well as in Fig. 6 introduced in the Sect. 3. However, they are not so extensive.

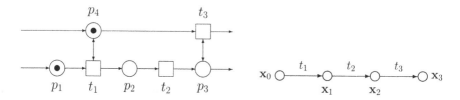

Fig. 2. The controlled segment of the IPN-based model (left) and its RG (right)

2.1   Illustrative Example

To illustrate the control process deeper, consider the more complicated configuration of the P/T PN model of a plant (without uncontrollable transitions and unmeasurable places) given in Fig. 3 left. Its reachability graph (RG) is in Fig. 3 right. The model parameters and the states reachable from the initial state (columns of $\mathbf{X}_{reach}^{PN_{pl}}$) are Control of the system can be modelled as it is shown in Fig. 4 left where the model of the controlled plant together with the controller are introduced. Corresponding RG is given in Fig. 4 right. The model parameters and the reachable states are as follows

$$\mathbf{F}^{PN_{pl}} = \begin{pmatrix} 1\,0\,0\,0\,0\,0 \\ 0\,1\,0\,0\,0\,0 \\ 0\,1\,0\,0\,0\,0 \\ 0\,0\,1\,0\,0\,0 \\ 0\,0\,0\,1\,0\,0 \\ 0\,0\,0\,0\,1\,0 \\ 0\,0\,0\,0\,0\,1 \end{pmatrix} ; \; \mathbf{G}^{PN_{pl}} = \begin{pmatrix} 0\,0\,0\,0\,0\,1\,1 \\ 1\,0\,0\,0\,0\,0\,0 \\ 0\,1\,0\,0\,0\,0\,0 \\ 0\,0\,1\,0\,0\,0\,0 \\ 0\,0\,0\,1\,0\,0\,0 \\ 0\,0\,0\,0\,1\,0\,0 \end{pmatrix} ; \; \mathbf{x}_0^{PN_{pl}} = \begin{pmatrix} 0 \\ 0 \\ 0 \\ 0 \\ 0 \\ 1 \\ 1 \end{pmatrix}$$

$$\mathbf{X}^{PN_{pl}}_{reach} = \begin{pmatrix} 0\;0\;0\;0\;0\;0\;0\;0\;0\;1 \\ 0\;0\;0\;1\;0\;0\;1\;0\;1\;0 \\ 0\;0\;0\;0\;0\;1\;0\;1\;1\;0 \\ 0\;1\;0\;0\;1\;0\;0\;1\;0\;0 \\ 0\;0\;1\;0\;1\;0\;1\;0\;0\;0 \\ 1\;0\;1\;0\;0\;1\;0\;0\;0\;0 \\ 1\;1\;0\;1\;0\;0\;0\;0\;0\;0 \end{pmatrix}$$

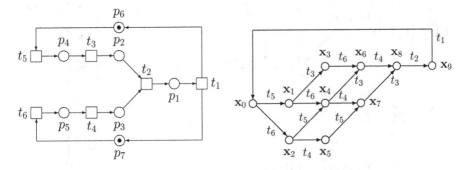

Fig. 3. The PN-based model of the plant (left) and its RG (right)

$$\mathbf{F}^{PN} = \left(\begin{array}{cccccc|cccc} 1 & 0 & 0 & 0 & 0 & 0 & 0 & 0 & 1 & 0 \\ 0 & 1 & 0 & 0 & 0 & 0 & 0 & 1 & 0 & 0 \\ 0 & 1 & 0 & 0 & 0 & 0 & 0 & 1 & 0 & 0 \\ 0 & 0 & 1 & 0 & 0 & 0 & 1 & 0 & 0 & 0 \\ 0 & 0 & 0 & 1 & 0 & 0 & 1 & 0 & 0 & 0 \\ 0 & 0 & 0 & 0 & 1 & 0 & 0 & 0 & 0 & 1 \\ 0 & 0 & 0 & 0 & 0 & 1 & 0 & 0 & 0 & 1 \\ \hline 0 & 0 & 0 & 0 & 1 & 1 & 1 & 0 & 0 & 0 \\ 0 & 0 & 1 & 1 & 0 & 0 & 0 & 1 & 0 & 0 \\ 0 & 1 & 0 & 0 & 0 & 0 & 0 & 0 & 1 & 0 \\ 1 & 0 & 0 & 0 & 0 & 0 & 0 & 0 & 0 & 1 \end{array}\right) ; \; \mathbf{G}^{PN} = \left(\begin{array}{ccccccc|cccc} 0 & 0 & 0 & 0 & 0 & 1 & 1 & 0 & 0 & 0 & 1 \\ 1 & 0 & 0 & 0 & 0 & 0 & 0 & 0 & 0 & 1 & 0 \\ 0 & 1 & 0 & 0 & 0 & 0 & 0 & 0 & 1 & 0 & 0 \\ 0 & 0 & 1 & 0 & 0 & 0 & 0 & 0 & 1 & 0 & 0 \\ 0 & 0 & 0 & 1 & 0 & 0 & 0 & 1 & 0 & 0 & 0 \\ 0 & 0 & 0 & 0 & 1 & 0 & 0 & 1 & 0 & 0 & 0 \\ \hline 0 & 0 & 0 & 1 & 1 & 0 & 0 & 0 & 1 & 0 & 0 \\ 0 & 1 & 1 & 0 & 0 & 0 & 0 & 0 & 0 & 1 & 0 \\ 1 & 0 & 0 & 0 & 0 & 0 & 0 & 0 & 0 & 0 & 1 \\ 0 & 0 & 0 & 0 & 0 & 1 & 1 & 1 & 0 & 0 & 0 \end{array}\right)$$

$$\mathbf{x}_0^{PN} = \begin{pmatrix} 0 \\ 0 \\ 0 \\ 0 \\ 0 \\ 1 \\ 1 \\ \hline 1 \\ 0 \\ 0 \\ 0 \end{pmatrix} ; \; \mathbf{X}^{PN}_{reach} = \left(\begin{array}{ccccccccccc} 0 & 0 & 0 & 0 & 0 & 0 & 0 & 0 & 0 & 1 & 1 & 0 \\ 0 & 0 & 0 & 0 & 0 & 1 & 0 & 1 & 1 & 0 & 0 & 0 \\ 0 & 0 & 0 & 0 & 0 & 0 & 1 & 1 & 1 & 0 & 0 & 0 \\ 0 & 1 & 0 & 1 & 1 & 0 & 1 & 0 & 0 & 0 & 0 & 0 \\ 0 & 0 & 1 & 1 & 1 & 1 & 0 & 0 & 0 & 0 & 0 & 0 \\ 1 & 0 & 1 & 0 & 0 & 0 & 0 & 0 & 0 & 0 & 0 & 1 \\ 1 & 1 & 0 & 0 & 0 & 0 & 0 & 0 & 0 & 0 & 0 & 1 \\ \hline 1 & 1 & 1 & 1 & 0 & 0 & 0 & 0 & 0 & 0 & 0 & 0 \\ 0 & 0 & 0 & 0 & 1 & 1 & 1 & 1 & 0 & 0 & 0 & 0 \\ 0 & 0 & 0 & 0 & 0 & 0 & 0 & 0 & 1 & 1 & 0 & 0 \\ 0 & 0 & 0 & 0 & 0 & 0 & 0 & 0 & 0 & 0 & 1 & 1 \end{array}\right)$$

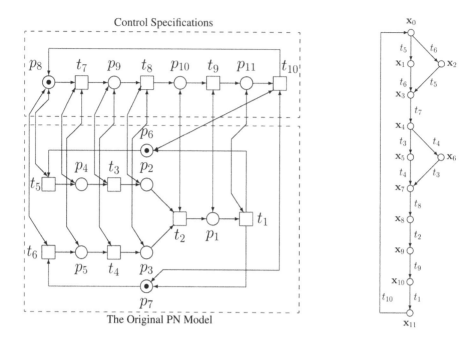

Fig. 4. The controlled PN-based model of the plant and its RG

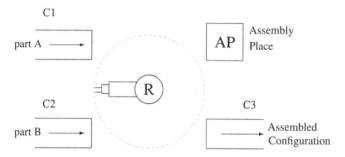

Fig. 5. The scheme of the robotized assembly cell

Comparing RGs introduced in Figs. 3 and 4 it is evident that the latter RG does not include complicated branching. Thus, the IPN-based approach can be successfully applied also on P/T PN without uncontrollable transitions and unmeasurable places.

3   Case Study on Robotized Assembly Cell

Consider the robotized assembly cell (RAC) displayed schematically in Fig. 5. The input conveyors C1 and C2 deliver to the RAC, respectively, parts A and parts B. The robot R takes the part A from C1 and inserts it into the assembly place

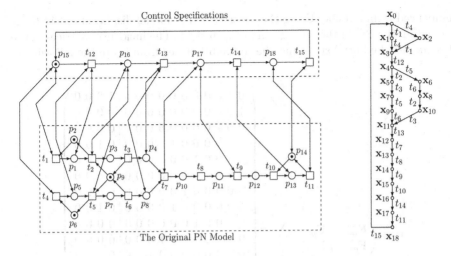

Fig. 6. The controlled IPN-based model of the plant (left) and its RG (right)

(AP). Then R takes the part B from C2 and inserts it into AP. In AP both parts are assembled into the final configuration. After finishing the assembly process, R takes the assembled configuration and put it on the output conveyor C3. Such the progress of work is repeated. The PN-based model of RAC is given in Fig. 6. The lower dashed box represents the PN model PN_{pl} of the plant, while the upper one expresses control specifications. The PN places represent the following activities: p_1 - C1 conveys the part A; p_2 - C1 is available; p_3 - R takes A from C1 and transfers it to AP; p_4 - R inserts A into AP; p_5 - C2 conveys the part B; p_6 - C2 is available; p_7 - R takes B from C2 and transfers it to AP; p_8 - R inserts B into AP; p_9 - it ensures the mutual exclusion, because R cannot take A from C1 and B from C2 simultaneously; p_{10} - the parts A, B are assembled in AP; p_{11} - R unloads the finished configuration from AP; p_{12} - R transfers the finished configuration from AP to C3; p_{13} - R put the finished configuration on C3; p_{14} - the free place on C3 is available.

In this model the transition t_8 is uncontrollable (i.e. PN turns to IPN). The RG of the uncontrolled PN model has 813 nodes and it is extensively branched. Consequently, it cannot be introduced here. When only one input part A and only one input part B are allowed the RG has 28 nodes. But it is also too big and branched.

3.1   Control of the Robotized Assembly Cell

Let us apply know the IPN-based approach to control the RAC with the uncontrollable transition t_8. The assembly process is spontaneous and it cannot be influenced from outside. Using the controlled IPN model displayed in Fig. 6 left we obtain the RG in the form given in Fig. 6 right. The incidence matrices of the PN model, the initial state and the reachable states from it are the following

$$
\mathbf{F}=\begin{pmatrix}\mathbf{F}_{pl} & \mathbf{F}_{pl\to cs}\\ \mathbf{F}_{cs\to pl} & \mathbf{F}_{cs}\end{pmatrix}=\left(\begin{array}{cccccccccc|cccc}
0&1&0&0&0&0&0&0&0&0&1&0&0&0\\
1&0&0&0&0&0&0&0&0&0&0&0&0&0\\
0&0&1&0&0&0&0&0&0&0&0&0&0&0\\
0&0&0&0&0&0&1&0&0&0&0&1&0&0\\
0&0&0&0&1&0&0&0&0&0&1&0&0&0\\
0&0&0&1&0&0&0&0&0&0&0&0&0&0\\
0&0&0&0&0&1&0&0&0&0&0&0&0&0\\
0&0&0&0&0&0&1&0&0&0&0&1&0&0\\
0&1&0&0&1&0&0&0&0&0&0&0&0&0\\
0&0&0&0&0&0&0&1&0&0&0&0&0&0\\
0&0&0&0&0&0&0&0&1&0&0&0&0&0\\
0&0&0&0&0&0&0&0&0&1&0&0&0&0\\
0&0&0&0&0&0&0&0&0&1&0&0&1&0\\
0&0&0&0&0&0&0&0&1&0&0&0&0&1\\
\hline
1&0&0&1&0&0&0&0&0&0&1&0&0&0\\
0&1&0&0&1&0&0&0&0&0&0&1&0&0\\
0&0&0&0&0&0&1&0&1&0&0&0&1&0\\
0&0&0&0&0&0&0&0&0&1&0&0&0&1
\end{array}\right)
$$

$$
\mathbf{G}^{T}=\begin{pmatrix}\mathbf{G}_{pl}^{T} & \mathbf{G}_{pl\to cs}^{T}\\ \mathbf{G}_{cs\to pl}^{T} & \mathbf{G}_{cs}^{T}\end{pmatrix}=\left(\begin{array}{cccccccccc|cccc}
1&0&0&0&0&0&0&0&0&0&1&0&0&0\\
0&1&0&0&0&0&0&0&0&0&0&0&0&0\\
0&1&0&0&0&0&0&0&0&0&0&0&0&0\\
0&0&1&0&0&0&0&0&0&0&0&1&0&0\\
0&0&0&1&0&0&0&0&0&0&1&0&0&0\\
0&0&0&0&1&0&0&0&0&0&0&0&0&0\\
0&0&0&0&1&0&0&0&0&0&0&0&0&0\\
0&0&0&0&0&1&0&0&0&0&0&1&0&0\\
0&0&1&0&0&1&0&0&0&0&0&0&0&0\\
0&0&0&0&0&0&1&0&0&0&0&0&0&0\\
0&0&0&0&0&0&0&1&0&0&0&0&0&0\\
0&0&0&0&0&0&0&0&1&0&0&0&0&0\\
0&0&0&0&0&0&0&0&1&0&0&0&1&0\\
0&0&0&0&0&0&0&0&0&1&0&0&0&1\\
\hline
1&0&0&1&0&0&0&0&0&0&0&0&0&1\\
0&1&0&0&1&0&0&0&0&0&1&0&0&0\\
0&0&0&0&0&0&1&0&1&0&0&1&0&0\\
0&0&0&0&0&0&0&0&0&1&0&0&1&0
\end{array}\right)
$$

$$\mathbf{x_0} = \begin{pmatrix} 0 \\ 1 \\ 0 \\ 0 \\ 0 \\ 1 \\ 0 \\ 0 \\ 1 \\ 0 \\ 0 \\ 0 \\ 0 \\ 0 \\ 1 \\ - \\ 1 \\ 0 \\ 0 \\ 0 \end{pmatrix} ; \ \mathbf{X}_{reach} = \left(\begin{array}{cccccccccccccccccc} 0&1&0&1&1&0&1&0&1&0&0&0&0&0&0&0&0&0 \\ 1&0&1&0&0&1&0&1&0&1&1&1&1&1&1&1&1&1 \\ 0&0&0&0&0&1&0&0&0&0&1&0&0&0&0&0&0&0 \\ 0&0&0&0&0&0&0&1&0&1&0&1&1&0&0&0&0&0 \\ 0&0&1&1&1&1&0&1&0&0&0&0&0&0&0&0&0&0 \\ 1&1&0&0&0&0&1&0&1&1&1&1&1&1&1&1&1&1 \\ 0&0&0&0&0&0&1&0&0&1&0&0&0&0&0&0&0&0 \\ 0&0&0&0&0&0&0&0&1&0&1&1&1&0&0&0&0&0 \\ 1&1&1&1&1&0&0&1&1&0&0&1&1&1&1&1&1&1 \\ 0&0&0&0&0&0&0&0&0&0&0&0&0&1&0&0&0&0 \\ 0&0&0&0&0&0&0&0&0&0&0&0&0&0&1&0&0&0 \\ 0&0&0&0&0&0&0&0&0&0&0&0&0&0&0&1&0&0 \\ 0&0&0&0&0&0&0&0&0&0&0&0&0&0&0&0&1&1&0 \\ 1&1&1&1&1&1&1&1&1&1&1&1&1&1&1&0&0&1 \\ \hline 1&1&1&1&0&0&0&0&0&0&0&0&0&0&0&0&0&0 \\ 0&0&0&0&1&1&1&1&1&1&1&0&0&0&0&0&0&0 \\ 0&0&0&0&0&0&0&0&0&0&0&1&1&1&1&1&0&0 \\ 0&0&0&0&0&0&0&0&0&0&0&0&0&0&0&0&1&1 \end{array}\right)$$

The first column of the reachability matrix $\mathbf{X}_{reach}$ represents the initial state $\mathbf{x_0}$ and other columns represent the states reachable from $\mathbf{x_0}$ - i.e. the feasible states. As it can be seen from Fig. 6, the PN model of the plant itself has no feedback - it is straightforward. The working cycle is closed by means of the loop of the control specifications - see the upper dashed box in Fig. 6.

4   Conclusion

Two main problems were investigated in this paper. Namely, (i) how to replace the P/T PN model of DES (in our case the assembly manufacturing system) by the IPN model in case when some uncontrollable transitions and unmeasurable places occur in the plant model; (ii) the control synthesis of DES at using the IPN model.

After introducing PN and P/T PN, IPN were introduced. Then, the creation of the IPN models was explained. To illustrate these techniques in detail, the simple demonstrative examples were introduced. Simultaneously, the attention was paid also to RGs. While in case of the P/T PN models RGs are usually extensively branched and contain a big number of nodes, the proposed approach yields RGs with fewer nodes and simpler structure. It was shown that the approach proposed for the IPN control can be applied also on the P/T PN (without the uncontrollable transitions and unmeasurable places). Thus, the simpler RGs of the P/T PN models can be found. As it is evident from the comparison of RGs displayed in Figs. 3 and 4, the latter RG does not include complicated branching - i.e. it is more straightforward. Consequently, the proposed IPN-based approach can be successfully applied also on P/T PN without uncontrollable transitions and unmeasurable places.

The main part of the paper concerns the case study on modelling and control of the robotized assembly cell. Here, the transition t_8 is uncontrollable, because its firing cannot be realized till the moment when the parts A, B are assembled (p_{10}). The assembly process cannot be influenced from outside, it depends only on the machine realizing the assembly. Namely, on its capability and efficiency. Moreover, some unexpected (accidental) influences may extend processing time of the assembly. Only, after finishing the assembly operation, t_8 is fired and the robot can unload the finished configuration from the machine p_{11} and the working progress of the RAC can continue.

The presented approach to controlling DES modelled by means of IPN seems to be useful for practice. Moreover, to a certain extent it is also comparatively general. It is able to deal with straightforward PN models without any problems. However, it is necessary to say that at systems, where PN models contain internal cycles, some problems can occur. Therefore, in future it is necessary to pay attention to the investigation of such problems in detail.

Acknowledgments. The author thanks the Slovak Grant Agency VEGA for the partial support under grant # 2/0039/13.

References

1. Čapkovič, F.: Agent based approach to modelling ATM network. In: Proceedings of the 2012 IEEE 6th International Conference on Intelligent Systems - IEEE IS 2012, Sofia, Bulgaria, 6–8 September 2012, vol. I, pp. 102–107. IEEE Press, Piscataway (2012)

2. Desel, J., Reisig, W.: Place/transition Petri nets. In: Reisig, W., Rozenberg, G. (eds.) APN 1998. LNCS, vol. 1491, pp. 122–173. Springer, Heidelberg (1998)

3. Dotoli, M., Fanti, M., Mangini, A.: On-line identification of discrete event systems by interpreted Petri nets. In: Proceedings of the IEEE International Conference on Systems, Man and Cybernetics - SMC 2006, Taipei, Taiwan, 8–11 October 2006, vol. 4, pp. 3040–3045. IEEE Press, Piscataway (2006)

4. Karatkevich, A.: Dynamic Analysis of Petri Net-Based Discrete Systems. Lecture Notes in Control and Information Sciences, vol. 356. Springer, Heidelberg (2007)

5. Lutz-Ley, A., Mellado, E.L.: Synthesis of fault recovery sequences in a class of controlled discrete event systems modelled with Petri nets. In: 2013 Iberoamerican Conference on Electronics Engineering and Computer Science, Procedia Technology, vol. 7, pp. 257–264 (2013)

6. Murata, T.: Petri nets: properties, analysis and applications. Proc. IEEE **77**, 541–580 (1989)

7. Peterson, J.L.: Petri nets. Comput. Surv. **9**, 223–252 (1977)

8. Peterson, J.L.: Petri Net Theory and the Modeling of Systems. Prentice-Hall Inc., Englewood Cliffs (1981)

9. Rivera-Rangel, I., Ramírez-Treviño, A., Aquirre-Salas, L.I., Ruiz-León, L.I.: Geometrical characterization of observability in interpreted Petri nets. Kybernetika **41**, 553–574 (2005)
10. Santoyo-Sanchez, A., Pérez-Martínez, M.A., De Jesús-Velásquez, V., Aguirre-Salas, L.I., Alvarez-Ureña, M.A.: Modeling methodology for NPC's using interpreted Petri nets and feedback control. In: Proceedings of the 2010 7th International Conference on Electrical Engineering, Computing Science and Automatic Control - CCE 2010, Tuxtla Gutiérrez, Chiapas, Mexico, 8–10 September 2010. CD-ROM, pp. 369–374. IEEE Press (2010)

Enhanced Guided Ejection Search for the Pickup and Delivery Problem with Time Windows

Jakub Nalepa[1(⊠)] and Miroslaw Blocho[1,2]

[1] Silesian University of Technology, Gliwice, Poland
jakub.nalepa@polsl.pl
[2] ABB IT, Krakow, Poland
miroslaw.blocho@pl.abb.com

Abstract. This paper presents an enhanced guided ejection search (GES) to minimize the number of vehicles in the NP-hard pickup and delivery problem with time windows. The proposed improvements decrease the convergence time of the GES, and boost the quality of results. Extensive experimental study on the benchmark set shows how the enhancements influence the GES capabilities. It is coupled with the statistical tests to verify the significance of the results. We give a guidance on how to select a proper algorithm variant based on test characteristics and objectives. We report one new world's best result obtained using the enhanced GES.

Keywords: Guided search · Ejection chain · Heuristics · PDPTW

1 Introduction

The pickup and delivery problem with time windows (PDPTW) is an NP-hard discrete optimization problem [1]. It consists in finding a feasible schedule for serving a set of transportation requests using homogeneous trucks. In each request (which must be served within one route), goods are picked up from one location, and delivered to another one (these locations must be served within the corresponding pickup and delivery time windows). The pickup location must be visited before the delivery location by the serving truck, and the capacity of all vehicles must not be exceeded in a feasible routing schedule. All vehicles start and finish their routes in a single depot which also specifies its own time window.

The PDPTW is a two-objective optimization problem [2]. The main objective is to minimize the fleet size, whereas the second one is to optimize the distance traveled by the vehicles. Due to numerous real-life applications of the routing problems (parcels delivery, rail distribution, food delivery, and many others) they attracted research attention [3,4]. Exact approaches tackling the PDPTW include—among others—branch-and-cut [5] and branch-and-price algorithms [6], dynamic programming, and column generation methods. They are however still not applicable for large-scale scenarios due to their execution times.

© Springer-Verlag Berlin Heidelberg 2016
N.T. Nguyen et al. (Eds.): ACIIDS 2016, Part I, LNAI 9621, pp. 388–398, 2016.
DOI: 10.1007/978-3-662-49381-6_37

Heuristics allow for obtaining high-quality solutions in a short time [7]. They usually tackle two objectives independently [4]. Approximate methods may be therefore developed to minimize the fleet size (in the first stage), and to minimize the traveled distance (in the second stage) [8]. Heuristics for minimizing the number of trucks are divided into construction and improvement techniques. The former methods build a solution from scratch, whereas the latter ones improve the initial solution. The approximate algorithms include tabu and neighborhood searches [9,10], guided ejection searches (GESes) [11], and other [2,12]. The GES became a local search framework and was applied to solve various optimization problems [13,14]. Thus, generic GES improvements can be incorporated into the algorithms tackling other optimization tasks. It is worth noting that the GES proved to be very flexible since it allows for building incomplete solutions.

In this paper, we present an enhanced GES for minimizing the number of vehicles serving customer requests in the PDPTW. The baseline version of the GES was presented in [11]. The proposed enhancements include randomizing the order of ejected requests during the re-insertion process, *squeezing* a temporarily infeasible solution (thus attempting to restore its feasiblity), and limiting the maximum computation time for requests ejected from a single route.

We perform an in-depth experimental analysis using Li and Lim's benchmark tests in order to assess the impact of the mentioned improvements on the quality of solutions and the algorithm behavior. We couple this study with the two-tailed Wilcoxon tests to verify the statistical significance of the results. We present a clear guidance on how to select an appropriate algorithm variant based on the test characteristics and the objective selected as the one of a higher importance. We report one new world's best result retrieved using our enhanced GES.

The paper is organized as follows. Section 2 formulates the PDPTW. The enhanced GES is discussed in Sect. 3. The experimental study complemented with the statistical analysis of the obtained results are reported in Sect. 4. Section 5 concludes the paper and highlights the directions of our future work.

2   Pickup and Delivery Problem with Time Windows

The PDPTW is defined on a directed graph $G = (V, E)$, with a set V of $C + 1$ vertices (v_i, $i \in \{1, \ldots, C\}$, vertices represent the customers, v_0 is the depot), and a set of edges $E = \{(v_i, v_{i+1}) | v_i, v_{i+1} \in V, v_i \neq v_{i+1}\}$ (travel connections). The travel costs $c_{i,j}$, $i, j \in \{0, 1, \ldots, C\}$, $i \neq j$, are equal to the distance between the corresponding travel points. Each request h_i, $i \in \{1, 2, \ldots, N\}$, where $N = C/2$, is a coupled pair of pickup (P) and delivery (D) customers—p_h and d_h, respectively, where $P \cap D = \emptyset$, and $P \cup D = V \setminus \{v_0\}$. For each h_i, the amount of delivered ($q^d(h_i)$) and picked up ($q^p(h_i)$) goods is defined, where $q^d(h_i) = -q^p(h_i)$. Each customer v_i defines its demand (delivery or pick up), service time s_i ($s_0 = 0$), and time window $[e_i, l_i]$ within which it should be visited. Since the fleet is homogeneous (let K denote its size), the capacity of each truck is Q. Each route $r = \langle v_0, v_1, \ldots, v_{n+1} \rangle$ in the solution σ, starts and finishes at v_0.

A solution σ (with K routes) is feasible if (i) Q is not exceeded for any vehicle (capacity constraint), (ii) the service of every request starts within its time window (time window constraint), (iii) every customer is served in one route, (iv) every vehicle leaves and returns to v_0 within $[e_0, l_0]$, and (v) each pickup is performed before the corresponding delivery (precedence constraint). The primary objective is to minimize K. Then, the total travel distance is minimized. Let σ_A and σ_B be two solutions. σ_A is then of a higher quality if $(K(\sigma_A) < K(\sigma_B))$ or $(K(\sigma_A) = K(\sigma_B)$ and $T(\sigma_A) < T(\sigma_B))$, and T is the total distance.

3   Enhanced Guided Ejection Search

3.1   Algorithm Outline

In the GES, the initial solution is iteratively improved to increase its quality—some solution components are removed from the solution in search of the partial solution of a higher quality. Then, the removed components are re-inserted into the partial solution. The GES involves the problem-specific guided local searches to exploit the knowledge about the underlying optimization task.

The GES for the PDPTW enhanced by our improvement techniques is given in Algorithm 1. The initial solution, in which every request is served in a separate route (line 1), is subject to the route minimization—the attempts to remove one route at a time are undertaken. A random route r is selected from σ, and its requests are inserted into the ejection pool (EP) containing unserved requests (line 3). The penalty counters p, indicating the re-insertion difficulty of each request, are initialized (line 4). A request h_{in} is popped from the EP (line 7). If there are several feasible positions to insert h_{in} into σ, then a random one is chosen (line 8)—$\mathcal{S}_{in}^{fe}(h_{in}, \sigma)$ is the set of feasible insertions of h_{in} into σ. This request is otherwise inserted so as it violates the constraints, and the solution is squeezed to restore its feasibility (line 9)—see Sect. 3.2. If it fails, then its p is increased (line 12), and at most k_m requests (with minimal sum of their p's) are ejected to insert h_{in} feasibly (lines 13–20). Then, σ is perturbed with local moves (out-relocate and out-exchange) to diversify the search (line 22).

The stopping conditions are verified—the GES may be terminated if its execution time τ surpasses τ_M, or the required number of routes has been retrieved (line 23). Additionally, we allow for breaking the inner loop aiming at removing a route r (lines 6–24), if the GES is stuck during the search (see Sect. 3.2). Finally, the best feasible solution (with the smallest K) is returned (line 29).

3.2   Improvements

In this paper, we introduce the improvements to the original version of the GES in order to boost the convergence capabilities of the algorithm, and to improve the quality of the final results. These enhancements include:

Algorithm 1. Enhanced guided ejection search.

1: Create an initial solution σ; $\sigma_B \leftarrow \sigma$; *finished* $\leftarrow$ **false**;
2: **while not** *finished* **do**
3:    Initialize EP with requests from a randomly removed route r;
4:    Initialize penalty counters for each request $p[i] \leftarrow 1$; $(i = 1, 2, \ldots, N)$;
5:    $c \leftarrow 0$; $\sigma_I \leftarrow \sigma$; $c_M \leftarrow N^2/K(\sigma)$;
6:    **while** (EP $\neq \emptyset$) **and** (**not** *stuck*) **do**
7:       Select and remove the request h_{in} from EP;
8:       **if** $\mathcal{S}_{in}^{fe}(h_{in}, \sigma) \neq \emptyset$ **then** $\sigma \leftarrow$ random solution $\sigma' \in \mathcal{S}_{in}^{fe}(h_{in}, \sigma)$;
9:       **else** $\sigma \leftarrow$ Squeeze(h_{in}, σ);
10:      **end if**
11:      **if** $h_{in} \notin \sigma$ **then**
12:         $p[h_{in}] \leftarrow p[h_{in}] + 1$;   ▷ Increase the request penalty counter
13:         **for** $k \leftarrow 1$ **to** k_m **do**
14:            Generate $\mathcal{S}_{ej}^{fe}(h_{in}, \sigma)$ with minimum $P_{sum} = p[h_{out}^1] + \cdots + p[h_{out}^k]$;
15:            **if** $\mathcal{S}_{ej}^{fe}(h_{in}, \sigma) \notin \emptyset$ **then**
16:               $\sigma \leftarrow$ random solution $\sigma' \in \mathcal{S}_{ej}^{fe}(h_{in}, \sigma)$;
17:               Add the ejected requests $\{h_{out}^1, \ldots, h_{out}^k\}$ to EP;
18:               **break**;
19:            **end if**
20:         **end for**
21:      **end if**
22:      Perturb σ; $c \leftarrow c + 1$;
23:      Verify termination conditions and update *finished*; *stuck* $\leftarrow c > c_M$;
24:   **end while**
25:   **if** EP $\neq \emptyset$ **then** $\sigma \leftarrow \sigma_I$;   ▷ Backtrack solution to the previous state
26:   **else** $\sigma_B \leftarrow \sigma$;
27:   **end if**
28: **end while**
29: **return** σ_B;

---

1. **Randomizing the EP.** In the original version of the GES, the requests which are being re-inserted into the partial solution are popped from the EP using the LIFO strategy. If the requests cannot be feasibly inserted into σ, the solution is backtracked to the previous state. Since the same route may be selected multiple times for the removal (line 3), it may lead to unnecessary attempts of re-inserting requests in the order which never leads to a feasible solution. Thus we propose to randomize the order of requests in the EP [15].

2. **Squeezing an Infeasible Solution.** Here, we adapt squeezing an infeasible solution (first introduced for VRPTW [14]) to the PDPTW (Algorithm 2). A temporarily infeasible solution undergoes local moves to restore its feasibility by decreasing the value of the penalty function $\mathcal{F}(\sigma) = \mathcal{F}_c(\sigma) + \mathcal{F}_{tw}(\sigma)$, where $\mathcal{F}_c(\sigma)$ and $\mathcal{F}_{tw}(\sigma)$ represent the sum of capacity exceeds in σ, and the sum of the time windows violations (lines 2–8). It thus reflects the severity of the constraint violations ($\mathcal{F}(\sigma) = 0$, if σ is feasible). These moves encompass out-relocate (a request is relocated to another route), and out-exchange

(a request from r is swapped with another one from another route) moves. The latter ones are executed only if the out-relocate moves do not improve the routing plan. These moves are applied to create $\mathcal{S}_r(\sigma_T)$, which denotes the set of the neighboring solutions resulted from performing local search moves affecting the route r. The solution σ' with the minimum value of $\mathcal{F}$ (after inserting h_{in}) is processed during the squeeze (line 1).

Algorithm 2. Squeezing an infeasible solution σ'.

1: Select σ' such that $\mathcal{F}(\sigma')$ is minimum; $\sigma_T \leftarrow \sigma'$;
2: **while** $\mathcal{F}(\sigma_T) \neq 0$ **do**
3:     Select a random infeasible route r;
4:     Find $\sigma'' \in \mathcal{S}_r(\sigma_T)$ such that $\mathcal{F}(\sigma'')$ is minimum;
5:     **if** $\mathcal{F}(\sigma'') < \mathcal{F}(\sigma_T)$ **then** $\sigma_T \leftarrow \sigma''$;
6:     **else break**;
7:     **end if**
8: **end while**
9: **if** $\mathcal{F}(\sigma_T) \neq 0$ **then** $\sigma_T \leftarrow \sigma'$;   ▷ Restore the initial solution
10: **end if**
11: **return** σ_T;

---

3. **Maximum Iteration Limit.** In the GES, one route at a time is selected randomly and its requests are inserted back from the EP into σ (see Sect. 3.1). Originally, the attempts of re-inserting the ejected requests from r were being undertaken until the maximum time of the GES elapsed. However, the selected route may contain "difficult" requests which cannot be re-inserted into the partial solution. We introduce the iteration limit c_M which bounds the analysis time of a single route. If r is "difficult" (the iteration counter c surpasses the limit, and the search is stuck), σ is backtracked to the previous state, and a new route is selected for deletion. Since removing a route becomes more challenging for lower K, c_M is adaptively increased during the search and is given as $c_M = N^2/K(\sigma)$, where N is the number of requests.

4   Experimental Results

The GES was implemented in C++, and run on an Intel Xeon 3.2 GHz computer (16 GB RAM). Its time limit was $\tau_M = 5$ min, and $k_m = 2$. To verify the impact of the enhancements, we analyzed 8 GES variants (Table 1). The GES was tested on 400-customer Li and Lim (LL)[1] tests. Six test classes (c1, c2, r1, r2, rc1 and rc2) reflect many scheduling factors: c1 and c2 encompass clustered customers, in r1 and r2 they are random, and rc1 and rc2 contain a mix of random and clustered customers. The classes c1, r1 and rc1 have smaller capacities and shorter time windows. Tests have unique names, lα_β_γ, where α is the class, β relates to the number of customers (4 for 400), and γ is the identifier ($\gamma = 1, 2, \ldots, 10$).

---
[1] See: http://www.sintef.no/projectweb/top/pdptw/li--lim-benchmark/.

Table 1. Investigated GES variants.

	GES(—)	GES(R)	GES(M)	GES(MR)	GES(S)	GES(SR)	GES(SM)	GES(SMR)
Squeeze	no	no	no	no	yes	yes	yes	yes
Max. iter	no	no	yes	yes	no	no	yes	yes
Random EP	no	yes	no	yes	no	yes	no	yes

4.1 Analysis and Discussion

The experimental results are given in Table 2. Each test (out of 60) was executed 5 times using each algorithm variant, and the outputs were averaged across the classes—the lowest K and T are indicated in boldface. GES(R) retrieved the best asymptotic results for most classes, when K is considered. This algorithm variant appeared stable, and gave the best schedules on average for instances with randomized customers with short time windows (r1 and rc1). On the other hand, GES(—) was able to converge to the best-quality results for clustered customers (c1 and c2) on average. Introducing the iteration limit turned out to be very beneficial for tests with relatively wide time windows and large truck capacities (thus requests are served in a lower number of routes)—GES(M) retrieved the best results for the rc2 class. It indicates that the intensive exploitation of a single route does not improve the final routing plan, and if the ejected requests cannot be re-inserted quickly, the other route should be selected for removal.

The results show that some algorithm variants can not only efficiently decrease K, but also optimize T (although it is not tackled directly in the GES)—see GES (SR) for c1, GES(R) for c2, GES(MR) for r1 and r2, GES(S) for rc1, and GES(M) for rc2. These GESes converged to small K's with significantly shortened T—the GES results are of a similar quality (with respect to K) to the world's best solutions (Table 3), obtained using *various* algorithms[2]. It is worth noting that two previous world's best solutions (lrc1_4_1 and lrc1_4_3) were infeasible[3]. Also, we report the new world's best solution (with a decreased number of trucks) for lrc2_4_2[4], obtained using GES(R). Clustered-customers tests appeared most challenging for all GESes. Additionally, instances with shorter time windows (c1, r1, and rc1) are easier to solve. Randomizing the EP was advantageous for all classes, and resulted in decreased K's.

The averaged convergence times (τ_C) are presented in Fig. 1. For most cases, converging to a low-quality solutions is very fast—the GES sticks in a local minimum and cannot improve the solution further. Applying the squeeze procedure increases τ_C for most LL classes, since it involves performing additional moves to restore the feasibility of a solution. However, there exist classes for which squeezing results in a faster convergence to high quality solutions (see e.g., GES(MR) compared with GES(SMR) for c2 and r2). Intermediate solutions in this case are

---

[2] The world's best solutions are available at: https://www.sintef.no/projectweb/top/pdptw/li--lim-benchmark/400-customers/; reference date: April 27, 2015.

[3] Our PDPTW feasibility checker is available at: http://sun.aei.polsl.pl/~jnalepa/PDPTW-checker/.

[4] See http://sun.aei.polsl.pl/~jnalepa/ACIIDS2016/ for the solution details.

Table 2. Comparison of the results obtained using different algorithm variants. The best results (across the GES variants) are rendered in boldface.

Variant↓			c1	c2	r1	r2	rc1	rc2
GES (−)	K	Min.	38.5	13	25.1	6.5	**28.6**	9
		Avg.	**38.82**	**13.14**	25.34	6.72	28.9	9.14
		Max.	39.3	**13.5**	**25.5**	**6.9**	29.1	9.3
	T	Min.	10115.95	8822.61	11742.58	13505.75	10563.75	13022.74
		Avg.	10697.55	9360.04	11997.83	14213.95	10780.72	13540
		Max.	11383.95	10378.49	12264.91	**14839.71**	10991.68	14225.02
GES(R)	K	Min.	**38.3**	**12.9**	**24.8**	6.4	**28.6**	9
		Avg.	38.88	13.18	**25.18**	6.7	**28.82**	9.14
		Max.	39.8	**13.5**	**25.5**	7	29.1	9.5
	T	Min.	10124.92	**8443.33**	11622.65	13237.41	10514.97	12861.2
		Avg.	10737.35	**9309.03**	11951.92	14106.34	10790.79	13323.24
		Max.	11619.54	**10250.63**	**12214.31**	15012.16	11055.56	13760.9
GES(M)	K	Min.	38.5	13	25	6.4	28.9	9
		Avg.	38.88	13.38	25.4	6.64	29.04	9
		Max.	39.3	13.6	25.7	7	29.3	9
	T	Min.	10282.62	8858.74	11762.1	13463.89	10485.03	**12722.11**
		Avg.	10810.73	9863.93	12109.18	14125.81	10795.29	**13301.3**
		Max.	11467.77	10714.03	12414.44	14916.2	11011.52	**13753.82**
GES(MR)	K	Min.	38.6	13	25	**6.3**	**28.6**	9
		Avg.	39.06	13.26	25.26	**6.6**	28.88	9.14
		Max.	39.4	13.7	25.6	7	**29**	9.3
	T	Min.	10374.72	8615.6	**11593.6**	**13072.05**	10434.08	12883.93
		Avg.	10959.36	9508.02	**11942.2**	**13868.02**	**10696.01**	13437.58
		Max.	11555.21	10676.07	12225.59	14930.55	10945.23	13898.37
GES(S)	K	Min.	38.6	13.1	25	6.8	28.7	9
		Avg.	38.88	13.44	25.4	6.92	29	9.24
		Max.	**39.1**	13.7	25.8	7.1	29.3	9.7
	T	Min.	10403.87	9103.78	11663.24	14223.21	**10422.58**	13006.71
		Avg.	10783.97	10086.76	12059.32	14629.37	10777.92	13642.39
		Max.	**11136.15**	10960.02	12450.51	15032.36	11036.97	14184.28
GES(SR)	K	Min.	38.5	13.1	24.9	6.7	**28.6**	9
		Avg.	38.86	13.32	25.38	6.94	28.88	9.34
		Max.	39.2	13.7	25.8	7.3	29.2	9.7
	T	Min.	**10107.87**	8844.77	11597.72	13868.99	10511.37	12788.72
		Avg.	**10665.17**	9672.94	11992.08	14639.32	10744.45	13551.67
		Max.	11215.17	10731.94	12372.72	15713.64	**10938.8**	14413.56
GES(SM)	K	Min.	38.8	13.2	25.4	6.7	29	9
		Avg.	39.06	13.48	25.58	6.96	29.26	9.28
		Max.	39.4	13.7	25.9	7.2	29.6	9.7
	T	Min.	10422.51	9153.72	11857.24	13796.16	10487.95	12766.84
		Avg.	10937.83	10087.59	12151.67	14680.42	10816.52	13575.71
		Max.	11472.97	10814.19	12535.17	15408.35	11023.34	14366.94
GES(SMR)	K	Min.	38.6	13.1	25.1	6.8	28.9	9.1
		Avg.	38.94	13.42	25.44	6.98	29.02	9.38
		Max.	39.2	13.9	25.6	7.4	29.2	9.7
	T	Min.	10270.51	8935.35	11700.42	13938.75	10454.8	12886.77
		Avg.	10824.38	9929.28	12050.66	14622.15	10751.32	13675.52
		Max.	11345.4	11353.65	12381.28	15669.95	10996.74	14474.46

Table 3. World's best currently-known results averaged for each class.

	c1	c2	r1	r2	rc1	rc2
K	37.1	12	24.2	5.8	27.8	8.5
T	7332.91	4019.16	8834.62	7888.37	7810.44	6080.93

often at the "feasibility border", and squeezing them help obtain well-optimized schedules faster, thus becomes beneficial. On the other hand, limiting the number of iterations leads to decreasing the average τ_C (see GES(—) compared with GES(M), GES(R) with GES(MR), and so on for rc1 and rc2). It helps avoid a too long exploitation of a partial solution with "difficult" requests in the EP.

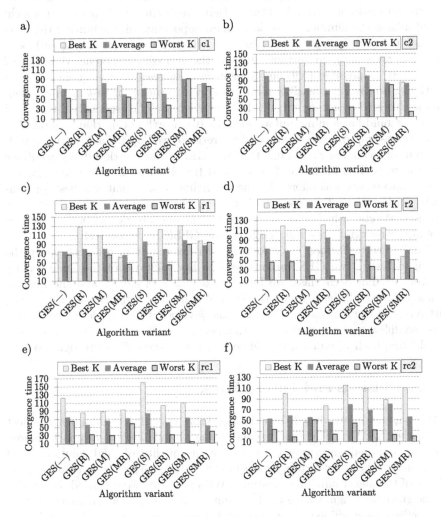

Fig. 1. Convergence time (in seconds) obtained using various GES variants.

Table 4. The guidelines for choosing the GES variant based on the test characteristics.

Class	Minimize K	Minimize T	Minimize K and T	Minimize time
c1	GES(R)	GES(SR)	GES(SR)	GES(R)
c2	GES(R)	GES(R)	GES(R)	GES(R)
r1	GES(R)	GES(MR)	GES(R)	GES(MR)
r2	GES(MR)	GES(MR)	GES(MR)	GES(R)
rc1	GES(R)	GES(MR)	GES(MR)	GES(R)
rc2	GES(M)	GES(M)	GES(M)	GES(M)

The guidelines of selecting an appropriate algorithm variant based on the test characteristics are given in Table 4. Here, we distinguish four cases in which: (i) optimizing the number of trucks is more important, (ii) optimizing the total distance traveled during the service is of a higher importance, and (iii) both objectives are equally important, and (iv) the average convergence time should be as short as possible (without a significant decrease in the quality of final solutions). Squeezing infeasible solutions appears crucial for test instances containing clustered customers with tight time windows (c1), especially if the travel distance should be as minimum as possible. It is worth noting that randomizing the EP helps improve the results and converge to well-optimized routing schedules in most cases. For r2 and rc2, the same GES variant retrieved best solutions with the respect to both the fleet size and traveled distance. Importantly, the fastest convergence was offered by the algorithm version that was best for minimizing either K or T across all LL classes. For some tests (c2 and rc2), the enhanced searches which allow for balancing the optimization of both objectives were also converging very fast (GES(R) and GES(M), respectively).

Finally, we performed the two-tailed Wilcoxon test in order to verify the null hypothesis: "two GES variants lead to the solutions of the same quality on average". For each class, we investigated the average K, and the average T independently. Table 5 shows the level of statistical significance—the greyed part refers to the analysis of T values (the p-values which are less than 0.05—thus the differences are statistically significant—are boldfaced). For most cases (considering both K and T), the differences across the GESes are significant.

5 Conclusions and Future Work

In this paper, we introduced the enhanced GES algorithm to minimize the fleet size in the PDPTW. The experiments on the Li and Lim's benchmark showed how the proposed improvements influence the quality of final solutions (and which are beneficial for certain classes of tests) along with the convergence capabilities of the algorithm. The two-tailed Wilcoxon tests were carried out to verify the statistical significance of the results. We gave a clear guidance on how to select a proper guided search variant for each class of scheduling problems, based on its characteristics and the importance of objectives (i.e., minimizing

Table 5. The level of statistical significance obtained using the two-tailed Wilcoxon test for each pair of the GES variants. The results which are statistically significant ($p < 0.05$) are rendered in boldface. The white part of the table refers to the analysis of K values, whereas the grey part—to the analysis of T values.

	GES(—)	GES(R)	GES(M)	GES(MR)	GES(S)	GES(SR)	GES(SM)	GES(SMR)
GES(—)	—	0.4473	0.215	0.7263	**0.0009**	**0.0071**	**<0.0001**	**0.0011**
GES(R)	0.0536	—	0.0658	0.2113	**0.0001**	**0.0002**	**<0.0001**	**0.0001**
GES(M)	0.5687	**0.0271**	—	0.4065	0.1615	0.3843	**<0.0001**	**0.0033**
GES(MR)	0.4122	0.6101	**0.0188**	—	**0.0366**	**0.0324**	**<0.0001**	**0.0003**
GES(S)	**0.0226**	**0.0003**	0.0688	**0.0019**	—	0.4473	**0.008**	0.1868
GES(SR)	0.3125	**0.0183**	1	**0.0477**	**0.0332**	—	**0.008**	**0.0366**
GES(SM)	**0.0039**	**0.0001**	**0.0012**	**<0.0001**	0.6312	**0.0143**	—	0.0673
GES(SMR)	**0.0203**	**0.0009**	0.1676	**0.0011**	0.9761	0.0751	0.3472	—

(i) the fleet size, (ii) traveled distance, (iii) both fleet size and travel distance, or (iv) the convergence time). Finally, we reported one new world's best routing schedule obtained using the enhanced GES.

Our ongoing research encompasses comparing the enhanced GES with other techniques in terms of their convergence capabilities. We aim at designing an adaptive algorithm which will dynamically update its parameters based on the search state. This will mitigate the necessity of tuning the parameters before the execution. We plan to validate our approach using the real-world data concerning the transportation network layout and other characteristics. Handling additional constraints would enable the enhanced GES to tackle other rich VRPs.

Acknowledgments. This research was supported by the National Science Centre under research Grant No. DEC-2013/09/N/ST6/03461, and performed using the infrastructure supported by the POIG.02.03.01-24-099/13 grant: "GeCONiI—Upper Silesian Center for Computational Science and Engineering".

References

1. Lau, H.C., Liang, Z.: Pickup and delivery with time windows: algorithms and test case generation. Int. J. Artif. Intell. Tools **11**(3), 455–472 (2002)
2. Kalina, P., Vokrinek, J.: Parallel solver for vehicle routing and pickup and delivery problems with time windows based on agent negotiation. In: Proceeding of the IEEE SMC, pp. 1558–1563 (2012)
3. Baldacci, R., Mingozzi, A., Roberti, R.: Recent exact algorithms for solving the vehicle routing problem under capacity and time window constraints. Eur. J. Op. Res. **218**(1), 1–6 (2012)
4. Nalepa, J., Blocho, M.: Adaptive memetic algorithm for minimizing distance in the vehicle routing problem with time windows. Soft Comput., 1–19 (2015). doi:10.1007/s00500-015-1642-4
5. Ropke, S., Cordeau, J.F., Laporte, G.: Models and branch-and-cut algorithms for pickup and delivery problems with time windows. Networks **49**(4), 258–272 (2007)
6. Bettinelli, A., Ceselli, A., Righini, G.: A branch-and-price algorithm for the multi-depot heterogeneous-fleet pickup and delivery problem with soft time windows. Math. Program. Comput. **6**(2), 171–197 (2014)

7. Parragh, S.N., Doerner, K.F., Hartl, R.F.: A survey on pickup and delivery problems. J. Betriebswirtschaft **58**(1), 21–51 (2008)

8. Bent, R., Van Hentenryck, P.: A two-stage hybrid algorithm for pickup and delivery vehicle routing problems with time windows. In: Rossi, F. (ed.) CP 2003. LNCS, vol. 2833, pp. 123–137. Springer, Heidelberg (2003)

9. Ropke, S., Pisinger, D.: An adaptive large neighborhood search heuristic for the pickup and delivery problem with time windows. Transp. Sc. **40**(4), 455–472 (2006)

10. Nanry, W.P., Barnes, J.W.: Solving the pickup and delivery problem with time windows using reactive tabu search. Transp. Res. **34**(2), 107–121 (2000)

11. Nagata, Y., Kobayashi, S.: Guided ejection search for the pickup and delivery problem with time windows. In: Cowling, P., Merz, P. (eds.) EvoCOP 2010. LNCS, vol. 6022, pp. 202–213. Springer, Heidelberg (2010)

12. Cherkesly, M., Desaulniers, G., Laporte, G.: A population-based metaheuristic for the pickup and delivery problem with time windows and LIFO loading. Comput. Oper. Res. **62**, 23–35 (2015)

13. Nagata, Y., Tojo, S.: Guided ejection search for the job shop scheduling problem. In: Cotta, C., Cowling, P. (eds.) EvoCOP 2009. LNCS, vol. 5482, pp. 168–179. Springer, Heidelberg (2009)

14. Nagata, Y., Bräysy, O.: A powerful route minimization heuristic for the vehicle routing problem with time windows. Op. Res. Lett. **37**(5), 333–338 (2009)

15. Blocho, M., Czech, Z.: A parallel memetic algorithm for the vehicle routing problem with time windows. In: Proceeding of the 3PGCIC, pp. 144–151 (2013)

How to Generate Benchmarks
for Rich Routing Problems?

Marcin Cwiek[1,2], Jakub Nalepa[1,2,3]([✉]), and Marcin Dublanski[1,2]

[1] Deadline24, Gliwice, Poland
{mcwiek,jnalepa,mdublanski}@deadline24.pl
[2] Future Processing, Gliwice, Poland
{mcwiek,jnalepa,mdublanski}@future-processing.com
[3] Silesian University of Technology, Gliwice, Poland
jakub.nalepa@polsl.pl

Abstract. In this paper, we show how to generate challenging benchmark tests for rich vehicle routing problems (VRPs) using a new heuristic algorithm (termed HeBeG—**He**uristic **Be**nchmark **G**enerator). We consider a modified VRP with time windows, in which the depot does not define its time window. Additionally, the taxicab metric is utilized to determine the distance between travel points, instead of a standard Euclidean metric. HeBeG was used to create a test set for the qualifying round of Deadline24—an international 24-hour programming marathon. Finally, we compare the best results submitted to the server during the qualifying round of the contest with the routing schedules elaborated using other algorithms, including a new heuristics proposed in this paper.

Keywords: Vehicle routing problem · Benchmark · Heuristics · VRPTW

1 Introduction

Solving the routing problems (VRPs) became a core issue in logistics and planning. There exist a plethora of different VRP formulations which reflect numerous real-life scheduling circumstances. Since these problems are inherently complex and have a wide practical applicability, they have attracted the research attention over the years. The applications of VRPs include the bus routing, cash delivery to ATMs, waste collection, food and parcel deliveries and many other [1].

In a standard traveling salesman problem (TSP), a single salesperson serves the customers scattered around the map, whereas in the multiple TSP (mTSP), a number of salesmen can be exploited. In practice, vehicles are always characterized by their maximum capacity, thus the capacitated VRP has been proposed to reflect this constraint. Also, each customer defines its own time window during which the service should be performed. The vehicle routing problem with time windows (VRPTW) incorporates this restriction and introduces the time windows that must not be violated in a feasible schedule. It is common that customers are divided into pickup and delivery ones. The former customers expect

© Springer-Verlag Berlin Heidelberg 2016
N.T. Nguyen et al. (Eds.): ACIIDS 2016, Part I, LNAI 9621, pp. 399–409, 2016.
DOI: 10.1007/978-3-662-49381-6_38

their goods to be delivered to the latter ones. This scenario—in turn—is reflected in the pickup and delivery problem with time windows (PDPTW) [2]. It is worth noting that these VRP variants are NP-hard [3,4]. Tackling the mentioned VRPs consists in finding a feasible schedule for serving a set of transportation requests (which may be the delivery requests in the VRPTW, or the pickup and delivery ones in the PDPTW) using a set homogeneous trucks, which start and finish their service in a single depot. It is common to define two objectives of the VRPs (therefore they are multi-objective): (i) to minimize the fleet size, and (ii) to minimize the total distance traveled in a routing schedule.

There are two streams of development towards solving difficult VRPs. The first aims at designing exact algorithms, which encompass branch-and-cut [5,6] and branch-and-price [7,8] techniques, dynamic programming, and other approaches [9,10]. These algorithms are however still not applicable to massive-scale real-world problems due to their execution times. Thus, the approximate methods are being actively developed to handle difficult routing problems in a reasonable time [11,12]. Heuristic algorithms for minimizing the fleet size are divided into construction and improvement techniques. The former methods build a feasible solution from scratch, whereas the latter algorithms aim at improving the initial solution. The approximate techniques (both sequential and parallel) include simulated annealing, tabu searches [13], neighborhood searches [14], agent-based approaches [2], guided ejection searches (GESes) [15], population-based metaheuristics [1,16,17], and other [12].

The emerging algorithms for the VRPTW are usually assessed and compared with other techniques based on the results obtained for two benchmark sets proposed by Solomon and Gehring and Homberger[1]. Since they contain tests with various characteristics (with different tightness of time windows, vehicle capacities and customer locations), they became a standard for evaluating new algorithms. These benchmarks include the instances with the maximum of 1000 customers, which is often well below the number of travel points in real-life problems. Also, the optimal solutions are not known for these tests. In this paper, we tackle the mentioned issues and propose a heuristics to generate difficult VRP tests, which may contain any number of customers. Since our algorithm determines a feasible solution during the creation of the test case, it is possible to determine an optimal (or nearly-optimal) schedules for the elaborated tests.

1.1   Contribution

In this paper, we propose a new heuristic algorithm (HeBeG) to generate difficult benchmark tests for the modified VRPTW. The elaborated tests can contain any number of customers. In HeBeG, we determine the nearly-optimal (when the number of vehicles is considered) solution during the generation of the corresponding test case. Additionally, we present a new simple heuristic algorithm (SIHA) for solving this variant of the VRPTW (however, SIHA can be easily adapted to handle other VRPs). It is worth noting that the instances generated

---

[1] https://www.sintef.no/projectweb/top/vrptw/.

using HeBeG were used during the qualifying round of Deadline24[2]—an international 24-hour programming marathon. Finally, we compare the solutions of these tests (i) found using the guided ejection search [1], (ii) elaborated with SIHA, and (iii) those submitted by the participants during the contest.

1.2   Paper Structure

Section 2 formulates the considered variant of the VRPTW. The heuristic algorithm for generating benchmark tests is discussed in detail in Sect. 3. In Sect. 4, we present a simple construction heuristics to solve the modified VRPTW. The experimental study is reported in Sect. 5. Section 6 concludes the paper.

2   Modified Vehicle Routing Problem with Time Windows

The VRPTW is defined on a directed graph $G = (V, E)$ with a set V of $N + 1$ vertices representing the customers and the depot, along with a set of edges $E = \{(v_i, v_{i+1}) | v_i, v_{i+1} \in V, v_i \neq v_{i+1}\}$ which represent the connections between travel points. Each vehicle starts and finishes its service at the depot (v_0). The travel costs are given as $c_{i,j}$, where $i \neq j$, $i, j \in \{0, 1, \ldots, N\}$. In a standard VRPTW, the travel cost between two travel points is equal to the Euclidean distance between them. In the modified VRPTW considered in this work, the cost between v_i and v_j (with the coordinates (x_i, y_i) and (x_j, y_j), respectively) is the distance between these travel points in the taxicab (Manhattan) metric:

$$c_{i,j} = |x_i - x_j| + |y_i - y_j| . \tag{1}$$

The non-negative customer demands d_i, $i \in \{0, 1, \ldots, N\}$, where $d_0 = 0$ and the time windows $[e_i, l_i]$, $i \in \{1, 2, \ldots, N\}$ are defined (note that the time window of the depot is infinite, which is in contrast to the VRPTW). The customers are characterized by their non-negative service times s_i, $i \in \{1, 2, \ldots, N\}$.

The fleet consists of K homogeneous vehicles (each of a capacity Q). The route is defined as a set of customers served by a single truck $\langle v_0, v_1, \ldots, v_{n+1} \rangle$, where $v_0 = v_{n+1}$ is the depot. A solution σ (being a set of routes) is feasible if: (i) the capacities of all vehicles are not exceeded, (ii) the service of v_i is started before the time window $[e_i, l_i]$ elapses, (iii) every customer is visited in exactly one route, and (iv) every vehicle starts its service from and returns to the depot.

The primary objective of the modified VRPTW is to minimize the fleet size K, where $K_{\min} = \lceil D/Q \rceil$, and $D = \sum_{i=1}^{N} d_i$. The secondary objective is to minimize the distance $T = \sum_{i=1}^{K} T_i$, where T_i is the distance traveled in the i-th route. Additionally, we introduce a new metric S for assessing feasible routing schedules (this metric was used during the qualifying round of Deadline24):

$$S = \frac{N}{K} + \frac{T_0}{T}, \tag{2}$$

where T_0 is the travel distance when $K = N$ (S is rounded to 3 decimal places).

---

[2] For details see: https://www.deadline24.pl/.

3   Heuristic Algorithm to Generate Benchmark Tests

In this section, we present the algorithm to generate benchmark instances (HeBeG) for the modified VRPTW. However, the proposed technique may be easily adapted to handle other scheduling constraints.

3.1   Benchmark Routes Approach

HeBeG generates benchmark tests along with their semi-optimal solutions. In this algorithm, we propose a reversed order approach—we determine a set of feasible routes at first. The order of visiting the clients should be reasonable, so as to keep the travel distance as short as possible. We assume that each truck (serving a single route) exploits its total capacity, and the visited clients are located one by another on a polygonal chain round a circle (a *benchmark route*).

Algorithm 1. Benchmark data generation for the modified VRPTW.

> **Input**: $size, R_{min}, R_{max}, Q, d_{max}, N, border, satellites$;
> 1: $\rho \leftarrow \emptyset$; $\sigma \leftarrow \emptyset$;
> 2: Generate the depot (randomly in a $(size/2 \times size/2)$ square centered on the map);
> 3: $K \leftarrow (N \cdot d_{max})/(2 \cdot Q)$;
> 4: **for** 1 to K **do**
> 5: $p \leftarrow \emptyset$;
> 6:   Split capacity Q randomly into C_r client demands;
> 7:   Put C_r client slots into p;
> 8:   Select random values for the middle point and the radius (make a base circle);
> 9:   Place clients on a base circle with the radius changes for each client;
> 10:   Randomize each client service time;
> 11:   Find the client (C_{min}^d) with the minimum distance to the depot;
> 12:   Make a benchmark route r starting from C_{min}^d and going round;
> 13:   Generate basic (narrow) time windows for each client using the sequence of r;
> 14:   Extend each client time window randomly;
> 15:   Calculate the final value of time needed for a truck to serve all clients;
> 16:   Adjoin clients p to the semi-optimal solution ρ; Add the route r to σ;
> 17: **end for**
> 18: **return** σ;

The process of generating a benchmark test starts with placing a depot on the map (Algorithm 1, line 2)—it is a random point in a $(size/2 \times size/2)$ square centered on the map. Then, we calculate K, denoting the number of benchmark routes to generate (line 3). This value is found based on the following estimates:

- The average client demand: $d_{avg} = d_{max}/2$,
- The average number of clients served by a single truck: $V_{avg} = Q/d_{avg}$.

Then, it is given as:

$$K = \frac{N}{V_{avg}} = \frac{N \cdot d_{max}}{2 \cdot Q}. \tag{3}$$

The main loop is repeated K times, each time generating a single benchmark route. First, we split the maximum truck capacity Q into customer demands (line 6). Each demand is a random value from the range $[1, d_{max}]$. This process continues until the client demands equals Q. It is worth noting that the number of clients of this route (C_r) is an immediate output of this HeBeG step. Then, we randomly generate the position of the middle point and the radius of a circle which will be the base for client locations (line 8). The radius is drawn from the range $[R_{min}, R_{max}]$ but might be decreased so as all the points of the circle fit in the map. Then, the client locations are generated.

C_r vertices of a regular polygon inscribed into a given circle constitute the initial locations of the clients. Each client distance from the middle of the circle might then be decreased to a random value between 60 % and 100 % of the initial circle radius (line 9). This operation creates the final location for a particular client (Fig. 1a). Additionally, we keep the order in which the clients should be served. This corresponds to the sequence of vertices of the regular polygon.

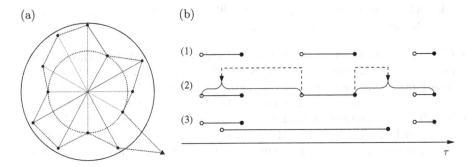

Fig. 1. (a) The depot (annotated as a black triangle in the bottom-right corner) and the clients (black dots) served in a benchmark route (dashed line). The radius of the dotted circle is 60 % of the radius of the base circle. (b) Possible ranges to extend a time window of a sample customer. Three exemplary client time windows are shown (1) before and (3) after the modification. (2) illustrates the possible ranges of extension (annotated as the braced lines). The empty dots represent the beginnings of time windows, the filled ones—the ends of them.

The next HeBeG steps involve generating the client service times (Algorithm 1, line 10)—a random value from the range 0–800 with the following probability distribution: 88 % – range $[1, 200]$; 8 % – range $[201, 800]$; 4 % – no service time, is drawn for each client. Determining the initial time windows (line 13) starts from finding the theoretical time that is needed to travel from the depot to the considered client starting from C_{min}^d (being the customer which is closest to the depot) and going round keeping the order of the previously saved service sequence. Here we consider the service time of already visited clients and possible waiting times at the travel points. The initial time window openings are drawn randomly with a narrow range (± 25 time units) around the theoretical arrival time to each client. The width of a time window is chosen randomly from the range [1, 80], and increased if the end of the time window would be before the arrival of the visiting truck. Furthermore, each client may have its time window extended (line 14). The extension may affect the beginning of the time window, the end of it, both or none. Whenever an extension is applied, the beginning of the time window can be moved somewhere between its current position and the beginning of the previous client time window within the route. The end of the time window can be moved as far as to the end of the time window of the next client (Fig. 1b). This approach allows for "obscuring" the benchmark routes so that they are not obvious to find with the reversed data engineering.

We register the total travel time and distance of the truck serving the clients along the considered route (Algorithm 1, line 15). Then, all generated values are stored (lines 16). The benchmark route finally becomes a part of the test which is being created. Additionally, we add this route to the semi-optimal solution.

3.2   Additional Modifications

In HeBeG, we propose two optional modifications for generating additional routes (the same reversed order approach is applied):

- **Corner Customers**—a single route which serves very distant travel points located in the corners of the map is added to the test case. Clients are randomly placed in the corners of the map, but the possible area for their coordinates contains only four map corners (squares of $size/10$ side length).
- **Satellites**—a number of randomly scattered customers with huge demands (between 90 % and 100 % of the capacity Q) are generated. Each of those clients is served in a separate route, containing this customer and the depot.

4   Simple Heuristic Algorithm for the Modified VRPTW

In this section, we discuss our construction heuristics for the modified VRPTW. This technique was used to solve the benchmark instances generated using the heuristic algorithm discussed in Sect. 3.

The proposed heuristics is given in Algorithm 2. At first, all customers are put into the vector A, which contains the unserved customers (line 2). Then, for each v_i, the neighborhood vector E_i is built (line 3). It contains all customers v_j (where $i \neq j$) which may be visited after serving v_i without violating the time window and capacity constraints. In SIHA, the routes are constructed separately, until all customers are served (line 4). The first customer v_c in a new route r is taken from A (line 6), and is removed from the vector of unserved customers (line 7). Then, r is being expanded if the feasible insertion positions (i.e., those which do not violate the constraints) exist (lines 9–13). For the current customer v_c, the neighborhood vector is analyzed, and the next feasible (unserved) customer is popped from E_c (line 10). Then, it is removed from A and appended to r (line 11). Finally, the route r is added to the (possibly partial) solution σ (line 14). Once all customers are put into σ, it is returned (line 16).

Algorithm 2. Simple construction heuristics (SIHA) for the modified VRPTW.

1: $\sigma \leftarrow \emptyset$;
2: Put all customers v_i $(i = 1, 2, \ldots, N)$ in $\boldsymbol{A}$;
3: Create N neighborhood vectors $\boldsymbol{E}_i$ $(i = 1, 2, \ldots, N)$;
4: **while** $\boldsymbol{A} \neq \emptyset$ **do**
5:     $r \leftarrow \emptyset$;
6:     $v_c \leftarrow$ first customer from $\boldsymbol{A}$;
7:     Remove v_c from $\boldsymbol{A}$;
8:     Add v_c to the route r;
9:     **while** (at least one customer can be appended to r) **do**
10:         $v_n \leftarrow$ first feasible customer from $\boldsymbol{E}_c$ such that $v_n \in \boldsymbol{A}$;
11:         Remove v_n from $\boldsymbol{A}$; Add v_n to the route r;
12:         $v_c \leftarrow v_n$;
13:     **end while**
14:     Add the route r to σ;
15: **end while**
16: **return** σ;

The performance of SIHA depends on the order of customers in $\boldsymbol{A}$ and the neighborhood vectors $\boldsymbol{E}_i$ (the order of ejecting the customers to be re-inserted into a partial solution affects the capabilities of various heuristic algorithms [1]). In this work, $\boldsymbol{A}$ is sorted descendingly by the distance from the depot (the closest customers have the lowest indices in the vector), whereas $\boldsymbol{E}_i$ by the distance from the customer v_i. Then—to diversify the search—these vectors are perturbed by the local swaps of neighboring vector cells. However, the impact of this perturb operation on the SIHA performance needs further investigation.

5   Experimental Results

In this section, we analyze the results obtained for 10 benchmark tests (distinguished by their unique names, r1–r10) generated using our algorithm[3]. The settings of HeBeG used for elaborating these tests are summarized in Table 1. Two exemplary tests are visualized in Fig. 2 (the dark red diamonds indicate the additional corner customers, and the customers with large demands served in separate routes). We compare the routing schedules obtained with the guided ejection search (GES) [1], our simple heuristic algorithm for solving the modified VRPTW (see Sect. 4 for details), and these submitted by the contestants during the qualifying round (which lasted 4 clock hours) of the Deadline24 marathon. The maximum time of the GES and SIHA was $\tau_m = 10$ min.

Table 2 presents the best results retrieved for all benchmark tests using the investigated techniques. We show the best results submitted by the Deadline24 contestants taking into account (a) the number of routes, (b) travel distance, and (c) the value of the metric given in Eq. 2. Additionally, we include the semi-optimal solutions generated using HeBeG. It is worth noting that the smaller

---

[3] For the details of these tests see: http://sun.aei.polsl.pl/~jnalepa/HeBeG/.

Table 1. The HeBeG settings used to generate the benchmark tests.

Id	size	R_{min}	R_{max}	Q	d_{max}	N	border	satellites
r1	200	10	40	400	200	30	yes	0
r2	300	10	20	200	100	50	yes	10
r3	1300	10	45	800	800	250	yes	5
r4	3000	2	40	800	600	1500	yes	0
r5	6000	5	30	900	400	2500	yes	4
r6	6000	1	30	6200	200	4500	no	5
r7	9000	3	40	500	50	5000	yes	10
r8	12000	10	20	500	400	1000	yes	20
r9	21000	1	20	5000	4000	1500	no	20
r10	48000	1	18	10000	10000	2000	no	50

numbers of routes in most cases result in longer travel distances, which is typical for many bi-criterion optimization problems (see e.g., (a) and (b)—K values are significantly larger in the latter case). The GES (which aims at minimizing K) retrieved the best fleet size in 8 cases (for r6 and r7, the best K was found by the contestants). Although SIHA is a very simple heuristics, it retrieved reasonable (albeit the worst) solutions to the generated tests.

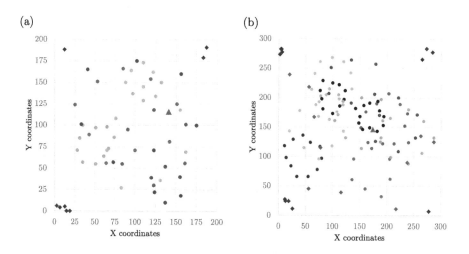

Fig. 2. Two HeBeG tests: (a) r1, (b) r2. The same-color customers are served in one route (excluding large-demand ones—the satellites), the depot is shown as a triangle (Color figure online).

In Table 3, we present the values of the quality metric used during the contest (the higher, the better). A trade-off between optimizing K and T is clearly visible in Table 3(c)—most solutions which were of a highest quality when this

Table 2. The results obtained for all tests: solutions submitted by the Deadline24 contestants and the best with respect to (a) the number of routes, (b) travel distance, and (c) the quality metric, (d) SIHA, (e) GES, (f) HeBeG. The best results (excluding HeBeG) are boldfaced.

Id↓	(a)		(b)		(c)		(d)		(e)		(f)	
	K	T	K	T	K	T	K	T	K	T	K	T
r1	9	**3248**	9	**3248**	9	**3248**	14	5936	**9**	4008	8	3714
r2	**23**	15124	24	**8898**	24	**8898**	30	14230	**23**	11062	23	10052
r3	135	343454	138	**214138**	138	**214138**	162	341836	**132**	256342	131	359166
r4	760	3403264	782	**2378960**	782	**2378960**	886	3440312	**758**	2757378	751	3842836
r5	646	6838672	695	**5143522**	657	5252520	693	6674556	**634**	5806848	630	6241540
r6	**102**	1758248	112	1177490	109	1228372	282	2670890	130	**1140568**	77	1107212
r7	**267**	4314380	322	3836172	278	3854464	423	6455720	269	**3574646**	261	4785040
r8	538	8038050	541	**8014922**	538	8038050	567	10809728	**524**	8697776	521	10040790
r9	812	23118748	815	**17442654**	815	**17442654**	923	24236594	**784**	18266808	770	20163786
r10	1089	**57030058**	1089	**57030058**	1089	**57030058**	1152	88746166	**1067**	62264388	1050	68854760

Table 3. The values of the quality solution metric for all investigated techniques (see Table 2 for acronym descriptions). The best results are boldfaced.

Id↓	(a) S	(b) S	(c) S	(d) S	(e) S	(f) S
r1	10.796	10.796	10.796	6.556	10.034	**11.139**
r2	7.577	**8.817**	**8.817**	6.431	8.348	8.636
r3	9.940	**11.786**	**11.786**	8.848	11.211	9.999
r4	12.105	**13.770**	**13.770**	10.958	13.15	11.698
r5	13.853	14.823	**15.172**	13.363	14.888	14.547
r6	58.396	60.563	60.926	24.88	55.41	**80.693**
r7	31.603	29.291	32.012	20.336	**33.607**	31.062
r8	**12.528**	12.506	**12.528**	10.723	12.281	11.621
r9	14.212	15.932	15.932	12.902	**15.958**	15.483
r10	**17.489**	**17.489**	**17.489**	14.017	16.98	16.403

metric is concerned are characterized by the shortest T. However, for r8, the solution did not have the smallest K nor T. It is interesting to notice that only two semi-optimal solutions (for r1 and r6 tests) appeared to be the highest-quality solutions according to our metric. This indicates that balancing both optimizations (of K and T values) was crucial to converge to the best solutions when two optimization objectives are transformed into a single objective.

6   Conclusions and Future Work

In this paper, we proposed a heuristic algorithm (HeBeG) to generate challenging, large-scale benchmark tests for a modified VRPTW. However, HeBeG can be easily modified to incorporate additional constraints of other routing problems. Also, we introduced a simple heuristics (SIHA) for solving this variant of the VRPTW. We compared the routing schedules for the generated tests retrieved using SIHA, the guided ejection search [1], and those submitted by the participants of the Deadline24 programming marathon. Finally, we investigated the quality of the solutions assessed by the metric utilized during the contest.

Our ongoing research includes generating a full benchmark set containing numerous test instances. Then, we plan to run multiple state-of-the-art algorithms on this set to establish the set of the world's best solutions of our tests. Finally, we will investigate adding new constraints to HeBeG and handling real network layouts. Thus, we will validate our approach with the real-world data.

Acknowledgments. This work was performed using the infrastructure supported by the POIG.02.03.01-24-099/13 grant: "GeCONiI—Upper Silesian Center for Computational Science and Engineering".

References

1. Nalepa, J., Blocho, M.: Adaptive memetic algorithm for minimizing distance in the vehicle routing problem with time windows. Soft Comput. 1–19 (2015). doi:10. 1007/s00500-015-1642-4
2. Kalina, P., Vokrinek, J.: Parallel solver for vehicle routing and pickup and delivery problems with time windows based on agent negotiation. In: Proceedings of IEEE SMC, pp. 1558–1563 (2012)
3. Lau, H.C., Liang, Z.: Pickup and delivery with time windows: algorithms and test case generation. Int. J. Artif. Intell. Tools **11**(03), 455–472 (2002)
4. Cordeau, J.F., Laporte, G., Savelsbergh, M.W., Vigo, D.: Chapter 6 vehicle routing. In: Barnhart, C., Laporte, G. (eds.) Transportation. Handbooks in Operations Research and Management Science, vol. 14, pp. 367–428. Elsevier (2007)
5. Bard, J.F., Kontoravdis, G., Yu, G.: A branch-and-cut procedure for the vehicle routing problem with time windows. Trans. Sci. **36**(2), 250–269 (2002)
6. Ropke, S., Cordeau, J.F., Laporte, G.: Models and branch-and-cut algorithms for pickup and delivery problems with time windows. Networks **49**(4), 258–272 (2007)
7. Abdallah, K.S., Jang, J.: An exact solution for vehicle routing problems with semi-hard resource constraints. Comp. Ind. Eng. **76**, 366–377 (2014)
8. Bettinelli, A., Ceselli, A., Righini, G.: A branch-and-price algorithm for the multi-depot heterogeneous-fleet pickup and delivery problem with soft time windows. Math. Program. Comput. **6**(2), 171–197 (2014)
9. Chabrier, A.: Vehicle routing problem with elementary shortest path based column generation. Comput. Oper. Res. **33**(10), 2972–2990 (2006)
10. Baldacci, R., Mingozzi, A., Roberti, R.: New route relaxation and pricing strategies for the vehicle routing problem. Oper. Res. **59**(5), 1269–1283 (2011)
11. Bent, R., Van Hentenryck, P.: A two-stage hybrid algorithm for pickup and delivery vehicle routing problems with time windows. In: Rossi, F. (ed.) CP 2003. LNCS, vol. 2833, pp. 123–137. Springer, Heidelberg (2003)
12. Parragh, S.N., Doerner, K.F., Hartl, R.F.: A survey on pickup and delivery problems. Journal fur Betriebswirtschaft **58**(1), 21–51 (2008)
13. Nanry, W.P., Barnes, J.W.: Solving the pickup and delivery problem with time windows using reactive tabu search. Transp. Res. **34**(2), 107–121 (2000)

14. Ropke, S., Pisinger, D.: An adaptive large neighborhood search heuristic for the pickup and delivery problem with time windows. Transp. Sci. **40**(4), 455–472 (2006)

15. Nagata, Y., Kobayashi, S.: Guided ejection search for the pickup and delivery problem with time windows. In: Cowling, P., Merz, P. (eds.) EvoCOP 2010. LNCS, vol. 6022, pp. 202–213. Springer, Heidelberg (2010)

16. Nalepa, J., Blocho, M.: Co-operation in the parallel memetic algorithm. Int. J. Parallel Program. **43**(5), 812–839 (2015)

17. Cherkesly, M., Desaulniers, G., Laporte, G.: A population-based metaheuristic for the pickup and delivery problem with time windows and LIFO loading. Comput. Oper. Res. **62**, 23–35 (2015)

Formal a Priori Power Analysis of Elements of a Communication Graph

Jacek Mercik[✉]

Wroclaw School of Banking and Gdansk School of Banking, Wrocław, Poland
jacek.mercik@wsb.wroclaw.pl

Abstract. This paper presents the idea of measuring the formal impact of elements of a communication graph structure consisting of nodes and arcs on its entirety or subparts. Arcs and nodes, depending on the context, can be assigned different interpretations. E.g. in game theory its nodes may represent the players, often referred to as policy makers and arcs symbolize the relationships between them. In another context, however, nodes and arcs of the graph represent elements of technical infrastructure, e.g. a computer. The graph representing the tested relationships is called the communication graph and the influence of the elements on the entire graph (or its subpart) is referred to as power of the element. Taking into account the power of nodes and connections creates so-called incidence-power matrix more completely than the one formerly describing the communication graph.

Keywords: Communication graph · Power index

1 Introduction

To fully understand the idea of power measurement of elements of a communication graph presented in the work, we will analyse an example transmission network, with a goal to transmit a signal between two selected points (Example 1).

Example 1. A signal is sent from A to D using the connections numbered from 1 to 5 (see Fig. 1). Each of these connections can be either functioning or non-functioning. In order to transmit the signal, there needs to be a sequence of functioning connections starting at A and finishing at D. Hence, for the signal to be transmitted

(i)   connection 3 must be functional,
(ii)   at least one of connections 1 and 2 must be functional,
(iii)   at least one of connections 4 and 5 must be functional.

It follows that at least three of the five connections must be functional in order to transmit the signal, but not all sets of connections satisfying this condition lead to the transmission of the signal. The sets of functional connections which lead to the signal being transmitted are as follows: {1,2,3,4,5}, {1,2,3,4}, {1,2,3,5}, {1,3,4,5}, {2,3,4,5}, {1,3,4}, {1,3,5}, {2,3,4}, {2,3,5}.

© Springer-Verlag Berlin Heidelberg 2016
N.T. Nguyen et al. (Eds.): ACIIDS 2016, Part I, LNAI 9621, pp. 410–419, 2016.
DOI: 10.1007/978-3-662-49381-6_39

Consider a simple voting game based on this scenario, in which the connections represent different channels of communication between voters (nodes), each voter has a single vote and the event "a vote is passed" corresponds to the signal being transferred in the situation described above. Note that unlikely in most assumptions in the literature we accept more than one connections between two adjacent nodes[1]. Note also the connection 6 is irrelevant to transmission of signal between nodes A and D.

Assuming the quota of 4, which is equivalent to a grand coalition {A, B, C, D} necessary to transmit the signal from node A to node D through nodes B and C. It is obvious all the nodes (players) {A, B, C, D} are equally valued and their power indices should be the same and node E has no power at all. Each voter of {A, B, C, D} has also the same veto power, i.e. can stop transmission of the signal and node E has no veto power at all. Moreover, since this is a first-degree veto, it cannot be overruled by the rest of players acting alone or together (Mercik 2011, 2015).

When one is analysing the channels of transmission we observe different situation. Assuming the quota of 3 and connections equally important with weight of 1 each. They have the following veto powers:

(a) connection 3 can successfully veto any coalition,
(b) connection 1 (or connection 2) can veto the coalition {3,4,5} (but e.g. connection 1 cannot veto the coalition {2,3,4}, neither can connection 2 veto the coalition {1,3,4}),
(c) connection 4 (or connection 5) can veto the coalition {1,2,3} (but e.g. connection 4 cannot veto the coalition {2,3,5}, neither can connection 5 veto the coalition {1,3,4}),
(d) connection 6 cannot stop action at all.

Hence, connection 3 has a veto of first degree (this veto cannot be overruled) and the remaining connections have a veto of second degree (this veto can be overruled). It is then evident that, assuming the connections outside a coalition use their veto (rather than abstain from it), then the set of winning coalitions coincides exactly with the sets of functional connections which lead to the signal being transmitted in the scenario above.

From the analysis of the above example we can draw the following conclusions:

(1) Communication between players is affected by both their power and the structure of the graph.
(2) Suitable power indices for various elements of the structure may differ if calculated individually.

The paper is set up as follows: After introduction, Sect. 2 presents preliminaries connected with the game theoretical language for modelling a simple voting game. The elements of graph theory describing communication between players are introduced in the following section. Section 3 presents a game model for games in which voters have

---

[1] It could be for example: traditional mail, e-mail, phone calls, personal meeting, etc. In such a case connections' weights can be modified by adding for example different values for to differentiate connections (and different weights of nodes).

unequal weights and there is restricted graph structure describing possible communication between players. The next section is devoted to concept of power index both for players and communication links between them. Finally, there are some conclusions and suggestions for future research.

2 Preliminaries

Let N be a finite set of committee's members, q be a quota, w_j be a voting weight of member $j \in N$.

A game on N is given by a map v: $2^N \rightarrow R$ with v(Ø) = 0. The space of all games on N is denoted by G. A coalition $T \in 2^N$ is called a carrier of v if $v(S) = v(S \cap T)$ for any $S \in 2^N$. The domain $SG \subset G$ of simple games on N consists of all $v \in G$ such that

(i) $v(S) \in \{0, 1\}$ for all $S \in 2^N$,
(ii) $(N) = 1$,
(iii)   of v is monotonic, i.e. if $S \subset T then v(S) \leq v(T)$.

A coalition S is said to be winning in $v \in SG$ f $v(S) = 1$ and losing otherwise. A simple game (N, v) is said to be proper, if and only if it is satisfied that for all $T \subset N$, if $v(T) = 1 then v(N \setminus T) = 0$.

We analyse only simple and proper games[2] where players may vote yes or no.

By $(N, q, w) = (N, q, w_1, w_2, \ldots, w_n)$ we shall denote a committee (weighted voting body) with member set N, quota q and weights $w_j, j \in N$. We shall assume that w_j are nonnegative integers. Let $t = \sum_{j=1}^{n} w_j$ be the total weight of the committee. Therefore, we described so called n-person cooperative game with transferable utility (TU game) introduced by Shapley (1953).

A flat graph G is an ordered pair $G = N, U$ wherein each u branch corresponds to at least one pair of ordered vertices, $x, y \in NxN$ such that $< x, u, y > \in NxUxNx, yy, xx, u, y \equiv y, u, x$. We assume that U is not an empty set, which means that the N set is not empty either. We also assume that in a graph G there are no so-called loops, or $< x, u, x > \notin NxUxN$.

In an analytic form the structure of communication between players, i.e. their ability to form a coalition, will be represented by a graph. Any graph with non-empty sets of nodes (N) and arcs (U), where all nodes and arcs are adequately numbered may be uniquely defined by incidence matrix $A(G) = [a_{ij}]_{nxm}$, where: $n = |N|$ represents cardinality of a set of nodes $N, m = |U|$ represents cardinality of a set of arcs $U, a_{ij} = 1 if$ there is relation u such that $< i, u, j > \in NxUxN$ (where for simplicity i and j denote x_i and y_j respectively). In Table 1 the example from Fig. 1 is presented in incidence matrix form. An incidence matrix may be generalized by assuming $a_{ij} > 0$ instead of $a_{ij} = 1$ if there is relation u such that $< i, u, j > \in NxUxN$ – which is why we presented an incidence matrix in an expanded form by simply showing all pairs of connected nodes.

---

[2] As we will see later, this condition is not necessary for games with graph representing communication between players.

Table 1. Incidence matrix for the transfer of signals from Example 1.

x	A	A	B	C	C	C
u	1	1	1	1	1	1
y	B	B	C	D	D	E

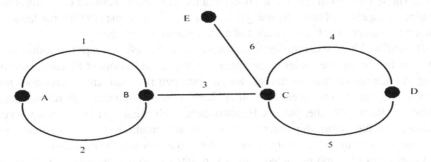

Fig. 1. Signal transfer between points A and D.

Definition. Path s_{ij} is a sequence of arcs and vertices joining vertices i and j, where each node and each arc can occur only once, i.e. $s_{ij} = \{i, u_{ik}, \dots, u_{lj}, j\}$ *dla* $k, l \neq i, j, \ k \neq l$.

3 Graph-Restricted Game Model

The introduction of Shapley values (1953) and Shapley-Shubik power index (1954) allowed for the introduction of cooperative relationships between the players to the TU (transferable utility) game description. It was assumed that any relationship between the players and thus each coalition is equally likely. According to Laplace's criterion, the lack of information on the existence or preferences regarding communication between the players denotes the probability of each of the possible coalitions, of which some are winning ones. The analysis of actually existing situations quickly showed that establishing a coalition of equal probability is unlikely. The first approach assuming the abandonment the establishment of equally likely coalitions (by indicating a possible pre-coalition) was proposed by Aumann and Dreze (1974). This allowed to capture possible similarities or contradictions between the players. For families of such TU games, Owen (1977) proposed a modification of the values in Shapley's game, and consequently a modification of the Shapley-Shubik power index. This modification initially distributes the total amount among the unions, according to the Shapley value, in the game played by the unions, then, once again uses the Shapley value within each union, taking into account their possibilities of joining other unions. Owen (1977) proposed the analogue modification for Banzhaf power index known as Banzhaf-Owen value.

Remember that in the proposed model, the winning coalitions are influenced by the weight of individual players and the nature of connections between them. It can be assumed that the strength of connection remains in some correlation with weights of players. This postulate is evident in a number of approaches to determining power of players. E.g. Myerson (1977) formulates two axioms, of which the first relates to expectations concerning equalling (for a game with a structure) the value of the sum of new index of the power of individual players with the value of the Shapley's game without structure – this axiom seems a natural expectation and we believe should be met by any proposal of a new index. The second axiom refers to the belief that the relationship between two players has equal value for either of them. By using these two axioms, Myerson (1977) formulates the claim on existence of only one solution of a TU game that meets them.

Hamiache (1999, 2012) verifies Myerson's axioms that refer to graph of relationships between the players. He denies the legitimacy of the second axiom of Myerson considering that the interruption of the relationship between two different players does not involve the same consequences if they have different amounts of relationships connecting them with other players. He introduces a value that describes how any given coalition sees the relationship described by the graph G, assuming that the coalition does not need to know the relationships between the players outside the coalition.

Other types of graph restricted games were introduced in Rosenthal (1988) where weights are attributed to the communication arcs, when these weights represent costs of communication or measures of trust or friendship. Vasques-Brage et al. (1996) present hybrid model which generalized the Owen value and the Myerson value when communication and coalitional structures are superimposed. Alonso-Meijide et al. (2009) consider the model with partial cooperation through a graph and an allocation of the total gains among players.

The model proposed in this paper assumes the only thing known is set of players with the structure of the connections between them (graph of communication) and the weighting attributed to each player. Much like Rosenthal, we attribute certain weight to connections without giving them any additional interpretation. These weights are designed to render validity of specific connections resulting only from the importance of the players, described in turn, by their weights (consistently with the second axiom of Myerson). The "weighing" of connections will be achieved through marking of the G graph.

We'll now mark the arcs. We assume that the weights of vertices are non-negative, which means that the marking also applies to the vertices with a weight equal 0, which in TU-cooperative game theory are referred to as empty players.

Definition. Let i and j be two neighbouring vertices, i.e. vertices connected by a relationship u_{ij}. If their respective weights are w_i and w_j, the relationship u_{ij} weights $w_{ij} = w_i + w_j$. If i and j vertices are not adjacent, yet connected by a path, s_{ij}, the weight of the relationship u_{ij} equals for $k, l \in s_{ij}, k \neq i, l \neq j, w_{ij} = w_{ik} + \sum\limits_{\substack{(k, l) \in s_{ij} \\ k \neq l}} w_{kl} + w_{lj} - \sum\limits_{\substack{k \in s_{ij} \\ k \neq i, j}} w_k$. There can be a

number of paths connecting the vertices i and j. In this case, each of these paths will be distinguished, s_{ij}^r, for $r = 1, 2, \ldots$ as well as w_{ij}^r. Moreover, the weight of each r paths will meet the inequality $w_{ij}^r \geq w_i + w_j$ and can be different, if necessary.

Definition. If for the path s_{ij} its weight w_{ij} meets the inequality $w_{ij} \geq q$, the path s_{ij} is referred to as a q-path. Note that q-paths correspond to winning coalitions in the game $[N, v, U]$ but such coalitions do not have to be minimal (i.e. Rae's postulate is not met). Moreover, they are not equivalent to coalitions present in the Shapley-Shubik power index calculation procedure. Furthermore, depending on the application of the proposed approach, the value q and, consequently, the q-path will change. For *Example I*, the value of q equals the weight of the shortest path between points A and D, i.e. the path whose weight is the smallest. In this example, using the value q lower than the one obtained from the shortest path will introduce a discussion of the analysis of the transmission of signals emitted from the node A, but not necessarily reaching the node D, and that's assuming that all the elements are efficient (there is no veto). Such a description allows for the analysis of e.g. information propagation in networks.

Definition. The G graph exclusively specifying the relationships between vertices may therefore be modified to graph G_w where any non-zero value a_{ij} that occurs in the matrix of incidence $A(G)$ is replaced by the corresponding value w_{ij}, thus forming a weighted incidence matrix $A_w(G)$.

Note that if the weight of one of the vertices is zero (i.e. empty player), the weight of the arch connecting it directly to another vertex with a non-zero weight will remain non-zero, this the incidence matrix $A_w(G)$ retains relationships between vertices. We interpret it in such a way that sometimes, for maintaining a route connecting two selected nodes, a node with zero weight must be enabled. In the TU cooperative game theory such vertex represents an empty player, which in our model, however, should be assigned a non-zero value of the power index (a winning coalition will not be possible without it, and thus it has power). It seems the approach to the problem of forming a coalition resulting from the graph of incidence allows such a possibility and distinguishes it from the previous a priori approach e.g. for Shapley's and Myerson's values.

If the graph describes a structure in which there is no start or end nodes, in such a graph we consider all the paths from such node. If in the analysis of a specific situation a node must be present (as the start or end of a path), all the analysed paths must contain it.

4   Power Indices of Graph-Structured Game

A power index is a mapping $\varphi: SG \rightarrow R^n$. For each $i \in N$ and $v \in SG$, the ith coordinate of $\varphi(v) \in R^n$, $\varphi(v)(i)$, is interpreted as the voting power of player i in the game v. In the literature there are two dominating power indices: the Shapley-Shubik (1954) power index and the Banzhaf (1965) power index. Both indices are based on the Shapley value concept (Shapley 1953) and belong to so called family of a priori power indices.

The Shapley-Shubik and Penrose-Banzhaf definitions of power indices are directly obtained from characteristic function games where a marginal value of power excess introduced into winning coalition is calculated[3].

Given the fact that relations between players are described by the set graph structure, we can now determine the power index of G graph element as follows.

---

[3] In Turnovec et al. (2008) one can find introduction of power indices without games theory but based on concept of permutations and their probability.

Let s_{ii} be a path consisting of only one player (excluding the possibility of loops in the graph describing the structure). Its length, by assumption, is equal to the weight value of the node, i.e. w_i.

Let S_i be the set of all paths s_i (including q-paths) passing through the i-th element (an arc or a node). Let S_i^q be the set of all paths s_i^q passing through the i-th element of the graph (an arc or a node) whose weight is not less than q, $S_i^q \subseteq S_i$.

Let S_i be the set of all paths s_i (including q-paths) extending from the i-th node. Let S_i^q be the set of all paths s_i^q extending from the i-th node, whose weight is not less than q, $S_i^q \subseteq S_i$. Note $S_A^{q=3}$ is the set of all $q(=3)$-path for example I.

$S = \cup_{i=1}^N S_i$ is the set of all paths in the graph described by the incidence matrix $A_w(G)$ and $S^q = \cup_{i=1}^N S_i^q$ is the set of all paths with length not less than q and starting with the $i = 1, 2, \ldots, N$ nodes.

Note that in a similar way we can define S_i i.e. a set of all paths ending in a i node. From the symmetricity of the definition $S^q = \cup_{i=1}^N S_i^q = \cup_{i=1}^N S_i^q$.

Thus, for the i-th node (a player) of a graph structure described by G graph an absolute power index is defined as follows[4]:

$$\varphi_i^p(v) = \frac{|S_i^q|}{|S|} = \frac{\left|S_i^q\right|}{\left|\bigcup_{i=1}^N S_i\right|}, \tag{1}$$

where p denotes power index of a player. By analogy, you can define an a priori power index for arcs (connections) $\varphi_i^c(v)$, where c stands for communication (and an arc between nodes for a given path).

The above definition also shows the calculation algorithm of both proposed indexes.

The proposed power index does not match the Shapley-Shubik power index, as every player present on a particular path is a pivotal player (their absence eliminates the path as a connection between the start and end nodes, or it may shorten the path so that it is no longer a type q-path for a given q value). Moreover, in axiomatics of both classical indices, i.e. Shapley-Shubik power index and Banzhaf power index there is a postulate (an axiom) of an empty player: an empty player has no power. Hence the proposed indices $\varphi_i^c(v)$ and $\varphi_i^p(v)$ as not meeting the postulate cannot be considered equal to Shapley-Shubik or Banzhaf indexes. The axiomatics of the proposed indices will be dealt with in subsequent papers.

Let's do relevant calculations, for the *Example 1*. Assuming that (to simplify the calculation) weights of all the nodes are equal to one. Hence the weights of all the arcs are also equal and amount to respectively $w_{ij} = 2$ for $\neq j$. Reaching designated end (D) from the designated beginning (A) in this case should require adopting $q = 4$. Thus, the appropriate four q-paths are: A1B3C4D, A1B3C5D, A2B3C4D and A2B3C5D. Each of weight equal 4, and thus equal to the set value q. This means that each of the nodes is equally important, because each occurs once in each q-path. Thus, the a priori value

---

[4] We use the | . | operator to denote the cardinality of a finite set.

of the power index of each node is $\varphi_i^p(v) = \frac{1}{4}$. Note that the introduction of the veto, which in this example is the inability of signal transmission through the node, does not change the value of the a priori index against the symmetry of the importance of individual nodes.

Also, please note that any fault on any of the connections of *Example 1* can be either eliminated through alternatives (connections 1, 2, 4, 5) or cannot be eliminated (connection 3) or simply not eliminated (connection 6). This leads to the conclusion that this fact should be reflected in the power indices also for the relationships. Proceeding as in the case of a power index for players we find that connections 1, 2, 3, 4, 5 and 6 exist in q-paths with respective frequencies: 2:2:4:2:2:0. Thus the a priori power index for connections without the possibility of veto, $\varphi_i^c(v)$ is respectively $\frac{1}{2}$, $\frac{1}{2}$, 1, $\frac{1}{2}$, $\frac{1}{2}$, 0. Note that the introduction of veto (inability of signal transmission through particular arc) makes the corresponding a priori index values again $\frac{1}{2}$, $\frac{1}{2}$, 1, $\frac{1}{2}$, $\frac{1}{2}$, 0 as no path between the vertices A and D is a q-path without connection 3, while connections 1, 2, 4, 5 are interchangeable (connection 6 belongs to no q-path in this example). The information about the a priori power of individual arcs can thus be added to a description of the graph and so-called incidence-power matrix can be created. Table 2 presents such a matrix for the Example 1.

Table 2. A priori incidence-power matrix for the transfer of signals from Example 1.

x	A	A	B	C	C	C
u	1/2	1/2	1	1/2	1/2	0
y	B	B	C	D	D	E

As it is seen from the *Example 1*, the node E is irrelevant and, what's more, this node is connected with 0 power connection. U values make also possible to recognise what kind of veto maybe associated with a given link. For $u = 1$ it is possible to introduce veto of first degree (cannot be overruled), for $0 < u < 1$ veto of second degree (can be overruled) and for $u = 0$ veto cannot be introduced at all.

Example 2. Let's analyse the previous example without defining what node is starting or ending nodes. Therefore there are twelve following q-paths for $q = 4$: A1B3C4D, A1B3C5D, A2B3C4D, A2B3C5D, A2B3C6E, A1B3C6E, D4C3B1A, D4C3B2A, D5C3B1A, D5C3B2A, E6C3B1A and E6C3B2A. The a priori power index for connections is presented in Table 3.

Table 3. A priori incidence-power matrix for Example 2.

x	A	A	B	C	C	C
u	1/2	1/2	1	1/2	1/2	1/3
y	B	B	C	D	D	E

Comparing results for Examples 1 and 2 one may notice that connection 6 has positive power value (1/3) but this connection is relatively weaker 1/3:1/2.

A priori incidence-power matrix contains not only local values associated with the connection between two given vertices but also describes the nature of this connection against the other relationships in the communication graph. Thus, we have modified the values for the relationship u so it describes not only the connection between given nodes, but also evaluates this connection globally for the entire communication graph, $u = [\varphi_i^c(v)]$ *for links between nodes*. We also believe that the method of marking arcs in graph G introduces, by incorporating the weights of players, information on the power of individual players to the features of their connecting arches.

Note that breaking the connection between the nodes and the arches which occurs in the marking process does not change our method although the problem of considering the power of nodes (players) remains open.

5 Conclusions

The presented way of measuring the impact of the structure of connections between elements of different weights allows to determine the relative importance of these elements. The introduced concept of q-path, in turn, allows to consider the impact of the graph nodes on its structure. The combination of these two approaches allows to create an incidence-power matrix that describes not only the connections in graph (including polyadic relationships), but also an analysis of possible changes associated with the change of weight and the strength of their influence.

The presented approach belongs to a class of a priori power indices different than those known to date; it allows multiple relationships between the players, and allows non-fulfillment of some of the axioms characteristic for power indices existent in literature. The future works should examine which of the axioms are met and assess the uniqueness of the solution.

The use of an incidence-power matrix for modeling of the information flow in networks is a separate issue. This could allow future studies on both network reliability (veto issue) as well as dissemination of information, including the problem of censorship and restrictions imposed on participants of such information exchange. In particular, the proposed approach can be used to model the flow of e.g. intellectual capital.

References

Alonso-Meijide, J.M., Alvares-Mozos, M., Fiestras-Janeiro, M.G.: Values of games with graph restricted communication and a priori unions. Math. Soc. Sci. **58**, 202–213 (2009)

Aumann, R.J., Dreze, J.: Cooperative games with coalitional structures. Int. J. Game Theory **3**, 217–237 (1974)

Banzhaf III, J.F.: Weighted voting doesn't work: and mathematical analysis. Rutgers Law Rev. **19**, 317–343 (1965)

Hamiache, G.: A value with incomplete communication. Games Econ. Behav. **26**, 59–78 (1999)

Hamiache, G.: A matrix approach to TU games with coalition and communication structures. Soc. Choice Welfares **38**, 85–100 (2012)

Mercik, J.: On a priori evaluation of power of veto. In: Herrera-Viedma, E., García-Lapresta, J.L., Kacprzyk, J., Fedrizzi, M., Nurmi, H., Zadrożny, S. (eds.) Consensual Processes. STUDFUZZ, vol. 267, pp. 145–156. Springer, Heidelberg (2011)

Mercik, J.: Classification of committees with vetoes and conditions for the stability of power indices. Neurocomputing **149**(Part C), 1143–1148 (2015)

Myerson, R.B.: Graphs and cooperation in games. Math. Oper. Res. **2**, 225–229 (1977)

Owen, G.: Values of games with a priori unions. In: Henn, R., Moeschlin, O. (eds.) Mathematical Economics and Game Theory. LNEMS, vol. 141, pp. 76–88. Springer, Heidelberg (1977)

Rosenthal, E.C.: Communication and its costs in graph-restricted games. Soc. Netw. **10**, 275–286 (1988)

Shapley, L.S., Shubik, M.: A method of evaluating the distribution of power in a committee system. Am. Polit. Sci. Rev. **48**(3), 787–792 (1954)

Shapley, L.S.: A value for n-person games. In: Kuhn, H.W., Tucker, A.W. (eds.) Contributions to the Theory of Games, vol. II. Annals of Mathematical Studies, vol. 28, pp. 307–317 (1953)

Turnovec, F., Mercik, J., Mazurkiewicz, M.: Power indices methodology: decisiveness, pivots, and swings. In: Braham, M., Steffen, F. (eds.) Power, Freedom, and Voting, pp. 23–37. Springer, Berlin (2008)

Vasques-Brage, M., Garcia-Jurado, I., Carreras, F.: The Owen value applied to games with graph-restricted communication. Games Econ. Behav. **12**, 42–53 (1996)

Angiogenic Switch - Mixed Spatial Evolutionary Game Approach

Michal Krzeslak, Damian Borys$^{(\boxtimes)}$, and Andrzej Swierniak

Department of Automatic Control, Silesian University of Technology,
Akademicka 16, 44-100 Gliwice, Poland
krzeslak.michal@gmail.com, {damian.borys,andrzej.swierniak}@polsl.pl

Abstract. The main goal of this paper is to study properties of a game theoretic model of an angiogenic switch using Mixed Spatial Evolutionary Games (MSEG). These games are played on multiple lattices corresponding to the possible phenotypes and give the possibility to simulate and investigate heterogeneity on the player-level in addition to the population-level. Furthermore, diverse polymorphic equilibrium points dependent on individuals reproduction, model parameters and their simulation are discussed. The analysis demonstrates the sensitivity properties of MSEGs and the possibility for further development of spatial games.

Keywords: Angiogenic switch · Intelligent information systems · Evolutionary games · Heterogeneity · Spatial population dynamics · Sensitivity

1 Introduction

One of the crucial steps and also necessary conditions for cancer invasiveness and motility is creation and development of an autonomous blood vessel network. During this process, called angiogenesis, new blood vessels are created basing on the existing network of vessels. Tumor cells, like any other cells, depend on nutrients and oxygen supplies, as well as they require an excretion of products of metabolic processes - toxic wastes and carbon dioxide. That is why, taking into account the excessive proliferation rates, without the tumor neovasculature the size of the tumor mass cannot exceed some limits related to penetration range of supplies from existing vasculature. Tumor angiogenesis is the proliferation of blood vessels network, which penetrates into cancerous growths, supplying nutrients and oxygen and removing waste products. It is released with cancerous tumor cells producing molecules sending signals to surrounding normal tissue. This signaling activates certain genes in the host tissue, which make proteins to promote a growth of new blood vessels. During tumor progression, an "angiogenic switch" is activated causing normally quiescent vasculature to sprout new vessels that help sustain expanding neoplastic growths [1,2]. The angiogenic switch is considered as a discrete event in the tumor development. The switch to the angiogenic phenotype involves a change in the local equilibrium between activators

© Springer-Verlag Berlin Heidelberg 2016
N.T. Nguyen et al. (Eds.): ACIIDS 2016, Part I, LNAI 9621, pp. 420–429, 2016.
DOI: 10.1007/978-3-662-49381-6_40

and inhibitors of angiogenesis and is a result of tilt towards pro-angiogenic regulators (Fig. 1). This signaling activates certain genes in the surrounding tissue that produce proteins that promote the growth and sprouting of blood vessels and thus tumor development. Tumor abnormal vasculature also alters tumor microenvironment and allows for growth and progression of a tumor. There are few mechanisms that cancer cells launch new vessels formation: sprouting from existing vessels (angiogenesis), recruitment of bone marrow-derived endothelial progenitor cells to form new vessels (vasculogenesis), and splitting a single vessel into two (splitting angiogenesis). Also cancer cells, to obtain nutrients for their growth, can intercept existing blood vessels or incorporate itself into the vessel wall. Despite numerous efforts to explain this process, it is still not clear why cancer so easily breaks down the regulatory pathways involved in the control of angiogenesis. One of the important questions is that about the intelligence of cancer cells. How much of this ability is genetically encoded and how much is related to a social intelligence of cancer cell population? Angiogenesis is a critical component of tumor metastasis, and highly vascular tumors have the potential to produce metastases [3]. Developing vascular network into the tumor mass, it provides an efficient route of exit for cancer cells from the primary site and enter the blood stream. This process eases the entry of cancer cells into the blood circulation by building the network of immature, loosely connected with slight basement membrane blood vessels that are highly permeable. Likewise the process of angiogenesis is very complex, being a well-orchestrated sequence of events involving endothelial cell migration, proliferation; degradation of tissue; new capillary vessel formation; loop formation and crucially, blood flow through the network. Once there is blood flow associated with the nascent network, the subsequent growth of the network evolves both temporally and spatially in response to the combined effects of angiogenic factors, migratory cues via extracellular matrix and perfusion-related haemodynamic forces. The spatial aspects are usually approximated by simple reaction-diffusion process, thus relating the change in some tumour cells to their diffusion in space, as well as their proliferation (see e.g. [4]). There exist many models which intend to fully reflect the complexity of the biological process and allow accurate simulations The list of mathematical tools includes partial differential equations (e.g. [5,6]), stochastic differential equations [7,8], random walk models [9], cellular automata [10,11], multi-scale-field models [12] and many others. Nevertheless, to answer some fundamental questions about the role of the interaction between tumor cells in the process of angiogenesis, it is also thoughtful to start the analysis with models under simplifying assumptions. One way to do this is to use machinery offered by the evolutionary games theory (EGT) proposed by John Maynard Smith and George Price [13] which combines the game theory (developed for the needs of the economy applications) with Darwinian fitness and species evolution. This combination seems to be especially tailored to answer the previously posed question about the role of evolution, on one hand, and social intelligence, on the other, in control of the angiogenic switch and its implications. Such combination allows to study the dynamics of changes due to interactions between different individuals described by different phenotypes (strategies, approaches) in a heterogeneous population. Individuals may compete or cooperate based on their encoded behaviour, instincts or evolutionary traits (phenotypes) without any rationality. The result of these interactions is a change of the degree of evolutionary adjustment by achieving access to food supplies, females, or life space. This change is called a payoff, and the evolutionary adjustment is termed fitness. The phenotypes frequency of occurrences in a population are changed and may achieve a stable monomorphic or polymorphic equilibrium due to different adjustments to the environment and due to different payoffs gained from interactions between individuals. Achieved phenotypic equilibrium is called the Evolutionary Stable Strategy (ESS) or Evolutionary Stable State. ESS is a phenotype that has been adapted by the majority of the population and cannot be repressed by other phenotypes. The opposite situation is possible, so ESS may coexist stably with other phenotypes or even suppress them and dominate in the population. The Evolutionary Stable State is the situation when the polymorphic population (more than one phenotype exists) is stable and resistant to the inflow of mutants or to environmental changes.

Fig. 1. The angiogenic switch.

2   Game Theoretic Models of Angiogenesis

Application of EGT allows to simulate and study the dynamic of changes and the final phenotypic structure of the population, which can be heterogeneous (polymorphic) or may be totally dominated by one phenotype (monomorphic). First approaches for application of evolutionary games in the modelling of interactions and communication between tumour cells were proposed in 1997 by Tomlinson and Bodmer [14]. Their work started a series of other papers describing different models containing different kinds of phenotypes and tumour populations (see [15,16] for surveys). Within one of these papers [17] the algorithm for spatial analysis of the phenomena of carcinogenesis, SEGT has been proposed by Bach and co-workers. SEGT suggests that each site (cell) on the spatial lattice reflects a single-strategy player. The local payoff for each player is the sum of payoffs due to interactions (according to the payoff matrix) with players in the neighbourhood. The player could be also treated as multi-strategic one, that can follow particular strategy with some probability. Thus, in the case of biological interpretations an individual, instead of homogenous, is heterogeneous and contains mixed phenotypes. Spatial games of this type proposed by us in [19] are called Mixed Spatial Evolutionary Games (MSEG). In fact, the game is performed on a multiple lattices (dependent on the number of defined phenotypes in the model), where each lattice (forming torus) represents a particular phenotype (as the frequency of occurrence) of the player. For the computation of the local adaptation, the sum of the payoffs between each phenotype (within two players) multiplied by their frequency of occurrence is calculated first. The second step is the summing of these values for each player in the neighbourhood. We find out that diversity of the phenotypes on the player level probably describes biological phenomena in more accurate way. However, for the sake of simplicity and following the way of reasoning from SEGT, the phenotypic composition of the population still will be analysed at the population level by measuring the overall contribution of the particular phenotype in the population. Both SEGT and MSEG algorithms contain 3 main steps: (1) payoff updating - the local fitness of each player is calculated taking into account the local neighbourhood; (2) site selection - 10 % of the sites from the lattice are chosen; (3) reproduction - players in the neighbourhood of the chosen site compete. In addition to deterministic (D) and probabilistic (P) reproductions proposed in [17], a different approach to the player interpretation (polyphenotypic description) allows to develop and use other reproductions: - the weighted mean of the best cells (C) - is computed for the players with the highest scores in accordance with the players payoffs; - the weighted mean of the best interval (I) - players are divided into intervals in accordance with their payoffs. It is computed only for the players from the best interval. When payoffs are equal, a tie is settled randomly for SEGT and for MSEG the average between phenotypes is computed. One of the problems considered in the seminal study of Tomlinson and Bodmer [14] is related to results of interactions between tumor cells in the context of production of proangiogenic growth factors. The model is the simplest possible. It contains only two phenotypes and two parameters in the payoff table. The phenotypes are: $A+$ cells produce proangiogenic growth factors (in paracrine fashion), $A-$ cells do not produce growth factors (baseline). The parameters are: i the cost of proangiogenic factor production, j the benefit of receiving growth factor. The payoff matrix (Table 1) should be read vertically. The main result from analysis of the model is that to reach stable dimorphism between the phenotypes cost of producing growth factors i should be smaller than benefit j. Resulting frequencies of occurrences are then dependent on the ratio of differences between the benefit and the cost. In the opposite situation A- is a strategy that is evolutionary stable and dominates the population. This model was extended in [17] where three directly interacting players are assumed. The additional requirement for the efficiency of the release of growth factor is a sort of synergy between players. The benefit j is applied only when at least two A+ cells interact at the same time. The results of this model are more differentiated than in the previous case. When $j < 2i$ then A- dominates population independently of initial frequencies, for $j = i$ and A+ less frequent than A- initially, the result is the same, but in the opposite case stable dimorphism may be reached with equal frequencies. If $j > 2i$ then two different scenarios may be observed. In the case of initial frequencies of A+ less than 30 %, the result is domination of A- in the population, in opposite case we are led to a stable dimorphic equilibrium. Then the authors consider the spatial version of the model. The number of parameters in the payoff table is increased by the introduction of different benefits of receiving growth factors by their producers and non-producers. The authors study results for different rules of site-selection, reproduction and different sizes of the neighbourhood. The results are compared with the mean-field model. The extension proposed in our study is to take into account heterogeneity on the single cell level which leads to MSEG model of inter-action between tumor cell. The line of reasoning follows study of hawk and dove game presented by us in [19]. Since the model contains two phenotypes therefore two layers (two lattices) are used to define the players phenotypic composition in MSEG. The sum of particular frequencies of occurrence equals 1 (exactly as for the mean-field model). For example, take two individuals: X (A+:0.6, A-:0.4) and Y (A+:0.3, A-:0.7). The local payoff for the interaction between X and Y is: $0.6 \cdot 0.3(1 - i - j) + 0.6 \cdot 0.7 * (1 - i - j) + 0.4 \cdot 0.3 \cdot (1 + j) + 0.4 \cdot 0.7$.

Table 1. Payoff matrix

Strategies	A+	A-
A+	1-i+j	1+j
A-	1-i+j	1

We assume four different schemes of reproduction: deterministic D, proba-bilistic P, the weighted mean of the best cells C, and weighted mean of the best intervals I. Different values of parameters of the payoff table and reproduction schemes are considered. The results are compared with the mean field case.

3   Results

The possibility of coexistence between all phenotypes can be studied by the Bishop-Canning theory [18] which states that the phenotypes mean payoff (1) in the population shall be equal for the stable polymorphism.

$$E(A^+) = 1 - i + j$$
$$E(A^-) = 1 + j \cdot A^+ \tag{1}$$

A+ and A- refers to frequencies of occurrence of the phenotypes in the popu-lation. Using the Bishop-Canning theory one can compute expected frequency of occurrences for cells that produce proangiogenic growth factors. Taking into account that the sum of A+ and A- equals 1 the expected frequencies of occur-rence are:

$$A^+ = \frac{j - i}{j} \quad and \quad A^- = \frac{i}{j} \tag{2}$$

Since A+ and A- shall be greater than 0 and less than 1 to achieve a stable poly-morphic, then $i < j$, otherwise the population is dominated by A-. The mean-field (not spatial) results for this model are independent of the initial frequencies of occurrence. Those results shall be used for a comparison with spatial games.

The first set of simulations is performed for $i < j$. It is a scenario where the population is polymorphic, and the frequency of occurrences conforms Eq. 2 for mean-field games (RD). Table 2 (left side) shows final frequencies of occurrences of A+ and A- for different reproduction types and RD game as well. Results for spatial games have been obtained by following the algorithm presented in Sect. 2. For the weighted mean reproductions, 3 cells and 3 intervals have been chosen for following simulations. Two general features may be seen. First one is that the MSEG results follow the equilibrium points from RD, so whenever A+ increases in the mean-field game (due to changing costs i), then A+ also increases for all reproductions in MSEG. The second one is that weighted mean reproductions promote A+ in all considered cases in comparison with another reproductions and with RD. Figures 2 and 3 presents final lattices for two sets of parameters. In both cases, for probabilistic and deterministic reproductions, some clusters and regular structures of the players with the same phenotypic composition are possible. However, those structures may be unstable, what can be seen on Fig. 2 where the changes of averages of the particular phenotypes through following generations are shown. For reproductions based on weighted mean, the entire lattice is smooth, and dynamics of changes are stable. The sec-

Table 2. Results for $i < j$ and $i > j$

| | $i < j$ | | | | | | | $i > j$ | | | | | | |
	i	j	RD	P	D	C	I	i	j	RD	P	D	C	I
A+	0.2	0.8	0.75	0.63	0.65	0.83	0.83	0.8	0.2	0	0	0	0.03	0
A-			0.25	0.37	0.35	0.17	0.17			1	1	1	0.97	1
A+	0.4	0.8	0.5	0.38	0.44	0.61	0.67	0.8	0.4	0	0	0	0.04	0.03
A-			0.5	0.62	0.56	0.39	0.33			1	1	1	0.96	0.97
A+	0.6	0.8	0.25	0.23	0.37	0.47	0.43	0.8	0.6	0	0.01	0.03	0.19	0.17
A-			0.75	0.77	0.63	0.53	0.57			1	0.99	0.97	0.81	0.83

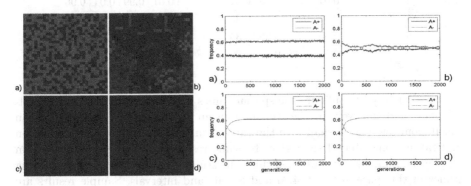

Fig. 2. MSEG, final lattices and dynamics for $i{=}0.4$, $j{=}0.8$ reproductions: (a) probabilistic; (b) deterministic; (c) weighted mean, best cells: 3; (d) weighted mean, intervals: 3.

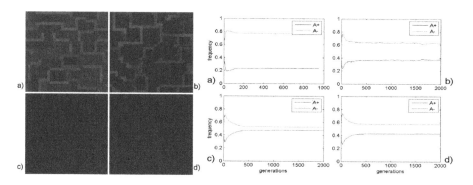

Fig. 3. MSEG, final lattices and dynamics for $i=0.6$, $j=0.8$ reproductions: (a) probabilistic; (b) deterministic; (c) weighted mean, best cells: 3; (d) weighted mean, intervals: 3

ond set of parameters describes a scenario when $i > j$, which for RD gives the total domination of the A- in population. The results are collected in Table 2 (right side). In almost all cases, the results are the same as for the mean-field model. Some small clusters of A+ can occur, however in the majority it is still close to the total domination of A- in the population. The difference can be visible for the last set of parameters ($i=0.8$, $j=0.6$). When j almost equals i then, especially for the weighted mean reproductions, some bigger fractions of A+ may survive in the population (Fig. 4).

Table 3. Results for weighted mean reproductions

	i	j	best cells		intervals		i	j	best cells		intervals	
			A+	A-	A+	A-			A+	A-	A+	A-
RD	0.4	0.8	0.5	0.5	0.5	0.5	0.8	0.4	0	1	0	1
2			0.61	0.39	0.63	0.37			0.01	0.99	0.04	0.96
4			0.64	0.36	0.61	0.39			0.09	0.91	0.02	0.98
6			0.65	0.35	0.65	0.35			0.13	0.87	0.01	0.99
8			0.57	0.43	0.64	0.36			0.3	0.7	0	1

Another way to configure the spatial games is to change the number of cells (players) and intervals in the weighted mean reproductions. Weighted mean taken from one player shall give the same result as deterministic reproduction and taken from interval shall give the same result as the weighted mean from all players. Simulations confirmed this. The studies were performed for following values of this parameter: 2, 4, 6 and 8 cells and intervals. Sample results are stored in Table 3. For the weighted mean from the best cells and when $i < j$, while number of cells increases the A+ frequency of occurrences also increases (Fig. 5) with one exception for 8 best cells. It may be related to the fact that

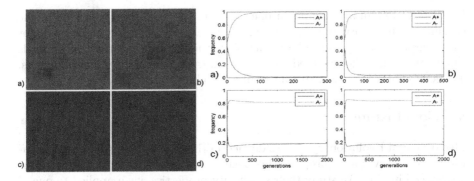

Fig. 4. MSEG, final lattices and dynamics for i=0.8, j=0.6 reproductions: (a) probabilistic; (b) deterministic; (c) weighted mean, best cells:3; (d) weighted mean, intervals:3

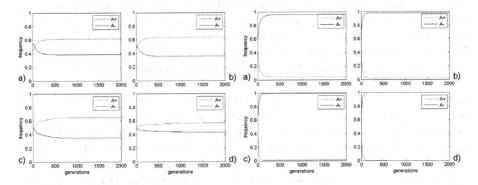

Fig. 5. Phenotypes' dynamics of changes for weighted mean from best cells (left) for i=0.4, j=0.8 and intervals (right) for i=0.8, j=0.4. Appropriately 2,4,6,8 (best or intervals) for a,b,c,d. Numerical values in Table 3.

almost all players in the neighbourhood are taken into account taking one more cell the result is nearly equal division between A+ and A- in the population. Also, a sudden increase of A+ between weighted mean from 1 (deterministic reproduction) and 2 cells can be observed. In a situation where $i > j$, then increasing the number of cells taken for the weighted mean also increases the A+ frequency. So depending on the different scenarios (meaning monomorphic or polymorphic population in case of the mean-field model) the increment of the parameter causes either promoting or repressing the particular phenotype. Moreover, for the 8 best cells the A+ frequency of occurrences is relatively much greater than for the other simulations. As previously, taking weighted mean from almost all cells in the neighbourhood may lead to such result. The weighted mean from the intervals seems to be more changeable in case of the relation between number of intervals and the frequencies of occurrences of the phenotypes. The reason is that number of intervals (apart from 1) does not indicate how many players are in one particular interval (Fig. 5). Everything depends on the ratio between maximum and minimum phenotype in the local neighbourhood of the certain site. In theory, the conclusion shall be exactly opposite to one made for the weighted mean from the best cells, since the increasing number of intervals should cause that the weighted mean is taken from a smaller number of players. Thats why for i=0.4 and j=0.8 all results are almost identical.

4 Conclusions

Mixed Spatial Evolutionary Games defined in [19] generalize in some sense, spatial evolutionary games proposed by Bach et al. [17], with some modifications introduced in [20]. We study their sensitivity properties when applied to one of the first game theoretical cancer models: growth factors production. In this approach, each cell expresses different traits and can use different strategies with some probability or on a certain level of efficiency. We have presented many, different results which show that in the case of spatial games A- are more promoted than in the case of mean-field games (based on the replicator dynamics). However, spatial games also give a chance for A+ cells when $i > j$ even if they are repressed from the population for the mean-field counterpart. We define sensitivity as different changes in the phenotypes frequency of occurrences and their dynamics through remaining generations obtained due to different parametrization of the game. Studies of different parameter values of payoff matrix, different types of reproductions and configuration parameter in case of weighted mean reproductions show that results are highly sensitive. Sensitivity occurs in plenty various results meaning different dynamics of changes, a different phenotypic composition of a population and different final spatial structures of cells. Additionally various initial lattices and the size of the neighbourhood may be studied enhancing the possibility of achieving different results. Though spatial games (both SEGT and MSEG) may provide more accurate simulation of interactions both in macro and micro environment, they may be also viewed as a complex tool that disrupts elegant simplicity ensured by mean-field games.

Acknowledgment. The study was supported by Polish National Centre of Sciences grants 2011/03/B/ST6/04384 in 2014 (MK, DB) and DEC-2014/13/B/ST7/00755 (AS) in 2015.

References

1. Hanahan, D., Folkman, J.: Patterns and emerging mechanisms of the angiogenic switch during tumorigenesis. Cell **86**, 353–364 (1996)
2. Bergers, G., Benjamin, L.E.: Tumorigenesis and the angiogenic switch. Nat. Rev. Cancer **3**, 401–410 (2003)
3. Zetter, B.R.: Angiogenesis and tumour metastasis. Annu. Rev. Med. **49**, 407–424 (1998)
4. Hahnfeldt, P., Panigraphy, D., Folkman, J., Hlatky, L.: Tumor development under angiogenic signaling: a dynamical theory of tumor growth, treatment response and postvascular dormancy. Cancer Res. **59**, 4770–4775 (1999)

5. McDougall, S.R., Anderson, A.R.A., Chaplain, M.A.J., Sherratt, J.A.: Mathematical modelling of flow through vascular networks: implications for tumour-induced angiogenesis and chemotherapy strategies. Bull. Math. Biol. **64**, 673–702 (2002)
6. Jackson, T., Zheng, X.: A cell-based model of endothelial cell migration, proliferation and maturation during corneal angiogenesis. Bull. Math. Biol. **72**, 830–868 (2010)
7. Sleeman, B.D., Hubbard, M., Jones, P.F.: The foundations of a unified approach to mathematical modelling of angiogenesis. Int. J. Adv. Eng. Sci. Appl. Math. **1**, 43–52 (2009)
8. Stokes, C.L., Lauffenburger, D.A.: Analysis of the roles of microvessel endothelial cell randommotility and chemotaxis in angiogenesis. J. Theor. Biol. **152**, 377–403 (1991)
9. Plank, M.J., Sleeman, B.D.: A reinforced random walk model of tumour angiogenesis and anti-angiogenic strategies. Math. Med. Biol. **20**, 135–181 (2003)
10. Alarcon, T., Byrne, H., Maini, P., Panovska, J.: Mathematical modelling of angiogenesis and vascular adaptation. Stud. Multidisciplinarity **3**, 369–387 (2006)
11. Anderson, A.R.A., Chaplain, M.A.J.: Continuous and discrete mathematical models of tumor-induced angiogenesis. Bull. Math. Biol. **60**, 857–900 (2003)
12. Travasso, R.D.M., Poire, E.C., Castro, M., Rodrguez-Manzaneque J.C., Hernandez-Machado, A.: Tumor angiogenesis and vascular patterning: a mathematical model. PLoS ONE 6, Article ID e19989 (2011)
13. Smith, J.M., Price, G.R.: The logic of animal conflict. Nature **246**, 16–18 (1973)
14. Tomlinson, I.P.M., Bodmer, W.F.: Modelling the consequences of interactions between tumour cells. British J. Cancer **75**, 157–160 (1997)
15. Basanta, D., Deutsch, A.: A game theoretical perspective on the somatic evolution of cancer. In: Bellomo, N., Chaplain, M., Angelis, E. (eds.) Selected Topics in Cancer Modeling: Genesis, Evolution, Immune Competition, and Therapy, pp. 1–16. Springer, New York (2008)
16. Swierniak, A., Krzeslak, M.: Application of evolutionary games to modeling carcinogenesis. Math. Biosci. Eng. **3**, 873–911 (2013)
17. Bach, L.A., Sumpter, D.J.T., Alsner, J., Loeschcke, V.: Spatial evolutionary games of interactions among generic cancer cell. J. Theor. Med. **5**, 47–58 (2003)
18. Bishop, D.T., Cannings, C.: A generalized war of attrition. J. Theor. Biol. **70**, 85–124 (1978)
19. Krzeslak, M., Swierniak, A.: Extended Spatial Evolutionary Games and Induced Bystander Effect. In: Piętka, E., Kawa, J., Wieclawek, W. (eds.) Information Technologies in Biomedicine, Volume 3. AISC, vol. 283, pp. 337–348. Springer, Heidelberg (2014)
20. Krzeslak, M., Swierniak, A.: Multidimensional extended spatial evolutionary games. Comput. Biol. Med. (2015). http://dx.org/10.1016/j.compbiomed.2015.08.003

Model Kidney Function in Stabilizing of Blood Pressure

Martin Augustynek[✉], Jan Kubicek, Martin Cerny,
and Marie Bachrata

Faculty of Electrical Engineering and Computer Science,
Department of Cybernetics and Biomedical Engineering,
VSB Technical University of Ostrava, Ostrava, Czech Republic
{martin.augustynek,jan.kubicek,
martin.cerny,marie.bachrata.st}@vsb.cz

Abstract. The aim of this work is to verify the model of kidney function by stabilizing blood pressure, which is implemented in Matlab Simulink. A user interface is also designed for educational purpose in the subject Biocybernetics thus enabling stage processes involved in the long-term regulation of blood pressure. At the end of the work was carried out physiological and functional verification model.

Keywords: Blood pressure regulation · Renal function · Heart · Hormones · Mean arterial pressure · Blood volume · Cardiac output · Peripheral resistance · Model · Matlab · Simuling

1 Introduction

The higher arterial pressure, the smaller tendency of kidneys to retain waste products appears, such as water, sodium, and chlorine. As a result the kidneys begin to excrete more urine (increasing U_0 value). Pressure control diagram is shown below (see Fig. 1).

This allows the reduced volume of $ECT(VECT)$ (Extracellular Fluid) and total blood volume (VB) in the bloodstream. Further both the venous return (VR) and heart filling pressure (PMS) decrease as a result of reduction in volumes of blood. Further the cardiac output (CO) decreases and finally the mean arterial pressure (PAS) and peripheral vascular resistance (RA) as well, which at the beginning along with the blood pressure increased. This regulation has in terms of time a long onset, but a long term duration, we can speak about hours or days [1, 2, 4].

When the arterial pressure is reduced, the mechanism works exactly the opposite. The kidneys thus seek to retain the largest possible number of ECT. The urinary output (U_0) decreases with the reduced arterial pressure (P_{AS}). This causes an increase in the body fluid volume (ECT), it means also in the blood volume (V_B), the mean systemic filling pressure (P_{MS}) and also in the cardiac output (CO).

The following table (Table 1) indicates the physiological values of the variables or their physiological range [1, 3, 8, 9].

© Springer-Verlag Berlin Heidelberg 2016
N.T. Nguyen et al. (Eds.): ACIIDS 2016, Part I, LNAI 9621, pp. 430–439, 2016.
DOI: 10.1007/978-3-662-49381-6_41

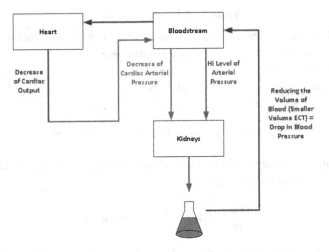

Fig. 1. Block diagram of the regulation of increased arterial pressure.

Table 1. Table of physiological variables [2].

Variables	Physiological values and their units	
P_{AS}	93.3–110	mmHg
P_{MS}	7	mmHg
CO	5000–5600	ml/min
V_B	4.5–5.5	L
R_A	1.08	mmHg*s/ml

2   The Mathematical Modeling

To create the model in The Simulink, it was needed to build the mathematical model. The Mean arterial pressure can be computed by using the values of systolic (ST) and diastolic (DT) values of pressure.

$$P_{AS} = DT + \frac{ST - DT}{3} \quad [mmHg] \tag{1}$$

The Renal function curve is a graphical curve which expresses the dependence of the volume of renal excretion (urine output $-U_0$) on different levels of the arterial pressure (P_{AS}). This curve was determined by increasing the arterial pressure to the isolated kidney, while the renal excretion was measured on different levels of blood pressure.

This dependence of U_0 on P_{AS} is modelled by the second-order polynomial. It is possible to simulate different types of hypertension that have their origin in altered kidney function, by changing the parameters (see Table 2) of the curve of renal excretion (a_{12}, a_{11}, a_{10}), see Eq. 2.

Table 2. Table of parameters.

Parameters for normotension	Parameters for Goldblatt hypertension
a12 = 0.003749	a12 = 0.00079365
a11 = −0.5999	a11 = −0.156
a10 = 23.96	a10 = 6.579

$$U_0 = a_{12} \cdot P_{AS}^2 + a_{11} \cdot P_{AS} + a_{10} \qquad [ml/min] \qquad (2)$$

The Changed volume of $ECT(V_{ECT})$ is determined by the difference between fluid intake (WS) and renal excretion (U_0).

$$\frac{dV_{ECT}}{dt} = WS - U_0 \qquad [ml/min] \qquad (3)$$

The Blood volume (VB) in the model is determined according to the volume $ECT(V_{ECT})$, the relationship is shown in Eq. 4. Nonlinear static dependency is replaced by quadratic functions with empirically obtained coefficients: a_{22}, a_{21}, a_{20}.

$$V_B = a_{22} \cdot V_{ECT}^2 + a_{21} \cdot V_{ECT} + a_{20} \qquad [L] \qquad (4)$$

where: $a_{22} = -0.01$
  $a_{21} = 0.6$
  $a_{20} = -1.75$

The Mean systemic filling pressure (P_{MS}) is then determined from Eq. 4, see Eq. 5.

$$P_{MS} = a_{32} \cdot V_B^2 + a_{31} \cdot V_B + a_{30} \qquad [ml/min] \qquad (5)$$

where: $a_{32} = 2.38$
  $a_{31} = -15.48$
  $a_{30} = 24.86$

The Venous return (VR) is determined by the difference of the mean systemic filling pressure (P_{MS}) and the pressure in the right atrium (P_{RA}) and is inversely proportional to the resistance of the venous return (VR).

$$VR = \frac{P_{MS} - P_{RA}}{P_{VR}} \qquad [ml/s] \qquad (6)$$

In the steady state, there is equality between the venous return and the cardiac output $(CO = VR)$. The Arterial pressure (P_{AS}) is determined from the value of the venous return (VR, CO) and the peripheral resistance R_A see Eq. 7.

$$P_{AS} = VR \cdot R_A \quad [mmHg] \tag{7}$$

The Peripheral resistance (R_A) is the sum of two components, see Eq. 10. The first component (R_{A1}) is determined by the difference in the cardiac output (CO, VR) and its standard value CO_0, see Eq. 8. The value of the second component is directly proportional to the arterial pressure (P_{AS}). This component is modelled by autoregulatory ability of blood vessels to regulate the increased arterial pressure by an increased resistance (vasoconstriction) and in this way to stabilize blood flow through tissues (called Bayliss effect), see Eq. 9.

$$\frac{dR_{A1}}{dt} = KR_P \cdot (CO - CO_0) \quad [mmHg/ml] \tag{8}$$

$$R_{A2} = KR_{PA} \cdot P_{AS} \quad [mmHg \cdot min \cdot ml^{-1}] \tag{9}$$

$$R_A = R_{A1} + R_{A2} \quad [mmHg \cdot min \cdot ml^{-1}] \tag{10}$$

where: $KR_P = 5 \cdot 10^{-9}$
$CO_0 = 5000\,ml/min$
$KR_{Pa} = 0.00011\,ml/min$

The Resistance of the venous return (R_{VR}) is modeled by a constant component R_{VR0} and direct proportion to the value of the peripheral resistance (R_A):

$$R_{VR} = R_{VR0} + KR_{VR} \cdot R_A \quad [mmHg \cdot min \cdot ml^{-1}] \tag{11}$$

where: $R_{VR0} = 0.00087984\,mmHg \cdot min/ml$
$KR_{VR} = 0.026$

The Right atrial pressure (P_{RA}), is determined from the obtained values of the venous return (VR, CO) and the coefficients a_{42}, a_{41}, a_{40}.

$$P_{RA} = a_{42} \cdot VR^2 + a_{41} \cdot VR + a_{40} \quad [Pa, kPa] \tag{12}$$

$$P_{RA} = \frac{P_{MS} \times \frac{a_{41}}{R_{VR}} - a_{40}}{K_{PRA} + \frac{a_{41}}{R_{VR}}} \tag{13}$$

where: $a_{40} = 2.3$
$a_{41} = 0.00058\,.$
$K_{PRA} = 1$

In the next picture (Fig. 2) is shown The Model of Renal Function in Stabilizing Blood Pressure in Matlab-Simulink [4, 7, 10, 12].

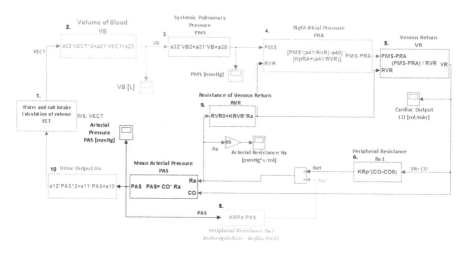

Fig. 2. Model in Matlab simuling.

3 Testing and Results

Model testing was conducted in two basic settings "normotensive" or "Goldblatt hypertension." Further testing were configured for three different values of the input variables WS (1, 3 and 6 ml /min). Pictures then show the waveforms of the individual output variables of tested model. For better view the initial waveforms are for individual physiological parameters insert always two graphs with time setting simulation for 5000 min and 60 min.

The figure below (see Fig. 3) shows the graph of the mean arterial pressure. The graph begins for all three values receiving an isotonic solution of sodium with water *(WS)*, 93.5 mmH . For parameter values receiving an isotonic solution of sodium water 1 and 3 ml/min, the values of mean arterial pressure are within physiological limits (93.3−110 mmHg). For the value $WS = 6$ ml/min, the value of P_{AS} stabilizes at a value that is outside of physiological ranges and ranges rather between hypertension (above 110 mmHg) [5, 6, 11].

Curves mean arterial pressure (P_{AS}) grow from the value of 93.5 mmHg due to increased fluid intake *WS*, thanks to which increases blood volume *(VB)*, systemic filling pressure (P_{MS}), cardiac output *(CO)* and peripheral vascular resistance (R_A). We can observe in all of these graphs that the values initially increase. After reaching the maximum value curve of mean arterial pressure (P_{AS}) continue falling because the control mechanism of kidneys began to function and kidneys began to excrete more urine. The curve *VB*, P_M, *CO* and R_A declines. After reaching the minimum value the curve begins to rise again slowly and because kidney function begins the E_{CT} levels in the body starts to consolidate and also stabilizes the mean arterial pressure P_{AS} and other physiological variables.

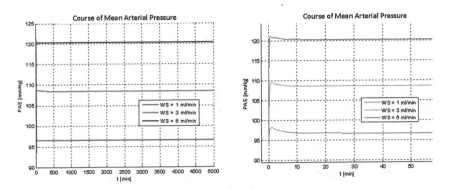

Fig. 3. On the left a graph during the mean arterial blood pressure values for three different intake isotonic sodium in water at 5000 min time (i.e., 3.5 days). On the right a zoomed graph of mean arterial pressure for three different values of reception isotonic sodium in water, 60 min.

The figure (see Fig. 4) shows the graph of the blood volume. For all three values WS graph starts at 5 L. The initial increase in value is due to the sudden increased fluid intake *WS*. Then, values are falling. That when the kidneys begin to operate as a regulatory mechanism of blood pressure. When the level of E_{CT} in the body due to kidneys parallels, we can observe the rise of value to their ultimate stabilization.

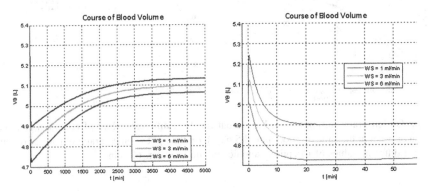

Fig. 4. On the left graph of blood volume for three different values of reception isotonic sodium in water at the time of 5000 min (i.e., 3.5 days). On the right zoomed graph of blood volume for three different values of reception isotonic sodium in water, 60 min.

The figure (see Fig. 5) displays a graph of systemic filling pressure. For all three values WS the graph starts at a value 6.96 *mmHg*.

The figure below (see Fig. 6) shows the graph of cardiac output in *ml/min*. For all three values *WS* graph starts at a value 4677 *ml/min*. Next image (see Fig. 7) represents

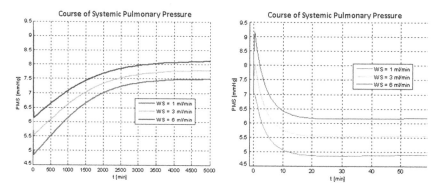

Fig. 5. On the left graph during systemic filling pressure for three different values of reception isotonic sodium in water at the time of 5000 min (3.5 days), on the right zoomed graph of systemic filling pressure for three different values of reception of an isotonic solution of sodium water-time 60 min.

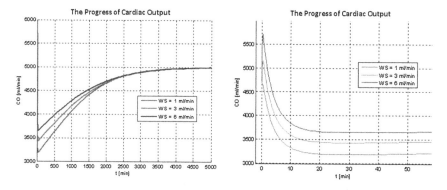

Fig. 6. On left graph during the cardiac output for three different values of reception isotonic sodium in water at the time of 5000 min (3.5 days), on the right zoomed graph of cardiac output for three different values of reception isotonic sodium in water, 60 min.

a graph of peripheral resistance vessels. For all three values *WS* the graph starts at a value of 1.2 *mmHg/ml*.

The waveform of peripheral resistance vessels (for all three values *WS*) is given by the course of blood pressure. When mean arterial pressure initially rises, also rise values of peripheral resistance, while subsequently P_{AS} curve declines, falls even curve R_A. This is so called myogenic effects. It is the reaction of small arteries and arterioles of the kidneys, through their activities, consisting of vasoconstriction and vasodilatation, ensuring a constant flow of blood glomeruli.

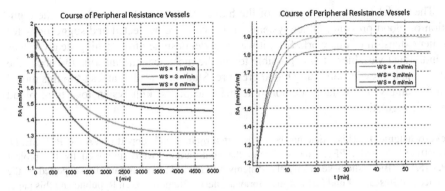

Fig. 7. On the left graph during the peripheral resistance of blood vessels for the three different values receiving an isotonic solution of sodium in water at the time of 5000 min (3.5 days), right zoomed graph of peripheral resistance vessels for three different values of reception of an isotonic solution of sodium water, time 60 min.

4 Conclusions

To verify the results obtained from the generated model, the data read from the individual graphs of the model compared with physiological values stated in the literature. Table (see Table 3) shows how the values which have been deducted from the resulting graphs and values that can be found in the literature. It can be observed that the values are not significantly different, minor variations may be occur due to rounding parameter values with which the model worked. The values read from the graphs are in the waveform moving within the physiological range given quantity.

Table 3. Table for comparing the values examined physiological variables when setting the fluid intake WS = 1 ml /min. Settling time values in the range of 5000–6000 min (3.5 days) [13, 14]

Physiological value	Initial value	Rise curve	Decline curve	Stable value	Physiological value	Units
P_{AS}	93.5	98.5	96.6	96.6	93.3–110 (normotenze)	mmHg
V_B	5	5.0092	4.7	5.06	4.5–5.5	L
P_{MS}	6.96	7.035	4.9	7.51	7	mmHg
CO	4700	4700	3183	5000	4500–6000	ml/min
R_A	1.2	1.82	1.16	1.16	1.08	mmHg*s/ml
WS	1	1	1	1	1.042	ml/min

Physiological validation was performed for stable values, which matched (or were within physiological limits) see Table 3 with the values shown in the literature.

The model is only one system of the human body, which can influence the regulation of blood pressure. Other systems in this case are neglected and suggestions for further extension of this work. Still, it was a model designed to watching steady values of individual variables, it was possible to get the most accurate information about the role of the kidney in stabilizing blood pressure. Studying steady values from this model can be simply and reliably demonstrate the role of the kidneys in long-term blood pressure regulation.

Acknowledgment. This article has been supported by financial support of TA ČR PRE SEED: TG01010137 GAMA PP1. The work and the contributions were supported by the project SP2015/179 'Biomedicínské inženýrské systémy XI' and This work is partially supported by the Science and Research Fund 2014 of the Moravia-Silesian Region, Czech Republic and this paper has been elaborated in the framework of the project "Support research and development in the Moravian-Silesian Region 2014 DT 1 - Research Teams" (RRC/07/2014). Financed from the budget of the Moravian-Silesian Region.

References

1. Despopoulos, A., Silbernagl, S.: Color atlas of physiology. Thieme, Stuttgart, New York (2003)
2. Kittnar, O.: Lékařská fyziologie (1. vyd.), 790 s. Grada, Praha. ISBN 978-802-4730-684 (2011)
3. van Brievingh, R.P.W., Möller, D.P.F.: Biomedical Modeling and Simulation on a PC: a Workbench for Physiology and Biomedical Engineering. Springer, New York (1993)
4. ECK, Vladimír a Miroslav RAZÍM. Biokybernetika. 1. Praha: ČVUT, Elektrotechnická fakulta, 155 s (1996)
5. Kubicek, J., Penhaker, M., Kahankova, R.: Design of a synthetic ECG signal based on the Fourier series. In: Proceedings of the 2014 International Conference on Advances in Computing, Communications and Informatics, ICACCI 2014, art. no. 6968312, pp. 1881–1885 (2014)
6. Cerny, M., Pokorny, M.: Circadian rhythm evaluation using fuzzy logic. Stud. Comput. Intell. **457**, 289–298 (2013)
7. Kubicek, J., Penhaker, M.: Fuzzy algorithm for segmentation of images in extraction of objects from MRI. In: Proceedings of the 2014 International Conference on Advances in Computing, Communications and Informatics, ICACCI 2014, art. no. 6968264, pp. 1422–1427 (2014)
8. Penhaker, M., Darebnikova, M., Cerny, M.: Sensor network for measurement and analysis on medical devices quality control. In: Yonazi, J.J., Sedoyeka, E., Ariwa, E., El-Qawasmeh, E. (eds.) ICeND 2011. CCIS, vol. 171, pp. 182–196. Springer, Heidelberg (2011)
9. Peter, L., Noury, N., Cerny, M.: A review of methods for non-invasive and continuous blood pressure monitoring: Pulse transit time method is promising? IRBM **35**, 271–282 (2014)
10. Kubicek, J., Penhaker, M., Bryjova, I., Kodaj, M.: Articular cartilage defect detection based on image segmentation with colour mapping. In: Hwang, D., Jung, J.J., Nguyen, N.-T. (eds.) ICCCI 2014. LNCS, vol. 8733, pp. 214–222. Springer, Heidelberg (2014)
11. Augustynek, M., Penhaker, M.: Non invasive measurement and visualizations of blood pressure. Elektronika Ir Elektrotechnika **116**, 55–58 (2011)

12. Peterek, T., Augustynek, M., Zurek, P., Penhaker, M.: Global courseware for visualization and processing biosignals. In: Dossel, O., Schlegel, W.C. (eds.) World Congress on Medical Physics and Biomedical Engineering, vol 25, pt 12, pp. 404–407 (2009)

13. Silbernagl, S., Despopoulos, A.: Atlas fyziologie člověka. 6. přeprac. a rozš. vyd. Grada, Praha, XII, 435 s. ISBN 80-247-0630-X (2004)

14. Trojan, S.: Lékařská fyziologie. 4. vyd. přepr. a dopl. Grada Publishing, Praha, 771 s. ISBN 80-247-0512-5 (2003)

Dynamic Diversity Population Based Flower Pollination Algorithm for Multimodal Optimization

Jeng-Shyang Pan[1]([⊠]), Thi-Kien Dao[2], Trong-The Nguyen[2],
Shu-Chuan Chu[3], and Tien-Szu Pan[2]

[1] College of Information Sciences and Engineering,
Fujian University of Technology, Fuzhou, China
jengshyangpan@gmail.com
[2] Department of Electronics Engineering,
National Kaohsiung University of Applied Sciences, Kaohsiung, Taiwan
jvnkien@gmail.com
[3] School of Computer Science, Engineering and Mathematics,
Flinders University, Adelaide, Australia

Abstract. Easy convergence to a local optimum, rather than global optimum could unexpectedly happen in practical multimodal optimization problems due to interference phenomena among physically constrained dimensions. In this paper, an altering strategy for dynamic diversity Flower pollination algorithm (FPA) is proposed for solving the multimodal optimization problems. In this proposed method, the population is divided into several small groups. Agents in these groups are exchanged frequently the evolved fitness information by using their own best historical information and the dynamic switching probability is to provide the diversity of searching process. A set of the benchmark functions is used to test the quality performance of the proposed method. The experimental result of the proposed method shows the better performance in comparison with others methods.

Keywords: Flower pollination algorithm · Dynamic diversity flower pollination algorithm · Multimodal optimization problems

1 Introduction

The nature-inspired algorithms have been applied to solve many NP-Hard optimization problems successfully in engineering, financial, and management fields [1, 2]. These algorithms have been developed based on the inspiration of the behaviors of biological systems [3]. For example, Genetic algorithms (GAs) were developed by drawing inspiration from the Darwinian evolution of biological systems [4]. Particle swarm optimization (PSO) was inspired from the swarm behavior of birds and fish [5]. Artificial bee colony algorithm (ABC) was emulated the intelligent foraging behavior of honey bee swarm [6]. Ant colony optimization (ACO) was mimicked the behavior of ants seeking a path between their colony and a source of food [7]. Flower pollination algorithm (FPA) was imitated by the pollination process of flowers [8]. All these algorithms have

© Springer-Verlag Berlin Heidelberg 2016
N.T. Nguyen et al. (Eds.): ACIIDS 2016, Part I, LNAI 9621, pp. 440–448, 2016.
DOI: 10.1007/978-3-662-49381-6_42

been applied to a wide range of applications. Moreover, the No Free Lunch (NFL) theorem [9] has proved logically that there is no meta-heuristic finest for solving all optimization problems nor others. Reliably, NFL makes the research of this field is highly active which results in proposing new meta-heuristics every year. Additionally, in real optimization problems, multiple global and local optima are often desirable to simultaneously find out of a given objective function [10]. Due to physical constraints of problems, the best results cannot always be realized. For interference phenomena among constrained dimensions, sometimes it is easy to converge to a local optimum for solving the multimodal optimization and complex constrained optimization problems. Enhancement of the diversity populations in the optimal algorithms is one of the solutions to this issue.

The idea of enhancing diversity agents by using neighborhood search strategies was introduced in the existing methods such as Ant colony system with communication strategies [11], and Diversity enhanced particle swarm optimization with neighborhood search [12]. More accuracy and extended global search capacity are proved the diversity artificial agents than the original structure in solving the complex problems [13]. Dynamic diversity method could be constructed by three main factors of the small size groups, neighborhood topology technique, and its own best historical information. The small size groups could be figured out by dividing the population into subpopulations or groups. The neighborhood topology technique could be implemented by applying some Niching techniques such as the crowding, the fitness sharing available resources and the speciation. The good information obtained by each group evolving optimization could be exchanged among the groups. The benefit of cooperation individuals is exploitation through local extremes to the global optimum. In this paper, the concept of the diversity is applied to FPA and an altering strategy for enhancing dynamic diversity populations FPA is proposed, namely dFPA.

Organization of this paper is as follows. A briefly review of FPA is given in Sect. 2. Methodology of the dynamic diversity FPA is presented in Sect. 3. The experimental results on the test functions and the comparison between original FPA and dFPA are discussed in Sect. 4. Finally, the conclusion is presented in Sect. 5.

2 Flower Pollination Algorithm

Flower Pollination Algorithm (FPA) [7] was developed by drawing inspiration from the characteristic of the biological flower pollination in flowering plant. The features of the flow pollination process of flowering plant are applied in this algorithm as follows.

Rule 1: The global pollination processes are biotic and cross-pollination. The pollens are transported by pollinators in a way that obeys Lévy flights.

Rule 2: Local pollination is viewed as abiotic and self-pollination.

Rule 3: Pollinators such as insects can develop flower constancy. Reproduction probability is considered as the proportional to the resemblance of the two flowers involved.

Rule 4: The switching probability $p \in [0, 1]$ can be used to control between the local and global pollination.

The aforementioned rules can be converted into updating equations. For instance, in the global pollination step, flower pollen gametes are carried by pollinators such as insects. Because insects can often fly and move in a much longer range, so pollen can travel over a long distance. Rule 1 and flower constancy can be applied to formulate mathematically as follows.

$$x_i^{t+1} = x_i^t + \gamma \times L(\lambda) \times (x_i^t - g^*) \tag{1}$$

where x_i is solution vector of the pollen i-th, and g^* is the current best solution found among all solutions at the current generation or iteration t. Here γ is a scaling factor to control the step size. $L(\lambda)$ is the parameter that corresponds to the strength of the pollination, and called the step size. Since insects may move over a long distance with various distance steps, a Lévy flight can be used to mimic this characteristic efficiently. Lévy distribution with L is positive as drawn equation.

$$L = \frac{\lambda \Gamma(\lambda) \times \sin(\frac{\pi \lambda}{2})}{\pi \times s^{i+\lambda}}, \ (s \gg s_0) \tag{2}$$

where $\Gamma(\lambda)$ is the standard gamma function, and this distribution is valid for large steps $s > 0$. The rule 2 and rule 3 can be used to model the local pollination.

$$x_i^{t+1} = x_i^t + u(x_j^t - x_k^t) \tag{3}$$

where x_j^t and x_k^t are pollen from different flowers of the same plant species. u is drawn from a uniform distribution in [0, 1]. If x_j^t and x_k^t comes from the same species or selected from the same population, this u becomes a local random walk. Flower pollination processes can occur at both local and global. In order to imitate this feature, the switch probability can be effectively used likely in the rule 4. The proximity probability p is to switch between common global pollination to intensive local pollination. To begin with, p can be set 0.55 as an initial value. A preliminary parametric showed that $p = 0.8$ might work better for some applications [14].

The basic steps of FPA are described as follows.

Step 1. Initialization: pollen population $x = (x_1, x_2, ..., x_d)$ is generated randomly. A switch probability $p \in$ arrange from 0 to 1. A stopping criterion is set.
Step 2. The best solution g^* is calculated with initial population, F_{min} is assigned to fitness at g^*.
Step 3. For each pollen in the population
  if rand $< p$,
    A step vector L is computed as Lévy distribution Eq. (2)
    Global pollination is updated via Eq. (1)
  else
    Draw u from a uniform distribution in [0,1]
    Local pollination is processed via Eq. (3)
  end if

Step 4. Evaluate new solutions, the function value F_{new} is assigned to fitness $(x(t+1))$. A new solution is accepted if the solution improves (F_{new} less than F_{min}), by updating the best solution $g*$ to $x(t+1)$ and assign the minimum function F_{min} to F_{new}.

Step 5. If the termination condition is not safety, go to Step 3.

Step 6. Output the best solution found.

3   Dynamic Diversity Populations FPA

In the diversity enhancement structure, several crowds are formed by dividing the population into subpopulations and the neighborhood topology is used to share the fitness available resources. The subpopulations could evolve themselves independently in regular iterations to locate for better area in the search space. The obtained information exchanges among subpopulations whenever the communication strategy is triggered. The benefit of cooperation and exploitation is achieved by communicating information. To adjust the proportion of the global and local searching processes, a dynamic switching probability strategy in the exchanging period is used to make the dynamic diversity. By setting the exchanging period of an activated schedule, a new configuration of small groups is started the searching the best global target.

The dynamic diversity FPA is designed based on original FPA optimization and a neighborhood based on Niching technique is used. There are two considered characters in neighborhood structure, small size and communicating. The former, small sized groups create the diversity in local search. FPA with small neighborhoods performs better on complex problems. The small sized groups could be employed by dividing the population in FPA into groups. Each group uses its own members to search for better area in the search space. The later, the better obtained information of evolving optimization in each group is exchanged among them to achieve the cooperative individuals and exploitation through local extremes to the global optimum. Since the small sized groups are searching using their own best historical information, they are easy to converge to a local optimum.

In FPA, a switching probability $p \in [0, 1]$ is used to control the proportion of local search and global search, and it is a constant value. Actively generate the dynamics for algorithm, the proportion of the global and local search processes should be changed in the exchanging period. Switching probability p can transform as following formula:

$$p = a - b \times \frac{R - mod(t, R)}{R} \tag{4}$$

where R and t are the exchanging period and current iteration, a, b are constants in arrange of $[0,1]$ and $a > b$. Because p is suggested arrange of $[0,1]$, a is set to 0.55 and b to 0.1 The fitness sharing available resources is one common used in the Niching techniques. The group of FPA has its own pollens as known the search agent and finest agents according to the fitness evaluation function. These finest agents among all the pollens in one group will be assigned to the poorer agents in other group, replace them or update agents for each crowd after running period.

Let m be number pollens of the group. It is called population size of the group. Let G_j be the group, where j is the index of the group. j is set to $0, 1, 2, ..., n-1$. n is number of groups. The top k fitness agents in the group $G(j)$ according to objection function evaluation will be copied to $G(j + 1)$ to replace the worst fitness agents with the same number of agents during run time with $t \cap R \neq \theta$. New configuration of the groups will be started searching in every R generations with dynamic switching probability triggered. The good information could be obtained by exchanging among the groups. Thus, simultaneously the diversity of the population could perform better on complex multimodal problems.

The steps of the diversity enhancement structure can be described as follows:

Step 1. **Initialization:** The $m \times n$ pollens are generate by initializing population size and dividing them into n groups randomly, with m individuals in each group G. Assign $R\text{-}th$ the exchanging period for executing X_{ijt} solutions, where $i = 0, 1, ..., m - 1; j = 0, 1, ..., n - 1$; t is the current iteration and set to 1.

Step 2. **Evaluation:** The value of $f(X_{ijt})$ for pollens is evaluated in $j\text{-}th$ group G_j.

Step 3. **Update:** The local and global solutions are updated by using Eqs. (1), and (3) and with p according to Eq. (4).

Step 4. **Communication Strategy:** Move k best pollens among G_{tj} to the $(j + 1)\text{-}th$ group G_{tj+1}, to replace k poorer pollens in that group and update all of the group in each R iterations.

Step 5. **Termination:** Go Step 2 if the predefined value of the function is not achieved or the maximum number of iterations has not been reached. The best value of the function $f(X_{ijt})$ and the best pollen solution among all the pollens X_{ijt} are recorded.

4 Experimental Results

A set of multimodal benchmark functions [15, 16] is used to test the quality performance of the proposed algorithm dFPA. The obtained evaluation output values of the test functions are averaged over 25 runs with different random seeds. The purpose of the optimization is to minimize the outcome of the test function of multimodal benchmarks. Let vector $X = \{x_1, x_2, ..., x_n\}$ be n-dimensional real-value.

The population size is set the same for all the algorithms of the proposed algorithm and original ones in the experiments. The detail of parameter settings of FPA can be found in [8]. The initial range, the dimension and total iteration number for all test functions are listed in Table 1.

For original FPA (oFPA), the setting parameters are as follows. The total population size N is set to 80. The initial probability p is set to 0.55. λ isset to 1.5, and the dimension d is set to 30. The parameters setting for dFPA is as follows. The initial probability p is according to Eq. (4). a and b are set to 0.55 and 0.1 respectively. λ isset to 1.5, the total population size $m \times n$ isset to 20×4, the exchanging period is set to 20 and the dimension d is set to 30. The full iterations of 1000 are set for each function and the different random seeds are over 25 runs. The final result is obtained by taking the average of the outcomes from all runs. The proposed method dFPA results are compared with the oFPA.

Table 1. The initial range and the total iteration of the multimodal benchmark functions

Test functions	Range	Dimension	Iteration
$f_1(x) = \sum_{i=1}^{n} -x_i \sin(\sqrt{|x_i|})$	± 500	30	1000
$f_2(x) = \sum_{i=1}^{n} [x_i^2 - 10\cos(2\pi x_i) + 10]$	± 5.12	30	1000
$f_3(x) = \sum_{i=1}^{n} \sin(x_i).(\sin(\frac{ix_i^2}{\pi}))^{2m}, \quad m = 10$	$0, \pi$	30	1000
$f_4(x) = [e^{-\sum_{i=1}^{n}(x_i/\beta)^{2m}} - 2e^{-\sum_{i=1}^{n}x_i^2}]\prod_{i=1}^{n}\cos^2 x_i, \quad m = 5$	± 20	30	1000
$f_5(x) = -\sum_{i=1}^{4} c_i \exp(-\sum_{j=1}^{6} a_{ij}(x_j - p_{ij})^2)$	$0, 10$	4	1000
$f_6(x) = -\sum_{i=1}^{5} [(X - a_i)(X - a_i)^T + c_i]^{-1}$	$0, 10$	4	1000

Table 2 compares the performance quality and running time for the multimodal optimization problems between dFPA and oFPA. The value in the comparison column in Table 2 is calculated percentage deviation of the proposed approach compared with the original method. Observation from Table 2, the results of the proposed algorithm on all of these cases of testing multimodal benchmark problems show that dFPA method almost increases higher than those obtained from original method. The maximum case of obtained from dFPA method increases higher than those obtained from the oFPA method is up to 66 %. However, the figure for the minimum case is only the increase 2 %. Thus, in general the proposed algorithm obtained the average cases of various tests multimodal optimization problems for the convergence, and accuracy increased more than those obtained from the oFPA method is 30 %.

Table 2. The quality performance evaluation and speed comparison of oFPA and dFPA for solving the multimodal optimization problems

Test functions	Time consumption (min)		Performance evaluation		Accuracy %
	oFPA	dFPA	oFPA	dFPA	Comparison
$f_1(x)$	0.837	0.8387	−6.100E + 03	−6.193E + 03	2 %
$f_2(x)$	0.8273	0.8502	2.083E + 02	1.287E + 02	62 %
$f_3(x)$	2.4793	2.4371	2.085E + 00	1.255E + 00	66 %
$f_4(x)$	0.7776	0.7734	1.600E-03	1.100E-03	45 %
$f_5(x)$	0.8918	0.9372	−3.240E + 00	−3.294E + 00	2 %
$f_6(x)$	0.9593	0.9793	−9.152E + 00	−9.724E + 00	6 %
Average	**1.1297**	**1.1307**	**−9.837E + 02**	**−1.013E + 03**	**30 %**

Figures 1, 2 and 3 show the experimental results of output obtained from dFPA and oFPA methods for the first three multimodal benchmark functions with the same iteration of 1000.

The above figures include four sub-figures in each of them, such as the multimodal function graph; Semilogy plot measures the performance through using a base 10 logarithmic scale with their index contains real numbers; Convergence plot measures

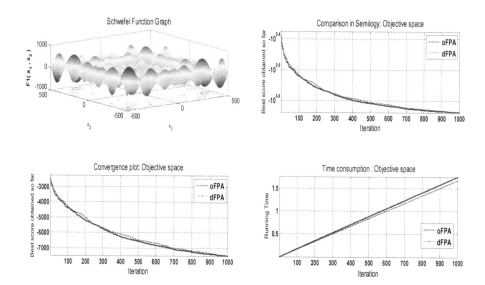

Fig. 1. The comparison the experimental results of the methods of oFPA and dFPA for function F1

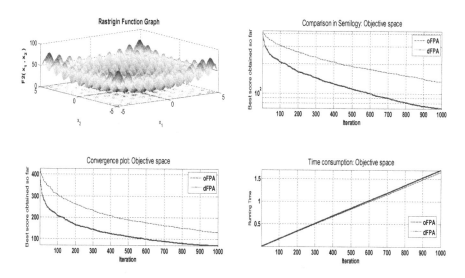

Fig. 2. The comparison the experimental results of the methods of oFPA and dFPA for function F2

the performance through using the convergence and the time consumptions of dFPA and oFPA methods for the multimodal optimizations. Clearly, all of these cases of testing multimodal benchmark functions for dFPA with performance quality is higher than those for oFPA in terms of the accuracy and convergence, even though the time consuming of two methods are equivalent.

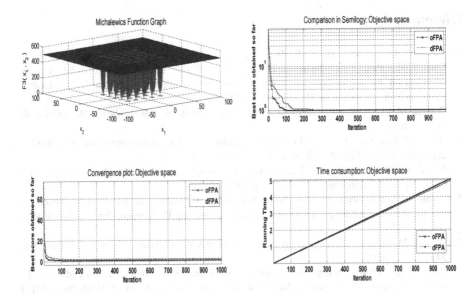

Fig. 3. The comparison the experimental results of the methods of oFPA and dFPA for function F3

5   Conclusion

A novel proposed method for the multimodal optimization problems was presented with the dynamic diversity population based flower pollination algorithm (dFPA) in this paper. The implementation of dynamic diversity population could have important significance for avoiding the loss of the global optimum nearby, rather easily converge to a local optimum of the optimal algorithms for solving the multimodal optimization and complex constrained optimization problems. In this new proposed algorithm, the pollens of flower pollination algorithm (FPA) are split into several independent groups based on the original structure of the FPA, and the neighborhood topology technique is used.

The communication strategy provides the information flow for the pollens to fly in different groups. By this way, the poorer pollens in the groups will be replaced with new better pollens from neighbor groups after running the exchanging period. The dynamic switching probability provides the diversity of searching process between the global and local search in each the exchanging period. A set of the test benchmark functions is used to test the quality performance, and the speed of the proposed method. According to the experimental results, the proposed method shows the better performance in comparison with the original method. The best case obtained from the proposed method increases higher than those obtained from the original method is up to 66 %.

However, the figure for the worst case is only the increase 2 %. Thus, in general the proposed algorithm dFPA obtained the average cases of various tests multimodal optimization problems for the convergence, and accuracy increased more than those obtained from the original method oFPA is 30 %.

References

1. Yang, X.-S.: Nature-Inspired Metaheuristic Algorithms. Luniver Press, Denmark (2010)
2. Dao, T.-K., Pan, T.-S., Nguyen, T.-T., Chu, S.-C.: A compact artificial bee colony optimization for topology control scheme in wireless sensor networks. J. Inf. Hiding Multimedia Sig. Process. **6**(2), 297–310 (2015)
3. Dorigo, M.: Optimization, learning and natural algorithms. Ph.D. thesis, Politecnico di Milano, Italy (1992)
4. Houck, C.R., Joines, J., Kay, M.G.: A genetic algorithm for function optimization: a Matlab implementation. NCSU-IE TR, vol. 95, no. 09 (1995)
5. Yuhui, S., Eberhart, R.: A modified particle swarm optimizer. In: IEEE World Congress on Computational Intelligence, pp. 69–73 (1998)
6. Karaboga, D.: An idea based on honey bee swarm for numerical optimization. Technical report-TR06, Engineering Faculty, Computer Engineering Department, Erciyes University, vol. T (2005)
7. Dorigo, M., Blum, C.: Ant colony optimization theory: a survey. Theoret. Comput. Sci. **344**(2), 243–278 (2005)
8. Yang, X.-S.: Flower pollination algorithm for global optimization. In: Durand-Lose, J., Jonoska, N. (eds.) UCNC 2012. LNCS, vol. 7445, pp. 240–249. Springer, Heidelberg (2012)
9. Wolpert, D.H., Macready, W.G.: No free lunch theorems for optimization. IEEE Trans. Evol. Comput. **1**(1), 67–82 (1997)
10. Qu, B.Y., Suganthan, P.N., Das, S.: A distance-based locally informed particle swarm model for multimodal optimization. IEEE Trans. Evol. Comput. **17**(3), 387–402 (2013)
11. Chu, S.C., Roddick, J.F., Pan, J.-S.: Ant colony system with communication strategies. Inf. Sci. **167**(1–4), 63–76 (2004)
12. Wang, H., Sun, H., Li, C., Rahnamayan, S., Pan, J.-S.: Diversity enhanced particle swarm optimization with neighborhood search. Inf. Sci. **223**, 119–135 (2013)
13. Qu, B.-Y., Suganthan, P.N.: Novel multimodal problems and differential evolution with ensemble of restricted tournament selection. In: IEEE Congress on Evolutionary Computation, pp. 1–7 (2010)
14. Raouf, O.A., El-henawy, I., Abdel-Baset, M.: A novel hybrid flower pollination algorithm with chaotic harmony search for solving Sudoku puzzles. Int. J. Modern Educ. Comput. Sci. **3**, 38–44 (2014)
15. Yao, X., Liu, Y., Lin, G.: Evolutionary programming made faster. IEEE Trans. Evol. Comput. **3**(2), 82–102 (1999)
16. Suganthan, P.N., Hansen, N., Liang, J.J., Deb, K., Chen, Y.-P., Auger, A., Tiwari, S.: Problem definitions and evaluation criteria for the CEC 2005 special session on real-parameter optimization, KanGAL report, vol. 05 (2005)

Hardware Implementation of Fuzzy Petri Nets with Lukasiewicz Norms for Modelling of Control Systems

Zbigniew Hajduk[1] and Jolanta Wojtowicz[2][✉]

[1] Faculty of Electrical and Computer Engineering,
Rzeszow University of Technology, Rzeszow, Poland
[2] Faculty of Mathematics and Nature, Chair of Computer Science,
University of Rzeszow, Rzeszow, Poland
jsmg@poczta.onet.pl

Abstract. In the paper an implementation of Fuzzy Petri Nets with Lukasiewicz norms is described based on FPGA integrated circuits. The proposed solution is used for modeling of control systems. The approach for the realization of fuzzy Petri net models is based on the synthesis method taking advantage of a fuzzy Petri net place module concept. The paper contains the description of a real-life control system, which is used to demonstrate new features of fuzzy Petri nets with Lukasiewicz norms. For the example control system, in order to analyze costs of implementation, a comparison is made between the fuzzy Petri net model with Lukasiewicz norms and the fuzzy Petri net model with triangular MIN/MAX norms.

1 Introduction

Rising requirements concerning capabilities of control systems for industrial processes necessitate the creation of new solutions both in the areas of hardware technology and control algorithms. Modern FPGAs allow implementation of complex control algorithms such as fuzzy systems or Petri nets, simultaneously offering the possibility of hardware optimization of a particular solution in terms of its performance and demand for resources [3]. Fuzzy Petri nets [5] are an extension of a classical Petri nets paradigm, which were implemented with success in FPGAs for the control of industrial processes [1,2,4]. The new model of fuzzy Petri nets allows overcoming some inadequacies of the classical Petri net model and its application in modelling of control systems makes it possible to improve control accuracy in comparison to analogous models built with classical Petri nets. In the new model Lukasiewicz norms are introduced as a replacement for classical MIN/MAX norms. Lukasiewicz norms constitute a natural generalization of OR and AND operators used to perform logical operations on binary values. The introduction of the new type of norms for fuzzy Petri nets allows designing of more accurate control systems [6] In the paper a method of fuzzy Petri nets implementation with FPGA integrated circuits taking advantage of

© Springer-Verlag Berlin Heidelberg 2016
N.T. Nguyen et al. (Eds.): ACIIDS 2016, Part I, LNAI 9621, pp. 449–458, 2016.
DOI: 10.1007/978-3-662-49381-6_43

the new type of norms is presented. The authors created a practical example of the control system in order to illustrate features of the new approach and to compare it with the model built with classical Zadeh norms.

2   Formal Description of Fuzzy Petri Nets

In the study a following class of fuzzy Petri net was implemented, which is defined as a n-tuple $FPN = (P, T, D, G, R, \Delta, \Gamma, \Theta, M_0)$, where: $P = \{p_1, ..., p_r\}$ is a non-empty, finite set of places; $T = \{t_1, ..., t_s\}$ is a non-empty, finite set of transitions; $D = \{d_1, ..., d_r\}$ is a non-empty, finite set of statements; $G = \{g_1, ..., g_s\}$ is a non-empty, finite set of conditions; P, T, D, G are pairwise disjoint; $R \subseteq (P \times T) \cup (T \times P)$ is the relation of incidence; $\Delta : P \rightarrow D$ is the statement binding function, which binds a statement to every place; $\Gamma : T \rightarrow G$ is the condition binding function, which binds a condition to every transition; $\Theta : T \rightarrow [0,1]$ is the truth degree function, which defines degrees of fulfillment for conditions associated with transitions; $M_0 : P \rightarrow [0,1]$ is the initial marking function. A pure Petri net with finite unit capacities of places and edges is considered. For every place in a net a number of input transitions is limited to 1.

In the model of fuzzy Petri net presented in the study Lukasiewicz norms are used. A pair of these norms is defined by the following relationships:

$$T_L(a, b) = max(0, a + b - 1) \tag{1}$$

$$S_L(a, b) = min(a + b, 1) \tag{2}$$

The dynamics of Petri net describes a method of marking in different places of the net and determines the conditions, when a given transition can be fired. Only an enabled transition can be fired. The transition $t \in T$ is enabled to be fired with marking M from the moment, when:

$$\forall p \in {}^\bullet t, M(p) = 1, \forall p \in t^\bullet, M(p) = 0, \tag{3}$$

to the moment, when:

$$\forall p \in {}^\bullet t, M(p) = 0, \forall p \in t^\bullet, M(p) = 1, \tag{4}$$

The determination of the new marking in fuzzy Petri net is accomplished in the following manner. If for a marking M a transition $t \in T$ is enabled for firing and the degree of fulfillment for the condition related to the transition defined as $\Theta(t) = \partial \in [0,1]$ changes by a value $\Delta\partial \neq 0$, then the new marking m' is determined with the following rule:

$$M'(p) = \begin{cases} M(p)T_L(1 - \Delta\partial) \Longleftrightarrow p \in {}^\circ t \backslash t^\circ \\ M(p)S_L\Delta\partial \Longleftrightarrow p \in {}^\circ t \backslash t^\circ \\ M(p) \Longleftrightarrow p \notin t \cup t^\circ \end{cases} \tag{5}$$

where T_L and S_L are fuzzy Lukasiewicz t-norm and t-conorm, respectively.

3   Implementation Method of the Proposed Fuzzy Petri Net Model

The implementation method is based on using two types of hardware modules, which are place modules and transition modules. Both types of modules are digital circuits hence the degree of fulfillment of the conditions associated with transitions, as well as values of the places' markings are elements of a discrete set Z_N:

$$Z_N = C_0 \cup Z \cup C_1 \tag{6}$$

$$C_0 = 0, Z = \{1, 2, ..., 2^N 2\}, C_1 = \{2^N - 1\} \tag{7}$$

where N is the number of bits.

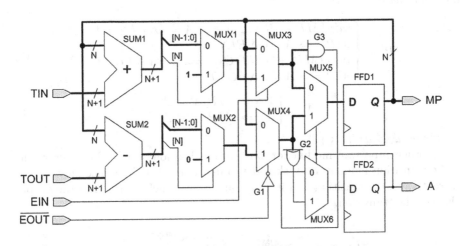

Fig. 1. The architecture of a fuzzy Petri net place module.

A block diagram of the implementation of a place module is shown in Fig. 1. The bold lines in the diagram denote multi-bit buses. The key elements of the module are two blocks performing arithmetic sum (SUM1) and difference (SUM2) of two operands, the first of which is a N-bit signal acquired from the module output (MP signal), and the other is a N+1 signal representing an input (TIN) or an output (TOUT) transition. The specific method of connecting multiplexers MUX1 and MUX2 assures maintaining of data processing outcome values to the ones comprised within the Z_N set. Selection inputs of these multiplexers are controlled by an oldest bit (denoted with number N), which value is a result of an arithmetic operation of the blocks SUM1 and SUM2 and to the first data inputs of multiplexers a value of selection of N least significant bits (numbered from N-1 to 0) is fed from the outputs of SUM1 and SUM2 blocks. To the other data inputs of multiplexers constant values are fed, denoted as C_1 and C_0. Multiplexers MUX3 and MUX4 are responsible for the activation of data processing from the input TIN (MUX3 multiplexer) and TOUT (MUX4). If appropriate inputs activating EIN and EOUT are in inactive states then the module's output remains unchanged, regardless of states of inputs TIN and TOUT. The MUX5 multiplexer is responsible for the selection of a function carried out by the place module, which can act either as the function of output or input place for a particular transition. A set of D type flip-flops (FFD1) stores a result of the operation. The selection input of MUX5 multiplexer is controlled by an output from the subsystem consisting of N input gates of type AND (G3) and OR (G2) and also of the MUX6 multiplexer and the FFD2 flip-flop. At the output of this subsystem a high state can be observed from the moment when the output of MP module achieves a value of C_1, to the moment when at this output a value C_0 is observed. The output A of the place module can be also used for binary control. The function carried out by the place module is described by the following equations:

$$MP' = \begin{cases} min\{MP+TIN, C_1\} \Longleftrightarrow A = 0 \wedge EIN = 1 \\ MP \Longleftrightarrow A = 0 \wedge EIN = 0 \\ max\{MP-TOUT, C_0\} \Longleftrightarrow A = 1 \wedge EOUT = 0 \\ MP \Longleftrightarrow A = 1 \wedge EOUT = 1 \end{cases} \tag{8}$$

$$A' = \begin{cases} 1 \Longleftrightarrow (MP, A) \in \Delta \\ 0 \Longleftrightarrow (MP, A) \notin \Delta \end{cases} \tag{9}$$

where: $\Delta(Z \cup C_1) \times \{1\}$

Another important module for hardware realization of fuzzy Petri net model is the transition module. Its task is determination of the difference between the current value of transition and its previous value, which is stored in the output place of the transition.

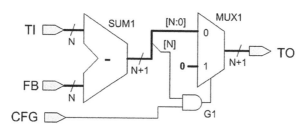

Fig. 2. The architecture of a fuzzy Petri net transition module.

The architecture of the transition module is shown in Fig. 2. The basic functionality of the module is realized by the system calculating difference SUM1. The part of the system consisting of the MUX1 multiplexer and the G1 gate is optional. Thanks to this subsystem, when a high state is observed in the configuration input CFG, the functionality of fuzzy Petri net with classic MIN/MAX norms can be obtained. The work of the transition module is described by the following equation:

$$TO = \begin{cases} TI - FB \Longleftrightarrow CFG = 0 \vee ((TI - FB) \geq 0 \wedge CFG = 1) \\ 0 \Longleftrightarrow (TI - FB) \leq 0 \wedge CFG = 1 \end{cases} \quad (10)$$

In Fig. 3 a generic fragment of a fuzzy Petri net is shown, and in Fig. 4 its hardware implementation is presented using place and transition modules described above.

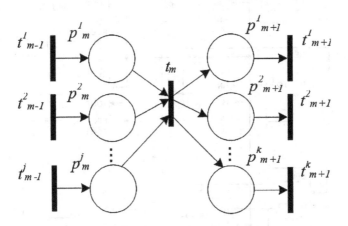

Fig. 3. Generic fragment of fuzzy Petri net.

The implementation of the t_m transition for the fragment of a fuzzy Petri net depicted in Fig. 2 requires the application of one transition module, for which TI input a value of the degree of condition fulfillment associated with the t_m transition is fed. The FB input of the module can be connected with any MP output of the place module for this transition. The TO output of the transition module should be connected with TOUT outputs of all place modules, which represent input places of the fuzzy Petri net and with TIN inputs for all modules representing output places of the fuzzy Petri net. The appropriate connection of A outputs of place modules with outputs EIN and OUT through multi-input AND/OR gates and also through inverters, ensures the realization of conditions for enabling of firing the t_m transition. Architectures of place and transition modules can be easily implemented in Hardware Description Language using a behavioral description. In turn, connections between these modules, which implement fragments of the fuzzy Petri net, can be described in HDL using a structural description. Additionally the description process for any fragment of the fuzzy Petri net can be automated with appropriate computer software, which generates ready program code in HDL language.

4   Description of the Control System

In order to demonstrate the characteristics of a fuzzy Petri net with operators using Lukasiewicz norms we consider the model of a chemical reactor control system. The functional diagram of the chemical reactor is presented in Fig. 5.

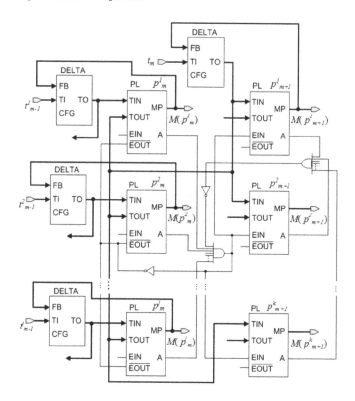

Fig. 4. Hardware implementation of fuzzy Petri net fragment.

The reactor consists of five containers. The three upper containers function as trays of ingredients 1 . . . 3. The middle container, labeled in Fig. 5 as scale can act as a device, which measures required amounts of ingredients. In the lower container, labeled as mixer, mixing of all ingredients supplied in appropriate proportions takes place. The sequence of operations, required to perform a given technological process, consists of the following phases:

– metering the desired quantity of component 1,
– emptying of the scale;
– metering the desired quantity of component 2,
– emptying of the scale;
– metering the desired quantity of component 3 (regulated through the length of valve V2 opening time),
– mixing of the ingredients;
– emptying of the mixer;

For the reactor control process to run at maximum capacity as many operations as possible should be performed simultaneously. For instance, during the ingredients' mixing in the mixer the next portion of ingredient 1 can be measured. The control algorithm implementing the above-mentioned sequence of

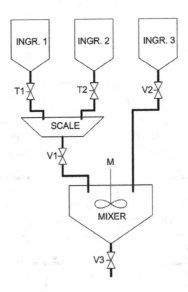

Fig. 5. Functional diagram of chemical reactor.

actions can be described by the model of fuzzy Petri net shown in Fig. 6. The following is a legend for the model's scheme:

P1 - supplying ingredient 1 on the scale by opening of the T1 valve,
t1 - obtainment of the required amount of ingredient 1,
P2 - obtainment of the required amount of ingredient 1 (no control action is being performed),
"1" - non conditional transition,
P4 - emptying of the scale by opening of the V1 valve (the measured quantity of ingredient 1 is passed to the mixer),
t2 - the scale is empty,
P5 - supplying ingredient 2 on the scale by opening of the T2 valve,
t3 - obtainment of the required amount of ingredient 2,
P6 - emptying of the scale by opening of the valve V1,
P11 - supplying ingredient 3 to the mixer by opening of the V2 valve,
t4 - obtainment of the required amount of ingredient 3,
P8 - obtainment of the required amount of ingredient 3 (no control action is being performed),
P7 - the scale is empty (no control action is being performed),
P9 - ingredients mixing (starting up of the mixer),
t5 - the process of ingredients mixing is finished,
P10 - emptying of the mixer by opening of the V3 valve,
t6 - the mixer is empty.

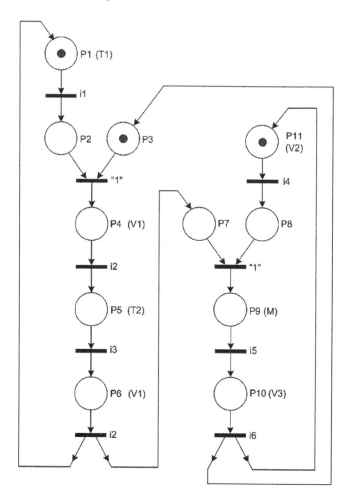

Fig. 6. Fuzzy Petri net model describing chemical reactor control algorithm.

5   Results

The analysis of implementation costs has been carried out for three versions of the model simulating the chemical reactor's work. The first version of the model implements fuzzy Petri net with Zadeh norms. The second version of the fuzzy Petri net model implements Lukasiewicz norms. In the third version of the model Lukasiewicz norms are implemented and additionally Petri net dynamics equations are modified. By modifying the change of the condition fulfillment operator (denoted as delta in the dynamics equation) we obtain for the fuzzy Petri net with Lukasiewicz norms a behavior analogous as in the case of fuzzy Petri net with Zadeh norms. The marking in the third version of the Petri net is updated, when the change of the degree of the condition fulfillment associated with the transition is greater than zero - that is, the fuzzy Petri net functions as the model with classic MIN/MAX norms. The results of Petri net implementation for the batch process of the chemical reactor are based on the Xiling ISE report for the FPGA family of Xilinx Spartan-6. Table 1 contains a comparison of the amounts of hardware resources required for the implementation of fuzzy Petri nets for the three versions of the model. These versions are denoted with letters A, B and C, which are respectively: A - the basic fuzzy Petri net with Zadeh norms, B - the version with Lukasiewicz norms, C - the version with Lukasiewicz norms and modified net dynamics equations, which functions exactly as the version A. In subsequent columns of the Table 1 included are the comparisons of hardware resources (divided explicitly into numbers of LUT tables and numbers of flip-flops) required for the realization of particular versions of the chemical reactor model and the comparisonsof maximal values of frequencies possible to achieve after the implementation of fuzzy Petri nets in FPGA.

Table 1. The results of implementation costs for the three versions of models simulating the batch process work.

Version	Number of LUT tables	Number of flip-flops	$F_{MAX}[MHz]$
A	332	99	153.8
B	481	99	120.5
C	552	99	117.6

On the basis of results collected in Table 1 it can be seen that the model of fuzzy Petri net denoted as A with classic MIN/MAX norms has the lowest resource requirements and has the highest maximum clock frequency. The model of fuzzy Petri net denoted as B with Lukasiewicz norms has higher hardware requirements and lower value of maximum clock frequency. However in comparison with the classic model its advantage is that it models token's reverse passing, which is not possible with the classic model. The net denoted as C has the highest hardware requirements of all the proposed versions and its maximum clock frequency is comparable with the version denoted as B.

6 Summary

The paper presents the hardware implementation of fuzzy Petri net with Lukasiewicz norms in application to industrial system control. The implemented model of fuzzy Petri net with Lukasiewicz norms has all the advantages of fuzzy Petri nets with classic standards MIN/MAX norms and through the modification of net dynamics equations it is possible to achieve new features such as reverse passing of a token for negative values of the change of the condition fulfillment degree associated with the particular transition. Such a situation is not possible for fuzzy Petri nets with triangular MIN/MAX norms. The feasibility of the implementation of fuzzy Petri nets with Lukasiewicz norms was shown for an actual case of the chemical reactor process control. Analysis of implementation cost was carried out for three versions of fuzzy Petri nets using both triangular MIN/MAX norms and Lukasiewicz norms. All the implemented versions of fuzzy Petri nets are more noise robust in comparison with binary Petri nets. The method of fuzzy Petri nets implementation proposed by the authors is universal because by the change of the state of one signal it is possible to obtain the classically functioning net (with MIN/MAX norms) or the net with Lukasiewicz norms. The results obtained in the course of research are encouraging and demonstrate the usefulness of applying Lukasiewicz norms in control systems. Further work will focus on the implementation of other types of fuzzy norms and checking their suitability for modeling problems of real control systems.

References

1. Gniewek, L., Kluska, J.: Hardware implementation of fuzzy Petri net as a controller. IEEE Trans. Syst., Man Cybern., Part B: Cybern. **34**(3), 1315–1324 (2004)
2. Gniewek, L.: Sequential control algorithm in the form of fuzzy interpreted Petri net. IEEE Trans. Syst. Man Cyber. Part A - Syst. Hum. **43**(2), 451–459 (2013)
3. Hajduk, Z.: An FPGA embedded microcontroller. Elsevier Microprocess. Microsyst. **38**(1), 1–8 (2014)
4. Kluska, J., Hajduk, Z.: Digital implementation of fuzzy Petri net based on asynchronous fuzzy RS flip-flop. In: Rutkowski, L., Siekmann, J.H., Tadeusiewicz, R., Zadeh, L.A. (eds.) ICAISC 2004. LNCS (LNAI), vol. 3070, pp. 314–319. Springer, Heidelberg (2004)
5. Looney, C.G.: Fuzzy Petri nets for rule-based decision-making. IEEE Trans. Syst. Man Cybern. **18**(1), 178–183 (1988)
6. Suraj, Z.: Parameterised fuzzy Petri nets for approximate reasoning in decision support systems. In: Hassanien, A.E., Salem, A.-B.M., Ramadan, R., Kim, T. (eds.) AMLTA 2012. CCIS, vol. 322, pp. 33–42. Springer, Heidelberg (2012)

ALMM Solver for Combinatorial and Discrete Optimization Problems – Idea of Problem Model Library

Ewa Dudek-Dyduch[1] and Sławomir Korzonek[2(✉)]

[1] Department of Automatics and Biomedical Engineering,
AGH University of Science and Technology, Kraków, Poland
edd@agh.edu.pl
[2] Advanced Technology Systems International, Zabierzów, Poland
Ademar@poczta.onet.pl

Abstract. The paper presents results of further research on a software tool named ALMM Solver. The objective of the ALMM Solver is to solve combinatorial and discrete optimization problems including NP-hard problems.

The solver utilizes a modeling paradigm named Algebraic Logical Meta Model of Multistage Decision Processes (ALMM of MDP) and its theory.

The ALMM of MDP enables a unified approach to creating discrete optimization problem models and representing knowledge about these problems. The models are stored in a Problem Model Library.

A new, extended modular structure of the ALMM Solver is presented together with a basic layout of the Problem Model Library.

Keywords: Solver · Optimizer · Algebraic-Logical Meta-Model (ALMM) · Multistage decision process · Discrete optimization problems · Scheduling problem · Software tool · Library of models

1 Introduction

The paper describes a software tool for solving combinatorial and discrete optimization problems named ALMM Solver.

Quite obviously, the literature presents a multitude of discrete optimization problems, general methods of solving them and dedicated algorithms [1, 21]. It is so vast an area that it is hard for a typical engineer to navigate, find a book or a paper tackling a specific question and get the needed algorithm. The situation may be improved by the development of software tools that let one solve certain kinds of problems by various known methods. The most popular such tools are based on the Constraint Logic Programming (CLP) approach, focusing on finding feasible solutions [19, 23]. The problems are modeled there by a finite, predetermined number of variables, their domains, and relations between them. However, wide range of actual optimization problems are not suitable for modeling through CLP. Another optimization software tools are based on multi agent approach [17, 24].

The first crucial issue when designing a solver is how to maintain the knowledge about the problem to be solved. In other words, how, in a formal way, should one enter

© Springer-Verlag Berlin Heidelberg 2016
N.T. Nguyen et al. (Eds.): ACIIDS 2016, Part I, LNAI 9621, pp. 459–469, 2016.
DOI: 10.1007/978-3-662-49381-6_44

into the system all the constraints and quality criteria which define the problem. The ALMM Solver operates on formal (mathematical) models of problems that belong to the class of models based on Algebraic-Logical Meta-Model of Multistage Decision Processes (ALMM of MDP). Using the ALMM of MDP it is possible to develop so-called algebraic logical (AL) models for a large class of discrete optimization problems as well as combinatorial ones.

The second basic issue is how to implement the AL problem models i.e. what software representation of AL models is to be like? AL models must be implemented in a way facilitating their use in implementation of solving methods and algorithms.

The third question is if there is a simple way to establish a software representations of AL models?

This paper contains answers to all the three questions.

The objective of the paper is dual, including:

- proposing a new, expanded structure of the ALMM Solver, including new module named Problem Model Library module,
- presenting a method of AL model implementation and component-oriented approach to their development.

The first, initiative structure of the ALMM Solver and the related requirements are presented in [15, 22].

2 Algebraic-Logical Meta-Model of Multistage Decision Processes

The algebraic-logical meta-model of multistage decision process (ALMM of MDP) is a general model development paradigm for deterministic problems, for which solutions that can be presented as a sequence (or a set) of decisions.

The idea of an ALMM of MDP paradigm was proposed and developed by Dudek-Dyduch [2, 3, 6 7] and recalled in many papers among others [8]. It is also put to use in multiple cases. Based on ALMM of MDP formal models (so-called AL models) may be established for a very broad class of discrete optimization problems from a variety of application areas, thus yielding to the meta-model designation.

In [2–5, 7, 9, 10, 12, 13] that lays the groundwork for the basic ALMM of MDP theory, the co-author herein has presented i. a. AL models for discrete manufacturing process control, for logistics problems, knapsack problem, and a model for a complex real-life problem encompassing both manufacturing and logistics. Basing on ALMM of MDP paradigm, new methods for solving discrete optimization problems have been created too [3, 11, 13, 14, 16, 18]. An overview of the new methods is given in [8].

ALMM of MDP (abbreviated as ALMM) provides a structured way of recording knowledge of the goal and all relevant restrictions that exist within the problems modelled. In line with this idea, all information is split into pre-defined basic components with appropriate links, defined both through algebraic and logical formulae.

Using this paradigm, the co-author has provided, i.a. in [1, 7] and recall in [8], the definition of two base classes for multistage decision processes: a common process, denoted here as cMDP and a dynamic process, denoted as MDDP. Let us review them briefly.

Definition 1. **Common multistage decision process** is a process that is specified by the sextuple cMDP = (U, X, x_0, f, X_N, X_G) where U is a set of decisions, X is a set of states, $f: U \times X \rightarrow X$ is a partial function called a transition function, (it does not have to be determined for all elements of the set $U \times X$), $x_0, X_N \subset X, X_G \subset X$ are respectively: an initial state, a set of not admissible states, and a set of goal states, i.e. the states in which we want the process to take place at the end. Subsets X_G and X_N are disjoint i.e. $S_G \cap S_N = \emptyset$.

Because not all decisions defined formally make sense in certain situations, the transition function f is defined as a partial one. As a result, all limitations concerning the decisions in a given state x can be defined in a convenient way by means of so-called sets of possible decisions $U_p(x)$, and defined as: $U_p(x) = \{u \in U: (u, x) \in Dom\, f\}$.

The formal definition of multistage dynamic decision process (MDDP) quoted below refers to dynamic decision processes, i.e. processes wherein both the constraints and the transition function (and in particular the possible decision sets) depend on time. Therefore, the concept of the so-called "generalized state" has been introduced, defined as a pair containing both the state and the time instant.

Definition 2. **Multistage dynamic decision process** is a process that is specified by the sextuple MDDP = (U, S, s_0, f, S_N, S_G) where U is a set of decisions, $S = X \times T$ is a set named a set of generalized states, X is a set of proper states, $T \subset \Re + \cup\{0\}$ is a subset of non negative real numbers representing the time instants, $f: U \times S \rightarrow S$ is a partial function called a transition function, (it does not have to be determined for all elements of the set $U \times S$), $s_0 = (x_0, t_0), S_N \subset S, S_G \subset S$ are respectively: an initial generalized state, a set of not admissible generalized states, and a set of goal generalized states, i.e. the states in which we want the process to take place at the end. Subsets S_G and S_N are disjoint i.e. $S_G \cap S_N = \emptyset$.

The transition function is defined by means of two functions, $f = (f_x, f_t)$ where $f_x: U \times X \times T \rightarrow X$ determines the next state, $f_t: U \times X \times T \rightarrow T$ determines the next time instant. It is assumed that the difference $\Delta t = f_t(u, x, t)-t$ has a value that is both finite and positive.

Just like in Definition 1, the transition function f is defined as a partial one. As a result, all limitations concerning the decisions in a given state s can be defined in a convenient way by means of so-called sets of possible decisions $U_p(s)$, and defined as: $U_p(s) = \{u \in U: (u, s) \in Dom\, f\}$.

For both defined types of the multistage decision processes, in the most general case, sets U and X may be presented as a Cartesian product $U = U^1 \times U^2 \times \cdots \times U^m$, $X = X^1 \times X^2 \times \cdots \times X^n$ i.e. $u = (u^1, u^2, ..., u^m), x = (x^1, x^2, ..., x^n)$. In particular, $u^i, i = 1, 2 ... m$ represent separate decisions that must or may be taken at the same time (for Definition 2) or at the same stage (for Definition 1) and relate to particular objects. There are no limitations imposed on the sets; in particular they do not have to be numerical. Thus values of particular co-ordinates of a state or a decision may be names of elements (symbols) as well as some objects (e.g. finite set, sequence etc.).

The sets X_N, S_N, X_G, S_G and U_p are formally defined with the use of both algebraic and logical formulae, **hence the algebraic-logic model descriptor**.

Based on the meta-model recalled herein, AL models may be created for individual problems consisting of seeking admissible or optimal solutions. In case of an admissible solution (admissible solution problem), an AL model is equivalent to a suitable multistage decision process, hence it is denoted as process P. An optimization problem then is denoted as a (P, Q) pair where Q, is a criterion. An optimization task (instance of the problem) is denoted as a (P, Q). A sequence of consecutive states from the initial state to a final state (goal or non-admissible), computed by transition function form a process trajectory.

3   Extended Structure of ALMM Solver

The task of the ALMM Solver is to provide solutions (exact or approximate) for combinatorial and discrete optimization problems or indicates that a solution hasn't been found. A solution has form of a sequence (or set) of decisions.

In the previous paper [15, 22], the initiative structure of the Solver has been given. Presented below (Fig. 1) is the new, extended structure, including the Problem Model Library. (Actually the modules presented in Fig. 1 only constitute a part of the real ALMM Solver structure, but they are sufficient for the purposes of this paper.)

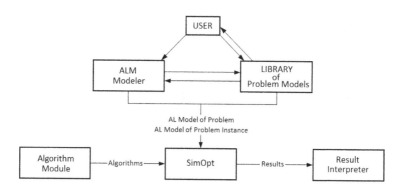

Fig. 1. Modified structure of ALMM Solver

Knowledge of problem and its instance is forwarded from the **ALM Modeler** to the main Solver module carrying out the computations. On this stage of ALMM Solver development, **SimOpt module** operates as its computation module. The module implements a class of algorithms based on state tree searches (e.g. backtrack method, dynamic programming, branch & bound method), and specially developed local optimization tasks (see [8]). SimOpt module task is to intelligently construct the state transition tree (either in its entirety, or only a part of it) by generating one or more trajectories (parts of trajectories). The solver database stores all generated states of the trajectories maintaining the structure of the tree. The SimOpt module are controlled using algorithms from the Algorithm Module. The structure of the SimOpt module is presented in Fig. 2 while its detailed description is given in [15, 22]. Due to the limited length of the paper it must be omitted here.

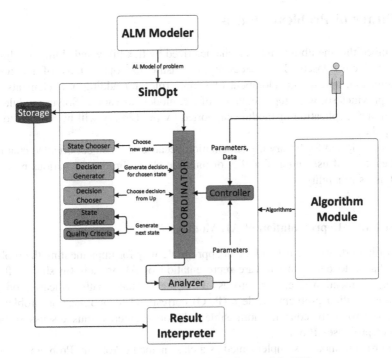

Fig. 2. Initial structure of SimOpt module [15]

The objective of ALM Modeler module is to initiate and forward information about the problem model and its instance to the SimOpt. module (the information can also come from the Problem Model Library). The model needs to take a form understandable to the SimOpt module, i.e. it should have a software form. (presented in chapter 4). The ALM Modeler module is linked to the User Interface, SimOpt module and Problem Model Library. Models can be created in a few ways, including:

- downloading a ready model from Problem Model Library
- building a model from Problem Model Library components
- developing a new model from scratch and implementing it in a proper programming language.

Results Visualization Interpreter gives the users the solution (one or many) or return an information that a solution was not found. In addition, another important feature of this module is that it contains the visualization of results. In the future, the module will also provide a mean to share the results of the analysis through external API. It will allow one to integrate the ALMM Solver with other systems, for example, with manufacturing planning systems, and to use Business Intelligence tools. Additionally, the module will provide the possibility to define various ways of analyzing results obtained from a series of computations for the same or different instances of a problem.

4 Library of Problem Models

It is assumed that the library must be characterized by flexibility and ability to adjust to certain problem subsets. It is necessary to ensure the opportunity of its gradual expansion, too, and the development of simple rules for adding new elements. The Library provides software representation of AL models to various Solver modules. At this point any application or module supported by the Library will be referred to as a client application.

C# was selected for library implementation because of its object-oriented character, capacities, ease of use and availability of multiple class libraries for various purposes, as well as its reliability.

4.1 Software Representation of AL Models

The most important issue is to adopt an appropriate way for implementing the problem model, that is to design a software representation of AL models (in short software models). As mentioned earlier, process **P** is the formal (mathematical) model of admissible solution problem, while a (**P**, **Q**) correspond to optimization problem. In this chapter we only consider admissible solution problems, thus discuss software models of processes **P** only.

The problem model is implemented as a class named cProblem. Problem instances will be matched by objects from that class.

The AL model indicates elements of the problem that need to be defined in the cProblem class, i.e. definitions of: decision and state structures, initial state, transition function, sets of possible decisions for given states, set of not admissible states, set of goal states and definition of problem data structures. The interface of the class that models a problem should provide methods related to the said elements:

- cState GetInitialState() – determining the initial state
- cState GetNextState(cState, cDecision) – transition function
- bool IsGoalState(cState) – verifying if a given state is a target state
- bool IsNonAdmissibleState(cState) – verifying if a given state is an inadmissible state
- bool IsPossibleDecision(cState, cDecision) – verifying if a given decision belongs to the set of possible decisions.

The C# code below is a declaration of interfaces matching individual elements of problem definitions and grouping them in a whole interface for a problem modeling class.

```
public interface iPossibleDecisionVerifier
{
    bool IsPossibleDecision( cState    CurrentState,
                             cDecision CurrentDecision );
}

public interface iStateInitializer
{
    cState GetInitialState();
}

public interface iStateTransformer
{
    cState GetNextState( cState    CurrentState,
                         cDecision CurrentDecision );
}

public interface iNonAdmissibleStateVerifier
{
    bool IsNonAdmissibleState( cState CurrentState );
}

public interface iGoalStateVerifier
{
    bool IsGoalState( cState CurrentState );
}

public interface iProblem : iPossibleDecisionVerifier,
                iStateInitializer, iStateTransformer,
        iNonAdmissibleStateVerifier, iGoalStateVerifier,
{

}
```

It should be noted that the process state and decision are elements of the method definitions that are included in interface of cProblem class. Each modification of a state or decision definition means a modification to the problem interface. In case of problems requiring e.g. different state definitions we are dealing with different software representation of models and different interfaces as well. The general principle of use of a modeling class interface remains the same, but the client application needs to process the data representing the process state in a different way.

4.2   Component Based Structure of Library

The developer has much freedom in designing and creating classes, using class inheritance and composition. In order to maintain clear structure enabling easy Library expansion it is important to categorize class sets constituting Library components. Dividing them into concrete classes, abstract classes, association classes and ones not matching any of these categories seems insufficient. The UML standard versions 2.0 and above [25] allows the definition of own types of the so-called generalization sets describing classification criteria for base and inherited classes.

For the reasons of availability to library clients and utility the following categorization of classes into distinct groups is proposed:

- utility (boundary) classes – the only classes available to the client application, to be used to build modeling objects, e.g. problem, state and decision instances. As such they cannot be abstract classes;
- atomic, elementary classes – responsible for exactly one component of the problem definition, these implement in a pre-defined way one or more methods related to that component. They constitute the basic components building other library classes,
- composite classes – classes with use similar to elementary classes, but combining atomic classes. They contain several components of respective elementary classes. This way it is possible to use ready implementation of atomic classes in order to create a new method made available by a given composite class (these classes are used because of no multiple inheritance available in C#),
- design classes – during implementation of solutions for complex problems these can be helpful in developing other classes. They can constitute base classes from which other classes may inherit. They can also be components of composite classes or design classes. They may be deemed links in the class chain of evolution leading from the most general class to a certain utility, elementary or composite one;

Once defined, the utility, atomic, composite and design classes cannot be modified by library users. In order to expand the library and provide support for new types of problems, the user should be able to create new classes and use already existing ones. Only utility classes are available to client applications. Only design classes can be abstract classes, but it is not necessary to define them in this way. Elementary and composite class objects can be components of other composite, design or utility classes.

It is assumed that the library of problems will be extended. Every class of problems being modeled corresponds to appropriate set of software classes. New classes must be added to the library when modeling new problems. It is desirable that the existing code is not duplicated. It is therefore necessary to check whether one can use ready-made classes to implement a new target. In view of the growing number of library classes it will be increasingly difficult to control the code. One possible way to handle this difficulty is to group classes according to certain categories. Software classes can be associated with a certain class of problems and it is the most general category of the division. A much more detailed category can be established by associating software classes with certain elements of problem definition, i.e. definition of goal state, forbidden state, etc. Dividing the library classes for atomic (elementary), boundary (utility), composite and design classes is an additional criterion and thus simplify the management of library items.

4.3 Atomic and Utility Classes

Atomic classes correspond to some parts of a reusable code.

The analysis of problem models for a given area such as task scheduling, graph problems and others shows that definitions of some AL model components are identical [2, 7, 9, 10]. As a result, in classes modeling different problems we may identify the same attributes and methods. That fact implies the idea of establishing utility class cProblem from relevant atomic classes. A design pattern "Strategy" [20] may be used to construct the class cProblem.

This way of building utility classes will be presented on the example of 5 consecutive job scheduling problems with a single machine, denoted here by $\alpha|\beta|\gamma$ notation [1], where α represent information on machines, β – restrictions referring to the jobs and γ determined criterion. The problems are: $1\|C_{max}$, $1|C_j \leq d_j|C_{max}$, $1|prec|C_{max}$, $1|r_j|C_{max}$, $1|\ C_{max} \leq \hat{C}|\Sigma C_j$, where C_j – completion time of the jth job, C_{max} – criterion of total completion time, ΣC_j – criterion of sum of job completion times, d_j, r_j denote restrictions referring to the job due dates and release time respectively, prec denotes existence of preceding constraints in the set of jobs. AL models of these problems have been developed by co-author and presented in [2, 4, 5, 7]. They are characterized by identical definitions of state and decision structures as well as the definition of initial state s_0. Definitions of not admissible state set S_N, final state set S_G, transition function f and sets of possible decisions $U_p(s)$ for problem $1\|C_{max}$ shall be assumed as default. Some of the definitions above can be used without modifications for the remaining 4 problems.

Table 1. Default or specific definitions of problem elements.

	S_N	S_G	f	$U_P(s)$
$1\|\ C_{max}$				
$1\|C_j \leq d_i\|\ C_{max}$	spec			
$1\|prec\|\ C_{max}$				spec
$1\|r_i\|\ C_{max}$			spec	
$1\|\ C_{max} \leq \hat{C}\|\ \Sigma c_i$	spec	spec		

In the Table 1 above empty fields mean definitions identical to formulas for problem $1\|C_{max}$. Fields marked with the "spec" tag mean a definition specific to a given problem. For problem $1|C_j \leq d_j|C_{max}$ it is necessary to develops a specific definition of S_N set. For problem $1|prec|C_{max}$ a new definition of sets $U_p(s)$ is necessary. In case of problem $1|r_j|C_{max}$ the transition function has to be redefined. Only for problem $1|\ C_{max} \leq \hat{C}|\Sigma c_j$ 2 definitions need to be changed: the definition of sets S_N

and S_G. The latter one will take a different form than the analogical definition for problem $1|C_j \leq d_j|C_{max}$. This means 3 definitions universal for all the problems need to be developed together with 4 default ones and 5 specific to individual problems.

Each problem component definition is matched by a single atomic class implementing one or more methods. Thus 12 elementary classes will emerge. Using these classes it is possible to develop the software models of all five given problems with reusable code.

5 Conclusion

The paper presents results of the new research concerning the ALMM Solver, the tool for solving various discrete optimization problems, especially NP-hard problems. The solver uses an Algebraic Logical Meta Model of Multistage Decision Process (ALMM of MDP) and its theory.

ALMM of MDP enables a unified approach to developing models of discrete optimization problems and recording problem-related knowledge. The models are stored in Problem Model Library.

A new, extended modular structure of ALMM Solver is described, a software representation of AL models is proposed and the basic structure of Problem Model Library is presented.

References

1. Błażewicz, J., Ecker, K., Pesch, E., Schmidt, G., Węglarz, J.: Handbook on Scheduling. Springer, Berlin (2007). ISBN 978-3-540-28046-0
2. Dudek-Dyduch, E.: Information Systems for Production Management (in Polish) Wyd. Poldex, Kraków (2002). ISBN 83-88979-12-4
3. Dudek-Dyduch, E.: Learning based algorithm in scheduling. J. Intell. Manufact. (JIM) **11**(2), 135–143 (2000). (accepted to be published 1998)
4. Dudek-Dyduch, E.: Discrete determinable processes - compact knowledge-based model. Notas de Matematica No 137, Universidad de Los Andes, Merida, Venezuela (1993)
5. Dudek-Dyduch, E.: Problems of knowledge representation in expert system aided control of DPP (in Polish) Inżynieria Wiedzy i Systemy Ekspertowe, Prace II Krajowej Konferencji, tom I, pp. 147–154. Politechnika Wrocławska, Wrocław (1993)
6. Dudek-Dyduch, E.: Control of discrete event processes - branch and bound method. In: Proceedings of IFAC/Ifors/Imacs Symposium Large Scale Systems: Theory and Applications, Chinese Association of Automation, vol. 2, pp. 573–578 (1992)
7. Dudek-Dyduch, E.: Formalization and analysis of problems of discrete manufacturing processes. Scientific Bulletin of AGH University, Automatics, vol. 54 (1990) (in Polish). ISSN 0454-4773
8. Dudek-Dyduch, E.: Algebraic logical meta-model of decision processes - new metaheuristics. In: Rutkowski, L., Korytkowski, M., Scherer, R., Tadeusiewicz, R., Zadeh, L.A., Zurada, J.M. (eds.) ICAISC 2015. LNCS, vol. 9119, pp. 541–554. Springer, Heidelberg (2015)

9. Dudek-Dyduch, E.: Modeling manufacturing processes with disturbances - a new method based on algebraic-logical meta-models. In: Rutkowski, L., Korytkowski, M., Scherer, R., Tadeusiewicz, R., Zadeh, L.A., Zurada, J.M. (eds.) ICAISC 2015. LNCS, vol. 9120, pp. 353–363. Springer, Heidelberg (2015)

10. Dudek-Dyduch, E.: Modeling manufacturing processes with disturbances – two-stage AL model transformation method. In: 20th International Conference on Methods and Models in Automation and Robotics, MMAR proceedings, pp. 782–787 (2015)

11. Dudek-Dyduch, E., Dutkiewicz, L.: Substitution tasks method for discrete optimization. In: Rutkowski, L., Korytkowski, M., Scherer, R., Tadeusiewicz, R., Zadeh, L.A., Zurada, J.M. (eds.) ICAISC 2013, Part II. LNCS, vol. 7895, pp. 419–430. Springer, Heidelberg (2013)

12. Dudek-Dyduch, E., Dyduch, T.: Formal approach to optimization of discrete manufacturing processes. In: Hamza, M.H. (ed.) Proceedings of the Twelfth IASTED International Conference Modelling, Identification and Control. Acta Press, Zurich (1993)

13. Dudek-Dyduch, E., Kucharska, E.: Learning method for co-operation. In: Jędrzejowicz, P., Nguyen, N.T., Hoang, K. (eds.) ICCCI 2011, Part II. LNCS, vol. 6923, pp. 290–300. Springer, Heidelberg (2011)

14. Dudek-Dyduch, E., Kucharska, E.: Optimization learning method for discrete process control. In: ICINCO 2011, vol. 1, pp. 24–33 (2011)

15. Dudek-Dyduch, E., Kucharska, E., Dutkiewicz, L., Rączka, K.: ALMM Solver - A Tool for Optimization Problems. In: Rutkowski, L., Korytkowski, M., Scherer, R., Tadeusiewicz, R., Zadeh, L.A., Zurada, J.M. (eds.) ICAISC 2014, Part II. LNCS, vol. 8468, pp. 328–338. Springer, Heidelberg (2014)

16. Dutkiewicz, L., Dudek-Dyduch, E.: Substitution tasks method for co-operation. In: Badica, A., Trawinski, B., Nguyen, N.T. (eds.) Recent Developments in Computational Collective Intelligence. SCI, vol. 513, pp. 103–113. Springer, Heidelberg (2014)

17. Jędrzejowicz, P., Ratajczak-Ropel, E.: Reinforcement learning strategy for solving the MRCPSP by a team of agents. In: Neves-Silva, R., Jain, L.C., Howlett, R.J. (eds.) IDT 2015. SIST, vol. 39, pp. 537–548. Springer, Heidelberg (2015)

18. Kucharska, E., Dudek-Dyduch, E.: Extended learning method for designation of co-operation. In: Nguyen, N.T. (ed.) TCCI XIV 2014. LNCS, vol. 8615, pp. 136–157. Springer, Heidelberg (2014)

19. Ligęza, A.: Improving efficiency in constraint logic programming through constraint modeling with rules and hypergraphs. In: Federated Conference on Computer Science and Information Systems, pp. 101–107. IEEE Computer Society Press (2012)

20. Metsker, S.J.: Design Patterns in C#. Addison Wesley, Boston (2004). ISBN 0-321-12697-1

21. Pinedo, M.L.: Scheduling: Theory, Algorithms and Systems. Springer, Berlin (2012)

22. Rączka, K., Dudek-Dyduch, E., Kucharska, E., Dutkiewicz, L.: ALMM solver: the idea and the architecture. In: Rutkowski, L., Korytkowski, M., Scherer, R., Tadeusiewicz, R., Zadeh, L.A., Zurada, J.M. (eds.) ICAISC 2015, Part II. LNCS, vol. 9120, pp. 504–514. Springer, Heidelberg (2015)

23. Rossi, F., Van Beek, P., Walsh, T.: Handbook of Constraint Programming. Elsevier, Amsterdam (2006)

24. Smith, R.E., Taylor, N.: A framework for evolutionary computation in agent-based systems. In: Looney, C., Castaing, J. (eds.) Proceedings of the 1998 International Conference on Intelligent Systems, pp. 221–224. ISCA Press (1998)

25. OMG: Unified Modeling Language Version 2.5. http://www.omg.org/spec/UML/Current

Integration of Collective Knowledge in Financial Decision Support System

Marcin Hernes[(⊠)] and Andrzej Bytniewski

Wrocław University of Economics,
ul. Komandorska 118/120, 53-345 Wrocław, Poland
{marcin.hernes,andrzej.bytniewski}@ue.wroc.pl

Abstract. Execution of a process supporting making financial decisions using the multiagent system entails the need of permanent cooperation between a human (humans) and agent (agents) collectives. Their knowledge is acquired from autonomous and distributed sources and they use different decision support methods therefore certain level of heterogeneity characterizes knowledge of collectives. In the decision-making process one, final decision is required therefore knowledge of individual members of the collective shall be automatically integrated. The aim of the paper is to develop consensus method in order to integrate knowledge of human-agent collectives in a multiagent financial decision support system built with the use of cognitive agent architecture. The first part shortly presents the state-of-the-art in the considered field; next a Multiagent Cognitive Financial Decision Support System has been characterized. The last part of paper presents the consensus algorithm for knowledge integration.

Keywords: Multiagent systems · Financial decision support systems · Human-agent collectives · Knowledge integration · Consensus methods

1 Introduction

Making financial decisions is a continuous process, it is connected with multivariance due to its multicriteria nature, and consecutive decision-related situations appear in a chronological order, in near real-time, which are why it has become necessary to employ systems supporting decision making processes, including multiagent systems. The systems enable automatic and fast access to information of adequate value, on the basis of which one can draw conclusions [4].

Execution of a process supporting making financial decisions using the multiagent system entails the need of permanent cooperation between a human (humans) and a program agent (agents). There may be various forms of such cooperation. One of them may include a situation when agents generate different variants of a decision, and human make the final decision. Cooperation may also consist of agents making final decisions automatically on the basis of criteria defined by people and specifying the level of his or her satisfaction from the decision (the criteria may include, for example the level of return rate, the level of risk). The form of cooperation may also be connected with making decisions concerning final decisions on the basis of variants created by a human

© Springer-Verlag Berlin Heidelberg 2016
N.T. Nguyen et al. (Eds.): ACIIDS 2016, Part I, LNAI 9621, pp. 470–479, 2016.
DOI: 10.1007/978-3-662-49381-6_45

(an expert) and variants generated by an agent (where a human and an agent are treated equally while making decisions).

Each of the forms of cooperation leads to the emergence of human-agent collectives (collectives, groups) characterized by the fact that they have knowledge from autonomous and distributed sources and they use different decision support methods. For example, decisions made by humans (people) may be made with the use of fundamental analysis, on the basis of experts' opinions, whereas decisions of an agent (agents) may be made using technical analysis, on the basis of various types of indicators. Additionally, one of the members of a collective may for example perform analysis of securities of a given group of companies, and another one may analyze securities of a different group of companies. Consequently, decision variants presented by individual members of the collective may differ. A certain level of heterogeneity characterizes knowledge of these collectives. Since, however, in the decision-making process one, final decision is required, knowledge of individual members of the collective shall be automatically integrated. It may be done, for example by using certain criteria or functions of assessing knowledge of individual members of the collective. However, in case of an inadequate or imprecise indication of the criteria or functions, the risk of selecting a variant which does not guarantee the desired level of satisfaction increases. The employment of consensus methods, which also enable integration of knowledge, seems to be more reliable. The consensus methods, however, assume that each party is taken into account, each party to a conflict "loses" as little as possible, each party contributes to the consensus, all parties accept the consensus, and it constitutes the representation of all parties to a conflict. Any decision made using the methods does not have to be a decision formulated by any of the members of a collective. It may only closely resemble one of such decisions. Thus the consensus enables integration of knowledge in real-time, and it guarantees reaching a satisfactory compromise at a lower level of risk, which consequently may lead to making decisions which bring satisfactory benefits to decision makers.

The aim of the paper is to develop consensus method in order to integrate knowledge of human-agent collectives in a multiagent financial decision support system built with the use of cognitive program agent architecture [2]. The integration of knowledge will consequently enable selection of final decisions presented by the system to users. A particular attention has been paid to the form of cooperation consisting of establishing final decisions on the basis of variants created by humans (experts) and variants generated by program agents.

This paper is organized as follows: the first part shortly presents the state-of-the-art in the considered field; next a Cognitive Multiagent Financial Decision Support System has been characterized. The last part of paper presents the consensus algorithm for knowledge integration.

2 Related Works

One of the first solutions of a multiagent financial decision support system has been suggested in paper [12]. The system presented in the paper facilitates cooperation of a user with many specialized agents which have access to various financial patterns.

The agents analyze the situation on a financial market taking into account criteria specified by a user. Paper [2] describes a system in which agents have been divided into two groups, however agents from the first group make decisions based on fundamental analysis methods, whereas agents from the second group make decisions based on technical analysis. Article [7] presents a multiagent system facilitating the process of investing on FOREX currency exchange market, and the method of assessing investment strategies of selected agents. The differences between these two approaches rely on more openness of the second one (e.g. behavioral agents can be also implemented). Paper [8], on the other hand, presents methods of passive and active learning by financial decisions making agents.

It needs to be stressed that more and more often, in practical solutions as well as in various sources on the subject, cognitive program agents are used to build multiagent systems e.g. [4, 11]. The agents play cognitive and decisive roles, the same as the ones taking place in the human brain, thanks to which they are capable of understanding the real meaning of observed business phenomena and processes taking place also on financial markets.

Aspects of human-agent cooperation have been extensively illustrated also in the paper by Jennings et al. [6]. The authors have concluded that the cooperation may be realized in different forms and methods, and that human imagination is the only limitation here.

Works on the use of the consensus method in order to integrate knowledge have been carried out by numerous authors e.g. [12, 13]. In papers [9, 10], a formal mathematical model of knowledge integration has been suggested. It uses the function of knowledge integration based on the consensus model. The methodology has been employed in various types of information systems to solve conflicts and inconsistencies of knowledge and to integrate knowledge.

The solutions which have been suggested so far however do not focus much on the problem of integrating knowledge of a collective in situations when in a multiagent financial decision support system the human-agent cooperation consists of establishing final decisions on the basis of variants created by humans (experts) and variants generated by agents. The problem has been undertaken and discussed in the paper. Further considerations will focus on characterizing the functional architecture of a cognitive multiagent financial decision support system.

3   A Cognitive Multiagent Financial Decision Support System

The aim of the Cognitive Multiagent Financial Decisions Support System (CMFDSS) is to support investing in the Stock Exchange or currency exchange markets by generating automatic decisions concerning creation of a securities portfolio (mainly stock portfolio) or a currency portfolio.

The Learning Intelligent Distribution Agent (LIDA), developed by Franklin [3], was used to build the CMFDSS. One of the advantages of the architecture is its emergent and symbolic nature, thanks to which it is possible to process both, structured (numerical and symbolic) knowledge as well as the unstructured one (recorded in the natural language) [1].

The CMFDSS system is made up of the following elements (Fig. 1):

1. Human-agent collectives consisting of several experts (people) and some LIDA cognitive program agents. The job of members of a collective is to analyze markets and to select (generate) decisions (securities portfolio or currency portfolio). Each expert/agent uses a different method supporting decisions (fundamental analysis methods as well as technical analysis methods are used). Each collective makes decisions concerning a different market (for example Collective 1 makes decisions concerning the stock market, Collective 3 concerning the currency exchange market).
2. Knowledge integration module which with the help of the consensus method is responsible for integrating knowledge possessed by individual members of a collective, and for selecting one, final decision which is then presented to users.
3. Users – people, financial investors, or program agents investing on behalf of a human. Users execute taken decisions (buy and sell) on financial markets.

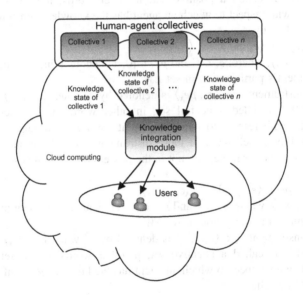

Fig. 1. Architecture of the CMFDSS.

One may notice that the CMFDSS can be viewed from a broader perspective. It is not just an information system as elements of sociological and social systems (experts-people groups) have been incorporated into its architecture. The level of satisfaction which the application of the consensus method to integrate knowledge should guarantee is specified by a user (decision maker), and it can be entered into a system in the form of parameters.

4   Integration of Knowledge in MCFDSS

Notice that each collective's knowledge state must be represented by using a concrete structure. Such structure was defined in previous work [5] as follows:

Definition 1. A knowledge structure representing decision P about finite set of financial instruments $E = \{e_1, e_2, ..., e_N\}$ is defined as a set:

$$P = \langle \{EW^+\}, \{EW^\pm\}, \{EW^-\}, Z, SP, DT \rangle$$

where:

(1) $EW^+ = \langle e_o, pe_o \rangle, \langle e_q, pe_q \rangle, ..., \langle e_p, pe_p \rangle$.
Couple $\langle e_x, pe_x \rangle$, where: $e_x \in E$ and $pe_x \in$ [0, 1] denote a financial instrument and this instrument's participation in set EW^+.
Financial instrument $e_x \in \langle e_x, pe_x \rangle$ is denoted by e_x^+ when $\langle e_x, pe_x \rangle \in EW^+$.
The set EW^+ is called a positive set; in other words, it is a set of financial instruments with respect to which an agent has the knowledge or information that they should be buy.

(2) $EW^\pm = \langle e_r, pe_r \rangle, \langle e_s, pe_s \rangle, ..., \langle e_t, pe_t \rangle$.
Couple $\langle e_x, pe_x \rangle$, where: $e_x \in E$ and $pe_x \in$ [0, 1] denote a financial instrument and this instrument's participation in set $EW^\pm$.
Financial instrument $e_x \in \langle e_x, pe_x \rangle$ is denoted by $e_x^\pm$ when $\langle e_x, pe_x \rangle \in EW^\pm$.
The set $EW^\pm$ is called a neutral set, in other words, it is a set of financial instruments, with respect to which an agent has no knowledge or information whether to buy or sell them. If these instruments are held by an investor, they should not be sold, or if they are not in the possession of the investor, they should not be bought.

(3) $EW^- = \langle e_u, pe_u \rangle, \langle e_v, pe_v \rangle, ..., \langle e_w, pe_w \rangle$.
Couple $\langle e_x, pe_x \rangle$, where: $e_x \in E$ and $pe_x \in$ [0, 1], denote a financial instrument and this instrument's participation in set EW^-.
Financial instrument $e_x \in \langle e_x, pe_x \rangle$ is denoted by e_x^- when $\langle e_x, pe_x \rangle \in EW^-$.
The set EW^- is called a negative set; in other words it is a set of financial instruments with respect to which an agent has the knowledge or information that they should be sell.

(4) $Z \in$ [0, 1] - decision rate of return forecast.

(5) $SP \in$ [0, 1] - degree of certainty of rate Z. It can be calculated on the basis of the level of risk related to the decision.

(6) DT- date of decision.

The percent of financial instrument's participation in positive, neutral or negative sets range <0, 1>. In our system, the financial decision consists of financial instruments, such as shares.

Integration of knowledge contained in human-agent collectives (realized in a module of knowledge integration) is performed in two stages. The concept of knowledge integration (Fig. 2) assumes that in the first stage a consensus is determined on the basis of knowledge status of all members of a collective, referred to as primary status of knowledge (primary profile in which the number of knowledge structures matches the number of all members of a given collective), and in the second stage an assessment of decisions of individual agents takes place.

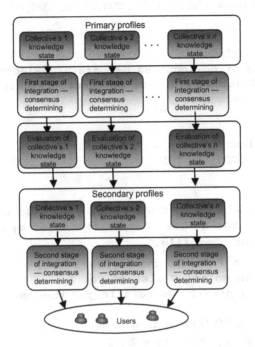

Fig. 2. A conception of knowledge integration process.

The assessment is performed by a separate agent performing evaluation on two levels:

- consistency of knowledge – various types of consistency evaluation functions are used; the evaluation is performed in such a way that decisions furthest from the consensus (in the sense of distances calculated according to different criteria) receive the worst score, and decision closest to the consensus receive the best score.
- efficiency (performance) of decisions made by members of a collective – a function of evaluation is used which takes into account performance and risk measurement indicators such as: return rate, number of profitable transactions, number of negative/loss transactions, costs of transactions, Sharpe ratio, etc.

A consensus algorithm for knowledge integration is as follows:

Algorithm 1.

Input: Profile $A = \{A^{(1)}, A^{(2)}, \ldots A^{(M)}\}$ consists of M structures of knowledge.
Result: Consensus $CON = \langle CON_+, CON_\pm, CON_-, CON_z, CON_{SP}, CON_{DT} \rangle$ according A.
BEGIN
1: Procedure primary.
2: Evaluation of collectives members` knowledge state and the elimination of lowest assessment knowledge states.
3: Procedure secondary.
END

Procedure primary:
1: Let $CON_+ = CON_\pm = CON_- = \emptyset$, $CON_z, CON_{SP}, CON_{DT} = 0$.
2: $j := 1$.
3: $i := +$.
4: If $t_i(j) > M/2$ then $CON_i := CON_i \cup \{e_j\}$. Go to: 6.
   // $t_i(j)$ - the number of occurrences of e_j element in sets EW^i of a profile.
5: If $i = +$ then $i := \pm$. If $i = \pm$ to $i := -$.
   If $i = -$ then go to: 6 else go to: 4.
6: If $j < Z$ then $j := j + 1$. Go to: 3. If $j \geq Z$ then go to: 7.
7: $i := DT$.
8: Determine $pr(i)$ //ascending order.
9: $k_i^1 = (M + 1) / 2$, $k_i^2 = (M + 2) / 2$.
10: $k_i^1 \leq CON_i \leq k_i^2$.
End procedure

Procedure secondary:
1: Determining a profile $B = \{B^{(1)}, B^{(2)}, \ldots B^{(M)}\}$ consist of N structures.
2: Let CON is a consensus determined in the same way as in Procedure primary but for the profile B.
3: $CON_z = \dfrac{1}{N} \sum\limits_{i=1}^{N} Z^i$.
4: $CON_{SP} = \dfrac{1}{N} \sum\limits_{i=1}^{N} SP^i$.
5: $CON_{DT} = \dfrac{1}{N} \sum\limits_{i=1}^{N} DT^i$, let $d := \sum\limits_{i=1}^{N} \left[\Psi\!\left(CON, A^{(i)}\right) \right]^2$ and $j := 1$.
6: If $e_j \in CON_+$ then

$$CON^{\backprime}:= \quad \langle CON_+ \backslash \{e_j\}, CON_\pm, CON_-, CON_z, CON_{SP}, CON_{DT}\rangle$$

Go to: 9, if $e_j \notin CON_+$ then go to: 7.

7: If $t_+(j) = 0$ then go to:10, If $t_+(j) > 0$ then go to:8.

8: If $e_j \cap CON \neq \emptyset$ and $e_j \in CON_\pm$ or $e_j \in CON_\pm$ then

$$CON^{\backprime}:= \quad \langle CON_+ \cup \{e_j\}, CON_\pm \backslash \{e_j\}, CON_- \backslash \{e_j\}, CON_z, CON_{SP}, CON_{DT}\rangle,$$

If $e_j \cap CON = \emptyset$ then

$$CON^{\backprime}:= \quad \langle CON_+ \cup \{e_j\}, CON_\pm, CON_-, CON_z, CON_{SP}, CON_{DT}\rangle. \text{ Go to: 9.}$$

9: If $\sum\limits_{i=1}^{N} \left[\Psi(CON^{\backprime}, A^{(i)})\right]^2 < d$ then $d := \sum\limits_{i=1}^{N} \left[\Psi(CON^{\backprime}, A^{(i)})\right]^2$ and

$CON := CON^{\backprime}$.

10: If $e_j \in CON_\pm$ then

$$CON^{\backprime}:= \quad \langle CON_+, CON_\pm \backslash \{e_j\}, CON_-, CON_z, CON_{SP}, CON_{DT}\rangle \text{ and go to:}$$

24, if $e_j \notin CON_\pm$ then go to: 11.

11: If $t_\pm(j) = 0$ then go to: 14, If $t_\pm(j) > 0$ then go to: 12.

12: If $e_j \cap CON \neq \emptyset$ and $e_j \in CON_+$ or $e_j \in CON_-$ then

$$CON^{\backprime}:= \langle CON_+ \backslash \{e_j\}, CON_\pm \cup \{e_j\}, CON_- \backslash \{e_j\}, CON_z, CON_{SP}, CON_{DT}\rangle,$$

If $e_j \cap CON = \emptyset$ then

$$CON^{\backprime}:= \quad \langle CON_+, CON_\pm \cup \{e_j\}, CON_-, CON_z, CON_{SP}, CON_{DT}\rangle.$$

13: If $\sum\limits_{i=1}^{M} \left[\Psi(CON^{\backprime}, A^{(i)})\right]^2 < d$ then $d := \sum\limits_{i=1}^{M} \left[\Psi(CON^{\backprime}, A^{(i)})\right]^2$ and

$CON := CON^{\backprime}$.

14: If $e_j \in CON_-$ then

$$CON^{\backprime}:= \quad \langle CON_+, CON_\pm, CON_- \backslash \{e_j\}, CON_z, CON_{SP}, CON_{DT}\rangle \text{ and}$$

go to: 17. If $e_j \notin CON_-$ then go to: 14.

15: If $t_-(j) = 0$ then go to: 18. If $t_-(j) > 0$ then go to: 16.

16: If $e_j \cap CON \neq \emptyset$ and $e_j \in CON_+$ or $e_j \in CON_\pm$ then

$$CON^{\backprime}:= \quad \langle CON_+ \backslash \{e_j\}, CON_\pm \backslash \{e_j\}, CON_- \cup \{e_j\}, CON_z, CON_{SP}, CON_{DT}\rangle.$$

If $e_j \cap CON = \emptyset$ then

$$CON^{\backprime}:= \quad \langle CON_+, CON_\pm, CON_- \cup \{e_j\}, CON_z, CON_{SP}, CON_{DT}\rangle.$$

17: If $\sum\limits_{i=1}^{N} \left[\Psi(CON^{\backprime}, A^{(i)})\right]^2 < d$ then $d := \sum\limits_{i=1}^{N} \left[\Psi(CON^{\backprime}, A^{(i)})\right]^2$ and

$CON := CON^{\backprime}$.

18: If $j < Z$ then $j := j+1$. Go to: 3.

End procedure.

Computational complexity of the algorithm equals: $O(N^2 M) + O(3NM)$.

It is worth noticing that in the second stage of knowledge integration determining a consensus is a NP-complete problem which is why the presented algorithm is a heuristic algorithm thanks to which its computational complexity is low.

The presented consensus determining algorithm allows agreeing on one decision presented by a system to a user taking into account evaluation of the status of knowledge of members of a collective. The algorithm is recalled automatically once all members of a collective have determined suggestions for decisions, and it is performed independently with respect to each collective.

On the basis of results of the preliminary research experiment performed by using 100 profiles it has been state, that in 92 cases consensus derived according to heuristic algorithm was in line with consensus derived by the optimal algorithm (compatibility level is 92 %). Consensus according to the optimal algorithm was calculated about 65 s, while the consensus heuristic algorithm in about 5 s. Therefore, the heuristic algorithm, developed in this paper, characterize a higher performance then performance of an optimal algorithm. Due to pages limitation of this paper, the wider research experiments will be presented in subsequent publication.

Knowledge integration allows for elimination of decisions generated by members of a collective whose knowledge status or condition has been assessed as being poor, which means that their decisions might not produce satisfactory benefits. Thanks to that, we are capable of eliminating the effect of such decisions on the final decision determined with the use of consensus methods and presented to a user. Additionally developed algorithm enables taking into account added knowledge of a collective as each individual status of knowledge of every member of a collective is taken into consideration.

5   Conclusions

Nowadays, in cognitive multiagent financial decision support systems, the cooperation of human-agent collectives is becoming more and more important. Authors of the paper have suggested implementing the collectives directly into the architecture of a system, thanks to which it is possible to automatically process collective knowledge. The problem of integration of knowledge of human-agent collectives, discussed in the paper, is also of great importance. Authors have pointed out that in order to solve the problem, consensus methods can be used. The developed consensus determining algorithm enables knowledge integration when there is cooperation between human and agents consisting of agreeing final decisions on the basis of variants created by experts (people) and variants generated by program agents. The algorithm includes the heterogenic nature of collective knowledge and enables generating added knowledge of a collective. In practical implementations it also allow also for decreasing risk level due to taking into consideration agents characterized by high level of knowledge. Consequently, it is possible to present to a user one satisfactory decision on the basis of which buy and sell transactions are made on financial markets. The developed consensus determining algorithm will also facilitate work of the creators of the multiagent financial decision support system as it can be directly implemented as a module of knowledge integration in the type of system. Consensus methods may also be used in order to integrate knowledge in decision support systems operating in others sectors (e.g. planning production, logistics, managing customers' relations). Since, however, knowledge structures differ with respect to each individual decision area, it is necessary to change the definition of the consensus algorithm.

Further research on the integration of knowledge of human-agent collectives shall focus, among other things, on verification of the effectiveness of the developed algorithm in systems functioning in practical environments, and on developing consistency evaluation function and on the assessment of knowledge of human-agent collectives in the cognitive multiagent financial decision support system.

References

1. Bytniewski, A., Chojnacka-Komorowska, A., Hernes, M., Matouk, K.: The implementation of the perceptual memory of cognitive agents in integrated management information system. In: Barbucha, D., Nguyen, N.T., Batubara, J. (eds.) New Trends in Intelligent Information and Database Systems. SCI, vol. 598, pp. 281–290. Springer, Heidelberg (2015)
2. Chiarella, C., Dieci, R., Gardini, L.: Asset price and wealth dynamics in a financial market with heterogeneous agents. J. Econ. Dyn. Control **30**, 9–10 (2011)
3. Franklin S., Patterson F.G.: The LIDA architecture: adding new modes of learning to an intelligent, autonomous, software agent. In: Proceedings of the International Conference on Integrated Design and Process Technology: Society for Design and Process Science, San Diego, CA (2006)
4. Goertzel, B., Wang, P.: Introduction: what is the matter here? In: Goertzel, B., Wang, P. (eds.) Theoretical Foundations of Artificial General Intelligence, pp. 1–9. Atlantis Press, New York (2012)
5. Hernes M., Sobieska-Karpińska J.: Application of the consensus method in a multiagent financial decision support system. Inf. Syst. e-Business Manag. Springer, Heidelberg (2015). doi:10.1007/s10257-015-0280-9
6. Jennings, N.R., Moreau, L., Nicholson, D., Ramchurn, S., Roberts, S., Rodden, T., Rogers, A.: Human-agent collectives. Commun. ACM **57**(12), 80–88 (2014)
7. Korczak J., Hernes M., Bac M.: Performance evaluation of decision-making agents' in the multi-agent system. In: Ganzha M., Maciaszek L., Paprzycki M. (eds.) Proceedings of Federated Conference Computer Science and Information Systems, Warszawa (2014)
8. LeBaron, B.: Active and passive learning in agent-based financial markets. East. Econ. J. **37**(1), 35–43 (2011)
9. Maleszka, M., Mianowska, B., Nguyen, N.T.: A method for collaborative recommendation using knowledge integration tools and hierarchical structure of user profiles. Knowl. Based Syst. **47**, 1–13 (2013)
10. Nguyen, N.T.: Advanced Methods for Inconsistent Knowledge Management. Springer, London (2008)
11. Rohrer, B.: An implemented architecture for feature creation and general reinforcement learning, workshop on self-programming in AGI systems. In: Fourth International Conference on Artificial General Intelligence, Mountain View, CA. http://www.sandia.gov/rohrer/doc/Rohrer11ImplementedArchitectureFeature.pdf. Accessed 11 September 2015
12. Sycara, K.P., Decker, K., Zeng, D.: Intelligent agents in portfolio management. In: Jennings, R.N., Wooldridge, M. (eds.) Agent Technology, pp. 267–281. Springer, Heidelberg (2002)
13. Zimniak, M., Getta, J.R., Benn, W.: Predicting database workloads through mining periodic patterns in database audit trails. Vietnam J. Comput. Sci. **2**(4), 201–211 (2015)

Framework for Product Innovation Using SOEKS and Decisional DNA

Mohammad Maqbool Waris[1(✉)], Cesar Sanin[1], and Edward Szczerbicki[2]

[1] The University of Newcastle, Callaghan, NSW, Australia
MohammadMaqbool.Waris@uon.edu.au, cesar.sanin@newcastle.edu.au
[2] Gdansk University of Technology, Gdansk, Poland
edward.szczerbicki@newcastle.edu.au

Abstract. Product innovation always requires a foundation based on both knowledge and experience. The production and innovation process of products is very similar to the evolution process of humans. The genetic information of humans is stored in genes, chromosomes and DNA. Similarly, the information about the products can be stored in a system having virtual genes, chromosomes and decisional DNA. The present paper proposes a framework for systematic approach for product innovation using a Smart Knowledge Management System comprising Set of Experience Knowledge Structure (SOEKS) and Decisional DNA. Through this system, entrepreneurs and organizations will be able to perform the product innovation process technically and quickly, as this framework will store knowledge in the form of experiences of the past innovative decisions taken. This proposed system is dynamic in nature as it updates itself every time a decision is taken.

Keywords: Product innovation · Product design · Set of experience · Decisional DNA · Innovation management

1 Introduction

Due to rapid changes in the dynamic environment, organizations involved in manufacturing products cannot grow through cost reduction alone. For the survival and prosperity of the manufacturing unit, entrepreneurs need to find out new ideas that can be implemented in the products leading to innovation. The reasons are frequent changes in the lifestyle of the users, rising costs of materials and energy, competition in the market at national and international level, and emerging technologies, among others. There are three types of approach in solving innovative problem [1]: A flash of genius, empiric path, and methodical path. Out of these the methodical path is a systematic approach for solving the innovative problem. By systematic analysis an optimal solution can be obtained quickly through this process.

The current study employs a systematic approach for product innovation. Innovation comes from the human minds like entrepreneurs who think and analyze the knowledge smartly [2]. Both knowledge and experience are essential attributes of an innovator that are necessary to find the optimal solution for the necessary changes. These changes

© Springer-Verlag Berlin Heidelberg 2016
N.T. Nguyen et al. (Eds.): ACIIDS 2016, Part I, LNAI 9621, pp. 480–489, 2016.
DOI: 10.1007/978-3-662-49381-6_46

are based on the innovative objectives that initiate the product innovation problem for finding the necessary changes to be implemented in the established product. But due to the enormous amount of ever increasing knowledge and rapid changes in the dynamic environment, the innovation process is difficult to practice. Innovators not only need to take proper decisions, they have to do it quickly and systematically so that the changes in the product may be implemented on time. Humans have the ability to store knowledge in their mind from the experiences they face during their life. This experience-based knowledge helps them in taking proper decisions for a relevant task by analyzing the knowledge smartly.

Erden et al. [3] pointed out that innovation is a collective work of a group of people or a team which draws the attention to "group tacit knowledge" and not an outcome of a single person. So, if the whole group is not available during the innovation process, it will be very difficult to find an optimal solution on time. If the Smart Knowledge Management System (SKMS) can be formulated for representing and analyzing data that is able to store experiences based on the decisions taken in the past, such a system will provide quick optimal solutions to a particular innovative problem. This system will act as a complete group of people required for finding solutions for query as it is using all the experiences from the past apart from other essential knowledge about the product. Moreover, the decision taken by this system will be quick due to fast computational ability of computers as compared to humans that also take more time for mutual coordination. Such a Knowledge Management (KM) system will help any innovative enterprise to survive in the dynamic environment.

Decision makers look back on the lessons learnt from the previous similar situations as a base for making current decisions [4]. However, the decision making process has some shortcomings like high response time, repetitions of decisions already made and inability to keep pace with the dynamic environment. This is due to inefficient knowledge administration and failure to capitalize the experiences within the organization. A smart knowledge management system, known as the Set of Experience Knowledge Structure (SOEKS) comprising Decisional DNA (DDNA), was presented by Sanin and Szczerbicki, see [5, 6]. Implementing this system in the process of product innovation will facilitate entrepreneurs and organizations to take proper decisions at appropriate time as this system stores information, knowledge as well as experiences of the formal decision events. In this way, the system grows and matures with time gaining more and more expertise in its domain.

The structure of the paper includes the background (Sect. 2) in which the concept of product innovation, SOEKS and DDNA are presented. Genetic structure of the product is explained in Sect. 3 along with the innovation process using SOEKS and the Decisional DNA. Finally the concluding remarks are presented in Sect. 4.

2 Background

2.1 Product Innovation

The most important questions encountered during innovation problem solving are: (i) when to innovate, and (ii) what to innovate? Most of the time, organizations fail to predict the proper time for analyzing and applying innovation. They start analyzing at a time when they should have been applying innovation. There must be some point, a particular time, at which the organization need to start analyzing the innovative objective. Once this point is recognized, the innovation process can be started for finding out the optimal solutions, so that the required innovative changes can be implemented into the product on time. There must be clear difference between the point of analyzing the innovative objective and the point of applying innovation. The time difference between these two will account for the complete innovation process, i.e. from analysis, innovative solution, design, manufacturing, and finally availability of the innovative product in the market. The alarming time for starting analyzing the innovation process is shown as a red circle in Fig. 1, called the critical zone. This is the point at which the sales are still increasing but the rate of increase in sales starts decreasing.

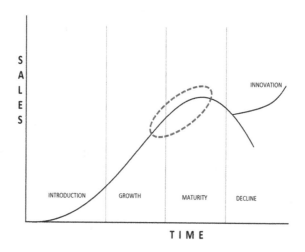

Fig. 1. The product life cycle with the introduction of innovation at later phases

Product innovation using set of experience knowledge structure and decisional DNA will make the innovation process systematic, smart and fast. After properly selecting the point of analyzing innovative objective and obtaining the optimal solution from the proposed system, the selected changes are designed and the product is manufactured accordingly. The product is then launched into the market at the point of applying innovation i.e. the time preferably at the end of maturity phase or at the beginning of decline phase and its sales start increasing again. The whole process can be repeated few times until the effect of innovative changes does not result in appreciable sales of the product or the product becomes obsolete [7].

2.2   Set of Experience Knowledge Structure and Decisional DNA

The Set of Experience Knowledge Structure (SOEKS) is a smart knowledge structure capable of storing explicitly formal decision events, see [8–10]. This smart knowledge based decision support tool stores and maintains experiential knowledge and uses such experiences in decision-making when a query is presented.

The SOEKS has four basic components: variables (V), functions (F), constraints (C) and rules (R) [4] as seen in Fig. 2. It comprises a series of mathematical concepts (logical element), together with a set of rules (ruled based element), and it is built upon a specific event of decision-making (frame element).

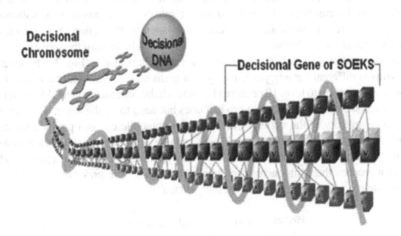

Fig. 2. Set of experience and decisional DNA [10]

Applications of SOEKS and DDNA have been successfully applied in various fields like industrial maintenance, semantic enhancement of virtual engineering applications, state-of-the-art digital control system of the geothermal and renewable energy, storing information and making periodic decisions in banking activities and supervision, e-decisional community, virtual organization, interactive TV, and decision support medical systems for Alzheimer's diagnosis to name a few. For details see Shafiq et al. [11]. Our research converges on the application of SOEKS in the development of systematic product innovation.

Variables are considered as the root of the structure as they are required to define other components. Functions are the relationship between a dependent variable and a set of input variables. Functions are used by the SOEKS for establishing links between variables and constructing multi-objective goals. Constraints are also functions that are used to set the limit to the feasible solutions and control system performance with respect to its goals. Rules, on the other hand, are the conditional relationships among the variables and are defined in terms of If-Then-Else statements.

A formal decision event is represented by a unique set of variables, functions, constraints and rules within the SOEKS. Groups of SOEKS are called chromosomes that represent a specific area within the organization and store decisional strategies for a category. Properly organized and grouped sets of chromosomes of the organization is collectively known as the Decisional DNA of the organization.

3   Product Innovation Using SOEKS and Decisional DNA

3.1   Genetic Structure of a Product

First the artifact, or the product, is structured in terms of a hierarchy of nested parts [12] as shown in Fig. 3. The product is divided into a number of systems performing some specific function of that product, which can be represented as subsystem level 1. Similarly, subsystem level 1 can be subdivided into further subsystems, represented as subsystem level 2 which are subassemblies associated with some sub-function that collectively perform the function at subsystem level 1. This nesting continues until the subsystem level reaches the component level. The four level hierarchy is shown in Fig. 3. The level of subsystems is different for any particular system performing a particular function of the product and can go up to level 10 or more [13] to reach the component level. Moreover the level of each subsystem in the same product does not need to be the same, as it depends upon the complexity of the system. Consider the automobile car as a product, it can be divided into subsystems level 1 like car body, engine, fuel system, suspension system, braking system, electrical system and so on. At subsystem level 2, it can be a piston and so on until it reaches the component level which can be a simple pressure ring under the engine system or a self-locking nut in car body system.

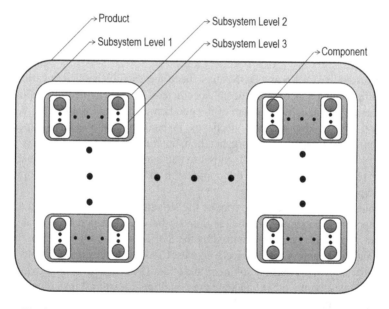

Fig. 3. Structure of a product representing its subsystems and components

There are inter-relationships among the systems, subsystems and components [14]. These relationships can be at the same level in the same system like piston and cylinder in the engine system, or it can be between subsystems at the same or different level under different systems like relationship between carburetor in fuel injection system and spark plug in engine system.

3.2 Innovation Process

Organizations involved in manufacturing products need to find out new ideas and innovate continuously to survive and prosper. They cannot grow only through cost reduction and improved engineering processes. The reasons are competition at international level, rising costs of material and energy, frequent changes in lifestyles, new technologies and automation. Innovation is defined as the process of making changes to something established by introducing something new that add value to users and contributes to the knowledge store of the organization [15]. Schumpeter [16] describes innovation as the use of an invention to create a new product or service resulting in the creation of new demand. He termed it as creative destruction as the introduction of a new product into the market destroys the demand of existing products and creates demand for new ones, and so on. There is a clear difference between an innovation and invention. Invention is the creation of something new and does not need to fulfil any customer need. Invention however can be exploited and transformed into a change that adds value to the customers; thus, becoming an innovation.

Manufacturers must respond quickly and effectively today to ever-growing demands from users for better products at lower costs. To survive, companies around the world must continuously pursue product innovation [14]. Most manufacturing organizations put their customers' satisfaction on top priority to improve their competitiveness [17]. A systematic and proper approach in product innovation can increase the life of the product. Based on innovative objectives, organizations can find out which features or functions of the product need to be upgraded, which ones may be excluded and which new features or functions may be added to the product. These features and functions are attributed to some systems of the product.

Innovative changes in the product can be performed by modifying one or more systems of the product. These modifications or changes can be at system, subsystem or component level. Accordingly, the required changes can be incorporated into the product to complete the innovation process. Product innovation is a continuous process; no organization, however big or successful, can continue to grow on the past achievements. Okpara [7] says that enterprises that rely exclusively on innovation will prosper until their products and services become obsolete and non-competitive.

3.3 Decisional DNA of Product Innovation

Organisms are created by nature through complex physical and chemical actions. Similarly products are also produced by manufacturers through physical and chemical actions [18]. Therefore, the essences of their origins are similar. DNA carries out the genetic information of the living organisms. It comprises four basic elements called nucleotides: adenine (A), thymine (T), guanine (G), and cytosine (C). Their combination represents a unique characteristic of this structure. A Gene is a portion of a DNA molecule that guides the operation of one particular component of an organism. Chromosome is a set of genes and multiple chromosomes represents the genetic code of the individual [19]. Similarly, the knowledge structure proposed by Sanin and Szczerbicki [4] can be used for storing and reusing the formal decisional events.

The unique combination of variables (V), functions (F), constraints (C), and rules (R) represents a formal decisional event or experience. This set of experience is considered as a part of long strand of DNA, that is, a gene or what we call here decisional gene of the product. A group of sets of experience of the same category is called as decisional chromosome. These chromosomes provide a schematic view of the knowledge. Finally the decisional DNA consists of stored experienced decision events, that is, experiential knowledge.

For example, a single decision associated with the fuel injection system of the car represents a set of experience, or decisional gene, of fuel injection system. Subsequently, many of these decisions, or sets of experience, associated with fuel injection will comprise a decisional chromosome of fuel injection system. Similarly many such types of chromosomes, like suspension chromosome, engine chromosome, car body chromosome, exhaust emission chromosome and so on, will comprise a Decisional DNA (DDNA) of the car. These chromosomes are never complete, as they keep growing with decisions added from time to time. In this way the decisional DNA continues to gain experiential knowledge which helps it to take decisions based on innovative objectives. This knowledge representation is dynamic in nature and behaves as an expert of the product from every perspective.

The architecture of product innovation DDNA is shown in Fig. 4. It is the knowledge representation of a product which is capable of capturing, storing, adding, improving, sharing as well as reusing knowledge in decision making in a way similar to an innovator or entrepreneur. The product innovation DDNA contains knowledge and experience of each important feature of a product. This information is stored in eight different modules of a product: Characteristics, Functionality, Requirements, Connections, Process, Systems, Usability, and Cost. The first five modules come under the decisional DNA of virtual engineering objects/virtual engineering process (VEO/VEP) that has been studied by Shafiq et al. [20, 21]. The information and experiential knowledge from VEO/VEP can be easily shared and used for innovation process.

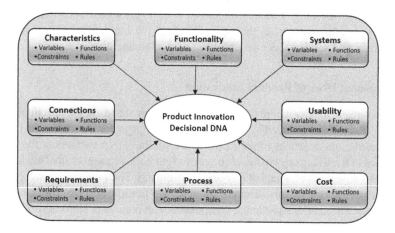

Fig. 4. Architecture of a product innovation DDNA

Characteristics represents the knowledge about dimensions, weight, appearance, etc. of the system, subsystems and components of the product as well as the possible concurrency attributes like versatility or ease of operation.

Functionality represents the knowledge about the basic working, input/output of the object (object collectively represents system, subsystem or a component of the product) and its operational principles. It also contains the operational knowledge of an object such as time consumed and outcome of the process that is performed.

Connections represents the knowledge about the relations between the VEOs in conjunction with the manufacturing scenario.

Requirements represents the knowledge about the necessities of the VEO required for its precise working. It includes the type and amount of power required, space requirements and the extent of user expertise necessary for operating a VEO.

Process represents the knowledge about the manufacturing process/process planning of the artifact having all shop floor level information including sequence and selection of operations, resources required for its manufacturing. This information helps in transforming a design model into a product in competent and economic way.

The above mentioned five modules can be extracted from the VEO/VEP DDNA studied by Shafiq et al. [20, 21]. Three more modules are introduced here;

Systems represents not only the knowledge about the relationships between various systems, subsystems and components like their hierarchy and dependability so as to represent a complete product but also stores the past history of every system, subsystem and component that were used for the same function. It also stores the possible alternative systems, subsystems or components that have the potential of replacing the current one. This module is continuously updated with the alternative systems used in advanced products as well as the new technological systems, inventions and advanced materials. This is the most important module for innovation process.

Usability represents the knowledge about the use of a particular system, subsystem or component of the product in other products. This will help in assessing its performance in other products. Information like which products have stopped using this system or component and in which products it has been introduced recently and its effect on the performance, popularity, sales or price of the product.

Cost represents the knowledge about the total cost of all the systems, subsystems and components. It will help in comparing and selecting a system, subsystem, optimum manufacturing process, component material, etc. on cost basis.

The query based on innovative objectives is fed into the SOEKS/DDNA system. This query is converted to a SOEKS containing a unique combination of variables, functions, constraints and rules as described in Sect. 2.2 above. The system will look for the most similar SOEKS for comparison and based on the similar experiences will provide proposed solutions. For example, the innovative objectives suggests possible changes in five functions or sub-functions. The system will relate these functions and sub-functions with some systems and subsystems of the product. Comparing the experiences from the past having some common innovative objectives, the system will provide the set number of possible solutions. At this point it may be noted that generally innovative changes are incremental and not modular i.e. changes are done only in few parts of the product not in all the systems and subsystem stated in innovative objectives.

So the system will compare the present query with the past queries having some common objectives. Based on the solutions of the past SOEKS, proposed solutions are obtained suggesting possible changes in some subsystems or components of the product.

The system will then compare the alternatives available in the systems and usability module. Based on the weights assigned to the attributes, list of solutions are presented by the system. The best solution is chosen and stored in the decisional DNA of the product innovation as a SOEKS that can be used for solving innovative problem in future. In this way the system also gains some experiential knowledge and with time it will behave as an expert innovator/entrepreneur having knowledge equivalent to a group of experts, capable of taking quick and smart decisions.

4   Conclusion

This paper applies Smart Knowledge Management System, comprising Set of Experience Knowledge Structure (SOEKS) and Decisional DNA, to solve the product innovation problem. The system is capable of storing knowledge as well as experiences of the past decisional events. The proposed framework for smart product innovation will allow the entrepreneurs and organizations to perform the innovation process quickly and technically sound. It stores the past decisional events or sets of experiences which help it in innovation process. The sets of experiences of the same category are grouped as a particular chromosome related to some function or system of the product and many such chromosomes represent a decisional DNA of the product. This proposed system is dynamic in nature as it updates itself every time the decision is taken. With time it will behave as an expert innovator/entrepreneur capable of taking quick and smart decisions.

References

1. Sheu, D.D., Lee, H.-K.: A proposed process for systematic innovation. Int. J. Prod. Res. **49**, 847–868 (2011)
2. Waris, M.M., Sanin, C., Szczerbicki, E.: Smart innovation management in product life cycle. In: 36th International Conference on Information Systems Architecture and Technology. Wroclaw University of Technology, Poland (2015)
3. Erden, Z., Krogh, G.V., Nonaka, I.: The quality of group tacit knowledge. J. Strateg. Inf. Syst. **17**, 4–18 (2008)
4. Sanin, C., Szczerbicki, E.: Set of experience: a knowledge structure for formal decision events. Found. Control Manage. Sci. **3**, 95–113 (2005)
5. Sanin, C., Szczerbicki, E.: A complete example of set of experience knowledge structure in XML. In: Szuwarzynski, A. (ed.) Knowledge Management: Selected Issues, pp. 99–112. Gdansk University Press, Gdansk (2005)
6. Sanin, C., Szczerbicki, E.: Decisional DNA and the smart knowledge management system: a process of transforming information into knowledge. In: Gunasekaran, A. (ed.) Techniques and Tool for the Design and Implementation of Enterprise Information Systems 2008, pp. 149–175. New York, USA (2008)
7. Okpara, F.O.: The value of creativity and innovation in entrepreneurship. J. Asia Entrepreneurship Sustain. **3**(2), 81–93 (2007)

8. Sanin, C., Szczerbicki, E.: Towards the construction of decisional DNA: a set of experience knowledge structure Java class within an ontology system. Cybern. Syst. **38**, 859–878 (2007)
9. Sanin, C., Szczerbicki, E.: Towards decisional DNA: developing a holistic set of experience knowledge structure. Found. Control Manage. Sci. **9**, 109–122 (2008)
10. Sanin, C., Toro, C., Haoxi, Z., Sanchez, E., Szczerbicki, E., Carrasco, E., Peng, W., Mancilla-Amaya, L.: Decisional DNA: a multi-technology shareable knowledge structure for decisional experience. Neurocomputing **88**, 42–53 (2012)
11. Shafiq, S.I., Sanin, C., Szczerbicki, E.: Set of experience knowledge structure (SOEKS) and decisional DNA (DDNA): past, present and future. Cybern. Syst. **45**(2), 200–215 (2014)
12. Murmann, J.P., Frenken, K.: Towards a systematic framework for research on dominant designs, technological innovations, and industrial change. Res. Policy **35**, 925–952 (2006)
13. Chen, K.Z., Feng, X.A., Chen, X.C.: Reverse deduction of virtual chromosomes of manufactured products for their gene-engineering-based innovative design. Comput. Aided Des. **37**, 1191–1203 (2005)
14. Chen, K.Z., Feng, X.A.: A gene-engineering-based design method for the innovation of manufactured products. J. Eng. Des. **20**, 175–193 (2009)
15. O'Sullivan, D., Dooley, L.: Applying Innovation. Sage Publications, London (2008)
16. Schumpeter, J.A.: The Theory of Economic Development. Harvard University Press, Cambridge (1934)
17. Ai, Q.S., Wang, Y., Liu, Q.: An intelligent method of product scheme design based on product gene. Adv. Mech. Eng. **2013**, 1–12 (2013)
18. Yong, C., Peien, F., Zhongqin, L.: A genetics-based approach for the principle conceptual design of mechanical products. Int. J. Adv. Manuf. Technol. **27**, 225–233 (2005)
19. Miller, K.R., Levin, J.: Biology. Prentice Hall, Saddle River (2002)
20. Shafiq, S.I., Sanin, C., Szczerbicki, E., Toro, C.: Virtual engineering objects/virtual engineering processes: a specialized form of cyber physical systems and Industry 4.0. Procedia Comput. Sci. **60**, 1146–1155 (2015)
21. Shafiq, S.I., Sanin, C., Toro, C., Szczerbicki, E.: Virtual engineering (VEO): towards experience-based design and manufacturing for industry 4.0. Cybern. Syst. **46**, 35–50 (2015)

Common-Knowledge and KP-Model

Takashi Matsuhisa[1,2(✉),3P]

[1] Karelia Research Centre, Institute of Applied Mathematical Research, RAS,
Pushkinskaya Ulitsa 11, Petrozavodsk, Karelia 185910, Russia
`takashi.matsuhisa@krc.karelia.ru`
[2] Faculty of Computational Mathematics and Cybernetics,
Lomonosov Moscow State University, Leninskie Gory, Moscow 119991, Russia
[3] Department of Advanced Course on Nursing,
Taisei Women Senior High School and College, Goken-cho 3-2-61,
Mito-shi, Ibaraki 310-0063, Japan
`takashimatsuhisa.mri.bsbh@gmail.com`

Abstract. This paper starts epistemic approaches of studying the Bayesian routing problem in the frame work of the network game introduced by Koutsoupias and Papadimitriou [LNCS 1563, pp.404–413. Springer (1999)]. It highlights the role of common-knowledge on the users' individual conjectures on the others' selections of channels in the network game. Especially two notions of equilibria are presented in the Bayesian extension of the network game; expected delay equilibrium and rational expectations equilibrium, such as each user minimizes own expectations of delay and social cost respectively. We show that the equilibria have the properties: If all users commonly know them, then the former equilibrium yields a Nash equilibrium in the based KP-model and the latter equilibrium yields a Nash equilibrium for social cost in the network game.

Keywords: Bayesian KP-model · Common-knowledge · Conjecture · Expected delay equilibrium · Information partition · Rational expectations equilibrium · Social cost

1 Introduction

This paper starts studying a Bayesian routing problem from the epistemic point of view. We focus on the role of common-knowledge on the users' individual conjectures on the others' selections of channels by giving Bayesian extension of KP-model, which is the network game introduced by Koutsoupias and Papadimitriou [7].

We propose two notions of equilibria, expected delay equilibrium and rational expectations equilibrium, which the former is the profiles of individual conjectures such as each user minimizes his/her own expectations of delay and the latter is the profiles of conjectures such as each user minimizes his/her own expectations of social cost respectively.

We will highlight the epistemic feature of these equilibriums, and we shall show the epistemic condition for the equilibria as below:

© Springer-Verlag Berlin Heidelberg 2016
N.T. Nguyen et al. (Eds.): ACIIDS 2016, Part I, LNAI 9621, pp. 490–499, 2016.
DOI: 10.1007/978-3-662-49381-6_47

Main Theorem. *If all users commonly know an expected delay equilibrium, then the equilibrium yields a Nash equilibrium in the based KP-model. If they commonly know a rational expectations equilibrium, then the equilibrium yields a Nash equilibrium for social cost in it.*

Garing et al. [5] is the first paper in which Bayesian Nash equilibrium is treated. They analysis Bayesian extension of routing game specified by the type-space model of Harsanyi [6] as information structure, and they collected several results: (1) the existence and computability of pure Nash equilibrium, (2) the property of the set of fully mixes Bayesian Nash equilibria and (3) the upper bound of the price of anarchy for specific types of social function associated with Bayesian Nash equilibria. In this paper we shall modify their model by adopting arbitrary partition structure following Aumann [1] instead of the type-space model. The merit of adopting information partition structure lies not only in getting the close connection to computational logic (Fagin et al. [4]) but also in increasing the range of its applications in various fields.

The paper is organized as follows: After reviewing KP-model in Sect. 2, we will present a Bayesian extension of the KP-models in Sect. 3. On highlighting the role of users' conjectures on the others' choices, we will propose the two equilibrium notions: expected delay equilibrium and rational expectations equilibrium. In Sect. 4, we will review the formal notion of common-knowledge, and will state the main theorem (Theorem 2). Further, we will make appraisal of the role of common-knowledge assumption in the main theorem. Section 5 concludes some remarks on agenda for further research.

2   KP-Model

In this paper we will treat only a simple type of the *KP-model* by Koutsoupias and Papadimitriou [7] as below.[1]

Let $m, n \in \mathbb{N}$ with $m, n \geqq 2$. The model consists of one *storage* S and n *users* (clients) $\{1, 2, \cdots, i, \cdots, n\}$ with which each has to use one of m *channels* (providers) to connect the storage. Each channel (or provider) $l = 1, 2, \cdots, m$ has a given capacity c_l. User i intends to send/receive information with volume w_i to/from the storage S through provider l_i. We call this structure a *KP-model* denoted by KP.

User i's actions are choices of the channels, so i's action set $L = \{l_i | l_i = 1, 2, \cdots, m\}$. This is common for all users. Let L^n denote $L \times L \times \cdots \times L$ (n times). We consider each member $l = (l_i)_{i \in N}$ of L as a *pure strategies profile*. Then the *delay* for a user i selecting a channel l_i is defined by

$$\lambda_i^{l_i} = \frac{1}{c_{l_i}}\{w_i + \sum_{k \in N \setminus i; l_k = l_i} w_k\}.$$

We consider a *social cost* $S(w, l)$ as a real valued function of pure strategies $l = (l_i)_{i \in N}$.

---

[1] Mazalov [9], Chap. 9 pp.314–351.

Example 1. The below are typical of social costs:

Linear cost: $LSC(w, l) = \sum_{i=1}^{n} \frac{1}{c_{l_i}} \left(\sum_{k;l_k=l_i} w_k \right)$;

Quadratic cost: $QSC(w, l) = \sum_{i=1}^{n} \frac{1}{c_{l_i}} \left(\sum_{k;l_k=l_i} w_k \right)^2$;

Maximal cost: $MSC(w, l) = \text{Max}_{i=1}^{n} \frac{1}{c_{l_i}} \left(\sum_{k;l_k=l_i} w_k \right)$. $\qquad\square$

Throughout this paper we shall assume that any volume bundle $w = (w_1, w_2, \cdots, w_n)$ is given a priori and each w_i is *indivisible*.

Let $\Delta(L)$ be the set of all probability distributions on L, and $\Delta(L^n) = \Delta(L)^n$. Each member $\sigma = (\sigma_i)_{i \in N}$ of $\Delta(L^n)$ is called a *mixed strategy*. The user i's *expected delay* of channel l_i for a mixed strategy $\sigma = (\sigma_i)_{i \in N}$ is

$$\lambda_i^{l_i}(\sigma) = \frac{1}{c_{l_i}} \{ w_i + \sum_{k \in N \setminus i} w_k \sigma_k(l_i) \}$$

Definition 1. A mixed strategy $\sigma = (\sigma_i)_{i \in N}$ is a *Nash equilibrium* for the KP-model if if for any client $i \in N$ and for any channel $l \in L$,

$$\lambda_i^l(\sigma) = \text{Min}_{j \in L} \lambda_i^j(\sigma) \text{ when } \sigma_i(l) \gneq 0 \text{ and}$$
$$\lambda_i^l(\sigma) \geqq \text{Min}_{j \in L} \lambda_i^j(\sigma) \text{ when } \sigma_i(l) = 0$$

Let NE(KP) be the set of all Nash equilibria.

By the *expected social cost* according to a mixed strategy $\sigma = (\sigma_i)_{i \in N}$ we mean

$$SC(w, \sigma) = \sum_{l=(l_k)_{k \in N} \in L^n} SC(w, l) \prod_{k \in N} \sigma_k(l_k).$$

Definition 2. A mixed strategy $\sigma = (\sigma_i)_{i \in N}$ is a *Nash equilibrium* for a social cost function $SC(w, l)$ if $SC(w, \sigma) \leqq SC(w, (l_i, \sigma_{-i}))$ for any user $i \in N$ and for any channel $l_i \in \text{Supp}(\sigma_i)$.

Let NE(SC) be the set of all Nash equilibria. According to Nash [10] the existence theorem for the above Nash equilibria is guaranteed:

Proposition 1. *Any KP-model KP always has its Nash equilibrium; i.e., NE(KP) $\neq \emptyset$, and furthermore it has also a Nash equilibrium for any social cost $SC(w, l)$; i.e.; NE(SC) $\neq \emptyset$.* $\qquad\square$

The *optimal social cost* for KP is defined by $\text{opt}(SC) = \inf_{\sigma \in \Delta(L^n)} SC(w, \sigma)$.

We can now obtain a characteristics of a network game KP: The *price of anarchy* for the network game KP is

$$PA(KP) = \sup_{\sigma \in \text{NE}(KP)} \frac{SC(w, \sigma)}{\text{opt}(SC)}$$

Example 2.[2] Let us consider the KP-model with one storage S and two users $N = \{1, 2\}$ sending/receiving their information to S of volume $w_1 = 1$ and $w_2 = 3$ through the two channels of the capacities $c_1 = c_2 = 1$ respectively. Denote $L_{(i)} = \{c_1, c_2\}$ for $i = 1, 2$. This KP-model has the unique mixed strategy Nash equilibrium $\sigma = (\frac{1}{2}c_1 + \frac{1}{2}c_2, \frac{1}{2}c_1 + \frac{1}{2}c_2)$. The linear social costs is $LSC(w, \sigma) = 4$.

3 Bayesian Approach

We will extend the above KP-model to the KP-model under uncertainty, called a *Bayesian KP-model*, and we treat only the network topology that *there is only one storage and every user connects directly to the storage'*.

3.1 Bayesian Extension

By *Bayesian KP-mode* we mean a KP-model equipped with a *partition information structure* $\langle \Omega, \mu, (\Pi_i)_{i \in N}, (l_i)_{i \in N} \rangle$ in which Ω is a non-empty *finite* set, called *state-space*, whose members are called *states*, whose subsets are *events*, μ is a probability measure on Ω with full support, and $(\Pi_i)_{i \in N}$ is a class of user i's information functions $\Pi_i : \Omega \to 2^\Omega$ together a profile of random variables (r.v.s) $l_i : \Omega \to L$, satisfying the below properties:

P1 $\{\Pi_i(\omega) | \omega \in \Omega\}$ is a partition of Ω;
P2 $\omega \in \Pi_i(\omega)$,
BP $\Pi_i(\omega) \subseteq [l]_i = [l_i = l]$ for any $l \in L$ and for any $\omega \in [l]_i$.[3]

We call also the above structure a *Bayesian extension* of a KP-model KP, and denote it by BKP. Furthermore, we shall denote by $\mathbf{B}(KP)$ the set of all the Bayesian extensions of a KP-model KP.

Example 3. The KP-model in Example 2 has a Bayesian extension by equipped with the below information partition:

- Two users $1, 2$;
- $\Omega = \{\omega_1, \omega_2, \omega_3, \omega_4\}$, with each state is interpreted as user 1 choses channel 1 at ω_1 and ω_2, and he choses channel 2 at ω_3 and ω_4. On the other hand, user 2 choses channel 1 at ω_1 and ω_3, and she choses channel 2 at ω_2 and ω_4. This is illustrated by

	c_1	c_2
c_1	ω_1	ω_2
c_2	ω_3	ω_4

---

[2] Example 9.6 in Mazalov [9] p. 324.
[3] Where $[l_i = l]$ is defined by $[l_i = l] = \{\omega \in \Omega | l_i(\omega) = l\}$. The last postulate **BP** means that 'user i knows absolutely his/her selection of channel l.

- μ is the equal probability measure on 2^{Ω}; i.e., $\mu(\omega) = \frac{1}{4}$:
- The information partition $(\Pi_i)_{i=1,2}$ is given as
  - The partition $\Pi_1 : \Omega \to 2^{\Omega}$:

$$\Pi_1(\omega) = \begin{cases} \{\omega_1, \omega_2\} \text{ for } \omega = \omega_1, \omega_2, \\ \{\omega_3, \omega_4\} \text{ for } \omega = \omega_3, \omega_4. \end{cases}$$

  - The partition $\Pi_2 : \Omega \to 2^{\Omega}$:

$$\Pi_2(\omega) = \begin{cases} \{\omega_1, \omega_3\} \text{ for } \omega = \omega_1, \omega_3, \\ \{\omega_2, \omega_4\} \text{ for } \omega = \omega_2, \omega_4. \end{cases}$$

- The r.v. $l_i : \Omega \to L_{(i)} = \{c_1, c_2\}$:

$$l_1(\omega) = \begin{cases} c_1 \text{ for } \omega = \omega_1, \omega_2, \\ c_2 \text{ for } \omega = \omega_3, \omega_4. \end{cases} \qquad l_2(\omega) = \begin{cases} c_1 \text{ for } \omega = \omega_1, \omega_3, \\ c_2 \text{ for } \omega = \omega_2, \omega_4. \end{cases}$$

3.2   Individual Conjecture

From Bayesian point of view we assume that each user i knows his/her choice of channel l, but he/she never know the other user's choices. The former assumption has already formulated in the above postulate **BP**. The latter requires to introduce i's *conjecture on the other user k's selection of channel*. By this we mean a probability distribution q_i on the others' section set $(L^n)_{-i}$ of channels $(l_k)_{k \in N \setminus i}$, i.e.; $q_i \in \Delta((L^n)_{-i})$. Denoting l by $l_{(k)}$ if user k selects l, the marginal probability $q_i(l_{(k)})$ on the other user k's section set L is the i's conjecture on k's selections. By using the random variable (r.v.):

$$\mathbf{q}_i([l]_k; \omega) = \mu([l]_k | \Pi_i(\omega))$$

we define the evens concerned of the conjecture q_i as follows:

$$[q_i(l_{(k)})] = \{\omega \in \Omega | \mathbf{q}_i([l]_k; \omega) = q_i(l_{(k)})\} \text{ and}$$
$$[q_i] = \cap_{(l_k)_{k \in N \setminus i} \in L^{n-1}} [q_i((l_k)_{k \in N \setminus i})]$$

3.3   Individual Expected Delay

A user i's *expectation of delay* λ_i^l for channel $l \in L$ according his/her conjecture q_i is defined by

$$\mathbf{E}_i[\lambda_i^l; q_i] = \frac{1}{c_l} \sum_{k \in N} w_k q_i(l_{(k)}).$$

It follows that for any $\omega \in [q_i(l_{(k)})]$, $\mathbf{E}_i[\lambda_i^l; q_i] = \frac{1}{c_l} \sum_{k \in N} w_k \mathbf{q}_i([l]_k; \omega)$. In viewing of **BP** it follows that $\mathbf{q}_i([l]_i; \omega) = 1$, and so

$$\mathbf{E}_i[\lambda_i^l; q_i] = \frac{1}{c_l}\{w_i + \sum_{k \in N \setminus i} w_k \mathbf{q}_i([l]_k; \omega)\} = \frac{1}{c_l}\{w_i + \sum_{k \in N \setminus i} w_k q_i(l_{(k)})\}.$$

Let us introduce an equilibrium for expected delay associated with individual conjecture:

Definition 3. i's conjectures $q_i \in \Delta((L^n)_{-i})(i \in N)$ is called an *rational expected delay* for BKP at state $\omega \in [q_i]$ if

$$\mathbf{E}_i[\lambda_i^l; (q_i)_{i \in N}] = \text{Min}_{j \in L}\mathbf{E}_i[\lambda_i^j; (q_i)_{i \in N}]$$

for any channel $l \in L$ adopted by $i \in N$.

We denote by $\mathrm{ED}^{\mathrm{BKP}}(q_i)$ the set of all the states in which q_i is a rational expected delay for $BKP \in \mathbf{B}(KP)$. Moreover,

Definition 4. A profile of all users' conjectures $q = (q_i)_{i \in N} \in (\Delta((L^n)_{-i}))^n$ is called an *expected delay equilibrium* (e.d.e.) for BKP at state $\omega \in [q_i]$ if

$$\omega \in \cap_{i \in N}\mathrm{ED}^{\mathrm{BKP}}(q_i)$$

We denote by $\mathrm{EDE}^{\mathrm{BKP}}(q)$ the set of all the states in which $q = (q_i)_{i \in N}$ is an e.d.e. for $BKP \in \mathbf{B}(KP)$.

3.4   Rational Expectation

A user i's *expectation of social cost* according to his/her conjecture q_i is

$$\mathbf{E}_i[SC(w, (l_i, 1_{-i})); q_i] = \sum_{\xi \in \Omega} SC(w, (l_i, 1_{-i}(\xi)))q_i(1_{-i}(\xi))$$

It follows that for $\omega \in [q_i]$, $q_i(l_{-i}) = \mathbf{q}_i([l_{-i}]; \omega)$ for any $l_{-i} \in (L^n)_{-i}$, and so

$$\mathbf{E}_i[SC(w, (l_i, 1_{-i})); q_i] = \sum_{l_{-i} \in (L^n)_{-i}} SC(w, (l_i, l_{-i}))\mathbf{q}_i([l_{-i}]; \omega)$$

$$= \sum_{l_{-i} \in (L^n)_{-i}} SC(w, (l_i, l_{-i}))q_i(l_{-i}).$$

Let us introduce an equilibrium for expected social cost associated with individual conjecture:

Definition 5. i's individual conjecture q_i is said to be *rational expectation* for a social function $SC(w, l)$ at $\omega \in \Omega$ if $\omega \in [q_i]$ provided with

$$\mathbf{E}_i[SC(w, (1_i(\omega), 1_{-i})); q_i] \leqq \mathbf{E}_i[SC(w, (l_i, 1_{-i})); q_i]$$

for any channel $l_i \in L$ adopted by i.

We denote by $\mathrm{RE}^{\mathrm{BKP}}(SC(w, l) : q_i)$ be the set of all the states in which $q_i(i \in N)$ is a rational expectation for a social cost $SC(w, l)$ for $BKP \in \mathbf{B}(KP)$. Furthermore,

Definition 6. A profile of all users' conjectures $q = (q_i)_{i \in N}$ is called an *rational expectations equilibrium* (r.e.e.) for a social function $SC(w, l)$ at $\omega \in \Omega$ if

$$\omega \in \cap_{i \in N} \mathrm{RE}^{\mathrm{BKP}}(SC(w, l) : q_i)$$

We denote by $\mathrm{REE}^{\mathrm{BKP}}(SC(w, l) : q)$ the set of all the states in which $q = (q_i)_{i \in N}$ is a r.e.e. for a social cost $SC(w, l)$ for $BKP \in \mathbf{B}(KP)$.

We can plainly observe that the below fundamental theorem follows from Proposition 1:

Theorem 1. *Any KP-model KP has a Bayesian extension $BKP \in \mathbf{B}(KP)$ such that it has simultaneously both an expected delay equilibrium $q' = (q'_i)_{i \in N}$ and a rational expectations equilibrium $q = (q_i)_{i \in N}$ for any social function $SC(w, l)$ at everywhere; i.e.:* $\mathrm{REE}^{\mathrm{BKP}}(SC(w, l) : q) = \mathrm{EDE}^{\mathrm{BKP}}(q') = \Omega.$

The converse is an interesting problem:

Problem 1. Given a Bayesian extension BKP of a KP-model together with an information structure, under what conditions can a rational expectations equilibrium for $SC(w, l)$ actually yield a Nash equilibrium for $SC(w, l)$? Or/and how about an expected delay equilibrium?

Example 4. Let us reconsider the Bayesian KP-model presented in Example 3. The r.v. $(\mathbf{q}_i)_{i=1,2}$ are given: For any $\omega \in \Omega$,

$$\mathbf{q}_1([l]_2; \omega) = \frac{1}{2} \text{ for } l \in L_{(1)} = \{1, 2\}; \quad \mathbf{q}([l]_1; \omega) = \frac{1}{2} \text{ for } l \in L_{(2)} = \{1, 2\}$$

Hence it can be plainly observed that there exists unique profile of conjectures $(q_i)_{i=1,2}$ with $[q_i] \neq \emptyset$. The conjectures are: $q_1(l) = q_2(l) = \frac{1}{2}$., and the profile of them gives the unique e.d.e. for this BKP, together with the unique r.e.e. for any social costs $SC(w, l)$.

4   Epistemic Approach

We will present an affirmative answer to Problem 1 as above by using common-knowledge following Aumann and Brandenburger [3].

4.1   Knowledge

We will give the formal model of knowledge.[4] Let us consider a Bayesian KP-model BKP based on a KP-model KP, and i be an arbitrary user. i's *knowledge operator* K_i on 2^Ω defined by

$$K_i E = \{\omega \mid \Pi_i(\omega) \subseteq E\}$$

The event $K_i E$ will be interpreted as the set of states of nature for which i knows E to be possible. We record the properties of i's knowledge operator, which actually characterize 'Knowledge.'[5] For every E, F of 2^Ω,

---

[4] This is called the *event-based approach* in Fagin et al. [4].

[5] According to these properties we can say the structure $\langle \Omega, (K_i)_{i \in N} \rangle$ is a model for the multi-modal logic **S5n**.

N $K_i \Omega = \Omega$, $K_i \emptyset = \emptyset$;   **K** $K_i(E \cap F) = K_i E \cap K_i F$;   **T** $K_i E \subseteq E$
4 $K_i E \subseteq K_i(K_i E)$;   **5** $\Omega \setminus K_i E \subseteq K_i(\Omega \setminus K_i E)$.

4.2   Common-Knowledge

The *mutual knowledge operator* $K_E : 2^\Omega \to 2^\Omega$ is the intersection of all individual knowledge operators: $K_E F = \cap_{i \in N} K_i F$ which is interpreted as that everyone knows E. By the composition of knowledge operators K_i, K_j we mean the operator $K_i K_j : 2^\Omega \to 2^\Omega$ defined by $K_i K_j(E) = K_i(K_j E)$. The *common-knowledge operator* $K_C : 2^\Omega \to \Omega$ is defined by

$$K_C E = \cap_{n \in \mathbb{N}} \cap_{\{i_1.i_2,\cdots,l_n\} \subseteq N} K_{i_1} K_{i_2} \cdots K_{i_n} E. \tag{1}$$

The intended interpretations are: An event E is *common-knowledge* at $\omega \in \Omega$ if $\omega \in K_C E$, and $K_C E$ is the event of common-knowledge of E. It can be plainly observed that K_C satisfies all the above properties **N**, **K**, **T**, **4** and **5**.

Example 5. Let us turn back to the Bayesian KP-model BKP presented in Example 3. The BKP has the profile of the conjectures $(q_i)_{i=1,2}$ that is simultaneously the unique e.d.e and the unique r.e.e. by Example 4. Here we should note that both q_1 and q_2 are common-knowledge among users 1 and 2 at everywhere, because $[q_1] = [q_2] = \Omega = K_C([q_1]) = K_C([q_2])$.

4.3   Epistemic Conditions for Equilibrium

The above observation will lead us to necessary conditions for these equilibrium in general case as the below theorem. For the proof, see Matsuhisa [8]

Theorem 2 (Matsuhisa [8]). *In a Bayesian KP-model, the following statements are true:*

(i) *If there is a state where for each $i \in N$, it is common-knowledge among N that i's conjecture q_i is a rational expectation for a social cost $SC(w, l)$, then the profile $(q_i)_{i \in N}$ yield a Nash equilibrium for $SC(w, l)$.*

(ii) *If there is a state where for each $i \in N$, it is common-knowledge among N that i's conjecture q_i is a rational expected delay, then the profile $(q_i)_{i \in N}$ of conjectures yield a Nash equilibrium for KP. Furthermore, i's expectation of delay $\mathbf{E}_i[\lambda_i^l; (q_i)_{i \in N}]$ coincides with i's expected delay $\lambda_i^l(\sigma)$.*

Remark 1. We shall make appraisal of the common-knowledge assumption in Theorem 2. In the case of two users, Theorem 2 is still true without the common-knowledge assumption. However, in the case of more than two users, the common-knowledge assumption plays crucial role in Theorem 2. In fact, a rational expectation equilibrium for a social cost $SC(w, l)$ can not always yield a Nash equilibrium for $SC(w, l)$ if the all users does not commonly know that one of the conjectures is a rational expectation, even though they mutually know that it is a rational expectation also. The example in Matsuhisa [8] shows the situation:

5   Discussions

This paper is the first time to treat a Bayesian routing game from the view point of the interaction of knowledge for users'. We have been keeping many important agenda left, so it will end this paper well by remarking some of them.

5.1   Expected Price of Anarchy

First of all, the most important agenda is to proceed with treating *social costs with its price of anarchy* from epistemic point of view. Let us extend them into our framework. By user i's *individual expected social cost* $\mathbf{E}_i[SC; q_i]$ according to q_i we shall mean the sum of the individual expected costs for all users; i.e.:

$$\mathbf{E}_i[SC; q_i] = \sum_{l_i \in L_{(i)}} \mathbf{E}_i[SC(w, (l_i, 1_{-i})); q_i]$$

and the *expected social cost* is $\mathbf{E}[SC; q] = \sum_{i \in N} \mathbf{E}_i[SC; q_i]$. The *expected price of anarchy* is

$$\mathbf{E}[PA]^{BKP}(SC) = \frac{\sup_{q \in \mathrm{REE}^{BKP}(SC(w,l))} \mathbf{E}[SC; q]}{\inf_{q \in (\Delta((L^n)_{-i}))^n} \mathbf{E}[SC; q]}$$

Then we have the below conjecture: Without loss of generality, assume $c_1 \gtrless c_2 \gtrless \cdots c_m \gneqq 0$.

Conjecture 1. Let $SC(w, l)$ be one of $LSC(W, l), QSC(w, l)$ and $MSC(w, l)$. Then we have a upper bound of the expected price of anarchy as

$$\mathbf{E}[PA]^{BKP}(SC) \lesseqgtr \begin{cases} \frac{c_1}{c_m} & for\ SC = LSC, MSC; \\ \left(\frac{c_1}{c_m}\right)^2 & for\ SC = QSC. \end{cases}$$

It is worthy noting that the conjecture implies a refinement of Theorem 9.4 in Mazalov [9] as below.

Corollary 1. *If BKP has homogeneous capacities (i.e.; $c_1 = c_2 = \cdots = c_m$) then* $\mathbf{E}[PA]^{BKP}(SC) = 1$ *for SC = LSC, MSC, QSC.*

5.2   Common-Knowledge

As we noted in Remark 1, the common-knowledge assumption plays essential role in Theorem 2. However it is actually very strong assumption, because common-knowledge is introduced by the infinite regress of interactions among individual knowledge as like as Eq. (1). So we would like to remove out it in our framework. There seems to be several ways to improving this point, here I would recommend adopting the communication process introduced by Parikh and Krasucki [11] replacing common-knowledge.

5.3 Indivisible Volumes

In this paper the volumes in BKP are considered as indivisible goods, but these should be consider as divisible ones when we shall consider KP models as the models of cloud computing system, because the volumes are considered as volumes of information. To treat KP models from this point of view the Bayesian KP-model shall be investigated as exchange economy under uncertainty, and it arises an interesting problem to investigate the relationship between the several core notions appeared and the equilibria presented in this paper.

5.4 General Information Partition

As mentioned previously, Garing et al. [5] is the first paper in which they analysis Bayesian extension of routing game specified by the type-space model of Harsanyi [6] as information structure, On the other hand, I modify their model by adopting arbitrary partition structure in this paper instead of the type-space model following Aumann [1]. The merit of adopting information partition structure lies in getting the close connection to computational logic (Fagin et al. [4]). The Bayesian Nash equilibrium in Garing et al. [5] seems to be very close to the correlated equilibrium of Aumann [2], so we have to introduce the notion of correlated equilibrium in BKP-model, and investigate the relationship between the equilibrium and the Bayesian Nash equilibrium in Garing et al. [5]. In this regards it is also important to reconstruct and refine the results obtained in Garing et al. [5] in our framework.

References

1. Aumann, R.J.: Agreeing to disagree. Ann. Stat. **4**, 1236–1239 (1976)
2. Aumann, R.J.: Subjectivity and correlation in randomized strategies. J. Math. Econ. **1**, 67–96 (1974)
3. Aumann, R.J., Brandenburger, A.: Epistemic conditions for mixed strategy Nash equilibrium. Econometrica **42**, 1161–1180 (1995)
4. Fagin, R., Halpern, J.Y., Moses, Y., Vardi, M.Y.: Reasoning About Knowledge. MIT Press, Cambridge (1995)
5. Garing, M., Monien, B., Tiemann, K.: Selfish routing with incomplete information. Theor. Compt. Syst. **42**, 91–130 (2008)
6. Harsanyi, J.C.: Games with incomplete information played by Bayesian players, I, II, III. Manag. Sci. **14**, 159–182, 320–332, 468–502 (1967)
7. Koutsoupias, E., Papadimitriou, C.: Worst-case equilibria. In: Meinel, C., Tison, S. (eds.) STACS 1999. LNCS, vol. 1563, p. 404. Springer, Heidelberg (1999)
8. Matsuhisa, T.: Selfish Routing with Common-knowledge. Working paper (2015)
9. Mazalov, V.: Mathematical Game Theory and Applications. Wiley (2014)
10. Nash, J.F.: Equilibrium points in n-person games. Proc. Natl. Acad. Sci. U. S. Am. **36**, 48–49 (1950)
11. Parikh, R., Krasucki, P.: Communication, consensus, and knowledge. J. Econ. Theor. **52**, 78–89 (1990)

Controllability of Semilinear Fractional Discrete Systems

Jerzy Klamka[✉]

Institute of Control Engineering, Silesian University of Technology, Gliwice, Poland
`jerzy.klamka@polsl.pl`

Abstract. In the present paper local constrained controllability problems for semilinear finite-dimensional discrete system with constant coefficients are formulated and discussed. Using some mapping theorems taken from functional analysis and linear approximation methods sufficient conditions for constrained controllability are derived and proved. The present paper extends the controllability conditions with unconstrained controls given in the literature to cover the semilinear discrete systems with constrained controls.

Keywords: Semilinear systems · Discrete systems · Controllability · Nonlinear operators

1 Introduction

Controllability is one of the fundamental concept in mathematical control theory [3, 4]. Roughly speaking, controllability generally means, that it is possible to steer system from an arbitrary initial state to an arbitrary final state using the set of admissible controls. In the present paper local constrained controllability problems for semilinear fractional discrete control system with constant coefficients are formulated and discussed. Semilinear fractional discrete systems [1, 7, 11] are subclass of general nonlinear discrete systems and contain both pure linear part and pure nonlinear in the right hand side of the difference state equation. Using some mapping theorems taken from functional analysis [12] and linear approximation methods [7] sufficient conditions for constrained controllability are derived and proved. The present paper extends in some sense the results given in papers [5–9] to cover the semilinear fractional discrete systems with constrained controls.

The research was done by the author as part of the projects funded by the National Science Centre in Poland granted according to the decisions: DEC-2012/07/B/ST7/01408.

N.T. Nguyen et al. (Eds.): ACIIDS 2016, Part I, LNAI 9621, pp. 500–507, 2016.
DOI: 10.1007/978-3-662-49381-6_48

2   Preliminaries

In this paper definition of the fractional difference of the form

$$\Delta^\alpha x_k = \sum_{j=0}^{k} (-1)^j \binom{\alpha}{j} x_{k-j}, \tag{1}$$

$$n - 1 < \alpha < n \in N = \{1, 2, \dots\}, \ k \in Z_+$$

will be used, where $\alpha \in R$ is the order of the fractional difference and the set of nonnegative integers is denoted by Z_+

$$\binom{\alpha}{j} = \begin{cases} 1 & \text{for } j = 0 \\ \frac{\alpha(\alpha-1)\cdots(\alpha-j+1)}{j!} & \text{for } j = 1, 2, \dots \end{cases} \tag{2}$$

Consider the fractional discrete semilinear control system, described by the difference state-space equations

$$\Delta_{k+1}^\alpha = f(x_k, u_k) + Cx_k + Du_k, \quad k \in Z_+ \tag{3}$$

with given initial condition x_0.

Where $x_k \in U \subset R^n$, $u_k \in R^m$ are the state and input vectors and A and B are given $n \times n$ and $n \times m$ constant matrices, respectively. Moreover, $f{:}R^n \times R^m \to R^n$, is a given function generally nonlinear and continuously differentiable near the origin. Moreover, we us assume that $f(0, 0) = 0$. In practical applications the function $f(x_k, u_k)$ may be treated as certain outside disturbances near the origin.

Taking into account the above assumption, let us introduce the notations for partial derivatives of the function $f(x, u)$ with respect to both vectors variables

$$G = f_x'(0,0), \qquad H = f_u'(0,0)$$

where G is $n \times n$-dimensional constant matrix, and H is $n \times m$-dimensional constant matrix.

In controllability investigations of semilinear fractional control system (3) the crucial role plays linear approximation of the system around the origin. Therefore, using standard methods, it is possible to construct linear approximation of the semilinear fractional discrete system (3). However, this linear approximation is valid only in some neighbourhood of the origin in the product space $R^n \times R^m$.

Approximated linear systems corresponding to semilinear fractional system (3) is given by the linear difference state Eq. (4)

$$\Delta^\alpha x_{k+1} = Ax_k + Bu_k, \quad k \in Z_+ \tag{4}$$

where matrices A, B are of the following form

$$A = C + G \quad and \quad B = D + H$$

In order to discuss controllability problem for semilinear fractional system (3) first of all it is necessary to find the solution for linear fractional difference state Eq. (4) from a given initial condition x_0. In order to do that let us consider the linear autonomous difference equation

$$\Delta^\alpha_{k+1} = Ax_k, \quad k \in Z_+$$

Following [1] the solution with initial condition x_0 has the form

$$x_{k+1} = S_k x_0$$

where $n \times n$ dimensional matrix S is a solution of difference equation

$$S_{k+1} = (A + \alpha I_n)S_k + \sum_{i=1}^{i=k+1} (-1)^{i+1} \binom{\alpha}{i} S_{k-i+1} \tag{5}$$

with initial condition

$$S_0 = I_n$$

where I_n is $n \times n$ dimensional identity matrix.

It should be mentioned, that for special case $\alpha = 1$ i.e., for standard system we have

$$S_k = A^k$$

Therefore, using matrix S the solution for the linear fractional Eq. (4) can be represented in the following compact form

$$x_k = S_k x_0 + \sum_{i=0}^{i=k-1} S_{k-i-1} Bu_k$$

In the literature there are many different controllability concepts, which strongly depends on the type of dynamical system and the given set of admissible controls. However, it is well known, (see e.g., [3–7]) that for linear, semilinear and generally nonlinear discrete dynamical control systems it is possible to define local controllability in a given number of steps and global controllability in a given number of steps.

In the sequel we shall concentrate on local controllability of semilinear fractional system (3) in a given number of steps i.e. in the interval $[0, q]$ and global controllability for linear fractional system (4) in the same interval $[0, q]$. For simplicity of consideration both cases will be discussed for zero initial condition $x_0 = 0$.

Definition 1. Linear fractional discrete system (4) is said to be globally U-controllable in a given interval $[0, q]$ if for zero initial condition $x_0 = 0$, and every vector $x' \in R^n$, there exists an admissible sequence of controls $u_p = \{u(i) \in U; 0 \le i < q\}$, such that the corresponding solution of the Eq. (4) satisfies condition $x(p) = x'$.

Definition 2. Semilinear fractional discrete system (3) is said to be locally U-controllable in a given interval *[0, q]* if for zero initial condition there exists neighbourhood of zero $D \subset R^n$, such that for every point x' $\in$ D there exists an admissible sequence of controls $u_p = \{u(i) \in U: 0 \leq i < p\}$, such that the corresponding solution of the Eq. (3) satisfies the final condition $x(p) = x'$.

There is an essential connection between local U-controllability of semilinear fractional system (3) and global U-controllability of linear fractional system (4) given in the next lemma.

Lemma 1. Global U-controllability of linear fractional discrete system (4) is a sufficient condition for local U-controllability near the origin for semilinear fractional discrete system (3).

From the above lemma follows two remarks.

Remark 1. Since global U-controllability of linear fractional system (4) is only sufficient but not necessary condition for local controllability near the origin for semilinear system (3), therefore, there are locally U-controllable semilinear fractional systems for which linear approximations are not globally U-controllable.

Remark 2. The crucial role in local U-controllability for semilinear fractional systems plays global U-controllability of corresponding linear fractional system.

3   Main Results

Taking into account Remark 2, first of all we shall prove necessary and sufficient condition for global controllability of linear fractional system (4) for unconstrained admissible controls i.e., $U = R^m$.

Theorem 1. The discrete fractional linear system (4) is controllable in q steps if and only if $n \times qm$ dimensional controllability matrix

$$M_q = [B, S_1 B, S_2 B, \dots, S_k B, \dots, S_{q-1} B] \tag{6}$$

has full row rank n.

Proof. Using solution of the linear fractional system given in (4) for $k = q$ and initial condition $x_0 = 0$ we obtain

$$x_k = x_q = \sum_{i=0}^{i=q-1} S_{q-i-1} B u_i = M_q \begin{bmatrix} u_{q-1} \\ u_{q-2} \\ \vdots \\ u_i \\ \vdots \\ u_1 \\ u_0 \end{bmatrix} \tag{7}$$

From Definition 1 it follows that linear fractional system (4) is controllable in q steps if and only if for every final state $x_q \in R^n$ there exists a input sequence $u_i \in U$, $i = 0, 1, \ldots, q - 1$ such that the equality (7) is satisfied. On the other hand it take place if and only if controllability matrix M_q has full row rank n. Hence Theorem 1 is proved.

Using the well-known properties of matrices from Theorem 1 follows directly the next Corollary.

Corollary 1. The fractional linear system (4) is globally controllable in q steps if and only $M_q M_q^T$ is invertible matrix, i.e. there exist matrix $\left(M_q M_q^T\right)^{-1}$, where M^T denotes transposition.

Taking into account the relationship between global controllability of linear fractional system (4) and local controllability of semilinear fractional system (3) we have the following sufficient condition for local controllability in q steps of semilinear fractional discrete system (3).

Corollary 2. The semilinear fractional discrete system (3) is locally controllable in q steps if $M_q M_q^T$ is invertible matrix.

Now let as consider constrained controllability problems for semilinear fractional discrete control system (3). Let $U \subset R^m$ be a given closed convex cone with vertex at zero. The sequence of controls $\{u_k, k = 0, 1, 2, \ldots\}$ is called an admissible sequence of controls.. In the sequel we shall also use the following notations: $\Omega^0 \subset R^m$ is a neighbourhood of zero, and $U^0 = U \cap \Omega^0$.

Therefore, in the next part of this section we shall formulate sufficient conditions of local U-controllability in a given interval $[0, p]$ and set U for the semilinear fractional discrete system (3).

Lemma 2. [12] Let $F: Z \to Y$ be a nonlinear operator from a Banach space Z into a Banach space Y which has the Frechet derivative $dF(0): Z \to Y$, whose image coincides with the whole space Y. Then the image of the nonlinear operator F will contain a neighbourhood of the point $F(0) \in Y$.

Lemma 3. [12] Let $F: Z \to Y$ be a nonlinear operator from a Banach space Z into a Banach space Y and suppose that $F(0) = 0$. Assume that the Frechet derivative $dF(0)$ maps a closed convex cone $C \subset Z$ with vertex at zero onto the whole space Y. Then there exists neighbourhoods $M_0 \subset Z$ about $0 \in Z$ and $N_0 \subset Y$ about $0 \in Y$ such that the equation $y = F(z)$ has for each $y \in N_0$ at least one solution $z \in M_0 \cap C$.

Now, we are in the position to formulate the main result on the local U-controllability for semilinear system (3).

Theorem 2. Let us suppose, that $U^c \subset R^m$ is a closed convex cone with vertex at zero. Then the semilinear fractional discrete system (3) is locally U^{c0}-controllable in the interval $[0, q]$ if its linear approximation near the origin given by the difference Eq. (4) is globally U^c-controllable in the same interval $[0, q]$.

Proof. Let us define for the nonlinear dynamical system (3) a nonlinear map g: $U_q \to R^n$ by

$$g(u) = x(q, u_q) = \sum_{i=0}^{i=q-1} S^{q-i-1}(f(x(i), u(i)) + Du(i))$$

where $u_q = \{u(0), u(1),...,u(k),...,u(q-1)\}$ is a sequence of admissible controls.

Similarly, for the associated linear dynamical system (4), we define a linear map

$$h: U_q^c \to R^n \, by \, hv = x(p, v).$$

By the assumption (iii) the linear dynamical system (4) is U^c-globally relative controllable in $[0, q]$. Therefore, by the Definition 2 the linear operator h is surjective i.e., it maps the cone U_q^c onto the whole space R^n. Furthermore, by Lemma 3 we have that $Dg(0) = h$.

Since U^c is a closed and convex cone, then the set of admissible controls U_q^c is also a closed and convex cone in the space U_q. Therefore, the nonlinear map g satisfies all the assumptions of the generalized open mapping theorem stated in the Lemma 3. Hence, the nonlinear map g transforms a conical neighborhood of zero in the set of admissible controls U_q^c onto some neighborhood of zero in the state space R^n. This is by previous definitions equivalent to the U^c-local controllability in $[0, q]$ of the semilinear fractional dynamical control system (3). Hence, our theorem follows.

Corollary 3. Under the assumptions stated in Theorem 2 the semilinear fractional system (3) is locally U^{c0}-controllable in the interval $[0, q]$ if the linear fractional system (4) is globally U^c-controllable in the interval $[0, q]$.

From Theorem 2 the linear system (4) is globally U^c-controllable in the interval $[0, q]$. Therefore, by Corollary 3 the semilinear fractional system (3) is locally U^{c0}-controllable in the same interval $[0, p]$. In the case when set U contains zero as an interior point we have the sufficient condition for local constrained controllability of semilinear fractional discrete systems (3).

Corollary 4. Let $0 \in int(U)$. Then the semilinear discrete system (3) is locally U-controllable in the interval $[0, q]$ if its linear approximation near origin given by the difference state Eq. (4) is locally U-controllable in the same interval $[0, q]$.

4   Example

Let us consider the following simple example, which illustrates theoretical considerations. Let us assume unconstrained admissible controls and let the semilinear fractional discrete dynamical control system defined on a given time interval $[0, q]$, has the following form

$$\Delta_{k+1}^{1,5} x_1(k+1) = x_1(k) + x_2(k) + \exp x_1(k) - \cos x_2(k) + \sin u(k)$$
$$\Delta_{k+1}^{1,5} x_2(k+1) = -x_1(k) + x_2(k) + \cos x_1(k) + \sin x_2(k) - \exp u(k)$$

(8)

Therefore,

$$n = 2, \; m = 1, \; x(t) = \left(x_1(t), x_2(t)\right)^{tr} \in R^2, \; U = R, \alpha = 1,5$$

and using the notations given in the previous sections matrices C and D and the nonlinear mapping F have the following form

$$C = \begin{bmatrix} 1 & 1 \\ -1 & 1 \end{bmatrix} \qquad D = \begin{bmatrix} 0 \\ 1 \end{bmatrix}$$

$$f(x(k), u(k)) = \begin{bmatrix} \exp x_1(k) - \cos_2(k) + \sin u(k) \\ \cos x_1(k) + \sin x_2(k) - \exp u(k) \end{bmatrix}$$

Moreover,

$$f(0,0) = \begin{bmatrix} 0 \\ 0 \end{bmatrix}$$

$$f_x(x(k), u(k)) = \begin{bmatrix} \exp x_1(k) & \sin x_2(k) \\ -\sin x_2(k) & \cos x_2(k) \end{bmatrix} \qquad f_x(0,0) = \begin{bmatrix} 1 & 0 \\ 0 & 1 \end{bmatrix}$$

$$f_u(x(k), u(k)) = \begin{bmatrix} \cos u(k) \\ -\exp u(k) \end{bmatrix} \qquad f_u(0,0) = \begin{bmatrix} 1 \\ -1 \end{bmatrix}$$

$$A = C + f_x(0,0) = \begin{bmatrix} 2 & 1 \\ -1 & 2 \end{bmatrix} \qquad B = D + f_u(0,0) = \begin{bmatrix} 0 \\ 1 \end{bmatrix} + \begin{bmatrix} 1 \\ -1 \end{bmatrix} = \begin{bmatrix} 1 \\ 0 \end{bmatrix}$$

Moreover, taking into account the equality (5) for $k = 0$ and $\alpha = 1, 5$ we have

$$S_{k+1} = (A + \alpha I_n)S_k + \sum_{i=1}^{i=k+1} (-1)^{i+1} \begin{pmatrix} \alpha \\ i \end{pmatrix} S_{k-i+1} =$$

$$= S_1 = (A + 1,5I_2)S_0 - \begin{pmatrix} 1,5 \\ 1 \end{pmatrix} S_0 = (A + 1,5I_2)I_2 - 1,5I_2$$

Thus substituting matrices A and I_2 we obtain

$$S_1 = \begin{bmatrix} 2 & 1 \\ -1 & 2 \end{bmatrix} + 1,5 \begin{bmatrix} 1 & 0 \\ 0 & 1 \end{bmatrix} = \begin{bmatrix} 3,5 & 1 \\ -1 & 3,5 \end{bmatrix}$$

Thus for $q = 2$ steps controllability matrix M_2 has the form

$$M_2 = \begin{bmatrix} B & S_1 B \end{bmatrix} = \begin{bmatrix} 1 & 3,5 \\ 0 & -1 \end{bmatrix}$$

Since *rank* $M_2 = 2 = n$ the linear fractional system is globally controllable in two steps and hence, the semilinear fractional discrete system (8) is locally controllable in two steps near origin.

5   Conclusions

In the present paper local unconstrained and constrained controllability problems for semilinear fractional finite-dimensional discrete control system with constant coefficients are formulated and discussed. Using some mapping theorems taken from functional analysis and linear approximation methods sufficient conditions for unconstrained and constrained controllability are derived and proved. The present paper extends the controllability conditions with unconstrained and constrained controls given in the literature for standard semilineaar discrete control systems to cover the semilinear fractional discrete systems with unconstrained and constrained controls.

References

1. Babiarz, A., Czornik, A., Klamka, J., Niezabitowski, M.: The selected problems of controllability of discrete-time switched linear systems with constrained switching rule. Bull. Pol. Acad. Sci. Tech. Sci. **63**(3), 657–666 (2015)
2. Kaczorek, T.: Reachability and controllability to zero of cone fractional linear systems. Arch. Control Sci. **17**(3), 357–367 (2007)
3. Klamka, J.: Controllability of Dynamical Systems. Kluwer Academic Publishers, Dordrecht (1991)
4. Klamka, J.: Controllability of dynamical systems - a survey. Arch. Control Sci. **2**(2), 281–307 (1993)
5. Klamka, J.: Constrained controllability of nonlinear systems. IMA J. Math. Control Inf. **12**(2), 245–252 (1995)
6. Klamka, J.: Constrained controllability of dynamics systems. Int. J. Appl. Math. Comput. Sci. **9**(2), 231–244 (1999)
7. Klamka, J.: Controllability of nonlinear discrete systems. Int. J. Appl. Math. Comput. Sci. **12**(2), 173–180 (2002)
8. Klamka, J.: Constrained controllability of semilinear systems with delayed controls. Bull. Pol. Acad. Sci. Tech. Sci. **56**(4), 333–337 (2008)
9. Klamka, J.: Constrained controllability of semilinear systems. Nonlinear Anal. Theory Methods Appl. **47**(5), 2939–2949 (2001)
10. Klamka, J., Czornik, A., Niezabitowski, M., Babiarz, A.: Controllability and minimum energy control of linear fractional discrete-time infinite-dimensional systems. In: Proceedings of the 11TH IEEE International Conference on Control and Automation (ICCA), pp. 1210–1214 (2014)
11. Klamka, J., Niezabitowski, M.: Trajectory controllability of semilinear systems with multiple variable delays in control. In: Proceedings of the 10th International Conference on Mathematical Problems in Engineering , Aerospace and Sciences (ICNPAA 2014), vol. 1637, pp. 498–503 (2014)
12. Robinson, S.: Stability theory for systems of inequalities. Differentiable nonlinear systems. SIAM J. Numer. Anal. **13**(6), 1261–1275 (1986)

Machine Learning and Data Mining

On Fast Randomly Generation of Population of Minimal Phase and Stable Biquad Sections for Evolutionary Digital Filters Design Methods

Adam Slowik[(✉)]

Department of Electronics and Computer Science,
Koszalin University of Technology, Sniadeckich 2 Street, 75-453 Koszalin, Poland
aslowik@ie.tu.koszalin.pl

Abstract. Evolutionary algorithms possesses many practical applications. One of the practical application of the evolutionary methods is digital filters design. Evolutionary techniques are very often used to design FIR (Finite Impulse Response) digital filters or IIR (Infinite Impulse Response) digital filters. IIR digital filters are very often practically realized as a cascade of biquad sections. The guarantee of stability of biquad sections is one of the most important element during IIR digital filter design process. If we want to obtain a stable IIR digital filter, the all poles of the transfer function for all biquad sections must be located into the unitary circle in the z-plane. Of course, if we want to have a minimal phase digital filter then all zeros of the transfer function for all biquad sections must be also located into the unitary circle in the z-plane. In many evolutionary algorithms which are dedicated to the IIR digital filter design the initial population (or re-initialized populations) of the filter coefficients are chosen randomly. Therefore, some of digital filters which are generated in population can be unstable (or/and the filters are not minimal phase). In this paper, we show how to randomly generate a population of stable and minimal phase biquad sections with very high efficiency. Due to our approach, we can also reduce a computational time which is required for evaluation of stability (or/and minimal phase property) of digital filter. The proposed approach has been compared with standard techniques which are used in evolutionary digital filter design methods.

1 Introduction

Evolutionary computation [1, 2] is a global optimization technique which possesses many practical applications [3–8]. Among evolutionary computation methods, we can mention evolutionary algorithms [2], genetic algorithms [1], differential evolution algorithms [9], particle swarm optimization algorithms [10], ant colony optimization algorithms [11], and many others nature based optimization techniques. The evolutionary algorithms are one of the first global optimization technique which is based on the nature. They are used very often in the optimization of NP-hard problems and multi-modal problems. One of the problem which possesses

© Springer-Verlag Berlin Heidelberg 2016
N.T. Nguyen et al. (Eds.): ACIIDS 2016, Part I, LNAI 9621, pp. 511–520, 2016.
DOI: 10.1007/978-3-662-49381-6_49

the multi-modal objective function is digital filters design problem [12]. Therefore, the evolutionary optimization methods are used in digital filters design. In paper [13] the differential evolution algorithm was proposed for IIR (Infinite Impulse Response) digital filter design. In the article [14] the finite word-length digital filter was designed using an annealing algorithm, in the paper [15] the finite word-length design for IIR digital filters based on the modified least-square criterion in the frequency domain was shown. In the paper [16], the differential evolution algorithm was used for design of IIR digital filters with non-standard amplitude characteristics, in work [17] the continuous ant colony optimization algorithm was used for the same problem. In the paper [18] the application of evolutionary algorithm to design of minimal phase and stable digital filters with non-standard amplitude characteristics and finite bits word length was shown. In the article [19] the hybridization of evolutionary algorithm with Yule Walker method to design minimal phase and stable digital filters with arbitrary amplitude characteristics was presented. In the algorithms described in papers [13–19], the initial population of individuals was created randomly. Therefore some of digital filters were non-stable and/or non-minimal phase. In practical applications, the IIR digital filters are realized using biquad sections. In this case for stability guarantee (or minimal phase guarantee) we can use the equations presented in [20]. These equations are true only for the case when the value of the coefficient $a_{k,0}$ and/or $b_{k,0}$ is equal to one (see Eq. (2)). The problem is more complicated if want to have the variable value of coefficient $a_{k,0}$ and/or $b_{k,0}$ into the continuous range $[-1; +1]$ (without zero value) or into the discrete range $[-1; 1 - 2^{-M}]$ (without zero value). The discrete range of coefficients value variability is especially important if our digital filter will be implemented into the hardware in given $Q.M$ fixed-point format. In this paper, we show how to randomly generate in evolutionary algorithm a population of stable and minimal phase biquad sections with very high efficiency (equal 100 % for continuous domain and higher then 99 % for discrete domain). The generation of stable and minimal phase biquad sections improve the quality of evolutionary algorithm search process. It is especially important in the algorithms were the re-initialization of population is occurred (as for example micro genetic algorithm [21]). It is significant to point out, that due to our approach, we can also reduce a computational time which is required for evaluation of stability (or/and minimal phase property) of digital filter.

2   IIR Digital Filter Biquad Section

The IIR digital filters can be realized in the hardware as for example the cascade of biquad sections. Then, the transfer function $H(z)$ of IIR digital filter realization with cascade of $k - th$ biquad sections is described as follows:

$$H(z) = H_1(z) \cdot H_2(z) \cdot H_3(z) \cdot ... \cdot H_{k-1}(z) \cdot H_k(z) \qquad (1)$$

where the $H_k(z)$ is the transfer function for the $k - th$ biquad section, and can be presented as follows:

$$H_k(z) = \frac{b_{k,0} + b_{k,1} \cdot z^{-1} + b_{k,2} \cdot z^{-2}}{a_{k,0} + a_{k,1} \cdot z^{-1} + a_{k,2} \cdot z^{-2}} = \frac{b_{k,0} \cdot z^2 + b_{k,1} \cdot z^1 + b_{k,2}}{a_{k,0} \cdot z^2 + a_{k,1} \cdot z^1 + a_{k,2}} \tag{2}$$

where the $b_{k,0}$, $b_{k,1}$, $b_{k,2}$ are the $k - th$ biquad section numerator coefficients, and $a_{k,0}$, $a_{k,1}$, $a_{k,2}$ are the $k - th$ biquad section denominator coefficients.

If we want to design the stable and minimal phase IIR digital filter then all zeros of the numerator and all poles of denumerator for all biquad sections must be located into the unitary circle in the z-plane. In other words if each transfer function $H_k(z)$ for $k - th$ biquad section will be minimal phase and stable then the transfer function $H(z)$ for IIR digital filter also will be minimal phase and stable.

3   Q.M Fixed-Point Format

In many DSP (Digital Signal Processing) system the numbers are represented by $Q.M$ fixed-point format. If we want to implement the designed digital filter into the $Q.M$ DSP system without any additional errors (i.e. filter coefficients rounding error) the value of $b_{k,0}$, $b_{k,1}$, $b_{k,2}$, $a_{k,0}$, $a_{k,1}$, $a_{k,2}$ coefficients must be taken from the predefined set X. The set X of potential values for these coefficients (in $Q.M$ fixed-point format) is defined as follows:

$$X = \left[\frac{(-1) \cdot 2^M}{2^M} ; \frac{2^M - 1}{2^M} \right] \tag{3}$$

In the 2's complement fractional representation, an $M + 1$ bits (in fixed-point format $Q.M$) binary word can represent 2^{M+1} equally space numbers from $\frac{(-1) \cdot 2^M}{2^M} = -1$ to $\frac{2^M - 1}{2^M} = 1 - 2^{-M}$. The binary word BW which consists of $M + 1$ bits (bw_i):

$$BW = [bw_M, bw_{M-1}, bw_{M-2}, ..., bw_1, bw_0] \tag{4}$$

we interpret as a fractional number x:

$$x = -(bw_M) + \sum_{i=0}^{M-1} \left(2^{i-M} \cdot bw_i \right) \tag{5}$$

If we assure that the value of filter coefficients $b_{k,0}$, $b_{k,1}$, $b_{k,2}$, $a_{k,0}$, $a_{k,1}$, $a_{k,2}$ will be the element from the set X then the digital filter will be resistive on rounding error after its implementation into $Q.M$ DSP system.

4   Biquad Section Stability and Minimal Phase Validation

Generally in the literature [20], we can find that the biquad section $H_k(z)$ with the values of $b_{k,0}$, and $a_{k,0}$ coefficients equal to one, and described as follows:

$$H_k\left(z\right) = \frac{1 + b_{k,1} \cdot z^{-1} + b_{k,2} \cdot z^{-2}}{1 + a_{k,1} \cdot z^{-1} + a_{k,2} \cdot z^{-2}} \tag{6}$$

is stable if and only if:

$$|a_{k,2}| < 1 \tag{7}$$

$$a_{k,1} < 1 + a_{k,2} \tag{8}$$

$$a_{k,1} > -1 - a_{k,2} \tag{9}$$

Of course the biquad section $H_k\left(z\right)$ will be minimal phase if and only if:

$$|b_{k,2}| < 1 \tag{10}$$

$$b_{k,1} < 1 + b_{k,2} \tag{11}$$

$$b_{k,1} > -1 - b_{k,2} \tag{12}$$

The Eqs. (7–12) are true only in the case when the value of $a_{k,0}$ or $b_{k,0}$ is equal to one. But what in the case when the all coefficient $a_{k,i}$ and $b_{k,i}$ (for i=0,1,2) can possesses continuous values from the range $[-1; 1]$ or discrete values from the set X (see Eq. 3). If we want to have a fast stability or/and minimal phase validation of the all biquad sections the Eqs. (7–12) are not sufficient. Moreover, in practical application (for given DSP system) the value of coefficients $a_{k,i}$ and $b_{k,i}$ (for i=0,1,2) are scaled into the range [-1; 1] and then they are quantized into the values from the one of $Q.M$ fixed-point format. Therefore, the Eqs. (7–12) must be replaced by another one, which can do a fast stability and minimal phase validation for any biquad section and in any $Q.M$ fixed-point format.

5   Proposed Approach

Let's consider the biquad section $H_k\left(z\right)$ described by Eq. (2). Assume that the values for all coefficients $a_{k,i}$ and $b_{k,i}$ (for i=0,1,2) are from the range $[-1; 1]$. Also, we must remember, that if the our biquad section will be implemented into DSP system with given $Q.M$ fixed-point format, then the values for all coefficients $a_{k,i}$ and $b_{k,i}$ (for i=0,1,2) must be from the range X (see Eq. 3). In the Fig. 1, we have present the stability region (white color) and non-stability region (black color) for different values of parameter $a_{k,0}$. The same regions will be for minimal phase (white color) and non-minimal phase (black color) biquad section for different values of parameter $b_{k,0}$. The horizontal axis represents the values of coefficient $a_{k,2}$ (or $b_{k,2}$), the vertical axis represents the values of coefficient $a_{k,1}$ (or $b_{k,1}$). The left-bottom corner is a point $(a_{k,2} = 1, a_{k,1} = -1)$ or $(b_{k,2} = 1, b_{k,1} = -1)$. The right-upper corner is a point $(a_{k,2} = -1, a_{k,1} = 1)$ or $(b_{k,2} = -1, b_{k,1} = 1)$.

Based on the dependencies presented in Fig. 1, we have elaborated an equations for the guarantee of stability for any biquad section. The biquad section with transfer function described by Eq. (2) is stable if and only if:

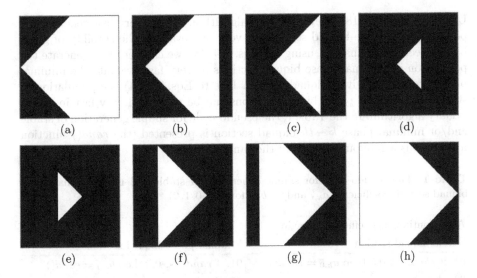

Fig. 1. The stability or minimal phase region (white color) and non-stability or non-minimal phase region (black color) for different values of parameter $a_{k,0}$ or $b_{k,0}$ equals to: -1 (a), -0.75 (b), -0.5 (c), -0.25 (d), 0.25 (e), 0.5 (f), 0.75 (g), 1 (h)

- for $a_{k,0} \in [+1; 0)$

$$(-1) \cdot a_{k,0} < a_{k,2} < a_{k,0} \tag{13}$$

$$(-1) < a_{k,1} < a_{k,2} + a_{k,0} \tag{14}$$

$$1 > a_{k,1} > (-1) \cdot a_{k,2} + (-1) \cdot a_{k,0} \tag{15}$$

- for $a_{k,0} \in (0; -1]$

$$(-1) \cdot a_{k,0} > a_{k,2} > a_{k,0} \tag{16}$$

$$1 > a_{k,1} > a_{k,2} + a_{k,0} \tag{17}$$

$$(-1) < a_{k,1} < (-1) \cdot a_{k,2} + (-1) \cdot a_{k,0} \tag{18}$$

Of course the biquad section $H_k(z)$ will be minimal phase if and only if:
- for $b_{k,0} \in [+1; 0)$

$$(-1) \cdot b_{k,0} < b_{k,2} < b_{k,0} \tag{19}$$

$$(-1) < b_{k,1} < b_{k,2} + b_{k,0} \tag{20}$$

$$1 > b_{k,1} > (-1) \cdot b_{k,2} + (-1) \cdot b_{k,0} \tag{21}$$

- for $b_{k,0} \in (0; -1]$

$$(-1) \cdot b_{k,2} > b_{k,2} > b_{k,0} \tag{22}$$

$$1 > b_{k,1} > a_{k,2} + b_{k,0} \tag{23}$$

$$(-1) < b_{k,1} < (-1) \cdot b_{k,2} + (-1) \cdot b_{k,0} \tag{24}$$

Using the Eqs. (13–18) in evolutionary algorithm, we can very fast generate the population of stable biquad sections or very fast evaluate the stability of given biquad section. In analogy, using the Eqs. (19–24), we can very fast generate the population of minimal phase biquad sections or very fast evaluate the minimal phase property of given biquad section. Due to Eqs. (13–24) the population of stable and minimal phase biquad sections can be generated very fast in evolutionary algorithm. In the Table 1, the pseudo-code for simple generation of stable and/or minimal phase $k - th$ biquad section is presented (the $rand()$ function in Table 1 is a random value from the range $(0; 1)$).

Table 1. The pseudo-code for simple generation of stable and minimal phase k-th biquad section coefficients $a_{k,i}$ and/or $b_{k,i}$ for $(i = 0, 1, 2)$

A. In continuous coefficient domain $[-1; 1]$	
Stable biquad section	Minimal phase biquad section
01. if $rand() < 0.5$ then $a_{k,0} = rand()$	01. if $rand() < 0.5$ then $b_{k,0} = rand()$
02. else $a_{k,0} = rand() - 1$;	02. else $b_{k,0} = rand() - 1$;
03. $a_{k,2} = 2 * rand() * a_{k,0} - a_{k,0}$;	03. $b_{k,2} = 2 * rand() * b_{k,0} - b_{k,0}$;
04. $tmp_1 = a_{k,2} + a_{k,0}$;	04. $tmp_1 = b_{k,2} + b_{k,0}$;
05. if $(tmp_1 > 1)$ then $tmp_1 = 1$; endif	05. if $(tmp_1 > 1)$ then $tmp_1 = 1$; endif
06. if $(tmp_1 < -1)$ then $tmp_1 = -1$; endif	06. if $(tmp_1 < -1)$ then $tmp_1 = -1$; endif
07. $a_{k,1} = 2 * rand() * tmp_1 - tmp_1$;	07. $b_{k,1} = 2 * rand() * tmp_1 - tmp_1$;
B. In discrete coefficient domain $[-1; 1 - 2^{-M}]$ (in given $Q.M$ fixed-point format)	
Stable biquad section	Minimal phase biquad section
01. if $rand() < 0.5$ then	01. if $rand() < 0.5$ then
02. $a_{k,0} = rand() * (1 - 2^{-M+1}) + 2^{-M}$;	02. $b_{k,0} = rand() * (1 - 2^{-M+1}) + 2^{-M}$;
02. else $a_{k,0} = rand() * (-2^{-M} + 1) - 1$;	02. else $b_{k,0} = rand() * (-2^{-M} + 1) - 1$;
03. $a_{k,2} = 2 * rand() * a_{k,0} - a_{k,0}$;	03. $b_{k,2} = 2 * rand() * b_{k,0} - b_{k,0}$;
04. $tmp_1 = a_{k,2} + a_{k,0}$;	04. $tmp_1 = b_{k,2} + b_{k,0}$;
05. if $(tmp_1 > (1 - 2^{-M}))$ then	05. if $(tmp_1 > (1 - 2^{-M}))$ then
06. $tmp_1 = 1 - 2^{-M}$; endif	06. $tmp_1 = 1 - 2^{-M}$; endif
07. if $(tmp_1 < -1)$ then $tmp_1 = -1$; endif	07. if $(tmp_1 < -1)$ then $tmp_1 = -1$; endif
08. $a_{k,1} = 2 * rand() * tmp_1 - tmp_1$;	08. $b_{k,1} = 2 * rand() * tmp_1 - tmp_1$;

If we want to generate (in evolutionary algorithm) a population of stable and/or minimal phase biquad sections with coefficients in given $Q.M$ fixed-point format, then the pseudo-code from the part B of Table 1 must be used and next the obtained values $a_{k,i}$ and $b_{k,i}$ (for $i = 0, 1, 2$) must be transformed into the discrete values from the range X (see Eq. 3). This transformation for $a_{k,i}$ coefficients (for $i=0,1,2$) can be done using following formula:

$$ind = round \left(\frac{1 - 2^{M+1}}{2 - 2^{-M}} \cdot (a_{k,i} + 1) + 2^{M+1} \right) \qquad (25)$$

where: $round(.)$ is a function which round the input argument to the nearest integer value, ind is a number of element in set X. The value of the element having index ind in set X is the new value for a given $a_{k,i}$ coefficient in $Q.M$ fixed-point format. The transformation for $b_{k,i}$ coefficients (for $i=0,1,2$) will be similar. We must only replace the symbol $a_{k,i}$ by the symbol $b_{k,i}$ in Eq. (25). If we have value of ind, we can also very fast compute, the new value of $a_{k,i}$ coefficient in given $Q.M$ fixed-point format ($a_{k,i}^{Q.M}$) using equation:

$$a_{k,i}^{Q.M} = 1 - ind \cdot 2^{-M} \qquad (26)$$

If we want to compute the new value of $b_{k,i}$ coefficient (in $Q.M$ fixed-point format) we must replace the symbol $a_{k,i}^{Q.M}$ by the symbol $b_{k,i}^{Q.M}$ in Eq. (26).

6   Description of Experiments

In order to test our approach, two procedures of randomly generation of population consists of $PopSize = 100$ individuals for evolutionary algorithm were created. Each individual consists of K stable and minimal phase biquad sections. In the first procedure (named "Our approach" in Table 2), the coefficient $a_{k,i}$ and $b_{k,i}$ (for $i=0, 1, 2$) values stored in individuals are generated randomly with the use of Eqs. (13-24). In the second procedure (named "Standard approach" in Table 2), the coefficient $a_{k,i}$ and $b_{k,i}$ (for $i=0, 1, 2$) values stored in individuals are generated randomly with uniform distribution. The experiments were performed using DELL Latitude E6420 computer and the FreeMat software [22] in version 4.2. The FreeMat is a free environment for rapid engineering and scientific prototyping and data processing [22]. It is similar to commercial system such as MATLAB [22].

In the first experiment, we have assume that the value of coefficients $a_{k,i}$ and $b_{k,i}$ (for $i=0, 1, 2$) are continuous from the range $[-1; 1]$. The pseudo-code from Table 1 (part A) has been applied. The symbols in Table 2 are as follows: K - the number of biquad sections in each individual in population, $Stab - Min$ - the average number (with standard deviation) of stable and minimal phase biquad sections in population. The average values (with standard deviations) were computed after 10-fold repetition of each procedure.

In Table 2 (part A), we can see that the biquad sections which were generated using our approach ("Our approach") are in all cases stable and minimal phase. The correctness of generated individuals in evolutionary algorithm population is equal to 100 %. In the case of procedure of standard individuals generation ("Standard approach") the correctness is from the range 8 % up to 10 %.

In the second experiment we want to show the computational time which is required for the generation of single biquad section using procedures: "Our approach" and "Standard approach". We have generate a population of 100 individuals (in one individual was stored 5 biquad sections) in 10-fold repetition. The average time (with standard deviation) for generation of one population of individuals was equal to: 109.50±15.1015 [ms] for "Standard approach" and

Table 2. The comparison of randomly generation of population of biquad sections with the values of $a_{k,i}$ and $b_{k,i}$ coefficients after 10-fold repetition of each procedure

| | A. The continuous domain $[-1; 1]$ | | B. The discrete domain $[-1; 1 - 2^{-15}]$ | |
| | Standard approach | Our approach | Standard approach | Our approach |
K	Stab-Min	Stab-Min	Stab-Min	Stab-Min
1	8.70±4.8546	100±0	9.40±2.5473	99.90±0.3162
2	17.90±2.8460	200±0	17.80±3.1552	199.70±0.6749
3	27.20±4.6857	300±0	26.60±5.9104	299.70±0.4830
4	32.50±6.4850	400±0	32.50±3.8369	399.60±0.5164
5	41.30±6.0378	500±0	41.20±6.7297	498.80±1.6193
100	868.70±26.2469	10000±0	869.40±28.9413	9989.40±12.6543

169.50±11.0277 [ms] for "Our approach". We can see that the "Our approach" procedure is about 50 % more time expensive than the "Standard approach" procedure.

In the third experiment, we have assumed that the biquad section will be implemented into the DSP system with Q.15 fixed-point format ($M = 15$). Therefore, the value of coefficients $a_{k,i}$ and $b_{k,i}$ (for $i=0$, 1, 2) are from the range $[-1; 1 - 2^{-15}]$. The pseudo-code from Table 1 (part B) has been applied and additionally, after generation of the values for biquad section coefficients, the generated coefficients were transformed into Q.15 fixed-point format using Eqs. (25) and (26). Before this experiments, the vector X (see Eq. 3) consists of $2^{M+1} = 2^{16} = 65536$ elements from the range $[-1; 1 - 2^{-15}]$ was created and stored in the computer memory. The value of the first element from the set X is equal to $1 - 2^{-15}$ ($X(1) = 1 - 2^{-15}$), and the value of the last element from the set X is equal to -1 ($X(65536) = -1$). The average values (with standard deviations) were computed after 10-fold repetition of each procedure.

In Table 2 (part B), we can see that the biquad sections which were generated using our approach ("Our approach") are in almost all cases stable and minimal phase. The correctness of generated individuals in evolutionary algorithm population is higher than 99 %. In the case of procedure of standard individuals generation ("Standard approach") the correctness is from the range 8 % up to about 9 %.

In the fourth experiment we want to show the computational time which is required for the generation of single biquad section using procedures: "Our approach" and "Standard approach" (for Q.15 fixed-point format). We have generate a population of 100 individuals (in one individual was stored 5 biquad sections) in 10-fold repetition. The average time (with standard deviation) for generation of one population of individuals was equal to: 2.75±0.0281 [s] for "Standard approach" and 2.03±0.0455 [s] for "Our approach". We can see that the "Our approach" procedure is about 30 % more time expensive than the "Standard approach" procedure.

In the fifth experiment, we have study the average computational time which is required to decide whether given biquad section is stable and minimal phase. To perform this study, the individuals in population (PopSize=100) have been generated randomly with uniform distribution (identical as in first experiment) in 10-fold repetition. Each individual consists of one biquad section. Each biquad section from the population was validate on fulfilling stability criteria and minimal phase criteria. We have used the Eqs. (13)–(24) for "Our approach", and $roots(.)$ and $abs(.)$ functions from FreeMat [22] software (in version 4.2) for "Standard method of validation". The average computational time which was required for evaluation of all individuals in population was equal to: 3.10 ± 0.1741 [s] for "Standard method of validation" and 1.84 ± 0.1718 [s] for "Our approach". We can see that the individual validation using "Our approach" procedure is about 40 % faster than the individual validation using "Standard method of validation".

7   Conclusions

Due to approach presented in this paper, we can generate the population of stable and minimal phase biquad sections for application in evolutionary algorithms (with 100 % of correctness for continuous values of biquad section coefficients and in over 99 % of correctness for discrete values of biquad section coefficients). The proposed approach can highly increase the efficiency of evolutionary methods for digital filters design problem. Especially, in the case when in the evolutionary algorithm, the block of re-initialization of individuals in population exists. Due to our approach we can efficiently generate the population of stable and minimal phase biquad sections and we can efficiently validate the stability criteria and minimal phase criteria for each individual in population. In our next study, we want to generalize presented approach to the efficient generation of stable biquad sections (both for continuous and discrete values of coefficients) with prescribed stability margin.

References

1. Goldberg, D.: Genetic Algorithms in Search, Optimization, and Machine Learning. Addison-Wesley Publishing Company Inc., Boston (1989)
2. Michalewicz, Z.: Genetic Algorithms + Data Structures = Evolution Programs. Springer-Verlag, Berlin (1992)
3. Vasicek, Z., Sekanina, L.: Evolutionary approach to approximate digital circuits design. IEEE Trans. Evol. Comput. **19**(3), 432–444 (2015)
4. Preen, R.J., Bull, L.: Toward the coevolution of novel vertical-axis wind turbines. IEEE Trans. Evol. Comput. **19**(2), 284–294 (2015)
5. Tersi, L., Fantozzi, S., Stagni, R.: Characterization of the performance of memetic algorithms for the automation of bone tracking with fluoroscopy. IEEE Trans. Evol. Comput. **19**(1), 19–30 (2015)
6. Graditi, G., Di Silvestre, M.L., Gallea, R., Riva Sanseverino, E.: Heuristic-based shiftable loads optimal management in smart micro-grids. IEEE Trans. Ind. Inform. **11**(1), 271–280 (2015)

7. Kim, J., Lee, J.: Trajectory optimization with particle swarm optimization for manipulator motion planning. IEEE Trans. Ind. Inform. **11**(3), 620–631 (2015)
8. Aghaei, J., Baharvandi, A., Rabiee, A., Akbari, M.A.: Probabilistic PMU placement in electric power networks: an MILP-based multiobjective model. IEEE Trans. Ind. Inform. **11**(2), 332–341 (2015)
9. Price, K.: An introduction to differential evolution. In: Corne, D., Dorigo, M., Glover, F. (eds.) New Ideas in Optimization. McGraw-Hill, London (1999)
10. Kennedy, J., Eberhart, R.C., Shi, Y.: Swarm Intelligence. Morgan Kaufmann Publishers, San Francisco (2001)
11. Socha, K., Doringo, M.: Ant colony optimization for continuous domains. Eur. J. Oper. Res. **185**(3), 1155–1173 (2008)
12. Erba, M., Rossi, R., Liberali, V., Tettamanzi, A.G.: Digital filter design through simulated evolution. In: Proceedings of ECCTD 2001, vol. 2, pp. 137–140 (2001)
13. Karaboga, N.: Digital IIR filter design using differential evolution algorithm. EURASIP J. Appl. Sig. Process. **2005**(8), 1269–1276 (2005)
14. Benvenuto, N., Marchesi, M., Orlandi, G., Piazza, F., Uncini, A.: Finite wordlength digital filter design using an annealing algorithm. In: International Conference on Acoustics, Speech, and Signal Processing, vol. 2, pp. 861–864 (1989)
15. Nakamoto, M., Yoshiya, T., Hinamoto, T.: Finite word length design for IIR digital filters based on the modified least-square criterion in the frequency domain. In: International Symposium on Intelligent Signal Processing and Communication Systems, ISPACS, pp. 462–465 (2007)
16. Slowik, A., Bialko, M.: Design of IIR digital filters with non-standard characteristics using differential evolution algorithm. Bull. Pol. Acad. Sci. Tech. Sci. **55**(4), 359–363 (2007)
17. Slowik, A., Bialko, M.: Design and optimization of IIR digital filters with non-standard characteristics using continuous ant colony optimization algorithm. In: Darzentas, J., Vouros, G.A., Vosinakis, S., Arnellos, A. (eds.) SETN 2008. LNCS (LNAI), vol. 5138, pp. 395–400. Springer, Heidelberg (2008). doi:10.1007/978-3-540-87881-0_39
18. Slowik, A.: Application of evolutionary algorithm to design of minimal phase digital filters with non-standard amplitude characteristics and finite bits word length. Bull. Pol. Acad. Sci. Tech. Sci. **59**(2), 125–135 (2011). doi:10.2478/v10175-011-0016-z
19. Slowik, A.: Hybridization of evolutionary algorithm with Yule Walker method to design minimal phase digital filters with arbitrary amplitude characteristics. In: Corchado, E., Kurzyński, M., Woźniak, M. (eds.) HAIS 2011, Part I. LNCS, vol. 6678, pp. 67–74. Springer, Heidelberg (2011)
20. STMicroelectronics, AN2874 Applications note, February 2009
21. Tiwari, S., Koch, P., Fadel, G., Deb, K.: Amga: an archive-based micro genetic algorithm for multi-objective optimization. In: Proceedings of the 10th Annual Genetic and Evolutionary Computation Conference, Atlanta, USA 12–16 July, pp. 729–736 (2008)
22. http://freemat.sourceforge.net/

Recursive Ensemble Land Cover Classification with Little Training Data and Many Classes

Yu Oya[⊠], Katsutoshi Kanamori, and Hayato Ohwada

Department of Industrial Administration,
Tokyo University of Science, Tokyo, Japan
7414606@ed.tus.ac.jp, {katsu, ohwada}@rs.tus.ac.jp

Abstract. Land-cover classification can construct a land-use map to analyze satellite images using machine learning. However, supervised machine learning requires a lot of training data since remote sensing data is of higher resolution that reveals many features. Therefore, this study proposed a method to generate self-training data from a small amount of training data. This method generates self-training, which is regarded as the correct class to consider various times and the surrounding land cover. As a result of self-training conducted using this method, the Kappa coefficient was 0.644 for 12 classification problems with one training data per class.

Keywords: Land cover classification · Ensemble learning · Self-training · Semi-supervised learning

1 Introduction

Many global and environmental applications require land use and land-cover information. One such application is utilizing land-cover change [1]. Updating land-use maps requires too much manpower, money, and time to conduct field research using traditional models. At present, land-cover classification is one of the most common remote sensing image analyses and constructs a land-use map [2, 3]. This is useful because it can make a land-use map automatically. The approaches utilized include the maximum likelihood method [4], the random forest [5], and the support vector machine [6]. These all involve supervised machine learning, which requires a large amount of training data. Therefore, it is not useful with unknown and uncontrolled forest because there is only old training data, or no data. In contrast, there are forms of unsupervised learning which require no training data. However, these can classify rough land cover only (e.g., water, urban, and vegetation) [7].

The objective of this study is to propose a new method that can classify many land covers on right and fine with little training data, as with an unknown and uncontrolled forest. This paper is organized as follows. Section 2 presents the theory of the proposed land-cover classification method that generates self-training data using time series remote-sensing data and land-cover information about the surrounding location, and which updates self-training data using a recursive ensemble algorithm. Section 3 describes how we experimented with this method and others to compare them.

© Springer-Verlag Berlin Heidelberg 2016
N.T. Nguyen et al. (Eds.): ACIIDS 2016, Part I, LNAI 9621, pp. 521–531, 2016.
DOI: 10.1007/978-3-662-49381-6_50

Section 4 reports the experiment results, Sect. 5 discusses comparative studies and methods, and Sect. 6 provides our conclusions.

2 Method

This section presents the proposed method of recursive ensemble land-cover classification. This method is an algorithm that generates and updates self-training data in cases where there is little or no training data. First, we describe the maximum likelihood method, which constitutes the bedrock of this classification approach. Second, we describe an algorithm to generate self-training data. Third, we provide an algorithm to update the self-training.

2.1 Maximum Likelihood Method

Land-cover classification determines the class at a given location using remote-sensing data. It then uses spectral reflectance characteristics. This spectrum includes several light wavelengths such as visible (e.g., blue, green, and red), near infrared, and middle infrared. A satellite image records reflectance as a particular bit of a wavelength band, and remote-sensing data includes several images. We call this the image band. In addition, a physical body has specific reflectance characteristics. Therefore, land-cover classification employs spectral reflectance characteristics. This classification thus looks for similar characteristics for water, loam, trees, and so on. This study classifies remote-sensing data using the maximum likelihood method (MLM), one of the most popular techniques for land-cover classification.

First, a reflectance vector $v(t, x, y)$ is defined by the expression below.

$$v(t, x, y) = (r_1(t, x, y), r_2(t, x, y), \ldots, r_B(t, x, y)) \tag{1}$$

Here, $r_b(t, x, y)$ is the reflectance at a point (x, y) in the band $b \in \{1, 2, \ldots, B\}$ for shooting time t of satellite images.

MLM is defined in the following equation [8].

$$f_k(v) = ln|\Sigma_k| + (v - m_k)^T \sum\nolimits_k^{-1} (v - m_k) \tag{2}$$

Equation (2) uses a discriminant function. MLM calculates the discriminant functions $f_k(v)$ and determines the class using k of the minimum $f_k(v)$. Here, the reflectance vector v is the observed data, constant $k \in \{1, 2, \ldots, K\}$ is the classification group, the variable m_k is a mean vector, and the variable Σ_k is a variance-covariance matrix of class k. The elements of the discriminant function are calculated as follows. Mean vectors m_k and the variance-covariance matrixes Σ_k are vector spaces. They are composed of training data v identified true answer class k.

2.2 Algorithm I - Generating Self-training Data

Algorithm I is a method for generating self-training data from few samples using ensemble learning of the classification results for every time series data set. Here, we define a model assumption that the self-training data has one and only land cover not to depend on classification results. It consists of a series of six steps.

1. $D = \emptyset$, (D: a generated self-training data set).
2. Let $V_0 = \{v_1, c_1, v_2, c_2, \ldots, v_n, c_n\}$ be a set of tuples of training data and class, where $v_i = v(t, x_i, y_i)$, c_i is the true class of v_i.
3. m_k and Σ_k in Eq. (2) are calculated for every class $k \in \{1, 2, \ldots, K\}$ using V_0.
4. Let $V = \{u_1, g_1, u_2, g_2, \ldots, u_m, g_m\}$ be a set of tuples of the test data and the predicted class, where $u_j = v(t, x_j, y_j)$, and $g_j = \mathrm{argmin}_k f_k(u_j)$.
5. It is considered that the true class of a point (x, y) does not depend on the observed time t, so we define $c(x, y) = g(t_1, x, y)$ if and only if it satisfies the following equations.

$$\forall t_i : g(t_i, x-1, y-1) = g(t_i, x, y-1) = g(t_i, x+1, y-1)$$
$$= g(t_i, x-1, y) = g(t_i, x, y) = g(t_i, x+1, y)$$
$$= g(t_i, x-1, y+1) = g(t_i, x, y+1) = g(t_i, x+1, y+1)$$

and

$$\forall t_i, t_j : g(t_i, x, y) = g(t_j, x, y)$$

where t_i is an element in the time series set $\{t_1, t_2, \ldots, t_T\}$, and $c(x, y)$ fulfills the given equation for every t_i. In other words, $c(x, y)$ is the class $g(t, x, y)$ when $9T$ predicted classes with around (x, y) every t are equivalent.

6. If $c(x, y)$ is obtained, then $D := D \cup \{W(x, y), c(x, y)\}$, where $W(x, y)$ is a reflectance vector defined by the expression below.

$$W(x, y) = \begin{matrix} (r_1(t_1, x, y), & r_2(t_1, x, y), & \ldots, & r_B(t_1, x, y), \\ r_1(t_2, x, y), & r_2(t_2, x, y), & \ldots, & r_B(t_2, x, y), \\ \vdots & \vdots & & \vdots \\ r_1(t_T, x, y), & r_2(t_T, x, y), & \ldots, & r_B(t_T, x, y)) \end{matrix}$$

2.3 Algorithm II - Updating Self-training Data

Algorithm II is a method for updating the generated self-training data set D by recursive and ensemble learning. There is a difference in classifier number between algorithm I and II, this is whether a number of time series T or any. Therefore, this method has a parameter of the number of the classifier α as the "any" value.

1. Let $\{W_1, c_1, W_2, c_2, \ldots, W_p, c_p\}$ be a set of tuples that has p elements in D, where $W_i = W(x_i, y_i)$, and their tuples are sampled D randomly. This p is a value more than $B \times T$, the number of features, and less than $|D| \div \alpha$, the number of elements in D divided by the number of the classifier, and it is decided randomly.

2. Let $V_s = \{w_1, c_1, w_2, c_2, \ldots, w_p, c_p\}$ be a set of tuples of training data and class, where w_i is a reflectance vector that has a subset of w_i or $w_i \subset W_i$, and their elements are sampled w_i randomly because of changing the number of features used. In addition, s is an index of the classifier between 1 and α.

3. Let $V = \{u_1(s), g_1(s), u_2(s), g_2(s), \ldots, u_q(s), g_q(s)\}$ be a set of tuples of the test data and the predicted class derived by classifier s, where u_j is a reflectance vector which has a subset of w_j but they are not in V_s. In addition, $g_j(s) = \mathrm{argmin}_k f_k(u_j(s))$.

4. Steps 1 to 3 are repeated if $s < \alpha$ then $s := s + 1$.

5. We define $c(x, y)$ as the class with the highest number in the class of $g(s, x, y)$, if it is greater than any threshold for the highest number in the class and if it satisfies the following equation.

$$\forall s : g(s, x-1, y-1) = g(s, x, y-1) = g(s, x+1, y-1)$$
$$= g(s, x-1, y) = g(s, x, y) = g(s, x+1, y)$$
$$= g(s, x-1, y+1) = g(s, x, y+1) = g(s, x+1, y+1)$$

6. If $c(x, y)$ is obtained, then $D := D \cup \{W(x, y), c(x, y)\}$, $s = 1$, and go to step 1.

7. If $c(x, y)$ is not obtained, then $D := D \cup \{W(x, y), d(x, y)\}$, where $d(x, y)$ is the class with the highest number in the class of $g(s, x, y)$.

8. We define D as the final result of this recursive ensemble land-cover classification.

3 Experiments

This section describes experiments conducted to demonstrate the usefulness of this method. First, we classified a large number of land-cover areas using this method and only one training data set per class. This is an adverse condition for machine learning. For comparison, we tried to classify the areas using both this method and other supervised learning methods and a large amount of training data on favorable terms.

3.1 Data Set

Practical applications should involve classifying unknown regions. This study looks into the usefulness of this method. Our target region was Kamakura City, Kanagawa Prefecture, Japan, which whose area is 102,724 km^2., and mMuch land cover information is known for its this region. The correct answer classifications may be found is in a land use map provided by an the environment ministry, and it this was used to evaluate an the accuracy of the results classified by of this method. This paper sets up as many classification groups as possible in order to make a fine vegetation map consisting of 12 categories.

We used remote- sensing data of from Landsat 8. It This is an artificial satellite launched in 2013, and its satellite images are not hyper- spectral images but coarser multi-spectral images. However, but they are public documents and t. Therefore we can use them at no charge, and it This data fulfills the requirements for time series data of this a algorithm I. In addition, Dube et al. evaluated the utility of Landsat 8 for quantifying the aboveground biomass in South Africa [9]. It denotes They determined that the land cover does not change to depend on the observed time within for the year, so we then selected four seasons satellite images shot on September 1, 2013 (summer), November 20, 2013 (fall), January 23, 2014 (winter), and May 22, 2014 (spring).

In addition, we selected bands affecting classification from Landsat 8. The Each satellite image, called a "bBand," records the reflectance ratios at various a particular wavelengths. For example, Landsat 8 has Operational Land Imager (OLI) and Thermal Infrared Sensor (TIRS) images consisting of nine spectral bands with a spatial resolution of 30 meters for bBands 1 to 7 and 9, and Thermal bBands 10 and 11 are useful in providing more accurate surface temperatures and are collected at 100-meter intervals. These bands have 16-bit pixel values. We then selected Bands 1, and 2 (blue), 3 (green), and 4 (red), and along with 5 and, 7 (near infrared).

3.2   Experiment Using This Method

In this study, there is only one training sample for each classification group, but this method cannot classify land cover because of a constraint of the maximum likelihood method (MLM). The number of training data sets should exceed the number of bands. Therefore, eight pixels in locations around the true training position were regarded as having the same land cover when algorithm I was run. In other words, a given category has training data for nine pixels, of which one is true data and eight are prospective data. The training data is randomly sampled from the correct answers, so the surrounding area may be of a different class with the training. However, the training data is being updated and more data is being added, so this error is decreasing.

In Algorithm I, we classified the land cover for each season using time series remote-sensing data, and generated new self-training data using the four results from spring to winter.

In Algorithm II, this method requires creating bundled time series data that has all of the features of reflectance because there are more training data than before. However, this is not sufficient, so this technique updates the self-training data using ensemble learning. In this study, we determined that the number of classifiers is 10. In addition, this algorithm repeats ensemble learning using 10 classifiers in MLM and when updating the self-training data. Figure 1 illustrates the classification flow using Algorithms I and II.

3.3   Comparison Experiment

Algorithm II is similar to the random forest method, one of the most famous and powerful ensemble learning techniques. The differences between the two lie in whether

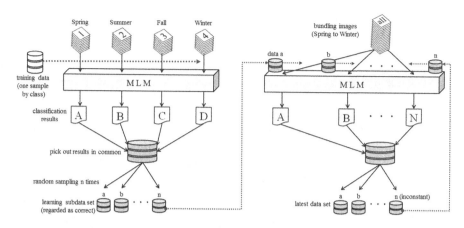

Fig. 1. Classification flow (Left: Algorithm I. Right: Algorithm II.)

the base method is the maximum likelihood method or a decision tree, semi-supervised or supervised learning, and recursive learning or not. Therefore, we classified land cover using a random forest to compare this method. In addition, we used the k-nearest neighbor algorithm (KNN). We tried two patterns in order to use both RF and KNN as per normal. Here, we used bundle data that has all of the features of reflectance.

1. To learn only one training data set (same number as this method).
2. To learn many training data set.

3.4   Evaluation Methodology

To evaluate the classification precision, we created a confusion matrix and worked out the Kappa coefficient during the verification process [10]. The Kappa coefficient is calculated using the following equation.

$$\kappa = \frac{\sum w_{ii} - \sum (w_{i*}w_{*i})}{1 - \sum (w_{i*}w_{*i})} \tag{3}$$

Here, w_{ki} represents the elements in the confusion matrix, and the weight of expected class i in observed class k. When the diagonal cells contain weights of 0 and all of the off-diagonal cells contain weights are 1, this formula produces the same value of kappa as the above calculation. The Kappa coefficient is generally considered to be a more robust measure than a simple percent agreement calculation since it takes into account the agreement occurring by chance. According to Landis et al. [11], a Kappa coefficient less than 0 indicates no agreement, 0 to 0.20 a slight agreement, 0.21 to 0.40 a fair agreement, 0.41 to 0.60 a moderate agreement, 0.61 to 0.80 a substantial agreement, and 0.81 to 1 an almost perfect agreement.

4 Results

This section presents the experiment results using this method, random forest, and KNN.

4.1 Result of This Method

Figure 2 presents the final classification result of this method. In addition, Table 3 represents the confusion matrix between the land use map depicted on the left in Fig. 2 and the classification result on the right.

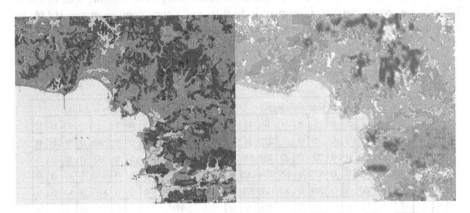

Fig. 2. Land cover map (Left: Correct answer. Right: Classification result.)

Table 1. Confusion matrix between land use map and this method

<table>
<thead>
<tr><th rowspan="2"></th><th rowspan="2"></th><th colspan="12">Expected class</th></tr>
<tr><th>1</th><th>2</th><th>3</th><th>4</th><th>5</th><th>6</th><th>7</th><th>8</th><th>9</th><th>10</th><th>11</th><th>12</th></tr>
</thead>
<tbody>
<tr><td rowspan="12">Observed class</td><td>1</td><td>39046</td><td>0</td><td>0</td><td>87</td><td>9</td><td>3</td><td>0</td><td>0</td><td>4</td><td>0</td><td>0</td><td>0</td></tr>
<tr><td>2</td><td>37</td><td>1227</td><td>0</td><td>0</td><td>23</td><td>138</td><td>238</td><td>13</td><td>31</td><td>129</td><td>3</td><td>51</td></tr>
<tr><td>3</td><td>431</td><td>0</td><td>24311</td><td>0</td><td>98</td><td>345</td><td>55</td><td>161</td><td>362</td><td>1313</td><td>131</td><td>436</td></tr>
<tr><td>4</td><td>26</td><td>0</td><td>0</td><td>4677</td><td>13</td><td>35</td><td>1</td><td>17</td><td>42</td><td>112</td><td>19</td><td>49</td></tr>
<tr><td>5</td><td>931</td><td>307</td><td>688</td><td>197</td><td>630</td><td>0</td><td>0</td><td>17</td><td>69</td><td>65</td><td>46</td><td>56</td></tr>
<tr><td>6</td><td>58</td><td>130</td><td>1086</td><td>400</td><td>0</td><td>846</td><td>0</td><td>32</td><td>31</td><td>200</td><td>21</td><td>29</td></tr>
<tr><td>7</td><td>45</td><td>82</td><td>656</td><td>446</td><td>0</td><td>0</td><td>460</td><td>59</td><td>41</td><td>250</td><td>50</td><td>114</td></tr>
<tr><td>8</td><td>252</td><td>469</td><td>3183</td><td>1392</td><td>93</td><td>361</td><td>78</td><td>3972</td><td>224</td><td>5053</td><td>461</td><td>1164</td></tr>
<tr><td>9</td><td>320</td><td>54</td><td>792</td><td>344</td><td>18</td><td>49</td><td>14</td><td>47</td><td>522</td><td>309</td><td>24</td><td>175</td></tr>
<tr><td>10</td><td>3</td><td>141</td><td>327</td><td>175</td><td>22</td><td>50</td><td>23</td><td>189</td><td>39</td><td>4035</td><td>237</td><td>456</td></tr>
<tr><td>11</td><td>9</td><td>71</td><td>243</td><td>113</td><td>14</td><td>31</td><td>51</td><td>199</td><td>17</td><td>1954</td><td>1654</td><td>829</td></tr>
<tr><td>12</td><td>3</td><td>49</td><td>327</td><td>63</td><td>22</td><td>40</td><td>18</td><td>122</td><td>24</td><td>1557</td><td>452</td><td>2268</td></tr>
</tbody>
</table>

In Table 3, the rows are the correct groups, the columns are the classification result groups, the elements are the number of pixels, and the diagonal elements are the correct classification. This method had the results described above. In the final classification result, the self-training data covered 82.1 % of the object region. The classification results are as follows: the simple percent agreement is 71.6 % and the Kappa coefficient is 0.644.

4.2   Results of Random Forest

Table 2 provides the confusion matrix of the random forest; on the left is the result using only one training data set and on the right is the result using many training sets. In left of Table 2, the simple percent agreement is 56.9 % and Kappa coefficient was 0.446. In right, the simple percent agreement is 60.0 % and Kappa coefficient was 0.476.

Table 2. Confusion matrix of random forest

	Expected class (one training)												Expected class (many training)											
	1	2	3	4	5	6	7	8	9	10	11	12	1	2	3	4	5	6	7	8	9	10	11	12
1	40170	172	379	50	0	0	21	0	7	221	51	0	39838	0	653	0	0	0	0	580	0	0	0	0
2	15	367	701	432	0	0	53	0	29	459	208	0	1	0	1531	0	0	0	0	732	0	0	0	0
3	91	5012	12134	3555	0	0	827	0	179	4214	1304	0	20	0	22232	0	0	0	0	5064	0	0	0	0
4	187	1798	2454	787	0	0	249	0	80	1438	573	0	163	0	5338	0	0	0	0	2065	0	0	0	0
5	26	61	161	129	0	0	17	0	8	105	68	0	22	0	380	0	0	0	0	173	0	0	0	0
6	37	420	238	172	0	0	163	0	26	295	185	0	46	0	877	0	0	0	0	613	0	0	0	0
7	6	183	28	19	0	0	90	0	17	139	128	0	4	0	309	0	0	0	0	297	0	0	0	0
8	16	280	87	51	0	0	26	0	61	594	556	0	1	0	350	0	0	0	0	1320	0	0	0	0
9	20	177	262	94	0	0	25	0	15	215	115	0	12	0	563	0	0	0	0	348	0	0	0	0
10	145	2279	646	226	0	0	127	0	753	4777	5185	0	1	0	1922	0	0	0	0	12215	0	0	0	0
11	49	268	70	77	0	0	25	0	175	310	1580	0	1	0	257	0	0	0	0	2296	0	0	0	0
12	131	517	244	103	0	0	33	0	166	1106	2799	0	0	0	744	0	0	0	0	4355	0	0	0	0

4.3   Result of KNN

Table 3 presents the confusion matrix for KNN.

On the left of Table 3, the simple percent agreement is 49.6 % and the Kappa coefficient is 0.368. On the right, the simple percent agreement is 56.1 % and the Kappa coefficient is 0.418.

Table 3. Confusion matrix of KNN

	Expected class (one training)												Expected class (many training)											
	1	2	3	4	5	6	7	8	9	10	11	12	1	2	3	4	5	6	7	8	9	10	11	12
1	40192	69	79	81	305	81	17	78	84	38	20	27	40015	1	374	48	213	2	7	341	65	0	5	0
2	132	178	259	367	459	202	27	140	181	141	83	95	218	19	851	229	315	32	6	504	46	1	43	0
3	1311	2185	5537	4943	2879	3938	242	1213	3055	786	669	558	1934	50	17101	2488	1293	119	44	3594	437	2	251	3
4	748	661	1190	1059	742	926	70	514	881	336	247	192	1093	2	4070	517	242	24	15	1422	120	0	59	2
5	44	40	61	74	122	61	16	23	52	22	27	33	67	0	212	43	97	10	0	126	6	0	14	0
6	163	168	162	191	165	151	38	109	137	93	86	73	212	16	666	72	82	26	17	409	15	0	21	0
7	57	48	28	49	71	52	19	55	99	37	45	50	87	2	251	13	18	1	1	214	9	0	14	0
8	121	139	45	49	78	40	41	173	254	325	184	222	284	0	213	20	29	1	3	1015	10	1	90	5
9	80	70	126	112	128	84	19	67	75	62	53	47	130	4	373	54	78	3	2	250	11	0	18	0
10	989	1245	395	251	426	265	304	1422	2204	2961	1734	1942	2448	3	1373	107	89	1	16	9190	97	1	786	27
11	123	144	57	38	71	44	31	178	432	431	401	604	315	0	161	29	36	7	2	1672	12	1	311	8
12	313	276	163	103	136	100	98	338	982	619	570	1401	622	0	507	41	54	1	2	3256	41	4	549	22

5 Discussion

Comparing this method, the random forest method, and KNN, the accuracy of this method is higher than the others. In fact, there are mostly correct classification results in the diagonal elements of Table 1. In contrast, there is a lot of misclassification in Table 2, and there are several classes having no predicted result (e.g., classes 5, 6, and 12). The propensity is strong for learning a lot of training data. This is obviously an invalid classification, since random forest does not have a beneficial effect on land-cover classification. KNN does not either, according to Table 3. In particular, KNN is the same as random classification when it learned only one training data set.

However, the principal cause is not the classification method but the conditions. This study has many classification groups, but there is little feature and training data. In addition, the remote-sensing data has coarse resolution. Just for reference, we compared a previous study by Kamagata et al. [12]. Table 4 suggests that our study had better accuracy than this previous work, although we had a larger area, coarser images, and more classes. This fact demonstrates that this method is highly useful and can be applied to unknown or uncontrolled forest.

Table 4. Comparison with previous work

	Kamagata et al.	This study
Target area	25 km^2	102,724 km^2
Training data	Too much	One per class
Spatial resolution	4 m	30 m
Number of class	7	12
Kappa coefficient	0.526	0.644

The objective of this study is to classify on right and fine and to require little training data. As a result, the described method has high potential in a hostile environment. However, the results still had many errors, meaning that the training data might not be robust enough to use the existing land-cover classification. This method offers the possibility of increasing the amount of self-training data until a full cover ratio is reached. The problem is the amount of time required for classification, because this method performs supervised learning over and over again. Therefore, this method should be optimized to decrease the amount of calculation.

6 Conclusion

This study proposed a new land-cover classification method utilizing time series data and surrounding land information. First, this method generated self-training data so as to require little true training and to predict whether the classification results are correct. Therefore, this required only one training data set. Second, this recursive ensemble learning updated the self-training data again and again. As a result, this study involved a greater number of classes, fewer features, less training data, a coarser satellite image, and higher accuracy than other supervised learning approaches and a previous study using a large amount of training data. This demonstrates that it is easy and popular to obtain geographical information on right and fine.

Today, many global and environmental applications require land-use and land-cover information. Land-cover classification requires a remote sensing image analysis and constructs a land-use map. However, supervised learning requires a lot of training data and at least the same amount of test data. However, it is difficult for unsupervised and semi-supervised learning to label many classification groups. As a result, there has been no algorithm to classify unknown and uncontrolled forest on right and fine. In traditional models, a great deal of manpower, money, and time are required to conduct field research. This study was able to classify the land cover over a large area. This paper thus demonstrated greater feasibility than previous methods. In future work, it will be necessary to optimize the parameters of this algorithm, for example the selected time series data. On that basis, we need to study utility factors in issues regarding land-use change.

References

1. Hansen, M., Loveland, T.: A review of large area monitoring of land cover change using Landsat data. Remote Sens. Environ. 122, 66–74 (2012)
2. Zhu, Z., Woodcock, C.: Continuous change detection and classification of land cover using all available Landsat data. Remote Sens. Environ. 114, 152–171 (2014)
3. Lu, D., Weng, Q.: A survey of image classification methods and techniques for improving classification performance. Int. J. Remote Sens. 28(5), 823–870 (2007)
4. Strahler, A.: The use of prior probabilities in maximum likelihood classification of remotely sensed data. Remote Sens. Environ. 10, 135–163 (1980)

5. Rodriguez-Galiano, V., Ghimire, B., Rogan, J., et al.: An assessment of the effectiveness of a random forest classifier for land cover classification. ISPRS J. Photogrammetry Remote Sens. **67**, 93–104 (2012)

6. Pal, M., Mather, P.M.: Support vector machines for classification in remote sensing. Int. J. Remote Sens. **26**(5), 1007–1011 (2005)

7. Banerjee, B., Bovolo, F., Bhattacharya, A., et al.: A new self-training based unsupervised satellite image classification technique using cluster ensemble strategy. Geosc. Remote Sens. Lett. **12**(4), 741–745 (2015)

8. Canty, M.: Image Analysis, Classification and Change Detection in Remote Sensing with Algorithms for ENVI/IDL. CRC Press, Boca Raton (2007)

9. Dube, T., Mutanga, O.: Evaluating the utility of the medium-spatial resolution Landsat 8 multispectral sensor in quantifying aboveground biomass in uMgeni catchment, South Africa. ISPRS J. Photogrammetry Remote Sens. **101**, 36–46 (2015)

10. Congalton, R.: A review of assessing the accuracy of classifications of remotely sensed data. Remote Sens. Environ. **37**, 35–46 (1991)

11. Landis, J., Koch, G.: The measurement of observer agreement for categorical data. Biometrics **33**, 159–174 (1977)

12. Kamagata, N., Hara, K.: Vegetation mapping by object-based image analysis, evaluation of classification accuracy and boundary extraction in a mountainous region of central Honshu, Japan. Veg. Sci. **27**, 83–94 (2010)

Treap Mining – A Comparison with Traditional Algorithm

H.S. Anand[1]([⊠]) and S.S. Vinodchandra[2]

[1] Department of Computer Science and Engineering,
Muthoot Institute of Technology and Science, Kochi, Kerala, India
coolhs03@gmail.com
[2] Computer Center, University of Kerala, Trivandrum, Kerala, India

Abstract. In this era of big data analysis, mining results hold a very important role. So, the data scientists need to be accurate enough with the tools, methods and procedures while performing rule mining. The major issues faced by these scientists are incremental mining and the huge amount of time that is virtually required to finish the mining task. In this context, we propose a new rule mining algorithm which mines the database in a priority based model for finding interesting relations. In this paper a new mining algorithm using the data structure Treap is explained along with its comparison with the traditional algorithms. The proposed algorithm finishes the task in O (n) in its best case analysis and in O (n log n) in its worst case analysis. The algorithm also considers less frequent high priority attributes for rule creation, thus making sure to create valid mining rules. Thus the major issues of traditional algorithms like creating invalid rules, longer running time and high memory utilization could be remedied by this new proposal. The algorithm was tested against various datasets and the results were evaluated and compared with the traditional algorithm. The results showed a peak performance improvement.

Keywords: Treap mining · Association rules · Rule mining · Apriori algorithm · Priority mining

1 Introduction

Association rules are if-then rules, which help to uncover the vast relationship between seemingly unrelated data [1]. It uses a combination of statistical analysis, machine learning and database management to exhaustively explore the data to reveal the complex relationships that exists. To do so, a complete study and analysis of the entire database is required. This work aims to provide all such details for analysis through valid rules by treap mining. The process is executed in a four stage subroutine call. The algorithm works in a priority model that even the least frequent high priority item is considered for rule creation. The algorithm is compared with the traditional mining algorithm like Apriori and FP growth. The performance of the algorithm is evaluated asymptotically and the result obtained shows a peak improvement in the validity of rules created. The algorithm was further tested with the mining datasets and a comparison was drawn with the help of a trend analysis graph. All results showed an enhancement in the

© Springer-Verlag Berlin Heidelberg 2016
N.T. Nguyen et al. (Eds.): ACIIDS 2016, Part I, LNAI 9621, pp. 532–543, 2016.
DOI: 10.1007/978-3-662-49381-6_51

system performance. The paper also points to the future enhancements like prioritizing cluster center of itemsets before rule creation.

2 Literature Survey

A lot of literature on association rule mining is available. Most of them discuss various algorithms, its advantages and working principle. Apriori algorithm put forward by Agrawal and Srikanth in 1993 was the most promising work in this area [2]. It quoted the various aspects of Apriori algorithm. Almost all association rule mining algorithms which were proposed after this holds the Apriori principle stated by Agrawal in some way or the other. There have been a lot of modified and advanced schemes proposed. Among them Zaki [3] introduced a new algorithm for fast discovery of association rules. The proposal was of great impact, but not all associated rules could be retrieved by this scheme. Association mining finds it application in various fields like rule mining in genomics [4], for finding synthetic data for testing market basket problem [5], for predictive analysis [6] etc. Apriori algorithm was also implemented using hash table technique [7, 8]. The new proposal was efficient but the time taken to complete the task was much higher. Hence while having a bulky dataset, the algorithm fails miserably.

Kryszkiewicz and Rybinski [9] during 2000 proposed a new algorithm which mines very large databases based on association rules generation. But the time taken for mining was considerably large and it was not found to be efficient. In 1999, Kosters [10] proposed a method to extract clusters based on the Apriori algorithm. The method considers the highest possible order association rules. It was a combined algorithm based on both the Apriori algorithm and rough set theory (initially proposed by Lin [11]). Frequent pattern growth was another mining algorithm. Christian Borgelt in his work [12] explains how mining task is carried out in the FP tree. The memory utilization was not perfect and the system failed while handling large database. There was an enhancement to the FP growth algorithm. Kuldeep Malik [13] implemented an Enhanced FP Growth Algorithm, which was found to be more improved on mining results. In the paper, Fast Set Operations Using Treaps by Margaret and Guy, details on the Treap data structure and its application in set theory [14]. Treap searching and various operation on Treap were detailed by Cecilia on her paper, Randomized Search Trees [15].

Even with large amount of literature available in this area, it is hard to find an algorithm which works in least time and space complexity. By the baseline principle of priority mining a dynamic method for finding itemsets with less frequency with high priority is yet to explore. This is where the scope of Treap mining comes into play.

3 Treap Mining

Treap is a data structure which has the properties of both tree and heap [8, 9]. A node in a Treap is like a node in a Binary Search Tree (BST). In BST, it has a data value, x, but in treap it has a unique numerical priority, p, in addition to, x. The nodes in a Treap maintains the heap property; that is, at every node *u*, *u.parent.p* < *u.p* (except root). By the way in which the root and its child are related, treap can be classified as Mini Treap (minimum priority) and Max Treap (maximum priority).

In this work, *Treap Mining* algorithm is presented with the help of four subroutines. As this algorithm works in a priority model, the first step is the priority calculation of each variable in the database. Subroutine *Priority* is called within the algorithm for this. After calculating the priority the Treap data structure need to be built, call to *Build-Treap* subroutine takes care of this task. The next step is to find out the frequent variables using a depth first treap traversal using the sub routine, *Traversal*. The rules are created by calling *GenerateRule* subroutine. These subroutines help in creating valid predictions and rules in various real time applications like statistical analysis of stock market, pattern predictions and fault prediction.

The basic input to the Treap mining algorithm is a set of items (variables) say *n*. The number of transactions and support is represented by *m* and *S* respectively. Major data structures used in this algorithm are Array and Treap. The pseudo code of treap mining algorithm is given in Algorithm 1.

Algorithm 1 Rule Mining - Treap

Input: Database *D*, with *n* itemsets and *m* transactions. Let the support be *S*
Output: Association Rules
Data structures Used: Array, Treap
Step 1: for each item set n ∈ D, calculate priority
Call Priority Procedure (D, S)
Step 2: for i=1 to n in Priority Array P[n]
Call BuildTreap Procedure
Step 3: Perform BSF traversal in Min Treap
Call Traversal Procedure
Step 4: for each mined item
Call GenerateRule Procedure
Step 5: Output rules R

Priority: In a database even the least frequent item can have a major impact in the rules generated. In all association algorithms such infrequent items are pruned off without much analysis. In this proposed work, the priority of all item sets is found and it is analyzed with the frequency. After this analysis, if the item set is still invalid it is eliminated or else it is considered for rule creation. For example, in case of serum analysis for myocardial infarction level of Troponins will be elevated. But in normal case its level in blood will be undetectable. All traditional mining algorithms prunes such items in the primary step itself, thus its level or presence may not be available in the final predicted rules.

The algorithm begins by scanning database *D* and frequency *f* is calculated for all items. Partial priority of items, which is the component of priority, is found out using the equation in Algorithm 2. The calculated partial priority is added up with normalized frequency to calculate the weight. Normalization is important since it is a variance maximizing exercise and it projects our original data onto directions which maximize the variance. Thus even if we have a big difference in the frequency range, it brings down the range to a favorable boundary. Here *Z-score* normalization technique is used [16, 17].

Normalized,

$$e_i = \frac{e_i - E'}{std(E)} \tag{1}$$

where

$$std(E) = \sqrt{\left\{ \frac{1}{n-1} \cdot \sum_{i=1}^{n} (e_i - E')^2 \right\}} \tag{2}$$

and

$$E' = \frac{1}{n} \sum_{i=1}^{n} e_i \tag{3}$$

Now the threshold is set and the priority of each itemset is found out. The weight added is the maximum allowed that can be given to any item set in that database D to maintain threshold.

The maximum possible weight (W_{mp}): Let Y be a *p-itemset*, and X is a superset of Y with the *k-itemset (p < k)*. The maximum possible weight for any *k-itemset* containing Y is defined as

$$W_{mp}(Y, k) = \frac{1}{k} \left(\sum_{i,j \in y, j=1}^{n} w_j + \sum_{l=1}^{k-n} w_l \right) \tag{4}$$

In Eq. 4, the first part is the average weight for the p-itemsets and the second part is the average weight of the *(k-p)* maximum remaining itemsets.

Algorithm 2 Procedure (D, S)

Input: Database D, with n itemsets and m transactions. Let S be the support of the given transaction.
Data structure used: Array
Output: Array P[n] - Priority Array with n itemset
Step 1: Scan D and find the frequency f of each itemset
Step 2: for i=1 to n, find partial priority p' for each itemset, $p_i' = S * \frac{fi}{\Sigma f}$
Step 3: Find the weight, $w = p_i' + fi$
Step 4: Find the mean of frequency, T and set it as threshold, $T = \frac{fmax - fmin}{n}$
Step 5: Priority, $pp = T + w$ (if frequency range is uneven, use normalized (F))

To illustrate the idea, consider the following database in Table 1, with five transactions and five different itemsets.

In a normal association mining algorithm, we take all frequent items above the threshold for rule formation. Thus there can be a situation where high priority non-frequent itemsets get pruned off in an early stage. Through this work, we provide additional weight to those itemsets and provide them with priorities. After running the dataset through our procedure priority, the following result is obtained (Table 2).

In traditional methods the itemset E would have been pruned off in an early stage if we set the threshold to be 50 % as its below par. Here, while providing a weighted priority we could observe that itemset E falls above the threshold and it there by makes itself available for rule creation. In situations where the frequency range is uneven, the above mentioned normalization method needs to be employed before calculating the priority.

BuildTreap: In this work we use the Min Treap data structure. This is a tree which satisfies the Min Heap Property. Basic property of Min Treap is, the root node will be having a smaller priority than its children. The *BuildTreap* procedure obtains its input from the Priority procedure. The algorithm (Algorithm 3) scans each element from the list and builds a Min Treap according to the priority. Whenever a new element is added the *BuildTreap* subroutine is recursively called in-order to maintain the Treap structure. The process is very similar to that of heap creation (Fig. 1).

By *MinTreap* procedure, treap for each transaction is build and the leaf node will be the one with highest priority. So, by employing a depth first search we get to the nodes with highest priority. For easy and quick retrieval of data, we always make use of tree like data structure; FP growth is one such algorithm. But one of the major drawbacks of the FP growth is the large amount of candidate prefix sub-tree that needs to be generated for each transaction. This can make the memory overflow leading to a system crash. In this work, only one treap per transaction is created. Figure 1 shows the corresponding Treap for Table 1.

Table 1. Sample database

Transaction ID	Itemset
100	B C D
200	C D E
300	A B D
400	B C D E
500	B D

Table 2. Priority calculation

Transaction ID	Itemset	f	p'	W	Pp
100	B C D	A = 1	A = 0.033	A = 1.033	A = 1.833
200	C D E	B = 4	B = 0.133	B = 4.133	B = 4.933
300	A B D	C = 3	C = 0.100	C = 3.100	C = 3.900
400	B C D E	D = 5	D = 0.166	D = 5.166	D = 5.966
500	B D	E = 2	E = 0.066	E = 2.066	E = 2.866

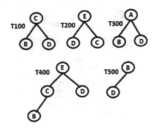

Fig. 1. Treap creation

Algorithm 3 BuildTreap

Input: Priority Array, P[n]
Data structure Used: Treap
Output: Minimum Valued Treap, MinTreap
Step 1: length of the priority array P is taken as treelength
treelength [A] ← length [P]
Step 2: for each node indexed at j
parent (j) ← return(j/2)
left (j) ← return (2j)
right (j) ← return (2j+1)
Step 3: for all nodes indexed at j, do
l ← left[j]
r ← right [j]

if(A[l] ≤ A[j]) then
        small ← l
else
        small ← j
else if(A[r] ≤ A[small]) then
        small ← r
else if(small ≠ j) then
        interchange A[j] ↔ A[small]
Step 4: Min Treap is obtained

Traversal: This is a depth first treap traversal where the algorithm tries to find the most frequent items with highest priority from the Treap (Algorithm 4). Search continues till the leave node and returns the priority value and the node value. If the priority is greater than the threshold, parent and the sibling of the node is returned. If there is no sibling, then the sibling of the parent is retuned by the subroutine. The search moves in a bottom up approach until the priority has reached its minimum threshold.

In the database discussed in our example, we can see that itemset B, D with root C and itemset D, C with root E are leaf nodes and the ones with maximum priority and frequency. By considering the level of confidence we need, we can extract the leaf nodes to that particular tree height. Figure 2 shows the DFS traversal in our example.

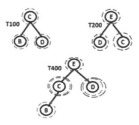

Fig. 2. DFS traversal

Algorithm 4 Traversal
_____
Input: Min Treap
Method Used: Treap Traversal
Output: Frequency Items with Maximum Priority, F
Step 1: Perform depth first traversal and return the left node, N.
Step 2: Return Parent (N) and Sibling (N)
Step 3: if Sibling (N) = NULL,
Return N, Parent (N), Sibling (Parent)

GenerateRule: This is the final step of the algorithm (Algorithm 5), here all those rules above the threshold T is calculated. Initially the rule set R is initialized to NULL. For each frequent item obtained from the treap traversal, the ratio of frequent item to its supporting item sets are found out. If it is greater than the threshold, the rule is added to R. This process is continued until all the frequent items are visited for rule creation. In our example database, we have two set of leaves as frequent. The support count for each set for all combination needs to be found out using the support count Eq. 5 or 6

$$Threshold = \frac{SC(IndivudualItems)}{SC(RuleItemsets)} \tag{5}$$

$$Threshold = \frac{SC(IndivudualItems)}{SC(ConclusionItemsets)} \tag{6}$$

Algorithm 5 GenerateRule
_____
Input: Frequent items with priority value, F; Threshold T
Output: Association Rules

Step 1: Rule R = NULL

Step 2: for i=1 to n in each Fi do
if(Support (Fi)/Support (Fn) ≥ T) then
        R = R U (Fi → Fn)

Step 3: Find R, until F = NULL

Table 3, shows the results of all those itemsets which has 100 % confidence. For getting more rules, we can reduce the threshold level and obtain more result. We are considering strong rules only. The itemsets which are below the support threshold is not considered here.

Table 3. Rules Generated

Set 1		
Rule	Support	Remark
B ∧ C → D	100 %	Accept
Set 2		
Rule	Support	Remark
D ∧ E → C	100 %	Accept
E ∧ C → D	100 %	Accept
E → D ∧ C	100 %	Accept

4 Comparison with Traditional Algorithm

When comparing Treap mining with the traditional algorithms, we could clearly see a distinction in the basic run time and space complexities. Time complexity of apriori algorithm [18, 20, 21] can be defined as the sum of time taken for generation of frequent items and time taken for rule generation. For n transactions with m itemsets, the time taken for frequent itemset generation is,

$$= n\, C_1^m + nC_2^m + \ldots + nC_m^m$$

$$= n \sum_{i=1}^{m} C_i^m \tag{7}$$

$$\cong O(e^n)\, as\, n \to \alpha$$

Similarly for rule generation the time complexity calculated as,

$$= C_1^k + C_2^k + \ldots + C_n^k$$

$$= \sum_{k=1, j=C_k^m}^{k=m} [\sum_{i=1}^{i=j} C_i^j]$$

$$= \sum_{k=1}^{k=m} m \sum_{i=1}^{i=j} C_i^j \tag{8}$$

$$\cong O(n)\, as\, n \to \alpha$$

Thus total time complexity is the sum of (7) and (8), which shows the Apriori algorithm grows exponential, i.e.,

$$\cong O(e^n) \, as \, n \rightarrow \alpha \tag{9}$$

whereas the treap performs the task in linear time order. We have four procedures called back to back, thus the total complexity will be the upper bound of these procedure calls. Priority calculation always takes a linear order as the amount of time taken for calculation is directly proportional to the number of input elements.

For the BuildTreap, we derive the time complexity by extending the complexity of heap sort. Time complexity is proportional to

$$T(n) = \sum_{j=0}^{h} j.2^{h-j} = \sum_{j=0}^{h} j \frac{2^h}{2^j} \tag{10a}$$

If the bottom most level has 2h nodes, where h is the height of the tree. Factor out the 2h term, we get

$$T(n) = 2^h \sum_{j=0}^{h} \frac{j}{2^j} \tag{10b}$$

We have,

$$\sum_{j=0}^{\infty} x^j = \frac{1}{1-x} \tag{11}$$

Taking derivative on both sides

$$\sum_{j=0}^{\infty} j.x^{j-1} = \frac{1}{(1-x^2)} \tag{12}$$

Multiplying x on both sides we get,

$$\sum_{j=0}^{\infty} j.x^j = \frac{x}{(1-x^2)} \tag{13}$$

when $x = \frac{1}{2}$, we get

$$\sum_{j=0}^{\infty} \frac{j}{2^j} = \frac{\frac{1}{2}}{(1-\frac{1}{2}^2)} = 2 \tag{14}$$

Substituting the value of Eq. 14 in Eq. 10a, we get

$$T(n) = 2^h \sum_{j=0}^{h} \frac{j}{2^j} \leq 2^h \sum_{j=0}^{\infty} \frac{j}{2^j} \leq 2^h.2 = 2^{h+1} \tag{15}$$

we already know, n takes the form of, $n = 2^{h+1} - 1$, Thus from Eq. (15) we can conclude,

$$T(n) \leq n + 1 \in O(n) \qquad (16)$$

Similarly for rule generation for n transactions with m itemsets can be given as,

$$= {}^k c_1 + {}^k c_2 + \ldots + {}^k c_n \qquad (17)$$

$$= \sum_{k=1, j=^m c_k}^{k=m} \sum_{i=1}^{j} {}^j c_i \qquad (18)$$

$$= \sum_{k=1}^{k=m} m \sum_{i=1}^{j} {}^j c_i \qquad (19)$$

$$\cong O(n) \; as \; n \to \alpha \qquad (20)$$

The analysis part clearly describes that the algorithm works in the linear order, O(n) in the best case scenario, and even keeps the algorithm steady in O(n log n) in worst case scenarios, which is far better than the competitor algorithms of the same genre.

5 Discussions

The proposed algorithm was tested against the traditional algorithms like FP growth and Apriori. The results obtained are shown in the Table 4. The standard databases available on the internet are used for the testing purposes. (Databases are available on this link: http://fimi.ua.ac.be/data/). Time is represented in milliseconds.

A graphical trend comparison of the algorithms with increased size of itemsets and records is provided in the graph (Fig. 3).

Table 4. Comparison of runtime against various databases

Minimum support	Database used	Apriori algorithm	FP growth algorithm	Treap mining algorithm
25 %	Chess	469	348	227
25 %	Accidents	42356	22588	16289
25 %	Mushroom	14432	4897	3189
25 %	Retail	7739	4230	3912
25 %	Webdocs	255650	225440	195230

The algorithms work with relative time complexity when the size of dataset is small. As the size increases, we can see Apriori gets to an indefinite working loop. After a particular record size, the algorithm continuously delivers the last generated output, thereby creating an invalid rule. Even FP growth algorithm shows indefinite halt signs during the last stage of cycle [19] (the flat part of the graph). The increase in sub-tree generation could be the major cause for this system halt while

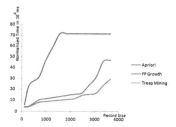

Fig. 3. Graphical comparison

working with large record size. Treap mining, on the other hand shows a very distinct performance improvement even with increases record size.

6 Conclusion

In this work a new association rule mining algorithm is put forward. It makes use of the priority model for building treap. From the treap the frequent items are mined and the rules are generated. The results are compared with the traditional mining algorithms like Apriori and FP growth. Algorithm was also tested by the datasets available in internet. Analysis of the algorithm in both time and space dimension was seen to be bound within O (n) and O (n log n), which is very much efficient than other mining algorithms. This proposal also helped to mine high priority non-frequent items for rule creation. Treap mining could be extended to market basket analysis, single variable prediction and even to practical applications like fault tolerance, estimation predictions and so on. A scheme for employing very large vertical database mining could be seen as a future scope of this treap mining. Also, prioritizing itemsets depending on the cluster centers can also be considered as an enhancement for this method.

References

1. Boney, L., Tewfik, A.H., Hamdy, K.N.: Minimum association rule in large database. In: Proceedings of the Third IEEE International Conference on Computing, pp. 12–16 (2006)
2. Agrawal, R., Srikant, R.: Fast algorithms for mining association rules. In: Proceedings of VLDB, pp. 487–499 (1994)
3. Zaki, M., Parthasarathy, S., Ogihara, M., Li. W.: New algorithms for fast discovery of association rules. In: Proceedings of the Third International Conference on Knowledge Discovery and Data Mining, vol. 2, pp. 283–296 (1997)
4. Anandhavalli, G.K.: Association rule mining in genomics. Int. J. Comput. Theory Eng. **1** (2007)
5. Cooper, C., Zito, M.: Realistic synthetic data for testing association rule mining algorithms for market basket databases. Knowl. Disc. Databases PKDD **9**, 398–405 (2007)
6. Varde, A.S., Takahashi, M., Rundensteiner, E.A., Ward, M.O., Maniruzzaman, M., Sisson, R.D.: Apriori algorithm and game of life for predictive analysis in materials science. Int. J. Knowl. Based Intell. Eng. Syst. **8**, 116–122 (2004)

7. Wu, H., Lu, Z., Pan, L., Xu, R., Jiang, W.: An improved apriori based algorithm for association rules mining. In: Proceedings of Sixth International Conference on Fuzzy Systems and Knowledge Discovery, pp. 51–55 (2009)
8. Bodon, F.: A fast apriori implementation. In: Proceedings of the IEEE ICDM Workshop on Frequent Item-set Mining Implementation, vol. 9 (2003)
9. Kryszkiewicz, M., Rybiñski, H.: Data mining in incomplete information systems from rough set perspective. Rough Set Methods Appl. **56**, 567–580 (2000)
10. Kosters, A.W., Marchiori, E., Oerlemans, A.J.: Mining clusters with association rules. In: Hand, D.J., Kok, J.N., Berthold, M.R. (eds.) IDA 1999. LNCS, vol. 1642, pp. 39–50. Springer, Heidelberg (1999)
11. Lin, T.Y.: Rough set theory in very large databases. In: Symposium on Modeling, Analysis and Simulation, vol. 2, pp. 936–941 (1996)
12. Borgelt, C.: An implementation of FP growth algorithm. In: Proceedings of the Workshop on Open ource Mining Software ACM (2005)
13. Malik, K., Raheja, N., Garg, P.: Enhance FP growth algorithm. Int. J. Comput. Eng. Manage. **12**, 54–56 (2011)
14. Guy, E.B., Margaret, R.M.: Fast set operations using treaps. In: Proceedings of the Tenth Annual ACM Symposium on Parallel Algorithms and Architectures, pp. 16–26 (1998)
15. Aragon Cecilia, R., Aragon, C.: Randomized search trees. Algorithmica **16**, 464–497 (1996)
16. Jane, A., Ross, A.: Score normalization in multimodal biometrics systems. Pattern Recogn. **38**, 270–285 (2005)
17. Ben Khalifa, A., Gazzah, S.: Adaptive score normalization: a novel approach for multimodal biometric systems. Int. J. Comput. Electr. Autom. Control Inf. Eng. **7**(3) (2013)
18. Vinodchandra, S.S., Anand, H.S.: Artificial Intelligence and Machine Learning. PHI Publishers, Delhi (2014). 368p. ISBN ISBN: 978-81-203-4934-6
19. Anand, H.S., Vinodchandra, S.S.: Mining association rules using improved frequent-pattern growth algorithm. Int. J. Appl. Eng. Res. **9**, 239–246 (2014). (ISSN: 0973-4562)
20. Anand, H.S., Vinodchandra, S.S.: Horizontal and vertical rule mining algorithms. In: ACCIS, Proceedings of Elsevier, pp. 26–28 (2014)
21. Anand, H.S., Vinodchandra, S.S.: Applying correlation threshold on apriori algorithm. In: IEEE ICE-CCN (2013)

SVM Based Lung Cancer Prediction Using microRNA Expression Profiling from NGS Data

Salim A.[1]([⊠]), Amjesh R.[2], and Vinod Chandra S. S.[3]

[1] College of Engineering Trivandrum, Thiruvananthapuram, Kerala, India
salim.mangad@gmail.com
[2] Center for Bioinformatics and Computational Biology, University of Kerala,
Thiruvananthapuram, India
[3] Computer Center, University of Kerala, Thiruvananthapuram, India

Abstract. microRNAs are single stranded non coding RNA sequences of 18 - 24 nucleotide length. They play an important role in post transcriptional regulation of gene expression. Last decade witnessed identification of hundreds of human microRNAs from genomic data. Experimental as well as computational identification of microRNA binding sites in messenger RNAs are also in progress. Evidences of microRNAs acting as promoter /suppressor of several diseases including cancer are being unveiled. The advancement of Next Generation Sequencing technologies with dramatic reduction in cost, opened endless applications and rapid advances in many fields related to biological science. microRNA expression profiling is a measure of relative abundance of microRNA sequences to the total number of sequences in a sample. Many experiments conducted in this kind of measure proved differential expression of microRNAs in diseased states. This paper discusses an algorithm for microRNA expression profiling, its normalization, and a Support Vector based machine learning approach to develop a Cancer Prediction System. The developed system classify samples with 97.6 % accuracy.

1 Introduction

A family of non coding RNA, around 22 nt long, found in many eukaryotes including humans are called microRNA. Around 1800 microRNAs are identified in human and have abundant evidences of its functionality in normal cell development, differentiation, growth control and human diseases [1]. microRNAs down regulate gene expression by repressing the process of protein translation or degradation of messenger RNA(mRNA)[2,3]. It is identified that several disease states are linked to altered microRNA expression. The regulatory role of microRNAs in Cancer, Heart diseases, Neurological diseases, Immune function disorders are proved with experimental evidences [4,5].

Next Generation Sequencing (NGS) is the latest techniques in DNA sequencing and are characterized by its unprecedented throughput and speed. The progress in NGS techniques with dramatic reduction in cost, attributed development to many applications in the fields related to biological science. microRNA

© Springer-Verlag Berlin Heidelberg 2016
N.T. Nguyen et al. (Eds.): ACIIDS 2016, Part I, LNAI 9621, pp. 544–553, 2016.
DOI: 10.1007/978-3-662-49381-6_52

expression profiling from NGS data, together with a high performance classifier could be utilized for the development of prediction systems, especially for Cancer. In this paper, the development of NGS data based Cancer Prediction System for lung cancer is discussed.

2   microRNA Profiling

The nucleic acid sequencing techniques find the exact order of nucleotides present in a given DNA sequence. Sanger Sequencing was the most popular method in DNA sequencing for around three decades, however recent development of Next Generation Sequencing (NGS) made a faster, and cheaper alternative. The Life Technologies : Ion Torrent Personal Genome Machine (PGM), Illumina : HiSeq, GAIIX, ABI : SOLiD, Roche: GS Flx+ or 454 are examples of NGS systems. Small RNA sequencing (RNA-Seq) is one the NGS library preparation method suitable for small non coding RNAs such as microRNAs [6]. The output of RNA-Seq experiments consist of millions of sequences called *reads*, which are the probable short RNAs present in a genome at a given instant of time. The length of sequences are less than 200 base pairs.

microRNA expression profiling is a measure of relative abundance of microRNA sequence to the total sequence output from a RNA-Seq experiment. microRNA profiling in wet lab is a difficult task due to the very low presence (0.01 %) of microRNA in total RNA mass, lack of common start or stop sequence and very short sequence length. Three different strategies are established despite these challenges - a hybridization based method (microarray, nCounter), Quantitative Reverse Transcription PCR (qRT-PCR), and High Throughput Sequencing [7]. Shirley et al. compared different profiling systems and concluded that NGS platform has highest detection sensitivity, highest differential expression analysis accuracy, and high level of technical re-productivity [8]. As a computational method, microRNA profiling is performed by proper sequence mapping between *reads* and a genome sub sequence. The term sequence map refers to the process of finding the most similar region where a short sequence could be attached to a comparatively long sequence. When profiling carried out between *healthy* and *diseased* states could be used in diagnostic applications.

3   microRNA Disease Association

microRNAs are proved as negative gene regulators due to the complementary sequence coupling with the target mRNA, which results in inhibition protein translation(the process of protein production from mRNA) or altered mRNA stability. The real outcome of this process depends on the target gene involved. Several microRNAs are acting as oncogenes, whereas another set as tumor suppressor. miR-15a, miR-16-1, miR 143, miR-145 and let-7 family members are examples of few early detected tumor suppressor microRNAs, whereas miR-21, mir 155 are proved as oncogenes [9]. Landi et al. reported that signature of five microRNAs (miR-25, miR-34c-5p, miR-191, let-7e and miR-34a) could be used to differentiate adenocarcinoma (AD) from squamous cell carcinoma (SCC), both sub types of lung cancer [10]. They used microRNA expression level as the signature. However, one cannot claim a global increase or decrease in microRNA expression level in case of cancer, but a variation in expression level is a reality. Presently, efforts are made by researchers to collect and publish microRNA disease association from published literature. MiRCancer, is one such database extracted from the literature, which contains 878 association between 79 human cancers and 236 microRNAs [11]. The PhenomiR is yet another manually curated database, where deregulation of microRNAs in diseases are investigated from 542 studies [12].

4   Materials and Methods

4.1   Data Collection

microRNA transcriptome data for lung cancer is used to build a classifier model to develop a Cancer Prediction System. Data samples are downloaded from NCBI Sequence Read Achieve (SRA). The downloaded lung cancer data set contains 41 samples, where 20 are of tumor and 21 are of normal tissues(SRP009408-microRNA expression profiles in lung cancer tissues versus adjacent lung tissues using next-gen sequencing). microRNA mature sequences database are downloaded from miRBase [13]. A list of microRNAs having direct link with lung cancer has been prepared by collecting data from miRCancer, PhenomiR, and from other published literature.

4.2   NGS Data Pre-processing

The downloaded data are in an archive format, which can be extracted with *SRA tool kit* from NCBI. The resultant sequences are in FASTQ format. A NGS sequence *read* may contain an adapter sequence or fragments of adapter sequence which are added during library preparation [14]. These sequence will be either at 3' end or 5' end, and needs to be removed before further processing of *reads*. If not, may lead to missed alignments, wrongful discarding of genuine match. A quality score is associated with each nucleotide of the sequence as a measure of error probability. The quality score $Q = -10\, log_{10} P$, where P is the probability of incorrect base call. When P=0.001, then quality score Q is 30, which ensure 99.9 % accuracy in base call, whereas quality score of 20 ensures 99.0 % accuracy only. Our algorithm is designed with an objective that every sequence read could complete adapter removal, quality trimming and sequence mapping in a single step. The quality threshold is fixed at 30 so that reads ensure 99.9 % accuracy.

4.3   Sequence Mapping

The primary objective of sequence mapping is to find the exact location where a given sequence gets aligned with another sequence. There are around 60 mapping tools with differing capabilities, majority of them developed after the Next

Generation Technologies came into existence. The read length supported by the mapping tool, accuracy, parallel processing of reads, computational efficiency when gaps/mismatches allowed are some of the front line characters on which these algorithms are designed [15].

Algorithm 1. Parallel processing of microRNA expression profiling

---

$NGSdata$: **Reads in FASTQ format**
$miRNAList$: **list of disease specific microRNAs**
procedure MICRORNAEXPRESSION
    for $i \leftarrow 1, Core\ in\ Parallel$ **do**
        Divide and assign NGSData to each core
    for $k \leftarrow 1, Core\ in\ Parallel$ **do**
        for each $M_j \in miRNAactiveList$ **do**
            $miRExp[j] \leftarrow 0$
        while *not end of NGSdata* **do**
            NGSRead to R_i
            Remove adapter(R_i)
            Rd = TrimQScore(R_i)
            if len $(Rd) \geq 17$ **then**
                for each $M_j \in miRNList$ **do**
                    $match = BMH(Rd, M_j)$
                if $match$ **then**
                    $miRExp[j] \leftarrow miRExp[j] + 1$
    return $miRExp$

To calculate read count, different approaches were employed. Kristina et al. in their study of microRNA expression profiles in Colorectal Cancer, reads mapped to mature microRNA sequence with a maximum of one mismatch is considered as a hit [16]. In another study, a hit can have a maximum of two mismatches between position 12 and 14 or three mismatches for longer reads [17]. Hang-Tai et al. did not allow a single mismatch to prevent *reads* mapped to paralogs of a given microRNA, and to avoid multiple ambiguous hits [18]. Considering the benefits of perfect match, our system limits only exact match between mature microRNA and *read* as a hit.

As we decided go for exact match for a *read hit*, a faster pattern matching algorithm is the best choice to map NGS reads to microRNAs. Let T be text string of length m and P be pattern of length n, the exact sequence match is to find the occurrences of P in T. The worst time complexity of naive approach to perform this match is $\Theta(m \times n)$. The improvements in search methods with additional pre-processing steps could reduce search time to $O(m + n)$. Boyer Moore introduced three techniques, namely, *Right to left scan*, *Bad character rule* and *Good suffix rule*. When the *right to left scan* being performed, and a mismatch occurred $x \neq y$, and $x \in T$ and $y \in P$, then *Bad character rule* says the pattern P could be shifted to the right most x belong to the pattern P. If such an x is not in the pattern, the shift distance could be the entire pattern length. The *Good Suffix Shift* is applicable when a substring t of the pattern P have match with the text T, and search halted due to mismatch at $y \in P$ to the left of t. If there is a $t' \in P$ and $t' = t$ with an $x \neq y$ to the left of t', then pattern can be shifted to the right so that x is below the current search pointer. Considering the implementation aspects, shift due to *Good suffix rule* is difficult and hence Horspool modified Boyer-Moore Algorithm suggesting that the shift due to *Bad character rule* alone is sufficient in all practical applications [19].

Algorithm 2. Boyer-Moore-Horspool Algorithm to map Reads to microRNA sequences

---

procedure BMH($R[n]$, $M[m]$)   ▷ $R[n]$ A *read* of length n
    $\Sigma = \{A,\ C,\ G,\ U\}$ ▷ $M[m]$ A microRNA sequence of length m

    for *each* $a \in \Sigma$ **do**   ▷ $D[]$ Shift distance array
        $D[a] = Max[j \mid M_j\ ==\ a]$
                                              ▷ right most occurrence of a
    $i \leftarrow m - 1$
    while $i \leq n - 1$ **do**
        $k \leftarrow 0$
        while $M[m - 1 - k] = R[i - k]$ *and* $k \leq m - 1$ **do**
            $k\ \leftarrow\ k\ +\ 1$
        if $k = m$ **then**
            return *true*
        else
            $i \leftarrow i + D[R[i]]$ ▷ increment i by shift distance

---

Sequence read input file consists of millions of *reads*, and our algorithm computes microRNA expression in parallel. We divide the input file into as many blocks as the number of processing units in the system and allocate one block to each unit. This strategy for microRNA expression profiling is illustrated in Algorithm 1. Each sequence *read* initially tested for adapter contamination, followed by a base quality check. Normally, read quality contamination is at the trailing end of *read* than the initial portion. Algorithm 3 keeps track of difference in quality score with that of a threshold value. When this accumulated difference falls below zero, remaining portion of *read* is trimmed off. If the trimmed sequence length is above 17, then search for a map with disease specific microRNA sequence using the Algorithm 2 is performed. The reason to fix length limit as above 17 is that the minimum sequence length of a microRNA is 18. The count of mapped *reads* from each processing units is integrated and returned as expression value.

4.4   Expression Normalization

In microRNA profiling, normalization is a significant step to find differently expressed genes. Several normalization methods were available, specifically

Algorithm 3. Algorithm to trim sequence reads based on Phread quality scores

---

   procedure TRIMQSCORE(R, $Qthreshold$])   ▷ R A Read sequence

                                           ▷ $Qthreshold$- Minimum quality score required

      $L = len(Read)$

      $i = 1$

      $Sum = 0$

      while $Sum \geq 0$ and $i \leq L$ **do**

         $Sum = Sum + Q(Read[i] - Qthreshold$

         $i = i + 1$

      if $i \geq 17$ **then**

         $Rd = trim(Read(i + 1, L)$

         return Rd

      else

         discard the Read

---

applicable to microArray analysis, Real time PCR, and Next Generation Sequencing with varying throughput, cost, and memory requirement [20]. In Next generation sequencing, relative count of microRNA can be calculated by normalizing *read count* against total number of *reads* or total number of microRNAs in the sample. Z-score normalization determines how an individual score value varies from the mean in units of standard deviation. Performance of Z-Score normalization is better when compared with Min-Max normalization, if prior knowledge about mean and standard deviation of the sample are available [21]. The normalized expression of microRNAs in this experiment is Z-score values of microRNA expression with respect to the total mapped microRNAs in a sample.

4.5 Classifier Model

Dimensionality Reduction: The initial list of mature microRNA sequences with respect to lung cancer consists of 82 sequences. Determining an optimal subset of differentially expressed microRNAs, will improve the performance of classifier, and reduce computational and storage requirements. Two different strategies can be used to select optimal set of features- filter method and wrapper method [22]. Filter method uses ranking function to select best attributes, whereas wrapper method tries to reduce feature set using the same learning algorithm used by the classifier. In the case of wrapper methods, if there are n variables, there are total of 2^n subsets, and hence exhaustive search is infeasible for large value of n. Wrapper method may follow sequential forward/ backward search or random search or heuristic search to find an optimal subset. In this experiment, wrapper method is applied and the most deciding 45 microRNAs are only used for classification.

Classification with Support Vector Machines (SVM): The data samples are classified using Support Vector Machine(SVM). SVM works by projecting training data in input space to a feature space of higher dimension. A linear classifier is based on discriminant function of the form $f(x) = \omega^T x + b$, where ω is weight vector and b is scalar value. $\omega^T x$ is dot product between two vectors, and $\omega^T x = \Sigma_i \omega_i x_i$. The set of all points with $\omega^T x = 0$ define a hyperplane, which separates input data into two classes. The bias b translate the hyperplane away from the origin. The closest point to he hyperplane among positive and negative samples define a *margin*. SVM minimizes the risk of misclassification by maximizing the margin between the data points. Thus, a maximum margin classifier becomes a constraint optimization problem

$$\text{minimize}_{\omega,b} \quad \frac{1}{2}\|\omega\|^2 + C\Sigma_i \,\epsilon_i$$

$$\text{subject to} \quad y_i(\omega^T x + b) \geq 1 - \Sigma_i \,\epsilon_i, \quad \epsilon_i \geq 0$$

The term, $C \Sigma_i \,\epsilon_i$ define penalty for margin errors. ϵ_i the margin error, it should be $0 \leq \epsilon_i \leq 1$

A non-linear SVM classifier is based on discriminant function of form $f(x) = \omega^T \phi(x) + b$, where ϕ is a non-linear function. To limit the size of feature space and thus memory and computational requirements, an efficient way of computation known as kernel trick k is employed, rather computing the mapping ϕ. Thus a polynomial kernel is

$$K(x,y) = (x^T y + 1)^d, \tag{1}$$

where d is degree of polynomial. A Gaussian kernel or Radial basis function (RBF) is defined by

$$K(x,y) = e^{(-\gamma \,\|x - y\|^2)} \tag{2}$$

the term γ determines the curvature of non linear decision boundary.

The Pearson VII kernel(PUK) is defined by

$$K(x,y) = \cfrac{1}{\left(1 + \left(\cfrac{2\,\sqrt{\|x - y\|^2}\,\sqrt{2^{(\frac{1}{\omega})-1}}}{\sigma}\right)^2\right)^{\omega}} \tag{3}$$

where ω and σ control half width and trailing factor of peak, respectively.

5   Results and Discussion

The classifier model was trained and tested with the data set prepared using the microRNA profiling followed by the normalization step of the experiment. Figure 1 shows the Z-score normalized microRNA expression values with respect to a normal and a tumor NGS data sample. Ranking of attributes based on wrapper method with SVM (RBF kernel) classifier unfold the top ranked microR-NAs that are differently expressed. List of a few top ranked microRNAs are - $hsa - miR - 21 - 5p$, $hsa - miR - 24 - 1 - 5p$, $hsa - miR - 200a - 3p$, $hsa - let - 7c - 3p$, $hsa - miR - 30c - 2 - 3p$, $hsa - let - 7d - 3p$, $hsa - miR - 210 - 3p$

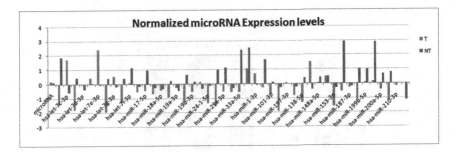

Fig. 1. Z-Score normalized microRNA expression levels corresponding to a Normal(NT) and a Tumor(T) sample

The evaluation method used in the experiment is 10 fold cross validation. 10 fold cross validation is a more preferable technique when number of data samples are less. In this method, the data samples are divided into 10 subsets, one randomly selected subset is used for testing and remaining 9 for training. This operation is repeated for 10 times. The performance of Support Vector Machine based classifier with different kernel functions are analyzed. The round of error(ϵ) is fixed at $1.0E - 12$ and complexity value (c) at 1.0. With RBF kernel, (γ) is set to 0.01 and with PUK kernel ω and σ are set to 1.0.

The analysis are performed based on the following measures:

$$precision = \frac{TP}{TP + FP}$$

$$True\ Positive\ Rate/Recall/Sensitivity = \frac{TP}{TP + FN}$$

$$True\ Negative\ Rate/Specificity = \frac{TN}{TN + FP}$$

$$False\ Positive\ Rate = \frac{FP}{TN + FP}$$

$$F - Measure = \frac{2 \times precision \times recall}{precision + recall}$$

$$Accuracy = \frac{TP + TN}{TP + TN + FP + FN}$$

The *Accuracy* alone could not be used as a measure of performance of a classifier, especially when number of samples in the positive and negative sets widely differ. The term *precision* indicates the relevancy of prediction. It represents out of all samples labeled as *class A*, what fraction actually belonged to the *class A*. *Recall* is the fraction of *class A* samples that the classifier picks up out of all samples that were originally belonged to *class A*. This is equivalent to *True Positive Rate*. In addition to this, *False Positive Rate* are also measured. *False Positive Rate* is proportion of samples which are classified as *class A*, but

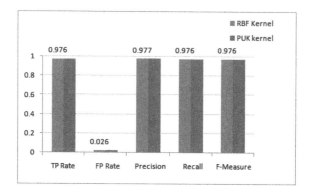

Fig. 2. Performance of cancer prediction system when SVM classifier with RBF Kernel and PUK Kernel were employed. SVM parameters were $c = 1$, $\epsilon = 1 \times 10^{-12}$, with RBF $\gamma = 0.01$, and with PUK $\omega = 1$ and $\sigma = 1$.

belong to a different class, among all samples that are not in *class A*. Figure 2 shows a comparison of performance cancer prediction system with different kernel functions. SVM with Pearson VII kernel(PUK) classifies with an accuracy 97.6 %. When the experiment is repeated with RBF Kernel accuracy is again 97.6 %, but γ, the curvature of non linear decision boundary increased from 0.01 to 0.1, the accuracy becomes 100 %.

6   Conclusion

The exact molecular mechanism behind gene expression regulation of microR-NAs is not yet unveiled completely. But, increasing evidences with experimental proofs are available to acknowledge the association between microRNAs and different diseases. Progress in Next Generation Sequencing added great momentum in microRNA research. Many studies related to differential expression of microR-NAs in specific diseases /cancer can be found in literature, but development of cancer prediction system using microRNA profiling is a novel approach. When experiment conducted with lung cancer samples and SVM classifiers, prediction accuracy was 97.6 %. We hope this method could be further extended to develop a more comprehensive cancer prediction system in future.

References

1. Bartel, D.P.: MicroRNAs: genomics, biogenesis, mechanism, and function. Cell **116**, 281–297 (2004)
2. Chandra, V.S., Reshmi, G., Achuthsankar, S.N., Sreenathan, S., Radhakrishna, M.P.: Mtar: A computational microrna target prediction architecture for human transcriptome. BMC Bioinform. **10**(S1), 1–9 (2010)
3. Salim, A., Chandra, V.S.: Computational prediction of microRNAs and their targets. J. Proteomics Bioinform. **7**(7), 193–202 (2014)

4. Li, Y., Kowdley, K.V.: MicroRNAs in common human diseases. Genomics Proteomics Bioinform. **10**, 246–253 (2012)
5. Esteller, M.: Non-coding RNAs in human disease. Nat. Rev. Genet. **12**, 861–874 (2011)
6. Ayman, G., Kate, W.: Next-generation sequencing: Methodology and application. Soc. Invest. Dermatol. **133**, e11 (2013)
7. Colin, C.P., Heather, H.C., Muneesh, T.: MicroRNA profiling: approaches and considerations. Nat. Rev. **13**, 358–369 (2012)
8. Shirley, T., de Richard, B., Ming-Sound, T., John, D.M.: Robust global microrna expression profiling using next-generation sequencing technologies. Lab. Invest. **94**, 350–358 (2013)
9. Esquela-Kerscher, A., Frank, J.S.: Oncomirs: microRNAs with a role in cancer. Nat. Rev. **6**, 259–269 (2006)
10. Teresa, M.L., Yingdong, Z., Melissa, R., Jill, K., Hui, L., Andrew, W.B., Maurizia, R., Alisa, M.G., Ilona, L., Francesco, M.M.: MicroRNA expression differentiates histology and predicts survival of lung cancer. Clin. Cancer Res. **16**(2), 430–441 (2010)
11. Boya, X., Ding, Q., Han, H., Wu, D.: MiRCancer: a microRNAcancer association database constructed by text mining on literature. BMC Bioinform. **29**, 638–644 (2013)
12. Andreas, R., Andreas, K., Daniel, S., Felix, B., Barbara, B., Irmtraud, D., Gisela, F., Goar, F., Corinna, M., Fabian, J.T.: Phenomir: A knowledge base for microrna expression in diseases and biological processes. Genome Biology
13. Kozomara, A., Griffiths-Jones, S.: MiRBase: Annotating high confidence micrornas using deep sequencing data. Nucleic Acids Res. **42**, D68–D73 (2014)
14. Seda, E., Danos, C.C., Francois, V., George, M.C., Seidman, J.G.: Quantification of microRNA expression with next-generation sequencing. Curr. Protoc. Mol. Biol. **4**(16), 1–20 (2013)
15. Nuno, A.F., Johan, R., Alvis, B., John C.M.: Tools for mapping high-throughput sequencing data. Bioinform. Adv. Access (2012)
16. Kristina, S., Susanne, L., Molton, M.W., Clara-Cecilie, G., Marit, H., Eivind, H., Oystein, F., Leonardo, A.M.Z., Kjersti, F.: Deep sequencing the microRNA transcriptome in colorectal cancer. PLOS-ONE **8**(6), 3169–3177 (2013)
17. Johannes, H.S., Tobias, M., Marcel, M., Philipp, R., Pieter, M., Stefanie, S., Theresa, T., Jo, V., Angelika, E., Stefan, S., Sven, R., Alexander, S.: Deep sequencing reveals differential expression of micrornas in favorable versus unfavorable neuroblastoma. Nucleic Acids Res. **38**(17), 5919–5928 (2010)
18. Hong-Tai, C., Sung-Chou, L., Meng-Ru, H., Hung-Wei, P., Luo-Ping, G., Ling-Yueh, H., Shou-Yu, Y., Wen-Hsiung, L., Kuo-Wang, T.: Comprehensive analysis of microRNAs in breast cancer. BMC Genomics **13**(6), s18 (2012)
19. Horspool, R.N.: Practical fast searching in strings. Softw. Pract. Experience **10**, 501506 (1980)
20. Meyer, S.U., Pfaffl, M.W., Ulbrich, S.E.: Normalization strategies for microrna profiling experiments: a normal way to a hidden layer of complexity? Biotechnol. Lett. **10**(1007), 1777–1788 (2010)
21. Anil, J., Karthik, N., Arun, R.: Score normalization in multimodal biometric systems. Pattern Recogn. **38**, 2270–2285 (2005)
22. Isabelle, G., Andre, E.: An introduction to variable and feature selection. J. Mach. Learn. Res. **3**, 1157–1182 (2003)

Forecasting the Magnitude of Dengue in Southern Vietnam

Tuan Q. Dinh[1,2(✉)], Hiep V. Le[1], Tru H. Cao[1,2], Quang C. Luong[3], and Hai T. Diep[3]

[1] Faculty of Computer Science and Engineering, Ho Chi Minh City University of Technology, Ho Chi Minh City, Vietnam
dqtuan10@gmail.com, lvhiep92@gmail.com,
tru@cse.hcmut.edu.vn
[2] John von Neumann Institute, Vietnam National University at Ho Chi Minh City, Ho Chi Minh City, Vietnam
[3] Department of Disease Control and Prevention, Pasteur Institute of Ho Chi Minh City, Ho Chi Minh City, Vietnam

Abstract. With recent rises of sophisticated and dangerous epidemics, there is a growing need for a system that could predict disease severity with high accuracy. In this paper, we address the problem of forecasting the magnitude of dengue in a short term period, i.e. one week ahead. We consider inputs as both statistics of historical cases and biological factors affecting the dengue virus, including the temperature, population and mosquito density. We propose a two-phase model simulating the disease transmission process, which are the local outbreak and then province transmission. The locality phase estimates the number of potential cases in each province independently in the following week. Then, in the transmission phase, an artificial neural network is used to predict the mobility of the dengue virus across provinces. Our proposed method obtains a higher accuracy than the conventional models of time series, linear regression, and ARIMA. Moreover, this provides the first research results about dengue prediction in Vietnam.

Keywords: Epidemics · Mosquito characteristics · Locality model · Transmission model · Artificial neural network

1 Introduction

Dengue [9], a mosquito-borne disease, is a serious epidemic in tropical areas, causing tremendous damages to people. This disease is still a major concern with which governments and scientists are trying to deal. Forecasting the magnitude of dengue will help the authorities to prevent and suppress the disease. Furthermore, we could save a large amount of resources for these actions thanks to accurate forecasting.

While there are works on long term prediction, i.e. more than one month ahead, which could give us a picture of a disease development in next four months or a half year [15], in this paper we focus on short term periods ranging from one week to one month [7]. In fact, during an epidemic period, the authority would take intervention to reduce and restrict the disease weekly or monthly. Therefore, a system that monitors

© Springer-Verlag Berlin Heidelberg 2016
N.T. Nguyen et al. (Eds.): ACIIDS 2016, Part I, LNAI 9621, pp. 554–563, 2016.
DOI: 10.1007/978-3-662-49381-6_53

and gives timely prediction every week could make a lot of help in reality. Here, to the best of our knowledge, this work is the first to predict for one week ahead the magnitude of dengue in Vietnam.

Conventional methods such as time series in [21] or linear regression in [12] still get limited accuracy. Meanwhile, an epidemic, especially one of mosquito-related diseases, is highly dependent on environmental factors such as the temperature, mosquito biology, virus, population, human behavior and shelter condition. Our model not only considers dengue situation in the past, but also examines other factors that contribute greatly to the number of dengue cases in the future.

We adapt the two-phase model in [2] for malaria transmission in China to forecast the magnitude of dengue in Vietnam. We note that, beside the disease difference (i.e., malaria vs. dengue), while [2] aimed at finding an underlying transmission network between provinces, our work here is to predict the number of disease cases in each province. Firstly, in the local outbreak phase, we employ Ross-MacDonald model [5] with dengue characteristics, instead of malaria's ones, to estimate the number of potential cases at each isolated province. The inputs for this phase consist of the number of dengue cases in a week, the average temperature that week, the mosquito density per person that month, and the population that year. Then, in the transmission phase, we construct a non-linear neural network receiving as inputs the outcomes from the first phase to forecast the final number of disease cases in each province.

The result of our proposed method is evaluated in comparison with the method in [2] modified for dengue, the Seasonal Auto Regressive Integrated Moving Average (SARIMA [18]), the Multivariate Linear Regression [1], and the Artificial Neural Network-Time Series (ANN-TS [19]) ones. The Root Mean Square Error (RMSE [4]) is used as the benchmark measure. It shows that our forecasting method, which takes into account environmental, biological and demographic features, has less error than the compared ones.

The rest of this paper is organized as follows. Section 2 reviews related works. Section 3 presents details of the proposed method. Performance evaluation in comparison with other methods is presented in Sect. 4. Finally, Sect. 5 draws some concluding remarks.

2   Related Works

About disease transmission, [2] determined the transmission rate of malaria among towns in Yunnan, China, by considering the impact of some factors, especially the temperature, which contributes to the evolution of malaria. The potential of malaria transmission was computed by a biological-inspired mathematical model with specific malaria parameters. They could be in turn combined together to infer a possible network of moving people that could reasonably find the hidden pattern of malaria transmission. Although focusing on simulation of a disease transmission network, the model could be adapted to predict the magnitude of a disease using biological and demographic factors.

Meanwhile, [6] showed that the percentage of current infected mosquitos in the population has a strong impact to the development of a malaria epidemic latter on.

The work also considered the vectorial capacity, which is computed by the current status of the mosquitos, population, temperature and some biological factors of both viruses and mosquitos, as a strong clue for determining how bad the epidemic could be in the next period. Additionally, it also introduced the entomological inoculation rate, which refers to the probability a person could be infected. By estimating that number, one can predict the potential of an epidemic in the following days.

In the recent work about monthly dengue prediction in Northeastern Thailand [21], the authors developed a temporal model using time series analysis. The used data set comprises historical dengue cases through 30 years.

Temperature is one of the most effective factors in a prediction model for dengue magnitude. In [14], the authors showed that a moderate fluctuation of temperature in daylight (diurnal temperature range - DTR) could result in a higher probability of mosquito survival, hence increasing the chance of infection to the community; otherwise a large temperature change could reduce the impact of the virus-carrying mosquito Aedes Aegypti, which causes dengue. Similarly, [13] proved a strong correlation between the temperature and the effect of mosquitos in dengue. It revealed that the best condition for mosquitos to grow was approximately 29 °C by the mean temperature, and the lower DTR, the better the condition was for mosquitos to spread the disease. That range of temperature is usual in a tropical country like Vietnam.

3   Proposed Method

We adapt the malaria transmission model in [2] for dengue transmission, which is divided into two phases, namely local outbreak and province transmission. However, for the locality model, we have to use the characteristics of the mosquito causing dengue instead. For the transmission model, we have to construct a new neural network for predicting the number of disease cases in each province in the following week, because the neural network in [2] was not supposed to do that but just to learn the transmission rate. Figure 1 shows the details of our proposed model.

At first, it considers each province independently. For each one of n ($n > 1$) provinces, the locality model (the first left rectangles) is built to estimate the potential

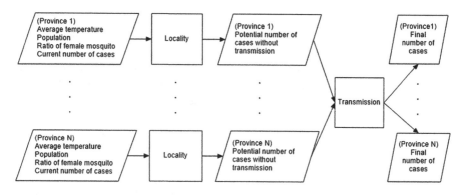

Fig. 1. The proposed two-phase model for prediction of dengue magnitude.

number of dengue cases from 4 inputs (the left parallelograms) in the current week. Since our research is about dengue that is caused by the mosquito type with different characteristics from malaria's one, we combine the locality model of malaria in [2] and the dengue's characteristics discovered in [13]. The output of this step (the middle parallelograms) is the potential number of disease cases.

In the following step, it considers the disease transmission among the provinces. We build a transmission model (the right rectangle) that receives as its inputs the outputs of all the locality models, and predicts the final number of dengue cases (the right parallelograms) in each province in the following week.

3.1   The Locality Model

The proposed locality model receives 4 inputs, namely, the average temperature, the population, the percentage of female mosquitos, and the number of dengue cases in the current week. It has two parts: (1) computing the infectious ability of dengue mosquitos, i.e. the vectorial capacity, from the temperature and the percentage of female mosquitos based on the equations given in [13] (Eqs. 4–9 below); and (2) computing the potential number of infected cases based on the obtained vectorial capacity using the equations given in [2] (Eqs. 1–3 below).

The potential number of infected cases at place i and time $t + 1$ is calculated as follows:

$$\delta_i(t+1) = \beta_i.P_i(t).h_i(t).EIR_i(t) \tag{1}$$

where $\delta_i(t+1)$ is the number of infected cases at place i and time $t + 1$, β_i is the intervention rate at place i, $P_i(t)$ is the place's population, $h_i(t)$ is the probability of an infected mosquito transmitting the virus to an uninfected person, and EIR (Entomological Incubation Rate) represents the number of infected bites each person receives per day at place i and time t.

The value of EIR at place i and time t is computed by:

$$EIR_i(t) = \frac{b_i(t).V_i(t).x_i(t)}{1 + b_i(t).a_i(t)x_i(t)/\mu_i(t)} \tag{2}$$

where $V_i(t)$ is the mosquito ability making infection at place i and time t, $b_i(t)$ is the probability of an infected person transmitting the virus to an uninfected mosquito, $a_i(t)$ is the biting rate, $\mu_i(t)$ is the mosquito mortal rate per day, and $x_i(t)$ is the percentage of infected people at place i and time t. With $y_i(t)$ being the number of infected cases, one has:

$$x_i(t) = \frac{y_i(t)}{P_i(t)} \tag{3}$$

For malaria, [2] adapted Ross-MacDonald model [5] with the specific parameters of the mosquito type causing malaria to compute the infectious ability $V_i(t)$ and other parameters in the above equations. For dengue, which is caused by another mosquito type, Ross-Macdonald model in [13] provides the following equation to compute the dengue mosquito infectious ability:

$$V_i(t) = \frac{m_i(t).a_i(t)^2.e^{-\mu_i(t).n_i(t)}}{\mu_i(t)} \tag{4}$$

where $n_i(t)$ represents the virus incubation and $m_i(t)$ is the percentage of female mosquitos.

The work [13] also provides the formulas to compute the other parameters for dengue as the functions of temperature $T_i(t)$ at place i and time t, except for β_i that has to be learned in the transmission phase as follows:

– The biting rate or the number of bites per day:

$$a_i(t) = 0.0043T_i(t) + 0.0943(21 \le T_i(t) \le 32) \tag{5}$$

– The probability of an infected mosquito transmitting the virus to an uninfected person:

$$h_i(t) = 0.001044T_i(t)(T_i(t) - 12.286)\sqrt{32.461 - T_i(t)} \tag{6}$$
$$(12.286 \le T_i(t) \le 32.461)$$

– The probability of an infected person transmitting the virus to an uninfected mosquito:

$$b_i(t) = \begin{cases} 0.0729T_i(t) - 0.9037(12.4 \le T_i(t) \le 26.1) \\ 1(26.1 < T_i(t) < 32.5) \end{cases} \tag{7}$$

– Incubation of the dengue virus:

$$n_i(t) = 4 + e^{5.15 - 0.123T_i(t)}(12 \le T_i(t) \le 36) \tag{8}$$

– The mosquito mortal rate:

$$\mu_i(t) = 0.8692 - 0.1590T_i(t) + 0.01116T_i(t)^2 - 3.408 * 10^{-4}T_i(t)^3 + $$
$$3.809 * 10^{-6}T_i(t)^4 \ (10.54 \le T_i(t) \le 33.41) \tag{9}$$

3.2   The Transmission Model

The proposed model is to simulate the spread of dengue viruses among nearby provinces in the cycle of one week. In reality, dengue disease spreads sequentially through adjacent provinces as far as possible. The final result of the model is the predicted number of dengue cases in each province after the spreading. On this basis, we build an artificial neural network to simulate this spreading. Due to different objectives, our network has significant differences from the one in [2].

Firstly, it is about the network structure as shown in Fig. 2. Every node of the network (a circle) represents a province. Each layer has 6 nodes corresponding to the number of provinces in the experimented data for Southern Vietnam. The first layer (the input layer) receives the potential numbers of infected cases factor in the provinces, obtained from the locality model. The last layer (the output layer) are the predicted numbers of infected cases in the provinces after the disease spreading in the cycle of one week.

The network has two hidden layers that are fully connected as in a typical neural network. Each hidden layer also has 6 nodes representing the 6 considered provinces. The weight matrix W represents the transmission rates between every pair of provinces. Each connection from an input node to the respective node in the first hidden layer has a coefficient β that represents the intervention factor of the authority. Such a coefficient takes into account the fact that the authority may have some intervention to reduce the local outbreak in each province when the disease happens. In [2], only the nodes that represent adjacent towns have connections to each other. Our argument here is that humans may travel and carry the disease from one place to a non-adjacent one.

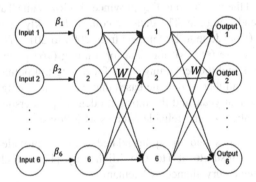

Fig. 2. Structure of the proposed neural network

Secondly, in [2] the authors used the linear activation function for each node, because their network was to learn the transmission rates between every pair of towns based on the movement of infected people, which was modeled as a linear problem. Meanwhile, since our network is to predict the number of infected cases in each province after the spreading cycle, we use the sigmoid function instead:

$$g(x) = \frac{1}{1 + e^{-x}} \qquad (10)$$

Then, the prediction output vector y is calculated by:

$$y = g((W')^d.diag(\beta).x) \tag{11}$$

where W' is the transpose of matrix W, $diag(\beta)$ is the main diagonal matrix generated from the intervention vector β, and the input vector x.

We use the Back Propagation algorithm [16] and the following RMSE as the evaluation function to train the network:

$$\text{RMSE} = \sqrt{\frac{1}{n}\sum_{i=1}^{n}(y_i - y_i')^2} \tag{12}$$

where y_i is a model value, y_i' is the corresponding actual value and n is the number of data points.

4 Evaluation

4.1 The Datasets

We conduct experiments on the data for 6 provinces in Vietnam, namely, Ho Chi Minh City as the center and the five surrounding provinces Ba Ria-Vung Tau, Dong Nai, Binh Duong, Tay Ninh, and Long An. The data are collected weekly in an around 28-week dengue rush period (from June to December) from 2004 to 2013. There are totally 272 week-data points. We use 60 % of the data for training (164 points), 20 % for cross validation (54 points), and the remaining 20 % for testing (54 points). Each data point has four features: the number of dengue cases that week, the average temperature that week, the population that year, and the mosquito density per person that month.

We collect the data from the reliable sources as follows:

- Dengue cases statistics and mosquito-related data are provided by the Dengue Center of Pasteur Institute at Ho Chi Minh City [17], which is the principal authority for dengue surveillance in Vietnam.
- Temperature data are obtained from MODIS (Moderate Resolution Imaging Spectro-radiometer) data crawled from IRI/LDEO Climate Data Library [11], with the radius of 1 km and 8-day circle. We then interpolate them by using the digital boundary of provinces provided by the Center for Developing Information Technology and Geographic Information System [3]. Finally, this dataset is tested against the 2-year official temperature dataset of Vietnam National Centre for Hydro meteorological Forecasting [20].
- The population of each province was obtained from the General Statistics Office of Vietnam [10].

4.2 The Experiments

We compare the prediction error using the RMSE measure of our proposed method with the following ones:

- *Benyun et al.'s method:* the network model in [2] is re-implemented by replacing the malaria locality model therein by the dengue one, and the output values of the last layer nodes in that neural network are used as predicted numbers of disease cases.
- *Artificial Neural Network-Time Series (ANN-TS):* this is a type of neural networks using time series data. Input data are the numbers of infected cases in some weeks prior to the prediction week, while the output is the number of dengue cases in each province in the prediction week. It was employed in [21] for prediction of dengue infection in Thailand.
- *Multivariate Linear Regression:* it produces a vector of prediction values for all provinces at a time by using the previous weeks' numbers of dengue cases.
- *Seasonal ARIMA:* this is a time series forecasting method based on auto-correlations within the data. It aims to catch the pattern in a period of time, which is a week in our case. The authors in [8] used this method to predict dengue infection in Brazil.

We use Weka to implement the ANN-TS, Multivariate Linear Regression, and Seasonal ARIMA models. In order to increase the training speed, we scale numbers of disease cases into values in [0, 1] by the factor of 1,000.

We set four criteria for the comparison experiments. Firstly, all experiments use the same dataset for the 6 provinces in Vietnam as presented above. Secondly, the training dataset is shuffled for each running time. Thirdly, the error of each method in a running time is the average one of those of the provinces. Fourthly, the final result is the average error of 10,000 running times. Table 1 shows the results of the compared methods. The improvement percentage of the proposed method over a compared one is computed by:

$$\text{Improvement} = \frac{\text{RMSE compared model} - \text{RMSE proposed model}}{\text{RMSE compared model}} \quad (13)$$

The results show that our proposed method has less error than the others, where Seasonal ARIMA has the worst prediction. Since our scale factor is 1,000, the RMSE of 0.02 of our proposed method roughly means the difference of 20 disease cases between the predicted number and the actual number. In addition, the mosquito-based models like ours and Benyun et al.'s one have smaller errors in comparison with the traditional models that rely only on historical numbers of disease cases.

Table 1. Comparison of the proposed method with others

Compared methods	RMSE	Improvement
Our proposed method	0.02	N/A
Benyun et al.'s method	0.023	13 %
ANN-TS	0.025	20 %
Multivariate Linear Regression	0.027	26 %
Seasonal ARIMA	0.046	57 %

5   Conclusion

This work is the first attempt to predict dengue spreading in Vietnam, in particular on real data for Ho Chi Minh City and its five surrounding provinces. Our proposed method combines machine learning with a biological model for the mosquito type causing dengue, and predicts the number of dengue cases in the following week. The selected features are the historical dengue cases, average temperature, population and mosquito density. The experimental results show that our proposed method outperforms the referred related works.

For the further work, we are extending the prediction period from one week to one month. Even though this period is still a short term, it could give the authority more time to prepare for actions. Also, we are researching more features for the built neural network, in order to further improve the prediction accuracy.

Acknowledgements. This work is funded by Vietnam National University at Ho Chi Minh City (VNU-HCMC) under the grant number B2015-42-02. We would also like to thank the anonymous reviewers for their constructive comments that help to make the final version of this paper.

References

1. Alvin, C.R., William, F.C.: Methods of Multivariate Analysis, 3rd edn. Wiley, New York (2012)
2. Benyun, S., et al.: Inferring Plasmodium vivax transmission networks from tempo-spatial surveillance data. PLOS Neglected Trop. Dis. **8**(2), e2682 (2014)
3. Center for Developing Information Technology and Geographic Information System (DITAGIS). http://www.ditagis.hcmut.edu.vn/
4. Chai, T., Draxler, R.R.: Root mean square error (RMSE) or mean absolute error (MAE)? – Arguments against avoiding RMSE in the literature. Geoscientific Model Dev. **7**, 1247–1250 (2014)
5. David, L.S., et al.: Ross, Macdonald, and a theory for the dynamics and control of mosquito-transmitted pathogens. PLoS Pathog. **8**(4), e1002588 (2012)
6. David, L.S., McKenzie, F.E.: Statics and dynamics of malaria infection in Anopheles Mosquitos. Malaria J. **3**, 13 (2004)
7. Dayama, P., Kameshwaran, S.: Predicting the dengue incidence in Singapore using univariate time series models. In: AMIA Annual Symposium Proceedings, pp. 285–292 (2013)
8. Edson, Z.M., Elisângela, A.S.S: Predicting the number of cases of dengue infection in Ribeirão Preto, São Paulo state, Brazil using a SARIMA model. Cadernos de Saúde Pública Reports in Public Health, Rio de Janeiro, pp. 1809–1818 (2011)
9. Felissa, R.L., Jerry, D.D.: Emerging Infectious Diseases, Trends and Issues, 2nd edn. Springer, New York (2007)
10. General Statistics Office of Vietnam. http://www.gso.gov.vn/
11. IRI/LDEO Climate Data Library. http://iridl.ldeo.columbia.edu/
12. Karim, M.N., et al.: Climatic factors influencing Dengue cases in Dhaka City: a model for Dengue prediction. Indian J. Med. Res. **136**(1), 32–39 (2012)

13. Liu-Helmersson, J., et al.: Vectorial capacity of Aedes Aegypti: effects of temperature and implications for global Dengue epidemic potential. PLoS ONE 9(3), e89783 (2014)
14. Louis, L.: Impact of daily temperature fluctuations on Dengue virus transmission by Aedes Aegypti. Proc. Natl. Acad. Sci. 108(18), 7460–7465 (2011)
15. Michael, C.W., et al.: A computer system for forecasting malaria epidemic risk using remotely sensed environmental data. In: Proceedings of the 2012 International Congress on Environmental Modelling and Software Managing Resources of a Limited Planet, pp. 482–489 (2012)
16. Michael, N.: Neural Networks and Deep Learning. Determination Press (2015)
17. Pasteur Institute at Ho Chi Minh City, Vietnam. http://www.pasteurhcm.gov.vn/
18. Robert, H.S., David, S.S.: Time Series Analysis and Its Applications. Springer, New York (2011)
19. Søren, B., Murat, K.: Time Series Analysis and Forecasting by Example, 1st edn. Wiley, Hoboken (2011)
20. Vietnam National Centre for Hydro meteorological Forecasting (NCHMF). http://www. nchmf.gov.vn/
21. Wongkoon, S., Jaroensutasinee, M., Jaroensutasinee, K.: Development of temporal modeling for prediction of Dengue infection in Northeastern Thailand. Asian Pac. J. Trop. Med. 5(3), 249–252 (2012)

Self-paced Learning for Imbalanced Data

Maciej Zięba[✉], Jakub M. Tomczak, and Jerzy Świątek

Department of Computer Science, Wroclaw University of Technology,
Wybrzeze Wyspianskiego 27, 50-370 Wroclaw, Poland
{maciej.zieba,jakub.tomczak,jerzy.swiatek}@pwr.edu.pl

Abstract. In this paper, we propose a novel training paradigm that combines two learning strategies: cost-sensitive and self-paced learning. This learning approach can be applied to the decision problems where highly imbalanced data is used during training process. The main idea behind the proposed method is to start the learning process by taking large number of minority examples and only the *easiest* majority objects and then gradually turning to more *difficult* cases. We examine the quality of this training paradigm comparing to other learning schemas for neural network model using a set of highly imbalanced benchmark datasets.

Keywords: Self-paced learning · Cost-sensitive learning · Imbalanced data

1 Introduction

The cognitive process of human or animal learning is highly organized and different tasks are scheduled so that less complicated concepts are presented first for further development of more difficult ideas. It has been noticed that providing a sequence of examples with an increasing level of difficulty can speed up animal and human learning process, a procedure known as *shaping* [6]. A natural question is whether the machine learning can also benefit from a similar training strategy. This issue was raised in [2] in the context of learning neural networks. It turned out that the procedure of formulating a curriculum that introduces different concepts at different times has indeed positive effect on learning in terms of increased speed of convergence and, in the case of a non-convex objective, it helps to obtain better quality of the local minima. The general procedure is known as *curriculum learning* (CL) [2].

The disadvantage of the curriculum learning is that it requires sequence of examples given in advance. This was the starting point for further development of automatic selection of *easy* examples to form a curriculum, a method called *self-paced learning* (SPL) [7]. The idea of the SPL is to subsequently solve optimization problems in which the *easy* examples are determined. The learner self-access the difficulty of examples through examining the loss function and picks those for which the loss function does not exceed some given value. Therefore,

© Springer-Verlag Berlin Heidelberg 2016
N.T. Nguyen et al. (Eds.): ACIIDS 2016, Part I, LNAI 9621, pp. 564–573, 2016.
DOI: 10.1007/978-3-662-49381-6_54

it is said that *the learner learns in its own pace*. The efficiency of the SPL was proved in various applications, such as object recognition, motif finding, document analysis, object localization [7]. Very recently it was successfully applied to matrix factorization [10].

The SPL was further developed to include other important information for learning. One idea was to choose easy examples in terms of the loss function but also *diverse* examples that are dissimilar from what has already been learned [4]. The SPL modified in this manner was proven to increase quality of learning in video analysis. A different approach aimed at combining the SPL with the CL by introducing an expert knowledge about *easy* examples [5]. In this method the curriculum was dynamically determined to adjust to the learning pace of the leaner. It has been argued that this corresponds to "instructor-student-collaborative" learning mode, as opposed to "instructor-driven" in the CL or "student-driven" in the SPL.

In this paper, we consider a different approach in which we modify the SPL to deal with the imbalanced data problem. The issue of imbalanced data is widely encountered in various domains, e.g., medical diagnosis [9] and credit scoring [8], where the number of examples from one class (e.g., healthy patients) is significantly higher than the number of examples from the other class (e.g., ill patients). Therefore, we aim to take advantage of the SPL to increase the quality of learning and allow learning from imbalanced datasets. For this purpose we re-formulate the learning objective of the SPL by introducing different misclassification costs. We apply the proposed approach to training neural networks and evaluate it on different highly imbalanced benchmark datasets.

2   Self-paced Learning for Imbalanced Data

We consider a supervised learning problem with a training dataset: $\mathbb{D}_N = \{(\mathbf{x}_n, y_n)\}_{n=1}^N$, where $\mathbf{x}_n \in \mathbb{R}^D$ and $y_n \in \{0,1\}$. Let $L(y_n, f(\mathbf{x}_n, \mathbf{w}))$ denote a loss function that represents the cost between the true label y_n and the label for $\mathbf{x}_n$ estimated by a model (a decision function) $f(\mathbf{x}_n, \mathbf{w})$ with parameters $\mathbf{w}$. The problem of training a model can be described as an optimization problem of minimizing a sum of loss functions over the examples in $\mathbb{D}_N$:

$$\underset{\mathbf{w}}{\text{minimize}} \; E(\mathbf{w}) = \sum_{n=1}^N L(y_n, f(\mathbf{x}_n, \mathbf{w})). \tag{1}$$

The structure of the loss function depends on the considered problem. Usually it is a squared error for classification and a cross-entropy for classification. The objective function is usually optimized with a gradient-based approach.

2.1   Cost-sensitive Learning

In the case of imbalanced data phenomenon the problem formulation in (1) causes a high risk of training a decision model biased toward the majority class.

Practically it means that the model tends to classify most of minority examples to the majority class. To overcome this issue the learning objective can be modified by introducing different misclassification costs:

$$E(\mathbf{w}; C_+, C_-) = C_+ \sum_{n \in \mathcal{N}_+} L(y_n, f(\mathbf{x}_n, \mathbf{w})) + C_- \sum_{n \in \mathcal{N}_-} L(y_n, f(\mathbf{x}_n, \mathbf{w})), \quad (2)$$

where C_+ and C_- are the costs of misclassification for minority (positive) and majority (negative) class $(C_+, C_- \geq 0)$, respectively.[1] Sets $\mathcal{N}_+$ and $\mathcal{N}_+$ are defined as the sets of positive and negative indexes of examples in training data, $\mathcal{N}_+ = \{n \in \{1, \ldots, N\} | y_n = 1\}$, $\mathcal{N}_- = \{n \in \{1, \ldots, N\} | y_n = 0\}$. It can be observed that in the context of solving the optimization problem of minimizing the criterion $E(\mathbf{w}; C_+, C_-)$ it is sufficient to operate on $\gamma = \frac{C_+}{C_-}$ ratio instead of considering C_+ and C_- separately. The cost-sensitive approach is applicable for imbalanced data if the mentioned ratio is much grater then one, $\gamma \gg 1$. Often, it is a good practise to set the ratio to $\gamma = \frac{N_-}{N_+}$, where N_+ and N_- are the total numbers of positive and negative examples in training data.

2.2 Self-paced Learning

The idea of self-paced learning is to take *easy* examples first, train the model and gradually increase the number of examples by taking more difficult cases. Contrary to curriculum learning the measure of *easiness* of an example is not given arbitrary and the model is used as an oracle to self-access which of the examples should be taken in current training iteration. Formally, the process of self-paced training can be described as the subsequent solution of the following optimization problem:

$$\underset{\mathbf{w}, \mathbf{v} \in \{0,1\}^N}{\text{minimize}} \; E(\mathbf{w}, \mathbf{v}; \lambda) = \sum_{n=1}^{N} v_n L(y_n, f(\mathbf{x}_n, \mathbf{w})) - \lambda \sum_{n=1}^{N} v_n, \quad (3)$$

where $\mathbf{v}$ is a vector of binary latent variables denoting whether nth datum should be considered as *easy* ($v_n = 1$) or not ($v_n = 0$) and λ is a parameter for controlling the learning pace. The learning objective given in the Eq. (1) can be seen as a sum of a modified learning objective in (1) that includes the latent variables and a regularizer of the form $r(\mathbf{v}) = ||v||_1 = \sum_{n=1}^{N} v_n$.[2]

Typically the optimization problem given by (3) is solved using Alternative Convex Search (ACS) [3]. It is an iterative method for biconvex optimization, in which the variables are divided into two disjoint blocks. In each iteration, a block of variables are optimized while keeping the other block fixed. Assuming fixed values of $\mathbf{w}$ the optimal values $\mathbf{v}^*$ can be determined as follows:

$$v_n^* = \begin{cases} 1, & \text{if } L(y_n, f(\mathbf{x}_n, \mathbf{w})) < \lambda, \\ 0, & \text{otherwise.} \end{cases} \quad (4)$$

---

[1] We consider two-class imbalanced data problem in which the minority class is assumed to be positive and the majority class is associated with the negative class.
[2] Absolute value can be omitted since $\mathbf{v} \in \{0, 1\}^n$.

The application of ACS procedure stays behind self-paced learning intuition. When $\mathbf{v}$ is optimized for fixed $\mathbf{w}$ the learner selects *easy* examples, i.e., if the cost function is lower than the threshold λ for considered example, it is assumed to be *easy*. Otherwise, the considered sample is *difficult* and the learner omits it during current stage of learning. For assigned vector of values $\mathbf{v}^*$ the optimization process is performed by minimizing (3) with respect to $\mathbf{w}$. This optimization step is equivalent to minimizing objective given by Eq. (1) considering only the examples with non-zero entries in $\mathbf{v}^*$.

2.3 Cost-sensitive Self-paced Learning

In this paper we aim at taking advantage of the SPL and the cost-sensitive learning to handle imbalanced data and obtain high quality model at the same time. It can be observed that the SPL alone fails if there are significantly more *easy* examples in one class than in the other. Moreover, if the imbalanced problem is observed, there is a high risk that no minority examples will be selected during one step of the SPL. The problem of the cost-sensitive self-paced learning can be formulated as a problem of minimizing the following objective function:

$$E(\mathbf{w}, \mathbf{v}; C_+, C_-) + r(\mathbf{v}; C_+, C_-) = \left(C_+ \sum_{n \in \mathcal{N}_+} v_n L(y_n, f(\mathbf{x}_n, \mathbf{w})) \right.$$

$$\left. + C_- \sum_{n \in \mathcal{N}_-} v_n L(y_n, f(\mathbf{x}_n, \mathbf{w})) \right) \qquad (5)$$

$$- \left(C_+^2 \sum_{n \in \mathcal{N}_+} v_n + C_-^2 \sum_{n \in \mathcal{N}_-} v_n \right)$$

where the first part of the sum is the cost-sensitive sum of loss functions and the second one is the regularization term. The value of C_+ is supposed to be higher than C_- so that examples from minority class $(y_n = 1)$ are more likely to be selected $(v_n = 1)$ than the majority examples $(y_n = 0)$. Additionally, the misclassification costs in the regularizer must be squared, otherwise they would have no effect during the stage of determining latent variables values. We replace λ by two parameters, C_+ and C_-, so we have two degrees of freedom, where one controls the pace of learning and the second takes care of ratio between C_+ and C_-. Assuming the constant values of $\mathbf{w}$ we obtain the following optimal solution for the latent variables $\mathbf{v}^*$:

$$v_n^* = \begin{cases} 1, & \text{if } L(y_n, f(\mathbf{x}_n, \mathbf{w})) < y_n C_+ + (1 - y_n)C_-, \\ 0, & \text{otherwise.} \end{cases} \qquad (6)$$

It can be observed that minority examples are much more likely to be selected than the majority cases (naturally, we assume, that $C_+ \gg C_-$), because of the higher threshold value of acceptance that is equal C_+ if $y_n = 1$ and C_- otherwise.

Algorithm 1. Training model using cost-sensitive self-paced learning.

Input : $\mathbb{D}_N$: training set, $\mathbb{V}_N$: validation set, $L(\cdot, \cdot)$: loss function,
$f(\mathbf{x}, \mathbf{w})$: decision model, C_+, C_-: misclassification costs, μ: step size,
$Q(\cdot, \cdot)$: quality criterion.

Output: $\mathbf{w}_{best}$: parameters for $f(\mathbf{x}, \mathbf{w})$ decision model.

1 Initialize: $\mathbf{v}^*$;
2 Initialize: $\mathbf{w}_{best} = \mathbf{0}$;
3 Initialize: $Q_{best} = 0$;
4 **while** *stopping criterion not reached* **do**
5 $\quad$ $\mathbf{w}^* = \arg\min_{\mathbf{w}} E(\mathbf{w}, \mathbf{v}^*; C_+, C_-) + r(\mathbf{v}; C_+, C_-)$ (see eq. (5));
6 $\quad$ **if** $Q_{best} < Q(\mathbb{V}_N, f(\mathbf{x}, \mathbf{w}^*))$ **then**
7 $\quad\quad$ $Q_{best} = Q(\mathbb{V}_N, f(\mathbf{x}, \mathbf{w}^*))$;
8 $\quad\quad$ $\mathbf{w}_{best} = \mathbf{w}^*$;
9 $\quad$ **end**
10 $\quad$ $\mathbf{v}^* = \arg\min_{\mathbf{v}} E(\mathbf{w}^*, \mathbf{v}; C_+, C_-) + r(\mathbf{v}; C_+, C_-)$ (see eq. (5));
11 $\quad$ **if** *updating criterion reached* **then**
12 $\quad\quad$ $C_+ = \frac{C_+}{C_-}(C_- + \mu)$;
13 $\quad\quad$ $C_- = C_- + \mu$;
14 $\quad$ **end**
15 **end**
16 **return** $\mathbf{w}_{best}$;

2.4   Algorithm for Cost-sensitive Self-paced Learning

The general procedure for the cost-sensitive self-paced learning is given in the Algorithm 1. In the first iteration the latent variables are initialized to start the learning procedure. The details of the latent variables values initialization is discussed in the next paragraph. Next, we initialize the vector of weights, $\mathbf{w}_{best}$. In step 3 we define the auxiliary variable, Q_{best}, that stores the highest value of the additional quality criterion that monitors progress of training, $Q(\cdot, \cdot)$ (see further paragraph for details). The quality criterion verifies training process using validation data $\mathbb{V}_N$ in order to avoid overfitting and biasing toward majority class.[3] In the step 5 the optimization process is performed on objective function given by the Eq. (5) with respect to parameters $\mathbf{w}$ of the model $f(\mathbf{x}, \mathbf{w})$. Practically, it means that $f(\mathbf{x}, \mathbf{w})$ is trained respecting different misclassification costs (C_+ and C_-) and taking into account only examples indicated by the non-zero entries of the vector $\mathbf{v}^*$. Next, the quality of the model is examined on validation data. If the quality is at the highest level comparing to previously examined model, the weights are stored to $\mathbf{w}_{best}$ and the auxiliary variable Q_{best} is updated. In step 10 the new set of examples for training the model is obtained by finding optimal vector $\mathbf{v}^*$ using (6) with fixed $\mathbf{w}^*$. The process of alternating optimization can be performed slowly, and may be stopped omitting large number of examples. Therefore, it is a good practice to increase the values of C_- and C_+ to collect more examples and speed up optimization process (steps 12 and 13). We propose

---

[3] We can always utilize about 10–20% data for validation.

to increase the more restrictive C_- by adding the step size μ and to modify C_+ by applying the equation presented in step 12 to keep the same ratio between C_+ and C_-.

Initialization of Latent Variables. The process of initializing the vector $\mathbf{v}^*$ plays crucial role in the SPL. This could be initialized randomly, however, wrong selection of values may cause the model totally untrainable using the SPL. To overcome this issue we propose to use an oracle model that can be a weaker learner than $f(\mathbf{x}, \mathbf{w})$ trained on entire data or the model $f(\mathbf{x}, \mathbf{w})$ that is trained on *easy* examples for randomly initialized weights $\mathbf{w}$. Both models applied to criterion (5) should result in selecting *easy* examples.

Stopping Criterion. The procedure should be stopped if the optimization criterion is optimal. Practically, the procedure can be stopped if the desired number of iteration is achieved, the monitored criterion on validation data did not increase for given number of iterations (early stopping) the change of optimized criterion is below given threshold value, or the convergence of $\mathbf{v}^*$ or $\mathbf{w}^*$ is very slow.

Quality Criterion. The quality criterion should be consistent with the optimized objective function. Formally, instead of finding parameters for which the quality criterion in (2) takes the highest value it is more recommended to take weights for minimal value of objective function in (2) for validation set. Practically, it is acceptable to consider other criterion like area under ROC curve (AUC), especially for the problem of imbalanced data.

Updating Criterion. The process of updating the values of C_+ and C_- prevents from stacking with insufficient number of examples in training process. Moreover, we start with low values of C_+ and C_- and increase them gradually monitoring the quality criterion. The misclassification costs can be updated under given number of iterations or if the change of $\mathbf{v}^*$ is not observed.

Setting Misclassification Costs. For two-class problems the cost function $L(\cdot, \cdot)$ for a single example can be often associated with the probability of belonging to the true class. Therefore, if $C_+ > 1$ all minority examples will be selected in training process and if $C_- > 1$ all majority instances will be chosen. Taking this fact into account the values above 1 for misclassification costs result in selecting all examples from the class. We notice that the values of C_+ and C_- are updated along with saving the constant ratio between them, the good practice for imbalanced data is to take C_- close to 0 and C_+ equals 1. It will result in one-class controlled training process in which the model will be trained on all minority cases and with an increasing number of the majority examples.

3   Experiments

To examine the quality of the proposed solution we used a set of 31 benchmark datasets available in the Keel data repository [1]. All datasets are characterized by the imbalance ratio higher than 9. The detailed description of the datasets used in the experiment is presented in Table 1.

The goal of the experiment is to compare the quality of prediction gained by selected decision model trained with the following approaches: minimization

Table 1. Detailed description of datasets that includes the information about number of attributes ((R)eal,(I)nteger, (N)ominal), number of instances and the ratio between cardinalities of majority and minority examples

ID	Dataset attributes (R/I/N)	Number of instances	Number of ratio	Imbalance
1	abalone-17_vs_7-8-9-10	8 (7/0/1)	2338	39.31
2	abalone-19_vs_10-11-12-13	8 (7/0/1)	1622	49.69
3	abalone-20_vs_8-9-10	8 (7/0/1)	1916	72.69
4	abalone-21_vs_8	8 (7/0/1)	581	40.5
5	abalone-3_vs_11	8 (7/0/1)	502	32.47
6	car-good	6 (0/0/6)	1728	24.04
7	car-vgood	6 (0/0/6)	1728	25.58
8	dermatology-6	34 (0/34/0)	358	16.9
9	flare-F	11 (0/0/11)	1066	23.79
10	kddcup-buffer_overflow_vs_back	41 (26/0/15)	2233	73.43
11	kddcup-guess_passwd_vs_satan	41 (26/0/15)	1642	29.98
12	kddcup-land_vs_portsweep	41 (26/0/15)	1061	49.52
13	kddcup-land_vs_satan	41 (26/0/15)	1610	75.67
14	kddcup-rootkit-imap_vs_back	41 (26/0/15)	2225	100.14
15	kr-vs-k-one_vs_fifteen	6 (0/0/6)	2244	27.77
16	kr-vs-k-three_vs_eleven	6 (0/0/6)	2935	35.23
17	kr-vs-k-zero-one_vs_draw	6 (0/0/6)	2901	26.63
18	kr-vs-k-zero_vs_eight	6 (0/0/6)	1460	53.07
19	kr-vs-k-zero_vs_fifteen	6 (0/0/6)	2193	80.22
20	poker-8-9_vs_5	10 (0/10/0)	2075	82
21	poker-8-9_vs_6	10 (0/10/0)	1485	58.4
22	poker-8_vs_6	10 (0/10/0)	1477	85.88
23	poker-9_vs_7	10 (0/10/0)	244	29.5
24	shuttle-2_vs_5	9 (0/9/0)	3316	66.67
25	shuttle-6_vs_2-3	9 (0/9/0)	230	22
26	winequality-red-3_vs_5	11 (11/0/0)	691	68.1
27	winequality-red-4	11 (11/0/0)	1599	29.17
28	winequality-red-8_vs_6	11 (11/0/0)	656	35.44
29	winequality-red-8_vs_6-7	11 (11/0/0)	855	46.5
30	winequality-white-3-9_vs_5	11 (11/0/0)	1482	58.28
31	winequality-white-3_vs_7	11 (11/0/0)	900	44

of the objective function given by the Eq. (1), cost-sensitive learning approach that trains the decision model by minimizing the objective given by the Eq. (2), self-paced learning approach that uses the objective given by the Eq. (3), cost-sensitive variant of the self-paced learning described in the Sect. 2.3.

The multilayer perceptron with one hidden layer (15 hidden units) was used as a decision model. The back-propagation algorithm was used for training the neural network. We set the learning rate equal 0.1, momentum term equal 0.01 and the number of epochs equal 220. For each of the neural networks the loss function for a single example was a squared difference between the true label and the probability of the label given by the model. Typically, the cross entropy is more suitable for the classification problem, however, the squared error could be better in terms of controlling the pace of learning because for a single observation it takes values between 0 and 1.

In the self-paced learning mode the parameters $\mathbf{v}$ are initialized using another neural network trained on the entire dataset (assuming $\mathbf{v} = 1$). As a stopping criterion we use given number of iterations equal 110. In each of the iterations we use number of epochs for training neural network assuming given $\mathbf{v}$ equal 2. The number of epochs for each of the self-paced iterations is relatively small, because only couple of examples are delivered in each iteration and the model is updated fast.

As a quality criterion for monitoring the training process on validation data we used area under ROC curve (AUC).[4] The same criterion was used to evaluate the prediction abilities of the considered models. For cost-sensitive approaches we set the following values of misclassification costs: $C_+ = 1$, $C_- = 0.01$. For non cost-sensitive self-paced learning we take $\lambda = 0.01$. We applied 5-fold cross validation with 10 repetitions for each of the datasets considered in the experiment.

In the experiment we took under consideration four methods of training the artificial neural network: typical back-propagation algorithm (**NN**), cost-sensitive back-propagation algorithm (**CSNN**), self-paced back-propagation algorithm (**SPNN**) and cost aware self-paced back-propagation (**CSSPNN**). The detailed results considering AUC as the evaluation criterion are presented in Table 2. In the Table 2 we present also the TPR values for each of the considered datasets. It can be observed that the self-paced learning fails completely if the imbalanced data is used for training. It is caused by the fact that in the case of imbalanced data all minority or majority examples are usually *hard to learn* if we take $\lambda = 0.01$. The neural network trained with ordinary back-propagation procedure also failed

---

[4] AUC is defined as the arithmetic mean of *True Positive Rate* (*TPR*, called *Sensitivity*)and the *True Negative Rate* (*TNR*, called *Specificity*), $AUC = \frac{TPR+TNR}{2}$. TP,TN,FP,FN are the elements of the confusion matrix, $TPR = \frac{TP}{TP+FN}$ and $TNR = \frac{TN}{TN+FP}$. We can represent the AUC value in such form if we consider classes, not probabilities while testing. In such case the ROC curve is represented by one point located in position (TPR,FPR). The area under ROC curve can be calculated using the procedure $AUC = \frac{1+TPR-FPR}{2}$. Making use of $TNR = 1 - FPR$ we have $AUC = \frac{1}{2}(TPR + TNR)$. In our opinion this method of calculating AUC is better for imbalanced data problems, because it evaluates true predictions instead of the ordering of data that is used for evaluation.

Table 2. AUC results for benchmark datasets considering various methods of training MLP.

Dataset	NN		SPNN		CSNN		CSSPNN	
ID	TPrate	AUC	TPrate	AUC	TPrate	AUC	TPrate	AUC
1	0.0702	0.5349	0.0000	0.5000	0.9825	0.7293	0.9474	0.6997
2	0.0000	0.5000	1.0000	0.5000	1.0000	0.5098	0.9355	0.5159
3	0.0000	0.5000	0.0000	0.5000	0.8800	0.8013	0.8400	0.8078
4	0.1538	0.5769	1.0000	0.5000	0.8462	0.6519	0.7692	0.7327
5	1.0000	1.0000	1.0000	0.5000	1.0000	0.9774	1.0000	1.0000
6	0.8824	0.9397	0.0000	0.5000	1.0000	0.9674	1.0000	0.9759
7	1.0000	0.9997	0.0000	0.5000	1.0000	0.9907	1.0000	0.9940
8	1.0000	1.0000	0.0000	0.5000	1.0000	1.0000	1.0000	1.0000
9	0.0238	0.5104	0.0000	0.5000	0.9524	0.8133	0.8810	0.8260
10	1.0000	1.0000	0.0000	0.5000	1.0000	1.0000	1.0000	1.0000
11	1.0000	0.9997	1.0000	0.5000	1.0000	0.9997	1.0000	0.9997
12	1.0000	1.0000	0.0000	0.5000	1.0000	0.9995	1.0000	0.9995
13	1.0000	1.0000	1.0000	0.5000	1.0000	1.0000	1.0000	1.0000
14	0.9524	0.9762	0.0000	0.5000	0.9545	0.9773	1.0000	1.0000
15	1.0000	1.0000	1.0000	0.5000	1.0000	1.0000	1.0000	1.0000
16	0.9500	0.9746	0.0000	0.5000	1.0000	0.9958	1.0000	0.9979
17	0.8365	0.9168	1.0000	0.5000	1.0000	0.9773	0.9905	0.9770
18	0.7308	0.8654	1.0000	0.5000	0.9231	0.9203	0.8462	0.9147
19	1.0000	1.0000	0.0000	0.5000	1.0000	1.0000	1.0000	1.0000
20	0.0000	0.5000	0.0000	0.5000	0.9167	0.5054	0.7083	0.5606
21	0.0000	0.5000	0.0000	0.5000	1.0000	0.5069	0.7917	0.7296
22	0.0000	0.5000	0.0000	0.5000	0.8750	0.4964	0.8750	0.6416
23	0.0000	0.5000	0.0000	0.5000	1.0000	0.5000	0.7143	0.6550
24	0.8750	0.9375	0.0000	0.5000	1.0000	0.9930	1.0000	0.9940
25	0.3333	0.6667	0.0000	0.5000	1.0000	0.5000	0.8889	0.9376
26	0.0000	0.5000	0.0000	0.5000	1.0000	0.5838	0.6667	0.6672
27	0.0000	0.5000	0.0000	0.5000	0.9615	0.4999	0.8462	0.5706
28	0.0000	0.5000	0.0000	0.5000	0.9412	0.6159	0.8333	0.7127
29	0.0000	0.5000	0.0000	0.5000	1.0000	0.5533	0.8889	0.7820
30	0.0000	0.5000	0.0000	0.5000	0.8750	0.5536	0.5000	0.6418
31	0.0000	0.5000	0.0000	0.5000	0.9474	0.5801	0.8421	0.6793
Avg.	**0.4777**	**0.7387**	**0.2581**	**0.5000**	**0.9695**	**0.7806**	**0.8956**	**0.8391**

comparing to the cost-sensitive ones. The average value of AUC for **CSSPNN** is over 0.05 higher than for **CSNN**. To investigate the significance of difference between results gained by this two approaches we applied signed rank Wilcoxson test. Assuming significance level equal 0.05 we can reject null median hypotheses, because p-value was equal 0.00016. We can conclude that the proposed approach gives significantly better results than the **CSNN**. It can be observed that model **CSNN** is biased toward minority class because it dominates over **CSSPNN** if TPR is taken under consideration.

4   Conclusion and Future Work

In this work we propose the extension of a typical self-paced learning approach that makes use of misclassification costs during training a decision model. The experimental results show that the cost-sensitive variant of the self-paced learning can significantly improve the quality of constructed model if the problem is spoiled by highly imbalanced data. For future work we propose to extend our studies to the deep neural networks and propose other criteria that can be used in the cost-sensitive self-paced learning.

Acknowledgments. The research conducted by the authors has been partially co-financed by the Ministry of Science and Higher Education, Republic of Poland, namely, Maciej Zięba: grant No. B50083/W8/K3, Jakub M. Tomczak: grant No. B50106/W8/K3.

References

1. Alcalá, J., Fernández, A., Luengo, J., Derrac, J., García, S., Sánchez, L., Herrera, F.: Keel data-mining software tool: Data set repository, integration of algorithms and experimental analysis framework. J. Multiple-Valued Log. Soft Comput. **17**(2–3), 255–287 (2010)
2. Bengio, Y., Louradour, J., Collobert, R., Weston, J.: Curriculum learning. In: ICML, pp. 41–48 (2009)
3. Gorski, J., Pfeuffer, F., Klamroth, K.: Biconvex sets and optimization with biconvex functions: a survey and extensions. Math. Methods Oper. Res. **66**(3), 373–407 (2007)
4. Jiang, L., Meng, D., Yu, S.I., Lan, Z., Shan, S., Hauptmann, A.: Self-paced learning with diversity. In: Advances in Neural Information Processing Systems, pp. 2078–2086 (2014)
5. Jiang, L., Meng, D., Zhao, Q., Shan, S., Hauptmann, A.G.: Self-paced curriculum learning. In: Twenty-Ninth AAAI Conference on Artificial Intelligence (2015)
6. Krueger, K.A., Dayan, P.: Flexible shaping: How learning in small steps helps. Cognition **110**(3), 380–394 (2009)
7. Kumar, M.P., Packer, B., Koller, D.: Self-paced learning for latent variable models. In: NIPS, pp. 1189–1197 (2010)
8. Tomczak, J.M., Zięba, M.: Classification restricted boltzmann machine for comprehensible credit scoring model. Expert Syst. Appl. **42**(4), 1789–1796 (2015)
9. Tomczak, J.M., Zięba, M.: Probabilistic combination of classification rules and its application to medical diagnosis. Mach. Learn. **101**(1–3), 105–135 (2015)
10. Zhao, Q., Meng, D., Jiang, L., Xie, Q., Xu, Z., Hauptmann, A.G.: Self-paced learning for matrix factorization. In: Twenty-Ninth AAAI Conference on Artificial Intelligence (2015)

A New Similarity Measure
for Intuitionistic Fuzzy Sets

Hoang Nguyen[(✉)]

Department of Engineering Sciences, Gdynia Maritime University,
Morska 83-87, 81-225 Gdynia, Poland
hoang@am.gdynia.pl

Abstract. Although there exist many similarity measures for intuitionistic fuzzy sets (IFSs), most of them can not satisfy the axioms of similarity measure or provide reasonable results. In this paper, a review of existing similarity measures for IFSs and their drawbacks is carried out. Then a new similarity measure between IFSs on the base of their knowledge measures is proposed. A comprehensive analysis of the performance features of the proposed measure is conducted in a comparative example. Finally, the proposed similarity measure is employed in application to the turbine fault diagnosis. We point out that the new proposed similarity measure overcomes the drawbacks of the existing similarity measures and gives reliable results in real world application.

Keywords: Intuitionistic fuzzy sets · Similarity measure · Knowledge measure · Fuzziness

1 Introduction

The intuitionistic fuzzy set (IFS for short) [1] was viewed as an alternative approach of ordinary fuzzy set to deal with imperfect information in solving various real-world problems. At present there are several applications of IFSs in many different fields, such as image processing [4], pattern recognitions [3, 6, 12, 15], fault diagnosis [14, 18, 23] or medical science [16]. Measuring similarity between IFSs has been also intensively explored for decades in both theory and application aspects. Among the most intensively explored and employed measures for IFSs, similarity measure is an essential tool to compare and determine degree of similarity between IFSs. Chen [5] first proposed some similarity measures between vague sets. Later on, Hong and Kim [7] pointed out by examples some unreasonable cases of Chen's measures and proposed a set of modified measures. Li and Chen [9] proposed some new similarity measures and their application in solving pattern recognition problems. However, Liu [13] showed that Li and Chen's methods have the same drawbacks as Chen's methods and proposed several new similarity measures between IFSs and between elements. Similarity measure is widely used in many applications of IFSs such as pattern recognition, classification and especially multiple attribute group decision making. Dengfeng and Chuntian [6] proposed a axiomatic definition of similarity measures between IFSs based on high membership and low membership functions and its application to pattern recognitions. Hung and Yang [8] adopted the Hausdorff distance and developed several similarity measures

© Springer-Verlag Berlin Heidelberg 2016
N.T. Nguyen et al. (Eds.): ACIIDS 2016, Part I, LNAI 9621, pp. 574–584, 2016.
DOI: 10.1007/978-3-662-49381-6_55

for linguistic evaluations. Ye [23] proposed cosine and weighted cosine similarity measures for IFSs and application to a small medical diagnosis problem.

Although there exist several similarity measures between IFSs, many unreasonable cases made by the such measures can be found as shown in [11, 20, 21]. Li et al. [11] showed that there always are counterintuitive examples in pattern recognition among existing similarity measures and pointed out a reason of this drawback as non-considering hesitancy degree in IFS. Szmidt and Kacprzyk [20] analyzed several possible geometric similarity between the IFSs and concluded that taking into account the symmetry of the complement elements in the description of the IFS element is necessary to attain intuitively reliable results. Tan and Chen [21] comprehensively analyzed most of published researches on similarity measures and proved that all existing similarity measures have counterintuitive cases.

In this paper, we present a new similarity measure between IFSs, based on the measure of amount of knowledge that makes it capable to evaluate differences between IFSs and provides reliable results. The performance evaluation of the proposed measure is shown in an comparative example and application to the fault diagnosis.

2 A Review of the Existing Similarity Measures Between IFSs

In 1986, Atanassov [1] generalized the concept of fuzzy sets by introducing an intuitionistic fuzzy set (IFS) defined as follows:

For any elements x of the finite universe of discourse X, an IFS A is an object having the form:

$$A = \{\langle x, \mu_A(x), \nu_A(x)\rangle | x \in X\} \tag{1}$$

where $\mu_A(x)$ denotes a degree of membership and $\nu_A(x)$ denotes a degree of non-membership of x to A, $\mu_A : X \to [0,1]$ and $\nu_A : X \to [0,1]$ such that

$$0 \le \mu_A(x) + \nu_A(x) \le 1, \forall x \in X. \tag{2}$$

To measure hesitancy degree of an element to an IFS, Atanassov introduced a third function given by:

$$\pi_A(x) = 1 - \mu_A(x) - \nu_A(x), \forall x \in X, 0 \le \pi_A(x) \le 1, \tag{3}$$

which is also called the intuitionistic fuzzy index or the hesitation margin of x to A.

If $\pi_A(x) = 0, \forall x \in X$, then $\mu_A(x) + \nu_A(x) = 1$ and the intuitionistic fuzzy set A is reduced to an ordinary fuzzy set.

The concept of a complement of an IFS A, denoted by A^c is defined as [1]:

$$A^c = \{\langle x, \nu_A(x), \mu_A(x)\rangle | x \in X\}. \tag{4}$$

For any two IFS A and B in $X = \{x_1, x_2, \ldots, x_n\}$ some commonly used relations between them are defined as follows [2]:

(R.1) $A \cup B = \{\langle x, \max\{\mu_A(x), \mu_B(x)\}, \min\{\nu_A(x), \nu_B(x)\}\rangle | x \in X\}$

(R.2) $A \cap B = \{\langle x, \min\{\mu_A(x), \mu_B(x)\}, \max\{\nu_A(x), \nu_B(x)\}\rangle | x \in X\}$
(R.3) $A \subseteq B$ iff $\mu_A(x) \geq \mu_B(x)$ and $\nu_A(x) \leq \nu_B(x)$
(R.4) $A = B$ iff $A \subseteq B$ and $B \subseteq A$
(R.5) $A \preceq B$ called A less fuzzy than B, i.e. for $\forall x \in X$,
   if $\mu_B(x) \leq \nu_B(x)$ then $\mu_A(x) \leq \mu_B(x)$ and $\nu_A(x) \geq \nu_B(x)$;
   if $\mu_B(x) \geq \nu_B(x)$ then $\mu_A(x) \geq \mu_B(x)$ and $\nu_A(x) \leq \nu_A(x)$.

A similarity measure between IFSs A and B is assumed to satisfy the following properties [8, 21]:
   (P.1) $0 \leq S(A, B) \leq 1$;
   (P.2) $S(A, B) = 1$ iff $A = B$;
   (P.3) $S(A, B) = S(B, A)$;
   (P.4) if $A \subseteq B \subseteq C$ then $S(A, B) \geq S(A, C)$ and $S(B, C) \geq S(A, C)$

Some existing similarity measures are recalled as follows:

(a) from Chen [5]:

$$S_C(A, B) = 1 - \frac{\sum_{i=1}^{n} |(\mu_A(x_i) - \nu_A(x_i)) - (\mu_B(x_i) - \nu_B(x_i))|}{2n}, \tag{5}$$

(b) from Hong and Kim [7]:

$$S_H(A, B) = 1 - \frac{\sum_{i=1}^{n} (|\mu_A(x_i) - \mu_B(x_i)| + |\nu_A(x_i) - \nu_B(x_i)|)}{2n}, \tag{6}$$

(c) from Li and Xu [10]:

$$S_L(A, B) = 1 - \frac{\sum_{i=1}^{n} |(\mu_A(x_i) - \nu_A(x_i)) - (\mu_B(x_i) - \nu_B(x_i))|}{4n} \\ + \frac{\sum_{i=1}^{n} (|\mu_A(x_i) - \mu_B(x_i)| + |\nu_A(x_i) - \nu_B(x_i)|)}{4n}, \tag{7}$$

(d) from Dengfeng and Chuntian [6]:

$$S_D(A, B) = 1 - \sqrt[p]{\frac{\sum_{i=1}^{n} |(\mu_A(x_i) + 1 - \nu_A(x_i))/2 - (\mu_B(x_i) + 1 - \nu_B(x_i))/2|^p}{n}}, \tag{8}$$

(e) from Mitchell [15]:

$$S_M(A, B) = \frac{1}{2} \left(1 - \sqrt[p]{\frac{\sum_{i=1}^{n} |\mu_A(x_i) - \mu_B(x_i)|^p}{n}} + 1 - \sqrt[p]{\frac{\sum_{i=1}^{n} |\nu_A(x_i) - \nu_B(x_i)|^p}{n}} \right), \tag{9}$$

(f) from Liang and Shi [11]:

$$S_e^p(A, B) = 1 - \sqrt[p]{\frac{\sum_{i=1}^{n} |(\mu_A(x_i) - \mu_B(x_i))/2| + |(\nu_A(x_i) - \nu_B(x_i))/2|^p}{n}}, \tag{10}$$

(g) from Hung and Yang [8]:

$$S_{HY}^1(A,B) = 1 - \frac{\sum_{i=1}^n max(|\mu_A(x_i) - \mu_B(x_i)|, |v_A(x_i) - v_B(x_i)|)}{n}, \tag{11}$$

(h) from Ye [23]:

$$C_{IFS}(A,B) = \frac{1}{n}\sum_{i=1}^n \frac{\mu_A(x_i)\mu_B(x_i) + v_A(x_i)v_B(x_i)}{\sqrt{(\mu_A(x_i))^2 + (v_B(x_i))^2}\sqrt{(\mu_B(x_i))^2 + (v_B(x_i))^2}}, \tag{12}$$

(i) from Boran and Akay [3]:

$$S_t^p(A,B) = 1 - \left(\sum_{i=1}^n \frac{1}{2n(1+p)}\right.$$

$$\left.\{|t(\mu_A(x_i) - \mu_B(x_i)) - (v_A(x_i) - v_B(x_i))|^p + |t(v_A(x_i) - v_B(x_i)) - (\mu_A(x_i) - \mu_B(x_i))|^p\}\right)^{1/p}, \tag{13}$$

(j) from Song et al. [19]:

$$S_Y(A,B) = \frac{1}{2n}\sum_{i=1}^n \left(\frac{\sqrt{\mu_A(x_i)\mu_B(x_i)} + 2\sqrt{v_A(x_i)v_B(x_i)} + \sqrt{\pi_A(x_i)\pi_B(x_i)} +}{\sqrt{(1 - v_A(x_i))(1 - v_B(x_i))}}\right). \tag{14}$$

In this paper, we find that Ye's similarity measure C_{IFS} [23] has the drawback of the "division by zero" problem. For example, let us consider three singleton IFSs $F = \langle x, 0, 0\rangle, M = \langle x, 1, 0\rangle$ and $A = \langle x, 0.3, 0.3\rangle$ in $X = \{x\}$, where $\mu_F(x) = v_F(x) = 0, \mu_M(x) = 1, v_M(x) = 0$ and $\mu_A(x) = 0.3, v_A(x) = 0.3$. Based on Eq. (12) we derive $C_{IFS}(F, M) = \frac{0}{0}$ and $C_{IFS}(F, A) = \frac{0}{0}$, which are the "division by zero" problem. Although the author assumed the similarity measure of these cases as 0, the remaining problem is, that C_{IFS} [23] between $F = \langle x, 0, 0\rangle$ and different singleton IFSs always gives the same value 0. Moreover, Ye's similarity measure [23] has also unreasonable case when $\mu_A(x) = v_A(x)$ and $\mu_B(x) = v_B(x)$, as based on Eq. (12) $C_{IFS}(A, B) = 1$. The same situations can be found for the Dengfeng and Chuntian's measure S_D [6], which according to Eq. (8) gives the same values equal to 1 when $\mu_A(x) = v_A(x)$ and $\mu_B(x) = \mu_B(x)$. We also find some unreasonable cases of the other existing similarity measures, e.g. Chen's S_C [5], Hong and Kim's S_H [7], Li and Xu's S_L [10], Dengfeng and Chuntian's S_D [6], Mitchell S_M [15], Liang and Shi's S_e^p [11], Hung and Yang's S_{HY}^1 [8], Boran and Akay's S_t^P [3] and Song et al.'s S_Y [19]. A comprehensive analysis of performance features of these measures is performed in comparative example of Sect. 4.

In the next section, inspired by these cases we propose a new similarity measure based on the knowledge measure for IFSs, which satisfies all axiomatic properties of IFSs and overcomes the abovementioned drawbacks.

3   A New Similarity Measure Between IFSs

The main reason of counterintuitive cases of the existing similarity measures is that there was no reliable measure on IFSs, which can be used to compare them. The similarity measures for IFSs based on distance measures between them have some problem with differentiating similarities between a fixed set and the complementary sets, respectively. It seems that, the most suitable and informative measure, which can be used to evaluate how similar the IFSs are, is the knowledge measure [17] of the information contained in IFSs.

Definition 1 [17]. Let A be an IFS in finite universe of discourse $X = \{x_1, x_2, \ldots, x_n\}$. The knowledge measure of A is defined as:

$$K_F(A) = \frac{1}{n\sqrt{2}} \sum_{i=1}^{n} \sqrt{(\mu_A(x_i))^2 + (\nu_A(x_i))^2 + (\mu_A(x_i) + \nu_A(x_i))^2}, \qquad (15)$$

where $0 \leq K_F(A) \leq 1, \forall x_i \in X$

The knowledge measure $K_F(A)$ evaluates quantity of information of an IFS A as its normalized Euclidean distance from the reference level 0 of information.

Theorem 1 [17]. Let IFSs(X) denotes the set of all IFSs in $X = \{x_1, x_2, \ldots, x_n\}$ and a mapping $K_F : IFSs(X) \longrightarrow [0, 1]$. $K_F(A)$ is said the knowledge measure of an IFS A if it satisfies the following axiomatic properties: $\forall x_i \in X$,

(A1.1) $K_F(A) = 1$ iff A is a crisp set;
(A1.2) $K_F(A) = 0$ iff $\pi_A(x_i) = 1$;
(A1.3) $0 \leq K_F(A) \leq 1$;
(A1.4) $K_F(A) = K_F(A^c)$, where A^c is the complement of A;
(A1.5) $A \preceq B$, i.e. A is less fuzzy than B iff $K_F(A) \geq K_F(B)$, for any A, B in $X = \{x_1, x_2, \ldots, x_n\}$.

Then, we propose a new similarity measure for IFSs, based on comparison of their amounts of knowledge, which is defined as follows:

Definition 2. For any two IFSs A and B in finite universe of discourse $X = \{x_1, x_2, \ldots, x_n\}$, a new similarity measure between IFSs A and B is defined as:

$$S_F(A, B) = 1 - |K_F(A) - K_F(B)|, \qquad (16)$$

where $K_F(A)$ and $K_F(B)$ are knowledge measures of A and B, respectively. Thus,

$$S_F(A, B) = 1 - \frac{1}{n\sqrt{2}} \left| \sum_{i=1}^{n} \sqrt{(\mu_A(x_i))^2 + (\nu_A(x_i))^2 + (\mu_A(x_i) + \nu_A(x_i))^2} \right.$$
$$\left. - \sum_{i=1}^{n} \sqrt{(\mu_B(x_i))^2 + (\nu_B(x_i))^2 + (\mu_B(x_i) + \nu_B(x_i))^2} \right|. \qquad (17)$$

The proposed in Definition 2 similarity measure S_F between IFSs A and B, satisfies the following axiomatic properties:

Theorem 2. For all IFSs A, B *and* C *in* X, the measure S_F is called a similarity measure between IFSs if satisfies the following properties:

(A2.1) $0 \leq S_F(A, B) \leq 1$

(A2.2) $S_F(A, B) = S_F(B, A)$

(A2.3) $S_F(A, B) = 1$ iff $A = B$

(A2.4) $A \preccurlyeq B \preccurlyeq C \Rightarrow S_F(A, C) \leq S_F(A, B)$ and $S_F(A, C) \leq S_F(B, C)$.

Proof: (A2.1) According to Definition 1, $K_F(A)$ and $K_F(B)$ are normalized Euclidean distances, i.e. $0 \leq K_F(A) \leq 1$ and $0 \leq K_F(B) \leq 1$, that implies $0 \leq |K_F(A) - K_F(B)| \leq 1$ and (A2.1) holds.

(A2.2) From Eq. (16) we have

$$S_F(A, B) = 1 - |K_F(A) - K_F(B)| = 1 - |K_F(B) - K_F(A)| = S_F(B, A)$$

(A2.3) Having in mind $\mu_A(x) + v_A(x) + \pi_A(x) = 1$ we have:

$$S_F(A, B) = 1 \Leftrightarrow |K_F(A) - K_F(B)| = 0 \Leftrightarrow$$
$$(\mu_A(x_i))^2 + (v_A(x_i))^2 + (1 - \pi_A(x_i))^2 = (\mu_B(x_i))^2 + (v_B(x_i))^2 + (1 - \pi_B(x_i))^2$$
$$\Leftrightarrow A = B \text{ or } A = B^C or B = A^C$$

(A2.4) $A \preccurlyeq B \preccurlyeq C \Rightarrow K_F(A) \geq K_F(B) \geq K_F(C)$ which implies

$$K_F(A) - K_F(C) \geq K_F(A) - K_F(B) \geq 0 \Rightarrow 1 - |K_F(A) - K_F(C)|$$
$$\leq 1 - |K_F(A) - K_F(B)| \Rightarrow S_F(A, C) \leq S_F(A, B)$$
$$\text{and } K_F(A) - K_F(C) \geq K_F(B) - K_F(C) \geq 0 \Rightarrow 1 - |K_F(A) - K_F(C)|$$
$$\leq 1 - |K_F(B) - K_F(C)| \Rightarrow S_F(A, C) \leq S_F(B, C)$$

Hence, the proof is completed.

The similarity measure $S_F(A, B) = 0$, when A and B are completely different (e.g. a crisp set vs. the most intuitionistic fuzzy set) and having in mind inequalities $0 \leq K_F(A) \leq 1$ and $0 \leq K_F(B) \leq 1$, amounts of knowledge of A and B have opposite values between 0 and 1, respectively. It means that we have the cases of a crisp set vs. the most intuitionistic fuzzy set. The similarity measure $S_F(A, B) = 1$, when amounts of knowledge of them are equal, which implies $A = B$. It indicates reliability and efficiency of our measure, which is consistent with property of complementary sets.

4   Comparative Example

In order to verify the proposed similarity measure, we compare its performance evaluation with some most recently used similarity measures, i.e. Chen's S_C [5], Hong and Kim's S_H [7], Li and Xu's S_L [10], Dengfeng and Chuntian's S_D [6], Mitchell S_M [15], Liang and Shi's S_e^p [11], Hung and Yang's S_{HY}^1 [8], Ye's C_{IFS} [23], Boran and Akay's S_t^p [3] and Song et al.'s S_Y [19]. In this example, the eight pairs of different singleton IFSs A and B were employed as test data. Table 1 presents a comprehensive comparison of these similarity measures between A and B.

The pairs of test IFSs are shown in the first two rows of Table 1. Each measure is expected to be capable to differentiate one pair from others. The unreasonable cases, where two (or more) different pairs are determined equally by similarity measure, are highlighted in bold type. From Table 1, we can see that the Chen's similarity measure S_C [5] has some unreasonable cases because does not differentiate the pairs of test IFSs number 1, 4 and 5 ($S_C(A, B) = 1$) and also the pairs number 7 and 8 ($S_C(A, B) = 0.7$). The another measures, i.e. Hong and Kim's S_H [7], Li and Xu's S_L [10], Dengfeng and Chuntian's S_D [6], Mitchell S_M [15], Liang and Shi's S_e^p [11], Hung and Yang's S_{HY}^1 [8] and Ye's C_{IFS} [23] have the same drawbacks for the pairs number 7 and 8.

We can see that only the last two measures do not have any unreasonable cases, i.e. Song et al.'s similarity measure S_Y [19] and our proposed similarity measure S_F. But let us consider pair 3 where $A = \langle x, 1, 0 \rangle$ is a crisp set and $B = \langle x, 0, 0 \rangle$ is the most fuzzy intuitionistic set, i.e. they are most different one from the other. The similarity measure between them should be 0, which is met only by our new proposed measure S_F. Thus, the newly proposed measure S_F is most reliable and accurate in measuring similarity between IFSs in comparison with other similarity measures.

Table 1. A comparison of the proposed similarity measures S_F with the existing similarity measures ($p = 1$ in S_D, S_M, S_e^p, S_t^p and $t = 2$ in S_t^p).

Similarity measures	Number of pairs of the test IFSs							
	1	2	3	4	5	6	7	8
A=	$\langle 0.5, 0.5 \rangle$	$\langle 0.3, 0.4 \rangle$	$\langle 1, 0 \rangle$	$\langle 0.5, 0.5 \rangle$	$\langle 0.4, 0.2 \rangle$	$\langle 0.4, 0.2 \rangle$	$\langle 0, 0.87 \rangle$	$\langle 0.6, 0.8$
B=	$\langle 0.4, 0.4 \rangle$	$\langle 0.4, 0.3 \rangle$	$\langle 0, 0 \rangle$	$\langle 0, 0 \rangle$	$\langle 0.5, 0.3 \rangle$	$\langle 0.5, 0.2 \rangle$	$\langle 0.28, 0.55 \rangle$	$\langle 0.28, 0$
S_C [5]	**1**	0.9	**0.5**	**1**	**1**	0.95	**0.7**	**0.7**
S_H [7]	**0.9**	**0.9**	**0.5**	**0.5**	**0.9**	0.95	**0.7**	**0.7**
S_L [10]	**0.95**	0.9	0.5	0.75	**0.95**	**0.95**	**0.7**	**0.7**
S_D [6]	**1**	0.8	0	**1**	**1**	0.95	**0.4**	**0.4**
S_M [15]	**0.9**	**0.9**	**0.5**	**0.5**	**0.9**	0.95	**0.7**	**0.7**
S_e^p [11]	**0.9**	**0.9**	**0.5**	**0.5**	**0.9**	0.95	**0.7**	**0.7**
S_{HY}^1 [8]	**0.9**	**0.9**	0	0.5	**0.9**	**0.9**	**0.68**	**0.68**
C_{IFS} [23]	**1**	0.96	**N/A**	**N/A**	**0.997**	**0.997**	0.891	0.779
S_t^p [3]	**0.967**	0.9	0.5	0.835	**0.967**	0.95	**0.7**	**0.7**
S_Y [19]	0.945	0.994	0.5	0.354	0.984	0.896	0.887	0.951
S_F	0.827	1	0	0.134	0.829	0.904	0.861	0.960

Note: Bold type denotes unreasonable results, "N/A" denotes the "division by zero" case.

5   An Application of the Proposed Similarity Measure Between IFSs in Turbine Fault Diagnosis

In this section, we apply the proposed similarity measure to the fault diagnosis with intuitionistic fuzzy information as follows. Suppose that there exist m known fault patterns, which are represented by IFSs $P_j = \left\{ \left\langle x_i, \mu_{P_j}(x_i), \nu_{P_j}(x_i) \right\rangle | x_i \in X \right\} (j = 1, 2, \ldots, m)$ in the finite universe of discourse $X = \{x_1, x_2, \ldots, x_n\}$ and that there is a fault-testing sample to be recognized, which is represented by an IFS $A = \{ \langle x_i, \mu_A(x_i), \nu_A(x_i) \rangle | x_i \in X \}$. The whole fault diagnosis process is developed in the following steps:

Step 1. Calculate the similarity measures $S_F(P_j, A)$ between P_j and A by Eq. (17).

Step 2. Select the largest one $S_F(P_{j_0}, A)$ from $S_F(P_j, A)$ $(j = 1, 2, \ldots, m)$: $S_F(P_{j_0}, A) = \max_{1 \le j \le m} \{S_F(P_j, A)\}$. Then we decide that the sample A should belong to the pattern P_{j_0} according to the principle of the maximum of similarity measure between IFSs.

The proposed method is employed to the fault diagnosis of steam turbine generator unit under intuitionistic fuzzy environment. The vibration of steam turbine generator unit suffers the influence of a lot of varying factors, such as mechanical load, vacuum degree, fluctuation of network load, temperature of lubricant oil and defects of mechanical structure as well. Interaction effects of these factors result in the vibration of the generator unit. Ten fault types in rotating machines is established as failure patterns, i.e. P_1- unbalance, P_2- pneumatic force couple, P_3- offset center, P_4- oil-membrane oscillation, P_5- radial impact friction of rotor, P_6- symbiosis looseness, P_7- damage of antithrust bearing, P_8- surge, P_9- looseness of bearing block and P_{10}- non-uniform bearing stiffness. The knowledge of fault types and detection samples (symptoms in the vibration frequencies of turbine) is adopted from Ye [22]. The vibration frequency of turbine (universe of discourse) is divided into nine different frequency ranges, in which the failure patterns are represented by IFSs as shown in Table 2.

Suppose that there are two fault-testing samples A and B expressed in IFSs as shown in Table 3. Our goal in the fault diagnosis analysis is to classify the fault-testing samples into one of the known fault patterns P_j, $(j = 1, 2, \ldots, 10)$, adopting the proposed similarity measure S_F from Eq. (17). The calculation results of similarity measures between the fault-testing samples and the known fault patterns are summarized in rows of Table 4.

According to the principle of maximum similarity measure, we can decide that the fault-testing sample A is most similar to the known fault pattern P_7- damage of antithrust bearing, which is consistent with the results obtained in [18, 22]. In the same manner we derive that the fault-testing sample B is most similar to the known fault pattern P_9- looseness of bearing block, which is also in agreement with the results obtained in [14, 22].

Table 2. The knowledge fault patterns in terms of IFSs [22].

Fault types	Frequency range (f- operating frequency)								
	0.01-0.39f	0.40-0.49f	0.50f	0.51-0.99f	f	2f	3-5f	Odd times of f	High freq > 5f
Unbalance	⟨0,1⟩	⟨0,1⟩	⟨0,1⟩	⟨0,1⟩	⟨0.85,0⟩	⟨0.04, 0.94⟩	⟨0.04, 0.93⟩	⟨0,1⟩	⟨0,1⟩
Pneumatic force couple	⟨0,1⟩	⟨0.28, 0.69⟩	⟨0.09, 0.88⟩	⟨0.55, 0.3⟩	⟨0,1⟩	⟨0,1⟩	⟨0,1⟩	⟨0,1⟩	⟨0.08, 0.83⟩
Offset center	⟨0,1⟩	⟨0,1⟩	⟨0,1⟩	⟨0,1⟩	⟨0.3, 0.42⟩	⟨0.4, 0.58⟩	⟨0.08, 0.83⟩	⟨0,1⟩	⟨0,1⟩
Oil-membrane oscillation	⟨0.09, 0.11⟩	⟨0.78, 0.18⟩	⟨0,1⟩	⟨0.08, 0.89⟩	⟨0,1⟩	⟨0,1⟩	⟨0,1⟩	⟨0,1⟩	⟨0,1⟩
Radial impact friction of rotor	⟨0.09, 0.88⟩	⟨0.09, 0.11⟩	⟨0.08, 0.88⟩	⟨0.09, 0.88⟩	⟨0.18, 0.79⟩	⟨0.08, 0.83⟩	⟨0.08, 0.83⟩	⟨0.08, 0.88⟩	⟨0.08, 0.88⟩
Symbiosis looseness	⟨0,1⟩	⟨0,1⟩	⟨0,1⟩	⟨0,1⟩	⟨0.18, 0.78⟩	⟨0.12, 0.83⟩	⟨0.37 0.55⟩	⟨0,1⟩	⟨0.22, 0.72⟩
Damage of antithrust bearing	⟨0,1⟩	⟨0,1⟩	⟨0.08, 0.88⟩	⟨0.86, 0.07⟩	⟨0,1⟩	⟨0,1⟩	⟨0,1⟩	⟨0,1⟩	⟨0,1⟩
Surge	⟨0,1⟩	⟨0.27, 0.68⟩	⟨0.08, 0.88⟩	⟨0.54, 0.38⟩	⟨0,1⟩	⟨0,1⟩	⟨0,1⟩	⟨0,1⟩	⟨0,1⟩
Looseness of bearing block	⟨0.85, 0.07⟩	⟨0,1⟩	⟨0,1⟩	⟨0,1⟩	⟨0,1⟩	⟨0,1⟩	⟨0,1⟩	⟨0.08, 0.88⟩	⟨0,1⟩
Non-uniform bearing stiffness	⟨0,1⟩	⟨0,1⟩	⟨0,1⟩	⟨0,1⟩	⟨0,1⟩	⟨0.77, 0.17⟩	⟨0.19, 0.77⟩	⟨0,1⟩	⟨0,1⟩

Table 3. The characteristic of the fault-testing samples presented by IFSs [22].

Fault-testing samples	Frequency range (f- operating frequency)								
	0.01- 0.39f	0.40-0.49f	0.50f	0.51- 0.99f	f	2f	3-5f	Odd times of f	High freq > 5f
A	⟨0,1⟩	⟨0,1⟩	⟨0.1, 0.9⟩	⟨0.9, 0.1⟩	⟨0,1⟩	⟨0,1⟩	⟨0,1⟩	⟨0,1⟩	⟨0,1⟩
B	⟨0.39, 0.61⟩	⟨0.07, 0.93⟩	⟨0,1⟩	⟨0.06, 0.94⟩	⟨0,1⟩	⟨0.13, 0.87⟩	⟨0,1⟩	⟨0,1⟩	⟨0.35, 0.65⟩

Table 4. The calculation results of similarity measures between the fault-testing samples and the known fault patterns

Similarity measures										
Fault-testing samples	P1	P2	P3	P4	P5	P6	P7	P8	P9	P10
A	0.963	0.948	0.903	0.973	0.930	0.927	0.990	0.963	0.969	0.962
B	0.941	0.930	0.885	0.959	0.939	0.936	0.945	0.925	0.963	0.944

6 Conclusion

Some similarity measures existing in literature are reviewed and a new similarity measure between IFSs is proposed. The proposed measure is based on a knowledge measure, which takes into account all information carried by IFSs, making it possible to evaluate fuzziness and intuitionism simultaneously. The main contribution of this paper is that the proposed similarity measure, though simple in concept and calculus, overcomes the drawbacks of the existing measures and that it is reliable and consistent with other measures in application to the fault diagnosis problem. The further researches should be focused on applications of the proposed measure to solve the real world problems, which would provide its advantages over the others.

References

1. Atanassov, K.T.: Intuitionistic Fuzzy Sets. Fuzzy Sets Systems **20**(1), 87–96 (1986)
2. Atanassov, K.T.: Intuitionistic Fuzzy Sets. Springer, New York (1999)
3. Boran, F.R., Akay, D.: A biparametric similarity measure on intuitionistic fuzzy sets with applications to pattern recognition. Inf. Sci. **255**, 45–57 (2014)
4. Bustince, H., Mohedano, V., Barrenechea, E., Pagola, M.: An algorithm for calculating the threshold of an image representing uncertainty through A-IFSs. In: IPMU 2006, pp. 2383–2390 (2006)
5. Chen, S.M.: Measures of similarity between vague sets. Fuzzy Sets Syst. **74**(2), 217–223 (1995)
6. Dengfeng, L., Chuntian, C.: New similarity measures of intuitionistic fuzzy sets and application to pattern recognitions. Pattern Recognit. Lett. **23**(1–3), 221–225 (2002)
7. Hong, D.H., Kim, C.: A note on similarity measures between vague sets and between elements. Inf. Sci. **115**(1–4), 83–96 (1999)
8. Hung, W.L., Yang, M.S.: Similarity measures of intuitionistic fuzzy sets based on the Hausdorff distance. Pattern Recognit. Lett. **25**(14), 1603–1611 (2004)
9. Li, D.F., Cheng, C.: New similarity measures of intuitionistic fuzzy sets and application to pattern recognition. Pattern Recognit. Lett. **23**, 221–225 (2002)
10. Li, F., Xu, Z.: Similarity measures between vague sets. J. Softw. **12**(6), 922–927 (2001)
11. Li, Y.H., Olson, D.L., Qin, Z.: Similarity measures between intuitionistic fuzzy (vague) sets: A comparative analysis. Pattern Recognit. Lett. **28**, 278–285 (2007)
12. Liang, Z., Shi, P.: Similarity measures on intuitionistic fuzzy sets. Pattern Recognit. Lett. **24**(15), 2687–2693 (2003)
13. Liu, H.W.: New similarity measures between intuitionistic fuzzy sets and between elements. Math. Comput. Model. **42**, 61–70 (2005)

14. Lu, Z.K., Ye, J.: Cosine similarity measure between vague sets and its application of fault diagnosis. Res. J. Appl. Sci. Eng. Technol. **6**(14), 2625–2629 (2013)
15. Mitchell, H.B.: On the Dengfeng-Chuntian similarity measure and its application to pattern recognitions. Pattern Recognit. Lett. **24**(16), 3101–3104 (2003)
16. Miaoying, T.: A new fuzzy similarity measure based on cotangent function for medical diagnosis. Adv. Model. Optim. **15**(2), 151–156 (2013)
17. Nguyen, H.: A new knowledge-based measure for intuitionistic fuzzy sets and its application in multiple attribute group decision making. Expert Syst. Appl. **42**, 8766–8774 (2015)
18. Shi, L.L., Ye, J.: Study on fault diagnosis of turbine using an improved cosine similarity measure of vague sets. J. Appl. Sci. **13**(10), 1781–1786 (2013)
19. Song, Y., Wang, X., Lei, L., Xue, A.: A new similarity measure between intuitionistic fuzzy sets and its application to pattern recognition. Appl. Intell. **42**, 252–261 (2015)
20. Szmidt, E., Kacprzyk, J.: Geometric similarity measures for the intuitionistic fuzzy sets. In: 8th Conference of the European Society for Fuzzy Logic and Technology (EUSFLAT 2013), pp. 840–847 (2013)
21. Tan, C., Chen, X.: Dynamic similarity measures between intuitionistic fuzzy sets and its application. Int. J. Fuzzy Syst. **16**(4), 511–519 (2014)
22. Ye, J.: Fault diagnosis of turbine based on fuzzy cross entropy of vague sets. Expert Syst. Appl. **36**, 8103–8106 (2009)
23. Ye, J.: Cosine similarity measures for intuitionistic fuzzy sets and their applications. Math. Comput. Model. **53**(1–2), 91–97 (2011)

Multiple Kernel Based Collaborative Fuzzy Clustering Algorithm

Trong Hop Dang[1,2], Long Thanh Ngo[1(✉)], and Wiltold Pedrycz[3]

[1] Department of Information Systems,
Le Quy Don Technical University,
Hanoi, Vietnam
dangtronghop@gmail.com, ngotlong@gmail.com
[2] Hanoi University of Industry, Bac Tu Liem, Hanoi, Vietnam
[3] Department of Electrical and Computer Engineering, University of Alberta,
Edmonton, AB, Canada
wpedrycz@ualberta.ca

Abstract. Cluster is found as one of the best useful tools for data analysis, data mining, and pattern recognition. The FCM algorithm and its variants algorithms has been extensively used in problems of clustering or collaborative clustering. In this paper, we present a novel method involving multiple kernel technique and FCM for collaborative clustering problem. These method endowed with multiple kernel technique which transform implicitly the feature space of input data into a higher dimensional via a non linear map, which increases greatly possibility of linear separability of the patterns when the data structure of input patterns is non-spherical and complex. To evaluate the proposed method, we use the criteria of fuzzy silhouette, a sum of squared error and classification rate to show the performance of the algorithms.

Keywords: Fuzzy clustering · Collaborative clustering · Fuzzy c-means · Multiple kernels

1 Introduction

Clustering is used to detect sound structures or patterns in the data set whose objects positioned within the same cluster exhibit a substantial level of similarity. This unsupervised technique has a long history in machine learning, pattern recognition, data mining, and many algorithms have been exploited in various applications. Clustering algorithms comes in numerous varieties including k-means and various improvements [1,2] and a family of Fuzzy C-Mean (FCM) [3].

Collaborative fuzzy c-means clustering was introduced by Pedrycz [4–6] as a vehicle to determine a structure and reveal similarity among separate data sets. There are two essential characteristics of collaborative data clustering. The first one is that the individual data cannot be shared. Second, we can only exchange findings about the structure, which are of a far higher level of generality than the original data. Through some interaction, the results obtained at one data site can impact clustering realized at other data site [4,5].

© Springer-Verlag Berlin Heidelberg 2016
N.T. Nguyen et al. (Eds.): ACIIDS 2016, Part I, LNAI 9621, pp. 585–594, 2016.
DOI: 10.1007/978-3-662-49381-6_56

In the sequel, Coletta et al. [7] extend Pedrycz method to optimize the parameters including the interaction level for all pairs of data sites and the number of clusters at each place. S. Mitra et al. 's research combine the advantages of fuzzy sets and rough sets for data collaboration clustering [16]. Falcon et al. is concerned with the application of multi-objective particle swarm optimization approaches to the framework of collaborative fuzzy clustering [15]. Tang and Cai focused on Horizontal Collaborative Fuzzy Clustering with spatial attributes data collection. In [12,13], the threshold level of membership function or entropy-based approach is used to identify of data site and regional data is involved in realizing collaboration.

In [10] when data sets described by multiple views, with each view having its own characterization of the data to be clustered take this advantage by applying Collaborative fuzzy c-means clustering so that we combine individual views coming from multiple clustering. Prasad [11] overcome some of the drawbacks of Pedrycz method by introducing preprocessing phase before running collaborative phase. Zhou et al. [9] proposed a novel collaborative clustering algorithm over a distributed P2P network. This algorithm searches the optimized clusters at each data site by collaborating only with prototypes of the neighbouring data site. The clustering solution could be improved by applying partial supervision, which involves a subset of labelled data augmented with their class membership. This knowledge-based hints have to be included into the objective function and reflect a fact that some patterns have been labeled [6–12]. Yu et al. [14] presented a new approach to implementing horizontal collaborative fuzzy clustering with the knowledge provided by the prototypes instead of partition matrixes.

The main challenge comes with the development of complexity of data. This complexity embraces various aspects, including the size of the data, the type of features, temporal aspects, and diversity of data, in general. Recently a number of approaches have been studied to solve these problems including kernel-based clustering [17–23], where data are transformed into some high-dimensional feature space increasing the probability of linear separability of the patterns within the transformed space and therefore simplifies the associated data structure. Multiple kernels or composite kernels [19,22–24] instead of a single fixed kernel gain more flexibility on kernel selections and also reflect the fact that practical learning problems often involve data from multiple heterogeneous or homogeneous sources. The paper presents new methods that apply multiple kernel technique to collaborative clustering with intent to cope with the complex and non-spherical structure of input data.

The paper is organized as follows: Sect. 2 offers a brief introduction to collaborative clustering. Section 3 proposes multiple kernel based collaborative fuzzy clustering. Section 4 show the experimental results. Conclusion and future studies are covered in Sect. 5.

2   Collaborative Fuzzy Clustering

Suppose there are P data sites $D[1], D[2], ..., D[P]$, which comprises of $N[1]$, $N[2]$, ..., $N[P]$ patterns data defined in the same feature space X. For each data site we group all patterns into c clusters. The objective function expressed for each data site when using the standard FCM algorithm comes in the well-known format $\sum_{k=1}^{N[ii]} \sum_{i=1}^{C} (u_{ik})^2 [ii](d_{ik})^2$ with $ii = 1, 2, ..., P$.

To accommodate the collaboration effect in the optimization process, the objective function is extended into the following form

$$Q_{[ii]} = \sum_{k=1}^{N[ii]} \sum_{i=1}^{C} (u_{ik})^2 [ii](d_{ik})^2 + \sum_{jj=1}^{P} \sum_{k=1}^{N[ii]} \sum_{i=1}^{C} \beta[ii|jj](u_{ik}[ii] - \tilde{u}_{ik}[ii|jj])^2 (d_{ik})^2 \tag{1}$$

In the above formula, the first part is the "standard" objective function of the FCM algorithm. The second part reflects the impact of structural clustering of data from other sites.

The collaborative clustering problem is converted to the optimization problems with the following membership constraints: $MinQ[ii]st.U[ii] \in U$ where U is a family of all fuzzy partition matrices, namely

$$U = \left\{ u_{ik}[ii] \in [0,1] | \sum_{i=1}^{c} u_{ik}[ii] = 1, \forall k, i \ \& \ 0 < \sum_{k=1}^{N[ii]} u_{ik}[ii] < N[ii] \right\}$$

Using the Lagrange method for optimization problems the objective function (1) we find the matrix $\mathbf{u}$ and $\mathbf{v}$ as the following:

$$u_{rs}[ii] = \frac{1}{\sum_{j=1}^{c} d_{rs}^2/d_{js}^2} \left[1 - \sum_{j=1}^{c} \sum_{jj=1, jj \neq ii}^{P} \frac{\beta[ii|jj]\tilde{u}_{js}[ii|jj]}{(1 + \beta[ii|jj](P-1)} \right]$$

$$+ \sum_{jj=1, jj \neq ii}^{P} \frac{\beta[ii|jj]\tilde{u}_{rs}[ii|jj]}{(1 + \beta[ii|jj](P-1)} \tag{2}$$

$$v_{rt}[ii] = \frac{\sum_{k=1}^{N[ii]} u_{rk}^2[ii]x_{kt} + \sum_{jj=1, jj \neq ii}^{P} \sum_{k=1}^{N[ii]} \beta[ii|jj](u_{rk}[ii] - \tilde{u}_{rk}[ii|jj])^2 x_{kt}}{\sum_{k=1}^{N[ii]} u_{rk}^2[ii] + \sum_{jj=1, jj \neq ii}^{P} \sum_{k=1}^{N[ii]} \beta[ii|jj](u_{rk}[ii] - \tilde{u}_{rk}[ii|jj])^2} \tag{3}$$

Among them $\tilde{u}_{ik}[ii|jj]$ is so-called the induced matrix caused by the impact of the data site ii on data site jj. The parameter $\beta[ii|jj]$ describes the level of collaboration between the data site ii and jj. The higher the value of β the stronger the level of collaboration among the sites is. The value of β can be acquired from experts given or calculated based on the similarity of the structure of the site. Detail about the $\tilde{u}_{ik}[ii|jj]$ and $\beta[ii|jj]$ can be referenced in [24]

3 Multiple Kernels Based Collaborative Fuzzy Clustering

MKCFC maps the data from the feature space of data into kernel space H by using transform functions $\psi = \{\psi_1, \psi_2, ..., \psi_M\}$ where

$$\psi_k(x_i)^T \psi_k(x_j) = K_k(x_i, x_j) \quad \text{and} \quad \psi_k(x_i)^T \psi_{k'}(x_j) = 0 | k \neq k' \tag{4}$$

In the paper, the prototypes v_i are constructed in the kernel space so we have the objective function as follow:

$$Q_{[ii]} = \sum_{k=1}^{N[ii]} \sum_{i=1}^{c} u_{ik}^2[ii](\Psi(x_k) - v_i)^2$$

$$+ \sum_{k=1}^{N[ii]} \sum_{jj=1,jj\neq ii}^{P} \beta[ii|jj] \sum_{i=1}^{c} (u_{ik} - \tilde{u}_{ik}[ii|jj])^2(\Psi(x_k) - v_i)^2 \quad (5)$$

In which

$$\Psi(x) = \omega_1\Psi_1(x) + \omega_2\Psi_2(x) + ... + \omega_M\Psi_M(x)$$

Subject to: $\omega_1 + \omega_2 + ... + \omega_M = 1$

and $\omega_k \geq 0, \forall k$ and $\sum_{j=1}^{c} u_{js}[ii] = 1, \forall s$ and $u_{js}[ii] \geq 0 \forall s, j$

where v_i is the center of the ith cluster in the kernel space, $(\omega_1, \omega_2, \cdots, \omega_M)$ is a vector of weights for each feature, respectively.

The distance d_{ik} concerns the kth data (pattern) in D[ii] and the ith prototype: $d_{ik}^2 = (\Psi(x_k) - v_i)^2$. Among them $\tilde{u}[ii|jj]$ is a so-called induced matrix caused by the impact of the data site ii on data site jj, calculated as:

$$\tilde{u}_{ik}[ii|jj] = \frac{1}{\sum_{j=1}^{c}\left(\frac{|x_k[ii]-v_i[jj]|}{|x_k[ii]-v_j[jj]|}\right)^2} = \frac{1}{\sum_{j=1}^{c}\frac{d_{ik}^2[ii|jj]}{d_{jk}^2[ii|jj]}} \quad (6)$$

Optimizing the objective function (5) by minimum condition are expressed as $\frac{\partial V}{\partial u_{rs}} = 0$ and $\frac{\partial V}{\partial v_i} = 0$

Where r = 1, 2, . . ., c; s = 1, 2, . . .,N[ii]. After computing the derivative with respect to the elements of the partition matrix, we obtain

$$u_{rs} = \frac{\sum_{jj=1,jj\neq ii}^{P}\beta[ii|jj]\tilde{u}_{rs}[ii|jj]}{(1 + \sum_{jj=1,jj\neq ii}^{P}\beta[ii|jj])} + \frac{1}{\sum_{j=1}^{c}\frac{d_{rs}^2}{d_{js}^2}}\left[1 - \sum_{j=1}^{c}\frac{\sum_{jj=1,jj\neq ii}^{P}\beta[ii|jj]\tilde{u}_{js}[ii|jj]}{(1 + \sum_{jj=1,jj\neq ii}^{P}\beta[ii|jj])}\right] \quad (7)$$

for $1 \leq r \leq c; 1 \leq s \leq N[ii]$;

Proceeding with the optimization of the objective function (5) with regard to the prototypes: $\frac{\partial Q_{[ii]}}{\partial v_i} = 0$ we have

$$v_i = \frac{\sum_{k=1}^{N[ii]} u_{ik}^2[ii]\Psi(x_k) + \sum_{k=1}^{N[ii]}\sum_{jj=1,jj\neq ii}^{P}\beta[ii|jj](u_{ik} - \tilde{u}_{ik}[ii|jj])^2\Psi(x_k)}{\sum_{k=1}^{N[ii]} u_{ik}^2[ii] + \sum_{k=1}^{N[ii]}\sum_{jj=1,jj\neq ii}^{P}\beta[ii|jj](u_{ik} - \tilde{u}_{ik}[ii|jj])^2} \quad (8)$$

for $1 \leq r \leq c; 1 \leq t \leq n$ (the number of attributes);

Now we can calculate the distance d_{ik} concerns the kth data (pattern) in D[ii] and the ith prototype as:

$$d_{ik}^2 = (\Psi(x_k) - v_i)^T (\Psi(x_k) - v_i) = \Psi(x_k)^T \Psi(x_k) - 2\Psi(x_k)^T v_i + v_i^T v_i$$

By replace the v_i in (8) to the above equation and using Eq. (4) we have

$$d_{ik}^2 = \sum_{t=1}^{M} \omega_t^2 K_t(x_k, x_k) - 2 \frac{P_{ik}}{\sum_{j=1}^{N[ii]} u_{ik}^2[ii] + \sum_{j=1}^{N[ii]} \sum_{jj=1, jj \neq ii}^{P} \beta[ii|jj](u_{ij} - \tilde{u}_{ij}[ii|jj])^2}$$

$$+ \frac{Q_{ik}}{\left(\sum_{j1=1}^{N[ii]} u_{ik}^2[ii] + \sum_{j1=1}^{N[ii]} \sum_{jj=1, jj \neq ii}^{P} \beta[ii|jj](u_{ij} - \tilde{u}_{ij}[ii|jj])^2 \right)^2}$$

$$\tag{9}$$

in which

$$P_{ik} = \sum_{j=1}^{N[ii]} u_{ij}^2[ii] \sum_{t=1}^{M} \omega_t^2 K_t(x_k, x_j)$$

$$+ \sum_{j=1}^{N[ii]} \sum_{jj=1, jj \neq ii}^{P} \beta[ii|jj](u_{ij} - \tilde{u}_{ij}[ii|jj])^2 \sum_{t=1}^{M} \omega_t^2 K_t(x_k, x_j)$$

and

$$Q_{ik} = \sum_{j1=1}^{N[ii]} \sum_{j2=1}^{N[ii]} u_{ij1}^2[ii] u_{ij2}^2[ii] \sum_{t=1}^{M} \omega_t^2 K_t(x_{j1}, x_{j2})$$

$$+ 2 \sum_{j1=1}^{N[ii]} \sum_{j2=1}^{N[ii]} \sum_{jj=1, jj \neq ii}^{P} \beta[ii|jj] u_{ij1}^2 (u_{ij2} - \tilde{u}_{ij2}[ii|jj])^2 \sum_{t=1}^{M} \omega_t^2 K_t(x_{j1}, x_{j2})$$

$$+ \sum_{jj=1, jj \neq ii}^{P} \sum_{j1=1}^{N[ii]} \sum_{j2=1}^{N[ii]} \beta^2[ii|jj](u_{ij1} - \tilde{u}_{ij1}[ii|jj])^2 (u_{ij2} - \tilde{u}_{ij2}[ii|jj])^2$$

$$\times \sum_{t=1}^{M} \omega_t^2 K_t(x_{j1}, x_{j2})$$

Above equation can be re-arranged as

$$d_{ik}^2 = \sum_{t=1}^{M} \alpha_{ikt} \omega_t^2 \tag{10}$$

where the coefficient α_{ict} can be written as

$$
\alpha_{ikt} = K_t(x_k, x_k)
$$

$$
-2\frac{\displaystyle\sum_{j=1}^{N[ii]} u_{ij}^2[ii]K_t(x_k, x_j) + \sum_{j=1}^{N[ii]} \sum_{jj=1, jj\neq ii}^{P} \beta[ii|jj](u_{ij} - \tilde{u}_{ij}[ii|jj])^2 K_t(x_k, x_j)}{\displaystyle\sum_{j=1}^{N[ii]} u_{ik}^2[ii] + \sum_{j=1}^{N[ii]} \sum_{jj=1, jj\neq ii}^{P} \beta[ii|jj](u_{ij} - \tilde{u}_{ij}[ii|jj])^2}
$$

$$
\cdot\ +\frac{Q_{ikt}}{\left(\displaystyle\sum_{j1=1}^{N[ii]} u_{ik}^2[ii] + \sum_{j1=1}^{N[ii]} \sum_{jj=1, jj\neq ii}^{P} \beta[ii|jj](u_{ij} - \tilde{u}_{ij}[ii|jj])^2\right)^2}
$$

$$(11)$$

in which

$$
Q_{ikt} = \sum_{j1=1}^{N[ii]} \sum_{j2=1}^{N[ii]} u_{ij1}^2[ii]u_{ij2}^2[ii]K_t(x_{j1}, x_{j2})
$$

$$
+2\sum_{j1=1}^{N[ii]} \sum_{j2=1}^{N[ii]} \sum_{jj=1, jj\neq ii}^{P} \beta[ii|jj]u_{ij1}^2(u_{ij2} - \tilde{u}_{ij2}[ii|jj])^2 K_t(x_{j1}, x_{j2})
$$

$$
+\sum_{jj=1, jj\neq ii}^{P} \sum_{j1=1}^{N[ii]} \sum_{j2=1}^{N[ii]} \beta^2[ii|jj](u_{ij1} - \tilde{u}_{ij1}[ii|jj])^2(u_{ij2} - \tilde{u}_{ij2}[ii|jj])^2 K_t(x_{j1}, x_{j2})
$$

Using above result we can rewrite the objective function as

$$
Q_{[ii]} = \sum_{k=1}^{N[ii]} \sum_{i=1}^{c} u_{ik}^2[ii] \sum_{t=1}^{M} \alpha_{ikt}\omega_t^2
$$

$$
+\sum_{k=1}^{N[ii]} \sum_{jj=1, jj\neq ii}^{P} \beta[ii|jj] \sum_{i=1}^{c} (u_{ik} - \tilde{u}_{ik}[ii|jj])^2 \sum_{t=1}^{M} \alpha_{ikt}\omega_t^2 \qquad (12)
$$

To optimization problem. we use a Lagrange multiplier.

$$
Q_{[ii]} = \sum_{k=1}^{N[ii]} \sum_{i=1}^{c} u_{ik}^2[ii] \sum_{t=1}^{M} \alpha_{ikt}\omega_t^2
$$

$$
+\sum_{k=1}^{N[ii]} \sum_{jj=1, jj\neq ii}^{P} \beta[ii|jj] \sum_{i=1}^{c} (u_{ik} - \tilde{u}_{ik}[ii|jj])^2 \sum_{t=1}^{M} \alpha_{ikt}\omega_t^2 - 2\lambda(\sum_{t=1}^{M} \omega_t - 1)
$$

with the constrain $\sum_{t=1}^{M} \omega_t = 1$ and after some mathematical transformations we have

$$
\omega_t = \frac{\dfrac{1}{\displaystyle\sum_{t=1}^{M} \dfrac{1}{\sum_{k=1}^{N[ii]} \sum_{i=1}^{c} u_{ik}^2[ii]\alpha_{ikt} + \sum_{k=1}^{N[ii]} \sum_{jj=1, jj\neq ii}^{P} \beta[ii|jj] \sum_{i=1}^{c} (u_{ik} - \tilde{u}_{ik}[ii|jj])^2\alpha_{ikt}}}}{\displaystyle\sum_{k=1}^{N[ii]} \sum_{i=1}^{c} u_{ik}^2[ii]\alpha_{ikt} + \sum_{k=1}^{N[ii]} \sum_{jj=1, jj\neq ii}^{P} \beta[ii|jj] \sum_{i=1}^{c} (u_{ik} - \tilde{u}_{ik}[ii|jj])^2\alpha_{ikt}} \qquad (13)
$$

Synthesis process variation on, we can compute α by (11), ω by (13) then partition matrix u can be calculated according to (7) and (10).

The essence of collaborative clustering is to collaboratively explore the structures of each data site through exchange of prototype. There are two main phases of the method: Initially, the FCM-type algorithm is performed independently at each data site with its optimization pursuits by focusing on the local data. Then the prototypes of each data site are broadcasted to all other data sites. In each collaborative step, the prototype and membership matrix at each data site are recalculated and optimized on a basis of interaction until some termination criterion has been satisfied.

MKCFC Algorithm

Input: the number of data site P, the number of item in each data site ii is $N[ii]$, the number of cluster in each data site is c, the number of attribute (features) of data is n, the data item in each data site $X[ii]$

Ouput: Membership matrix U and attributes weight.

Phase 1: locally clustering

Run IT2FCM for each data site

Phase 2: collaboration

Repeat

Communicate cluster prototypes among data sites

For each data site $D[ii]$

Compute induced partition matrices based on the formula (6).

Repeat Compute matrices α by the formula (11)

Compute attributes weight ω by the formula (13)

Update local partition matrix u by the formula (7, 10)

Until the objective function has been minimized

End for

Until Cluster prototypes do not significantly change in two consecutive iterations

4   Experiments

To evaluate the performance of the proposed algorithm, CFCM [5], CIT2FCM [24] and MKCFC clustering algorithms are chosen for comparative analysis and we used three measures FS, FSSE and CR for the quantitative assessment and following datasets are used in our examples: Canadian weather energy and engineering data sets and Iris UCCI data sets. To construct multiple kernel, we use Gaussian Kernel as k1 and Polynomial Kernel as k2

$k_1(x_i, x_j) = \exp\left(-\frac{||x_i-x_j||^2}{2\delta^2}\right)$, where $\delta > 0$

$k_2(x_i, x_j) = (x_i.x_j + \theta)^d$, where $c \geq 0, d \in N$

4.1   Validity Measures for Collaborative Fuzzy Cluster

Average Silhouette Width Criterion for collaborative clustering for all data site reads as follows [24]:

$$FS = \sum_{ii=1}^{P} FS[ii] = \sum_{ii=1}^{P} \frac{\sum_{j=1}^{N}(u_{rj} - u_{qj})S_j}{\sum_{j=1}^{N}(u_{rj} - u_{qj})} \tag{14}$$

The global FSSE criterion for all data site reads as follows [24]:

$$FSSE = \sum_{ii=1}^{P} FSSE[ii] = \sum_{ii=1}^{P} \sum_{k=1}^{C[ii]} \frac{1}{N_k} \sum_{\forall x_i \in C_k} u_{ik}[|x_i - v_k|^2 + |x_i - \tilde{v}_k|^2] \tag{15}$$

Classification Rate: The classification rate is the percentage of patterns that belong to a correctly labeled cluster [18]

4.2   Canadian Weather Energy and Engineering Data Sets

These data[1] concern hourly weather conditions occurring at 145 Canadian locations for up to 48 years of records, starting as early as 1953, and ending for most locations in 2001. The primary purpose of these files is to provide long term weather records for the use in urban planning, siting and designing of wind and solar renewable energy systems, and designing energy efficient buildings.

The study used Dry bulb temperature and Dew point temperature for clustering and each station data for one data site, 4 data sites and 4 stations, respectively are chosen: CowleyA, EdmontonStonyPlain, EdsonA and

Table 1. Collaborative Fuzzy Silhouette Criterion for weather data sets

	Data site 1	Data site 2	Data site 3	Data site 4	FS
CFCM	2.2285	3.4305	3.3156	3.3347	12.3093
CIT2FCM1	3.2102	2.9933	2.9779	3.8758	13.05732
MKCFC	3.3056	3.0504	3.1153	3.9724	13.4437

Table 2. Collaborative Fuzzy Sum of Squared Error for weather data sets

	Data site 1	Data site 2	Data site 3	Data site 4	FSSE
CFCM	96.24	143.35	100.26	150.52	490.37
CIT2FCM1	54.58	107.45	107.05	130.32	399.4
MKCFC	78.30	101.62	98.50	102.74	381.16

---

[1] http://climate.weather.gc.ca/.

FortChipewyanA. We used the algorithm in [7] to find out the number of cluster is 3 and used it for this experiment, set $\delta^2 = 4$ in kernel K1 and $\theta = 20$ and p=2 in kernel K2

Tables 1 and 2 show the evaluation results obtained for the different algorithms. The best efficient algorithm exhibiting largest FS value and smallest FSSE value is MKCFC and the second algorithm is CIT2FCM1.

4.3   Iris Data Sets

The second experiment, we use UCCI Iris Data sets, we randomly group 150 items (which are labeled in 3 categories) in this data sets to 3 data sites with the number of item in each data site are 49, 52, 49. The kernel K1 use $\delta^2 = 4$, kernel K2 use $\theta = 15$ and p=2. The Table 3 present the classification rate of MKCFC and compare with the CFCM [7] , CIT2FCM [24] and some others algorithms results in [18]. We can see that the MKCFC is 2nd best in all algorithms and it is better than FCM, other single kernel algorithms or other collaborative clustering algorithms.

Table 3. Clustering results for Iris Data Sets

	FCM	GK	KFCM-F	KFCM-K	KFCM-K	CIT2FCM	MKCFC
Classification rate %	84	95.3	84	85.3	88.7	86.7	90.7

5   Conclusion

In this paper, we reviewed and discussed some algorithms for collaborative fuzzy clustering and kernel technique. We have developed the idea of clustering collaboration by introducing multiple kernel which improved results of clustering and helped overcome the complexity of data where the data that are non spherical and can not separate linearly. The experimental results demonstrated that the FS and FSSE and classification rate yield better results.

References

1. Kanungo, T., Mount, D.M., Netanyahu, N.S., Piatko, C.D., Silverman, R., Wu, A.Y.: An efficient k-Means clustering algorithm: analysis and implementation. IEEE Trans. Pattern Anal. Mach. Intell. **24**(7), 881–893 (2002)
2. Zalik, K.R.: An efficient k-means clustering algorithm. Pattern Recogn. Lett. **29**, 1385–1391 (2008)
3. Bezdek, J.C., Ehrlich, R., Full, W.: FCM: the fuzzy c-Means clustering algorithm. Comput. Geosci. **10**(2–3), 191–203 (1984)
4. Pedrycz, W.: Collaborative fuzzy clustering. Pattern Recogn. Lett. **23**(14), 1675–1686 (2002)

5. Pedrycz, W.: Collaborative clustering with the use of fuzzy C-Means and its quantification. Fuzzy Sets Syst. **159**(18), 2399–2427 (2008)
6. Pedrycz, W.: Collaborative and knowledge based fuzzy clustering. Int. J. Innovative Comput. Inf. control **3**(1), 1–12 (2007)
7. Coletta, L.F.S., Vendramin, L., Hruschka, E.R., Campello, R.J.G.B., Pedrycz, W.: Collaborative fuzzy clustering algorithms: some refinements and design guidelines. IEEE Trans. Fuzzy Syst. **20**(3), 444–462 (2012)
8. Pedrycz, W., Reformat, M.: Evolutionary fuzzy modeling. IEEE Trans. Fuzzy Syst. **11**(5), 652–665 (2003)
9. Zhou, J., Chen, C.L.P., Chen, L., Li, H.X.: Collaborative fuzzy clustering algorithm in distributed network environments. IEEE Trans. Fuzzy Syst. **22**(6), 1443–1456 (2014)
10. Jiang, Y., Chung, F.L., Wang, S., Deng, Z., Wang, J., Qian, P.: Collaborative fuzzy clustering from multiple weighted views. IEEE Trans. Cybern. **45**(4), 688–701 (2015)
11. Prasad, M., Li, D.L., Liu, Y.T., Siana, L., Lin, C.T., Saxena, A.: A preprocessed induced partition matrix based collaborative fuzzy clustering for data analysis. In: IEEE International Conference on Fuzzy Systems, pp. 1553–1558 (2014)
12. Yu, F., Tang, J., Cai, R.: Partially horizontal collaborative fuzzy C-Means. Int. J. Fuzzy Syst. **9**(4), 198–204 (2007)
13. Yu, F., Tang, J., Cai, R.: A necessary preprocessing in horizontal collaborative fuzzy clustering. In: International Conference on Granular Computing, pp. 399–404 (2007)
14. Yu, F., Yu, S.: Prototypes-based horizontal collaborative fuzzy clustering. In: Web Mining and Web-based Application, pp. 66–69 (2009)
15. Falcon, R., Depaire, B., Vanhoof, K., Abraham, A.: Towards a suitable reconciliation of the findings in collaborative fuzzy clustering. In: ISDA International Conference, vol. 4, pp. 652–657 (2008)
16. Mitra, S., Banka, H., Pedrycz, W.: Roughfuzzy collaborative clustering. IEEE Trans.Syst. Man Cybern. Part B Cybern. **36**(4), 795–805 (2006)
17. Girolami, M.: Mercer kernel-based clustering in feature space. IEEE Trans. Neural Networks **13**, 780–784 (2002)
18. Graves, D., Pedrycz, W.: Kernel-based fuzzy clustering and fuzzy clustering: a comparative experimental study. Fuzzy Sets Syst. **161**, 522–543 (2010)
19. Sonnenburg, S., Ratsch, G., Schafer, C., Scholkopf, B.: Large scale multiple kernel learning. J. Mach. Learn. Res. **7**, 1531–1565 (2006)
20. Zhanga, D.-Q., Chen, S.-C.: A novel kernelized fuzzy C-means algorithm with application in medical image segmentation. Artif. Intell. Med. **32**(1), 37–50 (2004)
21. Yang, M.-S., Tsai, H.-S.: A gaussian kernel-based fuzzy c-means algorithm with a spatial bias correction. Pattern Recogn. Lett. **29**, 1713–1725 (2008)
22. Chen, L., Chen, C.L.P., Lu, M.: A multiple-kernel fuzzy C-Means algorithm for image segmentation. Syst. Man Cybern. Part B Cybern. **41**(5), 1263–1274 (2011)
23. Huang, H., Chuang, Y.Y., Chen, C.S.: Multiple kernel fuzzy clustering. IEEE Trans. on Fuzzy Syst. **20**(1), 120–134 (2011)
24. Dang, T.H., Ngo, L.T., Pedrycz, W.: Interval type-2 fuzzy C-Means approach to collaborative clustering. In: The IEEE International Conference on Fuzzy Systems (FUZZ-IEEE 2015) (2015)

Credit Risk Evaluation Using Cycle Reservoir Neural Networks with Support Vector Machines Readout

Ali Rodan[✉] and Hossam Faris

King Abdallah II School for Information Technology,
The University of Jordan, Amman 11942, Jordan
{a.rodan,hossam.faris}@ju.edu.jo

Abstract. Automated credit approval helps credit-granting institutions in reducing time and efforts in analyzing credit approval requests and to distinguish good customers from bad ones. Enhancing the automated process of credit approval by integrating it with a good business intelligence (BI) system puts financial institutions and banks in a better position compared to their competitors. In this paper, a novel hybrid approach based on neural network model called Cycle Reservoir with regular Jumps (CRJ) and Support Vector Machines (SVM) is proposed for classifying credit approval requests. In this approach, the readout learning of CRJ will be trained using SVM. Experiments results confirm that in comparison with other data mining techniques, CRJ with SVM readout gives superior classification results.

Keywords: Credit scoring · Reservoir computing · Echo state networks · Recurrent neural networks · Support vector machine

1 Introduction

Credit risk assessment is an important process in the financial risk management that is carried out by credit-granting institutions in order to ensure the willingness of the customer to return a loan with a given interest in a timely manner. The bank checks and study the credit history of the customer before approving loan, credit card, etc. [1]. Different aspects should be considered in order to assess the risk before approving a loan or credit. Such aspects include: the loan size, frequent of burrowing, length of commitment and credit worthiness. After completing a comprehensive process, the bank will be able to decide whether approving a giving a loan to a customer will be beneficial or it will be a risk.

In order to reduce the time and efforts in analyzing and studying credit requests, different data mining techniques were proposed in the literature for automatic evaluation of credit requests applied by customer. For example, in [2] Artificial Neural networks (ANN) with Backpropagation learning used for credit risk evaluation by applying it to the Australian credit approval dataset. Authors in [3] used a combination of ANN with decision trees model to overcome

© Springer-Verlag Berlin Heidelberg 2016
N.T. Nguyen et al. (Eds.): ACIIDS 2016, Part I, LNAI 9621, pp. 595–604, 2016.
DOI: 10.1007/978-3-662-49381-6_57

the problem of imbalanced credit data. Different data mining algorithms such as Radial Basis Functions (RBF), decision trees classifiers, and support-vector machines was applied to the problem of predicting consumer credit risk during the financial crisis of 2007-2009 is described in [5]. Genetic algorithms approaches were used also to build an efficient model for automatic credit approval application [6]. Moreover, a recent survey for credit scoring using different statistical and data mining techniques is presented in [7].

One of the most applied data mining algorithms for the automatic credit approval problem is the Support Vector Machines (SVM). SVM is very competitive machine learning algorithm which was first introduced by Vladimir Vapnik and his co-workers [12]. SVM implementation is based on the ideas of the structure risk minimization. For the credit risk evaluation and scoring problem, different variations of the SVM were proposed in the literature. Some of these variations include: Least Squared SVM [4,26], SVM based multiagent ensemble learning [27] and fuzzy SVM [28].

Reservoir Computing (RC) [8] is a framework that design and train recurrent neural networks (RNNs). RC tries to avoid the classical learning of RNNs by only fitting the readout (output) layer to the data. The state space which is random and fixed is called the *reservoir*. The reservoir is supposed to be complex so as to remember some aspects and features of the input data that can be represented by the readout mapping, Echo State Networks (ESNs) [9] is one of the simple and widely used models of RC. Rodan et al. [10,11] introduced simple deterministic reservoir networks to improve ESN performance, they proposed a simple deterministic cycle reservoir topology (SCR) [10] and extended this cycle topology using jumps (CRJ) [11]. They showed that the novel CRJ model has superior performance compared to the standard ESN on different benchmark datasets, and it can be used for several regression and classification problems.

In this paper, a novel SVM based classification approach is proposed for classifying credit approval requests. Instead of training the state space of CRJ network with linear regression (typical case), we will use the CRJ state space for training the SVM. In general, the goal of the new novel model is to help granting institutions and banks to take the right decision and accurately assess the risk associated with a consumer who applies for a credit or a loan.

The rest of this paper is organized as follows: Sect. 2 gives a background on Cycle reservoir with regular jumps and Support Vector Machines. In Sect. 3 the proposed CRJ with SVM-readout framework is presented. Credit approval datasets description is given in Sect. 4. Experiments and results are presented in Sect. 5. Finally, our work is concluded in Sect. 6.

2   Background

In this section we present Cycle Reservoir with Jumps (CRJ) and Support Vector Machines (SVM) which are the two main methods that are used to design and train our new proposed model.

2.1 Cycle Reservoir with Jumps (CRJ)

CRJ has a fixed deterministic simple regular model: the reservoir units are connected in a uni-directional ring (cycle) with bi-directional jumps (see Fig. 1). All the ring weights have the same value $r_c > 0$ and all jump weights share also the same value $r_j > 0$. In other words, non-zero values of the reservoir matrix $W_{[N \times N]}$, where N is the number of units in the hidden (reservoir) layer are [11]:

- the 'lower' sub-diagonal $W_{i+1,i} = r_c$, for $i = 1...N - 1$,
- the 'upper-right corner' $W_{1,N} = r_c$ and
- the jump values r_j. Consider the jump size $1 < \ell < \lfloor N/2 \rfloor$. If $(N \mod \ell) = 0$, then there are N/ℓ jumps, the first jump being from unit 1 to unit $1 + \ell$, the last one from unit $N + 1 - \ell$ to unit 1. If $(N \mod \ell) \neq 0$, then there are $\lfloor N/\ell \rfloor$ jumps, the last jump ending in unit $N + 1 - (N \mod \ell)$.

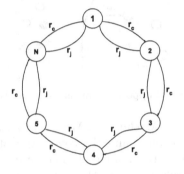

Fig. 1. Cycle reservoir with jumps (CRJ)

Suppose we have an CRJ with K input units, N hidden (internal) units and L output units. The activation of the input, internal, and output units at time step t are denoted by: $s(t) = (s_1(t), ..., s_K(t))^T$, $x(t) = (x_1(t), ..., x_N(t))^T$, and $y(t) = (y_1(t), ..., y_L(t))^T$, respectively. The connections between the input units and the internal units are given by an $N \times K$ weight matrix V, connections between the internal units are collected in an $N \times N$ weight matrix W, and connections from internal units to output units are represented by an $L \times N$ weight matrix U. The internal units is updated according to:

$$x(t + 1) = f(V s(t + 1) + W x(t)), \tag{1}$$

where f is the internal (hidden) layer activation function.

The readout is computed as:

$$y(t + 1) = g(x(t + 1)), \tag{2}$$

where g is the readout function, and can be either linear regression (normal case) or SVM (proposed case).

2.2 Support Vector Machine (SVM)

SVM is a supervised learning model used for regression and classification. Given a training examples, SVM builds a model that tries to map training data into points in a higher dimensional space using a nonlinear mapping function, so these points can be divided by a wide gap (largest distance) as possible [12,13]. SVM has many advantages over other traditional classification and prediction techniques such advantages include that the solution of the problem relies on a small subset of the dataset which gives SVM a great computational power. SVM seeks the minimization of the upper bound of generalization error instead of minimizing the training error. Figure 2 shows a hyperplane in SVM that separate two different datasets, where the vectors near the hyperplane are called support vectors (SV).

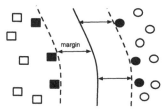

Fig. 2. Support vector machine (SVM)

Training Support Vector Machine (SVM) leads to a quadratic optimization problem with linear constraints, and the complexity of the solution in SVM depends on the complexity of the desired solution, rather than the dimensionality of the input space. The solution of SVM quadratic programming problem can be given as:

$$f(x) = \sum_{i=1}^{n}(\alpha_i - \alpha_i^*)K(x_i, x) + b \qquad (3)$$

where α_i and α_i^* are Lagrange multipliers which are subject to the following constraints:

$$\sum_{i=1}^{n}(\alpha_i - \alpha_i^*) = 0$$

$$0 \leq \alpha_i \leq C \quad i = 1, ..., n$$

$$0 \leq \alpha_i^* \leq C \quad i = 1, ..., n$$

$K(.)$ is the kernel function and its values are an inner product of two vectors x_i and x_j in the features space $\phi(x_i)$ and $\phi(x_j)$ and satisfies the Mercer's condition. Therefore, $K(x_i, x_j) = \phi(x_i).\phi(x_j)$.

The accuracy of SVM method depends on the selection of its parameters. Conventionally, SVM parameters like C (error/margin), ε (width) and kernel type are selected by cross validation.

3   CRJ with SVM-readout

In this section, we will propose the use of SVM algorithm for training the state space of Cycle reservoir with regular jumps (CRJ). Instead of using linear regression which is the classical way of training in reservoir computing community, we use SVM training that has the advantage of solving a quadratic optimization problem with linear conditions. In this work, SVM was selected as a basic classifier since it showed superior performance in the literature when applied for the credit risk problems. Moreover, the complexity of the SVM model only depends on the complexity of the desired solution, not on the space projection of the input stream [12,14,15]. In general, SVM has many advantages compared to other data mining methods such as the good generalization performance and the resistance to overfitting problems [16,17]. The proposed approach is applied and tested based on three different datasets for credit risk evaluation.

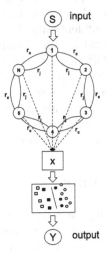

Fig. 3. CRJ with SVM training framework

CRJ can be trained with SVM readout by minimizing any loss function (see Fig. 3). Training can be done as follows:

1. Find CRJ parameters including reservoir weight r_c, jump weight r_j, and jump size ℓ using 10-fold cross validation.
2. Run CRJ on the dataset using the best CRJ parameters based on cross validation.
3. Dismiss data from initial *washout* period and collect remaining network states $(x(t))$ into a matrix X.
4. The matrix X which is the state space of CRJ, is presented as an input for SVM.
5. Find the best parameters for SVM using 10-fold cross validation.
6. The target values (examples) from the dataset is collected in a matrix y.
7. Compare each example in the dataset with the predicted class.
8. Compute the percentage of correctly predicted classes.

4   Credit Approval Datasets Description

Three popular datasets are selected to perform our experiments, The first two datasets are the German and the Australian datasets which are obtained from the UCI Repository of Machine Learning Databases [25]. The German dataset consists of 20 input variables with categorical and integer types, these variables include credit history, loan amount employment status, account balance, loan purpose, and some other personal information. It has 1000 instances with 700 records of approved credit and 300 whose credit should not be extended. The contributors of the data mentioned that several attributes of type 'ordered categorical' have been coded as integer.

The Australian credit card approval dataset [18] consists of 15 mixed attributes, these attributes are defining each customer with its class label that can be described as good or bad credit request. The types of attributes include 6 nominal, 5 integer and 3 real numerical features. The names of these attributes and their values were modified for confidentiality [18]. The dataset has 690 customer instances, with 383 records labeled as denied and 307 as approved.

The third dataset is called 'Give me some credit' provided by Kaggle competition[1]. The dataset includes 150,000 instances that represent anonymous borrowers with 59 attributes. Each instance is labeled as '0' or '1'. The dataset is highly imbalanced with a class ratio of about 13:1.

It is important to mention that all datasets are normalized in order to eliminate the effect of different scales of the features.

5   Experiments and Results

Before applying the developed approach on the selected datasets, the best CRJ parameters, including reservoir size of $N = 80$ units, input weight scale $v = 0.75$, cycle weight $r_c = 0.9$, jump weight $r_j = 0.45$, and jump size $\ell = 5$ are tuned using 10-fold cross validation. After that, we deploy the dataset of the CRJ network, and collect the state space X by representing it as an input for the SVM classifier. The RBF kernel is selected for SVM while the best SVM parameters including C, ε are chosen also by 10-fold cross validation.

In order to assess the performance of the classifiers in our experiments, we use the accuracy rate, specificity and sensitivity which are based on the confusion matrix shown in Table 1, where, $Accuracy = \frac{A+D}{A+B+C+D}$, $Specificity = \frac{A}{A+B}$ and $Sensitivity = \frac{D}{C+D}$.

Moreover, our new model (CRJ with SVM-readout) is also compared with the following popular machine learning algorithms: (**1**) Classical SVM, (**2**) Multilayer Perceptron Neural Network (MLP), (**3**) k-Nearest Neighbour (k-NN), (**4**) Naive Bayes, (**5**) Decision Trees C4.5, (**6**) Bagging, (**7**) AdaBoost and (**8**) Random Forest. In MLP, the number of neurons in the hidden layer is fixed to 6 neurons while the learning rate, momentum and epochs are set to 0.3, 0.2 and

---

[1] http://www.kaggle.com/c/GiveMeSomeCredit.

Table 1. Confusion matrix

	Predicted	
	Positive	Negative
Actual Positive	A	B
Actual Negative	C	D

Table 2. Results for the Australian dataset (Std. Dev. is shown in parentheses)

Model	Accuracy	Sensitivity	Specificity
SVM	88.7 (3.62)	89.8 (3.15)	83.5 (2.91)
MLP	84.3 (3.81)	84.6 (2.73)	86.7 (3.11)
k-nearest Neighbour	84.7 (3.24)	93.1 (3.22)	77.7 (1.27)
Naive Bayes	63.4 (3.77)	**100**	60.2 (2.73)
C4.5	85.5 (4.03)	93.7 (2.61)	72.8 (3.22)
Bagging	89.2 (2.78)	90.2 (3.17)	84.6 (2.21)
AdaBoost	90.7 (3.27)	91.1 (2.81)	85.1 (3.27)
Random Forest	87.6 (2.92)	94.8 (3.31)	78.1 (3.15)
CRJ with SVM readout	**93.6 (3.61)**	92.8 (2.13)	**88.1 (2.17)**

Table 3. Results for the German dataset (Std. Dev. is shown in parentheses)

Model	Accuracy	Sensitivity	Specificity
SVM	79.1 (3.64)	90.9 (3.21)	33.2 (2.36)
MLP	75.2 (5.21)	86.6 (2.81)	43.4 (1.62)
k-nearest Neighbour	73.8 (3.72)	**94.8 (3.52)**	18.1 (2.21)
Naive Bayes	61.5 (5.98)	87.8 (3.33)	38.1 (3.21)
C4.5	76.6 (4.27)	88.2 (2.71)	37.2 (2.95)
Bagging	79.2 (2.78)	91.2 (2.36)	35.2 (2.41)
AdaBoost	79.9 (3.72)	91.7 (3.31)	36.1 (3.23)
Random Forest	78.2 (2.61)	93.2 (2.97)	35.4 (2.72)
CRJ with SVM readout	**82.3 (3.21)**	91.1 (2.92)	**47.2 (1.72)**

500 respectively. For k-NN, the number of neighbours k is set to 10. The basic classifier for Bagging and AdaBoost is SVM.

For the German and Australian datasets, all algorithms are applied using 10-fold cross validation, where for each fold the original dataset is split into training and testing sets, and the final results are the average of results of 10 tests with their standard deviation. On the other side, since the Kaggle dataset has a large and sufficient number of examples 150,000, we divide the data to 80 % training and 20 % testing.

Evaluation results are presented in Tables 2, 3 and 4. It can be noticed that CRJ with SVM readout outperforms all other models achieving accuracy of 93.6 %, 82.32 % and 94.2 % for the Australian, German and Kaggle datasets, respectively. On the other side, our proposed model achieved the best specificity

Table 4. Results for the Kaggle dataset

Model	Accuracy	Sensitivity	Specificity
SVM	93.5	29.6	99.2
MLP	93.4	15.9	99.1
k-nearest Neighbour	92.9	14.8	98.7
Naive Bayes	92.4	**31.6**	96.5
C4.5	93.3	17.1	99.2
Bagging	93.7	30.2	99.4
AdaBoost	93.9	30.8	99.3
Random Forest	93.5	20.3	**99.6**
CRJ with SVM readout	**94.2**	31.2	99.1

Table 5. Related work based on the Australian and German datasets

Paper	Evaluation model	Australian dataset		German dataset	
		Acc.	Std. dev.	Acc.	Std. dev
Zhao et al. (2015) [19]	$SVM_{improved}$	**92.75**	-	**84.67**	-
Hens and Tiwari (2012) [20]	SVM+GA	86.76	3.78	76.84	3.82
	BPN	86.71	3.41	76.69	3.24
Wang et al. (2012) [21]	GP	86.83	3.96	77.26	4.06
	DT+RSFS	89.30	-	-	-
	LR+RSFS	87.50	-	-	-
	NN (RBF)+RSFS	88.40	-	-	-
Yu et al. (2011) [4]	LSSVM	**90.63**	-	78.64	-
Zhou et al. (2011) [22]	SVM_{linear}	84.78	-	71.10	
	SVM_{RBF}	85.65	-	71.10	
	$KASNP_{linear}$	85.81	-	70.98	
	$KASNP_{RBF}$	86.27	-	71.82	
Chen and Li (2010) [23]	SVM	84.34	5.69	75.40	5.96
	SVM+LDA	86.52	5.45	76.10	5.54
	SVM+DT	86.29	4.50	73.70	3.97
	SVM+RST	85.22	5.92	75.60	3.34
	$SVM+F_{score}$	85.10	4.83	76.70	6.07
Luo, Cheng, and Hsieh (2009) [24]	CLC	86.52	-	**84.80**	-
	MySVM	80.43	-	73.70	-
	SVM+GA	86.90	-	77.92	-
Our work	CRJ+SVM-readout	**93.60**	3.61	**82.32**	3.21

rates for the German and Australian datasets and the second best for the Kaggle dataset while maintaining very competitive sensitivity rates.

Moreover, in Table 5 we compare the obtained results of the CRJ with SVM readout to other models developed in the literature based on the Australian and German datasets. It can be noticed that CRJ with SVM readout obtained the best results for the Australian dataset while it is ranked third for the German dataset.

6 Conclusion

Since reducing the time and efforts in analyzing and studying credit request is a major issue for banks, many machine learning approaches used in the literature for automatic credit scoring systems. In this paper, we proposed an effective and high performance framework based on CRJ with SVM-readout to classify credit approval requests. Experiment results confirm that our novel model outperforms eight popular machine learning approaches and many other models proposed in the literature to solve the same classification problems.

References

1. Khashman, A.: Neural networks for credit risk evaluation: investigation of different neural models and learning schemes. Expert Syst. Appl. **37**(9), 6233–6239 (2010)
2. Khashman, A.: A neural network approach for credit risk evaluation. Int. J. Neural Syst. **19**(4), 285–294 (2009)
3. Bei, H.: Research on credit card approval models based on data mining technology. Comput. Eng. Des. **29**(11), 2989–2991 (2008)
4. Yu, L., Yao, X., Wang, S., Lai, K.: Credit risk evaluation using a weighted least squares SVM classifier with design of experiment for parameter selection. Expert Syst. Appl. **38**, 15392–15399 (2011)
5. Khandani, A., Kim, A., Lo, A.: Consumer credit-risk models via machine-learning algorithms. J. Bank. Finance **34**, 2767–2787 (2010)
6. Sakprasat, S., Sinclair, M.: Classification rule mining for automatic credit approval using genetic programming. In: IEEE Congress on Evolutionary Computation, CEC 2007, pp. 548–555 (2007)
7. Kraus, A.: Recent methods from statistics and machine learning for credit scoring, Ph.D. thesis, Ludwig-Maximilians-Universitat Munchen (2014)
8. Lukosevicius, M., Jaeger, H.: Reservoir computing approaches to recurrent neural network training. Comput. Sci. Rev. **3**(3), 127–149 (2009)
9. Jaeger, H.: The "echo state" approach to analysing and training recurrent neural networks, Technical report gmd report 148. German National Research Center for Information Technology, Technical report (2001)
10. Rodan, A., Tino, P.: Minimum complexity echo state network. IEEE Trans. Neural Netw. **22**(1), 131–144 (2011)
11. Rodan, A., Tino, P.: Simple deterministically constructed cycle reservoirs with regular jumps. Neural Comput. **24**(7), 1822–1852 (2012)
12. Vapnik, V.: The Nature of Statistical Learning Theory. Springer, New York (1995)

13. Vapnik, V.: An overview of statistical learning theory. IEEE Trans. Neural Netw. **5**, 988–999 (1999)
14. Cristianini, N., Shawe-Taylor, J.: An Introduction to Support Vector Machines and Other Kernel-Base Learning Methods. Cambridge University Press, Cambridge (2000)
15. Vapnik, V., Vashist, A.: A new learning paradigm: learning using privileged information. Neural Netw. **22**(5–6), 544–557 (2009)
16. Chan, W., Cheung, K., Harris, C.: On the modelling of nonlinear dynamic system using support vector neural networks. Eng. Appl. Artif. Intell. **14**, 105–113 (2001)
17. Zhu, G., Liu, S., Yu, J.: Support vector machine and its applications to function approximation. J. East China Univ. Sci. Technol. **5**, 555–559 (2002)
18. Marcano-Cedeno, A., Marin-de-la-Barcena, A., Jimenez-Trillo, J., Pinuela, J., Andina, D.: Artificial metaplasticity neural network applied to credit scoring. Int. J. Neural Syst. **21**, 311–317 (2011)
19. Zhao, Z., Xu, S., Kang, B.H., Kabir, M., Liu, Y., Wasinger, R.: Investigation and improvement of multi-layer perception neural networks for credit scoring. Expert Syst. Appl. **42**(7), 3508–3516 (2015)
20. Hens, A., Tiwari, M.: Computational time reduction for credit scoring: an integrated approach based on support vector machine and stratified sampling method. Expert Syst. Appl. **39**, 6774–6781 (2012)
21. Wang, J., Hedar, A., Wang, S., Mac, J.: Rough set and scatter search metaheuristic based feature selection for credit scoring. Expert Syst. Appl. **39**, 6123–6128 (2012)
22. Zhou, X., Jiang, W., Shi, Y., Tian, Y.: Credit risk evaluation with kernel-based affine subspace nearest points learning method. Expert Syst. Appl. **38**, 4272–4279 (2011)
23. Chen, F., Li, F.: Combination of feature selection approaches with SVM in credit scoring. Expert Syst. Appl. **37**, 4902–4909 (2010)
24. Luo, S., Cheng, B., Hsieh, C.: Prediction model building with clustering-launched classification and support vector machines in credit scoring. Expert Syst. Appl. **36**, 7562–7566 (2009)
25. Lichman, M.: UCI machine learning repository (2013)
26. Lai, K.K., Yu, L., Zhou, L., Wang, S.-Y.: Credit risk evaluation with least square support vector machine. In: Wang, G.-Y., Peters, J.F., Skowron, A., Yao, Y. (eds.) RSKT 2006. LNCS (LNAI), vol. 4062, pp. 490–495. Springer, Heidelberg (2006)
27. Yu, L., Yue, W., Wang, S., Lai, K.K.: Support vector machine based multiagent ensemble learning for credit risk evaluation. Expert Syst. Appl. **37**, 1351–1360 (2010)
28. Wang, Y., Wang, S., Lai, K.K.: A new fuzzy support vector machine to evaluate credit risk. IEEE Trans. Fuzzy Syst. **13**, 820–831 (2005)

Fuzzy-Based Feature and Instance Recovery

Shigang Liu(✉), Jun Zhang, Yu Wang, and Yang Xiang

School of Information Technology,
Deakin University, Melbourne, VIC 3125, Australia
{shigang,jun.zhang,y.wang,yang.xiang}@deakin.edu.au

Abstract. The severe class distribution shews the presence of under-represented data, which has great effects on the performance of learning algorithm, is still a challenge of data mining and machine learning. Lots of researches currently focus on experimental comparison of the existing re-sampling approaches. We believe it requires new ways of constructing better algorithms to further balance and analyse the data set. This paper presents a Fuzzy-based Information Decomposition oversampling (FIDoS) algorithm used for handling the imbalanced data. Generally speaking, this is a new way of addressing imbalanced learning problems from missing data perspective. First, we assume that there are missing instances in the minority class that result in the imbalanced dataset. Then the proposed algorithm which takes advantages of fuzzy membership function is used to transfer information to the missing minority class instances. Finally, the experimental results demonstrate that the proposed algorithm is more practical and applicable compared to sampling techniques.

Keywords: Imbalanced data · Information decomposition · Classification

1 Introduction

Class imbalance problem has been identified as one of the ten challenging problems in data mining research [1]. This issue occurs in two different types of data sets: binary and multiclass. For binary problem, the training data from the minority class or positive class are very small, and the rest which make up the majority class or negative class are very large. Nevertheless, for multiclass problems, each of the classes only contains a tiny fraction of samples. In these cases, standard classifiers generally fail to identify the minority class examples. However, minority class is often the most interesting and valuable for the data analyst [2]. Hence, a natural question in data mining research is how to improve the performance of classifiers when faced with imbalanced data?

So far, sampling (either oversampling or undersampling) [3] is a popular technique for imbalanced learning problem. Specifically, data sampling such as random oversampling (ROS), Synthetic Minority Oversampling Techniques (SMOTE) [5], or random undersampling etc., is a technique that either randomly

© Springer-Verlag Berlin Heidelberg 2016
N.T. Nguyen et al. (Eds.): ACIIDS 2016, Part I, LNAI 9621, pp. 605–615, 2016.
DOI: 10.1007/978-3-662-49381-6_58

or intelligently re-balances the distribution. However, each of them can be limited when applying to real-world problems. For example, random oversampling may cause over-fitting, while undersampling may cause some important information missing. Even though some 'smart' sampling techniques such as SMOTE, are developed to generate new instances, they may not suitable for all dataset. For example, if different features are independent to each other, these 'smart' techniques will be inefficient. In other words, in such scenarios these sampling techniques may fail to create useful minority class instances.

In this work, we address the class imbalance problem from a new perspective, in which imbalance means some data instances in the minority class are 'missing'. Therefore, we recover the 'missing data' in order to rebalance the dataset. In this regard, we propose a Fuzzy-based Information Decomposition oversampling (FIDoS) algorithm, which can recover the 'missing' instances feature by feature. We highlight our main contributions in the following.

- We formally redefine the class imbalance problem based on information distribution. In particular, we are the first to treat the imbalance problem as a special missing data problem.
- We propose a new fuzzy based information decomposition algorithm to solve the missing data problem. The algorithm has the capability to recover not only features but also instances.
- We carry out a large number of experiments to evaluate the proposed algorithm. In detail, the results on twenty-two different imbalanced datasets show that our algorithm significantly outperforms the five existing methods.

2   Related Work

This section presents a short review on five existing sampling techniques for the class imbalance problem, which are related to our work. For more information, please refer to the recent survey on data mining [4].

Random oversampling (ROS): The minority oversampling randomly select a training example from the minority class, and then duplicate it. This may cause over-fitting and longer training time during imbalance learning process.

Random undersampling (RUS): Majority undersampling draws a random subset from the majority class while discarding the rest instances. In doing so, some important information may be lost when examples are removed from the training data set randomly and especially when the dataset is small.

Synthetic Minority Over-sampling Techniques (SMOTE) [5]: This technique adds new artificial minority attribute examples by extrapolating instances from the k nearest neighbours (kNN) to the minority class instances. In experiments, the parameter is set to five. Because SMOTE generates numbers of synthetic data samples for each original minority class and without considering the original distribution, this may lead to more overlapping between classes.

Cluster-based undersampling (CBUS) [6]: The aim of Cluster based under-sampling approach is not to balance the data ratio of majority class of minority class into 1:1, but to reduce the gap between the numbers of minority class and minority class.

Majority Weighed Minority Oversampling Technique (MWM) [7]: This method first assign weights to the minority class according to their Euclidean distance from the nearest majority samples. Then, it 'intelligently' generate synthetic class examples from the weighted informative minority class samples using a modified hierarchical clustering algorithm.

3 Proposed Algorithm

As we discussed before, this paper advise a new way of addressing class imbalance problems. Before presenting the algorithm description, the knowledge of information distribution will be recalled in problem modelling part. We clarify the symbols used as follows: lower case **bold** Roman letters such as $\mathbf{x}$ and $\mathbf{x}_i$ means row vectors, and a superscript T means the transpose of a vector. $\mathbf{u}$ and $\mathbf{u}_i$ as the discrete universe of $\mathbf{x}$ and $\mathbf{x}_i$. X and $X_{minority}$ are matrices which denote training data and minority class, respectively. And A_i and A_{si} are intervals.

3.1 Problem Modelling

In this study, we make use of a linear distribution for transferring information from the observed data. In other words, FIDoS algorithm will employ linear distribution as one step to mine information for the 'missing' instances. Based on this, we recall the knowledge of information distribution [8], which can be regarded as a standard to choose linear distribution for our proposed algorithm.

Let $\mathbf{x} = (x_1, x_2, \cdots, x_n)^T$ be a sample observed from an experiment, and $\mathbf{u} = \{u_1, u_2, \cdots, u_m\}$ be the discrete universe of $\mathbf{x}$. A mapping from $\mathbf{x} \times \mathbf{u}$ to $[0, 1]$, $\mu : \mathbf{x} \times \mathbf{u} \to [0, 1]$, $(x_i, u_j) \to \mu(x_i, u_j)$ is called an information distribution of $\mathbf{x}$ on $\mathbf{u}$, if $\mu(x_i, u_j)$ has the following properties:

(1) Reflexive. $\forall x_i \in \mathbf{x}$, if $\exists u_j \in \mathbf{u}$, such that $x_i = u_j$, then $\mu(x_i, u_j) = 1$.
(2) Decreasing. For $x_i \in \mathbf{x}$, $\forall u_p, u_q \in \mathbf{u}$,
if $\| u_p - x_i \| \le \| u_q - x_i \|$, then $\mu(x_i, u_p) \ge \mu(x_i, u_q)$.
(3) Conserved. That is to say $\sum_{j=1}^{m} \mu(x_i, u_j) = 1$, $i = 1, 2, \cdots, n$.

For example, $X = (\mathbf{x}_1, \mathbf{x}_2, \cdots, \mathbf{x}_n)$ is a matrix such as imbalance dataset. $\mathbf{x}_i = (x_{1i}, x_{2i}, \cdots, x_{mi})^T, i = 1, 2, \cdots, n$ be a given feature/attribute, R is the universe of discourse of x_i, and $\mathbf{u}_i = \{u_{1i}, u_{2i}, \cdots, u_{ti}\}$ is the discrete universe of $\mathbf{x}_i$ where $u_{si} - u_{s-1,i} \equiv h_i, s = 2, 3, \cdots, t$. For $x_{ji} \in \mathbf{x}_i, j = 1, 2, \cdots, m$ and $u_{si} \in \mathbf{u}_i$, the following information distribution formula can meet all of the properties of the definition.

$$\mu(x_{ji}, u_{si}) = \begin{cases} 1 - \frac{\| x_{ji} - u_{si} \|}{h_i} & if \, \| x_{ji} - u_{si} \| \le h_i \\ 0 & if \, \| x_{ji} - u_{si} \| > h_i \end{cases} \tag{1}$$

where h_i is called step length and μ is called linear distribution.

Algorithm Design

1: INPUT: Renewed minority data $\tilde{X}_{minority}$, number of missing instances t.
2: OUTPUT: Re-balanced data set with minority class increased
/*Initialization*/
3: Data set prepare: Interpolate t minority instances to the minority class with 'NaN'
   as its features (labelled as minority class). In this step, we can obtain $\tilde{X}_{minority}$, let
   $\tilde{n}_{min}$ as the size of the renewed minority class
4: for $i = 1 : \tilde{n}_{min}$ do
5:     for $s = 1 : t$
6:         calculate a_i, b_i, h_i, A_{si}, A_{ti}, u_{si}, U_i
7:         use formula (5) obtain $\tilde{m}_{si}$
8:     end for
9: end for

3.2   Algorithm Description

Let $X_{minority} = (x_{ji})_{m \times (n+1)}, m \in N_+, n \in N_+$ be the minority class chosen from an imbalanced training dataset with missing instances inserted (NaN as its features) in the minority class, and the last column $x_{j,(n+1)}, j = 1, 2, \cdots, m$ is the classes label. In another way, we denote $X_{minority} = (\mathbf{x}_1, \mathbf{x}_2, \cdots, \mathbf{x}_n, \mathbf{x}_{n+1})$, and $\mathbf{x}_i = (x_{1i}, x_{2i}, \cdots, x_{mi})^T, i = 1, 2, \cdots, n+1$. Assume there are missing instances in the minority class, and the number of missing instances from minority class is t. Therefore, there are t missing features in $\mathbf{x}_i, i = 1, 2, \cdots, n$; denoted the missing feature as $\{\tilde{m}_{si} | s = 1, 2, \cdots, t\}$. Let $a_i = min\{x_{ji} | j = 1, 2, \cdots, m\}$; $b_i = max\{x_{ji} | j = 1, 2, \cdots, m\}$. Then we obtain an interval $[a_i, b_i]$. Let $h_i = (b_i - a_i)/t_i$, $A_i = \bigcup_{s=1}^{t} A_{si}$ and A_{si}, A_{ti} can be calculated from formulas (2) and (3). $\mathbf{u}_i = \{u_{1i}, u_{2i}, \cdots, u_{ti}\}$ is the discrete universe of x_i where $u_{si} - u_{s-1,i} \equiv h_i, s = 2, 3, \cdots, t$. And $u_{si} = (a_i + (s - 1) * h_i + a_i + s * h_i)/2$, that is to say u_{ij} is the center of A_{ji}. For $\mu(x_{ji}, u_{si})$ is chosen from formula (1), $x_{ji} \in x_i$, and $u_{si} \in U_i$, m_{jsi} obtained from formula (4) is called information decomposition (refer to [9] for an example) from x_{ji} to A_{ji}.

$$A_{si} = [a_i + (s - 1) * h_i, a_i + s * h_i), s = 1, 2, \cdots, t - 1 \tag{2}$$

$$A_{ti} = [a_i + (t - 1) * h_i, a_i + t * h_i] \tag{3}$$

$$m_{jsi} = \mu(x_{ji}, u_{si}) * x_{ji} \tag{4}$$

Then we calculate the sth missing feature in $\mathbf{x}_i$, which is $\tilde{m}_{si}$ from (5). Information decomposition algorithm framework is illustrated in Algorithm Design.

$$\tilde{m}_{si} = \begin{cases} \bar{\mathbf{x}}_i & if \sum_{j=1}^{m} \mu(x_{ji}, u_{si}) = 0 \\ \frac{\sum_{j=1}^{m} m_{jsi}}{\sum_{j=1}^{m} \mu(x_{ji}, u_{si})} & otherwise \end{cases} \tag{5}$$

where $\bar{\mathbf{x}}_i$ is the mean of 'non-NaN' values of $\mathbf{x}_i$.

4   Performance Evaluation

4.1   Datasets

The 22 datasets used in this paper are listed in Table 1. Data information include data name (first column), sample size (second column), minority data samples (third column), majority data samples (fourth column), the class attribute

Table 1. Dataset characteristics

Dataset	Instances(#)	Minority(#)	Majority(#)	Attributes	IR
MC2	125	44	81	39	1.84
Yeast1	1484	429	1055	8	2.46
vehicle2	846	218	628	18	2.88
vehicle1	846	217	629	18	2.9
vehicle3	846	212	634	18	2.99
G0-6	214	51	163	9	3.2
kc2	522	107	415	21	3.88
kc3	194	36	158	39	4.39
CM1	327	42	285	37	6.79
pc3	1077	134	943	37	7.04
pc4	1458	178	1280	38	7.19
Pb0	5472	560	4913	10	8.79
G016vs2	192	17	175	9	10.29
pc1	705	61	644	37	10.56
Glass4	214	13	201	9	15.47
Pb13vs4	472	115	357	10	15.86
G016vs5	184	9	175	9	19.44
Glass5	214	9	205	9	22.78
pc5	17186	516	16670	38	32.31
A17vs-10	2338	58	2280	8	39.31
MC1	1988	46	1942	38	42.22
pc2	745	16	729	36	45.56

(second last column), and IR(imbalance ratio: majority/minority). We can see from Table 1, the datasets cover a variety of sizes and imbalance level. Precisely, imbalance ratio varies from 1.84 (slightly imbalance) to 45.56 (highly imbalance). And the diversity of instances, varies from the smallest dataset with 125 examples to the largest dataset which contains 17186 observations. Moreover, these datasets are from PROMISE repository software engineering databases [11] and KEEL-dataset repository [10]. All these datasets can be downloaded from related website. In Table 1, we use the abbreviation version for some datasets. For example, G0-6 for Glass0123vs456, Pb0 short for Page-blocks0, G016vs2 for Glass016vs2, Pb13vs4 short for Page-blocks13vs4, G016vs5 for Glass016vs5, A17vs-10 for abalone-17_vs_7-8-9-10.

4.2   Measurement

Following previous study [12] in imbalanced learning, we employ F-measure (FM), G-mean (GM), Area under the ROC curve(AUC) and Kappa to evaluate the performance of the classifiers. Due to the space limitation, we only present the FM and Kappa values, however we outline the statistical ranking later based on the four performances.

In this paper, we use the minority class as the positive class and the majority class as the negative class. The confusion matrix values as: true positive (TP), false positive (FP), true negative (TN), and false negative (FN). The following formulas are used to calculate each metric.

$$FM = \frac{(1 + \beta^2) \cdot Recall \cdot Precision}{\beta^2 \cdot (Recall + Precision)} \qquad GM = \sqrt{Recall \cdot Precision}$$

$$AUC = \frac{1 + \frac{TP}{TP+FN} + \frac{FP}{FP+TN}}{2} \qquad Kappa = \frac{N \sum_{i=1}^{k} x_{ii} - \sum_{i=1}^{k} x_{i.}x_{.i}}{N^2 - \sum_{i=1}^{k} x_{i.}x_{.i}}$$

Where x_{ii} is the number of cases in the main diagonal of the confusion matrix, and N is the total examples, $x_{.i}$ and $x_{i.}$ are the number of column and row counts. And $Recall = (TP)/(TP + FN)$, $Precision = (TP)/(TP + FP)$.

4.3 Compared Methods

Recent study [13] has shown that simple methods such as ROS and RUS even perform better compared with elaborated approaches, and that Seiffert et al. outlines ROS, RUS, SMOTE are the most popular sampling techniques [14]. Therefore, RUS, ROS SMOTE are employed in our experiments. Moreover, other techniques such as CBUS and MWM are also considered in this paper. All of these techniques and learning processes are implemented with Matlab 2012B.

4.4 Experiment Strategy

In this study, 60 % of the original observation samples are randomly chosen as training data and the other 40 % as unseen test data.

We conduct 10 runs of HoldOut strategy under C4.5 [15] decision tree and k-nearest neighbours (KNN) [16] classification. Due to space limitation, only the overall average results of FM and Kappa are presented. All the sampling techniques are only used to training data rather than testing data.

For undersampling six parameter values, 20 %, 50 %, 70 %, 90 % are used, indicating the percentage of the majority class removed. For example, if a dataset contained 1000 majority class instances, 70 % means after sampling 700 samples will be removed. For oversampling approaches such as ROS, SMOTE oversampling rates are settled as 20 %, 50 %, 70 % and 90 %. For example, ROS% eathe minority examples 20 % of the minority class size are added. Besides, parameters in MWM are settled as $k1 = 5$, $k2 = 3$, $k3 = S_{min}/2$, $C_p = 3$, $C_f(th) = 5$, and $CMAX = 2$. Meanwhile, a balance ratio of 1:1 for techniques such SMOTE, ROS, and RUS are also considered in this paper. 1:1 means the same size of the minority class and majority class are equal to each other after using each sampling approach.

Finally, a statistical ranking time is collected and discussed, clearly show which technique has better ranking time.

4.5   Experiments Results

The results of our extensive suite of experiments will be presented in this section. We begin by analysing four metrics results, and then we focus on statistical ranking analysis. **Bold** text in the results mean better performance.

Results Based on Four Metrics: Tables 2 and 3 present the average FM, and Kappa values of each sampling technique under two classifiers. The first column is the classifier and the second column is the datasets. And the first row of each table lists 7 sampling techniques.

Precisely, for C4.5 decision tree classifier, both Tables 2 and 3 demonstrate that FIDoS performs better in most of cases in terms of FM and Kappa. For example, FIDoS achieves 0.511 of FM values on MC2 dataset, while none of other techniques obtain a FM value more than 0.5. Importantly, we can see that FIDoS also have the highest Kappa values on MC2, and this trend happens to many other datasets. As FM is used for measuring the performance of the classifier on minority class samples and Kappa represents the ability of the classifier in penalizing all-positive or all-negative predictions, we can conclude that the proposed FIDoS algorithm has a better performance regarding the minority class, and a good ability in penalizing all-positive or all-negative predictions.

When it comes to 1NN classifier, FIDoS also has an outstanding performance compared with the other sampling techniques. We can see that FIDoS achieves highest FM and Kappa values at least in 13 out of 22 cases.

Comparatively, Tables 2 and 3 clearly describe that other techniques such as CBUS and MWM have better performance in some cases under 1NN classifier. These results, however, are not stable. For example, MWM has better performance in 7 out 22 datasets based on FM under C4.5 classifier, however, its results are far worse when using 1NN classifier. Moreover, CBUS achieves the highest FM values in 5 out 22 datasets under 1NN classifier. Similar to MWM, its performance based on C4.5 classifier is even worse than MWM. Precisely, CBUS only results in better performance on kc2 and CM1 datasets in terms of FM metric, which means CBUS fails to work efficiently in classifying the minority class samples.

Statistical Ranking Results: Table 4 depicts the performance ranking in terms of two classifiers based on four performance measures, all datasets. The first column is sampling techniques, and the second row means the ranked number 1, 2, 3, 4, 5, 6, 7 of each classifier. A rank of one means that the sampling approach, for all the given datasets, a learner, and four performance measures, results in the highest value for four performance metrics. To be specific, for C4.5 classifier Table 4 demonstrates that FIDoS has the best ranking for 49 times, which is doubled of the second best ranking (MWM with 21). Moreover, RUS results in the worst performance with 35 times ranked 7. Even though CBUS achieve the third best ranking, we can see it produces the second worst performance for both classifiers.

Table 2. FM values

Classifier	Dataset	None	SMOTE	CBUS	ROS	RUS	MWM	FIDoS
C4.5	MC2	0.479	0.487	0.478	0.49	0.492	0.492	**0.511**
	Yeast1	0.480	0.500	0.434	0.498	0.478	0.504	**0.514**
	vehicle2	0.882	0.901	0.699	0.901	0.866	0.900	**0.904**
	vehicle1	0.507	0.531	0.484	0.524	0.492	**0.539**	0.527
	vehicle3	0.479	0.518	0.498	0.486	0.476	0.502	**0.53**
	G0-6	0.835	0.86	0.851	0.843	0.799	0.844	**0.863**
	kc2	0.474	0.491	**0.573**	0.457	0.486	0.479	0.509
	kc3	0.346	0.358	0.396	0.277	0.222	0.354	**0.409**
	CM1	0.167	0.261	**0.293**	0.241	0.199	0.224	0.285
	pc3	0.283	0.323	0.286	0.300	0.282	0.253	**0.328**
	pc4	0.488	0.520	0.505	0.504	0.459	0.506	**0.521**
	Pb0	0.848	0.835	0.742	0.823	0.794	**0.849**	0.834
	G016vs2	0.173	0.066	0.089	0.078	0.000	**0.189**	0.184
	pc1	0.294	0.307	0.321	0.287	0.281	0.274	**0.325**
	Glass4	0.529	0.489	0.324	0.649	0.000	**0.733**	0.607
	Pb13vs4	0.884	0.887	0.840	0.891	0.869	**0.905**	0.883
	G016vs5	0.671	0.679	0.338	0.734	0.034	0.631	**0.748**
	Glass5	0.502	0.560	0.307	0.556	0.000	0.680	**0.749**
	pc5	0.488	0.493	0.505	0.476	0.412	0.487	**0.542**
	A17vs78910	0.201	0.243	0.079	0.223	0.225	**0.335**	0.279
	MC1	0.183	0.181	0.199	0.267	0.070	0.239	**0.277**
	pc2	0.052	0.000	0.080	0.021	0.000	**0.088**	0.037
1NN	MC2	0.431	0.433	**0.504**	0.43	0.441	0.452	0.455
	Yeast1	0.482	0.505	0.400	0.486	0.473	0.492	**0.517**
	vehicle2	0.821	0.830	0.493	0.825	0.775	0.822	**0.833**
	vehicle1	0.451	**0.501**	0.480	0.495	0.465	0.453	0.421
	vehicle3	0.393	0.449	0.488	0.405	0.399	0.431	**0.498**
	G0-6	0.881	0.907	0.824	0.873	0.828	0.885	**0.913**
	kc2	0.419	0.457	**0.534**	0.426	0.462	0.469	0.451
	kc3	0.15	0.218	**0.281**	0.162	0.124	0.184	0.214
	CM1	0.211	0.194	**0.267**	0.221	0.183	0.201	0.211
	pc3	0.212	**0.245**	0.199	0.218	0.218	0.194	0.222
	pc4	0.216	0.255	**0.269**	0.233	0.215	0.216	0.256
	Pb0	0.783	0.793	0.559	0.793	0.751	0.798	**0.804**
	G016vs2	0.278	0.268	0.232	0.319	0.123	0.292	**0.320**
	pc1	0.104	0.141	0.170	0.116	0.094	0.123	**0.137**
	Glass4	0.626	0.690	0.390	0.677	0.000	0.728	**0.763**
	Pb13vs4	0.861	0.867	0.823	**0.873**	0.800	0.850	0.867
	G016vs5	0.515	0.556	0.284	0.649	0.070	0.579	**0.689**
	Glass5	0.568	0.616	0.289	0.548	0.106	0.464	**0.621**
	pc5	0.407	**0.450**	0.425	0.416	0.397	0.427	0.431
	A17vs-10	0.288	0.275	0.081	0.25	0.276	0.294	**0.302**
	MC1	0.191	0.198	0.188	0.182	0.066	0.189	**0.215**
	pc2	0.031	0.084	0.064	0.057	0.000	0.021	**0.109**

Table 3. Kappa values

Classifier	Dataset	None	SMOTE	CBUS	ROS	RUS	MWM	FIDoS
C4.5	MC2	0.220	0.202	0.148	0.220	0.219	0.218	**0.242**
	Yeast1	0.279	0.292	0.100	0.291	0.262	**0.312**	0.301
	vehicle2	0.842	0.865	0.542	0.866	0.819	0.866	**0.869**
	vehicle1	0.339	0.363	0.205	0.359	0.311	**0.389**	0.355
	vehicle3	0.312	0.354	0.238	0.325	0.303	0.340	**0.366**
	G0-6	0.783	0.815	0.799	0.798	0.726	0.793	**0.820**
	kc2	0.347	0.367	**0.448**	0.304	0.314	0.351	0.381
	kc3	0.217	0.213	0.223	0.108	0.152	0.185	**0.258**
	CM1	0.068	0.147	0.169	0.131	0.083	0.106	**0.176**
	pc3	0.182	0.221	0.186	0.203	0.175	0.156	**0.225**
	pc4	0.417	0.450	0.427	0.437	0.378	0.437	**0.452**
	Pb0	0.831	0.816	0.708	0.802	0.770	**0.831**	0.815
	G016vs2	0.101	0.091	-0.030	0.089	0.055	0.113	**0.166**
	pc1	0.229	0.235	0.241	0.224	0.209	0.209	**0.255**
	Glass4	0.506	0.545	0.253	0.627	0.181	**0.715**	0.583
	Pb13vs4	0.846	0.848	0.789	0.856	0.821	**0.875**	0.842
	G016vs5	0.653	0.663	0.290	0.724	0.228	0.623	**0.739**
	Glass5	0.484	0.612	0.263	0.541	0.227	0.673	**0.739**
	pc5	0.472	0.477	**0.490**	0.460	0.393	0.471	0.476
	A17vs-10	0.184	0.221	0.033	0.203	0.197	**0.317**	0.259
	MC1	0.167	0.181	0.181	0.249	0.111	0.224	**0.259**
	pc2	0.035	**0.089**	0.060	0.078	0.041	0.068	0.049
1NN	MC2	0.115	0.108	0.154	0.137	0.110	0.095	**0.158**
	Yeast1	0.278	0.291	0.049	0.284	0.260	0.285	**0.316**
	vehicle2	0.758	0.768	0.184	0.763	0.692	0.759	**0.774**
	vehicle1	0.273	0.296	0.211	0.283	0.278	0.267	**0.305**
	vehicle3	0.207	**0.263**	0.232	0.229	0.201	0.250	0.232
	G0-6	0.844	0.879	0.756	0.835	0.768	0.848	**0.886**
	kc2	0.286	0.321	**0.403**	0.295	0.283	0.331	0.318
	kc3	-0.030	0.030	**0.064**	-0.010	0.022	-0.008	0.043
	CM1	0.093	0.058	**0.137**	0.113	0.070	0.071	0.093
	pc3	0.105	**0.125**	0.088	0.113	0.095	0.077	0.108
	pc4	0.111	0.130	0.142	0.131	0.089	0.106	**0.157**
	Pb0	0.760	0.769	0.492	0.771	0.722	0.776	**0.782**
	G016vs2	0.218	0.196	0.151	0.254	0.189	0.225	**0.260**
	pc1	0.028	0.053	**0.082**	0.034	0.024	0.033	0.060
	Glass4	0.606	0.672	0.335	0.657	0.485	0.709	**0.747**
	Pb13vs4	0.814	0.822	0.759	**0.831**	0.718	0.801	0.824
	G016vs5	0.500	0.536	0.230	0.632	0.304	0.561	**0.673**
	Glass5	0.552	0.601	0.243	0.535	0.264	0.441	**0.618**
	pc5	0.389	**0.431**	0.407	0.399	0.376	0.409	0.413
	A17vs-10	0.273	0.256	0.036	0.236	0.252	0.279	**0.286**
	MC1	0.174	0.176	0.166	0.164	0.091	0.168	**0.198**
	pc2	0.035	0.064	0.037	0.033	0.047	-0.004	**0.087**

Table 4. Performance ranking based on four performance metrics

Technique	Ranked numbers(C4.5)							Ranked numbers(1NN)						
	1	2	3	4	5	6	7	1	2	3	4	5	6	7
None	0	5	7	14	26	26	10	0	0	8	18	27	19	16
SMOTE	2	22	41	12	6	4	1	13	32	16	13	8	3	3
CBUS	15	12	5	10	8	10	28	25	7	3	7	6	17	23
ROS	0	19	7	24	21	10	7	4	13	15	20	18	16	2
RUS	1	4	5	7	4	32	35	2	6	3	12	8	25	32
MWM	21	9	16	13	17	5	7	0	12	26	13	19	8	10
FIDoS	49	17	7	8	6	1	0	44	18	17	5	2	0	2

Similarly, for 1NN classifier, Table 4 also emphasizes that our proposed FIDoS algorithm has 44 times ranked the first which is nearly double of CBUS(25 times). It is worthwhile to mention that MWM has zero time ranked the first, which again supported our previous discussion.

Overall, the statistical ranking results from both tables outline that FIDoS ranks in the top three position for more than 75 % of its totally ranking times, and has the highest first ranking times, which again emphasized our previous experiment results. Therefore, the proposed algorithm is more applicable and practical.

5   Conclusions

This paper proposes a novel Fuzzy-based Information Decomposition oversampling (FIDoS) algorithm to re-distribute the data in different classes aiming to solve the class imbalance learning problems. Different from the previous sampling approaches, FIDoS introduces a new way of considering imbalanced data set issue. That is we assume there are missing instances in the minority class, which caused the dataset become imbalanced. Therefore, we recover the 'missing data' in order to rebalance the dataset. Finally, we have presented a comprehensive experimental analysis of learning from imbalanced data, using 7 sampling approaches with 22 real-world datasets. Experimental results demonstrate that FIDoS algorithm can achieve the best performance in most times which indicates that our proposed algorithm is more efficient in real-world problems.

In the future, a study to identify the influence of imbalance ratio on imbalanced learning problems will be an interesting work.

References

1. Yang, Q., Wu, X.: 10 challenging problems in data mining research. Int. J. Inf. Technol. Decis. Making **5**, 597–604 (2006)

2. Maratea, A., Petrosino, A., Manzo, M.: Adjusted F-measure and kernel scaling for imbalanced data learning. Inf. Sci. **257**, 331–341 (2014)
3. Dubey, R., et al.: Analysis of sampling techniques for imbalanced data: An n= 648 ADNI study. NeuroImage **87**, 220–241 (2014)
4. He, H., Garcia, E.: Learning from imbalanced data. IEEE Trans. Knowl. Data Eng. **21**(9), 1263–1284 (2009)
5. Chawla, N.V., Bowyer, K.W., Hall, L.O., Kegelmeyer, W.P.: SMOTE: synthetic minority over-sampling technique. J. Artif. Intell. Res. **16**, 321–357 (2012)
6. Rahman, M.M., Davis, D.: Cluster based under-sampling for unbalanced cardiovascular data. In: Proceedings of the World Congress on Engineering (2013)
7. Barua, S., et al.: MWMOTE-majority weighted minority oversampling technique for imbalanced data set learning. IEEE Trans. Knowl. Data Eng. **26**(2), 405–425 (2014)
8. Chongfu, H.: Demonstration of benefit of information distribution for probability estimation. Signal Process. **80**, 1037–1048 (2000)
9. Shigang, L., Honghua, D., Min, G.: Information-decomposition-model-based missing value estimation for not missing at random dataset. Int. J. Mach. Learn. Cybern. 1–11 (2015)
10. Alcala-Fdez, J., et al.: KEEL: a software tool to assess evolutionary algorithms for data mining problems. Soft Comput. **13**(3), 307–318 (2009)
11. Shirabad, J.S., Menzies, T.J.: The PROMISE Repository of Software Engineering Databases. School of Information Technology and Engineering, University of Ottawa, Canada (2005). http://promise.site.uottawa.ca/SERepository
12. Cano, A., Zafra, A., Ventura, S.: Weighted data gravitation classification for standard and imbalanced data. IEEE Trans. Cybern. **43**, 1672–1687 (2013)
13. Lin, M., Tang, K., Yao, X.: Dynamic sampling approach to training neural networks for multiclass imbalance classification. IEEE Trans. Neural Netw. Learn. Syst. **24**, 647–660 (2013)
14. Hulse, V., Jason, T.M., Khoshgoftaar, A.N.: Experimental perspectives on learning from imbalanced data. In: Proceedings of the 24th International Conference on Machine Learning. ACM (2007)
15. Quinlan, J.R.: C4.5: Programs for Machine Learning. Elsevier, New York (2014)
16. Mirkes, E.: KNN and Potential Energy (Applet). University of Leicester (2011)

An Enhanced Support Vector Machine for Faster Time Series Classification

Thapanan Janyalikit[1], Phongsakorn Sathianwiriyakhun[1], Haemwaan Sivaraks[2], and Chotirat Ann Ratanamahatana[1(✉)]

[1] Department of Computer Engineering, Chulalongkorn University, Phayathai Rd., Pathumwan, Bangkok 10330, Thailand
{Thapanan.Ja,Phongsakorn.Sa}@student.chula.ac.th, chotirat.r@chula.ac.th
[2] PTT ICT Solution Company Limited, Vidhavadi-Rangsit Rd., Chatuchak, Bangkok 10900, Thailand
haemwaan@gmail.com

Abstract. As time series have become a prevalent type of data, various data mining tasks are often performed to extract useful knowledge, including time series classification, a more commonly performed task. Recently, a support vector machine (SVM), one of the most powerful classifiers for general classification tasks, has been applied to time series classification as it has been shown to outperform one nearest neighbor in various domains. Recently, Dynamic Time Warping (DTW) distance measure has been used as a feature for SVM classifier. However, its main drawback is its exceedingly high time complexity in DTW computation, which degrades the SVM's performance. In this paper, we propose an enhanced SVM that utilizes a DTW's fast lower bound function as its feature to mitigate the problem. The experiment results on 47 UCR time series data mining datasets demonstrate that our proposed work can speed up the classification process by a large margin while maintaining high accuracies comparing with the state-of-the-art approach.

Keywords: Classification · Support vector machine (SVM) · Dynamic time warping (DTW) · Lower bound · LB_keogh

1 Introduction

Time series classification is widely used in many domains such as medicine, finance, and science [1–4]. One nearest neighbor (1-NN) classifier with Dynamic Time Warping (DTW) distance function is one of the most popular time series classifiers that has been demonstrated to work extremely well and is difficult to beat [5]. Although 1-NN with DTW is quite effective, it still has the known weaknesses in its high computational time and its sensitivities to noises in the training set. Figure 1 illustrates a 2-class problem with six training data sequences, one of which contains noises (candidate sequence 3). When we want to classify a new sequence (with the real class *B* label), the 1-NN with DTW could wrongfully classify it to class *A*, as it matches the best to sequence 3 that is contaminated by noises.

© Springer-Verlag Berlin Heidelberg 2016
N.T. Nguyen et al. (Eds.): ACIIDS 2016, Part I, LNAI 9621, pp. 616–625, 2016.
DOI: 10.1007/978-3-662-49381-6_59

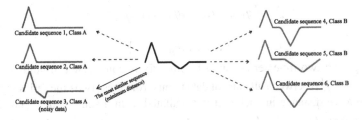

Fig. 1. One-nearest neighbor classification with noisy training data

To mitigate this problem, many approaches have been proposed, including Longest Common Subsequence (LCSS) [22] and variation of edit distances [23, 24]. However, they still suffer from high computational cost and difficulty in parameter selection. Some other researchers have proposed Support Vector Machine (SVM) that combined together the benefits of DTW by using DTW distance as the feature [6] to solve time series classification problems [7–9]. SVM is a powerful algorithm that can be applied in many domains because it builds a classifier from a hyperplane that separates features of input or exploits mapping of features into higher dimensional space through kernel function in non-linearly separable problems. It can also ignore some noisy data points that may lead to wrong classification. Recently, some research works have attempted to establish a kernel for DTW [7, 8] or have attempted to use other features together with DTW to increase the accuracy of SVM [9]. However, complexity and computational time for feature construction are still their weaknesses. In this work, we therefore mainly focus on reducing computational time, especially in the feature vector construction step, while being able to maintain high accuracies, comparing with the real DTW distance being used as a feature for SVM. Specifically, we take both the advantages of a lower bound function for DTW and robustness of SVM to build a faster yet accurate time series classifier.

The rest of the paper is organized as follows. The next section gives brief details on background and related works. Section 3 explains our proposed work with details on our experiment setup. Section 4 provides the experiment results and discussion, and the last section concludes our work.

2 Background and Related Works

In this section, we briefly give some details about Dynamic Time Warping (DTW), global constraint, Lower bound function for DTW, and Support Vector Machine (SVM).

2.1 Dynamic Time Warping (DTW)

Dynamic Time Warping (DTW) is a similarity measure for time series data that has gained its popularity over Euclidean distance metric due to its higher accuracy and more flexibility in non-linear alignments between two data sequences. Given two time series sequences, Q and C of length m and n, respectively, as follows.

$$Q = \{q_1, q_2, q_3, \ldots, q_i, \ldots, q_m\} \qquad (1)$$

$$C = \{c_1, c_2, c_3, \ldots, c_j, \ldots, c_n\} \qquad (2)$$

To align the two time series, an m by n matrix is constructed to hold cumulative distances ($\gamma(i^{th}, j^{th})$) between all pairs of data points from these two time series sequences. The cumulative distance in each cell is calculated from a sum of the distance of the current element, $d(q_i, c_j) = (q_i - c_j)^2$, and the minimum of the cumulative distance of the three adjacent elements:

$$\gamma(i,j) = d(q_i, c_j) + \min\{\gamma(i-1,j), \gamma(i,j-1), \gamma(i-1,j-1)\} \qquad (3)$$

Dynamic programming is used to obtain optimal alignment and cumulative distance from the elements $(1, 1)$ to (m, n). The optimal warping path is obtained by backtracking. However, in most time series mining tasks, we often are interested in only the distance value, contained in (m, n). So, backtracking can be omitted. The DTW distance is calculated by Eq. (4). Note that the squared root can also be omitted to speed up the calculation.

$$DTW(Q, C) = \sqrt{\gamma(q_m, c_n)} \qquad (4)$$

2.2 Global Constraint

To improve the performance of DTW, a global constraint is often applied to prevent unreasonable warping, which in turn increases the accuracy. While any shapes and sizes of the global constraints can be utilized, Sakoe-Chiba Band [10] is one of the simplest global constraints [11] that restricts the warping path to stay within specific boundary (r), as shown in Fig. 2. Equation (3) with the Sakoe-Chiba constraint can be rewritten as follows:

$$\gamma(i,j) = d(q_i, c_j) + \min\{\gamma(i-1,j), \gamma(i,j-1), \gamma(i-1,j-1)\} ; i+r \geq j \text{ and } i - r \leq j \qquad (5)$$

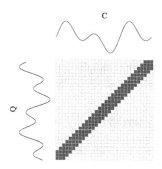

Fig. 2. The Sakoe-Chiba band restricts the warping path to stay within the boundary.

2.3 Lower Bound of DTW

Due to its $O(n^2)$ time complexity, DTW is considered too expensive for large time series data. In the past decade, various lower bound functions for DTW have been proposed to speed up the DTW calculation for data mining tasks [12–14]. The main idea of lower bound function is to approximate the real DTW distance with the crucial property that the computed lower bound will never exceed the real DTW distance. Such property will allow us to skip many expensive DTW calculations when the lower bound value is greater than the best-so-far distance. Regardless of many well proposed lower bounds [25, 26], we consider LB_Keogh [14] in this work due to its simplicity and tightness of the bound. However, LB_Keogh requires a DTW global constraint (r) to construct an upper (U) and lower (L) envelopes around any sequence Q, as follows:

$$U_i = max\left(q_{i-r} : q_{i+r}\right) \tag{6}$$

$$L_i = min\left(q_{i-r} : q_{i+r}\right) \tag{7}$$

Once the envelope is constructed, the lower bound value between a sequence C and the envelope is computed using Eq. (8). Any part of the sequence beyond the envelope (above or below) will be included as part of the lower bound, as shown in Fig. 3.

$$LB\_Keogh\left(Q, C\right) = \sum_{i=1}^{n}\begin{cases} \left(U_i - c_i\right)^2 & if\ c_i > U_i \\ \left(L_i - c_i\right)^2 & if\ c_i < L_i \\ 0\ otherwise \end{cases} \tag{8}$$

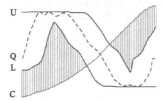

Fig. 3. An illustration of the upper and lower envelope in LB_Keogh function. The lower bound value is computed by the sum of the squared distance between the upper or lower envelope and the sequence C that are outside the envelope area.

2.4 Support Vector Machine (SVM)

SVM is a powerful classification algorithm that has been applied to a variety of problems and domains, including finance and medicine [15, 16]. SVM is based on the idea of constructing a hyperplane to separate the data that may be mapped to higher dimensional feature space using a kernel function. In the other words, if the data can be linearly separated, a hyperplane is constructed from input data, which gives a maximum margin called support vector as shown in Fig. 4. Otherwise, a kernel function is applied to map the input space into higher dimensional feature space that can be linearly separated by a hyperplane [17] as shown in Fig. 5.

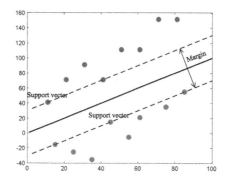

Fig. 4. Support vector machine with linearly separable data.

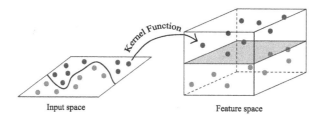

Fig. 5. Support vector machine uses a kernel function to map non-linearly separable data from input space into higher dimensional space that can be separated by a hyperplane.

2.5   Related Works

Some works have used DTW distance as an SVM's feature for time series classification problem [7–9]. Gudmundsson, S. et al. used DTW as a feature and concentrated on designing a complex SVM kernel so called ppfSVM [8], but the results turned out that it still could not beat 1-NN with DTW even though their SVM achieved better results over other existing SVM approaches. Most recently, Kate, R. J. found that employing a simple polynomial kernel function is sufficient with no need to design a complex kernel function [9]. Together with the use of other features such as Symbolic Aggregate approximation (SAX) [18], Euclidean distance, and DTW with global constraint, the results have showed to even outperform 1-NN using DTW with global constraint. However, regardless of its winning performance on accuracies, the time complexity is still its drawback. Therefore, this motivates us to aim at building a faster time series classifier while maintaining its robustness in achieving high classification accuracies.

3   Proposed Work and Experiment Setup

In this section, we provide details of our proposed work, which creates features based on LB_Keogh values that will be used in SVM classifier. Then, we also provide information about datasets and tools used in the experiments.

3.1   LB_Keogh Feature Construction

As discussed in Sect. 2.5, instead of using DTW with global constraint distance as a feature, which is extremely expensive, LB_Keogh with a global constraint is used to construct the feature. For each training data sequence, LB_Keogh feature vector is constructed by computing lower bound values from that data sequence to every single sequence in the training set. Testing is done very similarly; LB_Keogh feature vector is obtained from computing lower bound values from each of the test data to every training data sequence. More concretely, given a training dataset C with k time series sequences, a feature vector of any sequence Q that will be used in a classifier is defined as follows.

$$LB\_Keogh\_Feature\,(Q) = \{LB\_Keogh\,(Q, C_1)\,, LB\_Keogh\,(Q, C_2)\,, \dots, LB\_Keogh\,(Q, C_k)\} \qquad (9)$$

3.2   Making a Classifier

Once the LB_Keogh feature is obtained, LIBSVM [20] was used to build an SVM classifier with the following parameters: SVM type = C-SVC, Penalty parameter $C = 1$, and Kernel type = polynomial with coefficient and gamma = 1. We also perform 10-fold cross validation on each training set to determine the polynomial degree that fits the best to the problem. After the classifier is built for each dataset, its performance is tested using its corresponding unseen test data.

3.3   Datasets and Experiment Setup

In this work, we employ the whole 47 benchmark datasets from the UCR Time Series Classification Archive [19]. The information about the datasets (e.g., the number of classes, sequence length, train/test split) can be found on the archive. For each dataset, the DTW's global constraint parameter r is set according to what is reported on the UCR Time Series Mining webpage [19]. The performance is evaluated by comparing classification error rates as well as the computational time in feature vector construction of our proposed work (LB_Keogh_Feature) with the state-of-the-art SVM classifier with DTW-R distance as the feature proposed by [9]. All of the experiments were carried out using Weka (v.3.6.12) software [21] on core i7-3770 3.40 GHz CPU with 16 GB of RAM.

4   Experiment Results and Discussion

Figure 6 visually compares the classification error rates between our proposed SVM using LB_Keogh_Feature and the most recently proposed SVM using DTW-R_Feature.

Note that the points above the diagonal line denote the datasets on which the rival method DTW-R_Feature performs better (with lower error rate), and the points below the diagonal line denote the datasets on which our proposed LB_Keogh_Feature performs better, achieving lower error rate. The red triangular points (7 out of 47 datasets) denote the discrepancies in the error rates that are statistically significant at 5 % level. We can see that the error rates between the two approaches are mostly very comparable, with the exception of the following six datasets, Cricket_Y, FaceAll, FacesUCR, Lighting2, OSULeaf, and Two_Patterns, where DTW-R_Feature is better, and one dataset, CinC_ECG_torso, where DTW-R_Feature is worse.

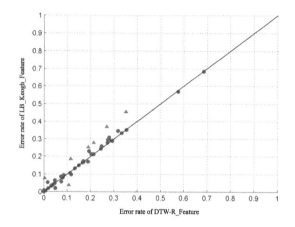

Fig. 6. A comparison between the classification error rates using our proposed LB_Keogh_Feature and the state-of-the-art DTW-R_Feature on all 47 datasets. The red triangular points indicate the discrepancies in the error rates that are statistically significant at 5 % level (Color figure online).

LB_Keogh_Feature generally works very well on most datasets. However, we have looked into the feature vector of the six datasets where it performed statistically worse than DTW-R_Feature. In particular, for each dataset, we sum up all the feature vectors in the training data and compare between the feature vectors from our proposed LB_Keogh_Feature and the rival DTW-R_Feature. Figure 7 shows some snippets of the feature vectors in each class of three datasets, i.e., Car, Cricket_Y, and Lighting2; the blue lines represent the LB_Keogh_Feature, and the red lines represent the DTW-R_Feature. Note that both algorithms have the same error rate in Car dataset, as it shows very similar feature vectors, whereas DTW-R_Feature outperforms in the other two datasets. In the latter case, we can see that since the features from LB_Keogh are essentially lower bounds or approximation to the real DTW-R distance, the blue lines show that the LB_Keogh_Feature generally underestimates the distance, i.e., achieving not very tight lower bounds, which could in essence affect the classification accuracy.

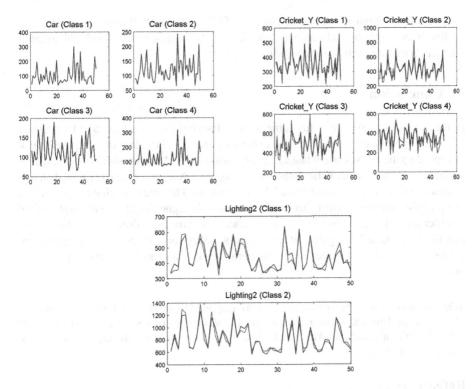

Fig. 7. Some snippets of the sum of the feature vectors within each class of the three training data, comparing between those from our proposed LB_Keogh_Feature in blue and those from the rival DTW-R_Feature in red (Color figure online).

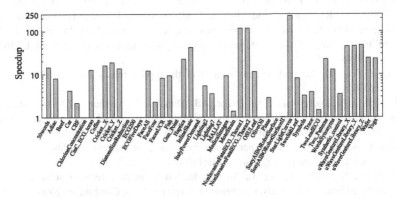

Fig. 8. The speedups of our proposed LB_Keogh_Feature over the rival method on all 47 datasets

The computational time in feature vector construction of both approaches are recorded. The speedups of our proposed LB_Keogh_Feature over the rival method are shown in Fig. 8. We can see that LB_Keogh_Feature vector is faster to be constructed in almost all datasets, except for 12 datasets whose global constraint values are zero (i.e., Euclidean distance is performed instead of DTW), where both approaches consume similar time complexity. The average speedup of all 47 datasets is 19.79 (26.24 speedup if 12 Euclidean datasets are excluded).

5   Conclusion

In this work, we propose a fast time series classification based on SVM that used LB_Keogh distance as a feature. The experiment results show that our proposed work can speed up the classification tasks by a large margin, while being able to maintain high accuracies, comparing with the state-of-the-art approach where the real DTW-R distance is used as a feature. It is also observed that our proposed feature can achieve comparable classification accuracies (little lower in some and little higher in some, with no statistical significance). However, in some datasets where the error rates drop down a little, statistically differing at 5 % significant level, good speedups on all the datasets are achieved, indicating the tradeoff between the classification accuracies and the running time it could save.

Acknowledgements. This research is partially supported by the Thailand Research Fund and Chulalongkorn University given through the Royal Golden Jubilee Ph.D. Program (PHD/ 0057/2557 to Thapanan Janyalikit) and CP Chulalongkorn Graduate Scholarship (to Phongsakorn Sathianwiriyakhun).

References

1. Sivaraks, H., Ratanamahatana, C.A.: Robust and accurate anomaly detection in ECG artifacts using time series motif discovery. Comput. Math. Methods Med. **2015**, 1–20 (2015)
2. Kurbalija, V., Radovanović, M., Ivanović, M., Schmidt, D., von Trzebiatowski, G.L., Burkhard, H.D., Hinrichs, C.: Time-series analysis in the medical domain: A study of Tacrolimus administration and influence on kidney graft function. Comput. Biol. Med. **50**, 19–31 (2014)
3. Zeng, Z., Yan, H.: Supervised classification of share price trends. Inf. Sci. **178**(20), 3943–3956 (2008)
4. Mahabal, A.A., Djorgovski, S.G., Drake, A.J., Donalek, C., Graham, M.J., Williams, R.D., Larson, S.: Discovery, classification, and scientific exploration of transient events from the Catalina Real-time Transient Survey. arXiv preprint (2011). arXiv:1111.0313
5. Bagnall, A., Lines, J.: An experimental evaluation of nearest neighbour time series classification. arXiv preprint (2014). arXiv:1406.4757
6. Cristianini, N., Shawe-Taylor, J.: An introduction to support vector machines and other kernel-based learning methods. Cambridge University Press, Cambridge (2000)
7. Jeong, Y.S., Jayaraman, R.: Support vector-based algorithms with weighted dynamic time warping kernel function for time series classification. Knowl.-Based Syst. **75**, 184–191 (2015)
8. Gudmundsson, S., Runarsson, T.P., Sigurdsson, S.: Support vector machines and dynamic time warping for time series. In: IEEE International Joint Conference on Neural Networks, IJCNN 2008, (IEEE World Congress on Computational Intelligence), pp. 2772–2776. IEEE (2008)

9. Kate, R. J.: Using dynamic time warping distances as features for improved time series classification. Data Min. Knowl. Discov., pp. 1–30 (2015)
10. Sakoe, H., Chiba, S.: Dynamic programming algorithm optimization for spoken word recognition. IEEE Trans. Acoust. Speech Signal Process. **26**(1), 43–49 (1978)
11. Niennattrakul, V., Ratanamahatana, C.A.: Learning DTW global constraint for time series classification. arXiv preprint (2009). arXiv:0903.0041
12. Park, S., Lee, D., Chu, W.W.: Fast retrieval of similar subsequences in long sequence databases. In: Proceedings of the Workshop on Knowledge and Data Engineering Exchange, (KDEX 1999), pp. 60–67. IEEE (1999)
13. Kim, S.W., Park, S., Chu, W.W.: An index-based approach for similarity search supporting time warping in large sequence databases. In: Proceedings of 17th International Conference on Data Engineering, pp. 607–614. IEEE (2001)
14. Keogh, E., Ratanamahatana, C.A.: Exact indexing of dynamic time warping. Knowl. Inf. Syst. **7**(3), 358–386 (2005)
15. Zhiqiang, G., Huaiqing, W., Quan, L.: Financial time series forecasting using LPP and SVM optimized by PSO. Soft. Comput. **17**(5), 805–818 (2013)
16. Labate, D., Palamara, I., Mammone, N., Morabito, G., La Foresta, F., Morabito, F.C.: SVM classification of epileptic EEG recordings through multiscale permutation entropy. In: The 2013 International Joint Conference on Neural Networks (IJCNN), pp. 1–5. IEEE (2013)
17. Awad, M., Khan, L., Bastani, F., Yen, I.: An effective support vector machines (SVMs) performance using hierarchical clustering. In: 16th IEEE International Conference on Tools with Artificial Intelligence, ICTAI 2004, pp. 663–667. IEEE (2004)
18. Lin, J., Keogh, E., Lonardi, S., Chiu, B.: A symbolic representation of time series, with implications for streaming algorithms. In: Proceedings of the 8th ACM SIGMOD Workshop on Research Issues in Data Mining and Knowledge Discovery, pp. 2–11. ACM (2003)
19. The UCR Time Series Classification Archive. http://www.cs.ucr.edu/~eamonn/time_series_data/
20. Chang, C.C., Lin, C.J.: LIBSVM: A library for support vector machines. ACM Trans. Intell. Syst. Technol (TIST) **2**(3), 27 (2011)
21. Hall, M., Frank, E., Holmes, G., Pfahringer, B., Reutemann, P., Witten, I.H.: The WEKA data mining software: an update. ACM SIGKDD Explor. Newsl. **11**(1), 10–18 (2009)
22. Das, G., Gunopulos, D., Mannila, H.: Finding similar time series. In: Komorowski, J., Żytkow, J.M. (eds.) PKDD 1997. LNCS, vol. 1263, pp. 88–100. Springer, Heidelberg (1997)
23. Marteau, P.F.: Time warp edit distance with stiffness adjustment for time series matching. IEEE Trans. Pattern Anal. Mach. Intell. **31**(2), 306–318 (2009)
24. Chen, L., Ng, R.: On the marriage of lp-norms and edit distance. In: Proceedings of the Thirtieth International Conference on Very Large Data Bases, vol. 30, pp. 792–803. VLDB Endowment (2004)
25. Sakurai, Y., Yoshikawa, M., Faloutsos, C.: FTW: fast similarity search under the time warping distance. In: Proceedings of the Twenty-Fourth ACM SIGMOD-SIGACT-SIGART Symposium on Principles of Database Systems, pp. 326–337. ACM (2005)
26. Rakthanmanon, T., Campana, B., Mueen, A., Batista, G., Westover, B., Zhu, Q., Keogh, E.: Searching and mining trillions of time series subsequences under dynamic time warping. In: Proceedings of the 18th ACM SIGKDD International Conference on Knowledge Discovery and Data Mining, pp. 262–270. ACM (2012)

Parallel Implementations of the Ant Colony Optimization Metaheuristic

Andrzej Siemiński[(✉)]

Faculty of Computer Science and Management,
Technical University of Wrocław, Wrocław, Poland
Andrzej.Sieminski@pwr.edu.pl

Abstract. The paper discusses different approaches to parallel implementation of the Ant Colony Optimization metaheuristic. The metaheuristic is applied to the well-known Travelling Salesman Problem. Although the Ant Colony Optimization approach is capable of delivering good quality solutions for the TSP it suffers from two factors: complexity and non-determinism. Overpopulating ants makes the ACO performance more predictable but increasing the number of ants makes the need for parallel processing even more apparent. The proposed Ant Colony Community (ACC) uses a coarse grained approach to parallelization. Two implementations using RMI and Sockets respectively are compared. Results of an experiment prove the ACC is capable of a substantial reduction of processing time.

Keywords: Ant colony optimization · Travelling salesman problem · Parallel implementations · RMI · Sockets

1 Introduction

The aim of the paper is to discuss the problem of parallel implementation of Ant Colony Optimization (ACO) metaheuristic used for the Travelling Salesman Problem (TSP). The ACO requires a large number of floating point operations and is therefore time consuming. Recent papers suggest that the increasing the number of ants has distinct advantages. Overpopulating of ants make the parallelization even more necessary.

The paper is organized as follows. The Sect. 2 briefly introduces the Travelling Salesman Problem and the version of the ACO used to solve it. The main contribution of the paper – the Ant Colony community (ACC) is introduced in the Sect. 3 and its implementations in the Sect. 4. The experiments, their results as well as the criteria used to evaluate the results are presented in the Sect. 5. The paper concludes with an overview future investigation areas.

2 Using Ant Colony Optimization Heuristic for TSP

From the theoretical point of view the TSP had been proven to be a NP-hard problem. It has become a touchstone in the area of AI. A recent comparison of metaheuristics used for TSP could be found in [1].

© Springer-Verlag Berlin Heidelberg 2016
N.T. Nguyen et al. (Eds.): ACIIDS 2016, Part I, LNAI 9621, pp. 626–635, 2016.
DOI: 10.1007/978-3-662-49381-6_60

This paper deals with the classical, static statement of the problem. The Dynamic TSP (DTSP) was introduced for the first time by Psarafits [2] are now subject of intensive study see e.g. [3–5]. The experiments reported in the paper were conducted on static distance matrix but there is strong indication that they could be useful for the DTSP.

2.1 Ant Colony Optimization

The Ant Colony Optimization (ACO) technique was first introduced by M. Dorigo in as early as in 1992 [6]. His extensive overview of the current research on the area is presented in [7]. An Ant Colony consists of ants which are extremely simple agents. They are capable only of moving from one node to another laying a pheromone trail on their way as well as detecting their current position, remembering already visited nodes, sensing the direct distances from its current position to other nodes and the amount of pheromone laid upon them. The colony works in iterations. At the start of each iteration the ants are placed randomly on the graph. Each ant works on its own, completing a route that connects all nodes. In each step of an iteration an ant, sensing the distance and pheromone levels placed on routes connecting nodes, selects the next node to visit. The iteration stops when all cities are included in an ants' route. The pheromone matrix represents the collective experience gained by all ants. The colony remembers the BSF (Best So Far) route, the current iteration number and the iterations' best route. The colony is responsible also for global pheromone updating.

The exact formulas necessary to operate an ACO are not presented here as they could be found in many other papers e.g. [8]. The key factors should be however stressed:

- The ACO metaheuristic requires a very large number of floating point operations. As a result the optimization process takes a few minutes on up-to-date computers even for moderate number of nodes.
- The process is no-deterministic, the same set of data and parameters could produce different solutions.
- There is no clear indication when to stop the optimizing process. As the number of iteration grows the computational effort is less and less productive.
- The optimization is controlled by a number of parameters. Various attempts to identify values offering better results were described e.g. in [9, 10] or [11] but they do not provide universal solution. There are also approaches that use fuzzy logic to facilitate the optimization process [12, 13].

The route length is not the only quality measure. The other one is execution time. In the study reported in this paper we deal with both of them. and in particular we study the correlation between the number of ants and the processing time and route lengths.

2.2 Increasing the Number of Ants

Overpopulating does not inevitably result in growing the execution time. The increased number of ants could be offset by the decreased number iterations. The Table 1 shows average values of the BSF for different sizes of colony and varying number of iterations. The numbers of iterations are shown in square brackets. Iteration numbers in the last column were selected in such a way as to preserve the same the computational complexity in each raw. The over populating of a single ant colony does not offer a significant reduction in route length but neither it prolongs execution.

Table 1. Observed average BSF route for varying colony populations, number of iterations, number of runs 30

Ant Colony Size	BSF fixed size	BSF normalized
30	1.82 [1000]	1.80 [1666]
50	1.81 [1000]	1.81 [1000]
80	1.80 [1000]	1.80 [625]
120	1.80 [1000]	1.85 [417]
150	1.79 [1000]	1.80 [333]
1000	1.78 [1000]	1.78 [50]

In the multi colony approach the ants are distributed over more than one colony. Their number in a colony could also exceed regular values as well. Using many colonies working in parallel brings up two problems:

- The necessary communication overhead must not diminish the advantages of speed-up due to parallelism.
- The lack or reduction of cooperation of ants from different colonies must have not much effect upon the quality of the resulting solution.

The theoretical analysis of the first problem is presented in [14]. The study on the second phenomena is for more complex and is delayed until the Sect. 5.

3   Ant Colony Community

The Ant Colony Community (ACC) consists of a number of colonies controlled by a server. The structure of the ACC is highly flexible. The colonies could run a single computer, many computers in a local network or on computers on internet servers. An exemplary structure with 4 computers and 5 Ant Colonies is presented on the Fig. 1.

The CS (Community Server) awaits for calls from an Ant Colony Client (ACC). When the CS receives a request from an ACC, it registers the client in its repository and creates a separate thread for handling data transmission. In the next step the CS passes parameters and data to the ACC. Depending on the type of CS the data can contain distance matrix, pheromone matrix and previously found routes. After the transmission has been completed the Community Server waits for a solution sent by one of the registered clients. An Ant Colony Client in its turn creates a local ant colony, feeds it with the data and parameters received from the server and starts the execution of the local colony. An ant colony produces a solution and passes it to ACC which in turn transmits it further to the Community server. The process of receiving parameters and sending solutions is repeated as long as necessary.

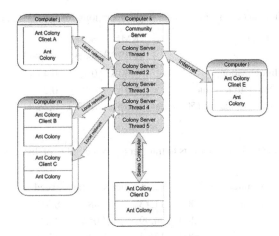

Fig. 1. An example of an ant colony community

The described above structure is relatively simple and at the same time both robust and flexible. The complex task of finding solutions is separated from the rest of the implementation. Replacing one type of an ant colony by another is relatively simple. The Community adopts itself automatically to the differences in processing power and/or connection transfer speeds. The "slower" clients, hosted e.g. on a remote Internet server are just less frequently assigned tasks by the Colony Server. Even dropping out of a computer from the Community does not disintegrate its operations.

4 Comparison of Corse Grain Implementations

The proposed implementation uses the Sockets mechanism. It keeps permanent connection between two processes until one of them closes it or stops operation. They communicate over network addresses. In the previous work the communication between the server and clients used the Java Remote Method Invocation which is a relatively high level mechanism [18]. It offers the developer many advantages. Once the connection has been established the code for handling local and remote objects is almost identical. This gives the developer full compiler support. The implementation and debugging of network programs is not much different from traditional programming. The complexity of organizing the data flow is handled by the compiler. Using the RMI is therefore most beneficial when the interaction pattern is complex.

The RMI enable us to have a parallel implementation with just one remote colony object that resides on a server and hosts the distance and pheromone matrixes. The individual ants could be located on client computers. The RMI could be easily used for fine-grained parallelization approaches.

The RMI looks attractive at first but it is actually less useful at all when it comes to an actual implementation. The time necessary for a remote procedure call to fetch a double value is approximately equal to 0.39 ms even when the client and the server reside on the same machine. Fetching the same value from a local object is equal to

Table 2. Time necessary to perform basic operations for the RMI and sockets.

Operation	RMI network	Sockets local	network
Initialization	1.30 s.	0.75 s.	0.45 s.
One double value	0.01 s.	0.01 s.	0.23 s.
Distance matrix (50 Nodes)	1.60 s.	0.03 s.	0,26 s.

Table 3. Time necessary for the Sockets to transfer a given number of floating point numbers (in milliseconds)

Connection	#	mean time	std. dev.	speed
Local	100	11.13	6.48	8.98
	1200	34.99	7.054	34.31
	4900	136.31	12.35	35.95
	19800	511.37	96.89	38.72
Network	100	228.75	12.55	0.44
	1200	261.27	39.37	4.59
	4900	342.85	31.13	14.29
	19800	726.55	330.26	27.25

0,001785 ms so it is two orders of magnitude faster. Calling a method with a single double parameter when an object is on a network computer is much slower – it takes 10 ms, see the Table 2. The passing of a pheromone matrix takes almost 2 s. In the case of the coarse grained parallel implementation of the Ant Colony Community the data flow pattern is straightforward and it could be implemented without using remote objects. All we need is pure data transfer. Such a functionality is offered by sockets.

The Sockets in an apparent contrast to the RMI do not provide remote objects and offers only means for data transfer. The programmer is solely responsible for ensuring the correctness of data flow. The communication between a server and a client requires many types of messages being transferred so this may sound prohibitive. Fortunately enough the JVM supports the transfer of any serializable object. This makes the task less daunting. Using serialized objects by the socket mechanism does not mean that we have access to objects methods. Only the data is transferred.

The data in the Table 2 shows the difference in performance between the RMI an Sockets. The RMI is optimized for parameter passing what is common for method evocation but are less efficient for passing large amounts of data. The sockets mechanism is optimized for transferring large blocks of data over the network as is clearly visible from the measurements from the Table 3. The transfer rate steadily increases with the size of data being transmitted and for the largest block it does not differ much from the rate achieved on a local server.

In the traditional client – server mode of operation thin clients call a server to perform complex calculations or obtain data from a centralized database. In the parallel implementation of the ACO the mode of operation is reversed. The bulk of computations is done by the clients.

5 Description of Experiments

The experiments were run on a network of 4 computers with codenames from Ca to Cd. The Ca computer was the fastest one: Intel i7-4700MQ 2.4 GHz 2.4 Ghz. The Cc computer equipped with Inter Core2Quad D6600 was the slowest one. The difference in computing power had an impact on the structure of the Communities.

Table 4. The efficiency of hosting many colonies on a single computer with SSD drive.

Number of Colonies	Total time	Time per Colony
1	81	81.0
2	83	41.5
3	86	28.7
4	89	22.2
5	100	20.0
6	115	19.8

All of the computers were powerful enough to host many colonies. Each colony has run as a separate process. The Table 4 shows the time used to run various number of an ant colony optimization tasks simultaneously on the computer Cc. Running simultaneously 5 colonies has increased the processing time by mere 25 % from 81 to 100 s while the time per colony has decreased fourfold from 81 to just 20 s. The Cc computer was equipped with a SSD drive. For computers with traditional hard disk drives the time per colony starts to increase for smaller number of colonies. The Cd computer was used as a server. The Cb computer, ranked in the middle according to processor power was used for reference purposes.

The structure of the tested communities is presented in the Table 5. The last column shows the computers and number of ant colonies that they hosted.

Table 5. Structure of ant communities

Code	Ant Colonies #	Computer /Colony number
TA	1	Cb/1
TB	3	Ca/3
TC	7	Ca/4; Cb/3
TD	12	Ca/6; Cb/4; Cc/2

The schema for ranking Ant Colony Communities respects the following principles:

- All communities use the same distance matrix. This does not eliminate non-determinism of their operation but they have at least identical task to solve.
- To evaluate a Community we run it several times using the same distance matrix.

- Instead of using mean of route lengths we used ranks of route lengths.
- There are two criteria for stopping the run:

  - Equal Complexity – the computational complexity measured by the number of node selection operations is the same for all Colonies. The time needed to find a solution could be different and depends on the architecture of the Community.
  - Equal Time – the clock time given to all Colonies is the same. The complexity of operation could be different and it is the sum of the complexities of individual Colonies that make up a Community.

Let *Len(Cx, Tk)* denote the length of the best route length found by the Community x in the k-th run. The process of ranking the communities starts the calculation of *PRV (Cx,Tk)* - Partial Ranking Value for each test run and each community. To rank the Communities we use the *RV(Cx)* which is sum their ranking values of their tasks.

$$PRV(C_x, T_k) = \sum_{i=1}^{N} \begin{cases} \dfrac{1 : if\ Len(C_x, T_k) < Len(C_i, T_i)}{0 : otherwize} \end{cases}$$

where N is the number of communities.

$$RV(C_x) = \sum_{i=1}^{M} PRV(C_x, T_i)$$

where M is the number of test runs.

The Equal Complexity criterion does not need much justification. It is well established in the computer science. It does not mean that it should be the only one used. Two Communities could have similar ranking values but could differ substantially in their processing time. The structure of a community does not have any influence upon the value of this criterion.

The Equal Time criterion is used to select a Community that is most likely to find the best solution in a given period of time. On many occasions the processing time is more important than the difference in route length. For Dynamic TSP's a community has to find solutions fast enough to adopt to changes in the route matrix. The proposed solution scales very easily and so adding more Colonies could sufficiently seed up the processing.

5.1 Equal Complexity Criterion

The Table 6 shows the ranking of Communities according the Equal Complexity criterion. The reference community (code name A) consisted of one colony with 50 ants and running 800 iterations, called in what follows the Standard Community Run. All the other communities have preserved the level of computational complexity: the increase in the number of colonies (the second column) and/or the number of ants in a colony was offset by the lower values of iteration numbers.

Table 6. Ranking of communities according the equal complexity criterion

Code	#	Ant #	Iter. #	1	2	3	4	5	6	7	8	9	10	RV
A	1	50	800	5	5	3	5	3	4	0	1	4	1	31
B	2	50	400	1	0	5	3	5	5	4	4	1	5	33
C	2	100	200	0	3	1	0	1	1	3	2	3	0	14
D	4	50	200	4	1	2	1	0	2	1	3	5	3	22
E	8	50	100	2	2	0	2	4	3	2	5	2	1	23
F	8	75	67	3	4	3	4	2	0	5	0	0	4	25

The good performance of the Standard Colony A comes as a no surprise. Its parameters were carefully chosen after running many experiments. We should not forget that it is much slower than the rest of colonies. The winner is the Community B which doubles the number of colonies and keeps relatively large number of iterations. It looks like the iteration number close to 400 is enough to achieve acceptable results.

5.2   Equal Time Criterion

In the Equal Time test the Communities have different structures. The selected time span was equal to the time necessary for the Cb computer to complete Standard Community Run. During the test all colonies were activated at the same time are were allowed to run for the mentioned above time span. After that they were stopped and the best found solution was recorded. Running more colonies on a single computer slows the execution and therefore in two cases the number of iterations was lowered from 400 to 350. This was to enable an ant colony to complete a task within the allowed time period.

The achieved results are presented in the Table 7. The Standard TA community is this time the looser. What is however worth noting is the performance of the TD community. It has a relatively small number of ants, very small amount of iterations and still it the occupies the second position in the ranking. This makes the community a good choice for dynamic TSP as discussed in Sect. 4.

Table 7. Ranking of communities according the equal time criteria

Code	Iter. #	Ant #	Colony	1	2	3	4	5	6	7	8	9	10	RV
TA	800	50	1	0	1	3	0	0	3	0	1	0	0	8
TB	350	50	3	1	3	4	3	3	1	4	4	2	4	29
TC	350	50	7	5	5	5	5	5	5	5	2	4	4	45
TD	100	50	12	3	4	1	4	4	3	3	3	5	3	32
TD	50	100	12	4	2	0	1	2	0	1	4	3	2	19
TD	100	100	12	2	0	1	2	1	2	2	0	1	1	12

6 Conclusions and Future Work

The paper presents initial studies in the performance of Ant Colony Communities. Such a community is a coarse grained parallel implementation of the Ant Colony Optimization algorithm. It has a very flexible structure. The server could coordinate the work of practically any number of individual colonies located upon one or many computers located in a local or wide area network. It uses low level socket mechanism which better suits the task than the previously proposed RMI mechanism.

The structure of the implementation is flexible, it could be adopted to the needs of changing environment. Computers with different processing power could cooperate easily. From the practical point of view it is important, that the approach enable us to harvest the spare computer power which is available for free on almost all computers. Computers working on a typical network have processor utilization less than 5 % most all the time.

The successful application of the ACC to other more refined versions of the ACO would be the final proof of its usefulness. The work on the area is currently underway.

References

1. Antosiewicz, M., Koloch, G., Kamiński, B.: Choice of best possible metaheuristic algorithm for the travelling salesman problem with limited computational time: quality, uncertainty and speed. Journal of Theoretical and Applied Computer Science 7(1), 46–55 (2013)
2. Psarafits, H.N.: Dynamic vehicle routing: Status and Prospects. National Technical Annals of Operations Research, University of Athens, Greece (1995)
3. Siemiński, A.: Using ACS for dynamic traveling salesman problem. In: Zgrzywa, A., Choroś, K., Siemiński, A. (eds.) New Research in Multimedia and Internet Systems. AISC, vol. 314, pp. 145–155. Springer, Heidelberg (2015)
4. Mavrovouniotis, M., Yang, S.: Ant colony optimization with immigrants schemes in dynamic environments. In: Schaefer, R., Cotta, C., Kołodziej, J., Rudolph, G. (eds.) PPSN XI. LNCS, vol. 6239, pp. 371–380. Springer, Heidelberg (2010)
5. Song, Y., Qin, Y.: Dynamic TSP optimization base on elastic adjustment. In: IEEE Fifth International Conference on Natural Computation, pp. 205–210 (2009)
6. Dorigo, M.,: Optimization, Learning and Natural Algorithms, Ph.D. thesis, Politecnico di Mila-no, Italie (1992)
7. Dorigo, M., Stuetzle, T.: Ant Colony Optimization: Overview and Recent Advances, IRIDIA - Technical Report Series, Technical Report No. TR/IRIDIA/2009-013, May 2009
8. Chirico, U.: A Java framework for ant colony systems. In: Forth International Work-shop on Ant Colony Optimization and Swarm Intelligence, Ants2004, Brussels (2004)
9. Siemiński, A.: TSP/ACO Parameter Optimization; Information Systems Architecture and Technology; System Analysis Approach to the Design, Control and Decision Support; pp. 151–161; Oficyna Wydawnicza Politechniki Wrocławskiej (2011)
10. Gaertner, D., Clark, K.L.: On optimal parameters for ant colony optimization algorithms. In: IC-AI, pp. 83–89 (2005)
11. Pedemonte, M., Nesmachnow, S., Cancela., H.: A survey on parallel ant colony optimization. Appl. Soft Comput. 11, 5181–5197 (2011)

12. Castillo, O., Lizßrraga, E., Soria, J., Melin, P., Valdez, F.: New approach using ant colony optimization with ant set partition for fuzzy control design applied to the ball and beam system. Inf. Sci. **294**, 203–215 (2015)

13. Castillo, O., Neyoy, H., Soria, J., Melin, P., Valdez, F.: A new approach for dynamic fuzzy logic parameter tuning in ant colony optimization and its application in fuzzy control of a mobile robot. Appl. Soft Comput. **28**, 150–159 (2015)

14. Siemiński, A.: Ant colony optimization parameter evaluation. In: Zgrzywa, A., Choroś, K., Siemiński, A. (eds.) Multimedia and Internet Systems: Theory and Practice. AISC, vol. 183, pp. 143–153. Springer, Heidelberg (2013). ISSN 2194-5357

15. Delévacq, A., Delisle, P., Gravel, M., Krajecki, M.: Parallel ant colony optimization on graphics processing units. J. Parallel Distrib. Comput. **73**, 52–61 (2013)

16. Randall, M., Lewis, A.: A parallel implementation of ant colony optimization. J. Parallel Distrib. Comput. **62**, 1421–1432 (2002). Academic Press Inc

17. Delisle, P., Gravel, M., Krajecki, M., Gagné, C., Price, W.L.: Comparing parallelization of an ACO: message passing vs. shared memory. In: Blesa, M.J., Blum, C., Roli, A., Sampels, M. (eds.) HM 2005. LNCS, vol. 3636, pp. 1–11. Springer, Heidelberg (2005)

18. Siemiński, A.: Potentials of hyper populated ant colonies. In: Nguyen, N.T., Trawiński, B., Kosala, R. (eds.) ACIIDS 2015. LNCS, vol. 9011, pp. 408–417. Springer, Heidelberg (2015)

A Segmented Artificial Bee Colony Algorithm Based on Synchronous Learning Factors

Yu Li[1], Jianxia Zhang[2], Dongsheng Zhou[1(✉)], and Qiang Zhang[1(✉)]

[1] Key Laboratory of Advanced Design and Intelligent Computing,
Dalian University, Ministry of Education, Dalian 116622, China
{donyson, zhangq26}@126.com
[2] School of Mechanical Engineering, Dalian University of Technology,
Dalian 116024, China

Abstract. In this paper, we propose a segmented ABC algorithm based on synchronous learning factors (SABC). For the problem of inferior local search ability and low convergence precision in the artificial bee colony (ABC) algorithm, we use the method of synchronous change learning factors for local search. Then under the guidance of the segmented thought, it updates the quality honey greedily. It improves the efficiency of nectar source updating, enhances the local search ability of artificial bee colony. The six standard test functions are chosen to do the simulation experiments. Compared with the other three experiments, the results show that SABC has a significant improvement in the convergence speed and searching optimal value.

Keywords: Artificial bee colony · Learning factors · Segmented thought · Local search

1 Introduction

ABC algorithm was proposed by Turkey scholar Karaboga in 2005 [1]. In nature, according to the different division of bees in the hive, the bees can be divided into three categories [2]: the queen, the drones and the worker bees. ABC algorithm also contains three parts: the employed bees, the onlooker bees and scout bees. In the process of finding the optimal solution, the role of the three bees is different. The employed bee is used to maintain the good solution. The onlooker bee is used to improve convergence speed. The scout bee is used to enhance the ability to escape from local optima [3]. ABC algorithm has been concerned by many scholars because of its simple principle, less control parameters, higher accuracy and better robustness. In solving function optimization problems, it shows strong vitality, which has become one of the hot spots in the field of swarm intelligence optimization algorithm [4]. However, the basic ABC algorithm is easy to fall into local optimum, premature convergence, slow convergence rate, and low convergence accuracy [5, 6]. Therefore, it has been improved from every aspect by many scholars. Banharnsakun [7] presented that the global optimal solution was used to replace the random selection of the neighborhood. Wu [8] introduced the chaos theory into ABC algorithm, and improved the global search ability of the algorithm. Inspired by the particle swarm optimization, Zhu [9] presented a new search

© Springer-Verlag Berlin Heidelberg 2016
N.T. Nguyen et al. (Eds.): ACIIDS 2016, Part I, LNAI 9621, pp. 636–643, 2016.
DOI: 10.1007/978-3-662-49381-6_61

operator to speed up the convergence rate of the algorithm. Bing Wang [10] presented that a search strategy based on the current local optimal solution is used in the onlooker bee stage. The population initialization adopts the general reverse learning strategy.

In order to improve the local search problems and the accuracy of ABC algorithm, this paper adopts synchronous change learning factors to improve the local search strategy of onlooker bee. At the same time, the segmented thought is used to update nectar source. Therefore, it is effective to improve the local search ability of ABC algorithm. And it improves the convergence precision and convergence speed of the algorithm.

2   The Basic ABC Algorithm

The process that bees finds food source is abstracted into the process of looking for the optimal solution of optimization problem [11–13]. Through the mutual cooperation among the three species of bees, the algorithm achieves the process of searching optimal source. The specific description of the algorithm is listed below:

Employed bees are in charge of the global search. Employed bees update each individual in turn. According to the formula (1), the new individual is created.

$$x'_{ij} = x_{ij} + R * (x_{ij} - x_{kj}) \tag{1}$$

Where x_{ij} is the j_{th} element of the individual x_i. x_{kj} is the j_{th} element of the individual x_k. j, k is a random selection, where $k \neq i$. R is a random number between $[-1, 1]$. It controls the range of the neighborhood of x_{ij}. x'_{ij} is a new j_{th} element. The new individual is recorded as x'_i.

The onlooker bees search local optima near the vicinity of some high-quality honey. According to the formula (2), we calculate the probability P_i. Select the nectar x_i which has better fitness from the population. It is executed the same mutation and selection operation as the employed bee.

$$P_i = fit_i / \sum_{j=1}^{SN} fit_j \tag{2}$$

Where fit_i is the fitness of nectar x_i.

Scout bees deal with the individuals in stagnation. If a solution is not improved after the specified limit cycles, employed bees that correspond to the solution turn into scout bees. Then we get a new solution according to the formula (3) as follows.

$$x_i = x_{min} + \phi * (x_{max} - x_{min}) \tag{3}$$

Where ϕ is a random number in $[0, 1]$. x_{max} and x_{min} are upper boundary and lower boundary in solution space.

3 A Segmented ABC Algorithm Based on Synchronous Learning Factors

3.1 Local Search Strategy of Onlooker Bee Based on Synchronous Learning Factors

In the basic ABC algorithm, onlooker bees use the same search strategy as employed bees, so onlooker bees also has strong global exploration ability. Its local search ability is relatively weak. In order to improve the local search ability of ABC algorithm, the improved onlooker bee looks for local optimal value under the guidance of synchronous learning factors. Therefore, inspired by the particle swarm optimization algorithm, this paper proposes a local search strategy of onlooker bee in formula (4) as follows.

$$x'_{ij} = x_{ij} + R * \varphi * (x_{ij} - x_{kj}) + R * \varphi * (x_{gj} - x_{ij}) \tag{4}$$

Where x'_{ij} is the j_{th} element of the new individual x'_i. R is a uniform random number in [- 1, 1]. j, k are random selections, where $k \neq i$. x_{ij} is the j_{th} element of current individual optimal solution x_i. x_{kj} is the j_{th} element of the random individual x_k. x_{jg} is the j_{th} element of the current global optimal solution x_g where $g \neq i$. Synchronous learning factor φ is given in formula (5) as follows.

$$\varphi = c_{max} - \frac{c_{max} - c_{min}}{MCN} * iter \tag{5}$$

Where $iter$ is the current iterations. MCN is the maximum iteration number. c_{max} and c_{min} are constants.

3.2 Local Search Strategy of Onlooker Bee Based on Segmented Thought

In the basic artificial colony algorithm, onlooker bees choose a component randomly to update it near the high quality nectar source. The fitness value of the new nectar source is uncertain after updating. If the new nectar source is better than before, we update it. So the probability of generating better source is small. The number of invalid search is more. The efficiency of local search declines. Therefore, the improved onlooker bees update nectar source greedily. To vividly describe the process of update, the specific idea is drawn in Fig. 1: onlooker x_i generates a new solution by formula (4). If we search a new nectar source x'_i, the line is divided into K equal segments from x'_i to x_{best}. In the opposite direction, we find K equal segments. We calculate the fitness values of nectar sources x_1, x_2, x'_i, x_3. The best fitness value of nectar source is updated. So it is easy to find a better nectar source near the current nectar source. It enhances the local search ability and avoids falling into local optimum. The probability of updating nectar sources x_i is high. We improve the search efficiency of onlooker bee and the performance of the algorithm. The segmented thought is described in Fig. 1 as follows.

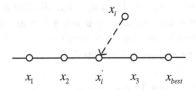

Fig. 1. The description of segmented thought

3.3 Scout Strategy Based on Synchronous Learning Factors

In the basic ABC algorithm, the scout bees are in charge of finding precocious individuals. When precocious individual is found, a random new solution replaces the precocious individual. This way cannot effectively avoid premature convergence phenomenon. So we combine the position of the current optimal solution and the evolution of the current algorithm to generate a new individual. We propose a scout strategy based on the current optimal solution. The formula (6) is given as follows:

$$x'_{ij} = x_{gj} + R * \varphi * x_{ij} \tag{6}$$

Where x'_{ij} is the j_{th} element of the new individual x'_i. x_{gj} is the j_{th} element of the current optimal solution x_g. x_{ij} is the j_{th} element of x_i. R is a uniform random number in [- 1, 1]. Synchronous learning factor φ is the same as formula (5).

Different from the basic ABC algorithm, the scout bees search around itself under the guidance of synchronous learning factors. In the early iterations, synchronous learning factor is large. It is effective to expand the range of searching, jump out of the local optimum, and prevent the premature phenomenon. With the increase of iterations, the range of the new solution decreases gradually. It is helpful to find the optimal solution quickly.

3.4 The Algorithm Process of Improved ABC

In this paper, the concrete steps of SABC are given as follows.

Step 1. Initialization parameter. The *SN* solutions are generated randomly. Every solution $x_i = (x_{i1}, x_{i2}, \cdots, x_{iD})$ is a D-dimensional vector. The maximal number of iteration is *MCN*. The segmentation point is K. The threshold parameter is lim *it*.

Step 2. Every employed bee x_i in population is searched by formula (1). If the new individual is better than the original individual, updates it; else, keep it x_i.

Step 3. The current global optimal solution x_g is selected after employed bee phase. The current global optimal solution will be applied to formula (4). Calculate the probability of each food source.

Step 4. According to roulette wheel, the onlooker bee chooses the better fitness nectar source. Every selected onlooker bee in population is updated by formula (4). The nectar source is updated greedily based on segmented thought.

Step 5. If the nectar source has not been updated afterlim *it* iterations, the scout bee generates a new nectar source to replace it using formula (4).

Step 6. If the algorithm achieves the maximum number of iteration *MCN*, output the optimal solution. Otherwise, go to the step 2.

4   Simulation Results and Discussion

In order to evaluate the performance of SABC algorithm, the six typical test functions [13] are used to test the performance of SABC algorithm, ABC algorithm, BABC algorithm [7] and HGABC algorithm [14].

4.1   Test Functions

The theoretical optimal values of six test functions are zero. They are listed as follows.

Table 1. The test results of six functions

Functions	Algorithms	Optimal values	Worst values	Average values	Standard deviations
Sphere	ABC	1.833e-05	3.071e-04	1.216e-04	8.080e-05
	BABC	1.147e-15	5.914e-11	4.733e-12	1.142e-11
	HGABC	7.325e-21	2.207e-17	1.115e-18	4.015e-18
	SABC	1.832e-22	5.792e-19	7.273e-20	1.206e-19
Step	ABC	0	0	0	0
	BABC	0	0	0	0
	HGABC	0	0	0	0
	SABC	0	0	0	0
Rastrigin	ABC	3.309	9.453	6.498	1.381
	BABC	5.684e-14	8.449e-08	3.611e-09	1.571e-08
	HGABC	0	0	0	0
	SABC	0	0	0	0
Ackley	ABC	5.536e-02	6.575e-01	2.398e-01	1.599e-01
	BABC	2.225e-07	3.534e-04	5.777e-05	7.885e-05
	HGABC	7.993e-15	1.509e-14	1.154e-14	2.798e-15
	SABC	8.881e-16	2.220e-14	7.283e-15	3.891e-15
Griewank	ABC	6.483e-04	6.366e-02	1.706e-02	1.581e-02
	BABC	0	1.742e-11	2.306e-12	3.903e-12
	HGABC	0	5.551e-16	1.110e-16	9.219e-17
	SABC	0	1.110e-16	1.073e-16	2.026e-17
Rosebrock	ABC	2.091e01	1.155e02	6.955e01	2.398e01
	BABC	1.410e-04	2.857e01	6.674	1.219e01
	HGABC	3.220e-04	0.2142	0.0239	0.0413
	SABC	1.597e-15	0.0165	0.0016	0.0036

Sphere: $f_1(x) = \sum_{i=1}^{n} x_i^2, [-50, 50]$

Step: $f_2(x) = \sum_{i=1}^{n} (\lfloor x_i + 0.5 \rfloor)^2, [-100, 100]$

Rastrigin: $f_3(x) = \sum_{i=1}^{n} [x_i^2 - 10 \cos 2\pi x_i + 10], [-5.12, 5.12]$

Ackley: $f_4(x) = -20 \exp(-0.2\sqrt{1/n * \sum_{i=1}^{n} x_i^2})$
$$- \exp(1/n * \sum_{i=1}^{n} \cos(2\pi x_i)) + 20 + e, [-32, 32]$$

Griewank: $f_5(x) = 1/4000 * \sum_{i=1}^{n} (x_i)^2 - \Pi_{i=1}^{n} \cos(x_i/\sqrt{i}) + 1, [-600, 600]$

Rosenbrock: $f_6(x) = \sum_{i=1}^{n-1} [100(x_{i+1} - x_i^2)^2 + (x_i - 1)^2], [-30, 30]$

4.2 Experimental Results and Analysis

Parameters of this paper include population size ($SN = 50$), segment number ($K = 2$), threshold parameter (lim $it = 10$), the dimension ($D = 30$) and maximum cycle times ($MCN = 1000$). The test results of six functions are list in Table 1 as follows.

Table 2. Iterations and time consumption

Functions	Algorithms	Precisions	Iterations	Time consumption /s
Sphere	ABC	1.3323e-14	2080	3.1824
	BABC	1.3323e-14	1125	2.5584
	HGABC	1.3323e-14	55	1.5600
	SABC	1.3323e-14	28	0.5772
Step	ABC	1.8328e-22	751	1.4352
	BABC	1.8328e-22	161	0.3432
	HGABC	1.8328e-22	20	0.6396
	SABC	1.8328e-22	3	0.0936
Rastrigin	ABC	1.7053e-13	3098	5.9436
	BABC	1.7053e-13	892	2.1060
	HGABC	1.7053e-13	97	3.2136
	SABC	1.7053e-13	34	0.8580
Ackley	ABC	8.3084e-14	4153	12.1213
	BABC	8.3084e-14	1885	6.7860
	HGABC	8.3084e-14	92	5.0388
	SABC	8.3084e-14	52	2.3088
Griewank	ABC	5.5511e-16	3022	6.1776
	BABC	5.5511e-16	1196	2.9172
	HGABC	5.5511e-16	78	2.7300
	SABC	5.5511e-16	43	1.1544
Rosenbrock	ABC	3.1205e-02	6094	11.4193
	BABC	3.1205e-02	2405	5.1012
	HGABC	3.1205e-02	449	14.2273
	SABC	3.1205e-02	46	1.1232

In Table 1, the data shows that the standard ABC algorithm is poor in stability, convergence speed and convergence accuracy. The BABC algorithm enhances the local search ability of the algorithm and improves the solution quality of the algorithm to some extent. The HGABC algorithm improves the exploration and exploitation procedure with balance quantity. The SABC algorithm performs local search based on synchronous learning factors. At the same time, onlooker bees apply the segmented thought to the local search strategy. The local search ability of the algorithm is improved dramatically. So the performance of the algorithm is obviously improved.

When the four algorithms achieve the same precision in the same function, iterations and time consumption data is listed in Table 2 as follows.

From Table 2, we can know that the segmented thought increases the amount of calculation. However, when the four functions achieve the same precision values, SABC need less Iterations and Time Consumption than other three functions. This fully shows that SABC improves the efficiency of nectar source updating.

The evolution curves of the average fitness value of the four algorithms are given from Figs. 2, 3, 4, 5, 6 and 7.

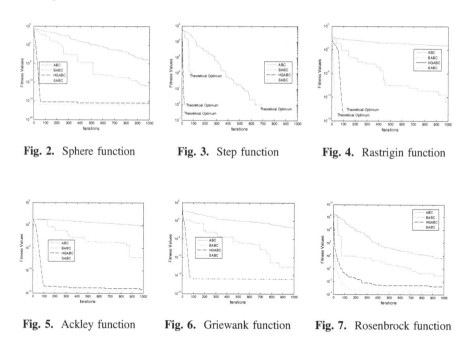

Fig. 2. Sphere function   **Fig. 3.** Step function   **Fig. 4.** Rastrigin function

Fig. 5. Ackley function   **Fig. 6.** Griewank function   **Fig. 7.** Rosenbrock function

From Figs. 2, 3, 4, 5, 6 and 7, SABC has a high precision and convergence speed. Especially in Figs. 3 and 4, SABC can find the theoretical optimum quickly. So SABC has a good performance of global searching and avoiding premature convergence.

5   Conclusion

On the basis of the algorithm analysis, this paper points out that basic ABC algorithm has flaws in onlooker bees and scout bees. This paper puts forward a specific improvement plan. Under the guidance of synchronous learning factors, the improved onlooker bee makes a local search around itself. The segmented thought improves the success rate of updating nectar source. The simulation shows that the improved ABC algorithm gets better results. The improved ABC algorithm is strong in robustness and can avoid the defect of reaching the part best value. So the improved ABC algorithm has more powerful optimizing ability precision and faster convergence ability.

References

1. Karaboga, D.: An idea based on honey bee swarm for numerical optimization. Technical Report-TR06. Erciyes University, Engineering Faculty, Computer Engineering Department (2005)
2. Gao, W.F., Liu, S.Y., Huang, L.L.: Enhancing artificial bee colony algorithm using more information-based search equations. Inf. Sci. **270**, 112–133 (2014)
3. Akay, B., Karaboga, D.: A modified artificial bee colony algorithm for real-parameter optimization. Inf. Sci. **192**, 120–142 (2012)
4. Karaboga, D., Akay, B.: A comparative study of artificial bee colony algorithm. Appl. Math. Comput. **214**, 108–132 (2009)
5. Gao, W.F., Liu, S.Y., Huang, L.L.: A novel artificial bee colony algorithm based on modified search equation and orthogonal learning. IEEE Trans. Cybern. **43**(3), 1011–1024 (2013)
6. Gao, W.F., Liu, S.Y.: A modified artificial bee colony algorithm. Comput. Oper. Res. **39**, 687–697 (2012)
7. Banharnsakun, A., Achalakul, T., Sirinaovakul, B.: The best-so-far selection in artificial bee colony algorithm. Appl. Soft Comput. **11**, 2888–2901 (2011)
8. Wu, B., Fan, S.-h.: Improved artificial bee colony algorithm with chaos. In: Yu, Y., Yu, Z., Zhao, J. (eds.) CSEEE 2011, Part I. CCIS, vol. 158, pp. 51–56. Springer, Heidelberg (2011)
9. Zhu, G., Kwong, S.: Gbest-guided artificial bee colony algorithm for numerical function optimization. Appl. Math. Comput. **217**, 3166–3173 (2010)
10. Wang, B.: Improved artificial bee colony algorithm based on local best solution. Appl. Res. Comput. **31**, 1023–1026 (2014)
11. Gao, W.F., Liu, S.Y.: Improved artificial bee colony algorithm for global optimization. Inf. Process. Lett. **111**, 871–882 (2011)
12. Ge, Y., Liang, J., Wang, X.P., Xie, X.C.: Improved artificial bee colony algorithms for function optimization. Comput. Sci. **40**, 252–257 (2013)
13. Karaboga, D., Basturk, B.: A powerful and efficient algorithm for numerical function optimization: artificial bee colony (ABC) algorithm. J. Global Optim. **39**, 459–471 (2007)
14. Shah, H., Herawan, T., Naseem, R., Ghazali, R.: Hybrid guided artificial bee colony algorithm for numerical function optimization. In: Tan, Y., Shi, Y., Coello, C.A. (eds.) ICSI 2014, Part I. LNCS, vol. 8794, pp. 197–206. Springer, Heidelberg (2014)

A Method for Query Top-K Rules from Class Association Rule Set

Loan T.T. Nguyen[1(✉)], Hai T. Nguyen[2], Bay Vo[3],
and Ngoc-Thanh Nguyen[4]

[1] Faculty of Information Technology, Ho Chi Minh City University of Foreign
Languages - Information Technology, Ho Chi Minh City, Vietnam
nttloan@huflit.edu.vn, nthithuyloan@gmail.com
[2] Faculty of Information Technology, VOV College,
Ho Chi Minh City, Vietnam
nguyenthihai@vov.edu.vn
[3] Faculty of Information Technology,
Ho Chi Minh City University of Technology, Ho Chi Minh City, Vietnam
bayvodinh@gmail.com
[4] Department of Information Systems,
Faculty of Computer Science and Management,
Wroclaw University of Technology, Wroclaw, Poland
Ngoc-Thanh.Nguyen@pwr.edu.pl

Abstract. Methods for mining/querying Top-k frequent patterns and Top-k association rules have been developed in recent years. However, methods for mining/querying Top-k rules from a set of class association rules have not been developed. In this paper, we propose a method for querying Top-k class association rules based on the support. From the set of mined class association rules that satisfy the minimum support and minimum confidence thresholds, we use an insertion-based method to query Top-k rules. Firstly, we insert k rules from the rule set to the result set. After that, for each rule in the rest, we insert it into the result rule set using the idea of insertion strategy if its support is greater than the support of the last rule in the result rule set. Experimental results show that the proposed method is more efficient than obtaining the result after sorting the whole rule set.

Keywords: Data mining · Class association rules · Top-k class association rules · Query Top-k rule query

1 Introduction

In 2012, Fournier-Viger et al. proposed TopKRules algorithm for mining Top-k association rules [23]. They used two thresholds: Minimum confidence (*minConf*) and k. The parameter *minConf* was used to remove the rules whose confidences do not satisfy *minConf* while k was used to mine Top-k association rules. The authors also used some techniques to prune the search space. This algorithm finds top-k rules with highest *minSup* values and satisfies the *minConf* threshold. At that time, Fournier-Viger and Tseng also proposed an algorithm for mining Top-k non-redundant association

© Springer-Verlag Berlin Heidelberg 2016
N.T. Nguyen et al. (Eds.): ACIIDS 2016, Part I, LNAI 9621, pp. 644–653, 2016.
DOI: 10.1007/978-3-662-49381-6_62

rules [24]. Besides, some non-redundant rules were ignored by their pruning scheme. Another method to filtered-top-k association was proposed in [9] or used constraints to mine association rules [8].

Deng and Fang proposed NTK algorithm to mine Top-rank-k frequent itemsets [6]. NTK used a divide-and-conquer scheme and an early pruning technique to mine Top-rank-k frequent itemsets. After that, Quyen et al. proposed an improved algorithm, named iNTK [12], to mine Top-rank-k frequent itemsets. iNTK used the subsume concept to quickly determine the itemsets with the same rank.

The above algorithms only focused on mining Top-rank-k frequent itemsets [22] or Top-k (non-redundant) general association rules. They cannot be used in the problem of mining Top-k class association rules. Mining top-rank-k frequent itemsets cannot be used for mining association rules and TopKRules cannot be used to mine Top-k class association rules because the right hand side of an association rule is any frequent itemset while the right hand side of a class association rule only contains the class label. Therefore, an effective solution for mining Top-k class association rules is necessary.

In this paper, we propose an algorithm for mining Top-k class association rules from the mined rule set. The proposed algorithm uses an insertion-like method for effective querying Top-k class association rules. It inserts each rule into a buffer and needs not to sort the whole rules. This strategy is very efficient compared to sorting the whole rule set.

The rest of the paper is organized as follows: Sect. 2 presents related works to mining CARs and Top-k frequent itemsets/association rules. The main contributions are described in Sect. 3. Section 4 presents the experiments and Sect. 5 presents the conclusions and suggestions for future work.

2 Related Works

2.1 Mining Class Association Rules

There are many methods for rule-based classification. Breiman et al. [2] proposed a binary tree for mining rules. CART (classification and regression tree) algorithm was proposed. CART chooses the attribute to split using the Gini measure. ID3 [25] and C4.5 [26], two decision tree based approaches, were proposed. ID3 used Information Gain to choose the attribute and C4.5 used Ratio Gain. ILA and ILA-2 [32, 33], another rule-based method for prediction was proposed. Unlike CART, ID3 and C4.5, ILA and ILA-2 do not build tree, they find the rule using maximum combination. In 2011, Parker analyzed seven ways to determine the best classifier in the set of classifiers [21]. This method can be applied in any classifier (not only focus on rule-based classifiers).

Associative classification (AC), which integrates association mining and classification [14, 30], is an efficient classification approach. A particular subset of association rules whose their right-hand side is restricted to the class attribute is mined. This subset of rules is denoted as CARs.

The first method for mining CARs was proposed in 1998 [1]. CBA algorithm has been developed in this work. CBA is based on Apriori method for mining CARs and uses a heuristic method to build classifier. In 2001, an FP-tree-based method for classification based on multiple association rules was proposed [13]. First of all, the authors modify the FP-tree for storing the single items with its class information. After that, they mine CARs from FP-tree and store them in CR-tree (Class Rule-tree). To build classifier, they use a database coverage threshold to select the rules. Classification based on predictive association rules, a statistical-based approach using CARs, was also proposed [27]. Thabtah et al. used multi-class, multi-label association classification to mine and predict class of new records [28, 29]. Later on, Thabtah and Cowling proposed a greedy method to build classifier and used the built classifier to predict the class of new records using multiple rules [31]. Class association rule mining based on equivalence class rule tree was proposed by Vo and Le [34]. The proposed algorithm (ECR-CARM) first scans dataset to build the first level of ECR-tree. After that, it expands ECR-tree to build child nodes using the parent nodes. CAR-Miner and CAR-Miner-Diff, two improved versions of ECR-CARM, were also developed in [16, 17]. Chen et al. proposed a principal association mining (PAM) method to improve the accuracy and the size of classifier [4]. Some efficient methods were also proposed to improve the accuracy such as: using CBA to handle class imbalance [3] and uncertain datasets [10], methods that uses interestingness measures [11, 27], a method that uses rule prioritization [5], and a method that uses closed sets [15].

None of the above techniques is designed for finding top-k class association rules.

2.2   Mining Top-rank-k Frequent Itemsets

Deng et al. proposed the NTK algorithm for mining Top-rank-k frequent itemsets [6]. NTK represented patterns by a node-list data structure. It uses t-patterns to form $(t + 1)$-patterns. By using Node-list, the algorithm needs not to rescan the whole dataset when computing the support of $(t + 1)$-patterns. Main ideas of NTK are as follows:

(1) NTK traverses the PPC-tree and generates a Node-list of 1-patterns. After that, it finds 1-patterns that satisfy Top-rank-k and inserts them into Top-rank-k table, this table contains frequent 1-patterns and their supports. All patterns with the same support are stored in the same entry. Therefore, the number of entries in this table is not greater than k.

(2) Using 1-patterns in Top-rank-k to generate candidate 2-patterns. NTK inserts candidate 2-patterns into Top-rank-k table if its support is not smaller than the smallest support of patterns in this table.

Go to step 2 by using t-patterns in Top-rank-k table to create candidate $(t + 1)$-patterns until there is no any candidate generated.

In [7], the authors developed an improved algorithm, called iNTK, based on NTK. iNTK is also using t-patterns to create candidate $(t + 1)$-patterns. By using N-list, it needs not to rescan the dataset to compute the support of candidate $(t + 1)$-patterns. Besides, the algorithm uses subsume concept to reduce the number of generated candidates compared to NTK and therefore, it saves time to generate candidates.

2.3   Mining Top-K Association Rules

Fournier-Viger et al. proposed the TopKRules algorithm for mining Top-k association rules from datasets [23]. This algorithm finds Top-k rules with highest *minSup* values and satisfies the *minConf* threshold. The change of the *minSup* value is dependent on the lowest support of itemsets. The TopKRules algorithm is based on the principle of extending rules and some methods for early eliminating rules that do not belong to Top-k rules. Fournier-Viger and Tseng also extended TopKRules for mining Top-k non-redundant rules [24] and Top-k sequential rules [22].

3   A Method for Mining Top-K Class Association Rules

3.1   Basic Concepts

Let D be the set of training data with n attributes $A_1, A_2, ..., A_n$ and $|D|$ objects (cases). Let $C = \{c_1, c_2,..., c_k\}$ be a list of class labels. A specific value of an attribute A_i and class C are denoted by the lower-case letters a and c, respectively [17].

Definition 1: An itemset is a set of some pairs of attributes and a specific value, denoted $\{(A_{i1}, a_{i1}), (A_{i2}, a_{i2}), ..., (A_{im}, a_{im})\}$.

Definition 2: A class-association rule r is of the form $\{(A_{i1}, a_{i1}), ..., (A_{im}, a_{im})\} \to c$, where $\{(A_{i1}, a_{i1}), ..., (A_{im}, a_{im})\}$ is an itemset, and $c \in C$ is a class label.

Definition 3: The actual occurrence *ActOcc(r)* of a rule r in D is the number of rows of D that match r's condition.

Definition 4: The support of a rule r, denoted *Sup(r)*, is the number of rows that match r's condition and belong to r's class.

Definition 5: The confidence of a rule r, denoted by *Conf(r)*, is defined as:

$$Conf(r) = \frac{Supp(r)}{ActOccr(r)}$$

For example (Table 1): Consider rule $r = \{< (A, a1) > \to y\}$. We have:

Table 1. An example of training dataset

ID	A	B	C	CLASS
1	a1	b1	c1	y
2	a1	b2	c1	n
3	a2	b2	c1	n
4	a3	b3	c1	y
5	a3	b1	c2	n
6	a3	b3	c1	y
7	a1	b3	c2	y
8	a2	b2	c2	n

$ActOccr(r) = 3$
$Sup(r) = 2$
$Conf(r) = \frac{Supp(r)}{ActOccr(r)} = \frac{2}{3}$.

Definition 6: Given a set of class association rules R and a rule $r \in R$, rank of r in R is defined as follows. $Rank(r) = |\{r_i \in R \mid Sup(r) > Sup(r_i)\}| + 1$.

Definition 7: (Top-k rules according to support): Given a set of class association rules R and a threshold k, mining Top-k class association rules is to find k best rules in R based on their supports, i.e.

$$\text{Top} - k(R) = \{r \in R | Rank(r) \geq k\}$$

Based on the two above definitions, the problem of mining Top-k class association rules simply filters out k rules whose supports are highest. There are two ways to solve the problem:

(i) Sorting class association rules and choosing k first rules.
(ii) Using the idea of insertion-based technique but only considering k rules.

Algorithm: Top-k-CARs-Insertion()
Input: A set of class association rules (R) and threshold k.
Output: Top-k class association rules in R

Begin
1. $RS = \varnothing$
2. *For* $(i = 1; i <= k \,\&\&\, i <= |R|; i++)$
   Begin
3. $j = i\text{-}1$
4.   *While* $(j > 0 \,\&\&\, Sup(R_i) > Sup(RS_j))$
          Begin
5. $RS_{j+1} = RS_j$
6. $j = j - 1$
          End
7. $RS_{j+1} = R_i$
   End
8. *For* $(i = k+1; i <= |R|; i++)$
   Begin
9.   If $Sup(R_i) > Sup(RS_k)$ then
          Begin
10. $j = k - 1$
11.   *While* $(j > 0 \,\&\&\, Sup(R_i) > Sup(RS_j))$
              Begin
12. $RS_{j+1} = RS_j$
13. $j = j - 1$
14. $RS_{j+1} = R_i$
              End
          End
   End
15. Return RS
End

Fig. 1. The proposed algorithm

3.2 Top-K-CARs-Insertion Algorithm

The proposed algorithm is shown in Fig. 1. The input is a set of class association rules R, and the output is Top-k rules with highest supports. Main steps of this algorithm as follows:

Step 1 (Line 1) is assigned null for RS.
Step 2 (Lines 2–7) is used for inserting k first rules into the rule set (RS).
Step 3 (Lines 8–14) is used for inserting the rest of rules into RS.
Step 4 (Line 15) returns RS that contains Top-k class association rules.

In contrast to the insertion sort, the algorithm in Fig. 1 only sort k first rules in the rule set. For the rest, it inserts each rule into the rule set RS if its support is greater than the support of the last rule in RS. In this way, the complexity of the proposed algorithm is $O(n*k)$ where n is the number of rules in R (typically, $k << n$). This complexity is smaller than using sorting ($O(n^2)$ in the worst case).

4 Experiments

4.1 Datasets and Testing Environment

The algorithms used in the experiments were coded with C# 2012, and run on a laptop with Windows 8.1 OS, CPU i5-4200U, 1.60 GHz, and 4 GB RAM.

Experimental datasets were downloaded from UCI Machine Learning Repository (http://mlearn.ics.uci.edu) and their charesteristics are showed in Table 2.

Table 2. Characteristics of experimental datasets

Dataset	#Attributes	#Classes	#Distinct items	#Records
Breast	12	2	737	699
German	21	2	1077	1000
Lymph	18	4	63	148

Table 3 shows the numbers of class association rules from experimental datasets with a given *minSup* for each dataset. Here, *minConf* is set to 50 %.

Table 3. Number of rules of each dataset

Dataset	*minSup* (%)	#Rules
Breast	0.3	13,870
German	5	19,343
Lymph	8	30,911

4.2 Mining Time

Figures 2, 3 and 4 compare the mining time of our proposed method based on an insertion scheme with that of the naïve method that sorts the whole class association rules.

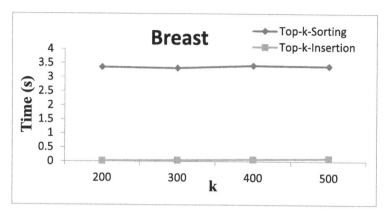

Fig. 2. Mining times of insertion-based and sorting-based methods in Breast dataset

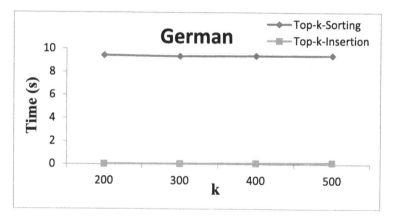

Fig. 3. Mining times of insertion-based and sorting-based methods in German dataset

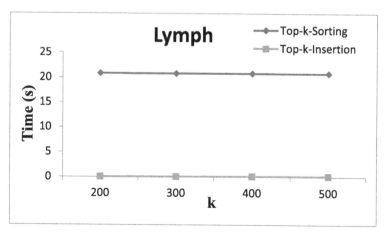

Fig. 4. Mining times of insertion-based and sorting-based methods in Lymph dataset

From these figures, we can see that the insertion-based algorithm is more efficient than the sorting-based method in all experiments. For example, considering Lymph dataset, for 30,911 class association rules with $minSup = 8\%$. The mining time for mining Top-k rules using sorting is around 20.8 (s) for k = 200, 300, 400, and 500. However, when we use the Insertion-based algorithm, the mining time is around 0.1 (s). We have the same results for Breast and German datasets.

5 Conclusions and Future Work

This paper has proposed a method for mining Top-k class association rules using Insertion-based algorithm. The contributions are as follows:

1. We define the problem of mining Top-k class association rules.
2. Our algorithm uses the idea of insertion to find Top-k class association rules but only retain k rules.
3. We develop an algorithm for fast mining Top-k class association rules from the rule set. The proposed algorithm has the complexity $O(n*k)$ where n is the number of rules. When k is smaller than n, the proposed algorithm is more efficient than method using sorting.

In future works, we will continue to study how to prune rules that cannot belong to Top-k to reduce the search space. Besides, we will expand our method for mining Top-k non-redundant class association rules. For very large datasets, we are also interested in some "anytime" algorithms such as [18–20] for further enhancing the performance as well as providing results under some resource constraints.

References

1. Agrawal, R., Imielinski, T., Swami, A.: Mining association rules between sets of items in large databases. In: Proceedings of the ACM SIGMOD Conference, Washington DC, US, pp. 207–216 (1993)
2. Breiman, L., Friedman, J.H., Olshen, R.A., Stone, C.J.: Classification and regression trees. In: Proceedings of Wadsworth, Belmont, CA, vol.1, pp. 14–23. CRC Press (1984)
3. Chen, W.C., Hsu, C.C., Hsu, J.N.: Adjusting and generalizing CBA algorithm to handling class imbalance. Expert Syst. Appl. **39**(5), 5907–5919 (2012)
4. Chen, F, Wang, Y., Li, M., Wu, H., Tian, J.: A Principal Association Mining An efficient classification approach. Knowl.-Based Syst. **67**, 16–25 (2014)
5. Chen, C.H., Chiang, R.D., Lee, C.M., Chen, C.Y.: Improving the performance of association classifiers by rule prioritization. Knowl.-Based Syst. **36**, 59–67 (2012)
6. Deng, Z.H., Fang, G.: Mining top-rank-k frequent patterns. In: ICMLC, pp. 851–856 (2007)
7. Deng, Z.H.: Fast mining top-rank-k frequent patterns by using Node-lists. Expert Syst. Appl. **41**(4), 1763–1768 (2014)
8. Duong, H.V., Truong, T.C.: An efficient method for mining associaiton rules based on minimum single constraints. Vietnam J. Comput. Sci. **2**(2), 67–83 (2015)
9. Webb, G.I.: Filtered-top-k association discovery. WIREs Data Min. Knowl. Discov. **1**, 183–192 (2011)

10. Hooshsadat, M., Zaïane, O.R.: An associative classifier for uncertain datasets. In: Tan, P.-N., Chawla, S., Ho, C.K., Bailey, J. (eds.) PAKDD 2012, Part I. LNCS, vol. 7301, pp. 342–353. Springer, Heidelberg (2012)

11. Lan, Y., Janssens, D., Chen, G., Wets, G.: Improving associative classification by incorporating novel interestingness measures. Expert Syst. Appl. **31**(1), 184–192 (2006)

12. Le, Q.H.T., Le, T., Vo, B., Le, B.: An efficient and effective algorithm for mining top-rank-k frequent patterns. Expert Syst. Appl. **42**(1), 156–164 (2015)

13. Li, W., Han, J., Pei, J.: CMAR: Accurate and efficient classification based on multiple class-association rules. In: Proceedings of the 1st IEEE International Conference on Data Mining, San Jose, California, USA, pp. 369–376 (2001)

14. Liu, B., Hsu, W., Ma, Y.: Integrating classification and association rule mining. In: Proceedings of the 4th International Conference on Knowledge Discovery and Data Mining, New York, USA, pp. 80–86 (1998)

15. Liu, H., Sun, J., Zhang, H.: Post-processing of associative classification rules using closed sets. Expert Syst. Appl. **36**(3), 6659–6667 (2009)

16. Nguyen, L.T.T., Vo, B., Hong, T.P., Thanh, H.C.: CAR-Miner: An efficient algorithm for mining class-association rules. Expert Syst. Appl. **40**(6), 2305–2311 (2013)

17. Nguyen, L.T.T., Nguyen, N.T.: An improved algorithm for mining class association rules using the difference of Obidsets. Expert Syst. Appl. **42**(9), 4361–4369 (2015)

18. Mai, S.T., He, X., Feng, J., Böhm, C.: Efficient anytime density-based clustering. In: Proceedings of SIAM International Conference on Data Mining (SDM 2013), pp. 112–120 (2013)

19. Mai, S.T., He, X., Feng, J., Plant, C., Böhm, C.: Anytime density-based clustering of Complex Data. Knowl. Inf. Syst. **45**, 319–355 (2015)

20. Mai, S.T., He, X., Hubig, N., Plant, C., Böhm, C.: Active density-based clustering. In: Proceedings of International Conference on Data Mining (ICDM 2013), pp. 508–517 (2013)

21. Parker, C.: An analysis of performance measures for binary classifiers. In: Proceedings of ICDM, pp. 517–526 (2011)

22. Fournier-Viger, P., Tseng, V.S.: Mining Top-K sequential rules. In: Tang, J., King, I., Chen, L., Wang, J. (eds.) ADMA 2011, Part II. LNCS, vol. 7121, pp. 180–194. Springer, Heidelberg (2011)

23. Fournier-Viger, P., Wu, C.-W., Tseng, V.S.: Mining Top-K association rules. In: Kosseim, L., Inkpen, D. (eds.) Canadian AI 2012. LNCS, vol. 7310, pp. 61–73. Springer, Heidelberg (2012)

24. Fournier-Viger, P., Tseng, V.S.: Mining Top-K non-redundant association rules. In: Chen, L., Felfernig, A., Liu, J., Raś, Z.W. (eds.) ISMIS 2012. LNCS, vol. 7661, pp. 31–40. Springer, Heidelberg (2012)

25. Quinlan, J.R.: Introduction of decision tree. Mach. Learn. **1**(1), 81–106 (1986)

26. Quinlan, J.R.: C4.5: program for machine learning. Morgan Kaufmann, San Francisco (1992)

27. Shaharanee, I.N.M., Hadzic, F., Dillon, T.S.: Interestingness measures for association rules based on statistical validity. Knowl.-Based Syst. **24**(3), 386–392 (2011)

28. Thabtah, F., Cowling, P., Peng, Y.: MMAC: A new multi-class, multi-label associative classification approach. In: Proceedings of the 4th IEEE International Conference on Data Mining, Brighton, UK, pp. 217–224 (2004)

29. Thabtah, F., Cowling, P., Peng, Y.: MCAR: Multi-class classification based on association rule. In: Proceedings of the 3rd ACS/IEEE International Conference on Computer Systems and Applications, Tunis, Tunisia, pp. 33–39 (2005)

30. Thabtah, F.A.: A review of associative classification mining. Knowl. Eng. Rev. **22**(1), 37–65 (2007)

31. Thabtah, F.A., Cowling, P.I.: A greedy classification algorithm based on association rule. Appl. Soft Comput. **7**(3), 1102–1111 (2007)
32. Tolun, M.R., Abu-Soud, S.M.: ILA: An inductive learning algorithm for production rule discovery. Expert Syst. Appl. **14**(3), 361–370 (1998)
33. Tolun, M.R., Sever, H., Uludag, M., Abu-Soud, S.M.: ILA-2: An inductive learning algorithm for knowledge discovery. Cybern. Syst. **30**(7), 609–628 (1998)
34. Vo, B., Le, B.: A novel classification algorithm based on association rules mining. In: Richards, D., Kang, B.-H. (eds.) PKAW 2008. LNCS, vol. 5465, pp. 61–75. Springer, Heidelberg (2009)

Hierarchy of Groups Evaluation Using Different F-Score Variants

Michał Spytkowski, Łukasz P. Olech$^{(\boxtimes)}$, and Halina Kwaśnicka

Department of Computational Intelligence,
Faculty of Computer Science and Management,
Wrocław University of Technology, Wrocław, Poland
{michal.spytkowski,lukasz.olech,halina.kwasnicka}@pwr.edu.pl

Abstract. The paper presents a cursory examination of clustering, focusing on a rarely explored field of hierarchy of clusters. Based on this, a short discussion of clustering quality measures is presented and the F-score measure is examined more deeply. As there are no attempts to assess the quality for hierarchies of clusters, three variants of the F-Score based index are presented: classic, hierarchical and partial order. The partial order index is the authors' approach to the subject. Conducted experiments show the properties of the considered measures. In conclusions, the strong and weak sides of each variant are presented.

Keywords: Clustering quality measures · F-score · Hierarchies of clusters

1 Introduction

Currently we are facing a time in which we are figuratively deluged with data. Thanks to the development and popularity of the Internet, almost every person can take an active part in the creation and distribution of new data. As this mass of data contain potentially useful information, this state presents a new, challenging tasks for scientists. It is no longer feasible to manually obtain this information, and as a result, the development of data mining algorithms has come into focus. Regarding the volume of the data and problems with labelling them, a special interest has turned to unsupervised methods such as cluster analysis, which tries to establish meaningful groups (called *clusters*) in a set of unlabelled data. This approach has been successfully used in many practical applications such as bioinformatics [1], social media [15], and audiovisual indexing [17]. In this paper we make a distinction between three types of clustering which are described below. For the purpose of this paper we only consider situations where a single data point belongs to only one cluster (hard clustering).

The first type is **flat clustering**, in which data are assigned to independent clusters. This type of clustering does not include relationships between groups. The primary goal of flat clustering is to build a model where data points within any given group are similar between themselves and dissimilar to data

© Springer-Verlag Berlin Heidelberg 2016
N.T. Nguyen et al. (Eds.): ACIIDS 2016, Part I, LNAI 9621, pp. 654–664, 2016.
DOI: 10.1007/978-3-662-49381-6_63

points of other groups. In this group two approaches should be mentioned: clusters are generated by iteratively relocating points between subsets, e.g., the *k-means* algorithm [7,8] or they are identified as areas of high density of data, e.g., DBSCAN [9]. Most flat clustering methods require to point the number of groups beforehand. The flat clustering methods are not considered in this paper.

Secondly, we consider **hierarchical clustering** methods. In contrast to flat clustering, they produce a hierarchy of partitions instead of one partition. Each partition differs in the number of clusters. Generally, there are two approaches to hierarchical clustering [3]: *agglomerative* and *divisive*. The agglomerative (*bottom-up* hierarhical clustering) approach starts with each data point in its own cluster and then, iteratively merges two clusters according to a distance function. It ends with all the points in a single cluster, having constructed a hierarchy in the process. The counterpart to the agglomerative approach is the *divisive* approach, also called *deagglomerative* (*top-down* hierarchical clustering). The key point in the divisive approach is a function indicating which cluster to split.

The results of hierarchical clustering methods are tree-like structures called *dendrograms* [5], where nodes represent clusters and the underlying hierarchy shows how the clustering process was performed. As data belong not only to the node it is assigned to, but also to its parents, a dendrogram represents a spectrum of possible clustering results – from one cluster containing all data to a number of clusters, each containing only one data point. In order to get a specific (flat) partition the dendrogram must be cut.

In this case the clusters are in a hierarchical relation (partial order), but the data is not. It is because the objects are only assigned to leaves. It remains true that data in a cluster should be maximally similar and that different clusters (at least those containing data) should be maximally dissimilar. Quality measures designed for flat clustering can be successfully applied to hierarchical clustering.

Finally we can consider **cluster hierarchies** [6,18]. In this class data can be assigned to any node in the hierarchy of clusters. As in hierarchical clustering there is a relation between groups. *Additionally*, as the data points can be not only in the leaves [14], there is also a hierarchical relation (partial order) between the objects assigned to the clusters. In this case maximum separation between clusters is not always desirable as clusters that are in relation to each other (and thus hold data that is in relations) should be less separated than unrelated clusters.

Methods generating cluster hierarchies include the Tree Structured Stick Breaking Process [6], which uses a Markov Chain to explore the space of possible partitions of data or the Bayesian Rose Tree [2] approach, which relies on a deterministic statistical process in order to build the tree. The properties of these clustering approaches appear useful in many fields of information retrieval and processing, especially in documents and images analysis. It is however not thoroughly researched if models generated by these methods can be validated in the same way as for more fully explored clustering analysis approaches.

An exhaustive information about clustering can be found in the available literature, for example, in [9,11,13,19]. From the perspective of this paper it is viable to divide existing clusterization techniques based on the characteristics of the result they produce i.e., whether it is a flat or hierarchical structure of clusters. This characteristic underlies our research.

The contribution of this paper is two-fold. Firstly, due to different paradigm of the methods able to build full hierarchies of clusters, we propose to use the name *generation of hierarchies of clusters* (groups) instead of *hierarchical clustering*. In our opinion, such distinction facilitates researchers to note that some new features of hierarchies of clusters are very important from practical point of view [18]. The second issue is connected with the evaluation process. The measures useful for evaluation of flat clustering may not be suitable for hierarchies of clusters. In order to at least partially fill this gap, we propose one new external measure, *Partial Order F-Score*. The additional two measures are also studied in experiments. The goal of experiments is to discover suitability of tested measures as well as their advantages and disadvantages.

The paper is organized as follows. Next section briefly describes problems with quality evaluation of hierarchies of clusters. Third section presents used F-Score based measures. Experiments and their results are described in Sect. 4. Section 5 concludes the paper.

2   Measuring Quality of Hierarchies of Clusters

The above presented problem raises the question if currently used clustering quality measures remain relevant for evaluating hierarchies of groups, and, if not, how can such structures be verified? In the case of traditional clustering the core principles [4] of quality verification can be summarized as data separation between clusters (for internal verification) and data purity within clusters (for external verification). These two approaches can be found at the root of all external and internal clustering measures. These principles fully translate into hierarchical clustering. For internal verification the separation between clusters is still paramount and can be applied on each level at which the hierarchy can be cut. For external verification the lack of data within intermediate clusters means that examining the leaf clusters for purity is all that is left. The data, which is stored in the leaves is not in relation to any data stored in different leaves unless examined level by level.

The perspective changes when we take into account hierarchies of clusters. When every cluster may contain elements the previously defined principles are no longer valid. Data separation becomes problematic as now relationships within clusters need to be considered. While both leaf and sibling clusters should remain maximally separated, the best case scenario for other clusters is not that obvious. As an example, a cluster should be less separated from its descendants and ancestors than it is from completely unrelated clusters. Moreover, for the purity principle, the relations between clusters also come into play. Taken in their classic form, these external measures are completely blind to clusters changing position within the hierarchy. As the classic measures used in cluster analysis no longer function as intended, the introduction of new measures, or adaptation of existing ones is an explorable field of research.

3   F-Score Based Measures

F-Score (also called as F-measure) is commonly used in the field of information retrieval [16]. It is the harmonic mean of the precision and recall measures. F-measure was also adapted for hierarchical clustering [12,14]. In this paper three different versions of the F-score quality measure are compared to each other. To describe these measures in a clear fashion a number of symbols are introduced:

X – set of all data points, or objects; X_c – set of all objects of class c;

x_i – i-th data point; ϵ – specific cluster;

$\mathbb{C}$ – set of all classes c; ϵ_{x_i} – cluster of object x_i;

c_{x_i} – class of object x_i; $\epsilon\epsilon_i$ – i-th child of cluster ϵ (if exists);

X_ϵ – set of all objects in cluster ϵ;

To define the hierarchy of groups a relationship between the ground truth classes must be established: $C_c = \{c\} \cup \bigcup_{i=1}^{n} C_{cc_i}$, where: C_c – set containing class c and all its descendant classes; C_{cc_i} – set containing class cc_i and all its descendant classes; n – number of children for class c.

A relationship between groups in the hierarchy must also be defined: $E_\epsilon = \{\epsilon\} \cup \bigcup_{i=1}^{m} E_{\epsilon\epsilon_i}$, where: E_ϵ – set containing node ϵ and all its descendant nodes; $E_{\epsilon\epsilon_i}$ – set containing node $\epsilon\epsilon_i$ and all its descendant nodes; m – number of children for node ϵ.

Using the above defined symbols the set of points belonging to a class (or cluster) and its descendants can be written:

$$X_{C_c} = \bigcup_{c' \in C_c} X_{c'}, \qquad X_{E_\epsilon} = \bigcup_{\epsilon' \in E_\epsilon} X_{\epsilon'}. \tag{1}$$

F-Score can be viewed as a statistical hypothesis. Thus, it is helpful to define a pair of relations based on pairs of points. In the classic and the partial order approach to F-Score, the statistical hypothesis is calculated based on the total of all possible pairs of different points. Here, the order of the points matters. In this way the condition is based only on the relationship of data points in the ground truth hierarchy of clusters, while the test is based only on the relationship of data points in the tested model. By taking pairs of points and keeping the relationships isolated there is no need to find a correlation between the classes and clusters:

$$G \subseteq X \times X, \qquad M \subseteq X \times X, \tag{2}$$

where: G – ground truth relation (condition); M – model relation (test outcome).

Given this, the four crucial statistical hypothesis values are as follows:

$$
\begin{aligned}
p_t &= |\{x_i, x_j \in X : i \neq j \wedge x_i G x_j \wedge x_i M x_j\}|, \\
p_f &= |\{x_i, x_j \in X : i \neq j \wedge \neg x_i G x_j \wedge x_i M x_j\}|, \\
n_t &= |\{x_i, x_j \in X : i \neq j \wedge \neg x_i G x_j \wedge \neg x_i M x_j\}|, \\
n_f &= |\{x_i, x_j \in X : i \neq j \wedge x_i G x_j \wedge \neg x_i M x_j\}|.
\end{aligned}
\tag{3}
$$

where: p_t – true positive; p_f – false positive; n_t – true negative, n_f – false negative.

3.1   Classic Clustering F-Score

In the classic F-Score measure for clustering, we rely on the principle of purity. The perfect clustering result has all of the points in any given cluster belonging to only one class and all of the points within a class belonging to only one cluster. This can be related to the above relations as class and cluster equality: $x_i G x_j \Leftrightarrow c_{x_i} = c_{x_j}$, $x_i M x_j \Leftrightarrow \epsilon_{x_i} = \epsilon_{x_j}$. Once these relations are defined the classic F-score measure can be calculated as:

$$
F_1 = 2p_t / (2p_t + n_f + p_f).
\tag{4}
$$

3.2   Hierarchical F-Score

This measure is an adaptation of F-Score aimed at hierarchical clustering. Initially it was used to measure the quality of clustering in [10], where the authors tried to create a topic hierarchy with related text documents. The proposed F-measure was used to validate the generated hierarchy as a whole, instead of finding and evaluating a single cut of the dendrogram. Because of this, hierarchical F-Score in its original form can be used in the field of hierarchy of groups.

For each class c it finds a cluster ϵ in the hierarchy with the maximal F-measure value:

$$
F_c = \max_{\epsilon \in \Theta} \frac{2|X_{E_\epsilon} \cap C_{C_c}|}{|X_{E_\epsilon}| + |X_{C_c}|},
\tag{5}
$$

where: $|X_{E_\epsilon} \cap C_{C_c}|$ – the number of points belonging to cluster ϵ and class c as well as their descendants.

When calculating F_c we assume that data in descendant clusters and classes belongs to their ancestors, hence X_{C_c} and X_{E_ϵ} is used. The final quality of a hierarchy is calculated as follows:

$$
F_1 = \frac{\sum_{c \in \mathbb{C}} |X_{C_c}| F_c}{\sum_{c \in \mathbb{C}} |X_{C_c}|}.
\tag{6}
$$

Because this version of the F-measure is a weighted average over all F_c, its value is greatly influenced by bigger classes. In [12,14] the calculation of F_1 is optimised because data points can belong to only one class. But when working with hierarchies of clusters this is no longer applicable.

3.3   Partial Order F-Score

The authors' version of F-Score is a new approach using partial order relation between points that are naturally formed in a hierarchy. This relation can be substituted for equivalence in the G and M relation definition: $x_i G x_j \Leftrightarrow C_{x_i} \subseteq C_{x_j}$, $x_i M x_j \Leftrightarrow E_{x_i} \subseteq E_{x_j}$, or alternatively: $x_i G x_j \Leftrightarrow C_{x_i} \supseteq C_{x_j}$, $x_i M x_j \Leftrightarrow E_{x_i} \supseteq E_{x_j}$.

Both of these are correct as all possible pairs of different points are considered when evaluating a hierarchy. There is a symmetry between the two alternate notations. By defining the relations between data in this way the measure becomes sensitive to not only the assignment of data from a single class to clusters, but also to the relative position of clusters in the tested model's hierarchy. As such, while it is calculated in the same way as the classic F-measure *(4)*, this variant fully makes use of the hierarchy of classes and of existing clusters.

4   Experiments

To verify the behaviour of the above three measures a series of experiments were conducted on generated data with known properties. The datasets used during verification were generated using the Tree Structured Stick Breaking Process [6] with eight different sets of parameters, each repeated thirty times and averaged. The data has been created by a generator which gives a points' features and class attribute. The values for the parameters used by the generator have been taken from the original publication [6] to showcase a number of different types of hierarchy structures. This selection offers hierarchies of varying width, depth and distributions of the data points among the nodes. The eight datasets used in experiments are as follows:

Name	s00	s01	s02	s03	s04	s05	s06	s07
α_0	1	1	1	5	5	5	25	25
λ	0.5	1	1	0.5	1	0.5	0.5	0.5
γ	0.2	0.2	1	0.2	0.2	1	0.2	1

The three important parameters in the data generation model are α_0, λ and γ. These parameters have a series of intuitions about the tree structure associated with them. The alpha function ($\alpha(\epsilon) = \alpha_0 \lambda^{|\epsilon|}$) controls the average density of data per level. Considering four cases for the parameters of this function:

- α_0 and λ values high (above 1) lead to extremely sparsely populated deep trees, which is not recommended;
- α_0 and λ values low (below 1) lead to densely populated shallow trees, data is either evenly distributed or grouped closer to the leaves;
- α_0 high, λ low leads to shallow trees, data more dense at lower levels of the tree;
- α_0 low, λ high leads to deep trees, the data generally grouped close to the top of the tree.

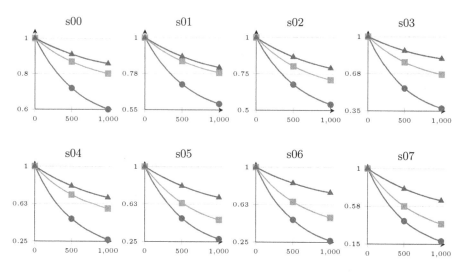

Fig. 1. Value of F-measures (vertical axis) depending on the number of random re-insertions (horizontal axis), sampled every 100 errors, for each of used datasets: red circles – Classic F-Score; blue triangles – Hierarchical F-Score; green squares – Partial Order F-Score. Note different scale on vertical axes (Color figure online).

When both parameters are close to 1 the general structure of the tree is hard to predict. The gamma (γ) parameter controls the average number of children a node can have. Hight values (generally above 1) lead to trees with more children per node. Lower values (below 1) lead to trees with less children per node on average. All three parameters interact together in the way, that alpha function controls the total data density per level of the hierarchy while γ splits the remaining density up between the children.

4.1 Random Error Introduction Tests

The first series of experiments carried out involved randomly re-inserting data within the cluster hierarchy. The results are in Fig. 1. The probability of a point of data being re-inserted was $\frac{1}{N}$ where N is the total number of data points. The re-insertion probability is the same as when generating a model using the TSSB distribution [6], the distribution parameters remain the same. The re-insertion process ignores data feature vectors and is based only on the structure of the hierarchy. The re-insertion results in data points changing the cluster they belong to, without changing their class, which the measures should perceive as an error.

The results of these experiments show that all measures similarly reacts negatively to the introduction of random errors. The drop rate of the measures appears proportional to the amount of random re-insertions, and none of the measures display any unusual behaviours during this experiment. All of the measures behave as intended in this situation. Since the values of the measures cannot be compared directly to each other it does not matter which variant of F-score achieves higher or lower results.

4.2   Reduction to a Single Cluster

In the second experiment all data points were moved into the root node. No random errors were considered during this round of testing. The tests results demonstrate that each of the measures reacts negatively to such extreme under-clustering (Table 1). It is notable however, that the results are no longer as

Table 1. The average values μ and standard deviations σ, all data points are in the root node, without errors introduced.

Set	Classic F-score		Partial order F-score		Hierarchical F-score	
	μ	σ	μ	σ	μ	σ
s00	0.6456	0.1616	0.8300	0.0857	0.8225	0.0719
s01	0.5973	0.2374	0.7977	0.1393	0.7501	0.0966
s02	0.5311	0.2600	0.7139	0.1854	0.6970	0.1062
s03	0.3729	0.1207	0.6891	0.0915	0.7598	0.0853
s04	0.1931	0.1389	0.5737	0.1384	0.5750	0.1089
s05	0.1952	0.0953	0.4253	0.1363	0.5680	0.0672
s06	0.2031	0.0872	0.4829	0.1368	0.6748	0.1049
s07	0.0611	0.0481	0.2528	0.0939	0.4723	0.0918

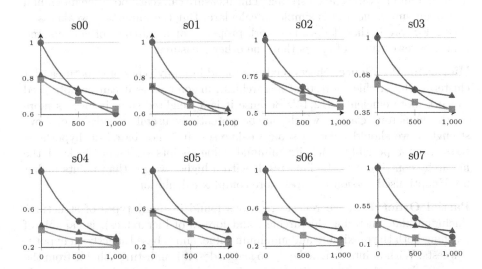

Fig. 2. Value of F-measures (vertical axis) depending on the number of random re-insertions (horizontal axis), sampled every 100 errors, for data with removed hierarchy: red circles – Classic F-Score; blue triangles – Hierarchical F-Score; green squares – Partial Order F-Score. Note different scale on vertical axes (Color figure online).

uniform and proportional. Notably the hierarchical F-Score measure stands out when compared with the other two, which show a correlation between their values and descending trend as the mass of data moves generally lower in the generated hierarchies. The value of the hierarchical F-Score however, does not follow this trend.

4.3   Removing the Cluster Hierarchy

The considered measures were tested using data with removed hierarchy. All clusters were rendered independently, as if the clustering was flat. This reduced the hierarchy to one level and essentially removed the hierarchical relationship between nodes. In these experiments errors were also introduced, but the structure of the cluster hierarchy always remained a single-level. It can be seen in Fig. 2 that the traditional F-measure is oblivious to these changes, as it was expected, because it only considers data points of the same class and cluster. However, the other two measures reacted negatively to this flattening.

5   Conclusions

Based on the results of the conducted experiments it is possible to comment the general behaviour of the three measures. Each of them possesses strengths and weaknesses. Ultimately the choice depends on the problem being evaluated.

Classic F-Score is based strictly on hypothesis tests and reflects relations found in the flat and hierarchical clustering. This measure can reach both the maximum and minimum value and is simple to calculate. But the important weakness of classic F-Score is that it does not work properly for hierarchies of clusters and it notices fewer types of errors than the other measures.

Hierarchical F-Score reflects relations found in many types of structures (flat clustering, hierarchies, forests of hierarchies), in some cases it can be optimised to work more efficiently [12,14]. Because it is a weighted sum, it focuses more on the numerous classes, which can be potentially useful. Besides the above strengths, we should mention some weaknesses. It is not based on hypothesis tests, cannot possibly reach its minimal value, points on lower levels of the hierarchy contribute to the final result with a higher weight than points higher up. Unoptimised version requires more complex calculation.

Partial Order F-Score reflects relations found in many types of structures (including flat clustering, hierarchies, and forests of hierarchies), is capable of reaching both the maximum and minimum value, and is based strictly on hypothesis tests. With points assigned only to leaf nodes, it is indistinguishable from the classic F-Score, therefore for flat clusters it can play the role of classic F-Score. It can be optimised to work as fast as classic F-score. However, when unoptimised, is more complex to calculate than classic F-Score, which should be mentioned as its weakness.

The paper offers some interesting insights into the adaptation of the F-Score measure to work with hierarchies of clusters. However, there are other common clustering measures that can be adapted in similar ways. Further research is required in order to judge the feasibility of other quality indices. We are continuing research on the generation of hierarchies of clusters as well as on the external and internal measures suitable to validate and compare their results.

Acknowledgements. The research was supported by the European Commission under the 7th Framework Programme, Coordination and Support Action, Grant Agreement Number 316097, ENGINE – European research centre of Network intelliGence for INnovation Enhancement (http://engine.pwr.wroc.pl/).

References

1. Andreopoulos, B., An, A., Wang, X., Schroeder, M.: A roadmap of clustering algorithms: finding a match for a biomedical application. Brief. Bioinform. **10**(3), 297–314 (2009)
2. Blundell, C., Teh, Y.W., Heller, K.A.: Bayesian rose trees. arXiv preprint. (2012). arxiv:1203.3468
3. Cimiano, P., Hotho, A., Staab, S.: Comparing conceptual, divise and agglomerative clustering for learning taxonomies from text. In: de Mántaras, R.L., Saitta, L. (eds.) Proceedings of the 16th Eureopean Conference on AI, Spain, pp. 435–439. IOS Press (2004)
4. Desgraupes, B.: Clustering indices (2013). https://cran.r-project.org/web/packages/clusterCrit/vignettes/clusterCrit.pdf
5. Everitt, B.S., Landau, S., Leese, M., Stahl, D.: Cluster Analysis. John Wiley and Sons Ltd, New York (2011)
6. Ghahramani, Z., Jordan, M.I., Adams, R.P.: Tree-structured stick breaking for hierarchical data. In: NIPS, pp. 19–27 (2010)
7. Hartigan, J.A., Wong, M.A.: Algorithm as 136: A k-means clustering algorithm. Appl. Stat. **28**(1), 100–108 (1979)
8. Jain, A.K.: Data clustering: 50 years beyond k-means. Pattern Recogn. Lett. **31**(8), 651–666 (2010)
9. Kogan, J., Nicholas, C.K., Teboulle, M.: Grouping Multidimensional Data: Recent Advances in Clustering. Springer, Heidelberg (2006)
10. Larsen, B., Aone, C.: Fast and effective text mining using linear-time document clustering. In: Fayyad, U.M., Chaudhuri, S., Madigan, D. (eds.) Proceedings of the 5th ACM SIGKDD International Conference on Knowledge Discovery and Data Mining, USA, pp. 16–22. ACM (1999)
11. Madhulatha, T.S.: An overview on clustering methods. CoRR abs/1205.1117 (2012)
12. Mirzaei, A., Rahmati, M., Ahmadi, M.: A new method for hierarchical clustering combination. Intell. Data Anal. **12**(6), 549–571 (2008)
13. Oded, M., Lior, R. (eds.): Data Mining and Knowledge Discovery Handbook. Springer, New York (2010)
14. Olech, L.P., Paradowski, M.: Hierarchical gaussian mixture model with objects attached to terminal and non-terminal dendrogram nodes. In: 9th International Conference on Computer Recognition Systems, Poland (2015)

15. Pohl, D., Bouchachia, A., Hellwagner, H.: Social media for crisis management: clustering approaches for sub-event detection. Multimed. Tools Appl. **74**(11), 3901–3932 (2015)
16. van Rijsbergen, C.J.: Information Retrieval. Butterworth, London (1979)
17. Sevillano, X., Valero, X., Alas, F.: Look, listen and find: A purely audiovisual approach to online videos geotagging. Inf. Sci. **295**, 558–572 (2015)
18. Spytkowski, M., Kwasnicka, H.: Hierarchical clustering through bayesian inference. In: Nguyen, N.-T., Hoang, K., Jędrzejowicz, P. (eds.) ICCCI 2012, Part I. LNCS, vol. 7653, pp. 515–524. Springer, Heidelberg (2012)
19. Xu, R., Wunsch, D.: Survey of clustering algorithms. Trans. Neur. Netw. **16**(3), 645–678 (2005)

Hierarchical Evolutionary Multi-biclustering
Hierarchical Structures of Biclusters Generation

Anna Maria Filipiak and Halina Kwasnicka$^{(\boxtimes)}$

Department of Computational Intelligence,
Faculty of Computer Science and Management,
Wroclaw University of Technology, Wroclaw, Poland
halina.kwasnicka@pwr.edu.pl
http://kio.pwr.edu.pl

Abstract. Biclustering is an important method of processing a big amount of data. In this paper, hierarchical structures of biclusters and their advantages are discussed. We propose the author's method called HEMBI (Hierarchical Evolutionary Multi-Biclustering) which creates this kind of structures. The HEMBI uses an Evolutionary Algorithm to split a data space into a restricted number of regions. The important feature of the method is ability to choice the optimal number of biclusters, which is restricted only to a maximum value. The conducted experiments and their results are presented and discussed.

Keywords: Bilustering · Hierarchical structures · Evolutionary algorithms

1 Biclustering – A Short Introduction

Clustering is an unsupervised learning, widely used for grouping observed data into such groups (clusters) that the data points within the group are more similar in a given sense than between groups. Each data is represented by the vector of its features. Clustering means a division of the data into a number of separate groups, and it is useful in many areas, e.g. pattern recognition [1], image analysis, information retrieval, statistics [2]. Increasing amount of data, as well as the demand for effective grouping it, cause that clustering remains growing area of interest. The Machine Learning methods have problem with high dimensional data – the problem is known as the *curse of dimensionality*. A **subspace clustering**, i.e., searching clusters of data points similar in some subspace instead of in full space of the data, is a possible remedy for the curse of dimensionality. There is three categories of subspace clustering [3]:

1. *Axis parallel*: the clustering is carried out in subspaces parallel to the axis of the feature vector.
2. *Pattern based*, called also *biclustering* or *co-clustering*: this approach uses the properties of the matrix, and searches in them certain specific patterns. It is focused on the search for areas with specific characteristics.

© Springer-Verlag Berlin Heidelberg 2016
N.T. Nguyen et al. (Eds.): ACIIDS 2016, Part I, LNAI 9621, pp. 665–676, 2016.
DOI: 10.1007/978-3-662-49381-6_64

3. *Correlation clustering*, called also *clustering in arbitrary oriented subspaces*: the search for such a subspace where the data points are correlated and in multidimensional space clearly create separate structures, e.g., a plane in the 3-dimensional space, or they come from a variety of the same distributions.

Subspace clustering is useful in many fields, e.g., gene expression analysis, [4–9], image analysis [10,11], Information Integration Systems [12], text mining [13], recommended systems [14]. The first definition of biclustering comes from [16], where this concept is defined as "interconnected cluster structures on both rows and columns as represented by their index sets I and J, respectively". The current meaning of biclustering was introduced by Cheng and Church in [15], it is perceived as a novel data mining technique useful in pattern recognition tasks.

Definition 1. *Biclustering is a division of data into groups of objects similar concurrently in rows and columns. Let A be a matrix $A = [a_{i,j}]$, consisting of R rows and C columns. Bicluster B is defined as a matrix $B(I, J)$ consisting of $I \subseteq R$ rows and $J \subseteq C$ columns: $B(I, J) : I \subseteq R, J \subseteq C$ such that $\forall i \in I$ and $\forall j \in J : a_{i,j} \in B(I, J)$.*

Thus, we are looking for data coherent with each other in terms of a given quality metrics, in a subset of objects and within a subset of their attributes.

Depending on the application, we can expect other properties of generated biclusters. This means that the quality of bicluster is closely related to its use. In this context, we distinguish four basic types of biclusters [17]:

1. constant biclusters;
2. constant rows or constant columns;
3. coherent values – additive or multiplicative model;
4. coherent evolution (trends) – on the rows or on the columns.

Constant biclusters means that each cluster's content must have the same value: $a_{ij} = const \; \forall i, j$. In reality we do not have perfect biclusters, so the variance of data belonging to biclusters is used as a quality measure of such biclusters. In *constant rows* (or *columns*) biclusters we expect constant values only in one dimension: $a_{ij} = a_i$ in i-th row (or $a_{ij} = a_j$ in j-th column). It is difficult to evaluate such biclusters, one can use the above measure by normalizing rows or columns, respectively. *Coherent values* present a regular change between the values of the respective rows or columns. The base value may be different, but the relationship between successive elements are fixed. Most often a combination of two conditions: MSR (*Mean Squared Residue*) (Eq. 1) and variance of rows (Eq. 2) is used in searching of this type of biclusters.

$$MSR(I, J) = \frac{1}{|I||J|} \sum_{i \in I, j \in J} (a_{ij} - a_{iJ} - a_{Ij} + a_{IJ})^2, a_{iJ} = \frac{1}{|J|} \sum_{j \in J} a_{ij},$$
$$a_{Ij} = \frac{1}{|I|} \sum_{i \in I} a_{ij}, a_{IJ} = \frac{1}{|J|} \sum_{i \in I, j \in J}, a_{ij} = \frac{1}{|I|} \sum_{i \in I}, a_{ij} = \frac{1}{|J|} \sum_{j \in J} a_{ij}. \quad (1)$$

Since the coherent values are a general case of the two previous, the variance of the rows is used as the support metrics to eliminate trivial biclusters (with the same values or consisting of the same rows). It is given by Eq. 2:

$$V(I, J) = \frac{1}{|J|} \sum_{j \in J} (a_{ij} - a_{Ij})^2 \quad (2)$$

Presented above MSR is quite universal; $MSR = 0$ for constant biclusters or containng constant rows or columns, or coherent values. However MSR is not resistant to outliers. Value of MSR of bicluster with one outlier can be lower than a bicluster containing noised values. Wang and Yang [18] proposed alternative measure, $pScore$: $pScore = |(a_{ij} - a_{il}) - (a_{kj} - a_{kl})|$, $pScore$ is checked for each 2×2 sub-array and the decision to accept or reject the bicluster is taken.

Other measure, ACV ($Average\ Correlation\ Value$) is proposed in [8]. It is assumed that a bicluster should be a subset of attributes from two dimensions that are strongly correlated. In opposite to the previous measures, the AVC is maximised – higher value is for the more correlated elements of the bicluster. Similar idea is presented in [19] – ASR ($Average\ Spearman's\ rho$) in which a correlation between two vectors: i-th and j-th rows, or columns, called Spearman's rank correlation is considered. ASR is in the range of $[-1, 1]$, the higher value indicates more correlated items.

$Coherent\ evolution$ (trends) biclusters indicate the same relationship between the values in rows, or more frequently, in columns. There is no one universal measure to find this type of biclusters, therefore other methods are used which are not based on optimization of a specific measure.

In addition to the criterion of the relationship between the values of the cluster, we can consider the structure of biclusters [17]: $Single\ bicluster$: we have only one bicluster in the data matrix. $Exclusive\ row\ and\ column$ biclusters: several biclusters are defined, but no row and column exists that is covered by more than one bicluster. $Checkerboard\ structure$: whole input matrix is covered by biclusters as a chess board. $Exclusive$-$rows$ (column) biclusters: all rows (columns) are covered by not more than single bilcuster. $Tree\ structure$ biclusters: is a results of application of hierarchical clustering for each dimension separately, it is the oldest ([20]). $Overlapping\ biclusters\ with\ hierarchy$ structure: the biclusters are disjoint or one embraces the other.

2 Some Approaches to Biclustering

Historically, there are three important approaches to biclustering, they are connected with the authors' approach. A significant basis for the biclustering gave Cheng and Church [15]. DBF – $Deterministic\ Biclustering\ with\ Frequent\ pattern\ mining\ algorithm$ [21], as one of the first introduces simultaneous search for multiple biclusters. Its idea coincides with QHB ($Quick\ Hierarchical\ Biclustering$) algorithm [9]. The last important approach is using a $Genetic\ Algorithm$ as an engine of searching for biclusters.

Cheng and Church in [15] introduced the metric capable identify clusters of coherent values, namely MSR, ($Mean\ Squared\ Residue$), end the concept of σ-$biclusters$: Given a matrix A, a sub-matrix A_{IJ} (where I is a subset of rows, and J is a subset of columns) is called the σ-$bicluster$ if the $MSR_{I,J} \leq \sigma$ for assumed value of $\sigma \leq 0$.

Their method relays on the searching for the largest σ-clusters, the starting point was a greedy algorithm using adding or deleting rows and columns.

More details one can find in [15]. The method was tested on real datasets *Yeast* and *Human*, the results have become a point of reference for many subsequent algorithms (http://arep.med.harvard.edu/biclustering).

DBF [21] consists of two phases: the generation of initial biclusters, and their extension in order to obtain the largest possible clusters. It uses the definition of MSR and σ-biclusters [15], and works in three phases. First phase is the generation of *Good Seeds* of possible biclusters, it consist of three steps: 1. the matrix data is *transformed to the matrix of relationship* between the values of successive columns; 2. the *frequent pattern mining* is applied on the transformed data with pre-determined minimum support count; 3. extraction of good seeds from the patterns. We assume that good seeds are these of frequent patterns, of which MSR is less then assumed σ value. Because a lot of frequent patterns can meet this assumption, the selection of best seeds is based on value of ratio of MSR to the size of the bicluster. BDF produces good biclusters taking into account their size and MSR.

SEBI (Sequential Evolutionary BIclustering), presented in [5], is an interesting evolutionary approach. It works in sequential way, similarly as that proposed by Cheng and Church. The aim of EBI was to find σ-biclusters with maximal size, relatively high row variance, and minimal overlapping among biclusters. In $SEBI$ this procedure is repeated sequentially. As a fitness function the authors use a linear combination of four elements [5]: MSR normalized by σ; the inverse of the row variance; empirically determined parameter aiming to balance the number of rows and columns in the bicluster; penalty for generation two identical or similar biclusters. The most important advantage of the genetic approach is a lack of hard threshold associated with σ.

3   Hierarchical Structures of Biclusters

Biclusters can be organized into hierarchical structures, but it is not frequent area of research, it has not yet been fully exploited. However it gives a broad spectrum of use in ontologies, analysis of social networks or biology.

QHB – Quick Hierarchical Biclustering was introduced in 2006 [9]. It searches biclusters using top-down approach, automatically generates a hierarchical structure. The idea of this algorithm is similar to the DBF, but in the third phase it uses a new measure of trend consistency of biclusters, namely MFD – *Mean Fluctuating Degree*. It informs how similar trends are. If the changes of sequential conditions are the same the $MFD = 0$, and for similar changes MFD is small. The resulting hierarchy is understood intuitively – child biclusters are included in the parent bicluster. The main advantage of QHB comparing with DBF is taking care about the cohesion of trends. For DBF, seeds are expanded while for QHB are reduced and clarified. DBF is based on the MSR what does not guarantee cohesion of trends. QHB guarantees compliance of trends, more, their similar value at a given level.

TreeBic Algorithm. Another interpretation of hierarchy is in [7], where *TreeBic* algorithm is presented. The method searches objects similar to each other for a some subset of conditions. The idea of *TreeBic* differs from others. It produces a probabilistic model, and then adjusts the bilcustered data. Integration of the model with data lies on the calculation of cumulative a posterior probability (data distribution). The advantage of the *TreeBic* method is the possibility of biological interpretation.

4 *HEMBI* – Hierarchical Evolutionary Multi-biclustering

HEMBI is an authors' method of generation of hierarchical structures of biclusters, based on a *Genetic Algorithm* (*GA*). *GA* is derived from the Darwinian theory of evolution, it uses the concept of survival of the fittest and real processes occurring in biological populations, such as the exchange of genes (*crossover*) or random changes of genes (*mutations*) [5]. In this paper we focus only on how we use *GA* in the *HEMBI* method for searching hierarchical structures of biclusters.

A chromosome. Chromosome consists of N rows (N is the maximal number of biclusters) and $|R| + |C|$ columns. Every row represents one bicluster. First R columns correspond to the indexes of rows of the input matrix, the next C columns correspond to the indexes of its columns. Value 1 on ij position of the chromosome means that i-th bicluster contains j-th rows of the input matrix if $j \leq |R|$ or $j - |R|$-th column of input matrix if $j \geq |R|$; 0 means opposite. Each bicluster can be *visible* or *hidden*, the size of the bicluster defines its state: a bicluster is *hidden* if its size is < 4.

Fitness function. Fitness function $F(X)$ of the chromosome X evaluates its quality, it determines which chromosomes in the population are the most promising and better meet the specified criteria. $F(X)$ consists of four components (Eq. 3), it uses MSR and it has to be minimised.

$$F(X) = w_{MSR} \times MSR(X) + w_{cov} \times Cov(X) + w_{mcov} \times mcov(X) + w_{cnt} \times cnt(X) \tag{3}$$

where X is a chromosome consisting of $M \leq N$ visible biclusters, w_k is a weight assigned with k-th component, ($MSR(X)$) – *Mean Squared Residue* of the chromosome X, is an arithmetic average of MSR of biclusters present in the chromosome X and normalized with respect to the MSR of the parent bicluster. The first element of $F(X)$ is defined by Eq. 4.

$$MSR(X) = \frac{\sum_{m}^{M} MSR(x_m)/M}{MSR(X_{RC})} \tag{4}$$

where M is a number of visible biclusters, $MSR(X_{RC})$ – MSR of the parent bicluster, $MSR(x_m)$ – MSR of m-th visual bicluster.

The total cover $Cov(X)$ is given as: $Cov(X) = 1 - cov(X)$, where $cov(X)$ is a fractional coverage of the matrix by biclusters. It is in range of $[0, 1]$.

Multiple coverage $mcov(X)$ refers to the situation when more than one bicluster covers a given data point in the matrix, i.e., the same pair row-column is defined in more than in one bicluster.

$$mcov(X) = \frac{1}{|I| \times |J|} \sum_{m=0}^{M-1} \sum_{l=m+1}^{M} dubCov(x_m, x_l) \qquad (5)$$

where $dubCov(x_m, x_l)$ is a shared part of biclusters x_m and x_l.

A number of biclusters $cnt(X)$ is calculated as: $cnt(X) = 1/M$ what reflects that more important is to add a new bicluster if there exists small number of biclusters than if this number is bigger. Initially each *weight* was set to 1.

Genetic operators. We apply the *tournament selection* of size 5. The best individual from 5 randomly chosen is selected. *Crossover* lies on the exchanging biclusters between two parent chromosomes. It is performed within the parts of chromosomes representing the indexes of rows and columns separately, with a given probability p_{cr}. Two children chromosomes are created by exchanging parts of parents' chromosomes. We apply classical *mutation* understood as a random change of single bit with assumed probability p_m.

The applied procedure uses a simple and intuitive method for iterative top-down depth-first search. The Genetic Algorithm running on a matrix generates a set of child biclusters. Next, on each of the resulting biclusters the procedure is repeated. The final result is a hierarchical structure arranged in a tree. Each node contains indexes of rows and columns belonging to the bicluster. The procedure ends according to the stop criterion, namely one of three conditions:

1. MSR of the bicluster is equal to 0 (the bicluster is fully coherent). In practice we assume maximal value of σ (we generate σ-biclusters).
2. Size of the bicluster is too small – we assume at least 2 rows and 2 columns.
3. Only one child bicluster is found.

A bicluster on which GA performs biclustarization is called a *parent bilcluster*. At the first iteration of the GA whole initial dataset (input matrix) is a parent bicluster. In each subsequent passing in depth, one of biclusters generated in previous iteration becomes a parent bicluster. Each node of the tree stores only a mask of the bicluster, i.e., indexes of rows and columns belonging to that bicluster.

5   Experiments with the $HEMBI$ Methods

The aim of this research is to verify the $HEMBI$ method in terms of existing solutions related to flat structures, and the correctness of generated hierarchical structures. This section presents the description of performed experiments and their results.

5.1 Planning of the Experiments

Research protocol. The main problem is the difficulty of comparison of the designed solution with existing ones, due to the different purposes and nature. However, we can explore some partial results and investigate their similarity to other in the literature. Therefore, firstly we have studied characteristics of flat biclustering obtained just after one iteration of the Genetic Algorithm, i.e., division on biclusters in a root of the tree. It is important because for the real-life data it allows for initial verification of the quality of generated biclusters, comparing with other methods of flat biclustering. This verification is done by examining coverage of data, rows, and columns, also the size of biclusters and their MSR. Examination of flat structures is also made on artificial data to verify whether the designed clusters are found. Remembering that the aim of the *HEMBI* is to achieve a hierarchical structure of biclusters, the main focal point of these experiments is a character of the structures: a depth of the tree, a number of nodes, and the structure itself. For real-life data, only expected lowering of the value of MSR of child biclusters comparing to the MSR of parent's bicluster indicates the correctness of the method. For artificial data it is possible to compare obtained results to structures that has been designed.

Used datasets. In experiments we have used three real datasets accessible in the Internet and popularly used by researchers, and five artificial datasets, generated especially for this study. Their characteristics are given below.

Yeast expression matrix [15] – the matrix consist of 2882 rows and 17 columns, available at *arep.med.harvard.edu/biclustering*.
Human B-cells expression data [15] – is based on real data [22], it contains 4026 rows and 96 columns, and is available at *arep.med.harvard.edu/biclustering*.
Arabidopsis thaliana [23] – this real dataset consists of 734 rows and 69 columns, is available at *www.eie.polyu.edu.hk/ nflaw/Biclustering*.
$1_{10 \times 10}$ – 10×10 artificial datadset, it is a small control matrix with very simple structure, it contains two biclusters, their $MSR = 0$. One bicluster is located in the left-top corner, the second in the right-bottom part, Fig. 1(a).
$2_{100 \times 100}$ – it is 100×100 artificial dataset, it does not contain a hierarchy. It was designed to test the ability of developed Genetic Algorithm generating flat biclusters. Biclusters partially overlap, Fig. 1(b).
$3_{6 \times 6}$ – 6×6 artificial hierarchical dataset, a small array with simple hierarchical structure with a depth equal to 2. Biclusters and their MSR are shown in Fig. 1(c).
$4_{100 \times 100}$ – 100×100 artificial hierarchical dataset, a test case of the hierarchical structure of a depth of 2 and of greater dimensions, Fig. 1(d)
$5_{300 \times 300}$ – 300×300 artificial hierarchical dataset, a more complicated test case with a hierarchical structure. The structure has a depth of 3 and biclusters at the lowest level have $MSR = 0$. Coverage on the first level of the hierarchy is 0.83, Fig. 1(e).

Creating test data with hierarchical structure requires a specific approach [15]. The expected value of the MSR of the matrix filled by uniformly generated random elements of range [a, b] is equal to $(b-a)^2/12$ and is independent of its size. The biclusters were created top-down, their areas were possibly disjoint, however some of them overlap in small part. Only the biclusters of the lowest level have not random values – their $MSR = 0$.

Parameters setting. Parameters of the Genetic Algorithm were chosen on the basis of a number of initial experiments, these parameters are: size of tournament n_t (from 2 to 8), population size N_Σ (from 60 to 140), crossover probability p_{cr} (from 0.3 to 0.8), and mutation probability p_m (from 0.0005 to 0.004). The method is more sensitive for the mutation probability and quite resistant for other parameters. Finally the following values of genetic parameters were chosen: $n_t = 5$, $N_\Sigma = 100$, $p_{cr} = 0.5$, $p_m = 0.002$.

For particular datasets it was necessary to chose maximal number of biclusters N_B, and weights present in the fitness function (Eq. 3). Parameters of experiments are collected in Table 1, other weights were set to 1.

Table 1. Summary of parameter values used in experiments for particular dataset

Data:	Yeast	Human	Arabid. thaliana	$1_{10\times10}$	$2_{100\times100}$	$3_{6\times6}$	$4_{100\times100}$	$5_{300\times300}$
N_B:	3,5,7	5,7	5,7	2,3,5,7	5,7	2	3,5	5,7
w_{MSR}:	1,2,3	1,2,3	1,2,3	1,2,5	1,2,5,10	1,2,3	1,2,3	1,2,3
w_{cov}:	1,2,3	1,2,3	1,2,3	1	1,2,3	1	2	1,2

5.2   Obtained Results

General observations. In the series of experiments one can notice some tendencies independently from a dataset used. The main assumption of a hierarchical structure of such biclusters is to get the child biclusters more precise than the parent bicluster. This means that the MSR of child biclusters should be lower than that of the parent bicluster. The main problem that arose during the study was to get the hierarchy of biclusters satisfying the above assumption. The problem is with arithmetic average of MSR in the finess function Eq. 4. The arithmetic average does not provide a better quality of all child biclusters. It is possible

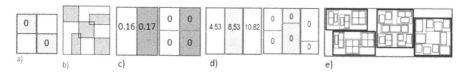

Fig. 1. Visual illustration of designed biclusters: (a) simple flat two biclusters; (b) flat biclusters partially overlapped; (c) simple hierarchical structures, MSR of the whole data is 0.92; (d) hierarchical structure of a depth of 2 and of greater dimensions; (e) hierarchical structure of a depth of 3 (rotated by $90°$ in a clockwise direction).

to obtain biclusters with significantly lower values of MSR and, at the same time, biclusters with MSR even higher than that of parent, however, due to the use of the arithmetic mean, this component of fitness value still achieves seemingly satisfactory value. Additionally, the results revealed good repeatability taking into account fitness value, data coverage, and MSR. Undesired behaviour is that in general, the created structures are not very deep and often highly unbalanced. It means that a tree is expanded only in a very few nodes, these with high values of MSR. Often a small bicluster with low MSR is achieved at the first level and such bicluster is not expanded further. To investigate different configurations of weights of the fitness function components, impact of enhancing of MSR component on general quality of the results was tested. Dramatic fall in the size of the clusters and thus little complete coverage of the data was the main observation. Increase the weight of the coverage component caused increase of coverage of the data, but the average arithmetic also increased. Thus, the depth of the structure and the number of its nodes have clearly increased. For example, for dataset *Yeast*, for weights of all components equal to 1, the $MSR = 451 \pm 92$ and depth of tree is 3, for $w_{cov} = 3$ a depth of tree is 12, but for $w_{MSR} = 3$ a depth of tree is 1.

Real datasets. In order to compare our results with those known from the literature, we focus on data coverage. Average coverage obtained by the best biclusters, in the nodes close to the root, fluctuated around $50 - 70\%$. On the deeper level of the structure these values were closer to 100%. It is because biclusters at lower levels of the tree are smaller, have lower value of MSR, and the search space for the Genetic Algorithm is smaller.

For *Yeast* dataset our method was able to reach the coverage of 75%. In relation to other methods, where biclusters covered more than 80% of the input matrix, our result is not satisfactory, especially in the context of an additional degree of generality – a hierarchical structure.

Coverage of *Human* dataset was similar to that obtained by other methods, it ranged around $35 - 40\%$ while original value is 36.4%. The method does not work fully satisfactorily, the problem was at the lowest levels of tree where coverage was far from the expected.

Concerning coverage of rows and coverage of columns, the method produces similar results as presented in [15]. Regarding the influence of N_B on the tree structure, we have observed a tendency to maximize the number of biclusters. In all runs the method created maximal number of biclusters N_B. In the nodes at the lower levels of tree a smaller number of biclusters were generated. Maximizing the number of biclusters at high levels is a simple way to reduce the value of the fitness function, and a greater number of biclusters helps to reduce the value of MSR. At the lower levels, if a parent bicluster is small, the visibility requirement can automatically enforce the formation of partially overlapping biclusters, what reduces the quality of the bicluster, although the coverage increases. The profit of MSR must be quite small, therefore even increase of the total quality of all other 3 components does not result in increase of the number of biclusters in the chromosome.

Artificial datasets. For $1_{10\times10}$ dataset all the obtained results have final value of the fitness function much lower than the expected 0.5. This means that in terms of the such defined fitness function the better solutions were found, but this solution is not the expected one. Detailed analysis shows that the normalization of MSR in relation to parent bicluster was a reason of that. The benefit from increased coverage was greater than relatively small loss on quality of MSR. It should be noted that the real datasets rarely contain biclusters with $MSR = 0$. Value of σ for *Human* and *Yeast* datasets was fixed respectively to 0.5 % and 2 % of MSR of the input matrices. Strengthening the MSR component in the fitness function caused that the proper solution was found almost in all runs.

For the $2_{100\times100}$ dataset the $HEMBI$ method was not able to produce the expected solution. $HEMBI$ found relatively easy biclusters with $MSR = 0$ but not all existing, in effect there was low coverage of the matrix, about $20 - 25\%$ to the real value of 44 %.

For simple hierarchical structure, $3_{6\times6}$ dataset, after tuning parameters of the Genetic Algorithm ($p_m = 0.01$, it was increased due to the small number of genes in the population), the method found two designed biclusters at the first level, and next, they were properly divided on biclusters with $MSR = 0$. For $4_{100\times100}$ hierarchical dataset the results are far from the designed, however some of their elements were present. Increasing the weight w_{MSR} in the fitness function caused that the produced structures were close to the expected one. The coverage was about 90 %. For $5_{300\times300}$ dataset the similar problems to those for flat structure were observed. During the first pass of the algorithm (on the full input matrix), the results were far from the designed top level biclusters. With each successive level of the hierarchy, generated biclusters are getting closer to the designed. Even biclusters with $MSR = 0$ were found. The generated structures were very broad and deep. In relation to broadness it partially corresponds to the designed structure, but the trees were too deep, what does not agree with the original.

Summing up the results of above experiments we can say that they were promising, but did not meet all our expectations. The main issue has proved to be a high value of MSR of some of the found biclusters, even higher than the value of their parent bicluster. This is due to the use of the arithmetic mean of a set of biclusters of the chromosome, where the rest of the biclusters clearly diminish the value of that component. Therefore, it is difficult to consider the results of experiments to be fully satisfactory, but it opened the way to further research that may improve the quality of the method.

6   Summary

The hierarchical biclustering is not a simple problem. A particular problem is connected with the multitude of its interpretation. Different types of biclusters are based on different assumptions, it depends of the goal of clustering. This means that depending on the goal, clustering methods can produce different clusters. Thus the comparison of results of hierarchical clustering is difficult and not clear. In addition, there is no accessible reference methods. The $HEMBI$

method, thanks to changes of weights in the fitness function, allows to tune the method and focus on the better biclusters according to MSR or more on better generalization and data coverage. The proposed method iteratively divides the flat space of the data on the predefined number of areas, but chooses the most advantageous, which could not be found in the accessible literature. $HEMBI$ in general works properly, but examining the results one can see undesirable behaviour of MSR of the child biclusters. The problem lies in a definition of fitness function, especially its MSR component.

Experiments with $HEMBI$ allow us to formulate some improvements. The serious problem lies in using the average of MSR of all child biclusters. The proposed solution to this problem is the introduction of the penalty function into the fitness function. This penalty should depend on the number of biclusters with value of MSR greater than the parent bicluster and on the value of sum of these differences. Additional possible modification of the fitness function is removing its component connected with the number of biclusters. It is due to observed behaviour that the method tried to generate maximal allowed number of biclusters. The above proposed modifications should at least reduce the adverse effects of higher values of MSR of child biclusters than MSR of the parent bicluster. It seems that the above modifications will not help in receiving better balanced trees. The remedy for this problem could be a variance of child biclusters. Clusters with similar MSR values should potentially give a similar hierarchy structures during further processing. MSR tells us how divergent from each other are values inside a cluster. So if the level of coherence of results is similar, there is a greater chance of getting a similar structure of child biclusters. A variance of visible biclusters in the chromosome should provide a level of similarity of their MSR. In subsequent iterations of the Genetic Algorithm, at the lower levels of tree, the mutation operator can produce too small changes in the population, because the size of chromosomes becomes smaller. Therefore it might be advisable to apply an adaptive probability of mutation, e.g., proportional to the size of the chromosome in the process of generation the hierarchy.

References

1. Fukunaga, K.: Introduction to Statistical Pattern Recognition. Academic Press Professional, San Diego (1990)
2. Murtagh, F.: A survey of recent advances in hierarchical clustering algorithms. Comput. J. **26**(4), 354–359 (1983)
3. Kriegel, H.P., Kröger, P., Zimek, A.: Clustering high-dimensional data: a survey on subspace clustering, pattern-based clustering, and correlation clustering. ACM Trans. Knowl. Discov. Data (TKDD) **3**, 1 (2009)
4. Alizedeh, A.A.: Distinct types of diffuse large B-cell lymphoma identified by gene expression profiling. Nat. **403**, 503–510 (2000)
5. Divina, F., Aguilar-Ruiz, J.S.: Biclustering of expression data with evolutionary. IEEE Trans. Knowl. Data Eng. **18**(5), 590–602 (2006)
6. Li, G., et al.: QUBIC: a qualitative biclustering algorithm for analyses of gene expression data. Nucleic Acids Res. **37**, e101 (2009)

7. Caldas, J., Kaski, S.: Hierarchical generative biclustering for MicroRNA expression analysis. In: Berger, B. (ed.) RECOMB 2010. LNCS, vol. 6044, pp. 65–79. Springer, Heidelberg (2010)
8. Teng, L., Chan, L.: Discovering biclusters by iteratively sorting with weighted correlation coefficient in gene expression data. J. Signal Process. Syst. **50**, 267–280 (2008)
9. Ji, L., Mock, K.W.L., Tan K.L.: Quick hierarchical biclustering on microarray gene expression data. In: Sixth IEEE Symposium on BioInformatics and BioEngineering, BIBE, VA, Arlington (2006)
10. Yang, A., et al.: Unsupervised segmentation of natural images via lossy data compression. Comput. Vis. Image Underst. **110**(2), 212–225 (2008)
11. Vidal, R., Tron, R., Hartley, R.: Multiframe motion segmentation with missing data using powerfactorization and GPCA. Int. J. Comput. Vis. **79**(1), 85–105 (2008)
12. Nie, Z., Kambhampati, S.: A frequency-based approach for mining coverage statistics in data integration. In: Proceedings of the 20th International Conference on Data Engineering, Toronto, Canada (2004)
13. de Castro, P.A.D., de França, F.O., Ferreira, H.M., Von Zuben, F.J.: Applying biclustering to text mining: an immune-inspired approach. In: de Castro, L.N., Von Zuben, F.J., Knidel, H. (eds.) ICARIS 2007. LNCS, vol. 4628, pp. 83–94. Springer, Heidelberg (2007)
14. Agarwal, N., Haque, E., Liu, H., Parsons, L.: Research paper recommender systems: a subspace clustering approach. In: Fan, W., Wu, Z., Yang, J. (eds.) WAIM 2005. LNCS, vol. 3739, pp. 475–491. Springer, Heidelberg (2005)
15. Cheng, Y., Church, G.: Biclustering of expression data. In: Proceedings of International Conference on Intelligent Systems for Molecular Biology (2000)
16. Mirkin, B.: Mathematical Classification and Clustering. Kluwer Academic Press, Boston, Dordrecht (1996)
17. Madeira, S.C., Oliveira, A.L.: Biclustering algorithms for biological data analysis: a survey. IEEE/ACM Trans. Comput. Biol. Bioinform. (TCBB) **1**, 24–45 (2004)
18. Wang, H., et al.: Clustering by pattern similarity in large data sets. In: Proceedings of the ACM SIGMOD International Conference on Management of Data, SIGMOD 2002, New York (2002)
19. Ayadi, W., Elloumi, M., Hao, J.-K.: A biclustering algorithm based on a Bicluster Enumeration Tree: application to DNA microarray data. BioData Mining, vol. 2(1) (2009). doi:10.1186/1756-0381-2-9
20. Hartigan, J.: Direct clustering of a data matrix. J. Am. Stat. Assoc. **67**(337), 123–129 (1972)
21. Zhang, Z., et al.: Mining deterministic biclusters in gene expression data. In: Proceedings of Fourth IEEE Symposium on Bioinformatics and Bioengineering, BIBE 2004 (2004)
22. Alizadeh, A.A., et al.: Distinct types of diffuse large B-cell lymphoma identified by gene expression profiling. Nat. **403**, 503–510 (2000)
23. Prelic, A., et al.: A systematic comparison and evaluation of biclustering methods for gene expression data. Bioinform. **22**(9), 1122–1129 (2006). (online access) (suppl. material)

Computer Vision Techniques

.

Feature Selection Based on Synchronization Analysis for Multiple fMRI Data

Ngoc Dung Bui[1], Hieu Cuong Nguyen[1(✉)], Sellappan Palaniappan[2], and Siew Ann Cheong[3]

[1] University of Transport and Communications, Hanoi, Vietnam
dnbui@utc.edu.vn, cuonggt@gmail.com
[2] Malaysia University of Science and Technology,
Petaling Jaya, Selangor, Malaysia
sell@must.edu.my
[3] Nanyang Technologycal University, Singapore, Singapore
cheongsa@ntu.edu.sg

Abstract. Functional magnetic resonance imaging (fMRI) can be used to predict the states of the human brain. However, solving the learning problem in multi-subjects is difficult, because of the inter-subject variability. In this paper, we use the synchronization of fMRI voxels when the brain responds to a stimulus in order to construct features for achieving better data representation and more efficient classification. With a simple definition of synchronization, the proposed method is insensitive to the reasonable choices over a broad range of thresholds. We also demonstrate a new unbiased method to compare multiple subjects by applying the singular value decomposition (SVD) to the discrimination matrix, which enumerates the different patterns. The method for analyzing the fMRI data works well for identifying the meaningful functional differences between subjects.

Keywords: fMRI · Synchronization · SVD

1 Introduction

Functional magnetic resonance imaging (fMRI) is a powerful tool to study the brain activity. The common method to analyze fMRI data is to find non-zero blood oxygen level dependent (BOLD) signal in a large cross section of voxels and apply statistical parametric models to create images of brain activation [1]. Recently, using patterns of the brain activity measured by fMRI data to predict the cognitive states of the subject have been receiving much attention of the neuroscience community [2, 3]. In this approach, one expects brain activation patterns to largely similar for lower-level brain functions, such as vision or motor responses in order to find a common activity model. Thereafter, all subjects are spatially normalized to a common template for creating the voxel correspondence [4]. However, empirical evidence reveals that the activation varies strongly from subject to subject [5, 6] and from group to group [7]. Therefore, simply normalizing data across subjects and pooling the normalized data into region of interest (ROI) super-voxels might not be the best option [8]. We believe that averaging multiple fMRI time series should only be performed after we understand the sources of variations in the data.

© Springer-Verlag Berlin Heidelberg 2016
N.T. Nguyen et al. (Eds.): ACIIDS 2016, Part I, LNAI 9621, pp. 679–687, 2016.
DOI: 10.1007/978-3-662-49381-6_65

When analyzing multiple subjects, a widely used method is to average data across subjects in the same task [8, 9]. However, for an inhomogeneous group of subjects, this method gives incorrect description of subject-to-subject variation. Most fMRI studies do not concentrate on diagnostic classification and utilize group averaging in order to differentiate subject classes, such as age or clinical conditions. Moreover, since the group-averaged activation profiles cannot be used for a high demanding problem of classification, there is simply no prediction accuracy. When a fMRI dataset becomes larger, the classification will be more severe, because some manually curated classifiers can be statistically meaningful, but other classifiers cannot. In general, these meaningful classifiers are not known beforehand. Thus, there is a need to develop automatic classification algorithms that can exploit the ever-growing of fMRI dataset for knowledge discovery.

The structure of the rest of this paper is as follows. In Sect. 2, we describe the synchronization method for analysis of the fMRI data and show how the synchronization patterns were constructed. Section 3 presents experimental results of the proposed method. Finally, we conclude the paper in Sect. 4.

2   Proposed Method

2.1   Synchronization Approach

An fMRI signal consists of activated signals and uninterested signals, such as physiology-related or motion-related signals. To extract the activation areas, we employ the independent component analysis (ICA) [10] to a group of Alzheimer's disease and normal subjects which we described in Sect. 3. The sensory-motor experiment suggests that the ROIs are those associated with visual processing and motor response. Next, we created a brain mask consisting of the regions using the Brodmann template. After applying the ICA, an activation map consisting of voxels, whose spatial map of highest correlation was selected. The ROIs which can be identified are the primary motor cortex (PMC), the supplementary motor area (SMA), the primary visual cortex (PVC), and the extrastriate visual cortical areas (EVC) (Fig. 1).

For a particular active region, many voxels must be activated simultaneously in order to make a strong response. This implies that a large cross section of voxels should be synchronized for reaction to a stimulus. These cross sections of voxels can be discovered through statistical clustering that is based on the magnitude of their responses to the experiments [11]. The similarity between voxel time series are commonly measured with linear techniques, such as the coherence [12] and the duration of coupling between a pair of neurophysiological processes [13].

Coherence measures the synchronization in the frequency domain, by comparing the average cross and power spectra of the two time series across the low-frequency band. At the same time, the duration of coupling between a pair of neurophysiological processes is the length of time that their band-pass filtered signals are in phase synchronization with each other. The color maps of the coherence and the duration of coupling between a pair of neurophysiological processes of four Brodmann areas are shown in Fig. 2.

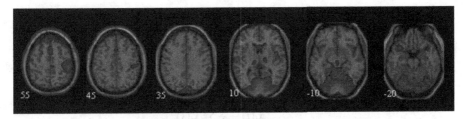

Fig. 1. Slices of the standardized anatomical brain showing the most strongly activated voxels obtained by ICA on the group of 27 subjects, over a brain mask comprising the visual processing and motor response Brodmann areas.

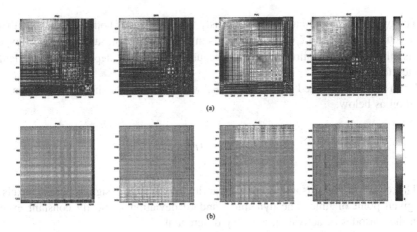

Fig. 2. Color maps of the (a) coherence matrices and (b) phase difference matrices between voxels within (left to right) the primary motor cortex (PMC, 1267 voxels), the supplementary motor area (SMA, 3601 voxels), the primary visual cortex (PVC, 1137 voxels), and the extrastriate visual cortical areas (EVC, 5949 voxels) of Alzheimer subject 2. The average coherences over all voxels over all times are 0.35 ± 0.29 (PMC), 0.32 ± 0.27 (SMA), 0.48 ± 0.31 (PVC), and 0.33 ± 0.29 (EVC). The average phase differences between all voxel pairs over all times are -0.06 ± 0.74 (PMC), 1.3 ± 1.5 (SMA), 0.02 ± 0.55 (PVC), and 0.01 ± 0.52 (EVC).

Although these approaches can measure the synchronization, they have some limitations. First, we have to choose a time window for measuring frequency spectrums. Moreover, no assumption is made for the synchronized cross section, i.e. as long as the synchronization is persistent in time, it can be detected even if only two of K voxels are synchronized. Another method, which applies a graph theory on resting state fMRI and uses graph measured as features for classification [14]. However, as we mentioned, physiologically meaningful synchronizations present large spatial cross sections. In this paper, we show how one can take advantage of this large spatial cross section of synchronized voxels in order to detect brief synchronizations.

Consider $x_i(t)$ and $x_j(t)$ are the BOLD signals from two voxels i and j. Their standardized fMRI activities can be defined as follows:

$$\zeta_i(t) = \frac{x_i(t) - \frac{1}{N}\sum_{s=1}^{N} x_i(s)}{\sqrt{\frac{1}{N-1}\left[\sum_{s=1}^{N} x_i^2(t) - \left(\frac{1}{N}\sum_{s=1}^{N} x_i(s)\right)^2\right]}},$$

$$\zeta_j(t) = \frac{x_j(t) - \frac{1}{N}\sum_{s=1}^{N} x_j(s)}{\sqrt{\frac{1}{N-1}\left[\sum_{s=1}^{N} x_j^2(t) - \left(\frac{1}{N}\sum_{s=1}^{N} x_j(s)\right)^2\right]}}.$$

The two voxels i and j are instantaneously synchronized if both of the standardized fMRI activities $\zeta_i(t)$ and $\zeta_j(t)$ achieve a given threshold at the same time t. In experiments, this synchronization is robust because it does not depend on a particular selected threshold.

From the standardized fMRI activities, we can calculate the dynamic standard deviation as below:

$$\sigma(t) = \frac{1}{N}\sum_{i=1}^{N} (\zeta_i(t) - \mu(t))^2.$$

This equation gives us a sense of the varying of the BOLD signals across voxels at any given point of time. The dynamic standard deviation is mostly constant except when the episodes of activation are very differential.

To ensure that only stimulation signals were picked up, we considered two voxels as being instantaneously synchronized only if they emerge together from a rejection band. The rejection band in $(-\sigma, +\sigma)$ for the rest of the paper is the time average of the dynamic standard deviation. It can be computed as follows:

$$\sigma = \frac{1}{T}\sum_{t=1}^{T} \sigma(t)$$

Next, we define the positive synchronization and negative synchronization fractions at time t:

$$\rho_+(t) = \frac{1}{N}\sum_{i=1}^{N} \theta(\zeta_i(t) - \sigma),$$

$$\rho_-(t) = \frac{1}{N}\sum_{i=1}^{N} \theta(-\zeta_i(t) - \sigma)$$

where the fractions of voxels whose standardized fMRI signals rise above or fall below $+\sigma$ $(-\sigma)$ and the unit step function can be computed as follows.

$$\theta(x) = \begin{cases} 1, & x > 0; \\ 0, & \text{otherwise} \end{cases}.$$

It is noted that $\rho_+(t)$ and $\rho_-(t)$ represent the spatial cross sections of positive and negative synchronizations.

In Fig. 3, the positive and negative synchronization fractions can be seen on the top and in the middle row. The relative strengths of the positive synchronization peaks coincide with troughs of the negative synchronization. It ensures the synchronization patterns observed are functionally meaningful, or at least as meaningful as the mean time course in measuring cognitive functions. These synchronization patterns in the four ROIs can be better visualized in a single color map (bottom row), where the blue area indicates strong negative synchronization and the red area indicates strong positive synchronization. In this color map, the green area indicates the absence of strong positive or negative synchronizations.

2.2 Feature Selection

We look for the described above differentiated responses in different ROIs in the brain. There are two ways for the responses of different ROIs to be discriminated: (1) synchronizing a ROI before synchronizing another ROI, (2) synchronizing a ROI is stronger than another ROI. In fMRI experiments, it may be difficult to see the first type of differentiation because of low time resolution, so we concentrate on looking for the second type of differentiated response, as shown in Fig. 3.

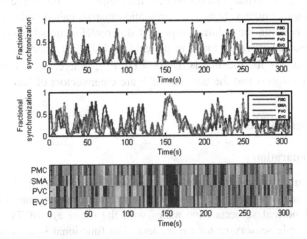

Fig. 3. Positive and negative synchronization patterns of four Brodmann areas: primary motor cortex (PMC) (red), supplementary motor area (SMA) (green), primary visual cortex (PVC) (blue), and extrastriate visual cortical areas (EVC) (cyan) in Alzheimer subject 2 (Color figure online).

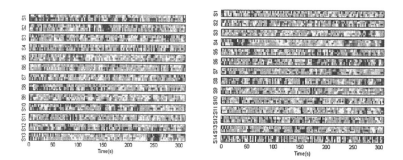

Fig. 4. Synchronization color maps of all ROIs in all 13 Alzheimer and 14 normal subjects

In Fig. 4, we show the synchronization fractions of all ROIs in 13 Alzheimer and 14 normal subjects. If the subjects were given the same task sequence, we would be able to directly compare functional differences in their responses to the tasks. In this figure, the task sequence varies from subject to subject. Therefore, though the synchronization patterns are interesting, the functional responses of different subjects can not be directly compared. For such fMRI data, we must perform functional comparison indirectly between different subjects. To this end, a discrimination matrix has been constructive.

Firstly, we go through the synchronization pattern of an individual subject and find an instance of a particular ordering of synchronization fractions. For example, the strongest synchronization can be found in the PMC, the next strongest synchronization in the SMA, followed by the PVC, and then the EVC areas for a particular stimulation episode. This is a functional pattern we may find in other subjects as well. Thus we search exhausting functional patterns and list the subjects we find these patterns in the form of a non-square matrix. In this discrimination matrix, called D the rows represent different subjects while the columns represent different functional patterns, such that $D_{ij} = 1$ if functional pattern j is found in subject i, and $D_{ij} = 0$ otherwise. After that, we utilize SVD analysis for the matrix D $(D = U\Sigma V^T)$, where the columns of U are eigenvectors of subjects and the columns of V are eigenvectors of functional patterns.

3 Experimental Results

3.1 Data Preparation

We used the publicly-known data of Washington University [15]: 13 subjects (six males with the mean age of 77.2 years) with very mild to Alzheimer's Disease conditions and 14 normal subjects (five males with the mean age of 74.9 years) were scanned in a simple sensory-motor experiment. The functional images were obtained using asymmetric spin-echo sequence sensitive to BOLD contrast with following parameters: TR = 2.68 s; 3.75 × 3.75 mm in-plane resolution; T2* evolution time = 50 ms (ms); alpha = 90°. Whole brain volumes were obtained using 16 contiguous 8-mm think slices with parallel to the plane of the anterior-posterior commissure. The raw data were received from the fMRI Data Center at Dartmouth College and preprocessed using SPM8 [16]. The images were motion corrected and normalized to coordinates of Talairach and Tournoux [17]. They were also smoothed with a 4 mm Gaussian kernel to decrease spatial noise.

3.2   Results

Referring to Fig. 4, we recognized that there is no easy way to directly compare the synchronization patterns of different subjects since the sequences of tasks that given to the subjects are different. Therefore, we constructed a discrimination matrix to look for hidden functional differences between subjects. As we are interested in the functional classification of subjects, we plot the weights of each subject along the first and second principal components of U. Clusters that appear in such a plot give a natural classification scheme (Fig. 5).

Alternatively, if we are interested in a natural classification scheme for the functional patterns, we can plot the weights of each functional pattern along the first and second principal components of V. Again clusters that might appear in such a plot would allow us to naturally classify functional patterns. We found that the Alzheimer and normal subjects are not differentiated, during positive or negative synchronizations. Thus, these two pieces of information in the form of a reordered version of the discrimination matrix can be combined. To reorder the discrimination matrix, we make use of the fact that the second principal component is generally associated with the greatest difference between subjects. In this second principal component, a subject with positive weight has similar functional patterns compared to another subject with positive weight, but dissimilar functional patterns from a subject with negative weight. Therefore, we reordered the subjects, so that those with positive weights come first, followed by those with negative weights.

We combined two pieces of information in the form of the reordered version of the discrimination matrix. In the second principal component, a subject with positive weight has similar functional patterns compared to another subject with positive weight, but

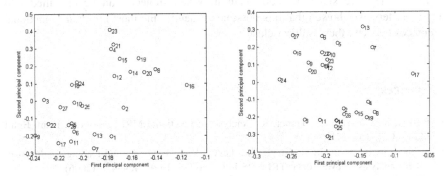

Fig. 5. Plots of weights of individual subjects (27 subjects, including 13 Alzheimer and 14 normal) of the first and second principal components of the discrimination matrices obtained from positive synchronization (left) and negative synchronization (right).

dissimilar functional patterns from a subject with negative weight. From reordered discrimination matrix, we found mostly nondiscriminatory patterns that appear in all subjects. However, we recognized discriminatory patterns and the two clusters of subjects are primarily discriminated by the positive synchronization patterns:

1 (PMC>SMA>PVC>EVC)	7 (PMC>EVC>PVC>SMA)
8 (SMA>PMC>PVC>EVC)	16 (PVC>PMC>EVC>SMA)
18 (PVC>SMA>EVC>PMC)	26 (EVC>PVC>PMC>SMA)
27 (EVC>PVC>SMA>PMC)	

From negative reordered matrix, it is shown that the two clusters of subjects are primarily discriminated by the negative synchronization patterns:

1 (PMC>SMA>PVC>EVC)	4 (PMC>PVC>SMA>EVC)
6 (PMC>EVC>SMA>PVC)	7 (PMC>EVC>PVC>SMA)
8 (SMA>PMC>PVC>EVC)	18 (PVC>SMA>EVC>PMC)
19 (PVC>EVC>PMC>SMA)	23 (EVC>SMA>PMC>PVC)
24 (EVC>SMA>PVC>PMC)	

Here we found the primary motor cortex (BA4) being most frequently the most negatively synchronized, followed by the primary visual cortex (BA17) and the extrastriate visual cortical area (BA18/BA19).

4   Conclusion

In this paper, we have presented a new and effective feature selection method for analyzing fMRI data to find the meaningful functional differences between subjects. Instead of looking for average activation profiles, we examined synchronization patterns in different parts of the human brain. Based on these patterns we constructed a discrimination matrix between subjects and between synchronization patterns, whose matrix elements tell us whether a given pair of subjects can be discriminated by a given positive or negative synchronization pattern. When subjects can be classified into natural clusters, we showed that it is possible to identify the most important functional differences between these subject clusters.

References

1. Friston, K.J., Jezzard, P., Turner, R.: Analysis of functional MRI time-series. Hum. Brain Mapp. **1**(2), 153–171 (1994)
2. Kenneth, A.N., Sean, M.P., Greg, J.D., James, V.H.: Beyond mind-reading: multi-voxel pattern analysis of fMRI data. TRENDS in Cogn. Sci. **10**(9), 424–430 (2006)
3. Mitchell, T.M., Hutchinson, R., Niculescu, R.S., Pereira, F., Wang, X., Just, M., Newman, S.: Learning to decode cognitive states from brain images. Mach. Learn. **57**, 145–175 (2004)

4. Friston, K.J., Holmes, A.P., Worsley, K.J., Poline, J.B., Frith, C.D., Frackowiak, R.S.J.: Statistical parametric maps in functional imaging: a general linear approach. Hum. Brain Mapp. **2**, 189–210 (1995)
5. Mechelli, A., Penny, W., Price, C., Gitelman, D., Friston, K.: Effective connectivity and inter-subject variability: using a multi-subject network to test differences and commonalities. NeuroImage **17**, 1459–1469 (2002)
6. Smith, S.M., Beckmann, C.F., Ramnani, N., Woolrich, M.W., Bannister, P.R., Jenkinson, M., Matthews, P.M., McGonigle, D.J.: Variability in fMRI- a re-examination of inter-session differences. Hum. Brain Mapp. **24**, 248–257 (2005)
7. Buckner, R.L., Snyder, A.A., Sanders, A.L., Raichle, M.E., Morris, J.C.: Functional brain imaging of young, nondemented, and demented older adult. J. Cogn. Neurosci. **12**, 24–34 (2000)
8. Wang, X., Hutchinson, R., Mitchell, T.M.: Training fMRI classifiers to discriminate cognitive state across multiple human subjects. In: Proceedings of the 18th Annual Conference on Neural Information Processing Systems. Vancouver, Canada (2004)
9. Schmithorst, V.J., Holland, S.K.: Comparison of three methods for generating group statistical inferences from independent component analysis of functional magnetic resonance imaging data. J. Magn. Reson. Imaging **19**, 365–368 (2004)
10. Calhoun, V.D., Adali, T.: Unmixing fMRI with independent component analysis. IEEE Eng. Med. Biol. Mag. **25**, 79–90 (2006)
11. Thirion, B., Faugeras, O.: Feature characterization in fMRI data: the information bottleneck approach. Med. Image Anal. **8**, 403–419 (2004)
12. Liu, D., Yan, C., Ren, J., Yao, L., Kiviniemi, V.J., Zang, Y.: Using coherence to measure regional homogeneity of resting-state fMRI signal. Front. Syst. Neurosci. **4**, 4–24 (2010)
13. Kitzbichler, MG, Smith ML, Christensen, SR, Bullmore E: Broadband criticality of human brain network synchronization. PLoS Computational Biology (2009)
14. Khazaee, A., Ebrahimzadeh, A., Babajani-Feremi, A.: Identifying patients with Alzheimer's disease using resting-state fMRI and graph theory. Clin. Neurophysiol. **126**(11), 2132–2141 (2015)
15. Buckner, R.L., Snyder, A.A., Sanders, A.L., Raichle, M.E., Morris, J.C.: Functional brain imaging of young, nondemented, and demented older adult. J. Cogn. Neurosci. **12**, 24–34 (2000)
16. SPM8, Statistical Parametric Mapping (2008). http://www.fil.ion.ucl.ac.uk/spm/
17. Talairach, J., Tournoux, P.: Co-Planar Stereotaxic Atlas of the Human Brain. Thieme. Stuttgart, Germany (1998)

Exploiting GPU for Large Scale Fingerprint Identification

Hong Hai Le$^{(\boxtimes)}$, Ngoc Hoa Nguyen, and Tri Thanh Nguyen

Vietnam National University of Hanoi, 144 Xuan Thuy, Hanoi, Vietnam
{hailh, hoann, ntthanh}@vnu.edu.vn

Abstract. Fingerprints are the most used biometrics features for identification. Although state-of-the-art algorithms are very accurate, but the need for fast processing speed for databases containing millions fingerprints is highly demanding. GPU devices are widely used in parallel computing tasks for its efficiency and low-cost. In this paper, we propose to adapt minutia cylinder-code (MCC) matching algorithm, an efficient algorithm in term of accuracy to GPU. The proposed method fits well with the architecture of the GPU that makes it easy to implement. The results of our experiments with a GTX- 680 device show that the proposed algorithm can perform 8.5 millions matches in a second that is suitable for real time identification systems having databases containing millions of fingerprints.

Keywords: Fingerprint identification · Matching · Minutiae · MCC · GPU · CUDA

1 Introduction

The fingerprint matching algorithms which compare two given fingerprints and return a degree of similarity are often classified into three types: correlation-based, minutiae-based, and ridge feature-based matching. The Fingerprint Verification Competitions (FVC) [2] shows that the minutiae-based matching is the most popular approach. Minutiae are the points where a ridge continuity breaks and it is typically represented as a triplet (x, y, θ); where x and y represent the point coordinates and θ is the ridge direction at that point. The task of the minutiae-based matching approach is to find the maximum number of matching minutiae pairs in the two given fingerprints. Figure 1 shows the matches between two fingerprints based on minutiae.

Minutia Cylinder-Code (MCC), a state-of-the-art matching algorithm in term of accuracy, takes 3 ms to perform a matching. So it takes 3 s to identify a fingerprint in a database of 1000 fingerprints. With a large database containing millions of fingerprints, the identification process will take a long time. This is the case of the huge civil identification systems being deployed [20].

There are two popular methods to increase the speed of fingerprint identification: reducing the total number of fingerprint comparisons (through fingerprint classification [16, 17], pre-filtering using indexing techniques [18, 19]), or using parallel architectures [14]).

© Springer-Verlag Berlin Heidelberg 2016
N.T. Nguyen et al. (Eds.): ACIIDS 2016, Part I, LNAI 9621, pp. 688–697, 2016.
DOI: 10.1007/978-3-662-49381-6_66

Graphics Processing Units (GPUs) have been proven to be very useful for accelerating the processing speed of computationally intensive algorithms. These devices introduce massive parallelism in the calculations and have been applied successfully in many fields such as artificial intelligence [21, 22], simulation [23] and bioinformatics [24]. Recently, a number of studies suggest using GPU in MCC fingerprint matching like the works of Gutierrez et al. [12], Capelli et al. [13]. In this paper, we propose a different approach to adapt MCC algorithm to GPU, the proposal fits well with GPU computing architecture, make it easy to be implemented.

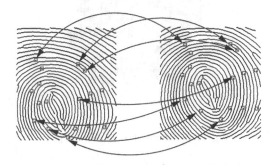

Fig. 1. Fingerprint matching based on minutiae.

The rest of the paper is organized as follows. First, we review the MCC fingerprint matching algorithm in Sect. 2. GPU programming model is briefly described in Sect. 3. Section 4 describes our adaption MCC to GPU. Finally, Sect. 5 details the experimental results over FVC 2002 DB database.

2   MCC Matching Algorithm

2.1   MCC Representation

Most minutiae-based matching algorithms consist of two steps: the first step performs a local structure matching, following by a consolidation step. The local structure matching allows quickly to find pairs of minutiae that can be matched locally and can be the candidate for aligning between the two fingerprints. The local structures are normally invariant to the fingerprint rotation and translation. Local structures of minutiae are typically represented by neighboring minutiae [3, 11], ridge [7, 8], orientation [9, 10], or combination of these [6]. Recently, Minutia Cylinder-Code (MCC) representation [4] shows a good performance in term of both accuracy and speed. In MCC presentation, each minutia is represented by a cylinder feature. This cylinder is centered at the minutia, has a fixed radius R, and a height of 2π. Each cylinder is divided into $N_s \times N_s \times N_d$ cells as shown in Fig. 2. N_s defines the resolution of the discretized 2D space around minutia m ($N_s \times N_s$) and N_d represents the number divisions applied to the height of the cylinder (2π) which represents angular distance.

The contribution of each minutia m_t to a cell (of the cylinder corresponding to a given minutia m_i), depends both on: spatial information (how much m_t is close to the center of the cell), and directional information (how much the directional difference between m_t and m_i), is similar to the directional difference associated to the section where the cell lies). In other words, the value of a cell represents the likelihood of finding minutiae that are close to the cell and whose directional difference with respect to m_i.

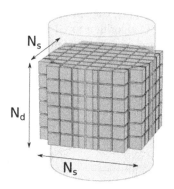

Fig. 2. Structure of a cylinder [12].

Once a cylinder c_i is built for minutia m_i, it can be simply treated as a single feature vector. With a negligible loss of accuracy [4], each element of the feature vector can be stored as a bit.

2.2 Similarity Score

A simple but effective similarity measure between two bit vectors of cylinders c_i and c_j is described in Formula 1 [4]

$$sim(c_i, c_j) = \begin{cases} 1 - \frac{\|v_i \oplus v_j\|}{\|v_i\| + \|v_j\|} & if \ d(\theta_i, \theta_j) \leq \delta_\theta \\ 0 & otherwise \end{cases} \qquad (1)$$

Where

- $\oplus$ represents the bitwise XOR operator;
- $\|.\|$ represents the Euclidean distance;
- $d(\theta_i, \theta_j)$ is the difference between the two angles of two minutiae m_i and m_j;
- δ_θ is the maximum rotation threshold allowed between two fingerprints.

With the cylinder set of the two fingerprints (T and T_q) to be matched, a local matching process is started. This computation is performed on every pair of cylinders and the results are stored in a matrix. In order to compare the similarity score s of two fingerprints T and T_q, there are two strategies: Local Similarity Sort (LSS) technique sorts all values of the matrix and computes the average of the top n_P values. A more accurate, but less efficient similarity measure is the Local Similarity Sort with Distortion-Tolerant Relaxation (LSS-DTR). LSS-DTR adds a consolidation step to

LSS, in order to obtain a score that modifies the local similarities to hold at global level. Figure 3 shows steps to calculate similarity score s using LSS, s is the average of the top n_P values chosen from the similarity matrix.

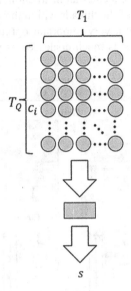

Fig. 3. Similarity score computing process using LSS [13].

3 Graphics Processing Units

The Compute Unified Device Architecture (CUDA) is one of the most widely-adopted frameworks for GPGPU. CUDA is a hardware and software architecture that enables NVIDIA GPUs to execute parallel kernels written in C/C ++. The physical architecture of CUDA-enabled GPUs consists of a set of Streaming Multiprocessors (SM), each containing 32 cores following SIMD (Single Instruction Multi Data) scheme. Threads are instances of a kernel and share the same code but each thread works on a different dataset. Threads are grouped into blocks. All threads of the same block are executed on the same SM and share the limited memory resources of that multiprocessor. The maximum number of threads in a block cannot be too large (1024 for the GPU device used in this work). However a kernel can be executed by multiple, equally-sized blocks, forming a grid: the total number of threads is then equal to the number of blocks times the number of threads per block (Fig. 4). Each SM schedules and executes threads in groups of 32 parallel threads (being 32 the number of cores in a SM) called warps. A warp executes one common instruction at a time, so full efficiency is realized when all 32 threads of a warp synchronize their execution path. If threads of the same warp take different paths (due to flow control instructions), they have to wait for each other. It is important to make GPU threads extremely lightweight.

CUDA threads have access to various memory types (Fig. 4): each thread has its registers, which are the fastest memory, and its private local memory (which is slower);

each block has a small shared memory, accessible to all threads of the block and with the same lifetime of the block; all threads have access to the global memory: the largest memory and slowest memory type, which is accessible by all threads of all blocks, so it is used for communication between different blocks and with the host (the program running on CPU). When a warp executes an instruction that accesses global memory, it coalesces the memory accesses of the threads within the warp into one or more of these memory transactions. Therefore a very important optimization in CUDA is ensuring that global memory accesses are as much coalesced as possible.

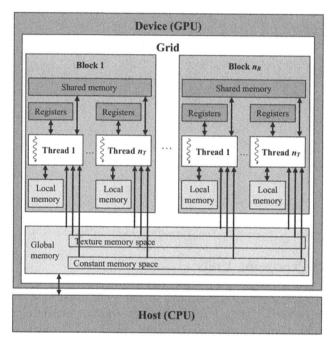

Fig. 4. CUDA: grid, blocks, threads, and the various memory spaces [13].

4   Adapting MCC Algorithm to GPU

For identifying a query fingerprint T_q in a database of N template fingerprints $\{T_1, \ldots, T_N\}$ using MCC matching algorithm, the first step is to calculate similarity matrices, after that similarity score set $S = \{S_1, S_2, .., S_N\}$ is calculated from the average of top n_P values of the similarity matrices (LSS approach). Figure 5 demonstrates the calculating steps of the MCC algorithm.

When adapting the algorithm to GPU, the aim is to maximize active threads. Threads are grouped by wraps, each of which contains 32 threads. Due to the varying minutiae number of fingerprints, to avoid divergence between threads in the warps, several approaches [12, 13] suggest to mainly divide the MCC matching algorithm into two separate kernel GPU calls. The first kernel GPU call is to calculate all similarity matrices. The second call is to calculate the match scores from similarity matrices.

When dividing the algorithm into separating calls, it is necessary to transfer data between kernel calls. Meanwhile, some advantages of the GPU architecture like share memory are not utilized. [13] uses a very careful design to adapt the algorithm and ad hoc technique to translate the similarity matrix to a fix size one.

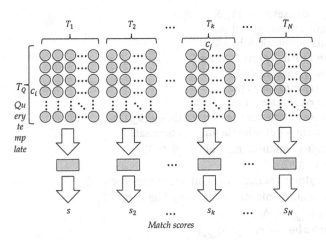

Fig. 5. Calculating steps for fingerprint identification process using MCC [13].

Our approach is based on the fact that using 32 minutiae for each fingerprint is enough for the matching process. From statistics on FVC 2002 fingerprint databases, the average number of minutiae of each fingerprint is 30, and the average number of matches for a genuine matching is only 6. In our algorithm, we use all minutiae for calculating cylinders of the fingerprint, after that, we choose 32 minutiae with cylinder having maximum number of 1 value. Minutiae with cylinder having small number of 1 value tend to be the outlines of a fingerprint. All minutiae are used to calculate cylinders so the bit vectors of cylinders are not affected.

By following our approach, all the similarity matrices in Fig. 5 have the same size 32×32. We use one GPU block for each matching between template fingerprint T_k and query fingerprint T_q, each block has 32 threads. The algorithm uses LSS technique to find the similarity score. Each thread of the block is used for calculating a column in the similarity matrix, and finding the maximum value in that column. The first thread of the block is used to calculate the similarity score from the average of maximum values found from 32 threads of the block. The detail algorithm is described in the Fig. 6, where:

- The GPU block with the index of b_{idx} is used to calculate the similarity score $S_{b_{idx}}$ between fingerprint template $T_{b_{idx}}$ of the fingerprint database with the query fingerprint T_q. Each thread with the index of t_{idx} of the block is used to calculate the maximum value of the matrix column and store found value into the array *maxValue*.
- All cylinders of the templates of the databases are loaded into the global memory of GPU before steps 1, 2 of the algorithm.

MCC Identification on GPU

Input:
− Template fingperprints $\{T_1, T_2, .., T_N\}$
− A query fingerprint T_q

Output:
−Matching score set $S = \{S_1, S_2, .., S_N\}$
Kernel execution configuration:
− 32 threads per block, N blocks

Algorithm for a kernel thread

Share memory $maxValue[32]$
Share memory $maxIndex[32]$
Share memory $maxMatching$
Block and thread index of the current thread b_{idx}, t_{idx}
1. $c_{t_{idx}}$: the cylinder of minutia $m_{t_{idx}}$ of template $T_{b_{idx}}$
2. θ_j: the angle of minutia $m_{t_{idx}}$ of template $T_{b_{idx}}$
3. For $i = 1$ to 32
4. c_i: the cylinder of minutia m_i of query fingerprint T_q
5. θ_i: the angle of minutia m_i of query fingerprint T_q
6.   If$(d(\theta_i, \theta_{t_{idx}}) \leq \delta_\theta)$
7. $temValue = sim(c_i, c_{t_{idx}})$
8.   updateMax$(maxValue[t_{idx}], tempValue)$
9.   End If
10. End For
11. _syncthreads()
12. If$(t_{idx} == 0)$
13. $S_{b_{idx}} = \sum_{t=1}^{32}(maxValue[t])/32$
}

Fig. 6. Proposal adapting MCC algorithm for GPU.

- Similarity value $sim(c_i, c_{t_{idx}})$ in line 7 is calculated by using Formula 1 (in Sect. 2.1).
- __syncthreads() function in line 11 is a barrier for all the threads of the block, after that all the results of the threads of the block are available for calculating.
- The similarity score $S_{b_{idx}}$ is calculated in line 13 by the first thread of the block using maximum values found by 32 threads of the block.

The proposed algorithm fits well with GPU computing architecture in which blocks contains the same number of threads, thus make it very easy to implement.

5 Experimental Results

In order to evaluate our proposed approaches, we used FVC 2002 DB fingerprint database to perform the experiments. In the minutiae extraction and MCC cylinders creation process, we used tools from Pérez et al. [25]. The generated MCC templates of fingerprints were stored on disk for the experiments.

For evaluating the accuracy of the proposed algorithm, the result of the proposed algorithm is compared with the result of the MCC baseline algorithm in which all minutiae are used for the matching process. Table 1 shows the results of the experiments.

Table 1. Experimental results on DB1 of FVC 2002

Algorithm	EER	FMR 100	FMR 1000	FMR Zero
MCC baseline	1.64 %	2.10 %	3.89 %	4.85 %
Our algorithm	1.76 %	2.29 %	4.32 %	5.46 %

We have achieved an EER of 1.76 % against a 1.64 % of MCC-baseline. These are certainly minor differences and can be accepted in real world applications.

For evaluating the speed of the proposed algorithm, we carried out all the experiments on an NVIDIA GeForce GTX 680 with 1536 CUDA cores, Kepler Architecture and 2 GB of memory. FVC 2002 DB was scaled to different database sizes (ranging from 10000 to 200 000) to study how the GPU based algorithm scaled with the database size. 10 input fingerprints were randomly selected to be identified. Table 2 shows the results of experiments with different database sizes.

Table 2. Execution time of the first ten queriess with different size databases

DB size	Time (ms)	Throughput (KMPS)
10000	14	7142
50000	61	8196
100000	119	8403
150000	850	8474
200000	1105	8510

At larger DB sizes, the throughput of the proposed algorithm is stable at 8.5 millions matches per second, no scalability issues were found. The result is higher than the result reported in [12], which gains 55.7 thousand matches per second using the same GeForce GTX 680 device, and comparable to state-of-the-art result [13] which gains 9 millions matches per second using Tesla C2075 GPU.

6 Conclusions

This paper proposed a simple approach of adapting MCC to GPU. Using all minutiae for calculating cylinders, then choosing 32 minutiae for matching, the approach actually fits well with the GPU computing architecture and can be easily implemented. The proposed method does not affect to the accuracy of the original algorithm. The speed of the adapting algorithm is comparable with the results of the state-of-the-art published algorithm. The proposed approach can be easily scaled-up. Thus, it is possible to implement a large-scale fingerprint identification system on inexpensive hardware.

In the future, we are planning to integrate this work with other modules: Query Module which maintains a queue of the queries received from the clients and sends the query to GPU, Refinement Module (RM) which gets score set S from GPU, considers top scores and compares them to Tq using MCC with LSS-DTR to provide the final result.

References

1. Maltoni, D., Maio, D., Jain, A.K., Prabhakar, S.: Handbook of Fingerprint Recognition. Springer, London (2009)
2. Cappelli, R., Maio, D., Maltoni, D., Wayman, J.L., Jain, A.K.: Performance evaluation of fingerprint verification systems. IEEE Trans. Pattern Anal. Mach. Intell. **28**, 3–18 (2006)
3. Chikkerur, S., Cartwright, A.N., Govindaraju, V.: K-plet and Coupled BFS: a graph based fingerprint representation and matching algorithm. In: Zhang, D., Jain, A.K. (eds.) ICB 2005. LNCS, vol. 3832, pp. 309–315. Springer, Heidelberg (2005)
4. Cappelli, R., Ferrara, M., Maltoni, D.: Minutia cylinder-code: A new representation and matching technique for fingerprint recognition. IEEE Trans. Pattern Anal. Mach. Intell. **32**, 2128–2141 (2010)
5. Xu, W., Chen, X., Feng, J.: A Robust Fingerprint Matching Approach: Growing and Fusing of Local Structures. In: Lee, S.-W., Li, S.Z. (eds.) ICB 2007. LNCS, vol. 4642, pp. 134–143. Springer, Heidelberg (2007)
6. Feng, J.: Combining minutiae descriptors for fingerprint matching. Pattern Recogn. **41**, 342–352 (2008)
7. Wang, X., Li, J., Niu, Y.: Fingerprint matching using orientation codes and polylines. Pattern Recogn. **40**, 3164–3177 (2007)
8. Feng, J., Ouyang, Z., Cai, A.: Fingerprint matching using ridges. Pattern Recogn. **39**, 2131–2140 (2006)
9. Qi, J., Yang, S., Wang, Y.: Fingerprint matching combining the global orientation field with minutia. Pattern Recogn. Lett. **26**, 2424–2430 (2005)
10. Tico, M., Kuosmanen, P.: Fingerprint matching using an orientation-based minutia descriptor. IEEE Trans. Pattern Anal. Mach. Intell. **25**, 1009–1014 (2003)
11. Medina-Pérez, M.A., García-Borroto, M., Gutierrez-Rodriguez, A.E., Altamirano-Robles, L.: Robust fingerprint verification using m-triplets. In: International Conference on Hand-Based Biometrics (ICHB 2011), Hong Kong, pp. 1–5 (2011)
12. Gutierrez, P.D., Lastra, M., Herrera, F., Benitez, J.M.: A high performance fingerprint matching system for large databases based on GPU. IEEE Trans. Inf. Forensics Secur. **9**(1), 62–71 (2014)
13. Cappelli, R., Ferrara, M., Maltoni, D.: Large-scale fingerprint identification on GPU. Inf. Sci. **306**, 1–20 (2015)
14. Peralta, D., Triguero, I., Sanchez-Reillo, R., Herrera, F., Benitez, J.M.: Fast fingerprint identification for large databases. Pattern Recogn. **47**(2), 588–602 (2014)
15. Luebke, D., et al.: GPGPU: general-purpose computation on graphics hardware. In: Proceedings of the 2006 ACM/IEEE Conference on Supercomputing, SC 2006 (2006)
16. Cappelli, R., Maio, D.: State-of-the-art in fingerprint classification. In: Ratha, N., Bolle, R. (eds.) Automatic Fingerprint Recognition Systems, pp. 183–205. Springer, New York (2004)

17. Hong, J.H., Min, J.K., Cho, U.K., Cho, S.B.: Fingerprint classification using one-vs-all support vector machines dynamically ordered with naive Bayes classifiers. Pattern Recogn. **41**(2), 662–671 (2008)
18. Cappelli, R., Ferrara, M., Maltoni, D.: Fingerprint indexing based on minutia cylinder code. IEEE Trans. Pattern Anal. Mach. Intell. **33**(5), 1051–1057 (2010)
19. Bhanu, B., Tan, X.: A triplet based approach for indexing of fingerprint database for identification. In: Bigun, J., Smeraldi, F. (eds.) AVBPA 2001. LNCS, vol. 2091, pp. 205–210. Springer, Heidelberg (2001)
20. Unique Identification Authority of India, Role of Biometric Technology in Aadhaar Enrollment (2012)
21. Krizhevsky, A., Sutskever, I., Hinton, G.E.: ImageNet classification with deep convolutional neural networks. In: NIPS 2012, pp. 1106–1114 (2012)
22. Zhang, Y., Yi, D., Wei, B., Zhuang, Y.: A GPU-accelerated non-negative sparse latent semantic analysis algorithm for social tagging data. Inform. Sci. **281**, 687–702 (2014)
23. Friedrichs, M., Eastman, P., Vaidyanathan, V., Houston, M., Legrand, S., Beberg, A., et al.: Accelerating molecular dynamic simulation on graphics processing units. J. Comput. Chem. **30**(6), 864–872 (2009)
24. Schatz, M., Trapnell, C., Delcher, A., Varshney, A.: High-throughput sequence alignment using graphics processing units. BMC Bioinformat. **8**, 474 (2007)
25. Medina-Pérez, M.A., Loyola-González, O., Gutierrez-Rodríguez, A.E., García-Borroto, M., Altamirano-Robles, L.: Introducing an experimental framework in C# for fingerprint recognition. In: Martínez-Trinidad, J.F., Carrasco-Ochoa, J.A., Olvera-Lopez, J.A., Salas-Rodríguez, J., Suen, C.Y. (eds.) MCPR 2014. LNCS, vol. 8495, pp. 132–141. Springer, Heidelberg (2014)

Extraction of Myocardial Fibrosis
Using Iterative Active Shape Method

Jan Kubicek[1(✉)], Iveta Bryjova[1], Marek Penhaker[1], Michal Kodaj[2],
and Martin Augustynek[1]

[1] FEECS, VSB–Technical University of Ostrava, K450, 17. Listopadu 15,
708 33 Ostrava–Poruba, Czech Republic
{jan.kubicek,iveta.bryjova,marek.penhaker,
martin.augustynek}@vsb.cz
[2] Podlesí Hospital, a.s., Konská 453, 739 61 Třinec, Czech Republic
michal.kodaj@gmail.cz

Abstract. The article deals with complex analysis of myocardial fibrosis. In clinical practice, myocardial fibrosis is commonly examined by MRI. This kind of disease is commonly assessed by human eyes. There isn't any diagnostic software alternative for evaluation of myocardial fibrosis features. The proposed method partially solves this problem. The main intention is automatic extraction of fibrosis area. This area is represented by closed curve which reflects shape of analyzed object. At the beginning of algorithm, initial circle is set on the fibrosis area. In iterative steps, this circle adopts shape of pathologic lesion. Before execution of segmentation process it is needed to specify region of interest (RoI) and image preprocessing which comprises especially low pass filtration. Filtration process suppresses unwanted adjacent objects. This step is quite important because active shape method could spread out of fibrosis borders and resulting curve has wouldn't reflect real shape of myocardial fibrosis.

Keywords: Magnetic resonance imagining · Cardiomyopathy · Myocardial fibrosis · Active shape method · Image segmentation

1 Introduction to Myocardial Fibrosis

Myocardial fibrosis is one of the most severe causes of cardiac insufficiency. Myocardial fibrosis which is caused by heart-attack takes the main cause of death worldwide. Heart-attack is common and frequent manifestation of ischemic heart diseases. During the heart-attack it goes to myocyte apoptosis with consequent ischemia which is result of unbalancing between delivery and demand of blood (blood supply). Myocardial fibrosis is defined as increased volume fraction of collagen in myocardium. In the dependence of pathological kind, saving of collagen is different. Therefore, it leads to deterioration of systolic and diastolic function of heart's chamber and to deterioration of myocardium elasticity. In severely affected myocardium, fibrosis may take up to 40 % myocardial space. Expect of ischemic diseases, myocardial fibrosis can manifest on rheumatic diseases as well [1–3]

© Springer-Verlag Berlin Heidelberg 2016
N.T. Nguyen et al. (Eds.): ACIIDS 2016, Part I, LNAI 9621, pp. 698–707, 2016.
DOI: 10.1007/978-3-662-49381-6_67

2 Myocardial Fibrosis Definition

According to etiology, we recognize three types of myocardial fibrosis:

1. Reactive interstitial fibrosis (RIF) is caused by diffusion of fibrosis distribution in interstitium with progressive beginning. Due of many facts, collagen is synthesized.
2. Infiltrative interstitial fibrosis (IIF) is caused by progressive saving of amyloid or glycosphingolipid in interstitium.
3. Substitution fibrosis (SF) is presented in the case of damaging or necrosis of myocytes. It leads to substitution of collagen I. degree, consequently scar is created. This type can be manifested locally (ischemic and hypertrophic cardiomyopathy, myocarditis, sarcoidosis) or it can goes to diffusion saving (Chronical renal insufficiency, toxic cardiomyopathy, inflammatory diseases). RIF and IIF in late stages come to SF [4, 5] (Fig. 1).

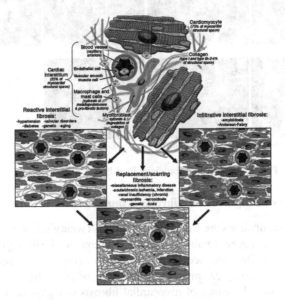

Fig. 1 Etiophysiopathology of myocardial fibrosis

3 Myocardial Fibrosis Diagnosis

During the diagnosis of myocardial fibrosis (presence, scale, prediction of further disease development) is indispensable examination on MRI. Local changes of myocardium are possible to detect and represent with imagining of late myocardium saturation by contrast agent. This imagining is based on the difference in signal intensity between fibrosis myocardium and normal myocardium after giving contrast agent. T1, T2 and T2* mapping techniques are used for imagining of myocardial fibrosis. By using sequence MOLLI (T1 mapping technique represents relaxation of myocardium) and shorten version shMOLLI, it is possible to set accurate values of T1 relaxing time. Those sequences are not commonly implemented to proprietary software on most MRI devices. The behaving of myocardial fibrosis is different on ischemic affection, where is typical subendocardial or holomyocardial fibrosis with pathological perfusion of this myocardial tissue. Contrarily, myocardial fibrosis is mainly appeared in central or subepicardial part of myocardium and doesn't exhibit change in myocardial perfusion studies [6, 7] (Fig. 2).

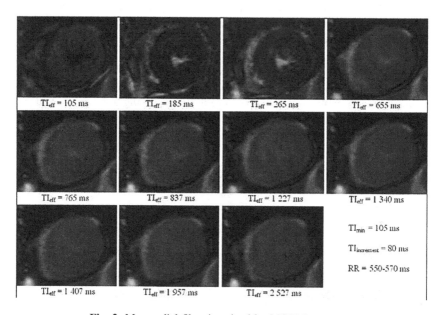

Fig. 2 Myocardial fibrosis gained by MOLLI sequence

3.1   The Process of Myocardial Fibrosis Examination

The performed examinations are mainly quantitative. Twenty minutes after application of double dose contrast agent (mainly Gadovist) on sequence PSIR, myocardial fibrosis quantity is assessed. On sequence PSIR, we try to choose inversion time same as inversion time of physiological myocardium. Signal of physiological myocardium is suppressed and contrarily signal of myocardial fibrosis is highlighted. In the case of inhomogeneous or global myocardium fibrotisation, it goes to repeated artifacts which don't allow accurate assessment of inversion time of myocardium. Sequence for assessment of inversion time physiological myocardium is closely related to sequences MOLLI and shMOLLI [1, 2, 6, 7].

4   The Proposed Methodology for Extraction of Myocardial Fibrosis

The main intention of proposed solution is automatic extraction of area myocardial fibrosis. This task is especially important in the field of clinical practice. Myocardial fibrosis is frequently solved task in the field of heart diseases. In the clinical practice, there is a significant problem with recognition area of myocardial fibrosis and quantifying geometrical parameters such fibrosis symmetry, perimeter and lesion area. The key benefit of proposed method is the fact that method approximates myocardial fibrosis by smooth curve. Consequently it is allowed to compute geometrical parameters which reflect clinical state of macular lesion. Currently, there is no any other suitable alternative for assessing myocardial fibrosis. Whole segmentation process is divided into three major parts. Firstly, it is necessary to specify region of interest (RoI) with interpolation. The problem is that myocardial fibrosis takes small part of input image area, myocardial fibrosis is extended by ROI. Another benefit is resizing of relatively small image part to larger area. If we performed only RoI, we would obtain image which is affected by worse contrast. Therefore, interpolation technique has been used. Interpolation procedure increases number of pixel and gives detailed image's information. Second step deals with filtration of RoI. For this task, low-pass filter has been used. Filtration partially suppresses adjacent structures. In the output of this procedure we obtain image, where higher frequencies are suppressed. This step is especially important for ensuring that active shape won't spread out of analyzed contour. After taking mentioned procedures segmentation is performed. The core of segmentation consist active geometrical model which iteratively adopts shape of myocardial fibrosis. It is necessary to set initial contour which is placed inside the object. Contour is consequently formed up to borders of fibrosis. Algorithm output gives continues and closed curve which should reflect area of myocardial fibrosis. Whole process of segmentation is illustrated on Fig. 3.

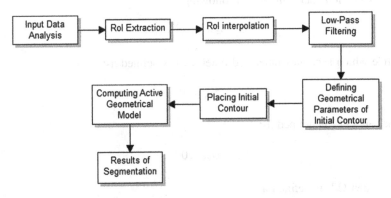

Fig. 3 The block diagram of myocardial fibrosis segmentation

4.1 Defining Active Geometrical Model

The active geometrical model is based on the level set method. Analyzed images are generated in two dimensional spaces. Therefore, following derivations are proposed in dimension Ω^2. Active model is given as: $\phi_{x,t}$, where $\mathbf{x} = (x,y)$ are coordinates of points and t denotes time. In given time, curve $\phi_{x,t}$ divides space into two parts. The development of active contour is show on Fig. 4.

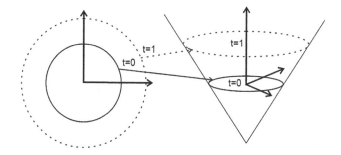

Fig. 4 The development of active geometrical model

For simplification, it is reasoning initial curve as unit circle K with center (0,0). This circle divides space into inner and outer area:

$$in\,K = \Omega^- = \{x|x| < 1\} \tag{1}$$

$$out\,K = \Omega^+ = \{x|x| > 1\} \tag{2}$$

With border:

$$d\Omega = \{x|x| = 1\} \tag{3}$$

Implicit expression such function is following:

$$\phi_{x,y} = x^2 + y^2 - 1 \tag{4}$$

The circle which separates inner and outer area is defined as:

$$\phi_{x,y} = 0 \tag{5}$$

Inner area Ω^- is defined as:

$$\phi_{x,y} < 0 \tag{6}$$

Outer area Ω^+ is defined as:

$$\phi_{x,y} > 0 \tag{7}$$

The Fig. 5 shows specification of initial contour. For our purposes circle with zero shift has been used.

Generally, active geometrical model is described as:

$$\Gamma_t = \{x|\phi_{x,t} = 0\} \tag{8}$$

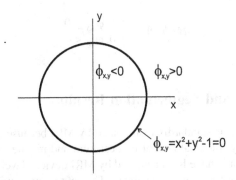

Fig. 5 The implicit circle representation with radius 1 and center (0,0)

For each point $\mathbf{x} = (x,y)$ in inner and outer area, distance is calculated by distance function d_x from nearest point $\mathbf{x_c} = (x,y)_c$ which is placed on boarder $d\Omega$.

$$d_x = min_{x_c \in d\Omega}(|x - x_c|) \qquad (9)$$

Value of each point $\mathbf{x}$ is given by sign distance function:

$$\phi_x = \begin{cases} -d_x, x \in \Omega^- \\ 0, x \in d\Omega \\ +d_x, x \in \Omega^+ \end{cases} \qquad (10)$$

Points which are placed in inner subspace have negative value of distance. Contrarily, points which are placed in outer subspace have positive values. Segmentation foreground from background is given on the base of minimization following functional:

$$E(c_1, c_2, \Gamma) = \alpha \int (n_{x,y} - c_1)^2 dxdy + \beta \int (n_{x,y} - c_2)^2 dxdy + \gamma \int |\Gamma_p| dp \qquad (11)$$

First and second part of integral equation describe deviation points intensities $n_{x,y}$ in inner space Ω^+ respectively in outer space Ω^- from middle point in these spaces c_1, c_2. The last part ensures smoothness of boarder Γ, where p denotes points which belong to boarder, α, β, γ are weighted coefficients.

These average values of foreground and background are represented by coefficients c_1, c_2.

$$c_1 = \frac{\int n_{x,y} \cdot (1 - H(\phi_{x,y})) dxdy}{\int (1 - H(\phi_{x,y})) dxdy} \qquad (12)$$

$$c_2 = \frac{\int n_{x,y} \cdot H(\phi_{x,y}) dxdy}{\int H(\phi_{x,y}) dxdy} \qquad (13)$$

$H(\phi_{x,y})$ denotes to Heaviside function, which is described follows: [8]

$$H(\phi_{x,y}) = \begin{cases} 1, \phi_{x,y} \geq 0 \\ 0, \phi_{x,y} < 0 \end{cases} \qquad (14)$$

5 Data Analysis and Segmentation Results

The myocardial fibrosis is standardly exanimated by MRI because of higher resolution and recognition of physiological part of myocardium and manifestation of myocardial lesions. Analyzed records have been acquired by MRI device. Twelve patient's records have been used for purposes of our analysis. The important requirement of segmentation process is separation foreground from background. Before taking segmentation process, it is needed to perform interpolation and filtering of selected RoI area. By those procedures we achieve smooth image and adjacent object should be suppressed. Segmentation procedure goes in iterative steps. Number of iterations is key parameter of segmentation procedure. If we choose smaller number of iterations, we would obtain resulting curve which doesn't reflect myocardial fibrosis shape. On the other hand computed time would have been reduced. For our purposes, we used 100 iterations. It increases computed time, but resulting curve relevantly reflects analyzed object. Experimental results of segmentation process are shown on Figs. 6 and 7 below.

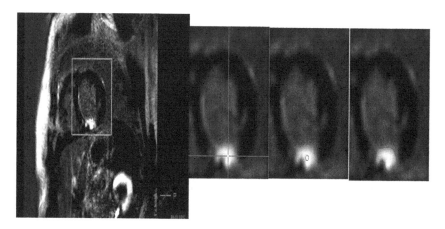

Fig. 6 Original MRI data with RoI (left), sequence of evolution active geometrical model (right)

After performing analysis, it is necessary to evaluate validation of proposed solution. Two alternatives have been used. We performed comparison with commercial software ImageJ. Software allows manually set reference points along the analyzed object. Consequently, individual points are fitted into continues curve. On the output, area of curve is calculated. Second alternative way deals with estimated results of curve's areas, calculated theoretically by MRI physicians.

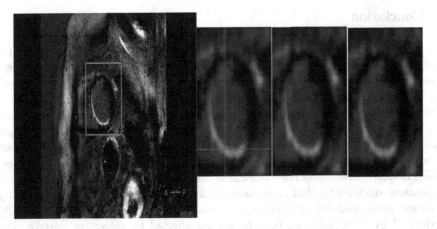

Fig. 7 Original MRI data with RoI (left), sequence of evolution active geometrical model (right)

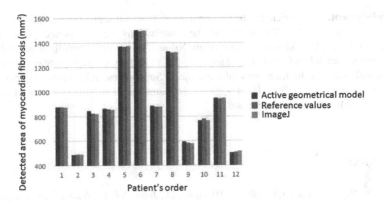

Fig. 8 Comparison of myocardial fibrosis area

The last part of analysis deals with comparison of proposed method with clinical results and manual fitting by software ImageJ. There is a statistical comparison, on the Fig. 8. Twelve patient's samples have been tested for this purpose. Mentioned methods exhibit identic results as used active geometrical model. This fact has been proofed by clinical experts from Department of Radiology. On the one hand software ImageJ gives nearly identic results as active shape method with minor observable differences, but on the other hand this software does not allow to automatic segmentation. It is significant problem because we are allowed set system of reference points around the object and those points are consequently fitted by smooth curve. It is obvious that this semi manual method is influenced by subjective error. Contrarily proposed active shape method adopts shape of macular fibrosis automatically. The significant benefit is that method is robust and reliable even in noisy environment.

6 Conclusion

The main intention of paper is design and implementation of complex methodology for analysis of myocardial fibrosis and evaluation of fibrotic area. The segmentation process is given by three essential steps: data analysis, image preprocessing and computation of segmentation model. After taking mentioned procedures, myocardial fibrosis is represented by closed smooth shape which reflects their geometrical features. There is a possible alternative for analysis of myocardial fibrosis by segmentation approach which uses color mapping of individual structures. This method allows clear separation of myocardial fibrosis from other structures. In result we would obtain area of object which is clearly observable but this method does not allow evaluation of geometrical parameters which are for clinical purposes crucial. Currently, there isn't any proprietary software for assessment of myocardial fibrosis. Fibrosis assessment is standardly performed by physicians and their results are affected by subjective mistake. The proposed way offers objective analysis and it is being currently tested in clinical practice.

Acknowledgment. This article has been supported by financial support of TA ČR PRE SEED: TG01010137 GAMA PP1. The work and the contributions were supported by the project SP2015/179 'Biomedicínské inženýrské systémy XI' and This work is partially supported by the Science and Research Fund 2014 of the Moravia-Silesian Region, Czech Republic and this paper has been elaborated in the framework of the project "Support research and development in the Moravian-Silesian Region 2014 DT 1 - Research Teams" (RRC/07/2014). Financed from the budget of the Moravian-Silesian Region.

References

1. Mewton, N., Ying, L.C., Croisille, P., Bluemke, D., João, A.C.L.: Assessment of myocardial fibrosis with cardiovascular magnetic resonance. J. Am. Coll. Cardiol. **57**(8), 891–903 (2011). doi:10.1016/j.jacc.2010.11.013
2. Flett, A.S., Hayward, M.P., Ashworth, M.T., Hansen, M.S., Taylor, A.M., Elliott, P.M., McGregor, C., Moon, J.C.: Equilibrium contrast cardiovascular magnetic resonance for the measurement of diffuse myocardial fibrosis: preliminary validation in humans. Circulation **122**(2), 138–144 (2010). doi:10.1161/CIRCULATIONAHA.109.930636. ISSN 0009-7322
3. Sado, D.M., Flett, A.S., Moon, J.C.: Novel imaging techniques for diffuse myocardial fibrosis. Future Cardiol. **7**(5), 643–650 (2011)
4. Kawel, N., Nacif, M., Zavodni, M., Jones, J., Liu, S., Sibley, Ch., Bluemke, D.: T1 mapping of the myocardium: Intra-individual assessment of the effect of field strength, cardiac cycle and variation by myocardial region. J. Cardiovasc. Magn. Reson. **14**(1), 27 (2012). doi:10.1186/1532-429X-14-27. ISSN 1532-429x
5. Kubicek, J., Penhaker, M., Bryjova, I., Kodaj, M.: Articular cartilage defect detection based on image segmentation with colour mapping. In: Hwang, D., Jung, J.J., Nguyen, N.-T. (eds.) ICCCI 2014. LNCS, vol. 8733, pp. 214–222. Springer, Heidelberg (2014). (including subseries Lecture Notes in Artificial Intelligence and Lecture Notes in Bioinfor-matics)

6. Kubicek, J., Penhaker, M.: Fuzzy algorithm for segmentation of images in extraction of objects from MRI. In: Proceedings of the 2014 International Conference on Advances in Computing, Communications and Informatics, ICACCI 2014, pp. 1422–1427 (2014). Article no. 6968264
7. Augustynek, M., Penhaker, M.: Non invasive measurement and visualizations of blood pressure. Elektron. Ir Elektrotechnika **116**(10), 55–58 (2011)
8. Augustynek, M., Penhaker, M.: Finger plethysmography classification by orthogonal transformatios. In: 2010 Second International Conference on Computer Engineering and Applications (ICCEA), pp. 173–177. IEEE (2010)

Increasing the Efficiency of GPU-Based HOG Algorithms Through Tile-Images

Darius Malysiak$^{(\boxtimes)}$ and Markus Markard

Computer Science Institute, Hochschule Ruhr West, Mülheim, Germany
darius.malysiak@hs-ruhrwest.de

Abstract. Object detection systems which operate on large data streams require an efficient scaling with available computation power. We analyze how the use of tile-images can increase the efficiency (i.e. execution speed) of distributed HOG-based object detectors. Furthermore we discuss the challenges of using our developed algorithms in practical large scale scenarios. We show with a structured evaluation that our approach can provide a speed-up of 30-180 % for existing architectures. Due to the its generic formulation it can be applied to a wide range of HOG-based (or similar) algorithms. In this context we also study the effects of applying our method to an existing detector and discuss a scalable strategy for distributing the computation among nodes in a cluster system.

Keywords: Image · Composed image · gpgpu · High performance computing · Histogram of oriented gradients · HOG · opencl · Cuda

1 Introduction and Previous Work

Since the pioneering work of [2] many architectures for object detection started to utilize histograms of oriented gradients (HOGs) as robust feature components. Sadly the computation of HOGs introduces a high computational complexity, yet the involved operations are highly parallelizable. The corresponding application of GPUs has been widely studied in e.g. [5,12] or [1], yet to our knowledge no detailed studies exist on the efficient distribution among multiple GPUs in a single node let alone multiple computers. In this paper we present a method which is capable of boosting the efficiency of existing HOG detectors without structurally modifying them. Furthermore we discuss a scalable strategy for distributing the computation among nodes in a cluster system.

Section 2 will briefly discuss the common approach to calculate HOG features on massively parallel architectures with a special focus on GPUs. The following Sect. 3 introduces our framework and discusses the challenges of practical application. We conclude this paper with Sects. 4 and 5 which present our results on standard image databases and an outlook for additional research, respectively.

© Springer-Verlag Berlin Heidelberg 2016
N.T. Nguyen et al. (Eds.): ACIIDS 2016, Part I, LNAI 9621, pp. 708–720, 2016.
DOI: 10.1007/978-3-662-49381-6_68

2   Histograms of Oriented Gradients

In order to understand our motivation for the later described tile-image approach one has to understand the general structure of a GPU implementation for HOG-based algorithms. Thus we will briefly explain the classic algorithm and use this description to introduce the common "tricks" of GPU-ports.

2.1   The Algorithmic Structure

The classical HOG algorithm by Dalal involves a loop which sequentially executes 1+3 major phases: shrink image, shift detection window over the image, extract a hog descriptor for each position, classify the descriptors. Where the last 3 phases are repeated until all possible window positions have been analyzed (depicted in Algorithm 1 where s, win_w, win_h, win_s represent scale factor, window width, window height and window stride, respectively).

Algorithm 1. Classic HOG

Require: HOG   parameters: $s, win_w, win_h,$
  $win_s, ...$
Ensure: l
1: $I_c = I, \tilde{s} = s, l = \emptyset$;
2: **while** Detection window fits into current image **do**
3:    Shrink($I_c, \tilde{s}$);
4:    Position the windows left upper corner at $(0, 0)$;
5:    **while** I_c not covered **do**
6:        Shift Window by one quantum win_s;
7:        Compute gradient image $I_G = (I_A, I_\phi)$ for window content;
8:        Calculate histogram $H = H(I_G)$;
9:        Classify H, add result to l;
10:    **end while**
11:    $I_c = I, \tilde{s} = \tilde{s} * s$;
12: **end while**
13: Group elements from l via Mean-Shift $l = MS(l)$;

The scaling itself with $\mathcal{O}(I_w \cdot I_h)$ weighs heavily on the overall complexity let alone the histogram calculation for each window position. The expression 'I_c not covered' refers to the check if there exists an image position which can be covered by the detection window if it is shifted in multiples of win_s (note that $win_s = (x_s, y_s)$ is a 2-tuple which defines a horizontal and a vertical shift quantum, i.e. shifting is done with two for-loops).

In order to derive an expression for the complexity of Algorithm 1 we state several facts in following

Theorem 1. *Let* $S_e = \min\{I_w/win_w, I_h/win_h\}$ *and* I_c *an image on scale level* c. *Furthermore let* $b_w, b_h, b_x, b_y, c_w, c_h, c_x, c_y, n$ *be the block width, block height, block stride x, block stride y, cell width, cell height, cell stride x, cell stride y and bin count respectively. The following statements are true*

- *The total amount of scaling steps is* $A := \lfloor \log(S_e)/\log(s) + 1.0 \rfloor$
- *Shrinking the image has a complexity of* $B := \mathcal{O}(I_w \cdot I_h)$.
- *There are* $C := \frac{I_{c,w} - win_w + x_s}{x_s} \cdot \frac{I_{c,h} - win_h + y_s}{y_s}$ *window positions in* I_c.

- *Computing I_G with two 1-dim convolution masks requires $\mathcal{O}(win_w \cdot win_h)$ operations.*
- *There are $D := \frac{win_w - b_w + b_x}{b_x} \cdot \frac{win_h - b_h + b_y}{b_y}$ blocks in each window.*
- *There are $E := \frac{b_w - c_w + c_x}{c_x} \cdot \frac{b_h - c_h + c_y}{c_y}$ cells in a block.*
- *Computing the histogram for a single window requires $\mathcal{O}(D \cdot E \cdot c_w \cdot c_h)$ operations.*
- *Each histogram has $F := D \cdot E \cdot n$ elements.*

The total complexity of a single HOG iteration for a single image is

$$\mathcal{O}(A \cdot (B + C \cdot (win_w \cdot win_h + D \cdot E \cdot c_w \cdot c_h + F))) \tag{1}$$

From the complexity expression is becomes clear that each HOG parameter plays an important role for the algorithms runtime, which is anything but small.

2.2 GPU Implementation

The usual structure of a GPU implementation is depicted in Algorithm 2. Since a modern GPU features several thousand execution units one attempts to delegate at least one operation onto each execution unit. For some algorithm steps there are more operations than execution units, for others there may be more execution units than operations. Yet GPUs follow the SIMT approach which enforces several restrictions onto the algorithms structure. Explaining these challenges would be beyond this papers scope, the interested reader might refer to e.g. [11]. The parallel for-loop in Algorithm 2 only works under a "trick" common to all implementations which attempt to reach state of the art detection speed; the block stride must equal the cell dimensions and the cells in a block must not overlap. This allows to precompute the gradient image for all window positions on a single scale. A complete analysis of this structure (e.g. in the PRAM [4] model) is beyond the scope of this paper. Thus we note at this point that parallelization can only modify the efficiency by a factor $c = c(\tau, \mathcal{W})$ with τ being the shader count and $\mathcal{W}$ the set of hog parameters (the $\mathcal{O}$ notation omits such factors). The comments in Algorithm 2 introduce such constants, the histogram normalization is usually achieved by an additional phase (due to the previously mentioned architecture restrictions), yet for the sake of simplicity we will regard it as one stage.

We point out that each constant c_i in Algorithm 2 has a different optimal shader count s_i for which c_i would equal 1. To illustrate this; $c_{shrink} = 1$ for $s_{shrink} = I_w \cdot I_h$ while $c_{classify} = 1$ for $s_{classify} =$ "histogram size". If a GPU would provide max s_i shaders and $\forall i, j : s_i = s_j$, the complexity of Algorithm 2 would be $\Theta(\text{scaleCount})$, which is unfeasible as memory latencies and the SIMT programming model must be considered as well, let alone the fact that current GPUs provide only a relatively small amount of shaders. One implication is that not all phases can be equally efficient on the GPU, which is an inherent aspect of most GPU-based multi-phase algorithms (e.g. [13]). It is difficult to counter this problem, usually it is omited for reasons of convenience or shadowed by arguments of a high speed up compared to a CPU implementation. Yet when it comes to practical (industrial) applications of such algorithms, it is often desired to save as much time as possible without overstepping a systems tolerance boundaries.

2.3 Efficiency Factors

Algorithm 2. GPU HOG

Require: HOG parameters: $s, win_w,$
1: $win_h, win_s, ...$
Ensure: l
2: $I_c = I, \tilde{s} = s^0, l = \emptyset;$
3: **while** Detection window fits into current image **do**
4:      Shrink$_{GPU}(I_c, \tilde{s});$ $\rightarrow c_{shrink}$
5:      Compute gradient image
6:      $I_G = (I_A, I_\phi)$ for I_C
7:      on GPU; $\rightarrow c_{grad}$
8:      **Par.-For** all Positions i in I_G **do**
         $\rightarrow c_{hist}$
9:         Calculate histograms
10:         $H_i = H_i(I_G);$
11:      **endParFor**
12:      **Par.-For** all Positions i in I_G **do**
         $\rightarrow c_{norm}$
13:         Normalize histograms
14:         $H_i = H_i(I_G);$
15:      **endParFor**
16:      **Par.-For** all Positions i in I_G **do**
         $\rightarrow c_{classify}$
17:         Classify H_i, add results to $l;$
18:      **endParFor**
19:      $I_c = I, \tilde{s} = \tilde{s} * s;$
20: **end while**
21: Group elements from l via
22: Mean-Shift $l = MS(l);$ $\rightarrow c_{group}$

Notice that by Theorem 1 the complexity is mainly influenced by the image size, since A, B and C directly depend on it. Several steps in Algorithm 2 work on global memory which exhibits very high latencies, for small amounts of schedules threads this results in large computation delays as shader units will have to wait for requested data. By scheduling a large number τ' of threads it is possible to mask these delays especially in combination with techniques as e.g. memory coalescing. Yet with increasing τ' the efficiency increase will stagnate as the overhead for scheduling will outweigh the gain of masking latencies. Note that due to the parallel approach in Algorithm 2 all elements within the while-loop are influenced by the image size. Thus we argue that an existing HOG implementations efficiency can be increased by varying the image size.

3   Cluster-Based Computation

The structure of Algorithm 1 gives rise to many different distribution strategies within computing clusters. One such method would be a strategy similar to pipelines in microprocessors, where each phase would be executed by a dedicated node. The state of the art total execution time for a classical HOG detection is roughly 70ms (GPU, image size 1600x1200, [7]), whereas a multicore CPU requires 180ms (CPU, image size 320x240, [8]). Thus, not considering inter-node

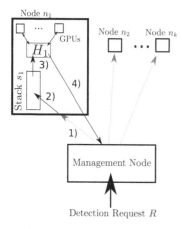

Fig. 1. A Beowulf cluster system for handling up to k parallel video streams. Each detection request is delegated by the the management node to the corresponding processing node (step 1). Each processing node contains a buffer stack whose size z_1 is optimized for the node-local detector H, each received image will be pushed onto the stack (step 2), which is essentially a tile image. Once the tile image has been filled it will be processes by H which utilizes multiple node-local GPUs (step 3). Afterwards the results for all z_1 images will be returned to the request source (step 4).

communication costs, it would require at least 10 CPUs to deliver the same performance. Motivated by these numbers we focused entirely on nodes equipped with multiple GPUs. This decision limits the distribution scheme to methods which compute entire HOG runs on each node, since communication latencies (GPU↔host and node↔node) outweigh the execution times for single phases.

3.1   Efficiency Through Tile Images

We will now introduce the concept of tile images

Definition 1. *Let $\mathcal{I}$ be a set of arbitrary images I_j. We define a tile image $I_{\mathcal{T}}$ to be an image which contains every image of $\mathcal{I}$ exactly once without overlap. The set $\mathcal{I}$ is called the base of $I_{\mathcal{T}}$ with each element being referred to as base image. The density $\delta_{\mathcal{T}}$ of $I_{\mathcal{T}}$ is defined as the area of $I_{\mathcal{T}}$ which is covered by any image from $\mathcal{I}$*

$$\delta_{\mathcal{T}} := \frac{1}{|I_{\mathcal{T}}|} \sum_{I \in \mathcal{I}} |I| \qquad (2)$$

Note that this definition imposes no restriction onto the maximal size of $I_{\mathcal{T}}$, yet since we are interested in minimizing the density $\delta_{\mathcal{T}}$ the maximal width and height are given by $\sum_{I \in \mathcal{I}} I_w$ and $\sum_{I \in \mathcal{I}} I_h$, respectively. This is a classic packing problem and corresponding strategies as well as algorithms have been studied by e.g. [9] or [3]. Yet in our context this problem becomes much more difficult due to time constraints.

Let us assume a HOG detector H requires n time units to process an image I and m units for an image J. We define the $I - J$ efficiency of H as

$$E_{I,J}(H) := \frac{J_w}{I_w} \cdot \frac{J_h}{I_h} \cdot n - m \qquad (3)$$

The first product represents the amount of times that J can be fitted into I, thus $E_{I,J}(H)$ states the amount of time units which are saved if one processes one image J instead of $\frac{J_w}{I_w} \cdot \frac{J_h}{I_h}$ images I. For tile images we generalize this to

Definition 2. *Let I_T be a tile image with base $\mathcal{I}$, H a HOG detector and $t_H(I)$ the processing time which H needs for I. The I_T efficiency of H is defined as*

$$E_{I_T}(H) := \left(\sum_{I \in \mathcal{I}} t_H(I) \right) - t_H(I_T) \qquad (4)$$

Remark 1. If $E_{I_T}(H) = 0$ then H performs equally fast as if being called with single images, if $E_{I_T}(H) > 0$ then H uses less time then for all single images. Note that the density δ_T has an implicit effect on $E_{I_T}(H)$, if e.g. $\mathcal{I}$ contains only one image I and $|I_T| = 50|I|$ it is very likely that $E_{I_T}(H) < 0$ whereas for $\delta_T = 1$ we obtain $E_{I_T}(H) = 0$. Thus it is desirable to aim for $\delta_T = 1$.

In case of $E_{I_T}(H) > 0$, i.e. the tile image yielded an efficiency gain, the computation of I_T should not take longer than $E_{I_T}(H)$, otherwise the gain would be canceled out.

3.2 Boundary Detections

A tile image will contain hard boundaries which can effect the detection results. In order to filter these false-positives we propose the use of tile grids.

Definition 3. *Let I_T be a tile image with base $\mathcal{I}$. The tile grid G_T is a set of $|\mathcal{I}|$ 4-tuples (x, y, w, h) where x, y represents the position of a base images top-left corner within I_T and w, h the base images size. Each tuple is called a base-rectangle.*

We propose the following approach to purge false-positive detections from a tile image. Let $\{d_i\}$ be the detections on I_T and I a base image with corresponding base-rectangle r_I, we associate a detection d_j with I iff $|d_{j,A} \cap r_{i,A}|/|d_{j,A}| = 1.0 \wedge |d_{j,A}| \leq |r_{i,A}|$, all unassociated detections are omitted ($d_{j,A}, r_{i,A}$ denote the corresponding areas).

3.3 Computing Tile Images

A restriction which was not mentioned so far is that the detector H remains unchanged in its implementation. This reduces the strategies for computing a tile image since the memory structure of each base image must be retained in the tile image. A simple memory copy into the linearized tile image is impossible as the tile images row stride differs from the base image. Thus one has to copy the images on a pixel- or row-base. Another observation comes from Algorithm 2, it is obvious that the image size effects the runtime, yet the algorithm does not favor any specific image proportion. Expressed differently; only the actual image size, i.e. $I_w \cdot I_h$ effects the performance. This leaves only the question of how many lines of base images should be used in the tile image in order to maximize its density.

In the case of identically sized base images one can sequentially enqueue (i.e. concatenate them in one line) as many as required in order to maximize the efficiency (this implies $\delta_{\mathcal{T}} = 1$). Using this strategy for differently sized images will a) force the image height to be equal to the largest image height within the image base and b) result in a density of

$$\delta_{\mathcal{T}} = \frac{|I_{\mathcal{T}}| - \sum_{I \in \mathcal{I}} I_w \cdot (I_{\mathcal{T},h} - I_h)}{|I_{\mathcal{T}}|} < 1 \qquad (5)$$

Each base images difference in height to $I_{\mathcal{T}}$ will reduce the density. We will not theoretically elaborate on this problem but rather discuss a solution for a concrete scenario.

3.4   Cluster Distribution

It was shown by [6] that large surveillance systems do not only generate huge amounts of data but also that it requires problem specific engineering in order to manage these amounts. In order to discuss a possible solution for the computation of tile images we will assume a system similar to the one described in [6]. Let us assume we have k videostreams where each one provides images of constant size, additionally we assume to have k detectors each with a dedicated computation device. Having an equal amount of detectors is a realistic assumption if one desires to perform realtime object detection, since each detection takes around $70ms$ (with a single GPU) one can process at most 14 frames per second. We propose a cluster structure as depicted in Fig. 1.

The management node represents a central hub which receives detection requests and redistributes them onto the designated nodes. A detection requests is a 2-tuple containing an image and meta data as e.g. sender address and image number. Since the image dimensions of each camera are known in advance it possible to optimize each node for a single camera. One such optimization is the size of each nodes image buffer which holds a pre-allocated tile image into which each received image is written. Once the buffer has been filled ($\delta_{\mathcal{T}} = 1$) the tile image will be send to the detector for processing.

In order to understand how the optimal buffer size is determined one has to analyze several system attributes. Let us assume a continuous stream of images to node n_i which contains a buffer of size z_1, the delay T until a client receives the results for the first batch of z_1 images is

$$T := (t_{prop,\uparrow} + t_{copy}) + (z_1 \cdot t_{tick} + t_{process,z_1} + t_{prop,\downarrow}) \qquad (6)$$

$t_{prop,\uparrow}$ is the time a request needs to arrive at the designated node, t_{copy} the time to copy an image into the tile image, $t_{process,z_1}$ the time to perform the detection process and $t_{prop,\downarrow}$ the time until the nodes response reaches the recipient. The first term is an initial latency T_1 which is carried throughout the remaining stream while the second term represents the time T_2 until z_1 image requests have been issued by the sender and the detection results have been received (t_{tick} represents the tick rate). The delay until the client receives the second batch is $z_1 \cdot t_{tick} + t_{process,z_1} + t_{prop,\downarrow}$ relative to the first batch. In order to harness as much computational power as possible a larger z_1 is desired, yet it will leave the sender waiting for T_2 time units until the next batch of results arrives, this in turn makes a smaller z_1 favorable.

We point out that this equation only holds as long as $T_1 \leq t_{tick}$, otherwise each received image would be delayed by an additional amount of $T_1 - t_{tick}$ time units. This implies that one can process higher image frequencies if the systems performance is large enough, i.e. one should be motivated to decrease T_1. Furthermore it holds that for larger images a smaller amount of images needs to be buffered.

Although T_2 increases with higher values for z_1 the efficiency $E_T(H)$ increases as well, thus a client may wait longer for receiving the next batch of detection results yet this time will be smaller compared to the situation where the client would obtain z_1 detections without using the tile image. As depicted in Fig. 1 a node might contain multiple GPUs here, independent of the workload distribution scheme (e.g. distributing single HOG phases or simply using multiple stacks) the GPUs performance will be determined by the amount of data which is available to the GPUs. In case of identical devices this might result in a linear scaling of T_2 with respect to the device count.

3.5   Algorithmic Details

In this section we explain the algorithms which determine t_{copy} and $t_{process,z_1}$. The remaining times $t_{prop,\uparrow}$ and $t_{prop,\downarrow}$ are determined mainly by the datastructures which are used in order to look up the correct node or sender, respectively. When a node receives an request it executes the steps in Algorithm 3. Since we assume a continuous image stream we have to ensure that any image which might arrive during an active detection are still enqueued on the node, i.e. at least two threads are active on Algorithm 3. This approach works of course only if the detection and copying of $z_1 - 1$ images from $\mathcal{B}$ into I_T requires less than $(z_1 - 1) \cdot t_{tick}$ time units, i.e. $(z_1 - 1) \cdot t_\mathcal{B} + t_{process,z_1} \leq (z_1 - 1) \cdot t_{tick}$. Since usually $t_\mathcal{B} = t_{copy}$ we get the following restrictions

$$(t_{prop,\uparrow} + t_{copy}) \leq t_{tick} \tag{7}$$

$$(z_1 - 1) \cdot t_{copy} + t_{process,z_1} \leq (z_1 - 1) \cdot t_{tick} \tag{8}$$

The second restriction should be fulfilled implicitly if $t_{prop,\uparrow} + t_{prop,\downarrow} \leq t_{copy} + t_{process,z_1}$. Should one be able to ensure that $t_{copy} + t_{process} \leq t_{tick}$ than the multithreaded approach can be omitted completely.

4   Evaluation

By using the SimpleHydra framework [10] we implemented the described cluster architecture with a total of 4 identically equipped processing nodes and a single management node. The processing nodes were equipped with one Radeon7970 GPU, a Core-i7 3.5 GHz CPU, 16GB RAM and ran ArchLinux with a 3.16 Kernel. The management node simulated the requests by using a local image database of images obtained from 4 different cameras on a small airport. Each camera sequence consisted of 1096 single images which were streamed to the detector in chronological order of their recording (streaming was done with a framerate of 10fps, which is the native speed of an AVT surveillance camera). We used a highly optimized HOG implementation which follows the original description by [2]. We studied the following aspects: How different does an already optimized detector H behave for a sequence of tile images? How does $E_T(H)$ develop? What values should be expected for $t_{copy}, t_{process,z_1}, t_{prop,\uparrow}, t_{prop,\downarrow}$? How low can t_{tick} get with respect to the latencies of our system? What is the behaviour of an generally optimized detector H for a sequence of tile images containing interleaved images of multiple cameras?

4.1   Results

Algorithm 3. Push on stack

Require: Request
  $R = (I, (senderID, imgNo))$
  if no detection runs **then**
      copy I into I_T;
      register I in G_T;
      if stack full **then**
          indicate running detection;
          execute detection on I_T;
          if B contains images **then**
              copy all images from B into
              I_T and register them in G_T;
          end if
          indicate no running detection;
      end if
  else
      buffer image in local buffer B;
  end if

Table 1a shows the speed-up for images of size 1600x1200, the efficiency increases continuously yet the increase begins to stagnate with more than 8 images (see the values of $\Delta E_T(H)$). Using our simple strategy one can except to save $\approx 30\,\%$ of computation time. The gain becomes even more significant for smaller images, Table 1b shows the results for using images of size 640x480. Using smaller images we can expect a speed-up of up to $\approx 180\,\%$. This difference in efficiency gain can be simply explained by the fact that a single small image will underutilize the large number of GPU shaders more than a large image, thus $E_T(H)$ increases more significantly in relation to this time. With even more shader units the same effect is to be expected for larger images.

Figure 2 shows that an example tile image and the corresponding detection, as indicated by that image there have been 0 detections on all image boundaries across all evaluated camera streams without even using the image grid G_T. This shows that conventional grouping methods as e.g. mean-shift are likely to be sufficient in order to prevent boundary detections. The actual amount of boundary detections is depicted in Fig. 3a, one can see that such detections indeed occur. If not filtered out before applying grouping procedures these detections may influence the final results.

In order to study the effect of boundary detections we evaluated the amount of additional detections with the following metric

$$\Delta D(I_T, I) := |H_I(I)| - |H_I(I_T)|, I \in \mathcal{I} \tag{9}$$

with $H_A(I)$ being the set of detections using detector H on image I and restricting the obtained results onto the area of image A. A positive value means that the

Table 1. (a) Processing times for differently sized tile images. The numbers in column T represent the stack size in units of images (each of size 1600x1200), t_{single} the time when single images would be used and $\Delta E_T(H)$ the efficiency increase in relation to the last table row. (b) Processing times for differently sized tile images. The numbers in column T represent the stack size in units of images (each of size 640x480), t_{single} the time when single images would be used and $\Delta E_T(H)$ the efficiency increase in relation to the last table row.

T	$t_{process,z_1}$	t_{single}	$E_T(H)$	$\Delta E_T(H)$	rel. speed-up	T	$t_{process,z_1}$	t_{single}	$E_T(H)$	$\Delta E_T(H)$	rel. speed-up
1	0.0429 s	0.0429 s	0.0000 s	-	-	1	0,0116 s s	0,0116 s	0,0000 s	-	-
2	0.0700 s	0.0858 s	0.0158 s	-	18.41 %	2	0,0147 s	0,0233 s	0,0086 s	-	58,86 %
3	0.0978 s	0.1287 s	0.0309 s	95.57 %	24.01 %	3	0,0190 s	0,0349 s	0,0160 s	85.25 %	84,35 %
4	0.1257 s	0.1716 s	0.0459 s	48.54 %	26.75 %	4	0,0227 s	0,0466 s	0,0239 s	49.44 %	105,27 %
5	0.1542 s	0.2145 s	0.0603 s	31.37 %	28.11 %	5	0,0266 s	0,0582 s	0,0316 s	32.34 %	118,80 %
6	0.1822 s	0.2574 s	0.0752 s	24.71 %	29.22 %	6	0,0309 s	0,0699 s	0,0390 s	23.29 %	126,16 %
7	0.2109 s	0.3003 s	0.0894 s	18.88 %	29.77 %	...					
8	0.2384 s	0.3432 s	0.1048 s	17.23 %	30.54 %	18	0,0732 s	0,2097 s	0,1365 s	5.96 %	186,39 %
9	0.2677 s	0.3861 s	0.1184 s	12.98 %	30.67 %	19	0,0767 s	0,2213 s	0,1446 s	5.95 %	188,42 %
10	0.2947 s	0.4290 s	0.1343 s	13.43 %	31.31 %	20	0,0809 s	0,2330 s	0,1520 s	5.17 %	187,93 %

a)   b)

Fig. 2. An example tile image which contains 10 sequential images from a camera stream. Using mean-shift grouping alone was sufficient to avoid false-positive results on the image boundaries. Note the changing boundary values in the last five images

tile image yielded less detections than the single image, a negative value indicates more results on the tile image while a 0 corresponds to an identical amount of detections on the tile image and corresponding base image. The results for one sequence are illustrated in Fig. 3b and c, the use of an image grid had little to no effect onto the results. In fact there are two reasons why the amount of detections can differ; The bilinear interpolation during the downscaling step will yield different values beyond the first base image in I_T. The other reason is the image grid itself, as without it the amount of pre-grouping detections is increased, which in turn will influence the grouping results. The same behaviour was observed on the remaining 3 sequences. Finally we evaluated the same aspect on an interleaved sequence of all 4 camera streams. Just as before there were no boundary detections without a pre-grouping filtering. The amount of detections differs by a at most 2 detections, the amount of identical detection counts was nearly identical. Thus applying the image grid yielded no significant difference in that aspect. Although our results indicate a small difference in detection count one should see H as a different detector when using tile images. The mere values of ΔD are just an indicator for judging if the actual recognition rate will change. As our result show one has to re-optimize the tile image detector for the specific image stream. We optimized our system in order to minimize the values of $t_{copy}, t_{prop,\uparrow}$ and $t_{prop,\downarrow}$, which were measured to be $t_{prop,\uparrow} \approx 4.767ms$, $t_{prop,\uparrow} \approx 4.201ms$ and $t_{copy} = 14.3ms$ (the major part of t_{copy} was owed to the high latency of host-device data transfers). The measured values might change for larger systems (more streams or higher framerate), yet our experiment indicate

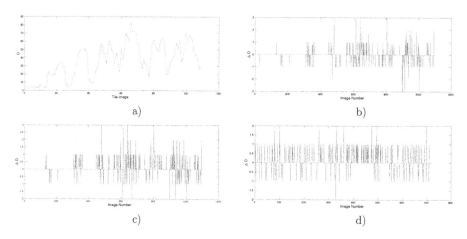

Fig. 3. (a) Example of boundary detections on the tile images for a single camera stream before grouping. (b) ΔD for a single sequence without filtering boundary detections before grouping, detections on I and the corresponding tile image region differ by maximal 3 detections. 84 % of all images yielded the same detection count. (c) ΔD for a single sequence with filtering boundary detections before grouping, detections on I and the corresponding tile image region differ by maximal 3 detections. 82 % of all images yielded the same detection count. (d) ΔD for the first 700 images within an interleaved sequence without filtering boundary detections before grouping, detections on I and the corresponding tile image region differ by maximal 2 detections. 78 % of all images yielded the same detection count.

that $t_{prop,\uparrow}$ and $t_{prop,\downarrow}$ are not likely to violate the previously stated restrictions. Note that the second restriction is not violated by any of our parameter sets from Table 1a or b.

5   Conclusion

We have analyzed the structure of existing GPU-based HOG implementations and identified the image size as a potential efficiency factor. Motivated by this we developed a framework of so called tile images, evaluated it in a small scale cluster system and provided a detailed analysis. Furthermore we provided a detailed analysis of the developed cluster architecture and identified critical time constraints. Our results correspond strongly with the theoretic analysis in section 2, since by varying the image size we obtained a speed-up of $\approx 30\,\%$ - 180 % (depending on the base image size). All conducted time measurements strongly indicate that our distribution scheme scales adequately for a larger amount of nodes since it is mainly limited by the efficiency of the management node's request processing. The tile image framework is applicable to most HOG-based systems, yet depending on the systems structure a different efficiency gain might be obtained. High performance object detection systems can benefit from our approach as the saved computation time might be used for additional reasoning steps without breaking any timing boundaries. Our results also show that tile images induce a relatively high amount of false-positives along base image boundaries. Yet their impact onto the detection count is significantly reduced by using mean-shift grouping alone, the use of an tile grid showed no benefit in this context. Future research should focus on how to efficiently embed images into tiles images, since this step represents the main challenge in situations where the image sizes fluctuate over time. An optimal packing algorithm should aim to minimize the overall computation time while maximizing the image density of the computed (fixed size) tile image. Furthermore it should be studied how the actual recognition rate develops, especially in situations where the images in a stream contain a strong variance to each other (e.g. through changing backgrounds in a PTZ camera).

References

1. Azmat, S., Wills, L., Wills, S.: Accelerating adaptive background modeling on low-power integrated gpus. In: 2012 41st International Conference on Parallel Processing Workshops (ICPPW), pp. 568–573. IEEE (2012)
2. Dalal, N., Triggs, B.: Histograms of oriented gradients for human detection. CVPR 1, 886–893 (2005)
3. Erdos, P., Graham, R.: On packing squares with equal squares. J. Comb. Theor. Ser. A 19(1), 119–123 (1975). http://www.sciencedirect.com/science/article/pii/0097316575900990
4. Gibbons, P.B.: A more practical pram model. In: Proceedings of the First Annual ACM Symposium on Parallel Algorithms and Architectures, pp. 158–168. ACM (1989)
5. Hirabayashi, M., Kato, S., Edahiro, M., Takeda, K., Kawano, T., Mita, S.: Gpu implementations of object detection using hog features and deformable models. In: 2013 IEEE 1st International Conference on Cyber-Physical Systems, Networks, and Applications (CPSNA), pp. 106–111. IEEE (2013)
6. Hommel, S., Malysiak, D., Grimm, M., Handmann, U.: Apfel - fast multi camera people tracking at airports, based on decentralized video indexing. Sci.2 - Saf. Secur. 2, 48–55 (2014)
7. Hommel, S., Malysiak, D., Handmann, U.: Model of human clothes based on saliency maps. In: 2013 IEEE 14th International Symposium on Computational Intelligence and Informatics (CINTI), pp. 551–556, November 2013
8. Kachouane, M., Sahki, S., Lakrouf, M., Ouadah, N.: Hog based fast human detection. In: 2012 24th International Conference on Microelectronics (ICM), pp. 1–4. IEEE (2012)
9. Lodi, A., Martello, S., Monaci, M.: Two-dimensional packing problems: A survey. Eur. J. Oper. Res. 141(2), 241–252 (2002). http://www.sciencedirect.com/science/article/pii/S0377221702001236
10. Malysiak, D., Handmann, U.: An efficient framework for distributed computing in heterogeneous beowulf clusters and cluster-management. In: 2014 IEEE 15th International Symposium on Computational Intelligence and Informatics (CINTI), November 2014

11. Prisacariu, V., Reid, I.: fasthog - a real-time gpu implementation of hog. Technical Report 2310/09. Department of Engineering Science, Oxford University
12. Sudowe, P., Leibe, B.: Efficient use of geometric constraints for sliding-window object detection in video. In: Crowley, J.L., Draper, B.A., Thonnat, M. (eds.) ICVS 2011. LNCS, vol. 6962, pp. 11–20. Springer, Heidelberg (2011)
13. Yudanov, D., Shaaban, M., Melton, R., Reznik, L.: Gpu-based simulation of spiking neural networks with real-time performance & high accuracy. In: The 2010 International Joint Conference on Neural Networks (IJCNN), pp. 1–8. IEEE (2010)

Gradient Depth Map Based Ground Plane Detection for Mobile Robot Applications

Dang Khanh Hoa, Pham The Cuong, and Nguyen Tien Dzung[(⊠)]

School of Electronics and Telecommunications (SET),
Hanoi University of Science and Technology (HUST), Hanoi, Vietnam
{hoa.dangkhanh,dzung.nguyentien}@hust.edu.vn,
phamcuong21478@gmail.com

Abstract. In the field of navigation and guidance for mobile robots utilizing stereo visual imagery, the main problem to be solved is to detect the ground plane in acquired images by a stereo camera system mounted on the mobile device. This paper focuses on effective detection of ground based on graphical analysis of the gradient depth map evaluated on the input depth map within a given window. The detected ground planes is further divided into blocks and then classified into ground or non-ground regions for elimination of false detected ground planes followed by smoothing in refinement process. This proposed approach also has been shown to be effective in detection of obstacles appearing in the ground plane too while the mobile device is moving. In addition, the algorithm is simple, reliable, feasible, and may be efficiently exploited for implementation in an embedded hardware with limited resources for real-time applications.

Keywords: Depth map · Gradient depth map · Ground plane · Mobile robot

1 Introduction

In the field of autonomous mobile robots moving on road which are mounted with a stereo camera, the most important problem is extraction of a ground plane and identification of obstacles. In recent years, there is a lot of work with the relatively diversified approach to solve this question. These include works [1–3] used the classic RANSAC algorithm to estimate the plane with high reliability. However, a large number of operations required in this method may lead to large elapsing time. Two approaches in [4, 5] apply an optical flow concepts and use input data that is a color video or multi-level gray image to get some very interesting results in environmental containing objects with special characteristics. In other cases, the ground plane contains complicated patterns, the detection accuracy would be reduced. For real-time application, the authors in [6] have presented an algorithm to simply find the ground plane by processing the 2D disparity map input data, however the input images were captured and tested in an indoor environment with a simple background. This limits the ability to solve problems in the real environment [7, 8]. The use of a combination of both classic Hough

© Springer-Verlag Berlin Heidelberg 2016
N.T. Nguyen et al. (Eds.): ACIIDS 2016, Part I, LNAI 9621, pp. 721–730, 2016.
DOI: 10.1007/978-3-662-49381-6_69

algorithm and RANSAC has brought results of identify objects very efficiently by promoting the advantages of each algorithm for each different specific circumstances [9]. But the volume of operations in the program is a problem not yet to be solved.

The process of selecting the simple data source to be applied by using a single camera [10–12] but it is evident that this solution could not extract the depth information of obstructions objects which could be appeared in the frame. This makes the movement of the mobile system not yet feasible in the context of land terrain.

Thus for real time application of mobile robots while moving, the mentioned methods may face up with high computational load and require to operate on a hardware platform with high configuration. In this article, we propose an enhanced approach to detect ground plane using approximation and grouping plane regions extracted from 2D depth map. The paper is structured in four sessions where Sect. 2 introduces some mathematical fundamentals in plane detection problem. Section 3 then illustrates the proposed algorithm. Finally Sect. 4 discusses the experimental results and performance evaluation.

2   The Basis Mathematics Theory of System

2.1   Mathematical Fundamentals

Assuming that the camera with focal length f is located at O with the height of h to the ground as shown in Fig. 1, where P_I and P_G are the image plane and ground plane, respectively. Let $O1$ and $O2$ be the projections of O on P_G and P_I, respectively; M_{I1} and M_{I2} given pixels lying on P_I and p the distance from O_2 to M_{I1}. Then, the depth for M_{I1} can be estimated as:

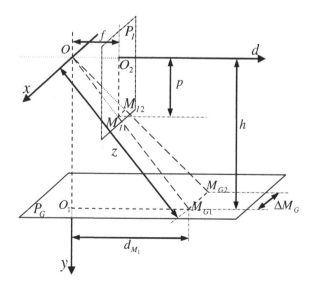

Fig. 1. Illustration of the depth difference determination

$$z = h\sqrt{1 + \frac{f^2}{p^2}} \tag{1}$$

Taking the differential both sides of (1), we can get:

$$dz = -hf^2 \frac{1}{p^2\sqrt{p^2 + f^2}} d\rho \tag{2}$$

It can be seen from Eq. (2) that the depth difference Δz_y may be determined from the height difference Δp in the vertical axis y.

In the other hand, also from Fig. 1, the depth difference Δz_x can be determined from the distance difference ΔM_G of M_{G1} and M_{G2} on the ground in the horizontal axis x as shown in Eq. (3).

$$\Delta z_x = \sqrt{\Delta M_G^2 + z^2} - z \tag{3}$$

The two Eqs. (2) and (3) may be utilized to calculate gradient depth map in y and x directions respectively. Assuming ΔM_I is one pixel distance, then $\Delta M_G = \frac{\Delta M_I}{OM_{I1}} \times z$, where the ratio $\frac{\Delta M_I}{OM_{I1}}$ is very small. Therefore $z >> \frac{\Delta M_I}{OM_{I1}} \times z$, and Δz_x calculated by Eq. (3) should be very small. Hence for simplicity in the proposed algorithm, Δz_x always is considered as 0 for pixels belonging to the ground plane. From graphical analysis of Eq. (2), it can be seen that for $\Delta z_x = 0$, Δz_y typically takes significant values above a given threshold T, which can be determined graphically by two-state approximation of the curve introduced by the ratio of $\frac{dz}{dp}$ in Eq. (2). This threshold value actually is the coordinate y at which $|\Delta z_y| = 1$. Since the depth of the ground pixels are normally quantized by 8 bits, therefore Δz_y is typically different from 0 for all $y < T$. Based on geometrical analysis regarding to ground plane properties, the proposed algorithm classify a pixel under consideration into ground or non-ground one as below:

1. $\Delta Z_x \neq 0$: Non-ground pixel
2. $\Delta Z_x = 0 \,\&\, \Delta Z_y \neq 0$: Ground pixel
3. $\Delta Z_x = 0 \,\&\, \Delta Z_y = 0$ & the pixel belongs to the bottom quarter of the input image: Also ground pixel, since the variation of ΔZ_y in this area.

2.2   Block Diagram of the Proposed System

The block diagram of the proposed method is presented in Fig. 2. The first stage calculates *gradient_x* and *gradient_y* for each pixel using the depth map as an input to construct the a gradient depth map. Then, the second stage groups the adjacent pixels that have a similar gradient into a range. The candidate ground plane is then formed by the ranges that meet the ground hypotheses. Since the candidate ground plane typically are affected by sporadic and random noise, the final stage will have to refine noise to construct the final ground plane by splitting the input image into blocks of size B.

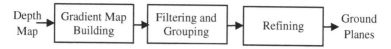

Fig. 2. A block diagram of the proposed ground plane detection.

3 Construction of Gradient Depth Map

The task of this stage is to create a map of depth difference, also called a gradient map from the depth map input by calculation of gradients in y and x directions using Eqs. (2) and (3) between two successive points, respectively. The resulting gradient depth map is further smoothed by consideration of depth difference of a point within a given window of size w, because of possible presence of noise in the input depth map.

4 Filtering and Grouping

The goal of this stage is to group the points having similar gradients in the gradient depth map into a homogeneous region called range, and then eliminate inappropriate regions which do not satisfy the following constraints of the ground plane:

- The number of pixels of the region must be greater than a predetermined threshold;
- $gradient\_x = 0$ and $gradient\_y \neq 0$; or if $gradient\_x = 0$ and $gradient\_y = 0$ then the region must be located completely in the quarter area from the bottom of the input image for higher accuracy in ground plane detection.

The grouping and elimination algorithms are illustrated in the following pseudo code.

```
//Algorithm: Grouping.
//Input: Image. //Output: Ranges
for each pixel do
  if this pixel is not in other Collection then
    Range add this pixel
    Range add all pixels satisfying Range's conditions
  end if
  if number pixel of Range > Range threshold and Range
  satisfies Ground Plane's conditions then
    Ranges Add Range
  end if
  Renew Range
end for
```

As the result, the ground plane from the acquired image would be roughly determined.

5 Refining

In order to extract more exact and smooth ground plane, this correction stage starts dividing the initial difference depth map into square blocks of size B and then estimates the ratio R of the ground pixels inside each block to the block size. This is an important parameter used to classify the blocks into ground or non-ground ones and then generate the final map which includes ground and non-ground regions. If R is greater than a given threshold θ, then the block is considered as ground, and vice versa. In order to evaluate the value of θ, the smallest rectangular bounding the detected ground regions is determined, and the ratio between the number of all ground pixels $P_{ground\_of\_ranges}$ over the square of the rectangular P_{rec} as depicted in Eq. (4):

$$\theta = \frac{\sum P_{ground\_of\_ranges}}{P_{rec}} \tag{4}$$

Obviously the non-ground areas which belong to obstacles appearing with large enough size would be detected. The following pseudo code illustrates the algorithm:

```
//Algorithm: Dividing Block.
//Input: Ranges.
//Output: Ground Plane
Divide gradient depth map into blocks of size B
Calculate threshold θ
for each block in Image do
  Ratio R of this block = number of ground pixel in a
  block / block size
  if R > θ then
    This block is assigned as Ground Plane
  else
    This block is eliminated
  end if
end for
```

6 Results and Discussion

The performance evaluation of the proposed method is carried out with 5 different images which are acquired by a stereo camera with and without obstacles as shown in Fig. 3.

The detected ground planes as illustrated in the last column are shown to be smoothly matched with the real scenes. For determination of the window size w in construction of the gradient depth map, the percentage of detected ground pixels in each image is compared for three different window sizes. The result in Fig. 4 indicates

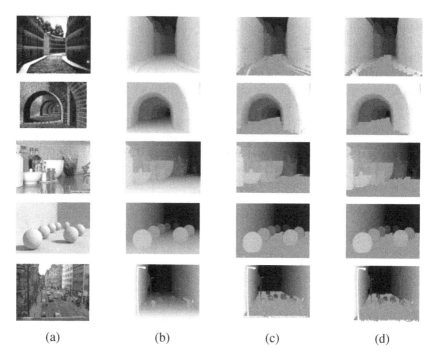

(a)   (b)   (c)   (d)

Fig. 3. The results of the tested images in the plenty of environment. (a) color images; (b) depth maps; (c) the rough ground planes; (d) refined ground planes. From top to bottom, the first row is the Street image [13], the Vaulted image [14], the Kitchen image [15], the Balls image [16], and the Canyon image [17], respectively;

that the window size of 5×5 generates highest detection rate above 90 % in most cases. The detected ground plane in the Canyon's image is relatively acceptable even though the street view and complex background are captured. Moreover, the obstacles have been successfully extracted in case of the last three images.

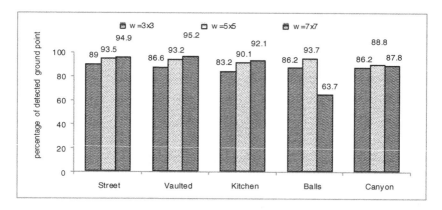

Fig. 4. The rate of detected ground pixels before correction process according to the window sizes w.

Figures 5, 6 and 7 illustrate the comparison of percentage of the detected ground pixels after refining with different block size B for each image. From these results, the block size of 8×8 outperforms the highest and stable detection rate of ground planes. The obstacles appearing in the images are kept detected.

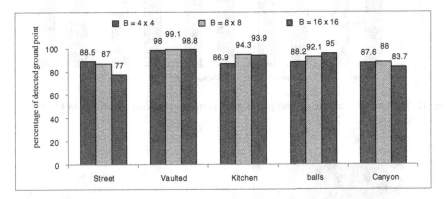

Fig. 5. The rate of detected ground pixels after correction process according to block's sizes with window size $w = 3 \times 3$

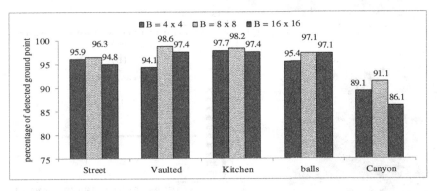

Fig. 6. The rate of detected ground pixels after correction process according to block's sizes with window size $w = 5 \times 5$

As mentioned above, the window size of $w = 5 \times 5$ and block size of $B = 8 \times 8$ have been utilized for best detection rate of ground plane pixels. Using these parameters, the next experiment demonstrates the variation curve of ratio R evaluated for all blocks lying on a row from the gradient depth map after the block division and the corresponding threshold θ. It can be seen from Fig. 8 that most of blocks belonging to ground planes have R greater than the determined threshold θ as discussed in paragraph 5 in refinement process for Canyon and Vaulted images, respectively. In case of

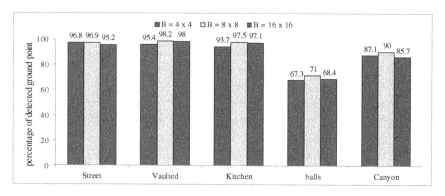

Fig. 7. The rate of detected ground pixels after correction process according to block's sizes with window size $w = 7 \times 7$

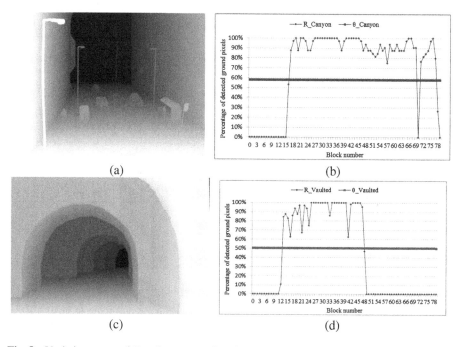

Fig. 8. Variation curve of R and corresponding threshold θ for a line in the gradient depth map after block division. (a) Depth map of Canyon; (b) corresponding R curve and θ calculated for row 45. (c) Depth map of Vaulted; (d) corresponding R curve and θ calculated for row 40

Canyon image, blocks 12 to blocks 69 are classified to ground ones because percentage of ground pixels is over threshold about $\theta = 60\%$. And in case of Vaulted image, the ground block numbers range from 12 to 48 which have percentage of ground pixels greater than threshold about $\theta = 50\%$.

7 Conclusions

In this paper a method for detection of ground plane including obstacles have been proposed which is based on the construction of gradient depth map to detect ground pixels, and then refine the resulting ground plane to obtain smoothing ground planes. The experimental result showed that the proposed algorithm could detect ground planes well in presence of obstacles. Based on the simple but effective algorithm, the proposed approach is appropriate for implementation in an embedded hardware. For future works, a mobile robot platform which is equipped with a stereo camera and running the proposed algorithm will be implemented for testing navigation and guidance ability of the robot.

References

1. Hast, A., Nysjö, J., Marchetti, A.: Optimal RANSAC – towards a repeatable algorithm for finding the optimal set. J. WSCG **21**(1), 21–30 (2013)
2. Hu, X., Rodriguez, F.S.A., Gepperth, A.: A multi-modal system for road detection and segmentation. In: Intelligent Vehicles Symposium Proceedings. IEEE, pp. 1365–1370 (2014)
3. Sakai, A., Tamura, Y., Kuroda, Y.: Visual odometry using feature point and ground plane for urban environment. In: 2010 41st International Symposium on Robotics (ISR), and 2010 6th German Conference on Robotics (ROBOTIK), Munich, Germany, pp. 1–8 (2010)
4. Wang, Z., Zhao, J.: Optical flow based plane detection for mobile robot navigation. In: Proceedings of the 8th World Congress on Intelligent Control and Automation, Taipei, Taiwan, pp. 1156–1160 (2011)
5. Jamal, A., Mishra, P., Rakshit, S., Singh, A.K., Kumar, M.: Real-time ground plane segmentation and obstacle detection for mobile robot navigation. In: Emerging Trends in Robotics and Communication Technologies (INTERACT), pp. 314–317 (2010)
6. Gong, K., Green, R.: Ground-plane detection using stereo depth values for wheelchair guidance. In: 24th International Conference Image and Vision Computing New Zealand (IVCNZ), pp. 97–101 (2009)
7. Haberdar, H., Shah, S.K.: Disparity map refinement for video based scene change detection using a mobile stereo camera platform. In: 20th International Conference on Pattern Recognition (ICPR), pp. 3890–3893 (2010)
8. Teoh, C., Tan, C., Tan, Y.C.: Ground plane detection for autonomous vehicle in rainforest terrain. In: IEEE Conference on Sustainable Utilization and Development in Engineering and Technology (STUDENT), pp. 7–12 (2010)
9. Fayez, T.-K., Landes, T., Grussenmeyer, P.: Hough-transform and extended ransac algorithms for automatic detection of 3d building roof planes from lidar data. In: ISPRS Workshop on Laser Scanning 2007 and SilviLaser 2007, vol. 36, pp. 407–412 (2007)
10. Arro'spide, J., Salgado, L., Nieto, M., Mohedano, R.: Homography-based ground plane detection using a single on-board camera. IET Intell. Transp. Syst. **4**(2), 149–160 (2010)
11. Mostof, N., Elhabiby, M., El-Sheimy, N.: Indoor localization and mapping using camera and inertial measurement unit (IMU). In: Position, Location and Navigation Symposium - PLANS 2014, 2014 IEEE/ION, pp. 1329–1335 (2014)

12. Mishra, P., Kishore, J.K., Shetty, R., Malhotra, A., Kukreja, R.: Department of electronics & communication engineering monocular vision based real-time exploration in autonomous rovers. In: 2013 International Conference on Advances in Computing, Communications and Informatics (ICACCI), pp. 42–46 (2013)
13. Yiningkarlli. http://blog.yiningkarlli.com/2010/11/city-street-playing-with-z-depth-and-ambient-occlusion.html
14. DeviantArt. http://mvramsey.deviantart.com/art/Vaulted-Cellar-depth-map-429569523
15. Evermotion. http://www.evermotion.org/tutorials/show/8320/tip-of-the-week-making-of-archmodels-vol-134-cover
16. The Foundry. http://community.thefoundry.co.uk/discussion/topic.aspx?f=36&t=48950
17. German Unix-AG Association. http://www.home.unix-ag.org/simon/files/street-canyon.html

Multiscale Car Detection Using Oriented Gradient Feature and Boosting Machine

Wahyono[1], Van-Dung Hoang[2], and Kang-Hyun Jo[1]([✉])

[1] The Graduate School of Electrical Engineering, University of Ulsan,
Ulsan 680749, Korea
wahyono@islab.ulsan.ac.kr, acejo@ulsan.ac.kr
[2] Quang Binh University, Dong Hoi City, Vietnam
dunghv@qbu.edu.vn

Abstract. In many car detection method, the candidate regions which have variation in size and aspect ratio are mostly resized into a fixed size in order to extract the same length dimensional feature. However, this process reduces local object information due to interpolation problem. Thus, this paper addresses a solution to solve a such problem by using Scalable Histogram of Oriented Gradient (SHOG). The SHOG enables to extract fixed-length dimensional features for any size of region without resizing the region to a fixed size. In addition, instead of using high dimensional features in training stage, our proposal divides the feature into several low-dimensional sub-features. Each sub-feature is trained using SVM, called weak classifier. Boosting strategy is applied for combining the weak classifier results for constructing a strong classifier. By conducting comprehensive experiments, it is found that the accuracy of SHOG is higher than standard HOG as much as 3 % and 4 %, without and with boosting, respectively.

Keywords: Car detection · Scalable Histogram of Oriented Gradient · Boosting machine · Support vector machine · Integral image · Weak classifier · Strong classifier

1 Introduction

Car detection is one of the most important modules for Advanced Driver Assistance System (ADAS). It assists the drivers see the preceding cars and road situation for preventing collision with them. In this system, the preceding cars are captured by mounting the camera on the car, and might have variation on shape, color, view, and scale. This is because local occlusion and illumination conditions take effect on capturing process. These problems challenge most of the state-of-the art car detection system [5–7]. Therefore, more effective car descriptor and detection method are required to develop a robust car detection system.

Histogram of Oriented Gradients (HOG) is one of the most effective descriptor for solving object detection system [1]. Many researchers have implemented

© Springer-Verlag Berlin Heidelberg 2016
N.T. Nguyen et al. (Eds.): ACIIDS 2016, Part I, LNAI 9621, pp. 731–740, 2016.
DOI: 10.1007/978-3-662-49381-6_70

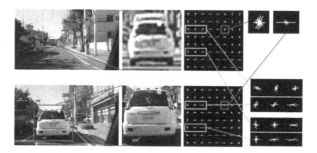

Fig. 1. HOG feature problem when two regions have different size. See text for details.

HOG with significantly results for many applications such as traffic sign detection [3,4], pedestrian detection [2], and car detection [5]. The HOG describes object shape and appearance based on the distribution of pixel gradients directions. It divides the image into small connected regions, called cells, and each cell compiles an histogram of gradient directions within neighbor pixels. The combination of these histograms generates the final descriptor.

Nonetheless, HOG has limitation for handling input region which have different scale and aspect ratio. In car detection system, the distance between camera and car might affect the size of the car region on the image. The common approach usually re-scales the object region into a fixed target size [1]. However, if the aspect ratio between the target and original sizes are totally different, the HOG might not represent the car object accurately. Figure 1 shows an example of HOG extraction for same car regions extracted from two different input images. When the car is located far away from the camera, as shown on the first row, it appears as small region on the image. Meanwhile, when the same car is close from camera, as shown on the second row, it appears as bigger region. For extracting same-length dimensional features, these regions should be resized to the same target size, e.g. 128x128. Ideally, as these regions represent same car, the HOG descriptor should have same value for corresponding block location. However, as shown on the last column, the HOG have different value for several same block locations. In consequence, the system may treat them as different object. Therefore, this work utilizes scalable histogram of oriented gradient (SHOG) for detecting car which produces a fixed-length dimensional feature from multi size object regions without resizing process.

Furthermore, the common approaches mostly extract high dimensional feature. They achieve good accuracy, but the highly dimensional features generate high computation cost. For solving such problem, the utilization of integral image is proposed. The integral image is very useful for faster gradient-based feature extraction. In addition, instead of using highly dimensional feature for training, our proposal divides the feature into several low-dimensional sub-features. Each sub-feature is trained using individual classifier SVM, called weak classifier. For final result, the weak classifier results are combined using boosting machine in order to obtain a strong classifier.

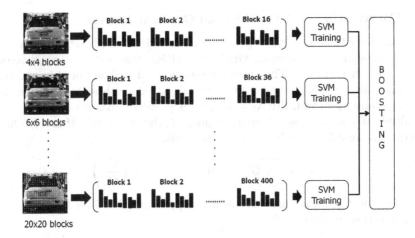

Fig. 2. Our proposed approach schema.

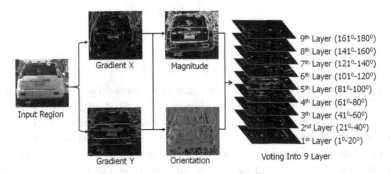

Fig. 3. Constructing 9 image layers based on gradient orientation.

2   The Proposed Approach

This section presents method for detecting multiscale car. As shown in Fig. 2, the schema of the proposed method is divided into three main processes: (1) feature extraction based on oriented gradient, (2) individual training using SVM for each block size which assigned as weak classifier, (3) boosting machine for combining several weak classifiers in order to obtain a strong classifier. Technically, input image is divided into several blocks according to block size as parameter. Different with standard HOG [1], the image does not need to be resized into a certain size. The oriented gradient features are extracted for each block region. Concatenation of the feature for all regions is then trained using support vector machine. Since the feature dimension of each block size is low, the training result may not be strong enough for classifying the data. It is called as weak classifier. Hence, for obtaining the highly accurate classifier, several weak classifiers are combined using boosting strategy.

2.1   Constructing Scalable Oriented Gradient Feature

This part describes an overview process for extracting fixed-length dimensional Scalable Histogram of Oriented Gradient (SHOG) feature from any different size and aspect ratio [12]. The process keeps local information of object as rich as possible and avoids resizing step which might reduce these information due to interpolation problem. SHOG descriptor is generated based on the gradient magnitude and orientation. Given input image I, the magnitude M and gradient directions θ are calculated by following formulas,

$$M = \sqrt{G_x^2 + G_y^2} \quad \text{and} \quad \theta = \arctan\left(\frac{G_y}{G_x}\right) \tag{1}$$

where G_x and G_y are intensity gradient by applying Sobel filter kernel [8] in x and y direction, respectively,

$$G_x = I \otimes \begin{bmatrix} -1 & 0 & +1 \\ -2 & 0 & +2 \\ -1 & 0 & +1 \end{bmatrix} \tag{2}$$

$$G_y = I \otimes \begin{bmatrix} +1 & +2 & +1 \\ 0 & 0 & 0 \\ -1 & -2 & -1 \end{bmatrix}. \tag{3}$$

As in standard HOG [1], 9 image layers are constructed by dividing the intensity gradient on basis of their orientation, as illustrated in Fig. 3. Each image layer collects the pixels with certain gradient orientations. For instance, the first image layer stores the magnitude value of pixel with orientation between 0 and 20 degrees, the second image layer stores the magnitude value of pixel with orientation between 21 and 40 degrees, and so on. The image layers are then divided into several block regions, and SHOG is extracted on each block. To be note that any block size may be used, and the pixel number on each block depend on its size. For example, if 128x128 pixels image is divided into 2x2 blocks, each block contains 64x64 pixels. However, if this image is divided into 4x4 blocks, each block consists of 32x32 pixels. For each block on each image layer, the average magnitude are computed. Since there are 9 image layers, each block gains 9-element vectors. The vector values are then normalized using the L_2-Norm [9]. The feature vector of all blocks are concatenated to form the final descriptor. Regarding to this step, it is possible to obtain the fixed-length dimensional descriptor from any size of images. However, the number of dimension is affected by the block sizes. For illustration, in Fig. 4, the input image is divided into 4x4 blocks. As result, the final descriptor contains 144-elements vector (4x4x9 = 144 elements). In the next step, SHOG descriptors are used to train the dataset using support vector machine.

2.2   Linear Support Vector Machine

Basically, to train SHOG descriptor, any machine learning method could be performed. However, in our implementation, support vector machine (SVM) is

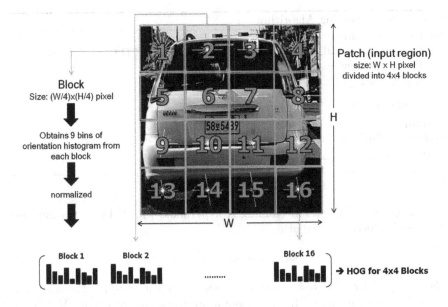

Fig. 4. Extracting SHOG descriptor for 4x4 block sizes.

used due to its empirically good performance in many applications [11]. SVM was initially developed from statistical learning theory and it deals with structural risk minimization which minimizes an upper bound on the Vapnik-Chervonenkis dimension. Mostly, SVM is used for solving classification problem in pattern recognition. In binary classification, the training data D are formulated as:

$$D = \{(x_i, y_i) | x_i \in R^d, y_i \in \{-1, 1\}, i = 1, \ldots, N\} \tag{4}$$

where x_i is d dimensional SHOG descriptor, y_i is either 1 or -1, representing the class to which point x_i belongs. In car detection problem, $y_i = 1$ is considered as car object, while $y = -1$ is non-car object.

The SVM is aimed to find the maximum-margin separating the points among two classes. By assuming the training data are linearly separable, two hyperplanes that separate the data and there is no point between them could be determined by the equations $w^T x_i + b = 1$ and $w^T x_i + b = -1$. The distance between these hyperplane is $\frac{2}{\|w\|}$, where $\|w\|$ is the Euclidean norm of w. To prevent data point from falling into the margin between hyperplanes, the constraint either $w^T x_i - b \geq 1$ for $y_i = 1$ or $w^T x_i - b \leq 1$ for $y_i = -1$ should be satisfied for each data. Mathematically, these two constraints can be rewritten as

$$y_i \left(w^T x_i + b \right) - 1 \geq 0 \quad \forall i. \tag{5}$$

So now, optimization problem is defined to find the maximum margin between hyperplanes. It is obtained by minimizing $\frac{\|w\|^2}{2}$ subject to constraints of (5).

Algorithm 1. Boosting Algorithm

---

for $t = 1, \ldots, T$ **do**

- Compute empirical error of week classifier t, $\epsilon_t = \sum_{i=1}^{N} D_t(i)[y_i \neq \Phi_t(x_i)]$
- If $\epsilon_t \geq 1/2$ then Stop
- Set $\omega_t = \frac{1}{2} \log \left(\frac{1+r_t}{1-r_t} \right)$, where $r_t = \sum_{i=1}^{N} D_t(i)\Phi_t(x_i)y_i$
- Update $D_{t+1}(i) = \frac{D_t(i)\exp(-\omega_t y_i \Phi_t(x_i))}{2\sqrt{\epsilon_t(1-\epsilon_t)}}$

end for

Output the final Classifier:

$H(x) = sign \left(\sum_{t=1}^{T} \omega_t \Phi_t(x) \right)$

The final output of the SVM training is hyperplane $\{w, b\}$. In evaluation, test data x is classified by calculating

$$\Phi(x) = sign(w^T x + b). \tag{6}$$

If $\Phi(x)$ is positive, the test data is considered as car object, otherwise it is classified as non-car object. In implementation, the standard linear SVM [13] is applied to learn SHOG descriptor for each block size as a weak classifier. Several weak classifiers are combined using boosting technique.

2.3 Boosting Machine

Boosting is a machine learning technique for constructing a highly accurate classifier by combining many weak and inaccurate classifiers. Here, SVM models for each block size are considered as weak classifiers. The weak classifier is represented by $m = \Phi_i(x, l)$, where x is SHOG descriptor for $l \times l$ block size. The combination of these weak classifiers is performed using boosting Support Vector Machine (SVM) [14] which is defined by the following formula

$$f(m) = \sum_{i=1}^{N} \omega_i \Phi_i(x, l) \tag{7}$$

where ω_i, $\omega_i > 0$ is the weight value of each weak classifier, and Φ_i is the output of SVM model, N is the number of block size. Our goal is discovering the weighting value using boosting machine.

There have been many boosting methods to find the weighting value of linear combination of weak classifiers. One of the simplest and the most efficient is Adaptive Boosting (AdaBoost) [15]. In our method, T weak classifiers are learned based on SHOG feature x of each block size. Given the training data $(x_1, y_1), \ldots, (x_N, y_N); x_i \in X, y_i \in \{-1, +1\}$, where x_i is SHOG descriptor, y_i is class where data i belong to, and N is the number of data. First, the weighting factor of each data is set to be $D_1(i) = 1/N$. The AdaBoost method is then performed as shown in Algorithm 1.

Fig. 5. Typical samples of car and non-car region used in implementation. First and second rows are car region samples; third and fourth rows are non-car region samples.

2.4   Integral Image of Intensity Gradient

As described in the previous section, the histogram of gradients might be repeatedly calculated many times within image regions. Therefore, the integral image method is used for reducing the number of computations in the feature extraction step. The integral image calculates the sum of intensity values (i.g. magnitude) in a given image effectively based on the dynamic programming strategy [10]. In technical implementation, it is necessary to create a Summed Area Table (SAT) T to store the cumulative sum of gradients on each image layer. T_k is SAT of the kth image layers, which is computed by

$$T_k(x,y) = M_k(x,y) + T_k(x-1,y) + T_k(x,y-1)$$
$$-T_k(x-1,y-1) \quad (8)$$

where $M_k(x,y)$ is the magnitude value at pixel location (x,y) corresponding to the kth image layer. The integral image S_k within block regions of the kth layer is calculated as:

$$S_k(p,q,w,h) = T_k(p+w-1,q+h-1) - T_k(p-1,q+h-1)$$
$$-T_k(p+w-1,q-1) + T_k(p-1,q-1) \quad (9)$$

where (p,q) is the coordinate of the left-top corner, while w and h are the width and height of the block, respectively. SHOG on each block consists of 9 elements which are computed on basis of 9 image layers. Therefore, extracting SHOG descriptor requires only 9 access operations. It can be confirmed that the complexity of SHOG extraction is almost constant $O(1)$ comparing to the conventional method that requires $O(wh)$ complexity.

3   Experiments

For proving the effectiveness of the proposed descriptor, the comprehensive experiment is conducted. The results would be compared with the state-of-the art histogram of oriented gradient. First, the dataset which consists of 1,525 car and 5,804 non-car sample was collected, treated as training data. This data was recorded on surrounding of University of Ulsan. The car samples are collected manually, while the non-car samples were automatically extracted from the non-car images. Figure 5 depicts several car and non-car samples used in our implementation. In verification stage, each extracted region was divided into several block sizes in order to find optimal block size. The maximum and minimum block size are defined as parameter in which our goal is to find the optimal value of the block size for the car candidate region. For each block size, the recognition is applied using the Support Vector Machines. Regarding to the verification results, it was found that parameter 4x4 to 20x20 block size gives the highest recognition rate of 99.06 %. Therefore, we set these parameters on the testing stage for evaluation purpose. In addition, for evaluation, we again collected 1,113 positive and 1,543 negative samples.

For evaluating the proposed descriptor, precision, recall and F-measure were used as measurement protocols. Precision is ratio between number of true positives and the number of all detected objects obtained from the experiment, while recall is ratio between the number of true positives and the number of true detected objects on the ground truth. Accuracy is defined as a harmonic mean between precision P and recall R with $A = \frac{2 \times P \times R}{P+R}$.

The comparison results of proposed descriptor and standard HOG are summarized on the Table 1. Our descriptor was evaluated using two different

Table 1. Comparison results between our strategy and standard HOG.

Properties	$SHOG^1$	$SHOG^1_{Boost}$	$SHOG^2$	$SHOG^2_{Boost}$	HOG^1	HOG^2
CAR SAMPLES						
True detected	1102	1104	1105	1107	1021	1057
False detected	11	9	8	6	92	56
Total	1113	1113	1113	1113	1113	1113
NON-CAR SAMPLES						
False detected	46	39	21	20	42	32
True detected	1497	1504	1522	1524	1501	1511
Total	1543	1543	1543	1543	1543	1543
PRECISION	95.99 %	96.59 %	98.13 %	96.05 %	96.05 %	97.06 %
RECALL	99.01 %	99.19 %	99.28 %	99.46 %	91.73 %	94.97 %
ACCURACY	97.47 %	97.87 %	98.7 %	97.72 %	93.84 %	96.00 %

$SHOG^1$ = 4x4 to 16x16 block sizes, 9 orientation bins (without boosting)

$SHOG^2$ = 4x4 to 20x20 block sizes, 9 orientation bins (without boosting)

$SHOG^1_{Boost}$ = 4x4 to 16x16 block sizes, 9 orientation bins (with boosting)

$SHOG^2_{Boost}$ = 4x4 to 20x20 block sizes, 9 orientation bins (with boosting)

HOG^1 = 56x56 pixels, 8x8 cell sizes, 2x2 block sizes

HOG^2 = 64x64 pixels, 8x8 cell sizes, 2x2 block sizes

strategies: without and with boosting approaches. Without applying boosting method, the features of all block sizes are concatenated into one highly dimensional feature vector and then are trained using SVM [12]. Without boosting, SHOG achieves an accuracy of about 3 % better than HOG features. However, when boosting strategy is implemented, it achieves better accuracy as much as 4 % higher than original HOG features. Even though Breiman et al. stated that a lot of weak classifiers are needed to construct boosting machine. However, in our case just using few weak classifiers is enough due to each single SHOG+SVM classifier achieves high accuracy.

4 Conclusion

This paper presents the solution for two challenges in car detection system: (1) extracting fixed-length dimensional feature of multi scale car regions, and (2) extracting car region candidates. First, the common method usually resizes the candidate region to a fixed size for extracting the same length dimensional features. However, it could cause problems if the aspect ratio between the original and target sizes are different. Our idea for solving this problem is introducing scalable histogram of oriented gradient (SHOG) feature for multi scale car detection. SHOG works on any size of the car region, but produces fixed-length dimensional features. The second challenge is that how to extract possible candidate regions from the image by improving sliding window method. The common sliding window needs a high computational cost due to the process try to find possible candidate regions in any position and scale on the image. Thus, integral image is performed to decrease the computational cost of sliding window. The boosting support vector machine is applied for both training and testing the proposed descriptor. The experiment results proved that the proposed method outperforms the standard histogram of oriented gradient approach.

References

1. Dalal, N., Triggs, B.: Histograms of oriented gradients for human detection. In: IEEE CVPR, vol. 1, pp. 886–893 (2005)
2. Zu, Q., Yeh, M.-C., Cheng, K.-T., Avidan, S.: Fast human detection using a cascade of histogram of oriented gradients. In: IEEE Conference on Computer Vision and Pattern Recognition, pp.1491–1498 (2006)
3. Wahyono, Kurnianggoro, L., Hariyono, J., Jo, K.-H.: Traffic sign recognition system for autonomous vehicle using cascade SVM classifier. In: Proceeding of the 40th Annual Conference of the IEEE Industrial Electronic Society, in: IECON (2014)
4. Greenhalgh, J., Mirmehdi, M.: Real-time detection and recognition of road traffic signs. IEEE Trans. ITS 13(4), 1484–1497 (2012)
5. Kim, M., Kim, C.-S.: A two-stage approach to people and vehicle detection with HOG-based SVM. IEEK Trans. Smart Process. Comput. 2(4), 189–196 (2013)
6. Kuo, C.-H., Nevatia, R.: Robust multi-view car detection using unsupervised subcategorization. In: Workshop of Application of Computer Vision, pp. 1–8, December 2009

7. Ramakrishnan, V., Prabhavathy, A.K., Devishree, J.: A survey on vehicle detection technique in aerial surveillance. Int. J. Comput. Appl. **55**(18), 43–47 (2012)
8. Sobel, I., Feldman, G.: A 3x3 isotropic gradient operator for image processing. Machine Vision for Three Dimensional Scenes, Freeman, H., Academic Pres, NY, 376-379 (1990)
9. Horn, R.A., Johnson, C.R.: Norms for Vectors and Matrices. Ch. 5 in Matrix Analysis. Cambridge University Press, Cambridge (1990)
10. Viola, P., Jones, M.: Robust real-time object detection. Int. J. Comput. Vis. **57**, 137–154 (2001)
11. Cortes, C., Vapnik, V.: Support-vector networks. Mach. Learn. **20**(3), 273 (1995)
12. Wahyono, Hoang, V.-D., Kurnianggoro, L., Jo, K.-H.: Scalable histogram of oriented gradients for multi-size car detection. In: Proceeding of the Mecatronics (2014)
13. Chih-Chung, C., Chih-Jen, L.: LIBSVM: A library for support vector machine. ACM Trans. Intell. Syst. Technol. **2**, 1–27 (2011)
14. Hoang, V.-D., Hernandez, D.C., Jo, K.-H.: Joint components based pedestrian detection in crowded scenes using extended feature descriptors, Neurocomputing (2015) in Press
15. Freund, Y., Schapire, R.E.: A decision-theoretic generalization of on-line learning and an application to boosting. J. Comput. Syst. Sci. **55**, 119–139 (1997)

Probabilistic Approach to Content-Based Indexing and Categorization of Temporally Aggregated Shots in News Videos

Kazimierz Choroś[(✉)]

Faculty of Computer Science and Management,
Wrocław University of Technology,
Wybrzeże Wyspiańskiego 27, 50-370 Wrocław, Poland
kazimierz.choros@pwr.edu.pl

Abstract. The most frequently stored and browsed videos in Web video collections, TV shows archives, documentary video archives, video-on-demand systems, personal video archives, etc. are broadcast news videos and sports news videos. Content-based indexing of news videos is based on the automatic detection of shots, i.e. of the main structural video units. Video shots can be of different categories such as intro and final animation, chart, diagram or table shots, anchor, reporter, statement, or interview shots, and finally the most informative report shots. The content analysis of a video shot is a very time-consuming process using specific strategy adequate for a given shot category. To analyse faster the content of videos it is desirable to reduce the video space analysed in time-consuming content-based indexing by using temporal aggregation. The temporal aggregation results in grouping of shots of the same event or the same category into scenes. Furthermore, the determination on the basis of time relations of the most likely category also reduces the analysis time enabling us to apply the adequate method of analysis as the first. The paper examines the usefulness of the time relations of shots to determine the most likely category of a shot and to optimize the order of applied strategies.

Keywords: Content-based video indexing · Video structures · News videos · Temporal aggregation · News shot categorization · Category probability · Temporal relations

1 Introduction

Content-based video indexing is still a process not easy to perform. Many solutions have been proposed for specific cases and for specific kinds of videos (see for example [1]). These solutions are based on different mathematical approaches and on different algorithms, most often very sophisticated but unfortunately not sufficiently effective. These methods are people face detection, foreground and background detection, detection of superimposed texts like names, posts, functions, institutions, cities, countries, etc., of talking people or of a place, detection of sports objects or of lines typical for a given playing field in sports news videos, recognition of people/audience

© Springer-Verlag Berlin Heidelberg 2016
N.T. Nguyen et al. (Eds.): ACIIDS 2016, Part I, LNAI 9621, pp. 741–750, 2016.
DOI: 10.1007/978-3-662-49381-6_71

emotions, and also detection of objects like buildings, historical monuments, or statues as well as natural landscapes typical for specific regions, etc. In a news video the main semantic units are: the intro animation, stories presented and commented by an anchorman, sequences of report shots optionally enhanced by reporter, politician's statement, interview, or chart, diagram, table shots. Similarly to the beginning the news is finished with a final graphical animation shot.

Because the content analysis of a video is extremely time consuming the indexing is usually performed only for the key-frames, i.e. for only one or for a few frames selected for every shot or for every video scene. A key-frame is a frame used to represent a shot or a scene and it should be the most meaningful frame in the scene.

The automatic detection of video structure [2] mainly the detection of shots, i.e. of the structural units of the lowest level, by the detection of transitions between shots in a digital video is performed using different temporal segmentation methods. It is nowadays very well done and effectively applied in practice. Different video genres may be edited in various style. The specific video montage technique has an significant influence on the performance of temporal segmentation algorithms. The segmentation parameters should be adjusted to the characteristics of analysed videos.

Unfortunately, the detection of scenes, i.e. of the second level of video structure units, is not effective. A scene is defined as a group of consecutive shots sharing similar visual properties and having a semantic correlation – following the classical rule of unity of time, place, and action. The detection of scenes can be achieved by shot grouping or by applying the temporal aggregation method [3]. The temporal aggregation method applied to sports news detects scenes with a great probability taking into account only shot lengths. A player scene presents a sports game, i.e. a scene recorded on the sports fields such as playgrounds, car racing trucks, cycling trucks, tennis courts, sports halls, swimming polls, ski jumps, etc. But also non-player shots and scenes are usually recorded mainly in a TV studio such as commentaries, interviews, charts, tables, announcements of future games, opinions on the decisions of sports federations, etc. They are called studio shots or studio scenes. These studio shots are not useful for sports shot categorization and therefore can be omitted. Because the studio scenes are two thirds or more of sports news such a rejection of non-player scenes at the beginning of the content analyses significantly reduces computing time and makes these analyses more efficient.

The main goal of content-based indexing of news videos is to segment videos both spatially and temporally into smaller and easier manageable structural units like shots and scenes and then to examine the content of the detected structural units. The analyses of contents would be more effective if video shots were first categorized. Then the method adequate for a given shot category can be applied. However, the categorization of news shots is normally carried out by analysing the content of all frames of a shot, or of key-frame or key-frames extracted from a shot. It would be desirable to initially categorize news shots but without using content analyses.

The analysis of a shot content using a specific method adequate to a given news shot category requires the knowledge on a shot category. How to detect shot category before the analysis of its content? A possible solution is to estimate the probability of the category of a given shot on the basis of time relations of shots in news videos.

In this paper a method of the estimation of probabilities of shot categories in temporally aggregated news video is presented. The paper is structured as follows.

The next section describes related work on automatic news shot categorization and indexing. The principles of the temporal aggregation method are reminded and the detection of pseudo-scenes using temporal aggregation is described in the third section. The fourth section presents the results of the tests demonstrating the usefulness of temporal analyses of shots in news videos to estimate the probability of shot category and to optimize the order of applied strategies. The final conclusions are discussed in the last fifth section.

2 Related Work

The process of content-based indexing is usually based on the video structure. The temporal segmentation process detects shots in a video. These shots can be of different categories. In [4] the following shot categories have been examined in news videos: anchor shot, animation (intro), black frames, commercial, communication (static image of reporter), interview, map, report (stories), reporter, studio (discussion with a guest), synthetic (tables, charts, diagrams), and weather. We can find in the literature many algorithms proposed to detect anchorperson shots. There are methods based on template matching, also on different specific properties of anchor shots, or on temporal analyses of shots. The most frequent speaker in news is the anchorman [5]. An anchor speaks several times during the news, so the anchorperson shots are distributed all along the programme timeline. Although, the speaker is most likely the anchorman, the shots with a reporter interviewing people or with political statements also frequently found in news can be wrongly identified as anchor shots.

The automatic detection and classification of shots in news videos proposed in [6] used a probabilistic framework based on the Hidden Markov Models and the Bayesian Networks paradigms. The system has been tested on news videos of two different Italian TV channels.

In the experiments described in [7] ten news videos were firstly segmented into shots by a four-threshold method and then the key frames were selected from each shot. The anchor detection was carried out from the key frames by using a clustering-based method based on a statistical distance of Pearson's correlation coefficient. The method presented in [8] was used for news videos with dynamic studio background and multiple anchors. This method is based on spatio-temporal slice analysis. The experiments conducted on news programs of seven different styles confirmed the effectiveness of this approach. The analyses of audio, frame and face information have been applied in [9] to identify the news content. These three elements were independently processed. The speaking anchor appears most often in the same scene (TV studio), so that fact can be used to identify the role played by the people detected in the video. Promising results have been obtained in the experiments conducted with broadcast news from eight different TV channels. Another method of anchor shot detection proposed in [10] finds an anchorperson by using skin colour and face detectors. In [11] a fast method of automatic detection of anchorperson shots as well as of reporter, interview, or any other statement shots has been presented.

Much attention has been paid to interview shots and scenes in [12]. It has been observed that an interviewer and an interviewee recursively appear in many interview scenes. A technique called interview clustering method based on face similarity has been proposed to merge these interview units.

3 Temporal Segmentation and Aggregation of Shots

The process of automatic content-based analysis and indexing of videos is carried out in several stages. The first stage is usually a temporal segmentation resulting in the segmentation of a video into small structural units like shots or scenes being groups of shots. The analysis of scenes instead of shots is more effective because the number of elements analysed during the very time-consuming content-based indexing is significantly lower. The detection of video scenes optimizes the indexing process. Furthermore, the automatic indexing of news videos will be less time consuming if the analysed video is reduced only to scenes the most informative for content-based analyses like player scenes in TV sports news or official political statements or interviews in TV news. The temporal aggregation method implemented in the Automatic Video Indexer AVI [13] is applied to detect a video structure and then to segment and aggregate shots. The shots are grouped into pseudo-scenes basing only on the length of the shots as a sufficient sole criterion. The AVI Indexer is an indexing system designed and implemented for testing new methods and algorithms of automatic content-based indexing and video retrieval, but first of all for testing the effectiveness of temporal segmentation of videos and temporal aggregation of shots.

The temporal aggregation method requires the setting of three parameters: minimum shot length as well as lower and upper limits representing the length range for the most informative shots. These values of parameters are determined taking into account characteristics of videos mainly montage style as well as its high-level structure. The temporal aggregation has been used in a sequence of experiments with TV sports news and TV news. In the case of sports news the parameters have such values that all shots shorter than only 15 frames have been joined to the next shot, then all consecutive shots shorter than 40 frames and similarly all shots longer than 160 frames have been aggregated. So, shots of the length of 40 to 160 frames have remained unchanged as the most probably belonging to player scenes. These shots are of the length of 2 to 6 s ± 10 frames of tolerance [4]. Scene recall was almost optimal and attained 94 %, whereas shot precision was also acceptable – more than 71 %. But what is the most important it was possible to reject about half of video and despite this almost all sports scenes reported in sports news would be indexed.

Whereas, for news videos other values of the parameters have been experimentally determined. Only shots of the lengths from 2 to 12 s ± 5 frames of tolerance, that is of the lengths not lower than 45 frames and not greater than 305 frames have been not aggregated in news videos [14].

Short shots including single frames are very frequent in news videos. Shots of one or several frames are usually detected in the case of dissolve effects applied by video editors but very often they are simply wrong detections. The causes of wrong detections in the sports news videos may be different. Most often it is the case of very dynamic movements of players during the game, very dynamic movements of objects just in front of a camera, changes (lights, content) in advertising banners near the player fields, very dynamic movements of a camera during the game, light flashes during games or interviews. To eliminate these cases very short shots detected in temporal segmentation process are joined with other shots in a video. The temporal aggregation of shots by reducing false cuts also improves the results of previous process of temporal segmentation.

The temporal aggregation method has two main advantages. First of all, it selects the most meaningful parts of news videos. And then, it groups numerous shots into scenes and significantly diminishes the number of units analysed in content-based indexing and permits to reduce indexing process only to scenes of chosen categories. For example in sports news videos the length of all player scenes is significantly lower than the length of all studio shots, although, the number of player shots is much greater than the number of anchor shots.

4 Shot Category Probabilities in Temporally Aggregated News

As in our previous experiments six editions of the TV News "Teleexpress" broadcasted in the first national Polish TV channel (TVP1) have been used in these new tests. The "Teleexpress" News is broadcasted every day and is of 15 min. Its montage style is very dynamic. The news program is conducted at a great pace with very quick anchor announcements and comments. This TV news program is mainly addressed to the young audience. The montage style of the "Teleexpress" News is comparable to that of TV sports news. However, one difference is easily observed that the number of political, economic, or social events reported in news is much greater than the number of sports events reported in sports news.

The tested videos have standard structure of news video. Every video starts with the intro animation, then several events are reported and commented by anchorman, reporter, politicians, or even casual observers. A sequence of report shots can be optionally illustrated by charts, tables, diagrams, or maps. A news video is always finished by a final graphical animation of several seconds usually with text imposed on the image. It should be also noticed that the lengths of the intro and of final animation varies, so the temporal segmentation detects different sequences of shots at the beginning as well as at the end of different news videos. It results from the fact that sometimes fade effects from black to the intro part are used and similarly the final animation often fades away to a black frame. Moreover, the first frame of the intro and similarly the last frame of the final animation are frozen for some short time, so the lengths of the intro and of the final animation of a news video are not constant. Nevertheless, in a temporally aggregated news video these parts of a video, i.e. intro and final animation, as well as headlines can be easily detected [15]. All shots before the first long anchor shot and similarly all shots after the last anchor shot have been eliminated from further analyses because their categories are already known. The main problem discussed in the paper is how to determine the most probable categories of the internal part with other shots of news. What kind of categories can be expected in this internal part? These are mainly such shots as: report shot, anchor shots, statement (interview) shots, or chart shots (Table 1).

Table 1. Characteristics of internal parts of six temporally aggregated news videos broadcasted in March 2014 (03-03, 03-05, 03-06, 03-08, 03-09, and 03-11).

Video	V1	V2	V3	V4	V5	V6	Average
Total number of shots	170	170	166	153	190	161	169
Number of anchor shots	11	14	13	13	13	12	13
The shortest anchor shot [in frames]	157	53	60	48	51	67	73
The longest anchor shot [in frames]	918	1122	768	913	828	1217	961
Average length of anchor shots [in frames]	455	440	375	495	359	437	427
Number of chart shots	0	0	0	0	2	1	0.5
The shortest chart shot [in frames]	0	0	0	0	121	225	173
The longest chart shot [in frames]	0	0	0	0	422	225	324
Average length of chart shots [in frames]	0	0	0	0	272	225	249
Number of statement shots	7	8	11	5	7	13	9
The shortest statement shot [in frames]	144	130	103	112	82	89	110
The longest statement shot [in frames]	342	409	327	261	273	368	330
Average length of statement shots [in frames]	220	225	236	203	163	230	213
Number of report shots	152	148	142	135	168	135	147
The shortest report shot [in frames]	45	45	45	45	45	45	45
The longest report shot [in frames]	347	248	432	354	281	379	340
Average length of report shots [in frames]	99	89	98	100	86	94	94

Tables 2–4 presents the statistics for the tested videos of the lengths of aggregated shots initially classified to anchor, chart, statement (interview), or report shot categories. These tables show the numbers of all shots of a given category (All) in the tested videos and the numbers of shots of a given category selected on the basis of shot lengths (Sel). Also the probabilities have been estimated that all shots of a given category have the lengths of a given range.

The first results (Table 2) are obtained for the following thresholds: 130, 210, and 290, that is it has been assumed that report shots are not longer than 130 frames, the lengths of statement shots are from 130 to 209, the lengths of chart shots are from 210 to 289, and anchor shots are longer than 290 frames.

Table 2. Selection of shots of a given category with following thresholds [in frames]: lengths of anchor shots $\geq$ 290, 210 $\leq$ lengths of chart shots $\leq$ 289, 130 $\leq$ lengths of chart shots $\leq$ 209, lengths of report shots < 130.

Video	V1		V2		V3		V4		V5		V6		Average	
	All	Sel	All	Sel	All	Sel	All	Sel	All	Sel	All	Sel	All	Sel
Anchor shots	11	10	14	11	13	11	13	11	13	10	12	10	12.7	10.5
Chart shots	0	0	0	0	0	0	0	0	2	0	1	1	1.5	0.5
Statement shots	7	4	8	5	11	3	5	1	7	4	13	4	8.5	3.5
Report shots	152	121	148	129	142	120	135	108	168	157	135	108	146.7	124.0

Because chart shots are rare the second test (Table 3) skips this category and uses two thresholds: 125 and 300 (5 and 12 s). Whereas, the Table 4 presents the results for the test with the same range for chart as well as for statement shots. This option is based on the suggestion that may be these two categories should be analysed in the same way (with the same ranges of lengths).

The estimated probabilities that all shots of a given category have the length of a given range are as follows: anchor shots – 0.83, chart shots – 0.33, statement shots – 0.41, and report shots – 0.85. Average is equal to 0.6049.

Table 3. Selection of shots of a given category with following thresholds [in frames]: lengths of anchor shots ≥ 300, $0 \leq$ lengths of chart shots ≤ 0, $126 \leq$ lengths of chart shots ≤ 299, lengths of report shots < 126.

Video	V1		V2		V3		V4		V5		V6		Average	
	All	Sel	All	Sel	All	Sel	All	Sel	All	Sel	All	Sel	All	Sel
Anchor shots	11	10	14	9	13	11	13	11	13	10	12	10	12.7	10.2
Chart shots	0	0	0	0	0	0	0	0	2	0	1	0	1.5	0.0
Statement shots	7	6	8	6	11	8	5	4	7	5	13	10	8.5	6.5
Report shots	152	118	148	128	142	118	135	106	168	153	135	108	146.7	121.8

The estimated probabilities that all shots of a given category have the length of a given range are as follows: anchor shots – 0.80, chart shots – 0.00, statement shots – 0.76, and report shots – 0.83. Average is equal to 0.5995. Average without taking into account chart shots is equal to 0.7993.

Table 4. Selection of shots of a given category with following thresholds [in frames]: lengths of anchor shots ≥ 300, $126 \leq$ lengths of chart shots ≤ 299, $126 \leq$ lengths of chart shots ≤ 299, lengths of report shots ≤ 125.

Video	V1		V2		V3		V4		V5		V6		Average	
	All	Sel	All	Sel	All	Sel	All	Sel	All	Sel	All	Sel	All	Sel
Anchor shots	11	10	14	9	13	11	13	11	13	10	12	10	12.7	10.2
Chart shots	0	0	0	0	0	0	0	0	2	0	1	1	1.5	0.5
Statement shots	7	6	8	6	11	8	5	4	7	5	13	10	8.5	6.5
Report shots	152	118	148	128	142	118	135	106	168	153	135	108	146.7	121.8

The estimated probabilities that all shots of a given category have the length of a given range are as follows: anchor shots – 0.80, chart shots – 0.33, statement shots – 0.76, and report shots – 0.83. Average is equal to 0.6828.

Now, let's estimate a probability that a shot of a given length is a shot of predicted, the most probable category (Table 5).

The statistical estimations of the probabilities that all shots of a given category have the length of a given range are as follows: anchor shots – 0.80, chart shots – 0.01, statement shots – 0.21, report shots – 0.97.

Table 5. Selection of shots of a given category with following thresholds [in frames]: lengths of anchor shots > 300, 126 < lengths of chart shots < 299, 126 < lengths of chart shots < 299, lengths of report shots < 125.

Video	V1		V2		V3		V4		V5		V6		Average	
	All	Sel	All	Sel	All	Sel	All	Sel	All	Sel	All	Sel	All	Sel
Anchor shots	12	10	11	9	16	11	13	11	11	10	13	10	12.7	10.2
Chart shots	40	0	28	0	29	0	32	0	20	0	37	1	93.0	0.5
Statement shots	40	6	28	6	29	8	32	4	20	5	37	10	31.0	6.5
Report shots	118	118	131	128	121	118	108	106	159	153	111	108	125.5	121.8

The tests have shown that anchor shots as well as report shots can be selected with very high probability. But the problem remains how to categorize shots of the medium length. Chart and statement shots are the most critical. Table 6 presents the probability of a given category if the length of a shot is neither adequate for anchor shots nor for report shots.

Table 6. Number of shots of the four particular categories and of the medium length from 126 to 299 frames, i.e. from 5 to 12 s.

	V1		V2		V3		V4		V5		V6		Average	
	All	Sel	All	Sel	All	Sel	All	Sel	All	Sel	All	Sel	All	Sel
Anchor shots	40	1	28	2	29	0	32	1	20	0	37	0	31,0	0,7
Chart shots	40	0	28	0	29	0	32	0	20	0	37	1	93,0	0,5
Statement shots	40	6	28	6	29	8	32	4	20	5	37	10	31,0	6,5
Report shots	40	33	28	20	29	21	32	27	20	15	37	26	31,0	23,7

The statistical estimations of the probabilities that shots of a given category have the length in the critical medium range are as follows: anchor shots – 0.02, chart shots – 0.01, statement shots – 0.21, report shots – 0.76.

The results of analyses of shots of the length from the medium range (Table 6) show that we can expect most probably report shots or less probably statement shots. Other categories are unlikely.

These estimations clearly suggest such a strategy that also for the shots with the lengths in this medium range the procedure of content-based analysis adequate for report shots should be first applied, next that for statements shots, and then that adequate for anchor shots. The special procedure for the detection of charts shots can be applied at the end (if at all).

It is interesting to compare the categorization of shots basing on these probabilities with the case of random order of categorization strategies. Random order means that the content analysis starts assuming that the analyzed shot is of a category randomly chosen but always the same for all shots. This comparison is presented in Table 7.

Table 7. Comparison of the probability of the best choice of the strategy for content-based video analysis based on time analyses with the probability in the case of random choice. The total number of shots in tested videos is 1010.

Shot category	Anchor shots	Chart shots	Statement shots	Report shots
Number of shots classified to a given shot category based on time analyses	76	186	186	748
Real number of shots of a given category	76	3	51	880
Number of shots correctly categorized	61	1	39	729
Probability of the good choice of a strategy based on time analyses	0.8026	0.0054	0.2097	0.9746
Probability that a random shot is of a given shot category	0.0752	0.0030	0.0505	0.8713

These analyses lead to the very practical conclusion that the order of strategies for content-based analyses can be optimized, i.e. for the long shots (more than 12 s) the anchor category is the most probable, whereas, for the short shots (less than 5 s) the report category is the most probable. Furthermore, for the shots of the length of a medium range (from 5 to 12 s) two categories are more probable than others, these are report shots and statement shots.

5 Conclusions

Most automatic methods of content-based video indexing are based on video structure. Also the methods proposed for news videos require the prior segmentation of news video into shots and if it is possible as well into scenes. The content analysis of a video shot is a very time-consuming process using specific strategy adequate for a given shot category. To analyse faster the content of videos it is desirable to reduce the number of video shots analysed in time-consuming content-based indexing by using temporal aggregation. The temporal aggregation mainly results in grouping of shots of the same category into scenes. Furthermore, the determination of the most likely category on the basis of time relations also reduces the analysis time enabling us to apply the adequate method of content-based indexing.

The temporal aggregation method is very efficient in detecting headlines, first welcome anchor shot, but also can be applied to optimize the decision which content-based indexing strategy should be used as the first.

The tests have shown that the analyses of time relations of shots can easily determine the most likely category of a shot in the news video and then to optimize the order of strategies applied for a given shot. If a shot is most likely anchor shot the detection of studio background or anchor face should be applied as the first. If the estimated probability suggests another category for an analysed shot the adequate strategy for such category should be launched.

References

1. Hu, W., Xie, N., Li, L., Zeng, X., Maybank, S.: A survey on visual content-based video indexing and retrieval. IEEE Trans. Syst. Man Cybern. Part C: Appl. Rev. **41**(6), 797–819 (2011)
2. Choroś, K.: Video structure analysis for content-based indexing and categorisation of TV sports news. Int. J. Intell. Inf. Database Syst. **6**(5), 451–465 (2012)
3. Choroś, K.: Temporal aggregation of video shots in TV sports news for detection and categorization of player scenes. In: Bǎdicǎ, C., Nguyen, N.T., Brezovan, M. (eds.) ICCCI 2013. LNCS, vol. 8083, pp. 487–497. Springer, Heidelberg (2013)
4. Valdés, V., Martínez, J.M.: On-line video abstract generation of multimedia news. Multimedia Tools Appl. **59**(3), 795–832 (2012)
5. Montagnuolo, M., Messina, A., Borgotallo, R.: Automatic segmentation, aggregation and indexing of multimodal news information from television and the Internet. Int. J. Inf. Stud. **1**(3), 200–211 (2010)
6. Colace, F., Foggia, P., Percannella, G.: A probabilistic framework for TV-news stories detection and classification. In: Proceedings of the IEEE International Conference on Multimedia and Expo ICME 2005, pp. 1350–1353. IEEE (2005)
7. Ji, P., Cao, L., Zhang, X., Zhang, L., Wu, W.: News videos anchor person detection by shot clustering. Neurocomputing **123**, 86–99 (2014)
8. Zheng, F., Li, S., Wu, H., Feng, J.: Anchor shot detection with diverse style backgrounds based on spatial-temporal slice analysis. In: Boll, S., Tian, Q., Zhang, L., Zhang, Z., Chen, Y.-P.P. (eds.) MMM 2010. LNCS, vol. 5916, pp. 676–682. Springer, Heidelberg (2010)
9. Broilo, M., Basso, A., De Natale, F.G.: Unsupervised anchorpersons differentiation in news video. In: Proceedings of the 9th International Workshop on Content-Based Multimedia Indexing (CBMI), pp. 115–120. IEEE (2011)
10. Lee, H., Yu, J., Im, Y., Gil, J.M., Park, D.: A unified scheme of shot boundary detection and anchor shot detection in news video story parsing. Multimedia Tools Appl. **51**(3), 1127–1145 (2011)
11. Choroś, K.: Automatic fast detection of anchorperson shots in temporally aggregated TV news videos. In: Nguyen, N.T., Trawiński, B., Kosala, R. (eds.) ACIIDS 2015. LNCS, vol. 9012, pp. 339–348. Springer, Heidelberg (2015)
12. Dong, Y., Qin, G., Xiao, G., Lian, S., Chang, X.: Advanced news video parsing via visual characteristics of anchorperson scenes. Telecommun. Syst. **54**(3), 247–263 (2013)
13. Choroś, K.: Video structure analysis and content-based indexing in the automatic video indexer AVI. In: Nguyen, N.T., Zgrzywa, A., Czyżewski, A. (eds.) Advances in Multimedia and Network Information System Technologies. AISC, vol. 80, pp. 79–90. Springer, Heidelberg (2010)
14. Choroś, K.: Automatic categorization of shots in news videos based on the temporal relations. In: Núñez, M., Nguyen, N.T., Camacho, D., Trawiński, B. (eds.) Computational Collective Intelligence. LNCS, vol. 9330, pp. 13–23. Springer, Heidelberg (2015)
15. Choroś, K.: Automatic detection of headlines in temporally aggregated TV sports news videos. In: Proceedings of the 8th International Symposium on Image and Signal Processing and Analysis (ISPA 2013), pp. 147–152. IEEE (2013)

A Method of Data Registration for 3D Point Clouds Combining with Motion Capture Technologies

Shicheng Zhang, Dongsheng Zhou[(✉)], and Qiang Zhang[(✉)]

Key Laboratory of Advanced Design and Intelligent Computing, Dalian
University, Ministry of Education, Dalian, China
{donyson, zhangq26}@126.com

Abstract. Data Registration is one of the key techniques in 3D scanning. The traditional methods for data registration have some disadvantages which always need many calibration markers or other accessories. Those will greatly reduce the convenience and usability for the scanning systems, and more markers will covered the limited useful surface of the measured object. This paper proposed a new method to overcome these shortcomings. In the method, the 3D scanner and the motion capture device, which have completely different elements, are effectively combined as a whole system. The position and posture of the measured object can be optionally changed as wish. Mocap system guides the spatial localization for the measured object which has a high flexibility and precision. Dynamic motion data and the static scan data can be obtained in real-time by using the Mocap system and 3D scanner, respectively. In the final, heterogeneous spatial data will be converted to a same 3D space, and the parts of point clouds will be spliced to a whole 3D model. The experiments show that the method is valid.

Keywords: 3D scan · Point clouds registration · Motion capture · Space transformation · System integration

1 Introduction

As the industry competition is more and more fierce and the consumer demand is more and more novel, the rapid and flexible market becomes an inevitable result. Reverse engineering [1], which can be applied to industrial production, as an advanced manufacturing technology, has become a significant subject in product design, development and innovation. Reverse engineering mainly includes data acquisition [2], data processing [3], surface reconstruction [4] and so on. Data processing [5] mainly covers data registration and other techniques [6]. Because the quality and the number of 3D point data have prodigious effect on the quality of latter data reconstruction [7], it is very essential to study on data registration. Due to the limitation of the current technologies, we cannot obtain the whole 3D point clouds data of the measured object in once time. A measured object can only be represented by a group of point clouds obtained by the canner. The several point-sets should be integrated to one uniform data.

© Springer-Verlag Berlin Heidelberg 2016
N.T. Nguyen et al. (Eds.): ACIIDS 2016, Part I, LNAI 9621, pp. 751–759, 2016.
DOI: 10.1007/978-3-662-49381-6_72

2 Related Works

In recent years, many scholars have done a lot of researches on the technology of point clouds registration. To registration, at present, general ways cover two steps: coarse registration and accurate registration. The coarse registration can transform two point sets in different coordinate systems to one point-set in one coordinate system. There are two main ways for coarse registration. One is the method based on aided device. Zhou et al. used the method based on rotating platform [8] to realize the data registration. The method calibrated the center axis of the rotating platform by point clouds in multi-view, so that they could obtain the relative position relationship between the scanning device and the rotating platform. The data for point clouds registration [9] could be calculated by above-mentioned relationship. Although the method has a high degree of automation, it can only scan the objects with small size. The accuracy of measurement equipment is required very high, which is not easy to ensure the positioning accuracy. Yuan et al. realized the highly precise registration of structure light scanning with the method based marked points [10], which effectively controlled the accumulation error caused by the increase of the registration times. This method needs the marked points stuck on the surface of the measured object, which results in the cover to the surface and the data loss of the surface. The other is the method based on surface features. This method requires the point clouds have obvious features, which were used to find the corresponding relationship between different point clouds. Based on the singular value decomposition (SVD) [11] method or quaternion method [12], they can calculate the coordinate transformation information between point clouds. In the literature [13], the author searched the corresponding point by the curvature of the observation point, which are calculated by the surface normal vector of itself and the points in its neighborhood. In order to overcome the multiple corresponding results caused by the existence of the points with similar features and calculation error, Xu et al. [14] used the points' mean curvature and gauss curvature as ties to set up the corresponding relation and avoid wrong match. In addition, Wang et al. [15] transformed the surface registration problem into the problem of searching the maximum weight clique, which represents the optimal corresponding relation of points. This method regards the uniform sampling points and Gaussian curvature points as candidate points. Its matching accuracy and efficiency need to be improved. In view of the shortcomings of mentioned methods and the characteristics of the current point clouds data, this paper aimed to propose a new method which can avoid the high cost and the loss of point clouds.

2.1 Rotating Platform Method

The object rotates on the Rotating Platform, which is equivalent to the coordinate transformation of the points [16]. If the rotating process of the object has been known, the rotation and translation matrix of the point cloud coordinate transformation can be obtained according to the changing process. The registration process of point clouds will be the reverse of the scanning process. It could be assumed that the coordinate of any points on the surface of the object are fixed on the rotating table, which could be described as $p'(x, y, z)$. The coordinate of the same point when the object has rotated by $-\omega$ degrees is

$$p_i' = Rp'$$ (1)

Where $p'(x, y, z, 1)^T$ is the coordinate of p' before the rotation and $p_i'(x, y, z, 1)^T$ is the coordinate of p' after the rotation; $i = 1, 2, 3. \ldots$ are the numbers of different field of views; R is the Rotation matrix. The formula (1) shows the coordinate conversion relationship between the point clouds before and after the rotation. To finish the point clouds registration, the inverse of the formula (1) could be used:

$$p' = R^{-1}p_i'$$ (2)

The problem of the transformation of the coordinate system is also related to the process of the registration, the Quaternion method could be used to solve the problem [17].

2.2 Method Based on Markers

Using markers is the traditional registration method. By identifying the marked points in the public area of different views [18], the transformation matrix between the two point sets could be calculated. Then, the connection of the multi-view data will be integrated.

3 Method of This Paper

3.1 Method Description

In this method, the 3D scanner and the motion capture device, which have completely different elements, are effectively combined as a whole system. The details are as follows:

Step 1: Calibrate the 3D scanner and the motion capture device. The 3D scanner and the motion capture device are kept on the unchanged position. There are marked points pasted on the bracket, while there are no markers on the measured objects.

Step 2: Acquire the dynamic motion data and the static scan data by using the Mocap system and 3D scanner, respectively.

Step 3: Calculate the rotation matrix and translation matrix.

Step 4: Transform motion capture data and 3D scanning data into a same 3D calibration system.

Step 5: Test and correct the registration of the point clouds.

3.2 Coordinate of Sphere Center

In the proposed methods, just three markers are needed which diameter is about 2 cm. The coordinate of the marker's center is necessary. We use the method of error equation to calculate the coordinate. If the radius of the sphere is R and the coordinate of a point on the surface is (x_i, y_i, z_i). As is well known, the equation is:

$$(x_0 - x_i)^2 + (y_0 - y_i)^2 + (z_0 - z_i)^2 = R^2 \tag{3}$$

Then, we got

$$\begin{aligned}
[aa]\delta_x + [ab]\delta_y + [ac]\delta_z + [al] &= 0 \\
[ab]\delta_x + [bb]\delta_y + [bc]\delta_z + [bl] &= 0 \\
[ac]\delta_x + [bc]\delta_y + [cc]\delta_z + [cl] &= 0
\end{aligned} \tag{4}$$

The value of $\delta_x, \delta_y, \delta_z$ is:

$$\begin{cases}
\delta_x = -[al]Q_{11} - [bl]Q_{12} - [cl]Q_{13} \\
\delta_y = -[al]Q_{12} - [bl]Q_{22} - [cl]Q_{23} \\
\delta_z = -[al]Q_{13} - [bl]Q_{23} - [cl]Q_{33}
\end{cases} \tag{5}$$

In the final, the center is

$$\begin{cases}
x = x_0 + \delta_x \\
y = y_0 + \delta_y \\
z = z_0 + \delta_z
\end{cases} \tag{6}$$

3.3   Rotation and Translation Matrix

The coordinates of the markers' center calculated in our experiments belong to different coordinate systems. If points $\{P_i, i = 1, \ldots, n\}$ in 3D Euclidean space are on the surface of a same rigid body, the shifting of them could be described by the rigid body motion. So, we can get the transformation equation:

$$P'_i = RP_i + t + \eta_i \tag{7}$$

In which, P'_i are the transferred corresponding points, R is the rotation matrix, t is the translation matrix and η_i is the additional noise which obeys the gauss distribution. So, the values of R and t could be figured out by calculate the minimum value of the following functions:

$$I(R, t) = \sum_{i=1}^{n} \left\| P'_i - (RP_i + t) \right\|^2 \tag{8}$$

Because the matrix R is an orthogonal matrix, it could be got:

$$\begin{cases} \displaystyle\sum_{i=1}^{n}(r_1^T P_i)P_i + \lambda_{11}r_1 + \lambda_{12}r_2 + \lambda_{13}r_3 = \sum_{i=1}^{n}P'_{i,1}P \\[2mm] \displaystyle\sum_{i=1}^{n}(r_2^T P_i)P_i + \lambda_{22}r_2 + \lambda_{23}r_3 + \lambda_{13}r_3 = \sum_{i=1}^{n}P'_{i,2}P \\[2mm] \displaystyle\sum_{i=1}^{n}(r_3^T P_i)P_i + \lambda_{33}r_3 + \lambda_{13}r_1 + \lambda_{23}r_2 = \sum_{i=1}^{n}P'_{i,3}P \end{cases} \qquad (9)$$

Notes that:

$$A = \sum_{i=1}^{n}P_i P_i^T, \qquad \Lambda = \begin{pmatrix} \lambda_{11} & \lambda_{12} & \lambda_{13} \\ \lambda_{21} & \lambda_{22} & \lambda_{23} \\ \lambda_{31} & \lambda_{32} & \lambda_{33} \end{pmatrix}, \qquad B = (b_1 \ b_2 \ b_3)$$

Where $b_k = \sum_{i=1}^{n}P'_{i,k}P_i$ and $\lambda_{ij} = \lambda_{ji}$, Λ is a symmetrical matrix. In the final, the rotation matrix R could be calculated by using the singular value decomposition of matrix, that is:

$$R = VU^T \qquad (10)$$

4   Experiments and Analysis

Experiments were performed to verify the effectiveness of the method. Konica Minolta vivid 9i 3D scanner and DIMS-9100 Mocap systems were used for the experiment, as shown in Fig. 1. The computing platform is based on MATLAB R2013a and an ordinary PC which processor is Intel(R) core(TM) i3-4160CPU@3.60 GHz, memory is 4G and operating system is Windows 7.

Fig. 1. The illustration of experiment system combined with 3D scanner and Mocap system

In experiments, three pieces of data were acquired, in which contained 19772, 19651 and 15896 points, respectively, as shown in Fig. 2. It can be easily perceived that the three point-sets with different colors represent the three different parts of the measured object. We combine the three point-sets to one complete data by using the calculated rotation and translation matrix. The complete data is presented in Fig. 3 which three pieces of point clouds have made up a whole. The fact means that our result was of high quality which we can hardly find a gap from the 3-D image in Fig. 3.

As the MATLAB software cannot fully display the quality of our work, so we adopted the widely used reverse software Geomagic Qualify. We got the CAD model (Fig. 4) of the original scanning data processed by the Qualify software as the reference. The comparison results were shown in Fig. 5 and Table 1.

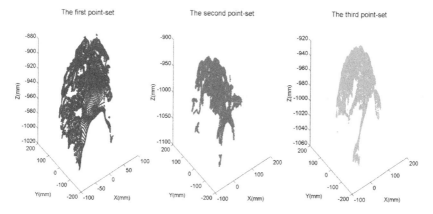

Fig. 2. The three initial point clouds

Fig. 3. The point clouds after registration

Figure 5 presents the comparison chromatogram map processed by Geomagic Qualify. According to the figure, most areas of the two models are nearly consistent with a green or yellow color. Table 1 is a random sample which shows the specific deviation of the point data. In this table, the reference columns are the 3D value of each

Fig. 4. The CAD model of the original data processed by the Qualify software

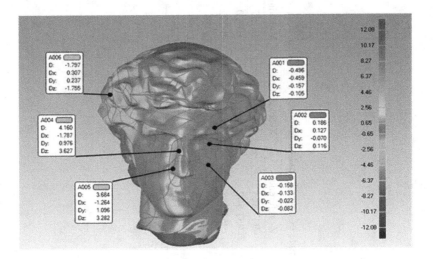

Fig. 5. Comparison chromatogram map

reference points. The deviation columns are the value of the deviations between the test and the reference points (The unit is mm). The average deviation of the results is 1.30 mm.

5 Conclusions

In this paper, a new method for point clouds registration combining motion capture technologies was presented. The method did not need highly precision rotating tables and either need paste a large number of markers on the surface of the measured object. The method is simple, time-saving and easy to be realized, and efficiently reduced the unnecessary loss of point clouds. The experiments show that the method has an acceptable precision of registration.

In order to achieve a better quality of registration, the future work will aim to improve the accuracy of the algorithm.

Table 1. The sample of several random points

Name	Deviation	X- reference	Y- reference	Z- reference	X-deviation	Y-deviation	Z-deviation
A001	-0.496	70.592	7.826	-947.865	-0.459	-0.157	-0.105
A002	0.186	61.083	-10.170	-951.502	0.127	-0.070	0.116
A003	-0.158	59.762	-33.322	-951.988	-0.133	-0.022	-0.082
A004	4.160	48.992	-17.424	-920.916	-1.787	0.976	3.627
A005	3.684	44.191	-36.469	-917.916	-1.264	1.096	3.282
A006	-1.797	-13.534	42.970	-886.719	0.307	0.237	-1.755

Acknowledgement. This work is supported by the National Natural Science Foundation of China (No. 61370141, 61300015), the Program for Liaoning Innovative Research Team in University (Nos. LT2015002), the Program for Science and Technology Research in New Jinzhou District (No. KJCX-ZTPY-2014-0012).

References

1. Sun, J.Z., Hu, Y., Ma, Y.Q.: An algorithm generate voronoi diagram based on delaunay triangulation. J. Comput. Appl. **1**(01), 75–77 (2010)
2. Lovato, C., Castellani, U., Zancanaro, C.: Automatic labelling of anatomical landmarks on 3D body scans. Graph. Models **76**(6), 648–657 (2014)
3. Han, F., Duan, X.F.: Research on the point cloud data processing method of 3D terrestrial laser scanning in existing railway track. Appl. Mech. Mater. **744**, 1298–1302 (2015)
4. Liu, M.M., Richard, H., Salzmann, M.: Mirror surface reconstruction from a single image. In: 2015 IEEE Transactions on Pattern Analysis and Machine Intelligence, pp. 760–773. IEEE Press (2015)
5. Zuo, C., Lu, M., Tan, Z.G.: A novel algorithm for registration of point clouds. Chin. J. Lasers **39**(12), 211–218 (2012)
6. Wu, S., Sun, W., Long, P., et al.: Quality-driven poisson-guided autoscanning. ACM Trans. Graph., **33**(6) (2014). Article No 203
7. Amenta, N., Choi, S., Dey, T.: A simple algorithm for homeomorphic surface reconstruction. Int. J. Comput. Geomet. Appl. **12**, 125–141 (2002)
8. Zhou, L.M., Zheng, S.Y., Huang, R.Y.: A registration algorithm for point clouds obtained by scanning objects on turntable. Acta Geodaetica Cartogr. Sin. **42**, 73–79 (2013)
9. Gressin, A., Mallet, C., Demantké, J.: Towards 3D lidar point cloud registration improvement using optimal neighborhood knowledge. ISPRS J. Photogrammetry and Remote Sens. **79**, 240–251 (2013)
10. Yuan, J.Y., Wang, Q., Li, Bailin.: Using multi-view network constraints among reference marker points to realize coarse registration in structured light system with high accuracy. J. Comput. Aided Des. Comput. Graph. 4,014 (2015)
11. Chen, Y., Medioni, G.: Object modeling by registration of multiple range images. In: 1991 IEEE International Conference on Robotics and Automation, pp. 2724–2729. IEEE Press (1991)
12. Dai, J.S.: Euler-rodrigues formula variations, quaternion conjugation and intrinsic connections. Mech. Mach. Theory **92**, 144–152 (2015)
13. Zhu, Y.J., Zhou, L., Zhang, L.Y.: Registration of scattered cloud data. J. Comput. Aided Des. Comput. Graph. **18**, 475–481 (2006)
14. Xu, J.T., Sun, Y.W., Liu, W.J.: Optimal localization of free-form shaped parts in precision inspection. Chin. J. Mech. Eng. **43**, 175–179 (2007)
15. Wang, J., Zhou, L.S.: Surface rough matching algorithm based on maximum weight clique. J. Comput. Aided Des. Comput. Graph. **20**(2), 167–173 (2008)
16. Long, X., Zhong, Y.X., Li, R.J.: 3-D surface integration in structured light 3-D scanning. J. Tsinghua Univ. (Sci. Technol.) **42**, 477–480 (2002)
17. Kim, J.Y., Kim, L.S., Hwang, S.H.: An advanced contrast enhancement using partially overlapped sub-block histogram equalization. Circuits Syst. Video Technol. **11**, 475–484 (2001)
18. Cheng, X.J., Wang, F.: Method of calculating sphere surface parameters by measuring several point coordinate on a sphere. Railway Invest. Surveying **32**, 1–2 (2007)

Detection and Recognition of Speed Limit Sign from Video

Lei Zhu[1]([⊠]), Chun-Sheng Yang[1], and Jeng-Shyang Pan[1,2]

[1] Shenzhen Graduate School, Innovative Information Industry Research Center, Harbin Institute of Technology, Shenzhen 518055, China
574371464@qq.com
[2] College of Information Science and Engineering, Fujian University of Technology, Fuzhou 350118, China

Abstract. The proper identification of speed limit traffic sighs can alarm the drivers the highest speed allowed and can effectively reduce the number of traffic accidents. In this paper, we put forward an efficient detection method for speed limit traffic signs based on the fast radial symmetry transform with new Sobel operator. when we detected the speed limit traffic sign, we need to segment the digits. Digit segmentation is achieved by cropping the candidate traffic sign from the traffic scene, making use of Otsu thresholding algorithm to binary it, and normalizing it to a uniform size. Finally we recognize and classify the signs using DAG-SVMs classifier which is trained for this purpose. In cloudy weather conditions and dusk illumination condition, we tested 10 videos about 28 min. The recognition rate of frames which contain speed limit sign is 90.48 %.

Keywords: Speed limit traffic sign · Fast radial symmetry transform · Otsu · DAG-SVMs

1 Introduction

Improving safety is an important goal in road vehicle technology. People set speed limit traffic signs to alert drivers pay attention to safety. Traffic signs usually possesses particular colors and shapes aiming to attract the drivers' attention against the natural environment. We ofen see the traffic signs whose colors are red and yellow, with white or black symbol in chinese road. The main shapes include rectangled, circles, and triangles. In this paper, we focus on the detection and classification of circular traffic signs, which posses a red circle with white background as shown in Fig. 1.

Recognition of traffic signs may involve several difficulties [1]: first of all, lighting conditions are uncontrollable. The scene could happen in all kinds of conditions and also different time. Second, some cars or shop signs can block the line of sight of a traffic sign. Besides, the long-term exposure to the sunlight also causes the color gradually fades. At the same time, the air and the paint

© Springer-Verlag Berlin Heidelberg 2016
N.T. Nguyen et al. (Eds.): ACIIDS 2016, Part I, LNAI 9621, pp. 760–769, 2016.
DOI: 10.1007/978-3-662-49381-6_73

Fig. 1. Speed-limit traffic sign (Color figure online)

produce chemical reaction in wet environment and causes the same result as above. As mentioned above, it is a huge challenge to recognize the traffic sign from complex environment.

2 Related Work

Traffic signs mainly possess two visual features, namely color and shape. People usually choose RGB color space because it is the most intuitive color space. Escalera et al. [2] segment the image by making use of the relations between different channels.

People choose Hough transform in [3,4]. Barnes and Zelinsky [5] propose the fast radial symmetry algorithm, which based on gradient. In [6], an efficient algorithm is proposed to detect regular polygons.

Rubel, Hasan and Moin [7] combine many methods which consists of three steps: Colour-based filtering, Circular Hough Transform and LOWESS regression technique. The accuracy of recognition was 98 %.

Torresen et al. [8] presented an algorithm which includes Colour-based filtering to detect speed limit signs in Norwegian. The experiment was tested on 198 images and the accuracy of recognition was 91 %.

Wu and Tsai [9] proposed an approach for detecting speed limit signs from video image. The experiment was tested on 123 images and the accuracy of recognition was 97 %.

Auranuch and Jackrit [10] proposed an approach which made use of Neural Network techniques to recognize traffic sign. Hoferlin and Zimmermann [11] proposed another approach which based on the application of Scale-Invariant Feature Transform (SIFT) local features.

3 System Overview

The detection system mainly includes two parts: detecting candidate sign region, and candidate of digit region segmentation. We decide to make use of the shape feature to avoid the impact of lighting conditions or weather on the color feature. The flow diagram of whole system is shown as Fig. 2.

3.1 Fast Radial Symmetry Transform

The fast radial symmetry is an accurate and efficient method which based on gradient to detect points of high symmetry. In this paper, we take an improved Sobel

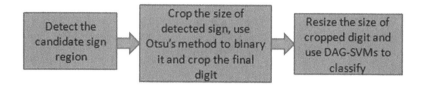

Fig. 2. The detection and recognition system

1	1	1
0	0	0
-1	-1	-1

(a)

-1	0	1
-1	0	1
-1	0	1

(b)

2	1	0
1	0	-1
0	-1	-2

(c)

0	1	2
-1	0	1
-2	-1	0

(d)

Fig. 3. The four Sobel templates. (a) The vertical template. (b) The horizontal template. (c) The 45° template. (d) The 135° template.

operator. To compensate for traditional Sobel's defect: to reduce the impact of noise, to describe the edge points more accurately, four directions of template T_x,T_y,T_{45},T_{135} are used as shown in Fig. 3.

For example, traditional Sobel operator is the partial derivative of $f(x,y)$ as the central computing 3×3 neighbourhood at y direction as Fig. 4.

f_{11}	f_{12}	f_{13}
0	0	0
$-f_{31}$	$-f_{32}$	$-f_{33}$

Fig. 4. The vertical template of Sobel operator

We can get the result of G_y:

$$G_y = |f_{11} + f_{12} + f_{13} - (f_{31} + f_{32} + f_{33})| \tag{1}$$

Let A, B replace part of formular:

$$A = f_{11} + f_{12} + f_{13} \tag{2}$$

$$B = f_{31} + f_{32} + f_{33} \tag{3}$$

Then we can denote G_y as:

$$G_y = |A - B| = \frac{|A - B|}{255} \times 255 \tag{4}$$

In this paper,we redefine G_y as:

$$G'_y = \frac{|A - B|}{A + B} \times 255 \qquad (5)$$

Here, the intensity gradient is the biggest output of gradient value of template.

At each radius n, an orientation projection image O_n are formed as shown in Fig. 5.

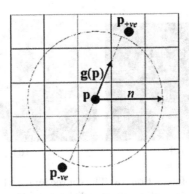

Fig. 5. The locations of pixels $p_{+ve}(p)$ and $p_{-ve}(p)$ affected by the gradient element $g(p)$ for a range of $n = 2$.

The coordinates of the positively-affected pixel are given by

$$p_{+ve}(p) = p + round(\frac{g(p)}{\|g(p)\|}n) \qquad (6)$$

while those of the negatively-affected pixel are

$$p_{-ve}(p) = p - round(\frac{g(p)}{\|g(p)\|}n) \qquad (7)$$

where "round" rounds each vector element to the nearest integer and $n \subseteq N$ is the radius (Fig. 6).

In the orientation projection image O_n, the point p_{+ve} is incremented by 1, while the point p_{-ve} is decremented by 1. That is,

$$O_n(p_{+ve}(p)) = O_n(p_{+ve}(p)) + 1 \qquad (8)$$

$$O_n(p_{-ve}(p)) = O_n(p_{-ve}(p)) - 1 \qquad (9)$$

Finally, we set an array to save the points which have more votes from different radius and make use of Quicksort algorithm to sort these points and just keep the five points, which have the maximum votes. If any two centre point's

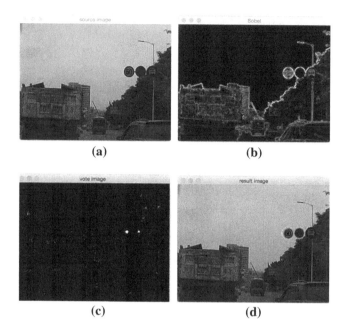

(a)   (b)

(c)   (d)

Fig. 6. The whole process of detection. (a) The source image. (b) The result of improved Sobel. (c) The orientation projection image. (d) The result of detection.

distance is less than their radii, we can judge that two circles are overlapped. If the circle's position is overlapped, then we compare the rate of votes to radii. We just keep the bigger one. Though this way, we confirm the signs position as shown in Fig. 7.

3.2 Digit Segmentation of Speed Limit Sign

When the speed limit traffic sign is detected, the following is to recognize the pictogram of this traffic sign and segment the digits. Inspired by [12], we make use of Otsu method to transform the ROI into a binary image when cropped the region of interest out of the scene.

However, we can't directly take this approach, because the target image which we captured from the source image is square not circle. It means that the part of background image maybe influences the effect of Otsu's method when the contrast ratio between the speed limit traffic sigh and background is obvious. To eliminate the bad effects, we have to crop the size of detected speed limit sign again. According to the standard of China road traffic signs, we decide to set a particular value as the target area's width and height, which is the four fifth of the radius. The advantage is that we can distinguish the foreground and background by just focusing on the speed limit sign itself. At the same time, we can crop part of sign's red rim. The comparison result is shown in Fig. 8.

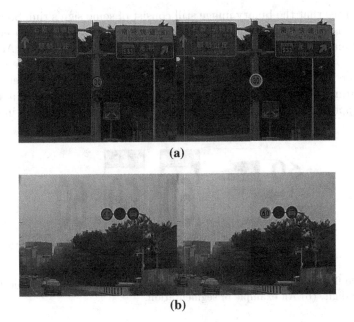

(a)

(b)

Fig. 7. The three detection results. (a) The detection result of 30 km/h. (b) The detection result of 60 km/h.

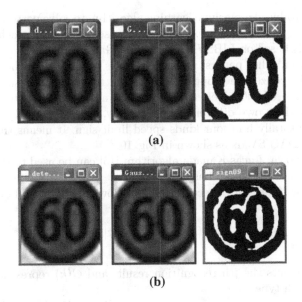

(a)

(b)

Fig. 8. The two kinds of Otsu's result. (a) Otsu's result after cropping the size of detected speed limit sign. (b) Otsu's result before cropping the size of detected speed limit sign (Color figure online).

We can see that the binary image still consist of some sign's rim. So the next step is to use maximal connected-component labeling algorithm to eliminate them. In binary image, we can see that there are several white regions. According to the coutours detection of white connected-component, we just keep the one which covers the maximum area and the others can be set as white pixels. Finally by calculating the black pixels to determine the location on the x-axis and y-axis of the cropped image where the target area starts and ends as shown in Fig. 9.

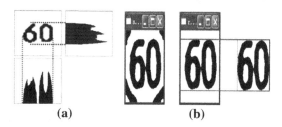

Fig. 9. The final output of detection. (a) The vertical and horizontal projection histograms of digit. (b) An example of segmentation.

3.3   Training Using DAG-SVMs

Vapnik [13] introduced a machine learning algorithm to solve pattern recognition problems, namely SVM. In this experiment, we use the Opencv machine learning library which is based on LIBSVM [14] with a Gaussian kernel.

Different from other papers, we view "60" as an integral rather than single "6" or "0". We resize the cropped ROI image to a standard size of 20×20 pixels. The value of 400 pixels can be viewed as input character. Each digit has 50 images which own different angle. Finally, we adopt DAG SVMs. In our experience, we totally find four kinds speed limit sign. It means that there are four classes in DAG SVMs as shown in Fig. 10.

As we all known, Lucas-Kanada algorithm [15] can be used to track. But we just keep its judgment method as following.

We can view the result of recognition as follows over a period of time n:

$$C(k) = \sum_{i=1}^{n} |x_i = k| \tag{10}$$

where, x_i represents the ith recognition result, and $C(k)$ represents the times recognized as kth type.

If $maxC(k)_k > th$ (th represents the recognition threshold), the output is the kth type of speed limit signs, otherwise, it would be passed.

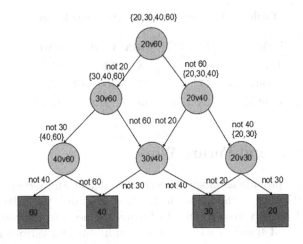

Fig. 10. The structure of DAG SVMs

4 Results and Analysis

We have tested the experiment under 10 videos about 28 min in ShenZhen. The frame rate of the video in this experiment is 33 frames per second, which has a resolution of 1920×1080 pixels. Speed limit signs: 20, 30, 40 and 60 km/h are used in the experiments. The result of the experiment is shown in Tables 1 and 2.

NV represents the Number of Video segment. NGT represents the reference number of speed limit signs appearing in the evaluation set. NRR is the number of the recognition result outputted by system. NTP is the number of Ground Truth which is recognized successfully. RR denotes the detection rate. FRR denotes the false recognition rate.

$$RR = \frac{NTP}{NGT}, FRR = 1 - \frac{NTP}{NRR} \tag{11}$$

NOA represents the Number of Output Alarm by system. NSA represents the Number of Succeed Alarm. CAR denotes the correct alarm rate. FAR denotes the false alarm rate.

$$CAR = \frac{NSA}{NV}, FAR = 1 - \frac{NSA}{NOA} \tag{12}$$

Table 1. The recogntion results of one frame

Scenes	NV	NGT	NRR	NTP	RR	FRR
cloudy+dusk	10	168	156	152	90.48 %	2.56 %
Cloudy [15]	17	341	329	321	94.13 %	2.43 %
Night [15]	23	284	264	253	89.08 %	4.17 %

Table 2. The recognition results of single sign

Scenes	NV	NOA	NSA	CAR	FAR
cloudy+dusk	10	8	8	100.00 %	0 %
Cloudy [15]	17	17	17	100.00 %	0 %
Night [15]	23	14	13	91.3 %	4.55 %

5 Conclusion and Future Work

In this paper, we provide an efficient method to detect and recognize the speed limit signs. With this method, the dream that alarming the drivers to notice the limit speed basiclly comes true. In future, we need to improve the recognition accuracy and speed in wide weather conditions and various illumination conditions.

References

1. Meng-Yin, F., Huang, Y.-S.: A survey of traffic sign recognition. In: 2010 International Conference on Wavelet Analysis and Pattern Recognition (ICWAPR), pp. 119–124. IEEE (2010)
2. De La Escalera, A., Moreno, L.E., Salichs, M.A., Armingol, J.M.: Road traffic sign detection and classification. IEEE Trans. Ind. Electron. **44**(6), 848–859 (1997)
3. Damavandi, B.Y., Mohammadi, K.: Speed limit traffic sign detection and recognition. In: 2004 IEEE Conference on Cybernetics and Intelligent Systems, vol. 2, pp. 797–802. IEEE (2004)
4. Escalera, S., Radeva, P., Pujol, O.: Traffic sign classification using error correcting techniques. In: VISAPP 2007, vol. 2, pp. 281–285 (2007)
5. Loy, G., Zelinsky, A.: A fast radial symmetry transform for detecting points of interest. In: Heyden, A., Sparr, G., Nielsen, M., Johansen, P. (eds.) ECCV 2002, Part I. LNCS, vol. 2350, pp. 358–368. Springer, Heidelberg (2002)
6. Barnes, N., Loy, G., Shaw, D.: The regular polygon detector. Pattern Recogn. **43**(3), 592–602 (2010)
7. Biswas, R., Fleyeh, H., Mostakim, M.: Detection and classification of speed limit traffic signs. In: 2014 World Congress on Computer Applications and Information Systems (WCCAIS), pp. 1–6. IEEE (2014)
8. Sekanina, L., Torresen, J.: Detection of norwegian speed limit signs. In: ESM, pp. 337–340 (2002)
9. Jianping, W., James, Y.: Tsai.: Real-time speed limit sign recognition based on locally adaptive thresholding and depth-first-search. Photogram. Eng. Remote Sens. **71**(4), 405–414 (2005)
10. Lorsakul, A., Suthakorn, J.: Traffic sign recognition for intelligent vehicle/driver assistance system using neural network on opencv. In: The 4th International Conference on Ubiquitous Robots and Ambient Intelligence (2007)
11. Höferlin, B., Zimmermann, K.: Towards reliable traffic sign recognition. In: 2009 IEEE Intelligent Vehicles Symposium, pp. 324–329. IEEE (2009)
12. Belongie, S., Malik, J., Puzicha, J.: Shape matching and object recognition using shape contexts. IEEE Trans. Pattern Anal. Mach. Intell. **24**(4), 509–522 (2002)

13. Vapnik, V.: The Nature of Statistical Learning Theory. Springer Science & Business Media, Heidelberg (2013)
14. Chang, C.-C., Lin, C.-J.: Libsvm: a library for support vector machines. ACM Trans. Intell. Syst. Technol. (TIST) **2**(3), 27 (2011)
15. Liu, W., Liu, Y., Hongfei, Y., Yuan, H., Zhao, H.: Real-time speed limit sign detection and recognition from image sequences. In: 2010 International Conference on Artificial Intelligence and Computational Intelligence (AICI), vol. 1, pp. 262–267. IEEE (2010)

FARTHEST: FormAl distRibuTed scHema to dEtect Suspicious arTefacts

Pablo C. Cañizares[✉], Mercedes G. Merayo, and Alberto Núñez

Departamento de Sistemas Informáticos y Computación,
Universidad Complutense de Madrid, Madrid, Spain
{pablocc,mlmgarci,albenune}@ucm.es

Abstract. Security breaches are a major concern by both governmental and corporative organisations. This is the case, among others, of airports and official buildings, where X-ray security scanners are deployed to detect elements representing a threat to the human life. In this paper we propose a formal distributed schema, formally specified and analysed, to detect suspicious artefacts. Our approach consists in the integration of several image detection algorithms in order to detect a wide spectrum of weapons, such as guns, knifes and bombs. Also, we present a case of study, where some performance experiments are carried out for analysing the scalability of this schema when it is deployed in current systems.

1 Introduction

After the 11/9 tragedy, security has become a national priority for governments around the world. Consequently, high security measures have been imposed in both governmental and corporative facilities to ensure the integrity of citizens. It is worth to emphasise the case of the airports, where only the U.S. government generated an annual $700 million market [10701] in the deployment of *Explosives Detection Systems* (EDS) [MR95], which are based on X-ray imaging for scanning baggage (see Fig. 1). The massive deployment of EDS has arisen the interest of the scientific community. Consequently, several detection techniques have been reported in the literature [WB12, Mer15a], like artificial neural networks [LW08], SVM [NPA08] and novel multiple-view approaches [Mer15b, US15]. All these techniques, especially the multiple-view ones, have high detection rates for processing all type of threat artefacts such as guns, weapons, bombs and knifes, which represent a beneficial contribution to protect the human life. However, these mentioned techniques are designed to be executed in a single machine, it being overexposed to risks such as computer attacks, lack of fault tolerance and replica.

Research partially supported by the Spanish MEC projects ESTuDIo and DArDOS (TIN2012-36812-C02-01 and TIN2015-65845-C3-1-R) and the Comunidad de Madrid project SICOMORo-CM (S2013/ICE-3006).

© Springer-Verlag Berlin Heidelberg 2016
N.T. Nguyen et al. (Eds.): ACIIDS 2016, Part I, LNAI 9621, pp. 770–779, 2016.
DOI: 10.1007/978-3-662-49381-6_74

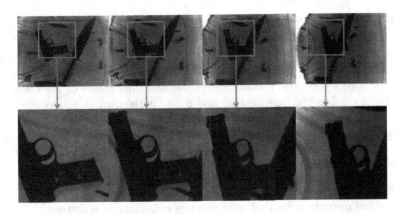

Fig. 1. X-ray image for detecting guns in baggage [Mer15b]

In order to avoid security risks, it is important to incorporate mechanisms to increase the confidence on the correctness of a system with respect to a specification. Thus, we consider that formal methods should be used to develop critical systems. Formal methods are techniques based on mathematics for modelling complex systems and to represent the specification, development, and verification of both software and hardware systems. It is important to mention that the combined use of formal methods and testing techniques [CHN15] allows us to ensure the fulfilment of a specific set of requirements and it is especially useful to detect unexpected behaviours.

In this paper we propose a distributed schema, called FARTHEST, to detect suspicious artefacts. We have used a formal approach to specify and analyse FARTHEST. In addition, we provide several specific set of communication requirements to ensure the correct behaviour of the proposed algorithm.

The rest of the paper is structured as follows. Section 2 presents the formal framework used in this paper. Next, in Sect. 3 we describe the proposed distributed scheme. Section 4 presents experimental results. Finally, in Sect. 5 we present the conclusions and some lines of future work.

2   Formal Framework Used in FARTHEST

In this section we review the framework used for specifying and testing complex systems [MN15] that has been used to model and specify FARTHEST. In addition, we introduce some extensions to the finite state machine model, that allows define and check the correctness of communications between components of the system.

2.1   Finite State Machines

Finite State Machines, in short FSM, are one of the formalism widely used to formally specify systems. We have chosen them to specify our system because they are well known are their definition and semantics are very simple.

Definition 1. *A Finite State Machine is a tuple* $M = (\mathcal{S}, s_{in}, \mathcal{I}, \mathcal{O}, \mathcal{T}_r)$ *where* $\mathcal{S}$ *is a finite set of states,* s_{in} *is the initial state of the machine,* $\mathcal{I}$ *is the set of input actions,* $\mathcal{O}$ *is the set of output actions, with* $\mathcal{I} \cap \mathcal{O} = \emptyset$, *and* $\mathcal{T}_r$ *is the set of transitions, with* $\mathcal{T}_r \subseteq \mathcal{S} \times \mathcal{S} \times \mathcal{I} \times \mathcal{O}$. *A transition belonging to* $\mathcal{T}_r$ *is a tuple* (s, s', i, o) *where* $s, s' \in \mathcal{S}$ *are the initial and final states of the transition respectively,* $i \in \mathcal{I}$ *is the input action and* $o \in \mathcal{O}$ *is the output action.*

We say that M is *deterministic* if for all state s and input i there exists at most one state s' and one output o' such that $(s, s', i, o) \in \mathcal{T}_r$. We say that M is *input-enabled* if for all state s and input i there exists at least one state s' and one output o' such that $(s, s', i, o) \in \mathcal{T}_r$. $\qquad\square$

In the following definition we introduce the concept of *trace*. A trace is a sequence of inputs and outputs pairs that captures the behaviour of a system.

Definition 2. *Let* $M = (\mathcal{S}, s_{in}, \mathcal{I}, \mathcal{O}, \mathcal{T}_r)$ *be a FSM. We say that* $< i_1/o_1, \ldots, i_n/o_n >$ *is a trace of* M *if there exist* n *states* $s_1, \ldots, s_n \in \mathcal{S}$ *such that*

$$(s_0, s_1, i_1, o_1), (s_1, s_2, i_2, o_2), \ldots, (s_{n-1}, s_n, i_n, o_n) \in \mathcal{T}_r$$

with $s_0 = s_{in}$. *We denote by* $<>$ *the empty trace and by* $\mathtt{trace}(M)$ *the set of all traces of M.* $\qquad\square$

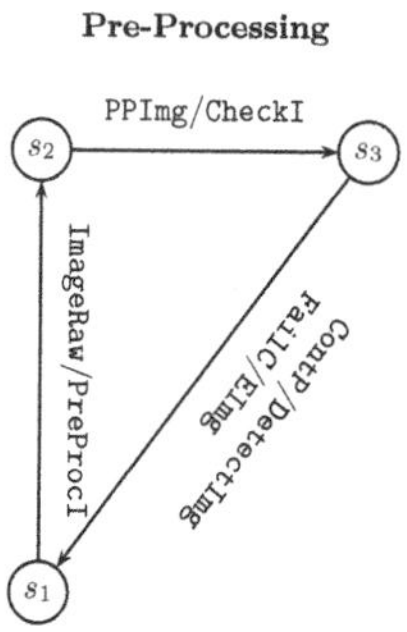

Fig. 2. Specification of an image pre-processing by using an FSM

Example 1. Let us consider the FSM depicted in Fig. 2 presenting a reduced version of the pre-processing image stage. It enhances the visual appearance and improves the manipulation of the image for later stages. The nodes represent the most relevant states of the algorithm, while the arcs represent the relevant transitions performed during the process. The initial state of the machine is s_1, corresponding to the point where the image to be pre-processed is received.

Let us consider the transition $(s_1, s_2, \mathtt{ImageRaw}, \mathtt{PreProcI})$. Intuitively, if the machine is the initial state s_1 and it receives an input $\mathtt{ImageRaw}$, then it produces the output $\mathtt{PreProcI}$ and the machine changes to state s_2. Also, we can observe that $(\mathtt{PPImg}/\mathtt{CheckI}, \mathtt{ContP}/\mathtt{DetectImg})$ is a trace of the system.

Next, we describe the set of steps required to perform the image pre-processing. At the initial state, the system receives an image `ImageRaw` and the process responsible to handle it `PreProcI` is invoked. Once all the pre-processing operations have been performed, the system checks the correctness of the generated image calling the `CheckI` process. Finally, the checking process returns a result which shows the diagnostic of the pre-processed image. If the checking process detects that the image has some faults, then the process will be interrupted.

2.2 Communicating Finite State Machines

In order to alleviate hard computational challenges, there is a whole new generation of systems. These systems are usually distributed along the nodes of a network. Thus, the communication between the components of this network becomes a critical factor for the overall system performance. Unfortunately, the behaviour of these systems cannot be represented by using classical finite state machines and, therefore, it is required to develop new methodologies that allow us both to represent properties related to communications and to establish its correctness.

Definition 3. A *Communicating Finite State Machine*, in short CFSM, is a FSM with a set of communication channels. A *Net Communicating Finite State Machines*, in short NETCOM, is a pair $\mathcal{N} = (\mathcal{M}, \mathcal{C})$, where $\mathcal{M} = \{M_1, \ldots, M_n\}$ is a set of CFSMs such that for all $1 \leq i \leq n$ we have that $M_i = (\mathcal{S}_i, s_{in}^i, \mathcal{I}_i, \mathcal{O}_i, \mathcal{T}_r^i,)$ and $\mathcal{C} = \{\mathcal{C}_i^a : i \leq n \wedge a \in \mathcal{I}_i\}$ represents the set of communication channels, where $\mathcal{C}_i^a$ means that M_i can receive the message a. We assume that $\mathcal{I}_1, \ldots \mathcal{I}_n$ are pairwise disjoint and for all $1 \leq i \leq n$ we have that $\mathcal{I}_i \cap \mathcal{O}_i = \emptyset$.

Given $\mathcal{I}_{\mathcal{N}} = \bigcup_{i=1}^n \mathcal{I}_i$ and $\mathcal{O}_{\mathcal{N}} = \bigcup_{i=1}^n \mathcal{O}_i$, we define the sets $Shared_{\mathcal{N}} = \mathcal{I}_{\mathcal{N}} \cap \mathcal{O}_{\mathcal{N}}$, $\text{envInput}_{\mathcal{N}} = \mathcal{I}_{\mathcal{N}} \setminus Shared_{\mathcal{N}}$ and $\text{envOutput}_{\mathcal{N}} = \mathcal{O}_{\mathcal{N}} \setminus Shared_{\mathcal{N}}$. □

A CFSM in a NETCOM can interact both with the environment and with another CFSM, by sending inputs and receiving output actions. Thus, two classes of transitions can be distinguished. On the one hand, *external transitions* are those labelled with input actions that are received from the environment. On the other hand, *internal transitions* are those that are triggered by an output produced by the execution of a transition in another CFSM.

The set $\text{shared}_{\mathcal{N}}$ contains those actions allowing the communication between two machines in a net. Those actions belong simultaneously to the set of *input* actions of a CFSM in the net and the set of *output* actions in another one. The set $\text{envInput}_{\mathcal{N}}$ ($\text{envOutput}_{\mathcal{N}}$) corresponds to the set of *not shared* input (output) actions appearing in $\mathcal{N}$, that is, the input (output) actions labelling external transitions.

Example 2. Let us consider the NETCOM depicted in Fig. 3. It can be seen as an evolution of Example 1 by including communication channels. In this case, the image pre-processing, previously performed by a single FSM has been separated into two different CFSM, *Pre-Processing* and *Checking*. Moreover, we have included another CFSM, called *Data Base*, that provides images to the image pre-processing system.

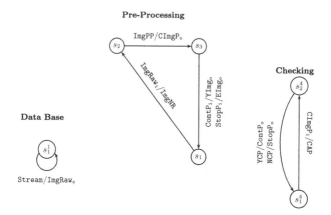

Fig. 3. Specification of pre-processing and checking image phases by using CFSM model

Next, we describe the set of steps required to perform the pre-processing and checking phases. At the initial state, the database machine provides the system with an image stream by sending the message ImageRaw$_o$ to the Pre-Processing machine. In this way, the Pre-Processing machine receives the message ImageRaw$_i$ and invokes the pre-processing operation ImgNR. Once all the pre-processing operations have been performed, the Pre-Processing machine sends the message CImgP$_o$ to Checking machine, that checks the correctness of the generated image. Finally, the checking process returns a result of the diagnostic of the pre-processed image. If the checking process detects that the image has some faults, the Checking machine sends a StopP$_0$. On the contrary, if the image is correct, it sends ContP$_0$.   □

In order to validate the correctness of a system by using a passive testing technique, we record and analyse the sequences of actions generated by the system under test. These sequences are checked against a certain set of properties, that we call *invariants*, representing the most relevant properties that the system must fulfill. Next, we introduce the notion of *communication invariants*, an extension of the usual notion of invariant used in a single FSM.

Definition 4. Let $\mathcal{N} = (\mathcal{M}, \mathcal{C})$ be a NETCOM. We say that a sequence ϑ is a *communicating invariant*, in short *c-invariant*, for the net $\mathcal{N}$, if ϑ is defined according to the following EBNF:

$$\vartheta ::= \vartheta_1 | \vartheta_2$$
$$\vartheta_1 ::= i/s, \vartheta_2 | i/s, \vartheta_3 | i \to S$$
$$\vartheta_2 ::= s/o, \vartheta_1 | s/o, \vartheta_3 | s/s', \vartheta_2 | s/s', \vartheta_3 | s \to O$$
$$\vartheta_3 ::= \star, \vartheta$$

In this expression we consider $s, s' \in$ shared$_\mathcal{N}$, $i \in$ envInput$_\mathcal{N}$, $o \in$ envOutput$_\mathcal{N}$, $S \subseteq$ shared$_\mathcal{N}$ and $O \subseteq$ envOutput$_\mathcal{N}$. The set of invariants for the net $\mathcal{N}$ is denoted by $\Omega_\mathcal{N}$, where we will omit the subindex if it can be deduced from the context.   □

The previous EBNF expresses that a c-invariant is a sequence of symbols where each component, but the last one, is either a pair with one of the elements being a shared action (s) and the other one an input (i) or an output (o) action, or the wildcard $\star$ that can replace a sequence of actions not containing the first input symbol that appears in the component of the c-invariant that follows it. Let us note that two consecutive pairs in the sequence $a/b, c/d$ must be *compatible*, that is, either $c = b \in$ shared$_{\mathcal{N}}$ or $b \in$ envOutput$_{\mathcal{N}}$, $c \in$ envInput$_{\mathcal{N}}$ and both a and c belong to the set of input actions of the same CFSM in $\mathcal{N}$. In addition, a c-invariant cannot contain two consecutive occurrences of $\star$. The last component is given by either the expression $i \rightarrow S$ or $s \rightarrow O$. The former corresponds to a input action followed by a set of shared actions and the latter represents a shared action followed by a set of output actions.

3 Distributed Schema to Detect Suspicious Artefacts

In this section we describe our proposed distributed schema for detecting suspicious artefacts, called FARTHEST. In order to show a detailed perspective of our approach, a formal specification of the different phases of this schema is provided. Moreover, we include a set of c-invariants to analyse the correctness of its behaviour.

FARTHEST can be divided into three different phases. The first phase corresponds to image pre-processing operations, the second phase ensures image integrity and the third phase performs image recognition based on majority voting process. Figures 4 and 5 show the formal specification of the NETCOM that represents the behaviour of FARTHEST, which have been developed by using the framework described in Sect. 2. First, an image to be processed (ImgRaw$_i$) is sent to the *pre-processing* phase. In this stage, the image is filtered and processed (ImgNR) with noise reduction algorithms to fix possible visual defects of the image. Next, the pre-processed image (ImgPP) is sent to the *checking* phase (CImgP$_o$), where the image (CImgP$_i$) is received and checked in order to detect format defects (CI). If some fault is detected, the checking phase emits a report error (StopP$_o$) and the execution is aborted (EImg$_o$). On the contrary, the image (VotImg$_o$) is sent to the *detection voting* phase. In this stage, a voting process is performed for determining the suspect nature of an element. Thus, the image received (VotImg$_i$) is sent to the detection algorithms (V1$_o$, V2$_o$, V3$_o$), where the image is processed by all of them. If the algorithm detects that the image matches with a suspicious artefact, it sends a positive vote (YA$_i$). On the contrary, it sends a negative vote (NA$_i$). Finally, if the absolute majority of the votes is positive, an alarm is triggered (WD$_i$) and the image is stored into a database (SaveDB$_i$).

In FARTHEST, an image stream flows through the different stages by following a pipeline model, where the output generated by a phase is considerer as the input of the next one. Inasmuch as each phase only can process one image at the same time, the execution of the phases can overlap, which allows processing multiple image detection simultaneously. Moreover, since the distributed design of FARTHEST allows to execute each phase in different physical machines, the resources provided by a distributed system, like an HPC cluster or a cloud computing system, can be exploited in parallel to increase the overall system performance.

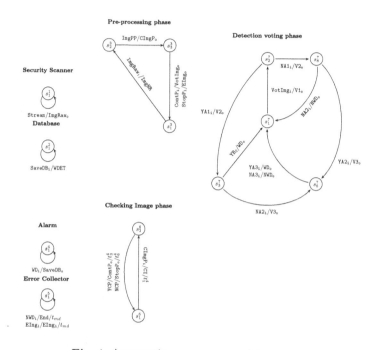

Fig. 4. Automatic weapon recognition system

We consider that the detection of elements that are a threat is critical. Thus, we provide a robust and extensible meta-detector suitable for different hazard environments, such that the detection of threat elements is performed through a voting process among several independent detection algorithms. In addition, we have included into **FARTHEST** *Online Stage* [Mer15b]. This is a detection algorithm based on the classification and analysis of the main keypoints extracted from an image.

Our algorithm can be divided into two main parts: *Monocular analysis* and *Multiple View Analysis*. Figure 5 shows the specification of this algorithm, that is represented by a **NETCOM** conformed by three **CFSM** . The first machine, *Online State*, describes the general behaviour of the algorithm, receiving an image ($V1_i$) and distributing the process among the other machines (MA_o, MV_o). The second machine describes the *Monocular analysis* process, performing operations such as segmentation (**SEG**), keypoints selection (**KEY**), classification and clustering (**CCS**). If the *Monocular Analysis* is correct (YMA_i), then the second phase of the algorithm is carried out by the machine *Multiple View Analysis*, performing operations such as data association (**DAS**) and data analysis (**DAN**). Finally, if

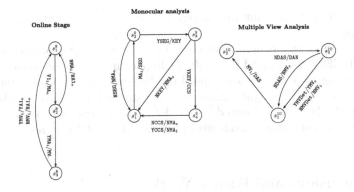

Fig. 5. Online stage algorithm

the algorithm detects a suspicious artefact, then it sends a positive vote ($\texttt{YA1}_o$). On the contrary, a negative vote is emitted ($\texttt{NA1}_o$).

Next, we include a set of communicating requirements, using c-invariants, that must be fulfilled by the implementation in order to ensure correctness.

$$\theta_1 = \underbrace{\texttt{ImgPP}}_{i} \rightarrow \underbrace{\{\texttt{VotImgB}_o, \texttt{CImgP}_o, \texttt{EImg}_o\}}_{S}$$

$$\theta_2 = \underbrace{\texttt{ImgRaw}_i / \texttt{ImgNR}, \star,}_{s/o} \underbrace{\texttt{EImg}_i}_{s} \rightarrow \underbrace{\{\texttt{SaveE}\}}_{O}$$

$$\theta_3 = \underbrace{\texttt{ImgRaw}_i / \texttt{ImgNR}, \star,}_{s/o} \underbrace{\texttt{WD}_i}_{s} \rightarrow \underbrace{\{\texttt{WDET}\}}_{O}$$

$$\theta_4 = \underbrace{\texttt{ImgRaw}_i / \texttt{ImgNR}, \star,}_{s/o} \underbrace{\texttt{NWD}_i}_{s} \rightarrow \underbrace{\{\texttt{End}\}}_{O}$$

4   Experiments

In this section we present several experiments to evaluate the scalability of FARTHEST when it is deployed in different cloud systems. These experiments have been conducted in a simulated environment by using the iCanCloud simulation platform [NVC+12].

Each modelled cloud contains 6 physical machines such that each phase of FARTHEST is executed in a dedicated machine and the other three machines are used for the voting process. In this model, there is one centralised database that contains the images to be processed. Each cloud where FARTHEST is deployed accesses this database through the Internet. In these experiments, FARTHEST has been deployed in 1, 2, 4 and 8 homogeneous cloud systems. Also, each experiment uses a different configuration for the physical machines, containing 1, 2, 4 and 8 CPU cores.

The results of our experiments show that increasing the number of clouds where FARTHEST is deployed has a direct impact in the overall system performance, leading to a performance speed-up. However, increasing the number of CPU cores per machine slightly increases the system performance. This is mainly caused by the bottleneck located in the data base system, which hampers the exploitation of computing parallelism by using all the CPU cores at the same time.

5   Conclusions and Future Work

In this paper we have presented FARTHEST, a formally specified and analysed distributed schema, to detect suspicious artefacts. FARTHEST has been specified by using a formal framework based on Finite State Machines. Also, a set of communicating requirements to check the correct behaviour of the proposed schema has been provided. In order to show the applicability of FARTHEST it has been deployed along several cloud systems in a simulated environment. The experiments of this paper have been conducted by using the iCanCloud simulation platform. The evaluation results show that FARTHEST provides an increasing in the overall system performance when it is deployed in different cloud systems. However, since all the images are stored in a centralised data base, the communication network to access the data base acts as a bottleneck, which leads to a performance loss.

A first line of future work consists in the inclusion of timed and probabilistic information in our models [HM09, HMN09, AMN12]. We would also like to use passive testing techniques to check the proposed schema by using more complex communicating requirements. A third line of work consists in studying optimizations to reduce energy consumption [CNLC13, CNNP15] and to increase performance due to parallelization [NFM13, NM14]. Finally, we would like to use learning techniques to improve the performance of our detection algorithms taking into account that an *attacker* might modify some of the components [LNRR02].

References

[10701] US PUBLIC LAW 107-71. Aviation and transportation security act. (2001)

[AMN12] Andrés, C., Merayo, M.G., Núñez, M.: Formal passive testing of timed systems: Theory and tools. Softw. Test. Verification Reliab. **22**(6), 365–405 (2012)

[CHN15] Cavalli, A.R., Higashino, T., Núñez, M.: A survey on formal active and passive testing with applications to the cloud. Ann. Telecommun. **70**(3–4), 85–93 (2015)

[CNLC13] Castañé, G., Núñez, A., Llopis, P., Carretero, J.: E-mc^2: A formal framework for energy modelling in cloud computing. Simul. Model. Pract. Theor. **39**, 56–75 (2013)

[CNNP15] Cañizares, P.C., Núñez, A., Núñez, M., Pardo, J.J.: A methodology for designing energy-aware systems for computational science. In: 15th International Conference on Computational Science, ICCS 2015, Procedia Computer Science 51, pp. 2804–2808. Elsevier (2015)

[HM09] Hierons, R.M., Merayo, M.G.: Mutation testing from probabilistic and stochastic finite state machines. J. Syst. Softw. **82**(11), 1804–1818 (2009)

[HMN09] Hierons, R.M., Merayo, M.G., Núñez, M.: Testing from a stochastic timed system with a fault model. J. Logic Algebraic Program. **78**(2), 98–115 (2009)

[LNRR02] López, N., Núñez, M., Rodríguez, I., Rubio, F.: Including malicious agents into a collaborative learning environment. In: Cerri, S.A., Gouardéres, G., Paraguaçu, F. (eds.) ITS 2002. LNCS, vol. 2363, pp. 51–60. Springer, Heidelberg (2002)

[LW08] Liu, D., Wang, Z.: A united classification system of x-ray image based on fuzzy rule and neural networks. In: 3rd International Conference on Intelligent System and Knowledge Engineering, ISKE 2008, vol. 1, pp. 717–722. IEEE Computer Society (2008)

[Mer15a] Mery, D.: Applications in X-ray testing. In: Computer Vision for X-Ray Testing, pp. 267–325. Springer, Heidelberg (2015)

[Mer15b] Mery, D.: Inspection of complex objects using multiple-X-ray views. IEEE/ASME Trans. Mechatron. **20**(1), 338–347 (2015)

[MN15] Merayo, M.G., Núñez, A.: Passive testing of communicating systems with timeouts. Inf. Softw. Technol. **64**, 19–35 (2015)

[MR95] Murray, N.C., Riordan, K.: Evaluation of automatic explosive detection systems. In: 29th Annual International Carnahan Conference on Security Technology, pp. 175–179. Institute of Electrical and Electronics Engineers (1995)

[NFM13] Núñez, A., Filgueira, R., Merayo, M.G.: SANComSim: A scalable, adaptive and non-intrusive framework to optimize performance in computational science applications. In: 13th International Conference on Computational Science, ICCS 2013, Procedia Computer Science 18, pp. 230–239. Elsevier (2013)

[NM14] Núñez, A., Merayo, M.G.: A formal framework to analyze cost and performance in Map-Reduce based applications. J. Comput. Sci. **5**(2), 106–118 (2014)

[NPA08] Nercessian, S., Panetta, K., Agaian, S.: Automatic detection of potential threat objects in x-ray luggage scan images. In: IEEE Conference on Technologies for Homeland Security, pp. 504–509. IEEE Computer Society (2008)

[NVC+12] Núñez, A., Vázquez-Poletti, J.L., Caminero, A.C., Castañé, G.G., Carretero, J., Llorente, I.M.: iCanCloud: A flexible and scalable cloud infrastructure simulator. J. Grid Comput. **10**(1), 185–209 (2012)

[US15] Uroukov, I., Speller, R.: A preliminary approach to intelligent X-ray imaging for baggage inspection at airports. Signal Process. Res. **4**, 1–11 (2015)

[WB12] Wells, K., Bradley, D.A.: A review of X-ray explosives detection techniques for checked baggage. Appl. Radiat. Isot. **70**(8), 1729–1746 (2012)

A Fast and Robust Image Watermarking Scheme Using Improved Singular Value Decomposition

Cao Thi Luyen, Nguyen Hieu Cuong[✉], and Pham Van At

Faculty of Information Technology,
University of Transport and Communications, Hanoi, Vietnam
caoluyengt@gmail.com, Cuonggt@gmail.com

Abstract. With the popularity of editing software and the Internet, digital content can easily be manipulated and distributed. Therefore, illegal reproduction of digital products became a real problem. Watermarking has been considered as an effective solution for copyright protection and authentication. However, watermarking schemes usually have encountered some difficulties, such as computational complexity, imperceptibility and robustness. In this paper, based on improved singular value decomposition (SVD), we proposed a new image watermarking scheme in order to reduce the computational complexity. To this end, we designed an algorithm to directly compute the largest eigenvalues and eigenvectors of the analyzed image segmented blocks. Moreover, an adaptive embedding technique was utilized to improve the robustness of the proposed scheme. Experimental results showed that the scheme is fast and good for digital image watermarking and it outperforms several widely used schemes in terms of robustness and imperceptibility.

Keywords: Digital watermarking · SVD · Eigenvalue · Eigenvector

1 Introduction

Due to the improvement of editing tools and the popularity of the Internet, illegal manipulations of digital objects became very popular. Among several approaches have been proposed for copyright protection and authentication, digital watermarking is commonly used. In every watermarking scheme, a watermark (some types of digital data, e.g. text, logos, labels etc.) is embedded into digital objects in order to present the authorship of the objects. In a fragile watermarking scheme, the embedded watermark could be distorted when any operation is applied to the watermarked objects. Conversely, in a robust watermarking scheme, the embedded watermark should be retained, even when some operations are applied to the watermarked objects. While fragile watermarking is applied for authentication, robust watermarking is mostly used for copyright protection. Basically, a watermarking scheme must satisfy some main requirements for robustness and imperceptibility. The first requirement is sufficed if the watermark is difficult to remove or distort, even though different illegal operations

© Springer-Verlag Berlin Heidelberg 2016
N.T. Nguyen et al. (Eds.): ACIIDS 2016, Part I, LNAI 9621, pp. 780–789, 2016.
DOI: 10.1007/978-3-662-49381-6_75

have been applied to the watermarked images. The second demand is fulfilled when the difference between the original image and the watermarked image is indistinguishable to human eyes.

In recent decades, various watermarking schemes have been proposed. In these schemes, different image processing operations and transformations were employed. Among them, the SVD is widely used in many watermarking schemes [1–13]. In the schemes, each segmented image block is decomposed into three matrices U, D, and V. After that, ones embed watermarks into different parts of the matrices: coefficients of the main diagonal of the matrix D [1, 9–13], the first column of the matrix U or the matrix V [3, 5], and both D and U [16]. Besides, some schemes employed a hybrid approach by combining SVD with other transformations, such as discrete Cosine transform (DCT), discrete wavelet transform (DWT), etc. [2, 4, 6–8].

Naturally, the most important problem of every SVD based watermarking scheme is the singular value decomposition of image blocks. This decomposition requires finding all eigenvalues and eigenvectors of the analyzed image blocks. Since the transform is complex and must be employed to every block, this step is rather time consuming. We realized that in various SVD based algorithms [2–9, 16], only the first coefficient of D and the first column of U (or V) were used. We denote these values as D(1,1), U(1) and V(1) respectively. In this work, we propose a new technique to directly compute the aforementioned values instead of conventionally analyzing SVD. Consequently, embedding process and extracting process are speed up. This improvement is essential for watermarking systems in practice, when ones often have to work with a large-scale image dataset.

We also conduct experiments to evaluate the robustness and imperceptibility of the proposed scheme and some widely used watermarking schemes. The results show that, all of the schemes are robust against popular attacks and the proposed scheme is slightly better than the others in terms of robustness and imperceptibility.

Moreover, based on a secret key, we employ an adaptive technique in order to increase the watermarking security. The watermark is not embedded in fixed positions, but can be embedded into different pairs of $\{U(1,1), U(2,1)\}$ or $\{V(1,1), V(2,1)\}$, depending on the key.

The rest of the paper has the structure as below. In Sect. 2 we briefly review essential background of the SVD. Next, we describe the proposed watermarking scheme. The experimental results are shown in Sect. 4. Lastly, we conclude the paper in Sect. 5.

2 Singular Value Decomposition (SVD)

The decomposition is not only be used in linear algebra, but also in many different applications, namely in image processing. The benefit of SVD is that it provides a robust method to decompose a large matrix to smaller and more manageable matrices. Concretely, every m × n matrix can be decomposed into three matrices:

$$A = U \times D \times V^T = U_1 D(1, 1) V_1^T + U_2 D(2, 2) V_2^T + \ldots + U_s D(s, s) V_s^T,$$

where $s = \min(m,n)$, $U \in \mathbb{R}^{m \times m}$, $V \in \mathbb{R}^{n \times n}$ are normalized orthogonal matrices and $D \in \mathbb{R}^{m \times n}$ is a diagonal matrix, containing the singular values of A (in the main diagonal of D): $D(1,1) \geq D(2,2) \geq \ldots \geq D(s,s) \geq 0$.

3 The Proposed Scheme

3.1 The Embedding Procedure

Input: An original image I, a watermark $W = (w_1, w_2, \ldots, w_t)$, a secret key $K = (k_1, k_2, \ldots, k_t)$ and some pre-defined thresholds (θ, τ).
  Output: The watermarked image I'.
  The embedding algorithm consists of following steps:

Step 1: Dividing the original image I into non-overlapped $m \times n$ blocks I_i.
Step 2: Inspired the method for finding the largest eigenvalue and eigenvector of a non-negative matrix [14], the first columns of U_i and V_i (i.e. U_i (1) and V_i (1)) and $D_i(1,1)$ can directly be computed. The values of U_i (1) and V_i (1) are the eigen-vectors corresponding to the largest eigenvalues of the non-negative matrices $I_i \times I_i^T$ and $I_i^T \times I_i$.
Step 3: Embedding the watermark bit w_i into the pair of coefficients $\{U_i(1,1), U_i(2,1)\}$ and $\{V_i(1,1), V_i(2,1)\}$.

Based on the secret key K, if $k_i = 0$ then the pair of $\{U_i(1,1), U_i(2,1)\}$ is selected for embedding bit w_i and in the other case, the pair of $\{V_i(1,1), V_i(2,1)\}$ is used. For embedding in $\{U_i(1,1), U_i(2,1)\}$, several steps are conducted as described below (similar to the $\{V_i(1,1), V_i(2,1)\}$).

- Computing x_i and l_i

$$x_i = U_i(1,1) + U_i(2,1),$$
$$l_i = \frac{x_i}{\theta}.$$

- Adjusting l_i in order to satisfy $(l_i + w_i) \bmod 2 = 0$. Then, we get l_i'.
- Computing x_i'

$$x_i' = l_i' \theta$$

- Changing $U_i(1,1)$, $U_i(2,1)$ to $U_i'(1,1)$ and $U_i'(2,1)$:

$$U_i'(1,1) = \frac{U_i(1,1) x_i'}{x_i}, U_i'(2,1) = \frac{U_i(2,1) x_i'}{x_i}$$

Step 4: Computing

$$I_i' = D_i(1,1) \times U_i'(1) \times V_i'^T(1) + I_i - D_i(1,1) \times U_i(1) \times V_i^T(1) \qquad (1)$$

At a result, I_i' can be computed without using the conventional SVD analysis. The correctness of Eq. 1 is proved as described follows:

According to the formular in [15, p. 448]:

$$I_i = D_i(1,1) \times U_i(1) \times V_i^T(1) + \sum_{k=2}^{s} D_i(k,k) \times U_i(k) \times V_i(k),$$

where s = min(m,n)

Therefore:

$$\sum_{k=2}^{s} D_i(k,k) \times U_i(k) \times V_i(k) = I_i - D_i(1,1) \times U_i(1) \times V_i^T(1) \qquad (2)$$

On the other hand:

$$I_i' = D_i'(1,1) \times U_i'(1) \times V_i'^T(1) + \sum_{k=2}^{s} D_i'(k,k) \times U_i'(k) \times V_i'(k) \qquad (3)$$

Since the values of $D_i(j,j), U_i(k)$ and $V_i(k)$ $(k \geq 2, j \geq 1)$ are retained after embedding, we have:
$D_i'(j,j) = D_i(j,j), U_i'(k) = U_i(k)$ and $V_i'(k) = V_i(k), j \geq 1, k \geq 2$.

Thus,

$$I_i' = D_i(1,1) \times U_i'(1) \times V_i'(1) + \sum_{k=2}^{s} D_i(k,k) \times U_i(k) \times V_i(k) \qquad (4)$$

Replace (Eq. 2) to (Eq. 4), we obtaine the equation (Eq. 1):

$$I_i' = D_i(1,1) \times U_i'(1) \times V_i'^T(1) + I_i - D_i(1,1) \times U_i(1) \times V_i^T(1).$$

3.2 The Extracting Procedure

It is noted that the watermarked image I' may be attacked by different operations. Subsequently, the receiver obtained an attacked image I^*, which is not quite the same as I'. The extracted watermark W^* (from I^*) can be compared to the original watermark W in order to evaluate the robustness of the watermarking scheme. Firstly, the attacked image I^* is partitioned into non-overlapped I_i^* blocks with the size of m × n. The watermark bits can be extracted from embedded blocks (selected blocks with high complexity) in following steps:

Step 1: Computing the matrices:

$$A_i^* = I_i^* \times I_i^{*^T} \in R^{m \times m}$$
$$B_i^* = I_i^{*^T} \times I_i^* \in R^{n \times n}$$

The eigenvectors $U_i^*(1) \in R^m$, $V_i^*(1) \in R^n$, corresponding to the largest eigenvalues of the non-negative matrices A_i^* and B_i^* were determined [14].

Step 2: w_i^* bit is extracted from P_i^* and Q_i^*:

$$\text{If } k_i = 0 \text{ then } x_i^* = U_i^*(2,1) + U_i^*(1,1)$$
$$\text{If } k_i = 1 \text{ then } x_i^* = V_i^*(2,2) + V_i^*(1,1)$$

Calculating w_i^*:

$$l_i^* = \left\lfloor \frac{x_i^* + \theta/2}{\theta} \right\rfloor$$
$$w_i^* = l_i^* \bmod 2$$

Step 3: Comparing the obtained watermark $W^* = \left(w_1^*, w_2^*, \ldots, w_t^*\right)$ with the original watermark $W = (w_1, w_2, \ldots, w_t)$ by using the error rate (ERR) between W^* and W.

The value of ERR can be computed as:

$$ERR = \frac{1}{t} \sum_{i=1}^{t} w_i XOR \; w_i^* \tag{5}$$

If the value of ERR is smaller than a predefined threshold τ then we conclude that the watermark W was embedded in I^* and I^* belongs to the owners of I'. The values of ERR is used to evaluate the robustness of the schemes in Tables 1, 2, 3, 4 and 5.

The author of [15] proved that the analysis of SVD for a matrix $A \in R^{m \times n}$ can be solved by computing all eigenvalues and eigenvectors of the matrix with size of $(m + n) \times (m + n)$. This problem is much more complex than finding only the largest eigenvalues and eigenvectors. Therefore, the computational complexity of the proposed scheme is lower than that of other SVD based schemes.

4   Experimental Results

In this section, several experiments were performed in order to verify the watermarking schemes. A set of standard uncompressed images of 256 × 256 pixels were used as host images (Fig. 1). A binary image with 32 × 32 bits was used as the watermark in our simulations (Fig. 2).

Firstly, we applied the embedding procedure of the proposed scheme to embed the watermark "Springer" to a host images. After that, the extracting procedure of the scheme was used to recover the embedded information from the watermarked images. We found that, the peak signal-to-noise ratio (PSNR) between the the original images and corresponding watermarked images were high (about 50 dB) and the diffrences between the original and the watermaked images are almost not distinguishable. The higher the PSNR, the better quality of the watermarked image The result implies that, the proposed scheme works well for watermarking when no attack is utilized.

Fig. 1. Four host images: (a) "Lena", (b) "Boat", (c) "Peppers", (d) "Cactus"

Fig. 2. The watermark image "Springer"

The experimental results for the imperceptual evaluated of the analyzed water-marking schemes are shown in Table 1. The results in Table 1 show that the evaluated watermarking schemes obtained high quality watermarked images with the PSNR values are about 50 dB and the proposed scheme is slightly better than the others.

Next, we evaluated the robustness of the proposed scheme and several widely used watermarking schemes of Sun et al. [1], Chung et al. [5], Lai [9]. To this end, after embedding a watermark into the host images, several attacks were applied to water-marked images. Then the watermark was extracted from the manipulated image. The error rates ERR (see Eq. 5) between the extracted watermarks and the embedded watermarks are used to measure the robustness of the watermarking schemes.

In our experiments, different attacks, including adding Gaussian noise (5 %), median filtering (3 × 3), Gaussian blurring (30 %), sharpening (30 %), and JPEG compression (50 %) were independently applied to watermarked images. For each case the error rate between the embedded watermark and the extracted watermark was computed. The value of the rate is in [0, 1] and a low error rate (about 0.1 or lower)

Table 1. PSNR of the watermarked images

Images	Threshold θ	Chung's [5]	Lai's [9]	Proposed
Lena	0.012	55.44	55.24	55.64
	0.020	55.20	54.89	55.47
	0.025	55.03	54.66	55.41
	0.040	54.61	53.99	55.18
Boat	0.012	54.21	53.19	54.88
	0.020	53.31	51.99	54.33
	0.025	52.83	51.41	53.94
	0.040	51.59	49.77	53.25
Peppers	0.012	54.46	53.99	55.27
	0.020	54.02	53.13	54.84
	0.025	53.74	52.71	54.71
	0.040	53.03	51.51	54.04
Cactus	0.012	54.16	53.19	55.02
	0.020	53.51	52.26	54.54
	0.025	53.14	51.78	54.29
	0.040	52.16	50.43	53.69
Average		53.77	52.75	**54.65**

means the difference between an original image and its watermarked image is small, in other words, the scheme is robust. The experimental results for the robustness of the evaluated schemes are shown in Tables 2, 3, 4, 5 and 6.

It can be seen that the proposed scheme and the scheme of Chung et al. are robust against Gaussian noise addition (Table 2). Although scheme of Chung et al. is robust against noise addition, it is so sensitive to median filtering. The schemes of Sun et al. and Lai are also sensitive, even they are more robust than Chung et al. In this case, only the proposed scheme is robust with the average error rate is about 0.06 (Table 3). Table 4 shows that all schemes are not robust against Gaussian blurring and the best scheme in this test case (Chung et al.) obtained an average error rate about 0.3. Table 5 shows that the schemes of Sun et al. and Lai are sensitive to sharpening. In this situation, the proposed technique seems robust with the average error rate is about 0.1. The scheme of Chung et al. is impressively robust against sharpening with the average error rate is only 0.0009 (Table 5). Table 6 shows that only the proposed scheme

Table 2. Results of the error rate of the extracted watermark after adding Gaussian noise

Images	Sun's [1]	Chung's [5]	Lai's [9]	Proposed
Lena	0.1589	0.0217	0.3930	0.0016
Boat	0.4900	0.0117	0.3462	0.0418
Peppers	0.4967	0.0535	0.2441	0.0870
Cactus	0.5167	0.0836	0.1890	0.1338
Average	0.4156	**0.0426**	0.2931	0.0660

Table 3. Results of the error rate of the extracted watermark after median filtering

Images	Sun's [1]	Chung's [5]	Lai's [9]	Proposed
Lena	0.0819	0.5334	0.2475	0.0067
Boat	0.3060	0.5284	0.1722	0.0151
Peppers	0.4448	0.5301	0.1355	0.0602
Cactus	0.4682	0.4348	0.1638	0.1639
Average	0.3252	0.5067	0.1798	**0.0615**

Table 4. Results of the error rate of the extracted watermark after Gaussian blurring

Images	Sun's [1]	Chung's [5]	Lai's [9]	Proposed
Lena	0.2475	0.0970	0.4114	0.2358
Boat	0.4732	0.2291	0.2876	0.2609
Peppers	0.4866	0.5301	0.3294	0.3545
Cactus	0.5050	0.3144	0.3562	0.3462
Average	0.4281	**0.2927**	0.3462	0.2994

Table 5. Results of the error rate of the extracted watermark after sharpening

Images	Sun's [1]	Chung's [5]	Lai's [9]	Proposed
Lena	0.0987	0	0.1839	0.0184
Boat	0.4218	0	0.1906	0.0535
Peppers	0.4599	0.0017	0.2040	0.1639
Cactus	0.4950	0.0017	0.1187	0.1856
Average	0.3689	**0.0009**	0.1743	0.1054

Table 6. Results of the error rate of the extracted watermark after JPEG compression

Images	Sun's [1]	Chung's [5]	Lai's [9]	Proposed
Lena	0	0.1355	0.3077	0.0535
Boat	0.1605	0.1706	0.3161	0.0769
Peppers	0.1271	0.2659	0.2642	0.0886
Cactus	0.5184	0.2308	0.3060	0.0886
Average	0.2015	0.2007	0.2985	**0.0769**

is robust against JPEG compression with the average error rate is about 0.07 and the other schemes are not so robust against this attack with the average error rates are about 0.2 and 0.3.

Besides, since the singular values of a matrix $A^{m \times n}$ are invariant to several geometric transformations [16, 17], we believe that the aforementioned SVD based schemes are also robust against rotation, scaling, cropping, etc.

5   Conclusion

In this paper, a fast and robust image watermarking scheme based on improved SVD is presented. Since the proposed scheme bypassed the step of conventional SVD analysis, the computational complexity was decreased. At a result, the procedures of embedding and extracting are speed up. This is significant when working with a large-scale image dataset in real life. Based on a secret key, we employed an adaptive embedding technique to improve the security of the scheme. Experimental results showed that the proposed scheme is not only fast but also robust against different types of image manipulations, such as median filtering and JPEG compression. Since embedding information in complex regions will be hard to recognize, in the future, we will design a method to compute the complexity for image blocks.

References

1. Sun, R., Sun, H., Yao, T.: A SVD and quantization based semi-fragile watermarking technique for image authentication. In: Proceedings of International Conference on Signal Processing, pp. 1952–1955 (2002)
2. Bhatnagar, G., Raman, B.: A new robust reference watermarking scheme based on DWT-SVD. Comput. Stand. Interfaces 31(5), 1002–1013 (2009)
3. Chang, C.C., Tsai, P., Lin, C.C.: SVD-based digital image watermarking scheme. Pattern Recogn. Lett. 26(10), 1577–1586 (2005)
4. Chen, H., Zhu, Y.: A robust watermarking algorithm based on QR factorization and DCT using quantization index modulation technique. J. Zhejiang Univ. 13(8), 573–584 (2012)
5. Chung, K.L., Yang, W.N., Huang, Y.H., Wu, S.T., Hsu, Y.C.: SVD-based watermarking algorithm. Appl. Math. Comput. 188(1), 54–57 (2007)
6. Fei, C., Kwong, R., Kundur, D.: Secure semi-fragile watermarking for image authentication. In: IEEE WIFS, pp. 141–145 (2009)
7. Arathi, C.: A semi fragile image watermarking technique using block based SVD. Int. J. Comput. Sci. Inf. Technol. 3(2), 3644–3647 (2012)
8. Gokhale, U.M., Joshi, Y.V.: A semi fragile watermarking algorithm based on SVD-IWT for image authentication. Int. J. Adv. Res. Comput. Commun. Eng. 1(4) (2012). ISSN-2278-1021
9. Lai, C.C.: An improved SVD-based watermarking scheme using visual characteristics. Optics Commun. 284(4), 938–944 (2012)
10. Liu, R., Tan, T.: An SVD-based watermarking scheme for protecting rightful ownership. IEEE Trans. Multimedia 4(1), 121–128 (2002)
11. Bao, P., Ma, X.: Image adaptive watermarking using wavelet domain SVD. IEEE Trans. Circ. Syst. Video Technol. 15(1), 96–102 (2005)
12. Sun, X., Liu, J., Sun, J., Zhang, Q., Ji, W.: A robust image watermarking scheme based on the relationship of SVD. In: Proceedings of International Conference on Intelligent Information Hiding and Multimedia Signal Processing (IIHMSP 2008), Harbin, pp. 731–734 (2008)
13. Zhu, X., Zhao, J., Xu, H.: A digital watermarking algorithm and implementation based on improved SVD. In: Proceedings of 18th International Conference on Pattern Recognition (ICPR 2006), vol. 3, Hong Kong, pp. 651–656 (2006)

14. At, P.V.: Determination of the largest characteristic number and the corresponding eigenvector of a nonnegative matrix. USSR Comput. Math. Math. Phys **21**(4), 1–19 (1982)
15. Golub, G.H., Loan, C.F.V.: Matrix Computations. Johns Hopkins University Press, Baltimore (1996)
16. Mohan, B.C., Kumar, S.S.: A robust image watermarking scheme using singular value decomposition. J. Multimedia **3**(1), 7–15 (2008)
17. Shieh, J.M., Lou, D.C., Chang, M.C.: A semi-blind digital watermarking scheme based on singular value decomposition. Comput. Stand. Interfaces **28**, 428–440 (2006)

Accelerative Object Classification Using Cascade Structure for Vision Based Security Monitoring Systems

Van-Dung Hoang[1(✉)] and Kang-Hyun Jo[2]

[1] Quang Binh University, Dong Hoi, Vietnam
dunghv@qbu.edu.vn
[2] University of Ulsan, Ulsan 680-749, Korea
acejo@ulsan.ac.kr

Abstract. Nowadays, object detection systems have achieved significant results, and applied in many important tasks such as security monitoring, surveillance systems, autonomous systems, human- machine interaction and so on. However, one of the most challenges is limitation of computational processing time. In order to deal with this task, a method for speed up processing time is investigated in this paper. The binary of cascaded structural model based detection method is applied for security monitoring systems (SMS). The classification based on cascade structure has been shown advance in extremely rapid discarding negative samples. The SMS is constructed based on two main techniques. First, a feature descriptor for representing data of image based on the modified Histograms of Oriented Gradients (HOG) method is applied. This feature description method allows extracting huge set of partial descriptors, then filtering to obtain only high-discriminated features on training set. Second, the cascade structure model based on the SVM kernel is used for rapidly binary classifying objects. In order taking advantage of optimal SVM classification, the local descriptor within each block is used to feed to SVM. The number of SVMs in each classifier is depended on the precision rate, which decided at the training step. The experimental results demonstrate the effectiveness of this method variety of dataset.

Keywords: Cascade structure · Local feature descriptor · Support vector machine

1 Introduction

In recent years, object detection systems have been developed and applied into several fields of intelligent systems such as security surveillance system, product inspection, automatic packing process, autonomous navigation, and other industry applications. However, there are many challenges in the detection procedures such as various skeletons, appearances of object and background, light conditions, occlusion, and consuming time for processing. In this paper, we are expected to deal with the problem of time processing consuming. The state of the art for real-time processing in object detection has rapidly improved. They can be roughly divided into several categories. The first group focuses on a parallel processing such as GPU/GPGPU based methods for high

© Springer-Verlag Berlin Heidelberg 2016
N.T. Nguyen et al. (Eds.): ACIIDS 2016, Part I, LNAI 9621, pp. 790–800, 2016.
DOI: 10.1007/978-3-662-49381-6_76

speed processing, such as [1–3]. Pascoe *et al.* in [1] presented the method using modern GPU applied into generating and texturing object data, which achieved from a LIDAR scanner and camera. The computational cost function is highly data parallel processing, and real-time performance on a GPU is achieved. In another way, Nammoto *et al.* in [2] presented the high speed and high accuracy visual servo-mechanizing system. Their method has improved to implement in practical applicable systems. In this task, operating speed is speed up by using GPGPU acceleration. That paper also illustrated some experimental results for providing effectiveness of their proposed framework. In order to parallelize the traditional algorithms to accelerate the processing, Schreiber *et al.* in [3] presented the hybrid methodology based on CPU-GPU. The experimental results illustrated the effectiveness framework when applied for fast directional Chamfer matching. That method was applied in human detection. Their proposed method achieved rapidly in computation processing with 32FPS in complex backgrounds and density of people. On the another task in object detection, Filonenko *et al.* in [4] described the smoke detection method based on GP/GPU technology applied for autonomous vehicles, which equipped with camera and LiDAR sensors. The results show that the smoke detection method is high performance in used both camera and LiDAR. In their proposal, some steps were processed in CUDA kernels and expected to supplement real-time robust fire detection.

The second group focuses on improving methods for speed up processing time [5–10]. Amount of them, the significant method for rapidly classification was presented by Viola *et al.* in [5]. Authors proposed the feature descriptor based on Haar wavelets method, called Haar like features. The numerous features are extracted by multiple scale of the box of Haar like feature. Then the boosting is used to training for selected just only high distinctive features, which feed to weak classifiers. The final classifier, strong classifier, is constructed based on the set of weak classifiers. The results of this research showed that the face detection system archives high accuracy and rapidly processing. In contrast, Zhang and Viola in [6] proposed method for learning efficient based on multiple- instance pruning for face detection. This proposed method was tested on than 20,000 positive samples collected from the web with roughly 10,000 faces and additional more than 2 billion negative examples. The processing procedure is ameliorated because the amount of negative samples is rejected in early step of cascade classification. The soft cascade based rejected threshold is used in multiple instance pruning. Pham and Lee in [11] improve the fruit defect detection system by using k-means clustering and graph-based algorithm. The experimental results showed that the system archives high accurance in the terms of human observation and in processing time. In another application, Cheng *et al.* in [7] proposed method for fast and accurate image matching based on the cascade hashing applied in scene reconstruction. The reconstruction based on a Cascade Hashing is used to speed up the image matching processing. Authors showed that the method accelerates image matching procedure faster more times when compared with brute force matching and traditional $K_{d\text{-tree}}$ matching method while obtained comparable precision to other methodologies. In the same direction, the cascade structure model is also considered to apply in people detection system for speed up processing time. In order to taking into account advantage of this method, a classification method based on combining of local descriptor HOG features with SVM kernel is considered application.

2   System Overview

A general ideal of classification method based on cascade structure model is that uses multiple layers classification, positive sample is passed all layers while each classification layer effort to discard negative samples as much as possible. This object classification is used as a module in our security monitoring system (SMS) work. An overview of the proposed method is presented in Fig. 1.

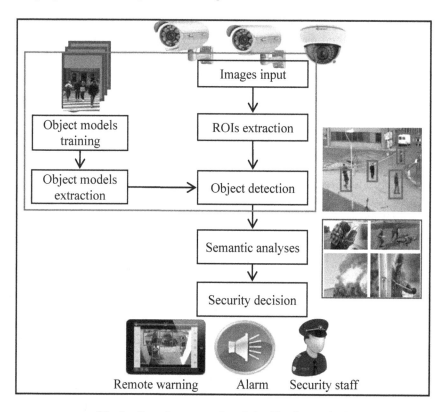

Fig. 1. Cascade structure based classification method

The SMS is constructed based on several main modules, which consists of learning object of interest (OOI) model, object detection, and sematic-based behaviors analyses. Data are retrieved from security cameras then extracts regions of interesting (ROI), which are fed to detection module to recognize the OOIs. The results of OOIs are continuous fed to semantic analyses module for prediction the behaviors of human. The final security decisions are propounded. In this presentation, we do not aim to present all techniques for constructing the fully system, instead of that we just only focus on object detection with the objective is speed up processing time, as illustrated within red rectangular of Fig. 1. In order to reduce the computational time, the classification stage is constructed based on the cascade structure using advantage of special local descriptor feed to SVM classifiers.

3   Preliminary Cascade Structure

A detection method based on cascade structure model is an ensemble model in machine learning, which is used for classification and regression. The basic idea is based on constructing of multiple layers of decision at the training step. The output prediction of each classification layer is combination of the set of weak classifiers for constructing strong classifier or just only a single classifier. However, it is different to other classification methods that mean it divides the detection system into multiple layers of classifications, as depicted in Fig. 2. In that ways, all most negative samples are discarded in early classification layers while all most all positive samples are retained. The method is different to the well-known random forest (RF). The basic RF based on multiple decision trees at the training step, the RF output is fed of all individual trees in forest. The trees are grown very deep tend to learn highly irregular patterns, which can made over-fitting the model with training data. On the contrary, the cascade structure model based classified method is a well-known technology, which allows to reduce the computational time while maintaining the accuracy rate [5]. Practice experiments showed that cascade structure rapidly discards almost negative examples after a few cascade layers using only small subset of features, as demonstrated in Fig. 4.

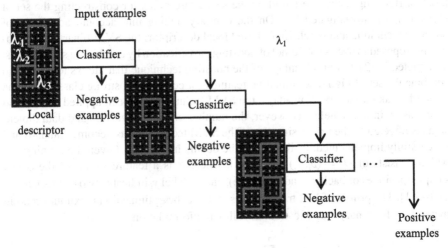

Fig. 2. Cascade structure based classification method

The cascade structure process is forced by a set of goals of classification and capability. The number of cascade level has to be sufficient to obtain high detection rate while spend low computation cost. The final classification archiving positive sample is formulated in (1).

$$H(x) \underset{L=Positive}{=} \prod_{i=1}^{N} C_i(x) \tag{1}$$

where N is the number of layers, $C_i(.)$ is classification at layer i^{th}.

In order to improve the accuracy and speed up of the proposed method based on the idea of using cascaded structural model, the set of SVMs is used for each layer of classified machine. Each SVM binary classification uses a block of HOG. Therefore, instead of the use whole features of sample fed to classification machine, the subset of local descriptors is used within a layer.

In this method, a strong classifier of each layer is instituted by combination of the set of weak classifier SVMs with a coefficient λ. The coefficient λ is depended on the correspondence SVM responded to training dataset. The classification machine of each layer is formulated in (2).

$$C(x) = sign \sum_{i=1}^{n} \lambda_i SVM_i(B_x) \tag{2}$$

where B_x is a feature vector, representing local features within each block of image x.

4   Kernel of Classification

This section briefly presents the SVM as kernel of weak classifiers using in classification process. Boosting technique based on the set of thresholds for constructing the set of classifiers was investigated [5]. On the contrary, taking into account advantage of boosting technique and block HOG-based local descriptor, the SVM binary classification is proposed to use as the kernel of boosting classification by combining set of SVMs, as depicted in (2). On of advantage of the boosting technique that allows to select and combine the set of high discriminative features for constituting strong classifier on the basic of weak classifiers. Nowadays, the SVM technique has been applied widely and successfully in many fields. However, the number of support vectors and data dimensions is affected to the processing time. The SVM technique has become standard and successfully implemented, as illustrated the details in [12–15]. Given the training set, which consists of $D = \{(v_i, y_i)|i = 1 \ldots n\}$, where v_i is a feature vector of the object sample (positive) or background (negative), and its label y_i indicates two class such that $y_i \in [-1,1]$. The primary SVM training tries to solve the optimization maximum-margin hyperplane for binary classification, as following formulation

$$\min_{w,b,\zeta} \quad \frac{1}{2} \|w\|^2 + C \sum_{i=1}^{n} \zeta_i \tag{3}$$
$$s.t. \quad y_i(w^T \phi(v_i) + b) \geq 1 - \zeta_i, \quad \zeta_i \geq 0, i - 1, \ldots, n$$

where $\phi(v_i)$ is mapping feature vector v_i into a higher-dimensional space for linear classification, and $C > 0$ is the regularization parameter assigning penalty to errors, with the optimal w satisfying $w = \sum_{i=1}^{ns} y_i \alpha_i \phi(v_i)$. The model parameters of machine are stored. The signed distance of feature vector v to the hyperplane margin of the SVM model is described as following formulation

$$h(v) = \sum_{i=1}^{ns} y_i \alpha_i k(\phi(v_i), \phi(v)) + b \tag{4}$$

In this experiment, instead of the use of binary classification, probability of SVM classifier is used for boosting a strong classifier. The SVM output is probability value, which is computed using the distance of hyperplane margin to input vector. The final SVM classification is formulated as follows

$$SVM(v) = \frac{1}{1 + \exp(-h(v))} \tag{5}$$

where v is a feature vector which represents the local features of HOG.

5 Experimental Results

For evaluation of the proposed method, the training and evaluation datasets were created by handmade processing. Image dataset is captured in special scenario of bird-view instead of side-view in traditional processing. In our application situation, security cameras are mounted on high, therefore the training dataset is also generated in the same situation.

There are more than 1,000 images, which are used to create roughly 10,000 positive samples and over 50,000 negative samples. They were generated from the same image set. Figure 3 shows some typical positive and negative samples, which were used for training the detection model. In training processing, the hard negative samples, which achieved from miss detection results, are used to re-train a model result with expected to reach higher accuracy.

Cascade structure is built up in similar method that was proposed by Viola *et al.* in [5]. In our experiment, the SVM is used as a kernel of weak classifier instead of scalar threshold in their proposed. Figure 4 illustrates that negative samples are rapidly removed only a few layers while retain all most of positive samples. Processing time for people detection is also validated by using different stride values tested on real data, as depicted in Fig. 5. The results imply that the cascade structure is significant improving processing time in real application. Figures 6 and 7 presented consuming time for training and detection with different the number layers used in cascade structure model. In the case of single layer, the classifier is equivalent with used non-cascade structure. Training time is proportionally increased with the number of layers while detection time is inversely proportional decreased with the number of layers.

Figure 8 shows detection results of the interface of SMS program. In this situation, we assume that there are four connecting to retrieve images with high resolution, 1280×1024 pixels. The detection system was also evaluated in real images, which retrieve from IP cameras free access on Internet with low resolution, as illustrated in Fig. 9. The images were achieved in resolution 640×480 pixels. Detection capability is depend on the resolution and the number of multiple scales. There is a tradeoff between detection precision and time consuming. The system setup with higher resolution and more the number of scales the system archives higher precision, but it should to pay more processing time and vice versa.

a) Some positive samples

b) Some negative samples

Fig. 3. Some samples used for training and evaluation.

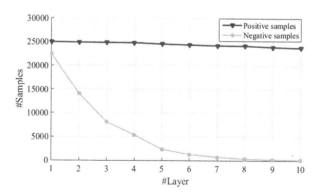

Fig. 4. The negative samples are discarded after each layer in cascade structure model

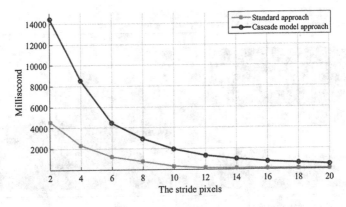

Fig. 5. The processing time with different kind of strides in window scans.

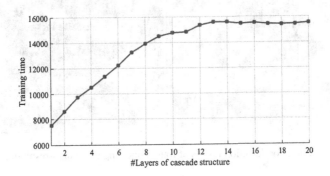

Fig. 6. The number of layers versus processing time in training task

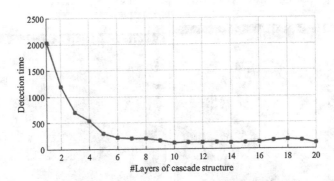

Fig. 7. The number of layers versus processing time in classification task

Fig. 8. Example for detection results of four connecting using high-resolution image

Fig. 9. Detection results using IP camera free access on internet with low-resolution

6 Conclusion

This paper presents the method for accelerating object detection using cascade structural model applying in security monitoring systems using vision sensors. Instead of the use single SVM classification, the cascade structural model using set of SVMs improves computational time. This model deals with two issues for speed up detection system. First, cascade structural model based classification showed advantage of rapidly discarded negative samples. In later steps of classification, positive samples and hard negative samples are only required to classify instead of both kinds of sample should be processed. Second, it is not necessarily to compute simultaneity entire features of a sample. A set of local features within block is requested for each layer processing. Therefore, all features are required to extract just only positive samples instead of the use simultaneously positive and negative samples like as in traditional methods. The number of SVMs in each classifier and the number of layers are depended on the predefined precision rate, which is expected the system result. The experimental results showed that the effectiveness of the proposed method for speed up classification processing.

References

1. Pascoe, G., Maddern, W., Stewart, A.D., Newman, P.: FARLAP: fast robust localisation using appearance priors. In: The International Conference on Robotics and Automation (ICRA), pp. 1–8 (2015)
2. Nammoto, T., Hashimoto, K., Kagami, S., Kosuge, K.: High speed/accuracy visual servoing based on virtual visual servoing with stereo cameras. In: 2013 IEEE/RSJ International Conference on Intelligent Robots and Systems (IROS), pp. 44–49. IEEE (2013)
3. Schreiber, D., Beleznai, C., Rauter, M.: Gpu-accelerated human detection using fast directional chamfer matching. In: IEEE Conference on Computer Vision and Pattern Recognition Workshops (CVPRW2013), pp. 614–621. IEEE (2013)
4. Filonenko, A., Hoang, V.-D., Jo, K.-H.: Smoke detection on roads for autonomous vehicles. In: IECON-40th Annual Conference of the IEEE Industrial Electronics Society, pp. 4063–4066. IEEE (2014)
5. Viola, P., Jones, M.J.: Robust real-time face detection. Int. J. Comput. Vis. **57**, 137–154 (2004)
6. Zhang, C., Viola, P.A.: Multiple-instance pruning for learning efficient cascade detectors. In: Advances in Neural Information Processing Systems, pp. 1681–1688 (2008)
7. Cheng, J., Leng, C., Wu, J., Cui, H., Lu, H.: Fast and accurate image matching with cascade hashing for 3d reconstruction. In: IEEE Conference on Computer Vision and Pattern Recognition (CVPR2014), pp. 1–8. IEEE (2014)
8. Hirschmüller, H.: Accurate and efficient stereo processing by semi-global matching and mutual information. In: IEEE Computer Society Conference on Computer Vision and Pattern Recognition (CVPR2005), pp. 807–814. IEEE (2005)
9. Hoang, V.-D., Vavilin, A., Jo, K.-H.: Fast human detection based on parallelogram haar-like feature. In: The 38th Annual Conference of the IEEE Industrial Electronics Society, pp. 4220–4225 (2012)
10. Hoang, V.-D., Le, M.-H., Jo, K.-H.: Hybrid cascade boosting machine using variant scale blocks based HOG features for pedestrian detection. Neurocomputing **135**, 357–366 (2014)

11. Pham, V., Lee, B.: An image segmentation approach for fruit defect detection using k-means clustering and graph-based algorithm. Vietnam J. Comput. Sci. **2**, 25–33 (2015)
12. Cristianini, N., Shawe-Taylor, J.: An Introduction to Support Vector Machines and other Kernel-Based Learning Methods. Cambridge University Press, Cambridge (2000)
13. Chih-Chung, C., Chih-Jen, L.: LIBSVM: a library for support vector machines. ACM Trans. Intell. Syst. Technol. **2**, 1–27 (2011)
14. Maji, S., Berg, A.C., Malik, J.: Efficient classification for additive kernel SVMs. IEEE Trans. Pattern Anal. Mach. Intell. **35**, 66–77 (2013)
15. Burges, C.C.: A tutorial on support vector machines for pattern recognition. Data Min. Knowl. Disc. **2**, 121–167 (1998)

Selections of Suitable UAV Imagery's Configurations for Regions Classification

Hai Vu[1][✉], Thi Lan Le[1], Van Giap Nguyen[2], and Tan Hung Dinh[3]

[1] International Research Institute MICA, HUST, Hanoi, Vietnam
{hai.vu,thi-lan.le}@mica.edu.vn
[2] School of Information and Communication Technology, Thai Nguyen University,
Thai Nguyen, Vietnam
giapnv.ictu@gmail.com
[3] Department of Aeronautical and Space Engineering, HUST, Hanoi, Vietnam
dinhtanhung@gmail.com

Abstract. Unmanned Aerial Vehicles (UAV) are used to conduct a variety of recognition as well as specific missions such as target tracking, safe landing. The transmitted image sequences to be interpreted at ground station usually face limited requirements of the data transmission. In this paper, on one hand, we handle a surveillance mission with segmenting a UAV video's content into semantic regions. We deploy a spatio-temporal framework that considers UAV videos specific characteristics for segmenting multi regions of interest. After post-processing steps on the segmentation results, a support vector machine classifier is used to recognize regions. In term of temporal feature, we combine the results from the previous frames by proposing to use a state transition formulating through a Markov model. On the other hand, this study also assesses the influences of data reduction techniques on the proposed techniques. The comparisons between the untreated configuration and control conditions under manipulations of the frame rate, spatial resolution, and compression ratio, demonstrate how these data reduction techniques adversely influence the algorithm's performance. The experiments also point out the optimal configuration in order to obtain a trade-off between the target performance and limitation of the data transmission.

Keywords: Image classification · Segmentation · Markov chain

1 Introduction

A central component of developing an UAV system is data transmission that will send imagery data to the ground stations for specific tasks. One of common missions could be identifying areas on ground classes by remote sensing data. This task provides operational commanders with real-time video of opposing forces, terrain factors, or safe landing. In fact some areas are spatially heterogeneous and with similar spectral response (i.e. artificial landscape). The artificial areas consist of several different structures like buildings, roads, gardens or other forestry areas. To handle these issues, on one hand, it required a robust multi-region segmentation approach for the UAV image sequences. On the other hand,

© Springer-Verlag Berlin Heidelberg 2016
N.T. Nguyen et al. (Eds.): ACIIDS 2016, Part I, LNAI 9621, pp. 801–810, 2016.
DOI: 10.1007/978-3-662-49381-6_77

data transmission techniques can be efficiently guard the limited capacity of the transmissions with causes of long range, transmitter/receiver energy consumptions. This study aims to partition each frame into multiple semantic regions and label them with the basic categories (such as building, road, sky, grass, tree). We also examine reduction techniques of data transmission while retaining sufficient performance of the specific task.

Image sequences taken from UAVs device meet several challenges such as camera translation, far distance from the ground plane. Moving of the UAV may also suffer other weather conditions such as wind throws, illuminations. Consequently, we need to deploy specific techniques to adapt the UAV video content. To detect building from aerial images, [1] starts with a seeded region growing algorithm to segment the entire image. Photometric and geometric features are extracted in [1] so that a classification is performed to differentiate the building and non-building. A three-stage framework is proposed in [2] to classify three difference objects such as buildings, ground and vegetation, in which LIDAR data as primary source. Recently, a hybrid classification technique is used in [4] to detect various artificial area like buildings, roads, gardens or other vegetation areas in remote sensing images. A survey in [3] reviews current practices, problems, and prospects of image classification based on remotely sensed data. According to [3], effective use of multiple features of remotely sensed data including spectral, spatial, multi-temporal information are especially significant for improving classification accuracy.

Although there are many approaches on remote sensing image segmentation and classification, choosing an reasonable tool for UAV video analysing is not always available. Most of the related works focus on spatial information, but the temporal information has been ignored. While the proposed algorithms on static image are adaptable on conventional approaches, we pay attention to use a temporal model so that the combination results are increased in term of the recognition rate. To ensure the proposed techniques could be well performed even the data transmission is limited, we examine various evaluations under different imagery configurations. To this end, a control group consists of image sequences generated under manipulating three parameters: frame rate, spatial resolution, and compression ratio. The optimal configuration was observed by comparing precision and recall rates between the untreated and control groups. This could be used as a starting point to define a minimal bandwidth that is required in designing an UAV system. While most of the related works focusing on specific missions, this work handles both tasks: proposing the appropriate algorithms and pointing out optimal imagery configurations for designing UAV's data transmission.

2   Regions Segmentation in UAV Image Sequences

2.1   Overview of the Proposed Algorithms

Given an image sequence, we classify four common regions that are: *sky, tree, construction (building), field (grass)*. Our algorithms try to use both spatial and temporal features to get a satisfying segmentation. The first step, we name

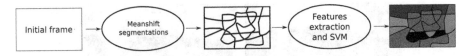

Fig. 1. The proposed method consists of a series of steps applied on every single frame.

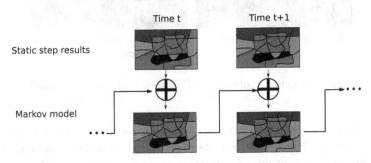

Fig. 2. The temporal model deployed in the proposed algorithms.

segmentation step, as shown in Fig. 1. On each frame, super pixels are extracted through a segmentation technique. Then each supper pixel is assigned a label by utilizing a classifier. In this step, we extract statistical descriptors of supper pixels to form the feature vector. The classifier is Support Vector Machine (SVM) algorithm in order to assign the class of each region. The details are given in Sect. 2.2. The second step is deployed through temporal features. Key idea of this step is to combine results of two consecutive frames through a Markov chain as shown in Fig. 2, as described in Sect. 2.4.

2.2 The Proposed Techniques for the Region Segmentations

We adapt mean-shift algorithm [5,6] to roughly segment regions from a static image. Given a frame F_k of size $M \times N$ pixels, simply using pixels intensity and without any procedure for tuning parameters, mean-shift algorithm gives results

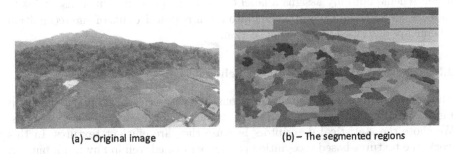

(a) – Original image   (b) – The segmented regions

Fig. 3. Segmentation using mean-shift algorithm. (a) Original image. (b) Segmentation result.

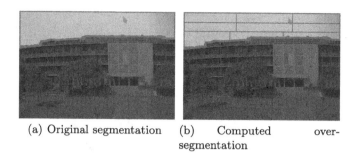

(a) Original segmentation   (b)   Computed   over-segmentation

Fig. 4. An example of object (sky) needs to be post-processed.

as shown in Fig. 3. Obviously, the segmentation results can be only geometric and not semantic. The results consists of scattered, non-connected components or particles without any meaning of the semantic regions. In fact, selection of the optimal parameters for this step is not an easy task. The main reasons are that size, the homogeneity, and border of components are very much depended on the objects. Therefore, a series of post-processing techniques is proposed. First, the small segmented regions should be automatically removed. Some cues help us deploy the post-processing steps. For example, if a region/component is too small, we can ignore this one. The fact that if an object consists of many sub-regions those are various colors, all the parts must be in the same object so that it is efficiently recognized. For example, the construction (e.g., building) can appear with many colors; and variety of colors cannot occur in other object like *Tree* or *Sky*. In this case, if regions are segmented by color features, such regions could not be classified well.

The post-processing steps also focus on eliminating over-segmentation of vast segments. Some contaminated object may appear in an uniform region. For example, as shown in Fig. 4, most of the parts of the sky are uniform color, however, the flag appear in the middle of the sky. To solve this issue, we divide the sky into many rectangular parts. Although the post-processing techniques make efforts to improve the segmentation results. This work still requires a strong solution to connect scattered components as well as to give their labels. A recognition scheme not only assigns a label C_i for a components, but it also helps to connect the same-label components into a full region. Details of the recognition scheme are described in following section.

2.3   The Proposed Recognition Scheme

The proposed recognition algorithm consists of a series of steps as shown in Fig. 5. Firstly, the image features are extracted for each segmented regions. We chose the statistical descriptors because they are efficient textures. In this work, the textures-based recognition can be performed well in any form, but not only a square region. Therefore, Following are the statistical features have been extracted:

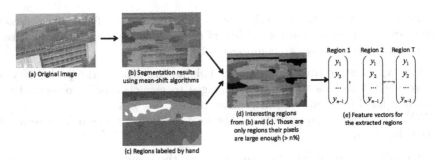

Fig. 5. Scheme of the semantic region segmentation: feature vectors extraction, learning SVM based on training data.

- Color histogram extraction: We examine color channels in RGB and HSV color spaces. The best results obtained when we extract only four bits from gray scale and four bits from Hue. Furthermore, statistical descriptors such as mean and variance over single channels of R, G, and B are calculated.
- For the texture extraction: We utilize Gabor filters responses. We examine mean of the each response over the all filter banks.

For each segmented frame, we classify each interested regions through a support vector machine (SVM). The SVM helps to predict a label corresponding to one of the classes. In this case, each segmented regions of the image labels with a certain class such as (*Sky, Tree, Construction/Building...*). SVM also provides a vector of probabilities. This vector is associated with each segmented regions: $Y_t = (y_{t,1}, y_{t,2}, \ldots, y_{t,k})$. The variable k is the number the interested classes/objects. $y_{t,i}$ is probability at the time t that belongs to the class C_i. In this study, we deploy a multiple classifier SVM, that adopt "One-against-one" scheme [7]. In a relevant work [8], it showns that the multi-classier SVM is a competitive method. For the deploying SVM, we use a SVM library such as libsvm [9]. Basing on "One-against-one" scheme, as 4 classes/objected are interested, there are six classifiers that are learnt. After applying these classifiers, we deploy a voting scheme to get the best class. Obviously, the best class archives 3 votes. Therefore, if we gets 3 votes for a class, a maximal probability α is assigned to this one. Then the another class is assigned a probability so that such value is proportional to its vote count. To avoid a null value, a minimal threshold ϵ is assigned to class that does not receive any vote.

2.4   The Proposed Spatio-Temporal Model

If the result of static step is examined carefully, it can be noticed that the classification output is not stable. If the camera is not moving quickly, the result should be stable. This observation gives us a model which is formed using a state-space model. According to form of a state-space model, the proposed temporal model is assumed as the first-order Markov chain. Consequently, the observations are followed the first-order Markov model. Basing on the form of a Markov chain, we can calculate the probability of each segmented regions through two consecutive frames using two components below:

- $P(X_t|X_{t-1})$: state-transition matrix. This matrix denotes the probability of a segmented regions that may be changed to new class or be preserved the current label between two consecutive frames.
- $P(Y_t|X_t)$: observation probability, the probability can be used from result of the classifiers.

According to above definition, and by using the assumption of the first order Markov chain, the probability of each state can be defined as below:

$$P(X_t|Y_{1:t}) \propto P(X_t|y_t)(\sum_{x_{t-1}} P(X_t|x_{t-1})P(x_{t-1}|y_{1:t-1})) \tag{1}$$

The above probability is recursively calculated; and the current results only depend on the results of the previous frame. It is notice that outputs of the multi classifiers give a probability vector. In this work, the elements of the state-transition matrix is simply defined by:

$$P(X_t = x_{t,i}|X_{t-1} = x_{t-1,j}) = \delta_{i,j}.\alpha + (1 - \delta_{i,j}).\frac{1-\alpha}{n-1} \tag{2}$$

According to above definition, a segmented region has a probability α in order to keep its label and $(1 - \delta_{i,j}).\frac{1-\alpha}{n-1}$ to change to other ones. The parameter α is an important parameter according to this scheme. Currently, this value has been selected through empirical study. However, more extensive work identifying the state-transition matrix is required.

3   Reduction Data Transmission Techniques

The task of semantic region segmentation can obtain the best performance when the UAV imagery is raw data. However, in fact size and the bit rate required for sending raw data are usually very large, while the data transmission capability is limited. Consequently, we have to reduce the data to fit the hardware configuration, also the imagery quality keeps enough quality to recognize the objects. We examine the proposed methods with different data reduction techniques such as: compression rate, frame rate, and spatial image resolution. Then we point out an optimal configuration of imagery data. This configuration suggests a trade-off between transmission limitations and performance of the mission.

In this study, the original image sequences are captured at a resolution of 2298×1294 pixels at 30 fps. Therefore, the data transmission requires a bandwidth of $2298 \times 1294 \times 24$ color bits per pixel $\times 30$ fps $= 2719$ kbyte per second (*kbps*). Such band width requirement is not feasible in designing a data link module of UAV system. By applying the data reduction techniques, the generated sequences name the control group. For each reduction technique, we adjust gradually its original values based on a scale factor. Consequently, various configurations (as shown in Table 1) are generated. We compare performances of untreated versus control groups in order to suggest an optimal configuration.

Table 1. Configurations of the control groups

Configurations	Compression ratio	Frame rate	Spatial resolution
R0(untreated)		30	2704 × 1524
F1..F4	no change	decreased by × 5	no change
R1..R5	no change	no change	resized by × 0.8
C1..C5	reduced by × 0.8	no change	no change

4   Experimental Results

4.1   Evaluation Results on Original Data

The material for evaluating consists of six video sequences captured by a consumer-graded camera attached in a UAV device. The videos are captured in various contexts such as urban, countryside and mountains. These videos have been captured with a spatial resolution of 2298 × 1294 pixels. The length of each images sequence is totally 1m22s. The image sequences are captured at 30 *fps*. For the training data, 66 frames are randomly selected from six videos. They have been labelled by hand. To evaluate performance with standard configuration, we select key frames in testing videos. Such key-frames are extracted by uniform sampling from the original videos. Figure 6 shows examples of the segmentation results.

For qualitative evaluation, precision and recall are usually used to evaluate the accuracy of an machine learning algorithm. Precision is the proportion of retrieved instances that are relevant, whereas recall is the proportion of relevant instances that are retrieved. The precisions and recalls are shown in Table 2. Moreover, to illustrate the roles of the proposed Markov model for temporal features, the output of the static step is compared with the results of the proposed frame-work that includes the static step and the temporal model. The results in Table 2 are achievable for the interested classes. Moreover, we also achieve the good performance on the proposed spatio-temporal model. In is noticed that the static step results are unstable for two classes that are *Tree* and *Grass/field*. However, it still keeps difficulty for recognizing *Construction*. The main reasons are that in some scenes trees and buildings are not clearly separated. Moreover, the appearance of the contractions or buildings have various colors and geometric/shape features. In these cases, the temporal model does not improve significantly the results but it should be really efficient at two different classes such as *Tree* and *Grass/field*. The results convince that the temporal model has a real impact to the final results. In these evaluations, we deployed the proposed method in a PC Intel Core i5 3.20 GHz. The computational time obtains at 1 fps.

4.2   Evaluating on the Control Groups

To easily compare results using control group, the averages of precision and recall rate of original image sequences with R0 configuration (untreated group)

Fig. 6. The recognition results on some examples. Left-to-right panels: original images, expected segmentations, the proposed model segmentations. Regions are marked in different colors: green (tree), red (construction), blue (sky), and yellow= (field/grass) (Color figure online).

are calculated. These results are averagely shown in Table 2. As shown in Table 1, for control groups, there are 4 configurations for frame rate reduction (F1 to F4); 5 configurations for spatial resolution reduction (R1 to R5); and 5 configurations for the compression ratio reduction (C1 to C5). The results of the control groups are given in Tables 3, 4 and 5, respectively.

Based on these evaluations, we select an optimal configuration for designing UAV imagery data transmission. Results in Table 3 show that the frame rate of video does not much affect the precision and the recall. However, the spatial resolution (Table 4) is an important factor impacts to the recognition rate. The performances significantly reduce when the resolutions are lower than $R2's$ configuration. Similarly, the compression ratio at $C3$ (Table 5) obtains equal performances to the original ones (Table 2). For compression ratio, we select values at $C3's$ configuration. Consequently, a set of the optimal parameters is shown in Table 6. As shown in Table 6, while an original video requires the transmission rate at 2719 kbps, by using the optimal configuration, it requires only 332 kbps. The transmission rate therefore is reduced to 87 % by comparing with original image sequences.

Table 2. The precision and recall results on the evaluated data.

| | Only static frame temporal model | | With temporal model | |
	Precision	Recall	Precision	Recall
Tree	0.56	0.39	0.94	0.78
Construction	0.51	0.75	0.56	0.76
Sky	0.96	0.99	0.90	0.99
Grass/field	0.51	0.48	0.65	0.79
Average	0.63	0.65	**0.76**	**0.84**

Table 3. Results of the frame rate variation experiments.

	Bitrate (kbps)	Frame rate	Average precision	Average recall
F1	5431	25:1	70.68	84.50
F2	4345	20:1	70.68	84.50
F3	3259	15:1	71.13	84.02
F4	1587	10:1	71.16	84.61
F5	**1086**	**5:1**	**71.16**	**84.61**

Table 4. Results of the resolution reduction experiments.

Config ID	Bitrate (kbps)	Reduction rate	Resolution	Average precision	Average recall
R1	4970	0.9	2434*1372	70.42	83.09
R2	**4030**	**0.8**	**2162*1218**	**60.79**	**76.53**
R3	3186	0.7	1892*1066	44.97	59.77
R4	2485	0.6	1622*914	43.25	58.83
R5	1809	0.5	1352*762	54.83	40.26

Table 5. Results of the compression ratio reduction experiments.

	Bitrate (kbps)	Compression ratio	Average precision	Average recall
C1	5205	1.24:1	71.02	84.50
C2	4131	1.56:1	71.20	84.74
C3	**3081**	**2.06:1**	**70.62**	**82.33**
C4	1972	3.15:1	70.95	78.45
C5	1741	3.54:1	70.40	72.49
C6	1587	3.85:1	70.04	72.21

Table 6. The optimal configuration for UAV imagery data transmission.

Bitrate	Resolution	Frame rate	Compression Ratio	Average precision	Average recall
332	2162*1218	10:1	5.87	70.06	82.09

5   Conclusion

In this paper, we have described an algorithm for multiple regions segmentation of UAV image sequences. The temporal features are efficiency to archive consistent segmentation and classification results. The proposed spatio-temporal model expanded a basic model to obtain the possibilities of the better segmentation results. We then evaluated the proposed method under manipulations of different data reduction techniques. We got satisfying results that reduced the required bandwidth transmission significantly from original configuration (87 %). This could suggest the down-link requirements in designing a UAV system. In further work, we are going to focus on examining different parameters and their impacts on the state-transition matrix. The proposed framework also needs to be evaluated on more larger data sets in order to examine the robust of the proposed frame work in term of the sensitivity to lighting, illumination, or weather conditions.

Acknowledgements. This work is supported by Hanoi University of Science and Technology (HUST), under grant reference number T2015-096.

References

1. Muller, S., Zaum, D.: Robust building detection in aerial images. In: Proceedings of ISPRS Workshop CMRT 2005, Vienna, pp. 143–148 (2005)
2. Forlani, G., Nardinocchi, C., Scaioni, M., Zingaretti, P.: Complete classification of raw LIDAR data and 3D reconstruction of buildings. Pattern Anal. Appl. **8**(4), 357–374 (2006)
3. Lu, D., Weng, Q.: A survey of image classification methods and techniques for improving classification performance. Int. J. Remote Sens. **28**(5), 823–870 (2007)
4. Malinverni, E.S., Tassetti, A.N., Mancini, A.: Hybrid object-based approach for land use/land cover mapping using high resolution imagery. Int. J. Geog. Inf. Sci. - IJGIS **25**(6), 1025–1043 (2011)
5. Fukunaga, K., Hostetler, L.: The estimation of the gradient of a density function, with applications in pattern recognition. IEEE Trans. Inf. Theory **21**(1), 32–40 (1975)
6. Comaniciu, D., Meer, P.: Mean shift: a robust approach toward feature space analysis. IEEE Trans. Pattern Anal. Mach. Intell. **24**(5), 603–619 (2002)
7. Knerr, S., Personnaz, L., Dreyfus, G.: Single-layer learning revisited: a stepwise procedure for building and training a neural network. In: Soulié, F.F. (ed.) Neurocomputing, pp. 41–50. Springer, Heidelberg (1990)
8. Hsu, C.-W., Lin, C.-J.: A comparison of methods for multiclass support vector machines. IEEE Trans. Neural Netw. **13**(2), 415–425 (2002)
9. Chih-Chung, C., Chih-Jen, L.: Libsvm: a library for support vector machines. ACM Trans. Intell. Syst. Technol. **2**(3), 27 (2011)

Author Index

Printed in the United States
By Bookmasters